What is Mathematics, Really?

By the same author

The Mathematical Experience
Descartes' Dream

What is Mathematics, Really?

Reuben Hersh

JONATHAN CAPE
LONDON

First published 1997

1 3 5 7 9 10 8 6 4 2

The illustration on p. vi "Arabesque XXIX" courtesy of Robert Longhurst. The sculpture depicts a "minimal surface" named after the German geometer A. Enneper. Longhurst made the sculpture from a photograph taken from a computer-generated movie produced by differential geometer David Hoffman and computer graphics virtuoso Jim Hoffman. Thanks to Nat Friedman for putting me in touch with Longhurst, and to Bob Osserman for mathematical instruction. The "Mathematical Notes and Comments" has a section with more information about minimal surfaces.

Figures 1 and 2 were derived from Ascher and Brooks, *Ethnomathematics*, Santa Rosa, CA.: Cole Publishing Co., 1991; and figures 6-17 from David, Hersh, and Marchisotto, *The Companion Guide to the Mathematical Experience*, Cambridge, Ma.: Birkauser, 1995

First published in the United Kingdom in 1997 by Jonathan Cape,
Random House, 20 Vauxhall Bridge Road, London SW1V 2SA

Random House Australia (Pty) Limited
20 Alfred Street, Milsons Point, Sydney,
New South Wales 2061, Australia

Random House New Zealand Limited
18 Poland Road, Glenfield,
Auckland 10, New Zealand

Random House South Africa (Pty) Limited
Endulini, 5A Jubilee Road, Parktown 2193, South Africa

Random House UK Limited Reg. No. 954009

A CIP catalogue record for this book is available from the British Library

Papers used by Random House UK Limited are natural,
recyclable products made from wood grown in sustainable forests.
The manufacturing processes conform to the environmental
regulations of the country of origin.

ISBN 0-224-04417-6

Printed and bound in Great Britain
by Mackays of Chatham PLC

to Veronka

Robert Longhurst, *Arabesque XXIX*

". . . , So long lives this, and this gives life to thee."

Shakespeare, *Sonnet 18*

Contents

Part Two

Summary and Recapitulation

Preface: Aims and Goals

Forty years ago, as a machinist's helper, with no thought that mathematics could become my life's work, I discovered the classic, *What Is Mathematics?* by Richard Courant and Herbert Robbins. They never answered their question; or rather, they answered it by *showing* what mathematics is, not by *telling* what it is. After devouring the book with wonder and delight, I was still left asking, "But what is mathematics, really?"

This book offers a radically different, unconventional answer to that question. Repudiating Platonism and formalism, while recognizing the reasons that make them (alternately) seem plausible, I show that *from the viewpoint of philosophy* mathematics must be understood as a human activity, a social phenomenon, part of human culture, historically evolved, and intelligible only in a social context. I call this viewpoint "humanist."

I use "humanism" to include all philosophies that see mathematics as a human activity, a product, and a characteristic of human culture and society. I use "social conceptualism" or "social-cultural-historic" or just "social-historic philosophy" for my specific views, as explained in this book.

This book is a subversive attack on traditional philosophies of mathematics. Its radicalism applies to philosophy of mathematics, not to mathematics itself. Mathematics comes first, then philosophizing about it, not the other way around. In attacking Platonism and formalism and neo-Fregeanism, I'm defending our right to do mathematics as we do. To be frank, this book is written out of love for mathematics and gratitude to its creators.

Of course it's obvious common knowledge that mathematics is a human activity carried out in society and developing historically. These simple observations are usually considered irrelevant to the philosophical question, what is mathematics? But without the social historical context, the problems of the philosophy of mathematics are intractable. In that context, they are subject to reasonable description and analysis.

The book has no mathematical or philosophical prerequisites. Formulas and calculations (mostly high-school algebra) are segregated into the final Mathematical Notes and Comments.

There's a suggestive parallel between philosophy of mathematics today and philosophy of science in the 1930s. Philosophy of science was then dominated by "logical empiricists" or "positivists" (Rudolf Carnap the most eminent). Positivists thought they had the proper methodology for all science to obey (see Chapter 10).

By the 1950s they noticed that scientists didn't obey their methodology. A few iconoclasts—Karl Popper, Thomas Kuhn, Imre Lakatos, Paul Feyerabend—proposed that philosophy of science look at what scientists actually do. They portrayed a science where change, growth, and controversy are fundamental. Philosophy of science was transformed.

This revolution left philosophy of mathematics unscratched. It's still dominated by its own dogmatism. "Neo-Fregeanism" is the name Philip Kitcher put on it. Neo-Fregeanism says set theory is the only part of mathematics that deserves philosophical consideration. It's a relic of the Frege-Russell-Brouwer-Hilbert foundationist philosophies that dominated philosophy of mathematics from about 1890 to about 1930. The search for indubitable foundations is forgotten, but it's still taken for granted that philosophy of mathematics is about—foundations!

Neo-Fregeanism is not based on views or practices of mathematicians. It's out of touch with mathematicians, users of mathematics, and teachers of mathematics. A few iconoclasts are working to bring in new ideas. P. J. Davis, J. Echeverria, P. Ernest, N. Goodman, P. Kitcher, S. Restivo, G.-C. Rota, B. Rotman, A. Sfard, M. Tiles, T. Tymoczko, H. Freudenthal, P. Henrici, R. Thomas, J. P. van Bendegem, and others. This book is a contribution to that effort.[1]

Mathematics Education

The United States suffers from "innumeracy" in its general population, "math avoidance" among high-school students, and 50 percent failure among college calculus students. Causes include starvation budgets in the schools, mental attrition by television, parents who don't like math.

There's another, unrecognized cause of failure: misconception of the nature of mathematics. This book doesn't report classroom experiments or make sug-

[1] This book is descended from my 1978 article, "Some Proposals for Reviving the Philosophy of Mathematics," published by Gian-Carlo Rota in *Advances in Mathematics*, and from Chapters 7 and 8 of *The Mathematical Experience*, co-authored with Philip J. Davis.

gestions for classroom practice. But it can assist educational reform, by helping mathematics teachers and educators understand what mathematics is.

There's discussion of teaching in Chapter 1, "The Plight of the Working Mathematician," and Chapter 13, "Teaching" and "Ideology."

Outline of Part 1

The book has two main parts. Part One, Chapters 1 through 5, is programmatic. Part Two, Chapters 6 through 12, is historical. The chapters are made of self-contained sections. Chapter 13 is a Summary and Recapitulation.

The book ends with Mathematical Notes and Comments. Often a mathematical concept mentioned in the main text receives more extended treatment in the Notes and Comments. I signal this by a double asterisk (**) in the main text.

Chapter 1 starts with a puzzle. How many parts has a four-dimensional cube? It's doubtful whether a four-dimensional cube exists. Yet as you read you'll figure out the number of its parts! After you've done so, the question returns. Does this thing exist? This is a paradigm for the main problem in philosophy of mathematics. In what sense do mathematical objects exist?

This beginning is followed by a quick overview of modern mathematics, and then a presentation of mathematical Platonism. Next comes the heart of the book: the social-historic philosophy of mathematics that I call humanism. Similar philosophies expounded by other recent authors are introduced in Chapters 11 and 12.

Chapter 2 evaluates criteria for evaluating a philosophy of mathematics. Some standard criteria are unimportant. Some neglected ones are essential. Later, in Chapter 13, I grade myself by these criteria. The first section of Chapter 3 exposes a scandal: Working mathematicians advocate two contradictory philosophies! The next section explains the front and the back of mathematics. Then we meet some mathematical myths, and knock them down with anecdotes from the back room. This chapter testifies that real-life mathematical experience supports humanism against Platonism or formalism.

Chapters 4 and 5 use the humanist point of view to reexamine familiar controversies:

proof
intuition
certainty
infinity
existence
meaning
object versus process
invention versus discovery

Why So Much History?

I advocate a historical understanding of mathematics. So it's natural to make an historical examination of different philosophies. We will find that foundationism and neo-Fregeanism are descendants of a centuries-old mating between mainstream philosophy of mathematics and religion/theology.

Raking up the past uncovers some surprises. René Descartes's famous Method is violated in the *Geometry* of René Descartes. Strange ideas about arithmetic and geometry were ardently held by George Berkeley and David Hume.

The history is told in two separate stories. First, starting with Pythagoras and Plato, we follow the idealists and absolutists, who see mathematics as superhuman or inhuman. I call them the Mainstream. Their story contains an unbroken thread of mutual support between idealist philosophy of mathematics and religion or theology.

Then, starting with Aristotle, we follow the thinkers who see mathematics as human activity. I call them "humanists and mavericks." ("Maverick" is taken from a fascinating article by Aspray and Kitcher.)

This unorthodox procedure finds some support in an interesting remark of Kurt Gödel: "I believe that the most fruitful principle for gaining an overall view of the possible world-views will be to divide them up according to the degree and the manner of their affinity to or, respectively, turning away from metaphysics (or religion.) In this way we immediately obtain a division into two groups, skepticism, materialism and positivism stand on one side, spiritualism, idealism and theology on the other. . . . Thus one would, for example, say that apriorism belongs in principle on the right and empiricism on the left side." (Gödel 1995, p. 375.)

"Dead White Males"

A glance at the index shows that nearly all the authors I cite are white males, many of them dead. Yet white males are a small fraction of the human race.

Why is this?

Art, music, poetry, botany, and architecture are available in some form to all peoples and both sexes. So are market-place arithmetic and architectural geometry. But disputing the meaning and nature of mathematics is ideological, not practical. Western society has been dominated by white males, and its ideologists have been white males.

Today this is no longer so true. I've been able to cite Juliet Floyd, Gila Hanna, Penelope Maddy, Anna Sfard, and Mary Tiles. I've been complimented in public for humanizing the male-hierarchical picture of mathematics.

Similar comments apply to the under-representation of persons of color.

In talking about "working mathematicians" or "academic philosophers" I lump the specimens I have met into some sort of statistic. Is it permissible to ignore differences? Not so many years ago national differences were thought interesting and important. Nowadays the differences thought significant are between males and females, and between advantaged and disadvantaged (white males and people of color).

Some people think female mathematicians and mathematicians of color see the nature of mathematics differently than do white male mathematicians. I'm not convinced such differences are present. If they are, I'm unqualified to write about them.

A defect of this book is neglect of non-Western mathematics. The crucial part of Arabic authors in restoring Greek science to Western Europe is well known. India and China sent important contributions to Europe. But compared with Greece, we hardly know the history of the philosophy of mathematics in Indo-America, Africa, or the Near and Far East. The literature on non-Western mathematics is valuable, but it's not philosophical. My report of Marcia Ascher's work in Chapter 12 goes beyond the Eurocentrism of the rest of the book. Sadly, I'm unprepared to translate archives in Mexico City or Beijing.

Did different religious philosophies in East and West result in different philosophies of mathematics? If so, did such differences affect mathematics itself? Future scholarship is sure to shed light on these fascinating questions.

Santa Fe, N.M. R. H.
December 1996

Acknowledgments

Thanks to V. John-Steiner for wise advice, for numerous encouragements, and especially for lessons on socio-cultural theory and practice.

Hao Wang was a famous computer scientist, logician, and philosopher. His *Beyond Analytic Philosophy* contains careful, thorough critiques of Carnap and Quine. Wang favored "doing justice to what we know." He saw philosophy related to life, not just an abstract exercise.

He started a new field of research by asking: Is there a set of tiles that tile the plane nonperiodically, but not periodically? The answer was yes. Then crystallographers found such tilings in nature. They're "quasi-crystals," a potential source of new technology.

Wang wrote one of the first programs to prove theorems automatically. In a few minutes it proved the first 150 theorems in Russell and Whitehead's *Principia Mathematica.*

He was one of very few who had many conversations with Kurt Gödel. His *Reflections on Kurt Gödel* has unique historical importance.

I was looking forward to his criticism of these chapters when I heard the sad news of his death.

Thanks: for financial support, to Sam Goldberg and the Sloan Foundation. For use of facilities, to The Rockefeller University, the Courant Institute of New York University, Brown University, and the University of New Mexico. For valuable illustrations, to Caroline Smith.

For conversations and letters, to Jose-Luis Abreu, Archie Bahm, Mike Baron, Jon Barwise, Agnes Berger, Bill Beyer, Gus Blaisdell, Lenore Blum, Marcelo de Caravalho Borba, John Brockman, Felix Browder, Mario Bunge, Dorie Bunting,

Ida and Misha Burdzelan, John Busanich, Mutiara Buys, Bruce Chandler, Paul Cohen, Necia Cooper, Richard Courant, Chan Davis, Martin Davis, Hadassah and Phil Davis, Jim Donaldson, Burton Dreben, Mary and Jim Dudley, Freeman Dyson, Ann and Sterling Edwards, Ed Edwards, Peter Eggenberger, Jim Ellison, Bernie Epstein, Dick Epstein, Paul Erdös, Paul Ernest, Florence and Dave Fanshel, Sol Feferman, Dennis Flanagan, Lois Folsom, Marilyn Frankenstein, Hans Freudenthal, Tibor Gallai, Tony Gardiner, Tony Gieri, Jay Ginsburg, Sam Gitler, Nancy Gonzalez, Nick Goodman, Russell Goodman, Luis Gorostiza, Russell Goward, Jack Gray, Cindy Greenwood, Genara and Richard Griego, Liang-Shin Hahn, Gila Hanna, Leon Henkin, Malke, Daniel, Eva and Phyllis Hersh, Josie and Abe Hillman, Moe Hirsch, Doug Hofstadter, John Horvath, Takashi Hosoda, Ih-Ching Hsu, Kirk Jensen, Fritz John, Chris Jones, Maria del Carmen Jorge, Mark Kac, Ann and Judd Kahn, Evelyn Keller, Joe Keller, Philip Kitcher, Morris Kline, Vladimir Korolyuk, Martin Kruskal, Tom Kyner, Marian and Larry Kugler, George Lakoff, Anneli and Peter Lax, Uri Leron, Ina Lindemann, Lee Lorch, Ray Lorch, Wilhelm Magnus, Penelope Maddy, Elena Anne Marchisotto, Charlotte and Carl Marzani, Deena Mersky, Ray Mines, Merle Mitchell, Cathleen Morawetz, Don Morrison, Joseph Muccio, Gen Nakamura, Susan Nett, John Neu, Otto Neugebauer, Bob Osserman, George Papanicolaou, Alice and Klaus Peters, Stan Philips, Joanna and Mark Pinsky, George Pólya, Louise Raphael, Fred Richman, Steve Rosencrans, Gian-Carlo Rota, Muriel and Henry Roth, Brian Rotman, Paul Ryl, Sandro Salimbeni, Joe Schatz, Andy Schoene, Susan Schulte, Anna Sfard, Abe Shenitzer, David Sherry, Jill and Neal Singer, Melissa Smeltzer, Joel Smoller, Vera Sos, Ian Stewart, Gabe Stolzenberg, David Swift, Anatol Swishchuk, Béla Sz.-Nagy, Robert Thomas, William Thurston, Mary Tiles, Uri Treisman, Tim Trucano, Tom Tymoczko, Francoise and Stan Ulam, Istvan Vincze, Cotten and Larry Wallen, Solveig and Burt Wendroff, Myra and Alvin White, Raymond Wilder, Carla Wofsy, and Steve Wollman.

What Is Mathematics, Really?

Dialogue with Laura

I was pecking at my word processor when twelve-year-old Laura came over.

L: What are you doing?

R: It's philosophy of mathematics.

L: What's that about?

R: What's the biggest number?

L: There isn't any!

R: Why not?

L: There just isn't! How could there be?

R: Very good. Then how many numbers must there be?

L: Infinite many, I guess.

R: Yes. And where are they all?

L: Where?

R: That's right. Where?

L: I don't know. Nowhere. In people's heads, I guess.

R: How many numbers are in your head, do you suppose?

L: I think a few million billion trillion.

R: Then maybe everybody has a few million billion trillion or so?

L: Probably they do.

R: How many people could there be living on this planet right now?

L: Don't know. Probably billions.

R: Right. Less than ten billion, would you say?

L: Okay.

R: If each one has a million billion trillion numbers or less in her head, we can count up all their numbers by multiplying ten billion times a million billion trillion. Is that right?

L: Sounds right to me.

R: Would that number be infinite?

L: Would be pretty close.

R: Then it would be the largest number, wouldn't it?

L: Wait a minute. You just asked me that, and I said there couldn't be a largest number!

R: So there actually has to be a number bigger than the biggest number in anybody's head?

L: Right.

R: Where is that number, if not in anybody's head?

L: Maybe it's how many grains of sand in the whole universe.

R: No. The smallest things in the universe are supposed to be electrons. Much smaller than grains of sand. Cosmologists say the number of electrons in the universe is less than a 1 with 23 zeroes after it. Now, ten billion times a million billion trillion is a 1 with

$$1 + 9 + 6 + 9 + 12$$

zeroes after it. That's a 1 with 37 zeroes after it, which is a hundred trillion times as much as a one with 23 zeroes it, which is more than the number of elementary particles in the universe, according to cosmologists.

L: Cosmologists are people who figure out stuff about the cosmos?

R: Right.

L: Awesome!

R: So there are way more numbers than there are elementary particles in the whole cosmos.

L: Pretty weird!

R: Never mind "where." Let's talk about "when." How long do you suppose numbers have been around?

L: A real long time.

R: Have they told you in school about the Big Bang?

L: I heard about it. It was like fifteen billion years ago. When the cosmos began.

R: Do you think there were numbers at the time of the big bang?

L: Yes, I think so. Just to count what was going on, you know.

R: And before that? Were there any numbers before the Big Bang? Even little ones, like 1, 2, 3?

L: Numbers before there was a universe?

R: What do you think?

L: Seems like there couldn't be anything before there was anything, you know what I mean? Yet it seems like there should always be numbers, even if there isn't a universe.

R: Take that number you just came up with, 1 with 37 zeroes after it, and call it a name, any name.

L: How about 'gazillion'?

R: Good. Can you imagine a gazillion of anything?

L: Heck no.

R: Could you or anyone you know ever count that high?

L: No. I bet a computer could.

R: No. The earth and the sun will vanish before the fastest computer ever built could count that high.

L: Wow!

R: Now, what is a gazillion and a gazillion?

L: Two gazillion. How easy!

R: How do you know?

L: Because one anything and another anything is two anything, no matter what.

R: How about one little mousie and one fierce tomcat? Or one female rabbit and one male rabbit?

L: You're kidding! That's not math, that's biology.

R: You never saw a gazillion or anything near it. How do

you know gazillions aren't like rabbits?

L: Numbers can't be like rabbits.

R: If I take a gazillion and add one, what do I get?

L: A gazillion and one, just like a thousand and one or a million and one.

R: Could there be some other number between a gazillion and a gazillion and one?

L: No, because a gazillion and one is the next number after a gazillion.

R: But how do you know when you get up that high the numbers don't crowd together and sneak in between each other?

L: They can't, they've got to go in steps, one step at a time.

R: But how do you know what they do way far out where you've never been?

L: Come on, you've got to be joking.

R: Maybe. What color is this pencil?

L: Blue.

R: Sure?

L: Sure I'm sure.

R: Maybe the light out here is peculiar and makes colors look wrong? Maybe in a different light you'd see a different color?

L: I don't think so.

R: No, you don't. But are you absolutely sure it's absolutely impossible?

L: No, not absolutely, I guess.

R: You've heard of being color blind, haven't you?

L: Yes, I have.

R: Could it be possible for a person to get some eye disease and become color blind without knowing it?

L: I don't know. Maybe it could be possible.

R: Could that person think this pencil was blue, when actually it's orange, because they had become color blind without knowing it?

L: Maybe they could. What of it? Who cares?

R: You see a blue pencil, but you aren't 100% sure it's really blue, only almost sure. Right?

L: Sure. Right.

R: Now, how about a gazillion and a gazillion equals two gazillion? Are you absolutely sure of that?

L: Yes I am.

R: No way that could be wrong?

L: No way.

R: You've never seen a gazillion. Yet you're more sure about gazillions than you are about pencils that you can see and touch and taste and smell. How do you get to know so much about gazillions?

L: Is that philosophy of mathematics?

R: That's the beginning of it.

Part One

one

Survey and Proposals

A Round Trip to the Fourth Dimension. Is There a 4-Cube?**

This section has two purposes. It's a worked exercise in Pólya's heuristic (see Chapter 11).

At the same time, it's an inquiry into mathematical existence. By guided induction and intelligent guessing, you'll count the parts of a 4-dimensional cube. Then you'll be asked, "Does your work make sense? What kind of sense does it make?"

You're familiar with two-dimensional cubes (squares) and three-dimensional cubes. Is there a four-dimensional cube?

To help you answer, here's a harder question:

"How many parts does a 4-cube have?"

I haven't explained the meaning of either "4-cube" or "part." What can you do? It's time for Pólya's problem-solving principle: *If you can't solve your problem, make up a related problem that you may be able to solve.*

In this case, what's an easier, related question? "How many parts has an *ordinary* cube, a 3-cube?"

What kinds of parts does it have?

A 3-cube has an interior (3-dimensional), some faces (2-dimensional), some edges (1-dimensional), and some vertices (0-dimensional). These are four different kinds of parts.

You can count each of these four kinds of parts.

Do it! (Hint: It has 12 edges.)

Write your four numbers in a row, from 0-dimensional to 3-dimensional. Add them up, and write the sum at the end of the row.

Answer: a 3-cube has 27 parts.

Ready for four dimensions?

Probably not yet.

What other related problem can you think of? Maybe a simpler one? (If you can't go up right away, try going down at first.)

How many kinds of parts has a *2-cube,* a square?

Count each kind. Write the numbers in a new row above the previous row, corresponding numbers above corresponding numbers. Add up the new numbers.

You get 4 + 4 + 1 = 9. Write 9 above the 27 from the 3-cube.

You went down from 4 dimensions to 3, and from 3 to 2. What should you understand by a "*1-cube*"? How many kinds of parts does it have? How many of each kind? And what's the sum?

The answer is 2 + 1 = 3.

Write these numbers above the corresponding numbers from the 2-cube and the 3-cube.

You have a table! It has three rows, one row for each "cube" from 1-dimensional to 3-dimensional.

The first row says 2, 1.

The middle row says 4, 4, 1.

The bottom row says 8, 12, 6, 1.

Your table has five columns. The first four columns give the number of parts of each dimension, from 0 to 3. The last column gives the sums.

We're trying to find the sum for the 4-cube. We have tabulated information for the 1-, 2-, and 3-cube, in three rows.

The 4-cube goes in the next row, the fourth! (It needs one additional space on the right, to count its four-dimensional part, the interior.)

You immediately see what to put first in row 4, below 2, 4, and 8. You've already found out how many vertices the 4-cube has! Stare at your table until you see what to put in the other places, for the one-, two- and three-dimensional parts of the 4-cube. Two of the diagonals follow obvious patterns. And there's a simple relation between every number in the table and a pair of numbers above it. Namely, each number equals the sum of the numbers diagonally above to the left plus double the number directly above. For instance, $6 = 4 + (2 \times 1)$ and $12 = 4 + (2 \times 4)$.

You've completed the fourth row! You know a 4-cube has 81 parts.**

BUT!!

BACK TO PHILOSOPHY:

Does a 4-cube really exist?

If yes, where is it? How does it exist? In what sense?

How do you *know* it exists? Could you be mistaken?

If no, how could you find out so much about it? If there is no 4-cube, what's the meaning of the numbers you found? Should other readers of this exercise get the same numbers? *Why* should they get the same numbers, if there's no such thing as a 4-cube?

For that matter, is there even such a thing as a 3-cube? You've seen and touched physical objects called cubes. They aspired to approximate a cube. But they couldn't *be* cubes. No ice cube or ebony cube or brass cube has 12 edges all *exactly* the same length, 8 corners all *perfectly* square.

Only a mathematical 3-cube is a perfect cube. So a mathematical 3-cube is like a 4-cube, in not being a physical object! Then what is it? Where is it? Is there a big difference between asking, "Does a 3-cube exist?" and asking, "Does a 4-cube exist?"

Answering these questions is the point of this book.

For experts, here are some exercises in the philosophy of mathematics that anticipate much of what follows.

How would these questions about 3-cube and 4-cube be answered by Gödel or Thom? (See below, "Must We Be Platonists?")

By Frege or Russell in his logicist period?

By Brouwer or Bishop?

By Hilbert or Bourbaki (Chapter 8)?

By Wittgenstein (Chapter 11)?

By Quine? By Putnam in his phase I, II, or III (Chapter 9)?

Quick Overview

Even without three years of graduate school, you can get a rough notion of modern mathematics. Here's a mini-sketch of its method and matter.

The method of mathematics is "conjecture and proof." You come to an inherited network of concepts and facts, properties and connections, called a "theory." (For instance, classical solid geometry, including the 3-cube.) This presently existing theory is the result of a historic evolution. It is the cooperative and competitive work of generations of mathematicians, associated by friendship and rivalry, by mutual criticism and correction, as leaders and followers, mentors and protégés.

Starting with the theory as you find it, you fill in gaps, connect to other theories, and spin out enlargements and continuations—like going up one dimension to dream of a "hypercube."

You just solved the hypercube problem. But you didn't solve it in isolation. You were handed the problem in the first place. Then you got helpful hints and encouragement as you went along. When you finally got the answer, you received confirmation that your answer was right.

Believe it or not, a mathematician has needs similar to yours. He/she needs to discover a problem connected to the existing mathematical culture. Then she needs reassurance and encouragement as she struggles with it. And in the end when she proposes a solution she needs agreement or criticism. No matter how isolated and self-sufficient a mathematician may be, the source and verification of his work goes back to the community of mathematicians.

Sometimes new theories seem to spin out of your head and the heads of your predecessors. Sometimes they're suggested by real-world subjects, like physics. Today the infinite-dimensional spaces of higher geometry are models for the elementary particles of physics.

Mathematical discovery rests on a validation called "proof," the analogue of experiment in physical science. A proof is a conclusive argument that a proposed result follows from accepted theory. "Follows" means the argument convinces qualified, skeptical mathematicians. Here I am giving an overtly social definition of "proof." Such a definition is unconventional, yet it is plainly true to life.

In logic texts and modern philosophy, "follows" is often given a much stricter sense, the sense of mechanical computation. No one says the proofs that mathematicians write actually *are* checkable by machine. But it's conventional to insist that there be *no doubt* they *could* be checked that way.

Such lofty rigor isn't found in all mathematics. From one specialty to another, from one mathematician to another, there's variation in strictness of proof and applicability of results. Mathematics that stresses results above proof is often called "applied mathematics." Mathematics that stresses proof above results is sometimes called "pure mathematics," more often just "mathematics." (Outsiders sometimes say "theoretical mathematics.")

A naive non-mathematician—perhaps a neo-Fregean analytic philosopher—looks into Euclid, or a more modern math text of formalist stripe, and observes that axioms come first. They're right on page one. He or she understandably concludes that in mathematics, axioms come first. First your assumptions, then your conclusions, no?

But anyone who has done mathematics knows what comes first—a problem. Mathematics is a vast network of interconnected problems and solutions. Sometimes a problem is called "a conjecture."

Sometimes a solution is a set of axioms!

I explain.

When a piece of mathematics gets big and complicated, we may want to systematize and organize it, for esthetics and for convenience. The way we do that is to axiomatize it. Thus a new type of problem (or "meta-problem") arises:

"Given some specific mathematical subject, to find an attractive set of axioms from which the facts of the subject can conveniently be derived."

Any proposed axiom set is a proposed solution to this problem. The solution will not be unique. There's a history of re-axiomatizations of Euclidean geometry, from Hilbert to Veblen to Birkhoff the Elder.

In developing and understanding a subject, axioms come late. Then in the formal presentations, they come early.

Sometimes someone tries to invent a new branch of mathematics by making up some axioms and going from there. Such efforts rarely achieve recognition or permanence. Examples, problems, and solutions come first. Later come axiom sets on which the already existing theory can be "based."

The view that mathematics is in essence derivations from axioms is backward. In fact, it's wrong.

An indispensable partner to proof is mathematical intuition. This tells us what to try to prove. We relied heavily on intuition in our hypercube exercise. It often gives true theorems, even with gappy proofs. We return to intuition and proof in Chapter 4.

So far I've described mathematics by its methods. What about its content? The dictionary says math is the science of number and figure ("figure" meaning the shapes or figures of geometry.) This definition might have been O.K. 200 years ago. Today, however, math includes the groups, rings, and fields of abstract algebra, the convergence structures of point-set topology, the random variables and martingales of probability and mathematical statistics, and much, much more. *Mathematical Reviews* lists 3,400 subfields of mathematics! No one could attempt even a brief presentation of all 3,400, let alone a philosophical investigation of them all. To identify a branch of study as part of mathematics, one is guided by its method more than its content.

Formalism: A First Look

Two principal views of the nature of mathematics are prevalent among mathematicians—Platonism and formalism. Platonism is dominant, but it's hard to talk about it in public. Formalism feels more respectable philosophically, but it's almost impossible for a working mathematician to really believe it.

The next section is about Platonism. Here I take a quick glance at formalism. I return to it in the section of Chapter 9 on David Hilbert. The third major school, intuitionism or constructivism, is also discussed in Chapter 9, in the sections on foundations, on L. E. J. Brouwer, and on Errett Bishop.

The formalist philosophy of mathematics is often condensed to a short slogan: "Mathematics is a meaningless game." ("Meaningless" and "game" remain undefined. Wittgenstein showed that games have no strict definition, only a family resemblance.)

What do formalists mean by "game" when they call mathematics a game? Perhaps they use "game" to mean something "played by the rules." (Now "play" and "rule" are undefined!)

For a game in that sense, two things are needed:

(2) people to play by the rules.
(1) rules.

Rule-making can be deliberate, as in Monopoly or Scrabble—or spontaneous, as in natural languages or elementary arithmetic.

In either case, the *making* of rules doesn't follow rules!

Wittgenstein and some others seem to think that since the making of rules doesn't follow rules, then the rules are arbitrary. They could just as well be any way at all. This is a gross error.

The rules of language and of mathematics are historically determined by the workings of society that evolve under pressure of the inner workings and interactions of social groups, and the physical and biological environment of earth. They are also simultaneously determined by the biological properties, especially the nervous systems, of individual humans. Those biological properties and nervous systems have permitted us to evolve and survive on earth, so of course they reflect somehow the physical and biological properties of this planet. Complicated, certainly. Mysterious, no doubt. Arbitrary, no.

People often make rules deliberately. Not only for games, but also for computer languages, for parliamentary procedure, for stopping at STOP signs, and for Orthodox weddings. These rule-making tasks don't follow rules. But that doesn't make them arbitrary. Rules are made for a purpose. To be played or accepted or performed by people, they have to be playable or acceptable by people. Tradition, taste, judgment, and consensus matter. Eccentricities of individual rule-makers matter. The resultant of such social and personal factors is what makes us make the rules we make. The outcome of rule-making isn't arbitrary. Neither is it rule governed.

Some details of a rule system may seem arbitrary or optional. In chess, for instance, the rule for castling might be varied without ruining the game.

Is there a sharp separation between playing by the rules and making the rules?

Some formalists in philosophy of mathematics say discovery is lawless—has no logic—while proof or justification is nothing but logic. If such a philosopher notices that real mathematical life isn't that way, the discrepancy seems like a scandal that must be kept out of the newspapers, or a crime calling for correction by Georg Kreisel's "logical hygiene."

In real life, in all games including mathematics (supposing for the moment that it is a game), the separation between playing the game and making the rules is imperfect, partial, incomplete.

Chess players don't change the rules of chess as they go along. Not in tournament chess, at any rate. Disputes are settled according to written procedures. But these procedures aren't *rules*. Settling disputes comes down to judgments and opinions. Big league baseball has plenty of rules. But the game would be impossible without umpires who use their judgment. In street stick ball, first base is supposed to be the left front fender on the closest car parked on the right side of the street. If no car is there, we improvise.

In real life there are no totally rule-governed activities. Only more or less rule-governed ones, with more or less definite procedures for disputes. The rules and procedures evolve, sometimes formally like amending the U.S. Constitution, sometimes informally, as street games evolve with time and mixing of cultures. Is there totally unruly or ruleless behavior? Perhaps not. Mathematics is in part a rule-governed game. But one can't overlook how the rules are made, how they evolve, and how disputes are resolved. That isn't rule governed, and can't be.

Computer proof is changing the way the game of mathematics is played. Wolfgang Haken thinks computer proof is permitted under the rules. Paul Halmos thinks it ought to be against the rules. Tom Tymoczko thinks it amounts to changing the rules. In the long run, what mathematicians publish, cite, and especially teach, will decide the rules. We have no French Academy to set rules, no cabal of team owners to say how to play our game. Our rules are set by our consensus, influenced and led by our most powerful or prestigious members (of course).

These considerations on games and rules in general show that one can't understand mathematics (or any other nontrivial human activity) by simply finding rules that it follows or ought to follow. Even if that could be done, it would lead to more interesting questions: Why and whence those rules?

The notion of strictly following rules without any need for judgment is a fiction. It has its use and interest. It's misleading to apply it literally to real life.

Must We Be Platonists?

Platonism, or realism as it's been called, is the most pervasive philosophy of mathematics. It has various variations. The standard version says mathematical entities exist outside space and time, outside thought and matter, in an abstract realm independent of any consciousness, individual or social. Today's mathematical Platonisms descend in a clear line from the doctrine of Ideas in Plato (see "Plato" in Chapter 6). Plato's philosophy of mathematics came from the Pythagoreans, so mathematical "Platonism" ought to be "Pythago-Platonism." I defer to custom and say "Platonism." (This debt of Plato is discussed by John Dewey in his 1929 Gifford lectures and by Bertrand Russell in Chapter 9.)

There are Platonisms of mathematicians and Platonisms of philosophers. I quote half a dozen eminent Platonists of past and present, mostly mathematicians. (Somerville and Everett are copied from Leslie White's article in *The World of Mathematics.*)

Edward Everett (1794–1865), the first American to receive a doctorate at Göttingen, an orator who shared the platform with Abraham Lincoln at Gettysburg, wrote: "In the pure mathematics we contemplate absolute truths which existed in the divine mind before the morning stars sang together, and which will continue to exist there when the last of their radiant host shall have fallen from heaven."

The scholar and mathematician Mary Somerville (1780–1872): "Nothing has afforded me so convincing a proof of the unity of the Deity as these purely mental conceptions of numerical and mathematical science which have been by slow degrees vouchsafed to man, and are still granted in these latter times by the Differential Calculus, now superseded by the Higher Algebra, all of which must have existed in that sublimely omniscient Mind from eternity."

G. H. Hardy, the leading English mathematician of the 1920s: "I have myself always thought of a mathematician as in the first instance an observer, who gazes at a distant range of mountains and notes down his observations. His object is simply to distinguish clearly and notify to others as many different peaks as he can. There are some peaks which he can distinguish easily, while others are less clear. He sees A sharply, while of B he can obtain only transitory glimpses. At last he makes out a ridge which leads from A and, following it to its end, he discovers that it culminates in B. B is now fixed in his vision, and from this point he can proceed to further discoveries. In other cases perhaps he can distinguish a ridge which vanishes in the distance, and conjectures that it leads to a peak in the clouds or below the horizon. But when he sees a peak, he believes that it is there simply because he sees it. If he wishes someone else to see it, he *points to it*, either directly or through the chain of summits which led him to recognize it himself. When his pupil also sees it, the research, the argument, the *proof* is finished" (1929, p. 18). Here the "chain of summits" is the chain of statements in a proof, connecting known facts (peaks) to new ones. Hardy uses a chain of summits to find a new peak. Once he sees the new peak, he believes in it because he sees it, no longer needing any chain.

The preeminent logician, Kurt Gödel: "Despite their remoteness from sense experience, we do have something like a perception also of the objects of set theory, as is seen from the fact that the axioms force themselves upon us as being true. I don't see any reason why we should have less confidence in this kind of perception, i.e., in mathematical intuition, than in sense perception. . . . This, too, may represent an aspect of objective reality."

The French geometer and Fields Medalist René Thom, father of catastrophe theory: "Mathematicians should have the courage of their most profound convictions and thus affirm that mathematical forms indeed have an existence that is independent of the mind considering them. . . . Yet, at any given moment, mathematicians have only an incomplete and fragmentary view of this world of ideas."

Thom's world of ideas is geometric; Gödel's is set-theoretic. They believe in an independent world of ideas—but not the same world!

Paul Erdös was a famous Hungarian mathematician who talked about "The Book." "The Book" contains all the most elegant mathematical proofs, the known and especially the unknown. It belongs to "the S. F."—"the Supreme Fascist"—Erdös's pet name for the Almighty. Occasionally the S. F. permits someone a quick glimpse into the Book.

The Book is a perfect metaphor for Platonism. But Erdös said he's not interested in philosophy. The Book and the S. F. are "only a joke."

However, in a film about Erdös (*N is a Number*, produced by Paul Csicsery) his friend and collaborator Fam Chung, says, "In Paul's mind there is only one reality, and that's mathematics."

Ron Graham, a well-known combinatorialist, collaborator friend of Erdös and husband of Chung, goes even further: "I personally feel that mathematics is the essence of what's driving the universe."

Another Erdös collaborator, Joel Spencer: "Where else do you have absolute truth? You have it in mathematics and you have it in religion, at least for some people. But in mathematics you can really argue that this is as close to absolute truth as you can get. When Euclid showed that there were an infinite number of primes, *that's it!*. There are an infinite number of primes, no ifs, ands, or buts! That's as close to absolute truth as I can see getting."

(As a small point of historical fidelity, Euclid never could have said there was an infinite number of anything. Proposition 20, Book IX, says, in Heath's translation, "Prime numbers are more than any assigned multitude of prime numbers"—there is no greatest prime. Heath immediately paraphrases this as "the important proposition that the number of prime numbers is infinite." Heath's and Spencer's formulation is natural in today's context of infinite sets. Not in Euclid's context.)

Why do mathematicians believe something so unscientific, so far-fetched as an independent immaterial timeless world of mathematical truth?

The mystery of mathematics is its objectivity, its seeming certainty or near-certainty, and its near-independence of persons, cultures, and historical epochs (see the section on Change in Chapter 5).

Platonism says mathematical objects are real and independent of our knowledge. Space-filling curves, uncountably infinite sets, infinite-dimensional manifolds—all the members of the mathematical zoo—are definite objects, with definite properties, known or unknown. These objects exist outside physical space and time. They were never created. They never change. By logic's law of the excluded middle, a meaningful question about any of them has an answer, whether we know it or not. According to Platonism a mathematician is an empirical scientist, like a botanist. He can't invent, because everything is already there. He can only discover. Our mathematical knowledge is objective and unchanging because it's knowledge of objects external to us, independent of us, which are indeed changeless.

An inarticulate, half-conscious Platonism is nearly universal among mathematicians. Research or problem-solving, even at the elementary level, generates a naive, uncritical Platonism. In math class, everybody has to get the same answer. Except for a few laggards, they *do* all get the same answer! That's what's special about math. *There are right answers.* Not right because that's what Teacher wants us to believe. Right because *they are* right.

That universality, that independence of individuals, makes mathematics seem immaterial, inhuman. Platonism of the ordinary mathematician or student is a recognition that the facts of mathematics are independent of her or his wishes. This is the quality that makes mathematics exceptional.

Yet most of this Platonism is half-hearted, shamefaced. We don't ask, How does this immaterial realm relate to material reality? How does it make contact with flesh and blood mathematicians? We refuse to face this embarrassment:

Ideal entities independent of human consciousness violate the empiricism of modern science. For Plato the Ideals, including numbers, are visible or tangible in Heaven, which we had to leave in order to be born. For Leibniz and Berkeley, abstractions like numbers are thoughts in the mind of God. That Divine Mind is still real for Somerville and Everett.

Heaven and the Mind of God are no longer heard of in academic discourse. Yet most mathematicians and philosophers of mathematics continue to believe in an independent, immaterial abstract world—a remnant of Plato's Heaven, attenuated, purified, bleached, with all entities but the mathematical expelled.

Platonism without God is like the grin on Lewis Carroll's Cheshire cat. The cat had a grin. Gradually the cat disappeared, until all was gone—except the grin. The grin remained without the cat.

MacLane is unusual in his unequivocal rejection of Platonism, without turning to formalism. "The platonic notion that there is somewhere the ideal realm of sets, not yet fully described, is a glorious illusion" (p. 385). He thinks there's no need to consider the question of existence of mathematical entities.

The Platonisms of philosophers are more sophisticated than those of mathematicians. One of them is logicism, once preached by Gottlob Frege and Bertrand Russell. Today's "most influential philosopher," W. V. O. Quine, has his own pragmatic-type Platonism (see Chapter 9). Here we talk mainly about "garden variety" or "generic" Platonism, Platonism among the broad mathematical masses.

The objections to Platonism are never answered: the strange parallel existence of two realities—physical and mathematical; and the impossibility of contact between the flesh-and-blood mathematician and the immaterial mathematical object. Platonism shares the fatal flaw of Cartesian dualism. To explain the existence and properties of mind and matter, Descartes postulated a different "substance" for each. But he couldn't plausibly explain how the two substances interact, as mind and body do interact. In similar fashion, Platonists explain mathematics by a separate universe of abstract objects, independent of the material universe. But how do the abstract and material universes interact? How do flesh-and-blood mathematicians acquire the knowledge of number?

To answer, you have to forget Platonism, and look in the socio-cultural past and present, in the history of mathematics, including the tragic life of Georg Cantor.

The set-theoretic universe constructed by Cantor and generally adopted by Platonists is believed to include all mathematics, past, present, and future. In it, the uncountable set of real numbers is just the beginning of uncountable chains of uncountables. The cardinality of this set universe is unspeakably greater than that of the material world. It dwarfs the material universe to a tiny speck. And it was all there before there was an earth, a moon, or a sun, even before the Big Bang. Yet this tremendous reality is unnoticed! Humanity dreams on, totally unaware of it—*except for us mathematicians.* We alone notice it. But only since Cantor revealed it in 1890. Is this plausible? Is this credible? Roger Penrose declares himself a Platonist, but draws the line at swallowing the whole set-theoretic hierarchy.

Platonists don't acknowledge the arguments against Platonism. They just re-avow Platonism.

Frege's point of view persists today among set-theoretic Platonists. It goes something like this:

1. Surely the empty set exists—we all have encountered it!
2. Starting from the empty set, perform a few natural operations, like forming the set of all subsets. Before long you have a magnificent structure in which you can embed the real numbers, complex numbers, quaternions, Hilbert spaces, infinite-dimensional differentiable manifolds, and anything else you like.
3. Therefore it's vain to talk of inventing or creating mathematics. In this all-encompassing, set-theoretic structure, everything we could ever want or dream of is already present.

Yet most advances in mainstream mathematics are made without reference to any set-theoretic embedding. Saying Hilbert space was already there in the set universe is like telling Rodin, "*The Thinker* is a nice piece of work, but all you did was get rid of the extra marble. The statue was there inside the marble quarry before you were born."

Rodin made *The Thinker* by removing marble. Hilbert, von Neumann, and the rest made the theory of Hilbert space by analyzing, generalizing, and rearranging mathematical ideas that were present in the mathematical atmosphere of their time.

A Way Out

What's the nature of mathematical objects?

The question is made difficult by a centuries-old assumption of Western philosophy: "There are two kinds of things in the world. What isn't physical is mental; what isn't mental is physical."

Mental is individual consciousness. It includes private thoughts—mathematical and philosophical, for example—before they're communicated to the world and become social—and also perception, fear, desire, despair, hope, and so on.

Physical is taking up space—having weight or energy. It's flesh and bones, sound waves, X-rays, galaxies.

Frege showed that mathematical objects are neither physical nor mental. He labeled them "abstract objects." What did he tell us about abstract objects? Only this: They're neither physical nor mental.

Are there other things besides numbers that aren't mental or physical?

Yes! Sonatas. Prices. Eviction notices. Declarations of war.

Not mental or physical, but not abstract either!

The U.S. Supreme Court exists. It can condemn you to death!

Is the Court physical? If the Court building were blown up and the justices moved to the Pentagon, the Court would go on. Is it mental? If all nine justices expired in a suicide cult, they'd be replaced. The Court would go on.

The Court isn't the stones of its building, nor is it anyone's minds and bodies. Physical and mental embodiment are necessary to it, but they're not *it. It's a social institution.* Mental and physical categories are insufficient to understand it. It's comprehensible only in the context of American society.

What matters to people nowadays?

Marriage, divorce, child care.
Advertising and shopping.
Jobs, salaries, money.
The news, and other television entertainment.
War and peace.

All these entities have mental and physical aspects, but none is a mental or a physical entity. Every one is a social entity.

Social reality distinct from physical and mental reality was explained by Émile Durkheim a century ago. These quotations are taken from an essay by L. White.

"Collective ways of acting and thinking have a reality outside the individuals who, at every moment of time, conform to it. These ways of thinking and acting exist in their own right. The individual finds them already formed, and he cannot act as if they did not exist or were different from how they are. . . . Of course, the individual plays a role in their genesis. But for a social fact to exist, several individuals, at the very least, must have contributed their action; and it is this combined action which has created a new product. Since this synthesis takes place outside each one of us (for a plurality of consciousness enters into it), its necessary effect is to fix, to institute outside us, certain ways of acting and certain judgments which do not depend on each particular will taken separately" (1938, p. 56).

"There are two classes of states of consciousness that differ from each other in origin and nature, and in the end toward which they aim. One class merely expresses our organisms and the object to which they are most directly related. Strictly individual, the states of consciousness of this class connect us only with ourselves, and we can no more detach them from us than we can detach ourselves from our bodies. The states of consciousness of the other class, on the contrary, come to us from society; they transfer society into us and connect us with something that surpasses us. Being collective, they are impersonal; they turn us toward ends that we hold in common with other men; it is through them and them alone that we can communicate with others. . . . In brief, this duality corresponds to the double existence that we lead concurrently: the one purely individual and rooted in our organism, the other social and nothing but an extension of society" (1964, p. 337).

Concepts have their own life, said Durkheim. "When once born they obey laws all their own. They attract each other, repel each other, unite, divide themselves and multiply" (1976, p. 424).

Mathematics consists of concepts. Not pencil or chalk marks, not physical triangles or physical sets, but concepts, which may be suggested or represented by physical objects.

In reviewing *The Mathematical Experience*, the mathematical expositor and journalist Martin Gardner made this objection: When two dinosaurs wandered to the water hole in the Jurassic era and met another pair of dinosaurs happily sloshing, there were four dinosaurs at the water hole, even though no human was present to think, "2 + 2 = 4." This shows, says Gardner, that 2 + 2 really is 4 in reality, not just in some cultural consciousness. 2 + 2 = 4 is a law of nature, he says, independent of human thought.

To untangle this knot, we must see that "2" plays two linguistic roles. Sometimes it's an adjective; sometimes it's a noun.

In "two dinosaurs," "two" is a *collective adjective.* "Two dinosaurs plus two dinosaurs equals four dinosaurs" is telling about dinosaurs. If I say "Two discrete, reasonably permanent, noninteracting objects collected with two others makes four such objects," I'm telling part of what's meant by discrete, reasonably permanent noninteracting objects. That is a statement in elementary physics.

John Stuart Mill pointed out that with regard to discrete, reasonably permanent non-interacting objects, experience tells us

$$2 + 2 = 4.$$

In contrast, "Two is prime but four is composite" is a statement about the pure numbers of elementary arithmetic. Now "two" and "four" are ***nouns***, not adjectives. They stand for pure numbers, which are concepts and objects. They are *conceptual objects*, shared by everyone who knows elementary arithmetic, described by familiar axioms and theorems.

The collective adjectives or "counting numbers" are finite. There's a limit to how high anyone will ever count. Yet there isn't any last counting number. If you counted up to, say, a billion, then you could count to a billion and one. In pure arithmetic, these two properties—finiteness, and not having a last—are contradictory. This shows that the counting numbers aren't the pure numbers.

Consider the pure number $10^{(10^{10})}$. We easily ascertain some of its properties, such as: "The only prime factors of $10^{(10^{10})}$ are 2 and 5." But we can't count that high. In that sense, there's no counting number equal to $10^{(10^{10})}$.

Körner made the same distinction, using uppercase for Counting Numbers (adjectives) and lowercase for "pure" natural numbers (nouns). Jacob Klein wrote that a related distinction was made by the Greeks, using their words "arithmos" and "logistiké."

So "two" and "four" have double meanings: as Counting Numbers or as pure numbers. The formula

$$2 + 2 = 4$$

has a double meaning. It's about counting—about how discrete, reasonably permanent, noninteracting objects behave. And it's a theorem in pure arithmetic (Peano arithmetic if you like). This linguistic ambiguity blurs the difference between Counting Numbers and pure natural numbers. But it's convenient. It's comparable to the ambiguity of nonmathematical words, such as "art" or "America."

The pure numbers rise out of the Counting Numbers. In a process related to Aristotle's abstraction, they disconnect from "real" objects, to exist as shared concepts in the mind/brains of people who know elementary arithmetic. In that realm of shared concepts, $2 + 2 = 4$ is a different fact, with a different meaning. And we can now show that it follows logically from other shared concepts, which we usually call axioms.

Platonist philosophy masks this social mode of existence with a myth of "abstract concepts."

From living experience we know two facts:

Fact 1: Mathematical objects are created by humans. Not arbitrarily, but from activity with existing mathematical objects, and from the needs of science and daily life.

Fact 2: Once created, mathematical objects can have properties that are difficult for us to discover. This is just saying there are mathematical problems which are difficult to solve. Example: Define x as the 200th digit in the decimal expansion of $23^{(45^{6789})}$. x is thereby determined. Yet I have no effective way to find it.

These two facts aren't theses waiting to be established! They're experiences needing to be understood. We need to "unpack" their philosophical consequences and their paradoxes.

Once created and communicated, mathematical objects are *there*. They detach from their originator and become part of human culture. We learn of them as external objects, with known properties and unknown properties. Of the unknown properties, there are some we are able to discover. Some we can't discover, even though they are our own creations. Does this sound paradoxical? If so, it's because of thinking that recognizes only two realities: the individual subject (the isolated interior life), and the exterior physical world. The existence of mathematics shows the inadequacy of those two categories. The customs, traditions, and institutions of our society are real, yet they are neither in the private inner nor the nonhuman outer world. They're a different reality, a social-cultural-historical reality. Mathematics is that third kind of reality—"inner" with respect to society at large, "outer" with respect to you or me individually.

To say mathematical objects are invented or created by humans makes them different from natural objects—rocks, X-rays, dinosaurs. Some philosophers (Stephen Körner, Hilary Putnam) argue that the subject matter of pure mathematics is the physical world—not its actualities, but its potentialities. "To exist in mathematics," they think, means "to exist potentially in the physical world." This interpretation is attractive, because it lets mathematics be meaningful. But it's unacceptable, because it tries to explain the clear by the obscure.

Consider this famous theorem of Georg Cantor: "If C is the set of points on the real line, and P is the set of all subsets of C, then it's impossible to put the points of C into 1–1 correspondence with the subsets of C—the elements of P." P can be regarded as the set of all functions of a real variable taking on the values 0 or 1. Nearly all these functions are nowhere continuous and nowhere measurable. We have no way to interpret them as physical possibilities.

The common sense of the working mathematician says this theorem is just a theorem of pure mathematics, not part of any physical interpretation. It's a human idea, recently invented. It wasn't timelessly or tenselessly existing, either as a Platonic idea or as a latent physical potentiality.

Why do these objects, our own creations, so often become useful in describing nature? To answer this in detail is a major task for the history of mathematics, and for a psychology of mathematical cognition that may be coming to birth in Piaget and Vygotsky. To answer it in general, however, is easy. Mathematics is part of human culture and history, which are rooted in our biological nature and our physical and biological surroundings. Our mathematical ideas in general match our world for the same reason that our lungs match earth's atmosphere.

Mathematical objects can have well-determined properties because mathematical problems can have well-determined answers. To explain this requires investigation, not speculation. The rough outline is visible to anyone who studies or teaches mathematics. To acquire the idea of counting, we handle coins or beans or pebbles. To acquire the idea of an angle, we draw lines that cross. In higher grades, mental pictures or simple calculations are *reified* (term of Anna Sfard) and become concrete bases for higher concepts. These shared activities—first physical manipulations, then paper and pencil calculations—have a common product—shared concepts.

Not everyone achieves the desired result. The student who doesn't catch on doesn't pass the course.Why can we converse about polynomials? We've been trained to, by a training evolved for that purpose. We do it without a definition of "polynomial." Even without a definition, polynomial is a shared notion of middle-school students and teachers. And polynomials are objective: They have certain properties, whether we know them or not. These are implicit in our common notion, "polynomial."

To unravel in detail how we attain this common, objective notion is a deep problem, comparable to the problem of language acquisition. No one understands clearly how children acquire rules of English or Navajo, which they follow without being able to state them. These implicit rules don't grow spontaneously in the brain. They come from the shared language-use of the community of speakers. The properties of mathematical objects, like the properties of English sentences, are properties of shared ideas.

The observable reality of mathematics is this: an evolving network of shared ideas with objective properties. These properties may be ascertained by many kinds of reasoning and argument. These valid reasonings are called "proofs." They differ from one epoch to another, and from one branch of mathematics to another.

Looking at this fact of experience, we find questions. How are mathematical objects invented? What's the interplay of mathematics with the ideas and needs of science? How does proof become refined as errors are uncovered? Does the network of mathematical reasoning have an integrity stronger than any link, so that the fracture of any link affects only the closest parts?

These questions can be studied by historians of mathematics. Thomas Kuhn showed the insight that the history of science can give to the philosophy of science. Such work is beginning in the history and philosophy of mathematics.

Generally speaking, before an answer is interesting or even makes sense, there has to be a question. This trivial remark applies to mathematics as well as to anything else. Mathematical statements, mathematical theorems, are answers to questions. Modern mathematics has been sarcastically described as "answers to questions that nobody asked." This is unfair. Most likely the mathematician who found the answer did first ask the question. And very likely he'll publish the answer without mentioning the question. To an unwary reader it can then look like a self-subsisting, self-justifying piece of information, a question-less answer.

The mystery of how mathematics grows is in part caused by looking at mathematics as answers without questions. That mistake is made only by people who have had no contact with mathematical life. It's the questions that drive mathematics. Solving problems and making up new ones is the essence of mathematical life. If mathematics is conceived apart from mathematical life, of course it seems—dead.

To learn how mathematics grows, study how mathematical problems are recognized, how they're attractive. It has to be both something somebody would *like* to do and something somebody might be *able* to do.

An adequate description of today's mathematics (or any other period's) has to include some problems that are considered interesting. That's one reason a formal axiomatic description is incomplete and misleading.

This is recognized by Kitcher in his *Nature of Mathematical Knowledge.* It's implicit in Lakatos's *Proofs and Refutations.* It's fatally absent in Frege, Russell, and their epigones.

Psychological and historical studies won't make mathematical truth indubitable. But why expect mathematical truth to be indubitable? Correcting errors by confronting them with experience is the essence of science. What's needed is explication of what mathematicians do—as part of general human culture, as well as in mathematical terms. The result will be a description of mathematics that mathematicians recognize—the kind of truth that's obvious once said.

Certain kinds of ideas (concepts, notions, conceptions, and so forth) have science-like quality. They have the rigidity, the reproducibility, of physical science. They yield reproducible results, independent of particular investigators. Such kinds of ideas are important enough to have a name.

Study of the lawful, predictable parts of the physical world has a name: "physics." Study of the lawful, predictable, parts of the social-conceptual world also has a name: "mathematics."

A world of ideas exists, created by human beings, existing in their shared consciousness. These ideas have objective properties, in the same sense that material objects have objective properties. The construction of proof and counterexample is the method of discovering the properties of these ideas. This branch of knowledge is called mathematics.

An Objection

There's a logical difficulty we have to look at.

I say the 3-cube or the 4-cube—any mathematical object you like—exists at the social-cultural-historic level, in the shared consciousness of people (including retrievable stored consciousness in writing). In an oversimplified formulation, "mathematical objects are a kind of shared thought or idea."

A mathematical 3-cube is just an idea we share.

This statement is open to an objection. If you turn it around, as by ordinary logic it seems you have a right to do, you get "A certain idea we share is a mathematical 3-cube."

That is, an idea has volume, and vertices, edges, and faces—all of which is nonsense. Probe my mind-brain anyway you like; you won't find inside it a cube or a hyper-cube.

What are we trying to say?

Things become clear if we turn to familiar material objects. We have an idea of a chair, but our idea of a chair isn't a chair. It's our mind-brain's representation of a chair, analogous to a photograph of a chair or to the definition of "chair" in Webster. We know little about the construction or functioning of ideas in the mind-brain. But there's no logical confusion between a chair and the idea of a chair.

Between a 4-cube and the idea of such, there is a confusion. Why? Because we have nowhere to point, to show a "real" 4-cube as distinct from the idea of a 4-cube.

There are two ways to go from here. One well-worn path is the Platonist way. "There *is* a real 4-cube. It's a transcendental immaterial inhuman abstraction. Our idea of a cube is a representation of this transcendental thing, parallel to our idea of chair being a representation of real chairs."

The other way is fictionalism. There is no more a "real" 4-cube than a "real" Mickey Mouse. Oedipus and Mickey Mouse exemplify shared ideas that don't represent anything real. They show that there can be representation without a represented.

Our mental picture of a 4-cube is only a picture, not a 4-cube. It doesn't have vertices or edges, but it does have representations of vertices and edges. It's different from a 4-cube, because it does exist (on the social-cultural-historic level) while the 4-cube, itself doesn't exist. Or, as I prefer to say, it exists only in its social and mental representations.

A 4-cube has 16 vertices. At each vertex, 4 edges meet at right angles. But there is no 4-cube! So *nothing* has 16 vertices at which 4 edges meet at right angles—except as we have a shared idea of such a thing, an idea so consistent, rigid, and reliable that we share each other's reasonings, and come to the same conclusions.

This may sound paradoxical. It's an honest account of the actual state of affairs.

It's a Futile Question

Some questions, which at first seem meaningful, are *futile*—to answer them is neither possible nor necessary.

Why are there rigid, reproducible concepts such as number or circle?

Why is there consciousness?

Why is there a cosmos?

We need not answer Kant's question, "How is mathematics possible?" any more than we need answer Heidegger's question, "Why should anything exist?"

I haven't heard about progress on either problem.

People who think up such questions may get compliments for asking amusing questions. But no physicist and few philosophers feel obliged to answer Heidegger's question. The existence of a world is the starting point from which we go forward.

Once upon a time an important question was, "How can the world be so simple, complicated, and beautiful unless Someone made it?" Now many would say that's a futile question.

Some of today's questions about cosmology, ethics, determinism, or cognition may be futile.

Kant answered his question, "How is mathematics possible?" If not because of the existence of external mathematical objects, then, he thought, our minds ("intuitions") must impose arithmetic and geometry universally.

Ethnology, comparative history, developmental psychology, the development of non-Euclidean geometry, and general relativity, all show that Euclidean geometry is not built into everyone's mind/brain. We think about space in more than one way. We reject Kant's answer. Must we still accept his question?

I counter Kant's question with a counter-question: "Why should your question have an answer?"

This much is clear: Mathematics *is* possible. It's the old saying, "What *is* happening *can* happen."

How does mathematics come about, in a daily, down-to-earth sense? That question belongs to psychology, to the history of thought, and to other disciplines of empirical science. It can't be answered by philosophy. Accept the *possibility* of mathematics as a fact of experience.

Major empirical discoveries about it are coming. Neuro-scientists are hunting for the brain structures we use in counting and spatial thinking. George Lakoff, George Johnson, Terry Regier, and others, using work of Antonio Damasio, Gerald Edelman, and others, may be approaching that goal.

When such discoveries come they'll have tremendous importance, both scientific and practical. But they won't decide philosophical controversies.

To see why not, consider a comparable question. Is what our eyes see really there? That is, is matter an illusion, as many brilliant idealists have said? Or, as Kant taught, is it impossible for us to know whether it's an illusion?

These questions have been of the highest concern to great philosophers.

Today, we realize that those philosophers had limited understanding of the workings of the eye and brain. We do know something about those workings. Maybe some day we'll understand them completely, for practical purposes. Would that understanding tell us whether the visible is real? No. Idealists and skeptics could find new distinctions, and go on being idealists or skeptics as long as they wished.

The reasons that apply to visual reality apply to mathematical reality. The philosophical issues around it will be influenced by empirical discovery, but not settled.

We can study how mathematics develops, in history, in society, and in the individual. We can study how mathematical theories give rise to one another. We can study how mathematics springs from and goes back to physics and other sciences. But the question, "How is mathematics possible?" tries to push mathematics into a pigeonhole: physical, mental, transcendental. None fits. I reject the question and its old alternatives.

Since Dedekind and Frege in the 1870s and 1880s, philosophy of mathematics has been stuck on a single problem—find a solid foundation to which all mathematics can be reduced, a ***foundation*** to make mathematics indubitable, free of uncertainty, free of any possible contradiction (see section in Chapter 8).

That goal is now admitted to be unattainable. Yet, with the exception of a few mavericks, philosophers continue to see "foundation" as the main interesting problem in philosophy of mathematics.

The key assumption in all three foundationist viewpoints is mathematics as a source of indubitable truth. Yet daily experience finds mathematical truth to be fallible and corrigible, like other kinds of truth.

None of the three can account for the existence of its rivals. If Platonism is right, the existence of formalism and constructivism is incomprehensible. If constructivism is right, the existence of Platonism and formalism is incomprehensible. If formalism is right, the existence of Platonism and constructivism is incomprehensible.

Humanism sees that constructivism, formalism, and Platonism each fetishizes one aspect of mathematics, insists that one limited aspect *is* mathematics.

This account of mathematics looks at what mathematicians do. The novelty is conscious effort to avoid falsifying or idealizing.

If we give up the obligation of mathematics to be a source of indubitable truths, we can accept it as a human activity. We give up age-old hopes, but gain a clearer idea of what we are doing, and why.

1. Mathematics is human. It's part of and fits into human culture.
2. Mathematical knowledge isn't infallible. Like science, mathematics can advance by making mistakes, correcting and recorrecting them. (This fallibilism is brilliantly argued in Lakatos's Proofs and Refutations.)
3. There are different versions of proof or rigor, depending on time, place, and other things. The use of computers in proofs is a nontraditional rigor. Empirical evidence, numerical experimentation, probabilistic proof all help us decide what to believe in mathematics. Aristotelian logic isn't always the only way to decide.
4. Mathematical objects are a distinct variety of social-historic objects. They're a special part of culture. Literature, religion, and banking are also special parts of culture. Each is radically different from the others.

Music is an instructive example. It isn't a biological or physical entity. Yet it can't exist apart from some biological or physical realization—a tune in your head, a page of sheet music, a high C produced by a soprano, a recording, or a radio broadcast. Music exists by some biological or physical manifestation, but it makes sense only as a mental and cultural entity.

What confusion would exist if philosophers could conceive only two possibilities for music—either a thought in the mind of an Ideal Musician, or a noise like the roar of a vacuum cleaner.

I have two concluding points.

Point 1 is that mathematics is a social-historic reality. This is not controversial. All that Platonists, formalists, intuitionists, and others can say against it is that it's irrelevant to their concept of philosophy.

Point 2 *is* controversial: There's no need to look for a hidden meaning or definition of mathematics beyond its social-historic-cultural meaning. Social-historic is all it needs to be. Forget foundations, forget immaterial, inhuman "reality."

two

Criteria for a Philosophy of Mathematics

Taking the Test

One way to test a philosophy of mathematics is to confront it with test questions:

What makes mathematics different?

What is mathematics about?

Why does mathematics achieve near-universal consensus?

How do we acquire knowledge of mathematics, apart from proof?

Why are mathematical results independent of time, place, race, nationality, and gender, in spite of the social nature of mathematics?

Does the infinite exist? How?

The social-historical approach gives better answers than the neo-Fregean, the intuitionist-constructivist, or any other proposed philosophy I know of.

Regarding the infinite, if mathematical objects are like stories we make up, we can make up fantastic weird stories if we want to, just so they fit together with other stories. N, an infinite set, is no harder to accept than $10^{(10^{10})}$. Both are socially validated inventions, real, not as physical objects, but as other socially validated inventions are real. Though they are our inventions, their properties are not arbitrary. They're forced to be what they are, by the purposes for which we invented them.

Guiding Principles

Evaluate a body of thought according to its own goals and presuppositions. Understand it historically, in the sense of history of ideas. Pay attention to its consequences, theoretical and practical. Beneficial consequences don't verify a doctrine. Harmful consequences don't falsify it. But consequences are as important as plausibility, consistency, or explanatory power.

I now list criteria for a philosophy of mathematics. No philosophy, including our own, is satisfactory by all criteria. The list is a vantage point from which to evaluate theories, including our own. We do so in Chapter 13.

1. Breadth

An adequate philosophy of mathematics would be aware at least of *some* active field of mainstream mathematical research (dynamical systems, say, or stochastic processes, or algebraic/differential geometry/topology). It would look at how mathematics is being used somewhere, whether in hydrodynamics, or meteorology, or geophysics. It would notice that theoretical physicists do mathematics differently from either "pure" or "applied" mathematics. It wouldn't ignore computing in mathematics today—real computing with real machines, not just idealized theory of ideal machines. It would be compatible with the history of mathematics and with how people learn mathematics.

Who dares to write philosophy of science without some acquaintance with quantum mechanics and relativity? But you must search long and far to find a philosopher of mathematics who claims a nodding acquaintance with functional analysis or differential topology.

This complaint can be parried by claiming that all mathematics "can be got down" to set theory and arithmetic (quote from Quine in Chapter 9). Such a claim depends on what you mean by "got down." It has a self-serving flavor. The argument, "What I don't know, *ipso facto* doesn't matter" isn't new. Age hasn't made it palatable.

2. Links to Epistemology and Philosophy of Science

Philosophy of mathematics should articulate with epistemology and philosophy of science. But virtually all writers on philosophy of mathematics treat it as an encapsulated entity, isolated, timeless, ahistorical, inhuman, connected to nothing else in the intellectual or material realms. Philosophy of mathematics is routinely done without reference to mind, science, or society. (We'll see exceptions in due course.)

Your view of mathematics should fit your view of physical science. Your view of mathematical knowledge should fit your view of knowledge in general. If you write philosophy of mathematics, you aren't expected simultaneously to write philosophy of science and general epistemology. To write on philosophy of mathematics alone is daunting enough. But to be adequate, it needs a connection with epistemology and with philosophy of science.

3. Valid against Practice

Philosophy of mathematics should be tested against five kinds of mathematical practice: research, application, teaching, history, computing. In all areas of mathematical practice, an essential role is played, one way or another, by something called mathematical intuition. There is an extended discussion of intuition in Chapter 4. Here I want to make clear that an adequate philosophy of mathematics must recognize and deal with mathematical intuition.

The need to check philosophy of mathematics against mathematical *research* doesn't require explication. Many important philosophers of mathematics were mathematical researchers: Pascal, Descartes, Leibniz, d'Alembert, Hilbert, Brouwer, Poincaré, Rényi, and Bishop come to mind.

Applied mathematics isn't illegitimate or marginal. Advances in mathematics for science and technology often are inseparable from advances in pure mathematics. Examples: Newton on universal gravitation and the infinitesimal calculus; Gauss on electromagnetism, astronomy, and geodesy (the last inspired that beautiful pure subject—differential geometry); Poincare on celestial mechanics; and von Neumann on quantum mechanics, fluid dynamics, computer design, numerical analysis, and nuclear explosions.

Not only did the same great mathematicians do both pure and applied mathematics, their pure and applied work often fertilized each other. This was explicit in Gauss and Poincaré. Nearer our time is Norbert Wiener. He was generally known for cybernetics, but his life work was mathematical analysis. His study of infinite-dimensional stochastic processes was guided by their physical interpretation as Brownian motion, illuminated by experiments of the French physicist, Jean Perrin (see Wiener, 1948). The standard mathematical model for Brownian motion is the Wiener process. His stochastic processes were useful in controlling anti-aircraft fire. When he renounced military research, he took up prosthetics for the blind—more applied work!

G. H. Hardy "famously" boasted: "I have never done anything 'useful'. No discovery of mine has made, or is likely to make, directly or indirectly, for good or ill, the least difference to the amenity of the world." Nevertheless, the Hardy-Weinberg law of genetics is better known than his profound contributions to analytic number theory. What's worse, cryptology is making number theory applicable. Hardy's contribution to that pure field may yet be useful.

Twenty years after the war, mathematical purism was revived, influenced by the famous French group "Bourbaki." That period is over. Today it's difficult to find a mathematician who'll say an unkind word about applied math.

Instead, we see the spectacular merger of elementary-particle physics and high-dimensional differential geometry. One famous practitioner, Ed Witten, is a physicist whose physical insight permits him unexpected discoveries in mathematics.

A philosophy of mathematics that ignores applied mathematics, or treats it as an afterthought, is out of date. The relationship between pure and applied mathematics is a central philosophical question. A philosophy of mathematics blind to this challenge is inadequate.

An interesting fact about mathematics is that it's taught and learned. (Even more interesting—sometimes taught but *not* learned!) A credible philosophy of mathematics must accord with the experience of ***teaching and learning*** mathematics. To a formalist or Platonist who presents an inhuman picture of mathematics, I ask, "If this were so, how could anyone learn it?"

We know a little about how mathematics is learned. It's not done by memorizing the times table or Peano's axioms. Mathematics is learned by computing, by solving problems, and by conversing, more than by reading and listening.

An account of the nature of mathematics incompatible with these facts is wrong. Piaget was one of the few writers on philosophy of mathematics to take teaching seriously. Today Thomas Tymoczko and Paul Ernest are doing so.

There is also ***the historical test***. Mathematics has seen enormous changes. Its story reaches from Babylonia to Maya, pre-colonial Africa, India, China, Japan. A philosophy about today's mathematics that leaves inexplicable the mathematics of 500 years ago is inadequate.

André Weil, a leading mathematician of the postwar period and now a historian of mathematics, wrote that there could hardly be two disciplines further apart than history of mathematics and philosophy of mathematics. Perhaps in this statement he was identifying philosophy of mathematics with foundations of mathematics.

One philosophical issue in which history is relevant is the reduction of all mathematics to set theory,** as in foundations textbooks, and as mentioned previously in point 1, breadth.

On the basis of this reduction, philosophers of mathematics generally limit their attention to set theory, logic, and arithmetic.

What does this assumption, that all mathematics is fundamentally set theory, do to Euclid, Archimedes, Newton, Leibniz, and Euler? No one dares to say they were thinking in terms of sets, hundreds of years before the set-theoretic reduction was invented. The only way out (implicit, never explicit) is that their own understanding of what they did must be ignored! We know better than they how to explicate their work!

That claim obscures history, and obscures the present, which is rooted in history.

An adequate philosophy of mathematics ***must*** be compatible with the history of mathematics. It ***should*** be capable of shedding light on that history. Why did the Greeks fail to develop mechanics, along the lines that they developed geometry? Why did mathematics lapse in Italy after Galileo, to leap ahead in England, France, and Germany? Why was non-Euclidean geometry not conceived until the nineteenth century, and then independently rediscovered three times? The philosopher of mathematics who is historically conscious can offer such questions to the historian. But if his philosophy makes these questions invisible, then instead of stimulating the history of mathematics, he stultifies it.

Computing is a major part of mathematical practice. The use of computing machines in mathematical proof is controversial. An adequate philosophy of mathematics should shed some light on this controversy.

Formalist and logicist philosophies, each in its own way, picture mathematics as essentially calculation. For the logicist, mathematical theorems are true tautologically, as logical identities. For the formalist, the undefined terms of mathematics are meaningless, so mathematical theorems are meaningless. Both say the essence of mathematics is proof in the sense of formal logic. Proof as a formal calculation in a formal language according to formal rules.

When these conceptions were developed early in the twentieth century, the possibility of realizing them was fantasy. The most famous attempt to formalize statements and proofs of mathematical theorems was the *Principia Mathematica* of Bertrand Russell and Alfred Whitehead. I'm told that finally on page 180 or so they prove 1 is different from 0.

Applied mathematicians have long used computers as a matter of course. Today computers are used more and more in pure mathematics research. Number theorists and algebraists use them to test and make up conjectures. With a computer you sometimes can finish a proof by a calculation impossible by hand. The four-color conjecture** is a famous example (see "Proof" in Chapter 5).

There are differences of opinion about computers. Clifford Truesdell, a leading authority on continuum mechanics, calls them a menace and an abomination. Some say the proof of the four-color theorem violates the traditional idea of proof, since it requires believing that computers work, which is not a mathematical belief. Such objections are ironic. For decades philosophers said valid mathematical proofs should be checkable by machine. Now, when part of a proof *is* done on a machine, some say, "That's not a proof!"

So far our criteria for a philosophy of mathematics have been external—how the philosophy relates to mathematics. There are criteria within the philosophical doctrine itself: consistency, elegance, economy, simplicity, comprehensibility, and precision.

4. *Elegance*

Elegance is more common in mathematical theories than in philosophical ones. Paul Cohen remarked that no one could accuse philosophy of being beautiful. He had in mind foundationist philosophy. Beauty or elegance is desirable, but not prerequisite. Otherwise, Aristotle and Kant would go out the window.

5. *Economy*

Good old Ockham's razor: Use what you need, nothing more. This principle justifies the set-theoretic reduction of mathematics. It's claimed that set membership suffices to define number systems, spaces, geometric figures, and all operations and operands in mathematics past, present, and future. That being so, there should be nothing in mathematics but sets, since there needn't be. But economy can conflict with other criteria such as comprehensibility and applicability. It may be fun to find minimal generators of a theory, even if they're neither unique nor convenient. Economy is like elegance: desirable, but optional.

6. Comprehensibility

Comprehensibility is valued by readers, not by all writers.

Philosophy students think that among professional philosophers, incomprehensibility gets "Brownie points," and comprehensibility gets demerits.

Unworthy suspicions aside, it's a question of comprehensibility *to whom*. What's impenetrable to you may be crystalline to the Heidegger expert.

This book aims to be easily comprehensible to anyone. If some allusion is obscure, skip it. It's inessential.

7. Precision

Should the philosophy of mathematics be precise? Analytic philosophers sometimes use pseudo-mathematical notation. Call a claim "Claim A" or "Hypothesis B," and obscure a conversation that would have gone better in some natural language, English, for example. If you notice that no other branch of philosophy even hopes for precision, the dispensability of precision in philosophy of mathematics becomes apparent.

Mathematics is precise; philosophy cannot be. Expecting philosophy of mathematics to be a branch of mathematics, with definitions and proofs, is like thinking philosophy of art can be a branch of art, with landscapes and still lives.

Art can be beautiful; philosophy of art cannot be beautiful.

Philosophy of politics cannot win at the ballot. Philosophy of law cannot fill the wallets of lawyers. Philosophical specialties, including the philosophy of mathematics, should be evaluated by philosophical standards, not the standards of the field they critique.

Is philosophy of mathematics, part of mathematics or part of philosophy? It's *about* mathematics, but it's *part* of philosophy. It happens that the creators of foundationist philosophy of mathematics were mathematicians (Hilbert, Brouwer) or mathematically trained (Husserl, Frege, Russell). This training may explain their bias. They sought to turn philosophical problems into mathematical problems, *to make them precise*. This bias was fruitful mathematically. Some of today's mathematical logic descended from the search for mathematical solutions to philosophical problems. But, even though mathematically fruitful, it was philosophically misguided. Today we can turn away from the philosophical failure of the foundationist schools. We can think of philosophy of mathematics, not as a branch of mathematics, but as a philosophical enterprise based on mathematical experience. Give up the illusion of mathematical precision. Aim for insight, enlightenment.

8. Simplicity

Both simplicity and precision are desirable, in science and in philosophy of science, especially in philosophy of mathematics.

But in science, philosophy of science, and philosophy of mathematics, there's another desideratum—truthfulness, faithfulness to the facts, simple honesty.

One wants all three: truthfulness, precision, and simplicity. But one can't usually maximize at once goal A and goal B. If you're not willing to pick one goal and ignore the others (maximum cash flow, for instance—reputation and legality be damned!) then you have to do some balancing or juggling. (Work on cash flow, but don't actually go to jail.)

Precision is easier to achieve in a simple situation than in a complicated one. Some phenomena are inherently imprecise.

Precision in philosophy of mathematics is sought by trying to mathematize it. Axiomatic set theory is a branch of mathematics. If the philosophy of mathematics were no more than a style of doing axiomatic set theory, it would attain mathematical precision. But it's impossible to talk about all interesting aspects of mathematics in the language of axiomatic set theory. And it's not necessary to do so! The notion that philosophy of mathematics is a branch of axiomatic set theory is no divine ordinance or self-enforcing decree. It's just one school, one trend, that hopes to borrow the prestige of mathematics by doing its nonmathematical thing ***more mathematico***. In this it's reminiscent of certain "mathematical" specialties in economics, sociology, and psychology.

Simplicity goes with single-mindedness. Where several factors interact to give a complex result, simplicity can be created by ignoring all factors but one. Different scholars may single out different factors. This kind of simplicity leads to fruitless controversy, like between Red Sox fans and White Sox fans. For example, both formalization and construction are central features of mathematics. But the *philosophies* of formalism and constructivism are long-standing rival schools. It would be more productive to see how formalization and construction interact than to choose one and reject the other.

The notions I advocate are less precise and less simple than familiar philosophies. This permits better faithfulness to experience. Some may think no loss of simplicity or precision is acceptable.

Putting simplicity and precision ahead of truthfulness is treating philosophy as a game, *art pour l'art*, like an exotic branch of algebra. Philosophy can be serious—no less than how and why to live. Being serious means putting truthfulness first. First get it right, then go for precision and simplicity.

My first assumption about mathematics is: It's something people do. An account of mathematics is unacceptable unless it's compatible with what people do, especially what mathematicians do.

9. Consistency

This is highly valued by logicians and logic-minded philosophers. Others downplay it. Ralph Waldo Emerson wrote, "A foolish consistency is the hobgoblin of little minds, adored by little statesmen and philosophers and divines."

Walt Whitman wrote:

> Do I contradict myself?
> Very well then I contradict myself.
> (I am large, I contain multitudes.)

Bourbaki wrote:

> Historically speaking, it is of course quite untrue that mathematics is free from contradiction. Non-contradiction appears as a goal to be achieved, not as a God-given quality that has been granted us once for all. There is no sharply drawn line between the contradictions which occur in the daily work of every mathematician, beginner or master of his craft, as the result of more or less easily detected mistakes, and the major paradoxes which provide food for logical thought for decades and sometimes centuries. (N. Bourbaki, 1949)

In a first course in logic, the teacher shows how, from any proposition A together with its negation you can deduce any other proposition B.**

From *any* contradiction, *all* propositions (and their negations) follow! Everything's both true and false! The theory collapses in ruins!

Outside of logic class, it isn't that way. The U.S. Constitution contains contradictions. Any courtroom prosecution or defense probably contains contradictions.

Quantum electrodynamics gives the most precise predictions of any physical theory. Yet physicists have known from its birth that it's self-contradictory. They make *ad hoc* rules for handling the inconsistency. Divergent series of divergent terms are manipulated and massaged. In a *Festschrift* for the famous physicist John Wheeler I found this praise: "He's never stopped by a formal contradiction" (i. e., a mathematical contradiction).

Players in this game know contradiction is seldom fatal. Richard Rorty quotes Aquinas: "Coherence is a matter of avoiding contradictions, and St. Thomas' advice, 'When you meet a contradiction, make a distinction,' makes that pretty easy. As far as I could see, philosophical talent was largely a matter of proliferating as many distinctions as were needed to wriggle out of a dialectical corner."

If A is B and also not B, make a distinction, A_1 and A_2. Example: "Mathematics is precise and imprecise." Whoops, contradiction! What to do? Distinguish between formal and informal mathematics. Formal is precise, informal is imprecise. Contradiction gone!

That contradiction is even useful! It calls attention to an interesting distinction.

Classical logic says all the consequences of a set of axioms exist—are derived—instantly, as soon as the axiom set is laid down. There are infinitely many consequences. This whole infinite theory, created instantly! Consistency holds or it's violated immediately, instantly.

In practice, consequences are derived step by step. At any time only a finite number have been derived. If a contradiction appears, some device is brought in to wall it off, to keep the rest of the theory from infection.

I once wrote that mathematicians hate contradiction. That's not accurate. We love it—like a duck hunter loves ducks. Nothing draws us to the chase like a contradiction in a famous theory.

In evaluating a mathematical theory, consistency is important. But it's less important than fruitfulness (inside and outside of mathematics), imaginative appeal, and linking new mathematical devices to old, respected problems. A contradiction can generally be fixed up one way or another.

As Bourbaki explained, "freedom from contradiction is attained in the process, not guaranteed in advance." They didn't notice that this fact discredits all standard foundationist or logicist theories about mathematics. In practice, we can't always prove in advance the consistency of all possible deductions. Instead, we develop a technique for preserving partial consistency—absence of contradiction up to the latest set of results. In that way we continue to forestall contradiction each time it raises its ugly head.

Frege announced that his building had collapsed. Then after its collapse he tried to patch it up, just as an ordinary, nonfoundationist mathematician would do. Mathematical buildings collapse—lose interest, are forgotten—not because of contradictions, but because their questions are no longer interesting, or because another theory answers them better.

10. Originality, Novelty

This quality is external; it relates to other philosophers and philosophies. It's rare for philosophical writing to be entirely novel. The basic ideas are in Plato and Aristotle. Nothing here is without some antecedent. Presentation, examples, arrangement, and flavor are mine. I hope to be provocative, even convincing to some, but not to all.

11. Certitude and Indubitability

Today the goals of certitude and indubitability are abandoned. The foundationist project has lost its philosophical rationale. "Foundations" (axiomatic set theory and related topics) is just one mathematical specialty of many. It compares axiomatic setups with respect to convenience, effectiveness, or strength. It doesn't pretend to give certainty or indubitability, the founding motive for "foundations."

An adequate philosophy of mathematics must account for the special role of proof in mathematics. Mark Kac, a famous probabilist, asked why mathematicians are obsessed with a need to prove everything. Many people, physicists among them, are willing to believe without proof. Yet mathematicians want proof. They even say, "Without a proof it's nothing." Is this the very nature of mathematics? Or is it one aspect among several of equal importance?

12. *Applicability*

This does not refer to mathematical applications, but to philosophical ones. Your philosophy may increase your feeling of being at home in the universe, or your ability to sleep with a clear conscience. But it should also be helpful in analyzing philosophical problems, perhaps even in solving one or two. If it's useless, who needs it? Chapters 4 and 5 below offer applications of humanist philosophy of mathematics.

13. Acceptability

This criterion is never explicitly demanded. Yet in practice it's the most important. It's why Cartesians prospered while Spinozists were damned (Chapters 6 and 7).

Mathematical theories "ahead of their time" have been ignored for decades, even centuries (Desargues, Grassmann.) In every field of learning, theory can't prosper if it too grossly violates current acceptability. In Europe until the seventeenth century, acceptability was conformity to the Holy Roman Catholic Church. In the Soviet Union, it was Marxism-Leninism-Stalinism, in genetics, linguistics, literature, and music. Here and now, no philosophy penetrates ***philosophia academica*** without bowing to that establishment's sine qua non.

The acceptability of this book—to mathematicians, philosophers, and the general public—remains to be seen.

Summing Up

Not all of these thirteen criteria are essential.

The first three are essential:

1. Recognize the scope and variety of mathematics.
2. Fit into general epistemology and philosophy of science.
3. Be compatible with mathematical practice—research, application, teaching, history, calculation, and mathematical intuition.

The next five are desirable:

4. Elegance
5. Economy
6. Comprehensibility
7. Precision
8. Simplicity

The next, 9. Consistency, is essential, but not as hard to attain as 1–3.

We reject 10. Novelty, Originality, as inessential and unattainable.

We reject 11. Certainty, Indubitability, as false and misleading.

We recognize 12. Applicability, as essential in practice. If you can't do anything with it, what good is it?

13. Acceptability can't be a goal, yet it can't be evaded.

Notice that in this list moral, ethical, and political considerations are excluded. I don't ask whether a theory is beneficial or harmful, progressive or reactionary, humane or inhumane. The preceding thirteen criteria suffice.

We return to these criteria in Chapter 13, and consider there the linkage between philosophical opinions and political opinions.

three

Myths/Mistakes/ Misunderstandings

This chapter is an impressionistic report of how mathematics looks to mathematicians today. I emphasize, but do not exaggerate, aspects ignored in conventional accounts. This picture of mathematical life is evidence that a humanist or social-historical account is truer to real life than traditional foundationist or neo-Fregean accounts.

Mathematics Has a Front and a Back

In *The Presentation of Self in Everyday Life* by the U.S. sociologist Erving Goffman, there's a chapter called "Regions and Region Behavior." There Goffman develops the concepts of "front" and "back." In a restaurant the dining area is the front; the kitchen is the back. In a theater, front stage is for the audience; backstage is for actors and stagehands. In front, actors (waiters) wear costumes (uniforms); in back, they change clothes or wear casual dress. In front the public is served; in back, professionals prepare to serve them. Front is where the public is admitted; back is where it's excluded.

Goffman's contribution is extending "front" and "back" from restaurants and theaters to all our institutions. In the university, classrooms and libraries are the front, where the public (students) is served. The chairman's and dean's offices are the back, where the products (classes and courses) are prepared.

This separation is a necessity. Goffman gives an example of distress from mixing front and back: a gas station where customers walk into the toolroom and help themselves to wrenches.

He quotes George Orwell: "It is an instructive sight to see a waiter going into a hotel dining room. As he passes the door a sudden change comes over him. The set of his shoulders alters; all the dirt and hurry and irritation have dropped off in an instant. He glides over the carpet, with a solemn, priest-like air . . . he entered the dining room and sailed across it, dish in hand, graceful as a swan" (p. 121).

The purpose of separating front from back isn't only to keep customers from interfering with the cooking. It's also to keep them from knowing too much about the cooking.

Everybody down front knows the leading lady wears powder and blusher. They don't know exactly how she looks without them.

Diners know what's supposed to go into the ragout. They don't know for sure what *does* go into it.

Traditional philosophy recognizes only the front of mathematics. But it's impossible to understand the front while ignoring the back.

The front and back of mathematics aren't physical locations like dining room and kitchen. They're its public and private aspects. The front is open to outsiders; the back is restricted to insiders. The front is mathematics in finished form—lectures, textbooks, journals. The back is mathematics among working mathematicians, told in offices or at cafe tables.

The front divides into subregions—first, second, third class. A restaurant may include banquet hall and snack bar. Theaters can have box, orchestra, and balcony. The mathematical public includes professionals, graduate students, and undergraduates.

The back also divides into subregions. In a restaurant, there are domains of salad chef, pastry chef, and dishwasher. Mathematicians divide into finite group theorists, numerical linear algebraists, nonstandard differential topologists, and so on and on.

Front mathematics is formal, precise, ordered, and abstract. It's broken into definitions, theorems, and remarks. Every question either is answered or is labeled: "open question." At the beginning of each chapter, a goal is stated. At the end of the chapter, it's attained.

Mathematics in back is fragmentary, informal, intuitive, tentative. We try this or that. We say "maybe," or "it looks like."

This distinction is described by George Pólya in his preface to *Mathematics and Plausible Reasoning*. "Finished mathematics presented in a finished form appears as purely demonstrative, consisting of proofs only. Yet mathematics in the making resembles any other human knowledge in the making. You have to guess a mathematical theorem before you prove it; you have to have the idea of the proof before you carry through the details. You have to combine observations and follow analogies; you have to try and try again."

Mainstream philosophy doesn't know that mathematics has a back. Finished, published mathematics—the front—is taken as a self-subsistent entity. Worse, mainstream philosophy doesn't look at the real front—real articles and treatises. It contemplates idealized texts as logicians would have them, not mathematical texts as they are and must be.

It would make as much sense for a restaurant critic not to know there are kitchens, or a theater critic not to know there is backstage.

Myths

The front/back separation makes possible a myth about the seasoning of the ragout or about the leading lady's complexion. This kind of myth takes the performance in front at face value, innocently unaware that it was concocted behind the scenes.

A myth isn't bad *per se*. It has allegorical or metaphorical power. It may increase the customer's enjoyment. It may be essential for the performance.

The myth of the divine right of kings was useful. So are the myths of Christmas, Easter, and those of other religions.

There's an unwritten criterion separating the professional from the amateur, the insider from the outsider: The outsider is taken in by the myths. The insider is not.

I will describe a few myths of mathematics. I limit myself to a few provocative comments. To present them in detail and refute them would fill a volume. The reader can dig deeper or extend the list, with help from the references.

The myth of Euclid: "Euclid's *Elements* contains truths about the universe which are clear and indubitable." Today advanced students of geometry know Euclid's proofs are incomplete and his axioms are unintelligible. Nevertheless, in watered-down versions that ignore his impressive solid geometry, Euclid's *Elements* is still upheld as a model of rigorous proof.

The plaster Newton fabricated in the eighteenth century is intact. Alexander Pope offered an epitaph:

> Nature and Nature's Laws lay hid in Night;
> GOD said, *Let Newton be!* and all was Light.

The strange, complex historical Newton is almost unknown, even among the mathematically literate.

The myths of Russell, Brouwer, and Bourbaki—of logicism, intuitionism, and formalism—are treated in *The Mathematical Experience*. I return to them in Part 2, in articles on foundationism, the foundationist philosophers, and Lakatos, their critic.

Now four more general myths.

1. *Unity*: There's only one mathematics, indivisible now and forever. Mathematics is a single inseparable whole.
2. *Universality*: The mathematics we know is the only mathematics there can be. If little green critters from Quasar X9 showed us their textbooks, we'd find again $A = \pi r^2$.
3. *Certainty*: Mathematics has a method, "rigorous proof," which yields absolutely certain conclusions, given truth of premises.

4. *Objectivity*: Mathematical truth is the same for everyone. It doesn't matter who discovers it. It's true whether or not anybody discovers it.

It would be easy to give citations to show that these four myths are generally accepted. Documentation would be pedantic.

Being a myth doesn't entail its truth or falsity. Myths validate and support institutions; their truth may not be determinable. Who can say the divine right of kings is false? Without a channel to the mind of God, this can't be proved or disproved. In its day it was credible and useful.

Myths 1–4 are almost universally accepted, but they aren't self-evident or self-proving. It's possible to question, doubt, or reject them. Some people do reject them. Standard and official though they are, they aren't taken literally or naively by backstage people.

Part of preparing mathematics for public presentation—in print or in person—is tying up loose ends. If there's disagreement whether a theorem has been proved, it's left out of the text. The standard exposition purges mathematics of the personal, the controversial, the tentative, leaving little trace of humanity in the creator or the consumer. This style is the "front" of mathematics.

Without it, the myths would lose their aura. If mathematics were presented in the style in which it's created, few would believe its universality, unity, certainty, or objectivity. These myths support the institution of mathematics. For mathematics is not only an art and a science, but also an institution, with budgets, administrations, rank, status, awards, and grants.

Let's examine 1–4 from backstage.

Myth 1 is unity. But at meetings of the American Mathematical Society, any contributed talk is understandable to only a small fraction of those present.

Sometimes pure and applied mathematicians interact. Most of the time they're oblivious to each other, working to different audiences, different standards, and different criteria. The purists sometimes even declare applied mathematics isn't mathematics. "Where are the definitions? Where are the theorems?"

The "unity" claimed in principle doesn't exist in practice

Myth 2 is universality. If there's intelligent life on Quasar X9, it may be blobs of plasma we can't recognize as life. What would it mean to talk about their literature, art, or mathematics? To ask if their mathematics is the same as ours requires a possibility of comparing. Comparing demands communication. The possibility of comparing isn't universal. It's conditional on their being enough like us to communicate with us.

Myth 3 is certainty. We're certain

$$2 + 2 = 4,$$

though we don't all mean the same thing by that equation. It's another matter to claim certainty for the theorems of contemporary mathematics. Many of these theorems have proofs that fill dozens of pages. They're usually built on top of other contemporary theorems, whose proofs weren't checked in detail by the mathematician who quotes them. The proofs of these theorems replace boring details with "it is easily seen" and "a standard argument yields" and "a calculation gives." Many papers have several coauthors, no one of whom thoroughly checked the whole paper. They may use machine calculations that none of the authors completely understands.

A mathematician's confidence in some theorem doesn't necessarily means she knows every step from the axioms of set theory up to the theorem she's interested in. It may include confidence in the word of fellow researchers, journals, and referees.

Certainty, like unity, can be claimed in principle—not in practice.

Myth 4 is objectivity. This myth is more plausible than the first three. Yes! There's amazing consensus in mathematics as to what's correct or accepted.

But just as important is what's interesting, important, deep, or elegant. Unlike correctness, these criteria vary from person to person, specialty to specialty, decade to decade. They're no more objective than esthetic judgments in art or music.

Mathematicians want to believe in unity, universality, certainty, and objectivity, as Americans want to believe in the Constitution and free enterprise, or other nations in their Gracious Queen or their Glorious Revolution. But while they believe, they know better.

To become a professional, you must move from front to back. You get a more sophisticated attitude to myth. Backstage, the leading lady washes off powder and blusher. She's seen with her everyday face.

The front-back codependence makes it hopeless to understand the front while ignoring the back. That's what Mainstream philosophy of mathematics tries to do. You can't understand a restaurant meal if you're unaware of the kitchen. Yet you *can* present yourself as a philosopher of mathematics, and be aware only of publications washed and ironed for public consumption.

Of the three historic schools, only intuitionism pays attention to the producer of mathematics. Formalists, logicists, and Platonists sit at a table in the dining room, discussing their ragout as a self-created, autonomous entity.

The Mathematician's Philosophical Dilemma

What is the mathematicians's dilemma?

How did it come about?

Can he escape?

Writers agree: The working mathematician is a Platonist on weekdays, a formalist on weekends. On weekdays, when doing mathematics, he's a Platonist, convinced he's dealing with an objective reality whose properties he's trying to determine. On weekends, if challenged to give a philosophical account of this reality, it's easiest to pretend he doesn't believe in it. He plays formalist, and pretends mathematics is a meaningless game.

I quote two famous mathematicians:

Jean Dieudonné: "We believe in the reality of mathematics but of course when philosophers attack us with their paradoxes we rush to hide behind formalism and say, 'Mathematics is just a combination of meaningless symbols,' and then we bring out Chapters 1 and 2 on set theory. Finally we are left in peace to go back to our mathematics and do it as we have always done, with the feeling each mathematician has that he is working with something real. This sensation is probably an illusion, but is very convenient. That is Bourbaki's attitude toward foundations."

(Bourbaki is a Parisian mathematical clique in which Dieudonné was a captain.)

Paul Cohen: "To the average mathematician who merely wants to know his work is securely based, the most appealing choice is to avoid difficulties by means of Hilbert's program. Here one regards mathematics as a formal game and one is only concerned with the question of consistency. . . . The Realist [Platonist] position is probably the one which most mathematicians would prefer to take. It is not until he becomes aware of some of the difficulties in set theory that he would even begin to question it. If these difficulties particularly upset him, he will rush to the shelter of Formalism, while his normal position will be somewhere between the two, trying to enjoy the best of two worlds."

(Cohen transformed logic and set theory in 1963 when he proved that the continuum hypothesis and the axiom of choice are independent of the other axioms of Zermelo-Frankel.)

This is shocking news! Most mathematicians hold contradictory views on the nature of their work.

Does it matter? Yes. Truth and meaning aren't recondite technical terms. They concern anyone who uses or teaches mathematics. Ignoring them leaves you captive to unexamined philosophical preconceptions. This has practical consequences.

Consequence 1: "What's interesting in mathematics?" is an urgent question for anyone doing research, or hiring or promoting researchers. There's no public discussion of this question. No vehicle for public discussion of it. No language or viewpoint that could be used for such a discussion.

Not to say there should be agreed-on standards of what's interesting. Precisely because tastes differ, we need discussion on taste. We have some common standards. That's proved by our identity as one profession, and our agreement that certain feats in mathematics deserve the highest rewards. Bringing out those standards for analysis and controversy would be important philosophical work. Our inability to sustain public discussion on values betrays philosophical unawareness and incompetence.

Consequence 2: In the middle of the twentieth century formalism was the philosophy of mathematics most advocated in public. (Recall the section introducing formalism in Chapter 1.) In that period, the style in mathematical journals, texts, and treatises was to insist on details of definitions and proofs and tell little or nothing of why a problem is interesting or why a method of proof is used that changed drastically in the last quarter century.

It would be difficult to document a connection between formalism in expository style and formalism in philosophical attitude. Still, ideas have consequences. What I think mathematics *is* affects how I present it.

Consequence 3: The unfortunate importation into primary and secondary schools, during the 1960s, of set-theoretic notation and axiomatics. This wasn't an inexplicable aberration. It was a predictable consequence of a philosophical doctrine: Mathematics *is* axiomatic systems expressed in set-theoretic language.

Critics of formalism in high school say "This is the wrong thing to teach, and the wrong way to teach." Such criticism leaves unchallenged the dogma that real mathematics *is* formal derivations from formally stated axioms. If this dogma rules, the critic of formalism is seen as asking for lower quality, to give students something "watered down" instead of the "real thing."

The fundamental question is, "What is mathematics?" Controversy about high-school teaching can't be resolved without controversy about the nature of mathematics. To discredit formalism in teaching, discredit the formalist picture of the nature of mathematics. The critic of formalism has to give a more satisfactory account of the nature of mathematics.

Mathematicians don't usually discuss philosophical issues. We think somebody else has taken that over—"the professionals." But with few exceptions, philosophers have little to say to us. Nearly all of them have only a remote notion of what we do. This isn't discreditable, in view of the technical prerequisites for understanding our work. But a mathematician who seeks enlightenment in philosophical books and journals is disappointed. Some philosophers are unfamiliar with anything beyond arithmetic and elementary geometry. Others are competent in logic or axiomatic set theory, but are as narrow technical specialists as we are.

There are a few philosophical comments by a few mathematicians whose interests go beyond set theory and logic. But philosophical discourse isn't well developed today among mathematicians. Philosophical issues as much as mathematical ones deserve careful argument, developed analysis, and consideration of objections. A bald statement of one's opinion is insufficient, even in philosophy.

The philosophical questions in the university math curriculum were raised by foundationists 50 years ago. "None of the three foundationisms succeeded," we tell students, "and there's no prospect of a 'solid' foundation. There's only one philosophical problem of interest—the foundation of the real number system—and that's intractable."

We can understand the working mathematician's oscillation between formalism and Platonism, if we look at her work experience and at the philosophical dogmas she inherited—Platonism and formalism. Both dogmas say mathematical truth must possess absolute certainty. Her experience in mathematics, on the other hand, offers plenty of uncertainty.

Emilio D. Roxin described Platonism: "Most mathematicians feel like a hunter in a jungle, where the theorems sit in the trees or fly around, where the definitions are like convenient ladders awaiting to be used in order to trap the theorems and corollaries which are there, even if nobody finds them." Pick a familiar theorem: Cauchy's integral formula, or Cantor's theorem on the uncountability of the continuum. Is it a true statement about the world? Does one discover such a theorem? Does such a discovery increase our knowledge? If our mathematician says yes, he's a Platonist for the moment.

The basis for Platonism is awareness that the problems and concepts of mathematics are independent of him as an individual. The roots of a polynomial are where they are, regardless of what he thinks or knows. It's easy to imagine that this objectivity is outside human consciousness as a whole. The working mathematician's Platonism expresses this aspect of his daily working experience. Yet, I repeat, it's a halfhearted, shamefaced Platonism, because it's incompatible with his general philosophy or worldview.

Platonism in the strong sense—belief in the existence of ideal entities, independent of or prior to human consciousness—was tenable with belief in a Divine Mind. Once mysticism is left behind, once scientific skepticism is focused on it, Platonism is hard to maintain,

The next question is: "To what objects or features of the world do such statements refer?" You don't meet complex integrals or uncountable sets walking down the street or flying in outer space. Outside our thoughts, where are complex integrals or uncountable sets? Perhaps such things aren't real!

The available alternative to Platonism is formalism. Instead of saying theorems are truths about eternal extra-human ideals, it says they're just transformations of symbols. If she retreats to this cautious disclaimer, she's a formalist for the moment. For the moment, she thinks mathematics doesn't mean anything, so she needn't worry about its meaning. "I cannot imagine that I shall ever return to the creed of the true Platonist, who sees the world of the actual infinite spread out before him and believes that he can comprehend the incomprehensible," said the American logician Abraham Robinson in declaring himself a formalist.

But formalism needs its own act of faith. How do we know that our latest theorem about diffusion on manifolds is formally deducible from Zermelo-Frankel set theory? No such formal deduction will ever be written down. If it were, the likelihood of error would be greater than in the usual informal or semi-formal mathematical proof.

Now another question: "How come these examples were known before their axioms were known? If a theorem is only a conclusion from axioms, then do you say Cauchy didn't know Cauchy's integral formula? Cantor didn't know Cantor's theorem?"

Formalism doesn't work! Back to Platonism.

We don't quit mathematics, of course. Just quit thinking about it. Just do it. That's roughly the situation of the mathematician's philosophy of mathematics.

Mathematicians mostly don't want to bother about philosophy. A bee gathers honey without wondering why. A salmon climbs up the river without wondering why. Mathematicians make conjectures and try to prove them, without wondering why. "Just to have fun and make a living," they say. No need to have a clue of what it's all about, any more than a bee or a salmon.

Yet more is possible for us. "The unexamined life isn't worth living," said Socrates. At least, the unexamined life is less worth living than the examined. If a mathematician thinks examining his life is worthy, he must examine the meaning of his obsession—mathematics.

Platonism and formalism, each in its own way, falsify part of daily experience. We talk formalism when compelled to face the mystical, antiscientific essence of Platonic idealism; we fall back to Platonism when we realize that the formalist description of mathematics has only a distant resemblance to our actual knowledge of mathematics. To abandon both, we must abandon absolute certainty, and develop a philosophy faithful to mathematical experience.

Mistakes

How is it possible that mistakes occur in mathematics?

René Descartes's Method was so clear, he said, a mistake could happen only by inadvertence. Yet, as we see in Chapter 5, his *Géométrie* contains conceptual mistakes about three-dimensional space.

Henri Poincaré said it was strange that mistakes happen in mathematics, since mathematics is just sound reasoning, such as anyone in his right mind follows. His explanation was memory lapse—there are only so many things we can keep in mind at once.

Wittgenstein said that mathematics could be characterized as the subject where it's possible to make mistakes. (Actually, it's not just possible, it's inevitable.) The very notion of a mistake presupposes that there is right and wrong independent of what we think, which is what makes mathematics mathematics. We mathematicians make mistakes, even important ones, even in famous papers that have been around for years.

Philip Davis displays an imposing collection of errors, with some famous names. His article shows that mistakes aren't uncommon. It shows that mathematical knowledge is fallible, like other knowledge.

Dreben, Andrews, and Anderaa show that lemmas in Jacques Herbrand's 1929 thesis are false. Herbrand used false lemmas to "prove" a theorem that has been influential for fifty years. Dreben, Andrews, and Anderaa prove his theorem, replacing false lemmas with correct ones. (Rohit Parikh informs me that for years Herbrand's thesis was physically inaccessible. Its errors could have been found sooner in normal circumstances.)

Some mistakes come from keeping old assumptions in a new context.

Infinite dimensional space is just like finite dimensional space—except for one or two properties, which are entirely different.

Kummer tried to prove Fermat's last theorem. He assumed falsely that every integral domain of algebraic numbers has the unique factorization property. To correct his blunder, he founded a new branch of algebraic number theory.

Riemann stated and used what he called "Dirichlet's principle" incorrectly.

Julius Koenig and David Hilbert each thought he had proved the continuum hypothesis. (Decades later, it was proved undecidable by Kurt Gödel and Paul Cohen.)

Sometimes mathematicians try to give a *complete classification* of an object of interest. It's a mistake to claim a complete classification while leaving out several cases. That's what happened, first to Descartes, then to Newton, in their attempts to classify cubic curves (Boyer).

Is a gap in a proof a mistake? Newton found the speed of a falling stone by dividing 0/0. Berkeley called him to account for bad algebra, but admitted Newton had the right answer (see Chapter 6). Mistake or not?

Euler had an art of working with divergent series. His answer was *usually* correct (see Pólya and Putnam in Chapters 9 and 10). Later a rigorous theory of divergent series showed in what sense Euler was right.

"The mistakes of a great mathematician are worth more than the correctness of a mediocrity." I've heard those words more than once. Explicating this thought would tell something about the nature of mathematics. For most academic philosophers of mathematics, this remark has nothing to do with mathematics or the philosophy of mathematics. Mathematics for them is indubitable—rigorous deduction from premises. If you made a mistake, your deduction wasn't rigorous. By definition, then, it wasn't mathematics!

So the brilliant, fruitful mistakes of Newton, Euler, and Riemann, weren't mathematics, and needn't be considered by the philosopher of mathematics.

Riemann's incorrect statement of Dirichlet's principle was corrected, implemented, and flowered into the calculus of variations. On the other hand, thousands of *correct* theorems are published every week. Most lead nowhere.

A famous oversight of Euclid and his students (don't call it a mistake) was neglecting the relation of "between-ness"** of points on a line. This relation was used *implicitly* by Euclid in 300 B.C. It was recognized *explicitly* by Moritz Pasch over 2,000 years later, in 1882. ***For two millennia, mathematicians and philosophers accepted reasoning that they later rejected.***

Can we be sure that we, unlike our predecessors, are not overlooking big gaps? We can't. Our mathematics can't be certain.

Gossip

A friend of mine was interested in some exciting work on "translation representations of linear operators" by our Professor Q. At another school, Professor Z worked on the same problem. When Z was in our neighborhood, my friend talked to him. She confided, "Neither one of them understands the other."

Can this be? Mathematics is straightforward. No secrets, just "If A then B." Yet here are top experts, neither understanding the other, ***neither admitting he didn't understand the other.***

Maybe their use of different methods makes it hard for them to understand each other. When Q meets a difficulty, he turns to his usual device E. He may have trouble forgetting E to think about Z's favorite device F.

In a philosophy journal it's no surprise if author X says author Y doesn't understand him. The charge of not understanding is popular with philosophers. Popper is one of the clearest philosophical writers, yet he said he'd been understood only 2 or 3 times. Russell was Wittgenstein's mentor, and the world champion philosopher of science. But Wittgenstein said Russell didn't understand his *Tractatus.*

Mathematicians are different. You never see mathematicians accusing each other in print of not understanding, even when they ***don't*** understand, and it's apparent that they don't. Mathematics changes. The whole point of view can change in a few years. In the early part of this century mathematics was "set-theorized"—reformulated in Cantor's set theory. In the 1930s, led by Bourbaki, "structure" was the word; examples were buried in the exercises. The algebraization of geometry and topology is a third example.

When these transformations occur, a few outstanding mathematicians lead the revolution. Then more and more fresh PhDs are trained to think the new way. Meanwhile, some of the older generation continue in the old style. Among them are brilliant veterans of the previous revolution. If they don't master the new methods, that says something about mathematics. If it were simply correct reasoning from arbitrary premises, good mathematicians couldn't fail to understand good mathematics.

Solomon Lefschetz was a gray eminence of modern topology. According to Gian-Carlo Rota, Lefschetz's theorems were always correct, his proofs never correct.

My friend W. collaborated with B. (not Lefschetz.). "B is the most amazing mathematician I ever knew," W said. "His theorems are always right, his proofs are always wrong."

Professors D and K are probabilists, creators and shapers of their field. D uses measure-theoretic arguments. K works "analytically" or "classically." D doesn't use K's methods. K doesn't use D's methods. S, a student of both D and K, combined their methods to go further than either had gone separately.

In several branches of mathematics you hear: "Our deepest theorems were found by Professor Z. No one but he ever understood his methods. By now we can get his results with methods anyone can follow. With one or two exceptions, all his formulas and theorems were correct." The same story is told by probabilists, partial differential equators, algebraists, and topologists. The name of the hero changes.

Professor Z's knowledge without proof is inexplicable in the formalist account of mathematics.

At an International Congress of Mathematicians, a famous analyst reports his new results. They're not quite certain, he says, because there hasn't been time for other famous analysts to check them. *Until you check with other people, you don't know if you've overlooked something*, he tells us. A well-known example of this was Wiles's original "proof" of Fermat's last theorem

What about Ramanujan, the brilliant self-taught Indian mathematician? Dozens of amazing results that he did not see the need to prove, or didn't know how to. American mathematicians Ed Witten and Bill Thurston today are honored for fascinating results that, it is said, they left in part for others to prove.

I can't omit this reminiscence of J. J. Sylvester, one of the supreme algebraists of the nineteenth century. "Sylvester's *Methods*! He had none. Statements like the following were not unfrequent in his lectures: 'I haven't proved this, but I am as sure as I can be of anything that it must be so. From this it will follow, etc.' At the next lecture it turned out that what he was so sure of was false. Never mind, he kept on forever guessing and trying, and presently a wonderful discovery followed, then another and another. Afterward he would go back and work it all over again, and surprise us with all sorts of side lights. He then made another leap in the dark, more treasures were discovered, and so on forever" (E. W. Davis, 1890).

Sylvester's partner, Arthur Cayley, had the opposite style. Always careful and responsible, never saying more than he knew.

One of my honored teachers (a world class mathematician) astonished me by saying he didn't read mathematical papers. When something interesting happens, somebody tells him.

The same was reported of David Hilbert, in his day the world champion of mathematics. Didn't read.

Every young math PhD has had this experience: You've done something good. You talk about it. Your seminar or colloquium has an audience of a dozen or two. You feel successful if two or three show they understood.

Facing these ugly facts, the easy choice is hypocrisy. *Pretend you don't notice the gap between preaching and practice.* If you don't choose hypocrisy, you must give up either myth or reality. You can hold onto the myth, by saying mathematics as practiced by mathematicians isn't what it ought to be. You can hold onto reality, by admitting that mathematical proof isn't a mechanical procedure, even "in principle."

Isn't there a middle way? How about, "We're not as careful as we should be, but that doesn't detract from the ideal." Yes, let's try not to make mistakes!

But if you think mathematics would be infallible if we wrote our proofs to be checkable by computer, you're mistaken. As you already know, if you've done much debugging. A proof doesn't become certain by being formalized. To the doubtfulness of the proof are added the doubtfulness of the coding, the programming, the logical design of the machine, and the physical functioning of its components. In real computers and real computations, the correctness of a formal proof (a code, in computer lingo) has to be verified by a human being. Her ***understanding*** of the meaning and purpose of the program permits her to check its formal correctness. Programs don't work until they're checked and corrected by people. And even then, if they're big enough, they're still "buggy." We have no formal definition of "understanding." Nevertheless, understanding is required to verify formal computations, as well as the other way around.

Sometimes mathematicians make mistakes. Sometimes we disagree whether a proof is correct. There isn't absolute certainty in mathematics. There's ***virtual*** certainty, as in other areas of life.

four

Intuition/Proof/Certainty

There's an old joke about a theory so perfectly general it had no possible application. Humanist philosophy is applicable.

From the humanist point of view, how would one investigate such knotty problems of the philosophy of mathematics as mathematical proof, mathematical intuition, mathematical certainty? It would be good to compare the humanist method with methods suggested by a Platonist or formalist philosophy, but I am really not aware how either of those views would lead to a method of philosophical investigation. A humanist sees mathematics as a social-cultural-historic activity. In that case it's clear that one can actually look, go to mathematical life and see how proof and intuition and certainty are seen or not seen there.

What is proof? What should it be? Old and difficult questions.

Rather than, in the time-honored way of philosophers, choose a priori the right definitions and axioms for proof, I attempt to ***look carefully, with an open mind***. What do mathematicians (including myself) do that we call proving? And what do we mean when we talk about proving? Immediately I observe that "proof" has different meanings in mathematical practice and in logico-philosophical analysis. What's worse, this discrepancy is not acknowledged, especially not in teaching or in textbooks. How can this be? What does it mean? How are the two "proofs" related, in theory and in practice?

The articles on intuition and certainty, and half a dozen more in the next chapter, are in the same spirit.

"Call this philosophy? Isn't it sociology?"

The sociologist studying mathematics comes as an outsider—presumably an objective one. I rely, not on the outsider's objectivity, but on my insider's experience and know-how, as well as on what my fellow mathematicians do, say, and write. I watch what's going on and I understand it as a participant.

This is not sociology. Neither is it introspection, that long discredited way to find truth by looking inside your own head. My experience and know-how aren't isolated or solipsistic. They're part of the web of mutual understanding of the mathematical community.

There used to be a kind of sociology called "Verstehen." That was perhaps close to what I've been talking about. But my looking and listening with understanding isn't an end in itself. It's a preparation for analysis, criticism, and connection with the rest of mathematics, philosophy, and science. That's why I call it philosophy.

Proof

The old, colloquial meaning of "prove" is: *Test, try out, determine the true state of affairs* (as in Aberdeen *Proving* Ground, galley *proof*, "the *proof* of the pudding," "the exception that proves the rule," and so forth.) How is the mathematical "prove" related to the old, colloquial "prove"?

We accuse students of the high crime of "not even knowing what a proof is." Yet we, the math teachers, don't know it either, if "know" means give a coherent, factual explanation. (Of course, we know how to "give a proof" in our own specialty.)

The trouble is, "mathematical proof" has two meanings. In practice, it's one thing. In principle, it's another. We *show* students what proof is in practice. We *tell* them what it is in principle. The two meanings aren't identical. That's O.K. But *we never acknowledge the discrepancy*. How can that be O.K.?

Meaning number 1, the *practical* meaning, is informal, imprecise. *Practical mathematical proof is what we do to make each other believe our theorems.* It's argument that convinces the qualified, skeptical expert. It's done in Euclid and in *The International Archive Journal of Absolutely Pure Homology*. But what is it, *exactly*? No one can say.

Meaning number 2, theoretical mathematical proof, is formal. Aristotle helped make it. So did Boole, Peirce, Frege, Russell, Hilbert, and Gödel. It's transformation of certain symbol sequences (formal sentences) according to certain rules of logic (modus ponens, etc). A sequence of steps, each a strict logical deduction, or readily expanded to a strict logical deduction. This is supposed to be a "formalization, idealization, rational reconstruction of the idea of proof" (P. Ernest, private communication).

Problem A: *What does meaning number 1 have to do with meaning number 2?*
Problem B: *How come so few notice Problem A?* Is it uninteresting? Embarrassing?
Problem C: *Does it matter?*

Problem C is easier than A and B. It matters, morally, psychologically, and philosophically.

When you're a student, professors and books claim to prove things. But they don't say what's meant by "prove." You have to catch on. Watch what the professor does, then do the same thing.

Then you become a professor, and pass on the same "know-how" without "knowing what" that your professor taught you.

There is an official, standard viewpoint. It makes two interesting assertions.

Assertion 1. Logicians don't tell mathematicians what to do. They make a theory out of what mathematicians actually do. Logicians supposedly study us the way fluid dynamicists study water waves. Fluid dynamicists don't tell water how to wave. They just make a mathematical model of it.

A fluid dynamicist studying water waves usually does two things. In advance, seek to derive a model from known principles. After the fact, check the behavior of the model against real-world data.

Is there a published analysis of a sample of practical proofs that derives "rigorous" proof as a model from the properties of that sample? Are there case studies of practical proof in comparison with theoretical proof? I haven't heard of them. The claim that theoretical proof models practical proof is an assertion of belief. It rests on intuition. Gut feeling.

You might say there's no need for such testing of the logic model against practice. If a mathematician doesn't follow the rules of logic, her reasoning is simply wrong. I take the opposite standpoint: What mathematicians at large sanction and accept *is* correct mathematics. Their work is the touchstone of mathematical proof, not *vice versa*. However that may be, anyone who takes a look must acknowledge that formal logic has little visible resemblance to what's done day to day in mathematics.

It commonly happens in mathematics that we believe something, even without possessing a complete proof. It also sometimes happens that we *don't* believe, even in the *presence* of complete proof. Edmund Landau, one of the most powerful number theorists, discovered a remarkable fact about analytic functions called the two constants theorem.** He proved it, but couldn't believe it. He hid it in his desk drawer for years, until his intuition was able to accept it (Epstein and Hahn, p. 396n).

There's a famous result of Banach and Tarski,** which very few can believe, though all agree, it has been proved.

Professor Robert Osserman of Stanford University once startled me by reporting that he had heard a mathematician comment, "After all, Riemann never proved anything." Bernhard Riemann, German (1826—1866), is universally admired as one of the greatest mathematicians. His influence is deeply felt to this day in many parts of mathematics. I asked Osserman to explain. He did:

"Clearly, it's not literally true that Riemann never proved anything, but what is true is that his fame derives less from things that he proved than from his other contributions. I would classify those in a number of categories. The first consists of conjectures, the most famous being the Riemann Hypothesis about the zeros of the zeta function. The second is a number of definitions, such as the Riemann integral, the definition of an analytic function in terms of the Cauchv-Riemann equations (rather than via an 'analytic' expression), the Riemann curvature tensor and sectional curvature in Riemannian geometry. The third, related to the second, but somewhat different, consists of new concepts, such as a Riemann surface and a Riemannian manifold, which involve radical rethinking of fundamental notions, as well as the extended complex plane, or Riemann sphere, and the fundamental notions of algebraic curve. The fourth is the construction of important new examples, such as Riemann's complete periodic embedded minimal surface and what we now call hyperbolic space, via the metric now referred to as the 'Poincaré metric', which Riemann wrote down explicitly. And finally, there is the Riemann mapping theorem, whose generality is simply breathtaking, and again required a depth of insight, but whose proof, as given by Riemann, was defective, and was not fully established until many years later. Perhaps one could combine them by saying that Riemann was a deep mathematical thinker whose vision had a profound impact on the future of mathematics."

Many readers have not heard of all these contributions of Riemann. But this example is enough to refute the catch-phrases. "A mathematician is someone who proves theorems" and "Mathematics is nothing without proof."

As a matter of principle there can't be a strict demonstration that Meaning 2, formal proof, is what mathematicians do in practice. This is a universal limitation of mathematical models. In principle it's impossible to give a mathematical proof that a mathematical model is faithful to reality. All you can do is test the model against experience.

The second official assertion about proof is:

Assertion 2. Any correct practical proof can be filled in to be a correct theoretical proof.

"If you can do it, then do it!"

"It would take too long. And then it would be so deadly boring, no one would read it."

Assertion B is commonly accepted. Yet I've seen no practical or theoretical argument for it, other than absence of counter-examples. It may be true. It's a matter of faith.

Take a mathematically accepted proof and undertake to fill in the gaps, turn it into a formal proof. If you meet no obstacles, very well! And if you do meet an obstacle? That is, a mathematically accepted step that you can't break down into successive *modus ponens*? You've discovered an implicit assumption, a hidden lemma! Join it to the hypotheses of the theorem and go merrily forward.

If this is what's meant by formalization of proof, it can always be done. There's a preassigned way around any obstacle! There remains one little worry. Is your enlarged set of assumptions consistent? This can be hard to answer. You may say, if there's doubt about their consistency, it was wrong to claim the original proof was correct. On the other hand, we don't even know if the hypotheses of our number system are consistent. We presume it's O.K. on pragmatic grounds.

No one has proved that definition 2, the logic model of proof, is wrong. If a counter-example were found, logic would adapt to accommodate it.

By "proof" we mean correct proof, complete proof. By the standards of formal logic, ordinary mathematical proofs are incomplete. When an ordinary mathematical proof is offered to a referee, she may say, "More detail is needed." So more detail is supplied, and then the proof is accepted. But this acceptable version is still incomplete as a formal proof! The original proof was mathematically incomplete, the final proof is mathematically complete, but ***both are formally incomplete.*** The formal concept of proof is irrelevant at just the point of concern to the mathematician!

We prefer a beautiful proof with a serious gap over a boring hyper-correct one. If the *idea* is beautiful, we think it will attain valid mathematical expression. We even change the meaning of a concept to get a more beautiful theory. Projective geometry is a classic example. Euclidean geometry uses the familiar axiom, "Two points determine a line, and two lines determine a point, ***unless the lines are parallel.***" Projective geometry brings in ideal points at infinity—one point for each family of parallel lines. The axiom becomes: "Two points determine a line, and two lines determine a point." This is "right." The Euclidean axiom by comparison is awkward and clumsy.

When a mathematician submits work to the critical eyes of her colleagues, it's being tested, or "proved" in the old sense. With few exceptions, mathematicians have no other way to test or "prove" their work—invite whoever's interested to have a shot at it. Then the mathematical "Meaning number 1" of "proof" agrees with the old meaning. The *proof* of the pudding is in the eating. The *proof* of the theorem (in that sense of being tested) is in the refereeing.

Philosophers call this social validation "warranting."

Computers and Proof

In the late 1970s or early 1980s, while visiting a respected engineering school, I was told that the dean was vexed with his mathematicians. Other professors used his computer center; mathematicians didn't. Today no one's complaining that mathematicians don't compute. We were just 10 or 15 years late.

The issue about proof today is the impact of computers on mathematical proof. The effects are complex and manifold. I describe 2 trends, 3 examples, 2 critics, and 5 reasons.

First, two trends: (Trend A) Computers are being used as aids in proving theorems; and (Trend B) computers are encroaching on the central role we give proof in mathematics.

Three examples of trend A:

Example 1. In 1933, before general-purpose computers were known, Derrick Henry Lehmer built a computer to study prime numbers. It collected number-theoretic data and examples, from which he formulated conjectures. This was a mechanization of Lagrange and Gauss, who conjectured the prime number theorem from tables of prime numbers.

Example 2. Wolfgang Haken and Kenneth Appel's proof that a plane map needs at most four colors. The conjecture had been in place since 1852. Appel and Haken turned it into a huge computation. Then their computer did the computation. Thus they proved the conjecture.

Example 3. In the Feigenbaum-Lanford discovery of a universal critical point for doubling of bifurcations, computers played two distinct essential roles. Mitchell Feigenbaum discovered the universal critical point by computer exploration, comparable to Lehmer's number-theoretic explorations. Later, rigorous proofs were given by Oscar Lanford III (9, 10, 11) and by M. Campanino, H. Epstein, and D. Ruelle. At that stage the computer played a role like its role in the four-color problem: assistant in completing the proof.

Now the proof wasn't algebra but analysis. A certain derivative had to be estimated. This was accomplished by the computer, using approximation methods from numerical analysis, with a rigorous error estimate.

Example 1, Lehmer's exploratory use, didn't challenge the standard notion of proof. That notion doesn't care how a conjecture is made. I heard a distinguished professor tell his class, "It doesn't matter if you find the answer lying in a mud puddle. If you prove it's the answer, it makes no difference how you found it."

Examples 2 and 3 are different. Now the machine computation is *part of the proof.* For some people, such a proof violates time-honored doctrines about the difference between empirical knowledge and mathematical knowledge. According to Plato, Kant, and many others, common knowledge and scientific knowledge are a posteriori. They come from observation of the material world. Mathematical knowledge is a priori—independent of contingent facts about the material world. Daily experience and physical experiment could conceivably be other than what they are, but mathematical truths will hold in every possible world, or so thought Plato and Kant.

The operation of computers depends on properties of copper and silicon, on electrodynamics and quantum mechanics. Confidence in computers comes from confidence in physical facts and theories. These are not a priori. We learn the laws of physics and the electrical properties of silicon and copper from experience. It seems, then, that while old-fashioned theorems proved "by hand" are a priori, computer-assisted theorems are a posteriori! (Philip Kitcher thoroughly goes into the issue of a priori knowledge in mathematics.)

But despite the illusions of idealist philosophy, old-fashioned person-made proofs are also a posteriori. They depend on credence in the world of experience, the material world. We believe our scribbled notes don't change from hour to hour, that our thinking apparatus—our brain—is reliable more often than not, that our books and journals are what they pretend to be. We sometimes believe a proof without checking every line, every reference, and every line of every reference. Why? We depend on the integrity and competence of certain human beings—our colleagues. But human beings are *less* reliable at long computations than a Cray or a Sun! And still less in collaborations by large groups, as in the classification theorem of simple finite groups. Recognizing this demolishes the dream of a priori knowledge in advanced mathematics, and in general.

Most mathematicians see the difference between computer-assisted proofs and traditional proofs as a difference of degree, not of kind. But Paul Halmos does object to the Haken-Appel proof. "I do not find it easy to say what we learned from all that. We are still far from having a good proof of the Four Color Theorem. I hope as an article of faith that the computer missed the right concept and the right approach. 100 years from now the map theorem will be, I think, an exercise in a first-year graduate course, provable in a couple of pages by means of the appropriate concepts, which will be completely familiar by then. The present proof relies in effect on an Oracle, and I say down with Oracles! They are not mathematics."

Why does Halmos call the computer an Oracle? He can't know every step in the calculation. Indeed, the physical processes that make computers work aren't fully understood. Believing a computer is like believing a successful, well-reputed fortune teller. Should we regard the four-color conjecture as true? Should our confidence in it be increased by the Appel-Haken proof? Halmos doesn't say.

Halmos dislikes the Appel-Haken proof because it uses an Oracle, and because, he thinks, we can't learn anything from it. His criticism isn't for logical defects—incompleteness, incorrectness, or inaccuracy. It's esthetic and epistemological. This is normal in real-life mathematics. It would be senseless in formalized or logicized mathematics.

Views like Halmos were also vented by Daniel Cohen. "Our pursuit is not the accumulation of facts about the world or even facts about mathematical objects. The mission of mathematics is understanding. The Appel and Haken work on the Four Color Problem amounts to a confirmation that a map-maker with only four paint pots will not be driven out of business. This is not really what mathematicians were worried about in the first place. Admitting the computer shenanigans of Appel and Haken to the ranks of mathematics would only leave us intellectually unfulfilled."

The eloquence of Halmos and Cohen won't deter mathematicians who hope a computer will help on their problem. Computers in pure mathematics will increase for at least five reasons.

Reason 1. Access to more powerful computers spreads ever more widely.

Reason 2. As oldsters are replaced by youngsters schooled after the computer revolution, the proportion of us at home with computers rises.

Reason 3. Despite the scolding of Halmos, Cohen, and sympathizers, success like that of Haken-Appel and Lanford inspires emulation.

Reason 4. Scientists and engineers were computerized long ago. You can't imagine a mathematician interacting with a scientist or engineer today without a computer.

Reason 5. Reasons 1, 2, 3, and 4 stimulate those branches of mathematics that use computation, leaving the others at a disadvantage. This reinforces Reasons 1, 2, 3, and 4.

Some will still reject computer proofs. They'll have as much effect as old King Canute. (By royal order, he commanded the tide to turn back.)

In fields like chaos, dynamical systems, and high Reynolds number fluid dynamics, we come to trend B. Here, as in number theory, the computer is an explorer, a scout. It goes much deeper into unknown regions than rigorous analysis can. In these fields we no longer think of computer-obtained knowledge as tentative—pending proper proof. We prove what we can, compute what we can. This part of mathematical reality rests on computation and analysis closely yoked together. The computer finds "such and such." We believe it—not as indubitable, but as believable. Proof of "such and such" then would help us see *why* it's so, though not perhaps increase our conviction it *is* so.

This effect of computers, Trend B, is profound. It erodes the time-honored understanding of what mathematics is. Machine computation as part of proof is radical, but still more radical is machine computation accepted as ***empirical evidence of mathematical truth, virtually a weak form of proof.*** Such acceptance contradicts the official line that mathematics *is* deductions from axioms. It makes mathematics more like an empirical science.

An interesting proposal to control this disturbing new tendency was made in the *Bulletin of the American Mathematical Society* by the distinguished mathematical physicist Arthur Jaffe and the distinguished topologist Frank Quinn. They want to save rigorous proof, the chief distinction of modern mathematics, from promiscuous contamination with mere machine calculations. Yet they acknowledge that "empirical" or "experimental" computer mathematics will not go away, and may have a place in the world. Their solution is—compulsory rigid distinction between the two. Genuine mathematics stays as it should be. Experimental or numerical or empirical mathematics is labeled appropriately. Analogous to kosher chickens versus non-kosher. I don't have the impression that this proposal met with general acclaim.

A different departure from traditional proof was found by Miller (1976), Rabin (1976), Davis (1977), and Schwartz (1980). There's now a way to say of an integer n whose primality or compositeness is unknown, "On the basis of available information, the probability that n is prime is p." If n is really prime, you can make p arbitrarily close to 1. Yet the primality of a given n is *not* a random variable. n either is prime or it isn't. But if n is large, finding its primality by a deterministic method is so laborious that random errors must be expected in the computation. Rabin showed that if n is very large, the probability of error in the deterministic calculation is greater than the probability p in his fast probabilistic method!

How will we adjust to such variation in proof? We can think of proofs as having variable quality. Instead of "proved," label them either "proved by hand" or "proved by machine." Even provide an estimate for the reliability of the machine calculation used in a proof (Swart, 1980).

There are two precedents for this situation. Between the world wars, some mathematicians made it a practice to state explicitly where they used the axiom of choice. And in 1972 Errett Bishop said the clash between constructivists and classicists would end if classicists stated explicitly where they used the law of the excluded middle (L. E. M.). (Constructivists reject the L. E. M. with respect to infinite sets.) These issues didn't involve computing machines. They involved disagreement about proof.

If experience with the axiom of choice and the L. E. M. is indicative, Swart's proposal is unpromising. Nobody worries any more about the axiom of choice, and few worry about the L. E. M. Perhaps we don't care much about distinctions of quality or certainty in proof. Perhaps few care today if proofs are handmade or computer-made.

There's a separate trend in computer proof, that so far has had more resonance in computer science than in main-line mathematics. In this trend, the idea is not to use the computer to supply a missing piece of a mathematician's proof. Rather, the computer is a logic assistant. Give it some axioms, and send it out to find interesting theorems. Of course, you have to give it a way to measure "interesting." Or tell it what you want proved—a conjecture you haven't proved yourself, or a proved theorem, to see what different proof it might find.

Since the computer is expected to follow the rules of logic, the relation to definition 2 is apparent. Since it really does find proofs that convince mathematicians, it conforms to definition 1. People who think mathematicians will become obsolete have this kind of thing in mind—computer proofs of real theorems.

Professor Wos writes movingly, "We have beaten the odds, done the impossible, automated reasoning so effectively that our programs have even answered open questions. Yet skeptics still exist, funding is not abundant, and recognition of our achievements is inappropriately small and not sufficiently widespread."

This work can be regarded as practical justification of the formal logic notion of proof. To the extent that computers following only the rules of formal logic do reproduce discoveries of live mathematicians, they show that formal logic is an adequate model of real live proof. But "technical limitations" restrict automated proof to relatively simple theorems.

There are centers of this research in Austin, Texas, and at Argonne National Laboratory. There is a *Journal of Automated Reasoning*. Bledsoe and Loveland is an instructive and readable review.

I've said nothing about applied mathematics. In applied mathematics, infiltration and domination by computers has long been a fait accompli. Under the influence of computers, pure mathematics is becoming more like applied mathematics.

Fallibility

Philosophical discussions of mathematical proof usually talk about it only as it's seen in journals and textbooks. There, proof functions as the last judgment, the final word before a problem is put to bed. But the essential mathematical activity is *finding* the proof, not checking after the fact that indeed it is a proof.

How does formal proof differ from real live proof?

Real-life proof is informal, in whole or in part. A piece of formal argument—a calculation—is meaningful only to complete or verify some informal reasoning. The formal-logic picture of proof is a topic for study in logic rather than a truthful picture of real-life mathematics.

Formal proof exists only in a formalized theory, cleaned and purged of all associations and connotations. It uses a formal vocabulary, formal axioms, and formal inference rules.

The passage from informal to formalized theory must entail loss of meaning or change of meaning. The informal has connotations and alternative interpretations not in the formalized theory. Consequently, anything proved formally can be challenged: "How faithful are this statement and proof to the informal concept we're actually interested in?"

For some investigations, formalization and complete formal proof would take time and persistence beyond human capability, beyond any foreseeable computer.

Mathematicians say that any "Theorem and Proof" in a pure mathematics journal must be formalizable *in principle*. A glance into any mathematical *Archive* or *Bulletin* or *Journale* or *Zeitschrift* shows a great proportion of the text in natural language. Even pages of solid calculation turn out, on inspection, not to be formalized. In presenting calculations for publication, we include steps that we consider nonroutine, which should be explained to fellow specialists. And in every calculation there are routine steps that needn't be explained to fellow experts. These, naturally, we leave out. (If we include them, the editor throws them out.)

So the published proof is incomplete. The reader accepts the result on faith, or fills in the steps herself.

Even in a graduate math class proof isn't completely formalized. The professor leaves out as a matter of course what she considers routine or trivial. A nonroutine step may be assigned as homework.

Practically none of the mathematical literature is formalized (except for computer programs, which are another story.) Yet these far-from-formalized proofs are accepted by mathematicians as "formalizable in principle."

Why are they accepted? Because they're convincing to the experts, who'd be only too happy to find a serious error or gap.

We accept incomplete, natural-language proofs on the basis of our experience, our know-how in looking for the weakest link. And also on the author's reputation. A known bungler, or just an unknown, or an authority of proved accomplishment?

Peano hoped that his formal language would guarantee proofs to be correct. In today's language, that proofs could be checked by computer. But trying to check proofs by computer may introduce new errors. There is random error by physical fluctuations of the machine, and there is human error in designing and producing hardware and software (logic and programming.) For most nontrivial mathematics, the vision of formal proof is still visionary. And when a machine does part of a proof, as in the four-color theorem of Appel and Haken, some mathematicians reject it because the details of machine computation are inevitably hidden. More than *whether* a conjecture is correct, we ask *why* it is correct. We want to understand the proof, not just be told it exists.

The issue of machine error is more than just electrical engineering. It's the difference between computation in principle (infallible) and computation in practice (fallible). A simple calculation shows that no matter how small the chance of error in one step, if the calculation or formal proof is long enough, it's almost sure to contain errors. But we can have other grounds for believing the conclusion of a proof, apart from the claimed certainty of step-by-step reasoning. Examples and special cases, analogy with other results, expected symmetry, unexpected elegance, even an inexplicable feeling of rightness. These illogical logics may say, "It's true!" If you have something you hope is a proof, such nonrigorous reasons can make you sure of the conclusion, even while you know the "proof" is sure to contain uncorrected errors. Such intuition is fallible in principle. Attempted rigorous proof is fallible in practice.

Beyond the certainty of error in long calculations, you face the fact that calculation is finite, mathematics infinite. There's a limit to the biggest computer, how tightly it can be packed, how fast it can run, how long it will operate. The life of the human race is a limit. Put these limits together, and you have a bound on how much anyone will ever compute. If you can't *know* anything in mathematics except by formal proof (a particular kind of computation), you've set a bound on how much mathematics you'll ever know. The physical bound on computation implies a bound on the number and length of theorems that will ever be proved. The only way out would be a faster, less certain way to mathematical knowledge (Knuth, 1976).

Meyer (1974, p. 481) quotes a theorem, proved with L. J. Stockmeyer: "If we choose sentences of length 616 in the decidability theory of WSIS (weak monadic second-order theory of the successor function on the nonnegative integers) and code these sentences into $6 \times 616 = 3,696$ binary digits, then any logical network with 3,696 inputs which decides truth of these sentences contains 10^{123} operations." (WSIS is much weaker than ordinary arithmetic. The conclusion applies a fortiori to sentences longer than 616 digits.)

"We remind the reader," Meyer writes, "that the radius of a proton is approximately 10^{-13} cm., and the radius of the known universe is approximately 10^{28} cm. Thus for sentences of length 616, a network whose atomic operations were performed by transistors the size of a proton connected by infinitely thin wires would densely fill the entire universe." No decision procedure for sentences of length 616 in WSIS can be physically realized. Yet WSIS is "decidable": One says that a decision procedure "exists" for sentences of any length.

Our notion of rigorous proof isn't carved in granite. We'll modify it. We'll allow machine computation, numerical evidence, probabilistic algorithms, if we find them advantageous. We mislead our pupils if we make "rigorous proof" a shibboleth in class.

In Class

The role of proof in class isn't the same as in research. In research, it's to *convince*. In class, students are all too easily convinced! Two special cases do it. In a first course in abstract algebra, proof of the fundamental theorem of algebra is often omitted. Students believe it anyway.

The student needs proof to *explain*, to give insight why a theorem's true. Not proof in the sense of formal logic. As the graduate student said to the Ideal Mathematician in *The Mathematical Experience*, she never saw such a proof in class. In class informal or semiformal proofs are presented in natural language. They include calculations, which are formal subproofs inside the overall informal proof.

Some instructors think, "If it's a math class, you prove. If you don't prove anything, it isn't math." That makes a kind of sense. If proof is math and math is proof, then in math class you're duty bound to prove. The more you prove, the more honest and rigorous you feel your class is.

Exposure to proof can be more emotional than intellectual. If the instructor gives no better reason for proof than "That's math!," the student knows she saw a proof, but not wherefore, except: "That's math!"

I call this view "absolutist," despite that word's unfortunate associations (absolute monarchy, absolute zero, etc.). If mathematics is a system of absolute truths, independent of human construction or knowledge—then mathematical proofs are external and eternal. They're to admire. The absolutist teacher wants to tell only what he intends to prove (or order the students to prove). He'll usually try for the shortest proof or the most general one. The main purpose of proof isn't explanation. The purpose is certification: admission into the catalog of absolute truths.

The view I favor is humanism. To the humanist, mathematics is *ours*—our tool, our plaything.

Proof is complete explanation. Give it when complete explanation is appropriate, rather than incomplete explanation or no explanation.

The humanist math teacher looks for enlightening proofs, not necessarily the most general or the shortest. Some proofs don't explain much. They're called "tricky," "pulling a rabbit out of a hat." Give that kind of proof when you want your students to see a rabbit pulled out of a hat. But in general, give proofs that explain. And if the only proof you can find is unmotivated and tricky, if your students won't learn much from it, must you do it "to stay honest"? That "honesty" is a figment, a self-imposed burden. Better try to be clear, well-motivated, even inspiring.

This attitude disturbs people who think proof is the be-all and end-all of mathematics—who say "a mathematician is someone who proves theorems" and "without proof, there's no mathematics." From that viewpoint, a mathematics in which proof is less than absolute is heresy.

For the humanist, the purpose of proof, as of all teaching, is understanding. Whether to give a proof as is, elaborate it, or abbreviate it, depends on what he thinks will increase the student's understanding of concepts, methods, and applications.

This policy uses the notion of "understanding," which isn't precise or likely to be made precise. Do we understand what it means "to understand"? No. Can we teach to foster understanding? Yes. We recognize understanding, though we can't say precisely what it is.

In a stimulating article, Uri Leron (1983) borrowed an idea from computing—"structured proof." A structured proof is like a structured program. Instead of starting with little lemmas whose significance appears at the end, start by breaking the proof-task into chunks. Then break each chunk into subchunks. The little lemmas come at the end, where you see why you need them.

In the general classroom, the motto is: "Proof is a tool in service of teacher and class, not a shackle to restrain them."

In teaching future mathematicians, "Proof is a tool in service of research, not a shackle on the mathematician's imagination."

Proof can convince, and it can explain. In research, convincing is primary. In high-school or undergraduate class, explaining is primary.

Intuition

If we look at mathematical practice, the intuitive is everywhere. We consider intuition in the mathematical literature and in mathematical discovery.

A famous example was the letter from Ramanujan to Hardy, containing astonishing formulas for infinite sums, products, fractions, and roots. The letter had gone to Baker and to Hobson. They ignored it. Hardy didn't ignore it.

Ramanujan's formulas prove there is mathematical intuition, for they're correct, even though Ramanujan didn't prove them, and in some cases had hardly an idea what a proof would be. But what about Hardy? He also made a correct judgment without proof—the judgment that Ramanujan's formulas were true, and that Ramanujan was a genius. How did he do that? Not by checking his formulas with complete proofs. By some mental faculty associated with mathematics, Hardy made a sound judgment of Ramanujan's formulas, without proofs. Was Hardy's judgment of Ramanujan's letter a mathematical judgment? Of course it was, in any reasonable understanding of the word mathematics. It was an exceptional event, yet not essentially different from mathematical judgments made every day by reviewers and referees, by teachers and paper-graders, by search committees and admission committees. The faculty called on in these judgments is mathematical intuition. It's reliable mathematical belief without the slightest dream of being formalized.

Since intuition is an essential part of mathematics, no adequate philosophy of mathematics can ignore intuition.

The word intuition, as mathematicians use it, carries a heavy load of mystery and ambiguity. Sometimes it's a dangerous, illegitimate substitute for rigorous proof. Sometimes it's a flash of insight that tells the happy few what others learn with great effort. As a first step to explore this slippery concept, consider this list of the meanings and uses we give this word.

1. Intuitive is the opposite of rigorous. This usage is not completely clear, for the meaning of "rigorous" is never given precisely. We might say that in this usage intuitive means lacking in rigor, yet the concept of rigor is defined intuitively, not rigorously.

2. Intuitive means visual. Intuitive topology or geometry differs from rigorous topology or geometry in two ways. On one hand, the intuitive version has a meaning, a referent in the domain of visualized curves and surfaces, which is absent from the rigorous formal or abstract version. In this the intuitive is superior; it has a valuable quality the rigorous version lacks. On the other hand, visualization may mislead us to think obvious or self-evident statements that are

dubious or false. The article by Hahn, "The Crises in Intuition" is a beautiful collection of such statements.

3. Intuitive means plausible, or convincing in the absence of proof. A related meaning is, "what you might expect to be true in this kind of situation, on the basis of experience with similar situations." "Intuitively plausible" means reasonable as a conjecture, i.e., as a candidate for proof.

4. Intuitive means incomplete. If you take a limit under the integral sign without using Lebesgue's theorem, if you expand a function in a power series without checking that it's analytic, you acknowledge the logical gap by calling the argument intuitive.

5. Intuitive means based on a physical model or on some special examples. This is close to "heuristic."

6. Intuitive means holistic or integrative as opposed to detailed or analytic. When we think of a theory in the large, when we're sure of something because it fits everything else we know, we're thinking intuitively. Rigor requires a chain of reasoning where the first step is known and the last step is the conjecture. If the chain is very long, rigorous proof may leave doubt and misgiving. It may actually be less convincing than an intuitive argument that you grasp as a whole, which uses your faith that mathematics is coherent.

In all these usages intuition is vague. It changes from one usage to another. One author takes pride in avoiding the "merely" intuitive—the use of figures and diagrams as aids to proof. Another takes pride in emphasizing the intuitive—showing visual and physical significance of a theory, or giving heuristic derivations, not just formal post hoc verification.

With any of these interpretations, the intuitive is to some degree extraneous. It has desirable and undesirable aspects. It's optional, like seasoning on a salad. It's possible to teach mathematics or to write papers without thinking about intuition.

However, if you're not doing mathematics, but watching people do mathematics and trying to understand what they're doing, dealing with intuition becomes unavoidable.

I maintain that:

1. All the standard philosophical viewpoints rely on some notion of intuition.
2. None of them explain the nature of the intuition that they postulate.
3. Consideration of intuition as actually experienced leads to a notion that is difficult and complex, but not inexplicable.
4. A realistic analysis of mathematical intuition should be a central goal of the philosophy of mathematics.

Let's elaborate these points. By the main philosophies, I mean as usual constructivism, Platonism, and formalism. For the present we don't need refined distinctions among versions of the three. It's sufficient to characterize each crudely with one sentence.

The constructivist regards the natural numbers as the fundamental datum of mathematics, which neither requires nor is capable of reduction to a more basic notion, and from which all meaningful mathematics must be constructed.

The Platonist regards mathematical objects as already existing, once and for all, in some ideal and timeless (or tenseless) sense. We don't create, we discover what's already there, including infinites of a complexity yet to be conceived by mind of mathematician.

The formalist rejects both the restrictions of the constructivist and the theology of the Platonist. All that matters are inference rules by which he transforms one formula to another. Any meaning such formulas have is nonmathematical and beside the point.

What does each of these three philosophies need from the intuition? The most obvious difficulty is that besetting the Platonist. If mathematical objects constitute an ideal nonmaterial world, how does the human mind/brain establish contact with this world? Consider the continuum hypothesis. Gödel and Cohen proved that it can neither be proved nor disproved from the set axioms of contemporary mathematics. The Platonist believes this is a sign of ignorance. The continuum is a definite thing, independent of the human mind. It either does or doesn't contain an infinite subset equivalent neither to the set of integers, nor to the set of real numbers. Our *intuition* must be developed to tell us which is the case. The Platonist needs intuition to connect human awareness and mathematical reality. But his intuition is elusive. He doesn't describe it, let alone analyze it. How is it acquired? It varies from person to person, from one mathematical genius to another mathematical genius. It has to be developed and refined. By whom, by what criteria, does one develop it? Does it directly perceive an ideal reality, as our eyes perceive visible reality? Then intuition would be a second ideal entity, the subjective counterpart of Platonic mathematical reality. We have traded one mystery for two: first, the mysterious relation between timeless, immaterial ideas and the mundane reality of change and flux; and second, the mysterious relation between the flesh and blood mathematician and his intuition, which directly perceives the timeless and eternal. These difficulties make Platonism hard for a scientifically oriented person to defend.

Mathematical Platonists simply disregard them. For them, intuition is something unanalyzable but indispensable. Like the soul in modern Protestantism, the intuition is there but no questions can be asked about it.

The constructivist, as a conscious descendant of Kant, knows he relies on intuition. The natural numbers are given intuitively. This doesn't seem problematical. Yet Brouwer's followers have disagreed on how to be a constructivist. Of course, every philosophical school has that experience. But it creates a difficulty for a school that claims to base itself on a universal intuition.

The dogma that the intuition of the natural numbers is universal violates historical, pedagogical, and anthropological experience. The natural number system seems an innate intuition to mathematicians so sophisticated they can't remember or imagine before they acquired it; and so isolated they never meet people who haven't internalized arithmetic and made it intuitive (the majority of the human race!).

What about the formalist? Does intuition vanish along with meaning and truth? You can avoid intuition as long as you consider mathematics to be no more than formal deductions from formal axioms. A. Lichnerowicz wrote, "Our demands on ourselves have become infinitely larger; the demonstrations of our predecessors no longer satisfy us but the mathematical facts that they discovered remain and we prove them by methods that are infinitely more rigorous and precise, methods from which geometric intuition with its character of badly analyzed evidence has been totally banned."

Geometry was pronounced dead as an autonomous subject; it was no more than the study of certain particular algebraic-topological structures. The formalist eliminates intuition by concentrating on refinement of proof and dreaming of an irrefutable final presentation. To the natural question, Why should we be interested in these superprecise, superreliable theorems?, formalism turns a deaf ear. Obviously, their interest derives from their meaning. But the all-out formalist throws out meaning as nonmathematical. Then how did our predecessors find correct theorems by incorrect reasoning? He has no answer but "Intuition."

Surely Cauchy knew Cauchy's integral theorem, even though (in the formalist's sense of knowing the formal set-theoretic definition) he didn't know the meaning of any term in the theorem. He didn't know what is a complex number, what is an integral, what is a curve; yet he found the complex number represented by the integral over this curve! How can this be? Cauchy had great intuition.

"But during my last night, the 22–23 of March, 1882—which I spent sitting on the sofa because of asthma—at about 3:30 there suddenly arose before me the Central Theorem, as it has been prefigured by me through the figure of the 14-gon in (Ges. Abh., vol. 3, p. 126). The next afternoon, in the mail-coach (which then ran from Norden to Emden) I thought through what I had found, in all its details. Then I knew I had a great theorem. . . . The proof was in fact very difficult. I never doubted that the method of proof was correct, but everywhere I ran into gaps in my knowledge of function theory or in function theory itself. I could only postulate the resolution of these difficulties, which were in fact completely resolved only 30 years later (in 1921) by Koebe" (F. Klein).

But what *is* this intuition? The Platonist believes in real objects (ideal, to be sure), which we "intuit." The formalist believes no such things exist. So what is there to intuit? The only answer is, unconscious formalizing. Cauchy subconsciously knew a correct proof of his theorem, which means knowing the correct definitions of all the terms in the theorem.

This answer is interesting to the many mathematicians who've made correct conjectures they couldn't prove. If their intuitive conjecture was the result of unconscious reasoning, then: (a) Either the unconscious has a secret method of reasoning that is better than any known method; or (b) the proof is there in my head, I just can't get it out!

Formalists willing to consider the problem of discovery and the historical development of mathematics need intuition to account for the gap between their account of mathematics (a game played by the rules) and the real experience of mathematics, where more is sometimes accomplished by breaking rules than obeying them.

Accounting for intuitive "knowledge" in mathematics is the basic problem of mathematical epistemology. What do we believe, and why do we believe it? To answer this question we ask another question: what do we teach, and how do we teach it? Or what do we try to teach, and how do we find it necessary to teach it? We try to teach mathematical concepts, not formally (memorizing definitions) but intuitively—by examples, problems, developing an ability to think, which is the expression of having successfully internalized something. What? An intuitive mathematical idea. The fundamental intuition of the natural numbers is a shared concept, an idea held in common after manipulating coins, bricks, buttons, pebbles. We can tell by the student's answers to our questions that he gets the idea of a huge bin of buttons that never runs out.

Intuition isn't direct perception of something external. It's the effect in the mind/brain of manipulating concrete objects—at a later stage, of making marks on paper, and still later, manipulating mental images. This experience leaves a trace, an effect, in the mind/brain. That trace of manipulative experience is your representation of the natural numbers. Your representation is equivalent to mine in the sense that we both give the same answer to any question you ask. Or if we get different answers, we compare notes and figure out who's right. We can do this, not because we have been explicitly taught a set of algebraic rules, but because our mental pictures match. If they don't, since I'm the teacher and my mental picture matches the one all the other teachers have, you get a bad mark.

We have intuition because we have mental representations of mathematical objects. We acquire these representations, not mainly by memorizing formulas, but by repeated experiences (on the elementary level, experience of manipulating physical objects; on the advanced level, experiences of doing problems and discovering things for ourselves). These mental representations are checked for veracity by our teachers and fellow students. If we don't get the right answer, we flunk the course. Different people's representations are always being rubbed against each other to make sure they're congruent. We don't know how these representations are held in the mind/brain. We don't know how *any* thought or knowledge is held in the mind/brain. The point is that as shared concepts, as mutually congruent mental representations, they're real objects whose existence is just as "objective" as mother love and race prejudice, as the price of tea or the fear of God.

How do we distinguish mathematics from other humanistic studies? There's a fundamental difference between mathematics and literary criticism. While mathematics is a humanistic study with respect to its subject matter—human ideas—it's science-like in its objectivity. Those results about the physical world that are reproducible—which come out the same way every time you ask—are called scientific. Those subjects that have reproducible results are called natural sciences. In the realm of ideas, of mental objects, those ideas whose properties are reproducible are called mathematical objects, and ***the study of mental objects with reproducible properties is called mathematics.*** Intuition is the faculty by which we consider or examine these internal, mental objects.

There's always some discrepancy between my intuition and yours. Mutual adjustment to keep agreement is going on all the time. As new questions are asked, new parts of the structure come into sight. Sometimes a question has no answer. The continuum hypothesis doesn't have to be true or false.

We know that with physical objects we may ask questions that are inappropriate, which have no answer. What are the exact velocity and position of an electron? How many trees are there growing at this moment in Minnesota? For mental objects as for physical ones, what seems at first an appropriate question is sometimes discovered, perhaps with great difficulty, to be inappropriate. This doesn't contradict the existence of the particular mental or physical object. There are questions that *are* appropriate, to which reliable answers can be given.

The difficulty in seeing what intuition is arises because of the expectation that mathematics is infallible. Both formalism and Platonism want a superhuman mathematics. To get it, each of them falsifies the nature of mathematics in human life and in history, creating needless confusion and mystery.

Certainty

Even if it's granted that the need for certainty is inherited from the ancient past, and is religiously motivated, its validity is independent of its history and its motivation. The question remains: is mathematical knowledge indubitable?

Set aside history and motivation. Look at samples of mathematical knowledge and ask: Is this indubitable?

We take three examples. First, good old

$2 + 2 = 4.$**

Second, familiar to all former high-school students,

"The angle sum of any triangle equals two right angles." Finally, a more sophisticated example: a convergent infinite series.

Label the first example

Formula A: $2 + 2 = 4.$

Everyone knows Formula A is a mathematical truth. Everyone knows it's indubitable. *Ergo*, at least one mathematical truth is indubitable.

There's Russell's cavil: indubitable, but no great truth.

$2 + 2$ by definition means $(1 + 1) + (1 + 1)$
4 by definition means $1 + (1 + (1 + 1))$

By the associative law of addition, Formula A then is:

$$1 + 1 + 1 + 1 = 1 + 1 + 1 + 1.$$

Indubitable but unimpressive! As Frege said, it is an analytic a priori truth, not a synthetic one. (See Chapters 7 and 8 about Kant and Frege.)

Bertrand Russell thought *every* mathematical truth is a tautology like $2 + 2 = 4$ —trivially indubitable. He said that a mathematical theorem says no more than "the great truth that there are three feet in a yard." This view is regarded by mathematicians as absurd and without merit.

To probe into Formula A, we must ask, what is 2? What is 4? What is +? What is =? Trivial as these questions may seem, they serve to distinguish the different schools (logicist, formalist, intuitionist, empiricist, conventionalist).

The most elementary answer is the empiricist one. "$2 + 2 = 4$" means "Put two buttons in a jar, put in two more, and you have four buttons in the jar." John Stuart Mill is the classic advocate of this interpretation. For him, formula A is *not* indubitable. It's about buttons. Buttons are material objects, which never can be known with certainty (Heraclitus et al., as expounded, for instance, in Russell's *History of Western Philosophy*, 1945). Who knows if some exotic chemical reaction might give

two buttons + two buttons = zero buttons

or

two buttons + two buttons = five buttons.

For indubitability, forget buttons.

Another answer, along formalist or logicist lines, might be given by a graduate student of mathematics. "1, +, = are symbols defined by the Dedekind-Peano axioms.**

"2 is short for $1 + 1$,
"3 is short for $2 + 1$,
"4 is short for $3 + 1$.
"Now Formula A can be proved."

(Our reduction above of Formula A to

$$1 + 1 + 1 + 1 = 1 + 1 + 1 + 1$$

is a sketch of the formal proof in the Mathematical Notes and Comments.)

So then is Formula A doubtable?

Before I worry about doubtability, how sure am I that the proof is even correct? It *seems* to me that it is correct. Could I be mistaken? Overlooked something staring me in the face? I make mistakes in math. I think I'm sure the proof is O.K., but am I *really* sure that I'm *totally* certain?

A second worry is more substantial. How do I know Peano's axioms produce the same number system 1,2,3, . . . that I had in mind (or Dedekind had in mind) in the first place? They seem to work. How certain can I be that they'll always work? The numbers 1,2. . . . with which we start (before anybody gives us axioms) are an informal, "intuitively given " system. For that reason, it's impossible to prove formally that they correspond to any formal model. Such a proof is possible only between two formal models. Does the formal model correspond to the original intuitive idea? That question can never by answered by a rigorous, formal proof! It must rest on informal, intuitive reasoning that has no claim to be rigorous, let alone indubitable.

I have no doubt that Peano's axioms actually do describe the intuitive natural numbers 1,2,. . . . But I can't claim this belief as indubitable. Consider my knowledge that I'm now writing on a yellow pad with a blue pen sitting at a round table, and so forth. I can barely conceive that this might be false. But by the standards of Heraclitus, Plato, and others, such knowledge is doubtable or dubitable. If so, I can more readily believe Peano's seemingly convincing axioms might be doubtable. If they are, then every theorem in Peano arithmetic is doubtable. Even Formula A,

$$2 + 2 = 4 \text{ !!}$$

Another often-cited distinction between sensory knowledge and mathematical knowledge is that sensory knowledge *could conceivably be* other than it is. It's *conceivable* that my blue pen is really yellow and my yellow pad really blue.

Mathematical knowledge, on the other hand, such as Formula A, is supposed to be not only indubitably true, it's supposed to be inconceivable that it could be false. We supposedly can't imagine

$$2 + 2 = 3$$

as we supposedly can imagine a blue pad that looks yellow. This observation has been held to demonstrate that Formula A is indubitable.

But whether the denial of Formula A is inconceivable can't be judged until the import and significance of Formula A is explicated. It has one meaning as a *report of a property of buttons*, coins, or other such discrete objects. It has a second meaning as *a theorem in an axiom system* with +, =, and 1 as undefined terms. But we've just seen that as a formula about buttons, Formula A is doubtable. And as a theorem in an abstract axiom system, it's doubtable still. Whether in terms of buttons or of Peano arithmetic, Formula A is doubtable. Its negation is conceivable.

Let's consider the angle-sum theorem in Euclidean plane geometry.** It was Baruch Spinoza's favorite example of an absolutely certain statement. We'll follow proper terminology and call it

Formula B: Angle A + Angle B + Angle C = 2 right angles.

The angles A, B, and C on the left side of the equation are the internal angles of a triangle, any triangle at all.

Trite as this example seems to us today, it was already the twentieth century when Gottlob Frege berated David Hilbert for not understanding that there can be only one real, true geometry (Euclid's, of course).

We have theorem and proof. They have withstood scrutiny for 2,000 years. Isn't this indubitable?

Consider what is meant by the term "angle." Some say the theorem tells what will happen if you find a triangle lying somewhere and measure its angles. The sum should be 180 degrees.

If you actually do it, you find the sum not exactly 180 degrees. You retreat. It ***would*** be 180 degrees if you could draw perfect straight lines and measure perfectly. But you can't. You claim only that the sum will be *close* to 180 degrees, and even closer if you remeasure more accurately. That's ***not*** what Euclid said. It's a modern reinterpretation based on our post-Newtonian idea of "limit" or "approximation." Euclid's theorem is about the angles themselves, not about measurements or approximations.

A more sophisticated interpretation says Euclid's talking about ideal triangles, not triangles you actually draw or measure—an idealization that our mind/brain creates or discovers after seeing lots of real triangles. The indubitability of the theorem flows from the indubitability of the argument and the indubitability of the axioms with respect to the ideal points and lines.

The most important of the axioms about lines is the parallel postulate, "Euclid's Fifth." It can be stated in many equivalent forms. Most popular is Playfair's: "Through any given point, parallel to any given line, can be drawn exactly one line." For thousands of years the fifth axiom was accepted as intuitively obvious. Using it, we obtain Euclidean geometry, including the "indubitable" theorem about the angle sum of a triangle. But it was too complicated for an axiom, some thought. It ought to be a theorem.

In the nineteenth century Carl Friedrich Gauss, Janos Bolyai, a Hungarian army officer, and Nicolai Ivanovich Lobachevsky, professor and later rector at the University of Kazan, all without knowing of each other's work, had the same idea: Suppose the fifth postulate is *false*, and see what happens. Others had made the same supposition, in search of a contradiction that would prove the fifth postulate. Gauss, Bolyai and Lobatchevsky recognized that they got, not a contradiction, but a new geometry!

There is more information about non-Euclidean geometry in the article about Kant and in the mathematical notes and comments. Here we see how this discovery affects the certitude of the angle sum theorem. Suppose we replace Euclid's fifth by the non-Euclidean "anti-Playfair" postulate: "Through any point there pass more than one line parallel to any given line." From this strange axiom it follows that the angle sum of any triangle is *less* than 180 degrees. And if we try a different non-Euclidean fifth postulate—assume there are *no* parallel lines—then the angle sums of triangles are *greater* than 180 degrees.

We can't claim the angle-sum theorem is indubitable unless Euclid's parallel postulate is indubitable, for if we alter that postulate, the angle sum theorem is falsified. It's tempting just to declare that obviously or intuitively Euclid is correct. This wasn't believed by Gauss. There's a story that he tried to settle the question by measuring angles of a triangle whose vertices were three mountain tops. The larger the triangle, the likelier would be a measurable deviation from Euclideanness. According to the legend, the measurement was indecisive.

Some hundred years later, Einstein's general relativity depended on Riemannian geometry, a far-reaching generalization of non-Euclidean geometry. In relativity texts non-Euclidean geometry represents relativistic velocity vectors. So physics gives no license to favor Euclid over non-Euclid. Our prescientific, intuitive notions of space are learned on a small scale, relative to the universe at large. Locally, "in the small," Euclidean and non-Euclidean geometries are indistinguishable. Local intuition can't tell which is true in the large. Belief in the Euclidean angle sum theorem as indubitable was based on belief in an infallible spatial intuition. That belief has been refuted by non-Euclidean geometry and the following development of Riemannian geometry with its application in relativity.

Then where are we? Surely the angle-sum theorem is true in some sense? Of course. It's true as a theorem in Euclidean geometry. It follows from the axioms—not from Euclid's axioms alone, however. They need correction and completion. Hilbert took care of that, so we finally have a correct proof of the angle sum theorem, from a corrected and completed Euclidean axiom set.

We're in the same position as in the interpretation of

$$2 + 2 = 4$$

by Peano's axioms. We can't be completely sure which axioms describe the triangle we had in mind. By the nature of the case, such a thing can't be proved. Geometry is worse off than arithmetic, for its axioms are more subtle and elaborate. No one says Hilbert's axioms for Euclidean plane geometry are self-evident or indubitable.

Formula C is our last:

$$1 + \frac{1}{4} + \frac{1}{9} + \frac{1}{16} + \frac{1}{25} + \ldots = \frac{\pi^2}{6}$$

This formula was first written down by Euler. On the left is an infinite series. It says, "Add all terms of the form $1/n^2$, where n is any positive whole number." It asserts that the result of such an addition is the same as if you take π (the ratio of the circumference of a circle to its radius), square it, and then divide by 6.

The novice should be staggered. How can he ever add infinitely many numbers? And why would the answer have anything to do with a circle? The proof is easy when you know how to expand functions in series of sines and cosines (Fourier series).**

Formula C is abstruse and remote compared to Formulas A an B. You check

$$2 + 2 = 4$$

on your fingers. You check the angle sum theorem by measuring angles. What can you do here? Well, if you have a computer, or a calculator plus patience, you add up the series—as much as you have patience for. Then press the π button on your calculator, square and divide by 6.

A reader in Johannesburg sent these numbers (found after 30 hours on a hand calculator):

$$\sum \frac{1}{n^2} = 1.644914943$$

$$\frac{\pi^2}{6} = 1.644934067$$

What should you conclude from this?

If you're an optimist, and the error after 30 hours of calculation is only 0.000019124, you expect it would be even less after 40 hours. The formula is probably right. This is hope, not certainty.

No amount of addition could yield certainty, because the formula is about the *limit*—infinitely many terms! Any computation is finite.

If you want indubitable truth for this formula, you need *proof.* Fortunately, from the calculation of Euler or Fourier, which yields Formula C, it takes only a few more steps to such a proof.

The proof relies on the real number system. That has served long and well, but Gödel told us we can't prove it's consistent. Intuitionist and constructivist mathematicians don't rely on it without radical trimming. For practical purposes, we confidently write

$$\sum \frac{1}{n^2} = \frac{\pi^2}{6}$$

At the same time, if the angle-sum theorem and the formula "2 + 2 = 4" are subject to possible doubt, then this formula, involving limits, irrationals, and infinites, is more doubtable.

five

Five Classical Puzzles

"Vacancy in a schedule" uses a doctor's appointment to drive home the real existence of nonphysical, nonmental, nontranscendental entities. "Creating/discovering" is a new solution to the old dilemma, based on real math-talk. The other sections develop related ramifications of the humanist philosophy of mathematics.

Vacancy in a Schedule

I phone Dr. Caldwell for an appointment. He's booked way ahead, but his receptionist finds me an opening.

It's fortunate that the opening in Caldwell's schedule exists. That existence may mean he'll see me in time to detect a life-threatening illness.

The opening in his schedule can't be weighed, or measured, or analyzed chemically. It's not a physical object. The suggestion is absurd.

Is it a mental object? A thought in the receptionist's mind/brain? No. If she overlooked it, the opening would still be "there." A thought in Dr. Caldwell's mind/brain? No. When you ask about his schedule, he asks the receptionist.

The opening in Dr. Caldwell's schedule really exists. It's definite; it's recognizable. It has the right to be called "an object." But neither physical, mental, nor transcendental object.

It exists in a web of social arrangements, which includes doctor, receptionist, and patients. It's a social-cultural-historic object, and real as can be.

Creating-Discovering

Is mathematics created or discovered? This old chestnut has been argued forever. The argument is a front in the eternal battle between Platonists and anti-Platonists.

Platonists think mathematical entities can't be created. They already exist, whether we know them or not. We can discover them, but we can't create what's already there.

Formalists and intuitionists, on the contrary, know that mathematics is created by people (mainly mathematicians). It can't be discovered, because nothing's there to discover until we create it.

Each position is internally consistent. They seem incompatible.

Let's not replow this well-trodden ground. Instead, let's listen impartially for "create" and "discover" in nonphilosophical mathematical conversation. Why do both words—"create" and "discover"—seem plausible?

Think about some simple school problem.

"Find the area of a polygon, with vertices such and such and such."

Each student chooses her own method to get the area. Yet their different methods must all yield the same number. If someone gets a different answer, he made a mistake. After the mistakes are found and corrected (and any new mistakes made in the correction process also corrected), the whole class ends up with the same number.

This is the canonical, paradigmatic, fundamental experience of mathematical problem solving. That's why we say, "discover." We're not free to answer according to our fancy! The answer is the answer, like it or not. So we're convinced the answer is somewhere *there*.

If a problem is clearly and definitely formulated, we take it for granted that it has a solution.

(Sometimes the solution is a proof that no solution of the sought-for type is possible.)

When we solve the problem, we say the solution is "found" or "discovered." Not *created*—because the solution was already determined by the statement of the problem, and the known properties of the mathematical objects on which the solution depends.

The curvature of a circle of radius 7 is completely determined by the notion of "circle" and the notion of "curvature." We know there's a unique answer before we know it's 1/7.

When a problem is solved, the solution is brought from below the surface. It moves from the implicit to the explicit, from the hidden to the accessible. Before being discovered or *un*covered, it lay hidden, like an image on exposed, undeveloped photographic film. Once it's found—*developed*—we see where it lay hidden in the known mathematical context. And everybody agrees, this is the solution.

But solving well-stated problems isn't the only way mathematics advances. We must also invent concepts and create theories. Indeed, our greatest praise goes to those like Gauss, Riemann, Euler, who *created* new fields of mathematics.

A well-known classification of mathematicians is problem-solvers and theory builders. When speaking of a *theory*—Galois's theory of algebraic number fields,

Cantor's theory of infinite sets, Robinson's theory of nonstandard analysis, Schwartz's theory of generalized functions—we don't say it was "discovered." The theory is ***in part*** predetermined by existing knowledge, and ***in part*** a free creation of its inventor. We perceive an intellectual leap, as in a great novel or a great symphony.

In the 1930s and 1940s a need was recognized for a generalized calculus that would include singular functions like Dirac's delta function** (see "Change" in this chapter). Theories of generalized functions were created by Solomon Bochner, Sobolev, and Kurt Friedrichs. Meant to serve similar purposes, they were not identical in the way that different answers to a calculus problem are identical. Sobolev spaces and the "distribution theory" of Laurent Schwartz won general acceptance.

When several mathematicians solve a well-stated problem, their answers are identical. They all ***discover*** that answer. But when they create theories to fulfill some need, their theories aren't identical. They ***create*** different theories.

Such was the case with Gibbs's vector analysis versus Hamilton's quaternions (see section on change).

The distinction between inventing and discovering is the distinction between two kinds of mathematical advance. Discovering seems to be completely determined. Inventing seems to come from an idea that just wasn't there before its inventor thought of it. But then, after you ***invent*** a new theory, you must ***discover*** its properties, by solving precisely formulated mathematical questions. So inventing leads to discovering.

Invention doesn't happen only in theory-building. In a well-stated problem, the answers must be the same, but the methods of different problem-solvers may be different. It may be necessary to do something new, to bring in some new trick. You may have to ***invent*** a new trick to ***discover*** the solution. Again, inventing is part of discovering.

Maybe your new trick will help other people to solve their problems. Then it will receive a name, and be studied for its own sake. To ***discover*** its properties, ***inventing*** still other new tricks may be necessary.

For example, Fourier invented Fourier analysis to solve the linear equations of heat and vibration. Then some natural-seeming questions about Fourier series turned out to be very difficult. So Fourier analysis became a field of research on its own. The needs of Fourier analysis (and other parts of mathematics) led to infinite-dimensional linear spaces (Hilbert space and spaces of generalized functions). These new spaces themselves are now big fields of research, which need new ideas in linear operator theory and functional analysis. And so on, and on, and on.

Were the natural numbers 1, 2, 3 . . . discovered or invented? One can't help recalling the diktat of Ludwig Kronecker: "The integers were created by God; all else is the work of man." Since Kronecker was a believer, it's possible he meant this literally. But when mathematicians quote it nowadays, "God" is a figure of speech. We interpret it to mean: "The integers are discovered, all else is invented." Such a statement is an avowal of Platonism, at least as regards the integers or the natural numbers.

How do I as a humanist answer it?

I recall the distinction between Counting Numbers—adjectives applied to collections of physical objects—and pure numbers—objects, ideas in the shared consciousness of a portion of humanity. (See Chapter 1, "A Way Out.")

Counting numbers are discovered. Pure numbers were invented.

Is mathematics created or discovered? Both, in a dialectical interaction and alternation. This is not a compromise; it is a reinterpretation and synthesis.

Finite/Infinite

All the numbers calculated since the formation of the earth are less than $10^{(10^{(10^{10})})}$ (or some higher iterate of iterates.) In other words, they're all finite.

Yet mathematics is full of the infinite. The line R^1 is infinite; the space R^3 is infinite; N, the set of natural numbers, is infinite. There are infinitely many infinite series. There are points "at infinity" on the real line, in the complex plane, in projective space, and, of course, Cantor's hierarchies of infinite sets, infinite ordinal numbers, infinite cardinal numbers.

Where do these infinites come from? Not from observation and not from physical experience. If you don't believe in a separate spiritual or transcendental universe, they must be born in human mind/brains.

Poincaré said arithmetic is based on "the mind's" conviction that what it has done once, it can do again. You needn't search far to find minds holding no such conviction. That "conviction" is trivially false. Any thought or action can be repeated only finitely many times. The person (or animal or machine) that's repeating the thought or action will wear out or die.

The brain is a finite object. It can't contain anything infinite. But we do have ideas of the infinite. It's not the infinite that our mind/brains generate, but *notions* of the infinite.

Logic doesn't force us to bring infinity into mathematics. Euclid had finite line segments, never an infinite line.

In set theory it's the axiom of infinity that provides an infinite set. Without adopting that axiom, Frege and Russell would have had only finite sets.

Sometimes we consciously exclude the infinite. A convergent infinite series is interpreted as a sequence of finite partial sums. We still *call* the series infinite, but we're really interested in the finite partial sums. Yet the "meaningless" intuition of summing infinitely many terms is still the core meaning of "infinite series."

People ask what kind of mathematics would be produced by an alien intelligence (big green critters from Sirius). Maybe they would have no infinity, since it comes out of our heads, not from physical reality.

The Hungarian logician Rózsa Péter wrote a survey of mathematics called *Playing with Infinity*. For her, playing with infinity was the essence of mathematics.

Donald Knuth is an accomplished mathematician and a world champion computer scientist. His *Art of Computer Programming* is called a "bible." Knuth makes vivid a point often scored against constructivists or intuitionists. From Brouwer to Bishop, they can't stomach infinite sets as completed objects, yet they balk at no finite set, no matter how huge. Knuth shows us finite sets so vast that they're just as obscure as the infinite.

The reader may have noticed that nearly all finite sets are vast and huge. To check this obvious fact, just pick a huge number, call it M. How many sets are smaller than M? Very many, but still, finitely many. How many sets are bigger than M? Infinitely many. Infinitely many is *way more* than finitely many, so nearly all finite sets are bigger than M, no matter how big M may be. If you pick a finite set at random, it's almost sure to be bigger than M, no matter how big M might be. Its finitude is small comfort.

In fact, infinite sets are often brought in because they're *simpler* than the given finite sets. Integrating is usually simpler than summing a huge finite number of terms. Differential equations are usually easier than the corresponding finite difference equations.

Moral: If you allow *all* finite sets, you have no excuse to refuse the infinite. The infinite is unintuitive and metaphysical? So are nearly all finite sets. If you want all math to be concrete and intuitive, stick with small finite numbers and small finite sets.

How small? Hard to say, because if n is small, so's $n + 1$. There's no sharp line between small and large, no smallest big number or biggest small number. You could use the M you just picked, and increase Peano's five axioms of arithmetic** by a sixth:

"For all n in N, $n < M$ or $n = M$."

Sad to say, this set of six axioms is inconsistent. To restore consistency, you could modify the first five. For instance,

"If $M/2 < A < M$, then $A + A$ is out of bounds."

Redefine $A + A$ any way you please, or leave it undefined (it doesn't "exist"). Either way, the old arithmetic is out the window. Perhaps we're better off with our infinite number system.

The huge mystery of infinity is an artifact of Platonism. In some transcendental realm, do infinite sets exist? This is the wrong question. Number systems are invented for the convenience of human beings. The appropriate questions about infinity are:

Is it good for anything?

Is it interesting?

Mathematicians have long since answered, yes!

Object/Process

This section strays far from the philosophy of mathematics. In the end it returns to the main subject, with an insight picked up along the way.

Frege said numbers are abstract *objects*. Plato's Ideas, including numbers, were objects of some sort. But intuitionists and formalists deny that numbers are objects.

Kreisel and Putnam say what's needed aren't objects, but objectivity: Numbers are *objective*. We needn't trouble whether they're objects.

Some people say, "Numbers exist, that's plain as the chair I'm sitting on."

Others say, "It's obvious that numbers don't exist in the sense that this chair exists."

The two statements aren't contradictory. Both are true.

If I have no prior explication of "exist," my knowledge of numbers isn't increased if you tell me they exist or don't exist. In his great polemic on the foundations of arithmetic, Frege refrains from explaining the meaning of "exist" or the meaning of "object."

Is a cloud an object? Sort of yes.

A roaring fire on the prairie? No, more like a process.

My grandchild's temper fit? Definitely a process. $10^{(10^{10})}$? An object a la Frege, or just "something objective" a la Kreisel-Putnam?

Does "object" mean "something like a rock"? Something with definite shape and volume? Then only solids would be objects. Melt a piece of ice, and you turn object into nonobject. So stringent an objecthood is too special, too transient, too conditional for philosophy.

Sometimes "object" seems to mean any physical entity, with volume and shape or without. Then atoms are objects. So are electrons, photons, and quarks. But at this level, the distinction between object and process vanishes. Depending on how they're observed, electrons, protons, and photons have particle-like (object) or wavelike (process) behavior.

A wider meaning of "object" is, "independent of my consciousness." Maybe this is what Kreisel-Putnam mean by objective. Does it mean independent of *anyone's* consciousness?

Are you and I objects to each other, but subjects to ourselves? Are we both just objects? Both just subjects? Or objective subjects? Subjective objects?

A related meaning of "object" is, "anything that can affect me." Such a definition was used by Paul Benacerraf in a widely referenced paper. A tree could affect me if I drove into it, and a germ could affect me by giving me the sniffles. An attack of paranoia could make me very sick. Would that attack be an object?

Still another meaning is by opposition to "process." "'Object' is noun, 'process' is verb. Object acts or is acted on. Process is the action."

In all these explications, objects are independent of individual consciousness. They have relatively permanent qualities. They can be observed or experienced by anyone with the appropriate sense organ, the appropriate training of eye and brain, the appropriate scientific instrument.

Niagara Falls is an example of the dialectic interplay of object and process. Niagara Falls is the outlet of Lake Ontario. It's been there for thousands of years. It's popular for honeymoons. To a travel agent, it's an object.

But from the viewpoint of a droplet passing through, it's a process, an adventure:

> over the cliff!
> fall free!
> hit bottom!
> flow on!

Seen in the large, an object; felt in the small, a process. (Prof. Robert Thomas informs me that the Falls move a few feet or so, roughly every thousand years.)

Movies show vividly two opposite transformations:

A. Speeding up time turns an object into a process.
B. Slowing down time turns a process into an object.

To accomplish (A), use time-lapse photography. Set your camera in a quiet meadow. Take a picture every half hour. Compose those stills into a movie.

> Plant stalks leap out of the ground, blossom, and fall away!
> Clouds fly past at hurricane speed!
> Seasons come and go in a quarter of an hour!
> Speeding up time transforms meadow-object to meadow-process.

To accomplish (B), use high-speed photography to freeze the instant.

A milk drop splashed on a table is transformed to a diadem—a circlet carrying spikes, topped by tiny spheres. By slowing down time, splash-process is transformed to splash-object.

A human body is ordinarily a recognizable, well-defined *object*. But physiologists tell us our tissues are flowing rivers. The molecules pass in and out of flesh and bone. Your friend returns after a year's absence. You recognize her, yet no particle of her now was in her when she left. Large-scale, object—small-scale, process.

In the social-cultural-historic domain, the continuity between object and process is blatant, even though some institutions, beliefs, and practices seem eternal. All institutions change. If they change slowly, over centuries—slavery, piracy, royalty, private property, female subjection—they are thought of as objects. If they change daily—clothing fashions, stock market prices, opinion polls—they are thought of as processes.

A nation defends or alters borders, signs treaties, makes war. It claims to be an object. Yet it's also a horde of individual persons being born, dying, immigrating, and emigrating. A few individuals decide the fate of nations, and nations decide the fate of many individuals. A nation is like a waterfall: an object in the large, a process in the small.

In computing machines, the difference between object (hardware) and process (software) is almost arbitrary. The designer decides which functions to embody in hardware, which in software.

Some features of mind/brain are object-like, life-long. Some change by the minute. A psychotherapist tries to overcome a patient's belief that his neurotic symptom is an object. A step toward cure is convincing him that it is a process.

Before Galileo turned his telescope to the sun, philosophy thought earth was change and process, and the sun was a changeless object. We learned that the "changeless" sun is a bonfire. In time it must fade and die, or explode. The dead moon too had a birth and a history.

For the geologist, using a long time scale, earth is a process. Looking at the huge earth from a short distance, we ordinarily don't see it as a whole, as an object. But when astronauts stood on the moon, they saw earth from a long distance for a short time. They saw an object.

In Einstein's equation

$$E = mc^2$$

where m is mass: ***object***, and E is energy: ***process***; and c is the speed of light in vacuum. (This tremendous number must be ***squared***!)

This equation tells how at Hiroshima and Nagasaki mass-object-bombs transformed into energy-process holocausts. In stars as in nuclear explosions, mass is converted to energy. It's equally sensible or senseless to say a star is an object or a process.

When Edwin Hubble photographed his red shift, and later when Arno Penzias picked up his 15 billion year-old echo, we learned that the cosmos passes through history, from a Big Bang to a Big Crunch or a Cold Death. Cosmology became historical, like biology and geology.

Presently, elementary particles called hadrons and bosons seem to compose the universe. But those particles exist only during one stage of cosmic evolution, when the configuration of the Cosmos as a whole permits that existence. Physics itself takes on a historical, evolutionary aspect.

High-speed and time-lapse photography show that the object-process polarities are ends of a continuum. Any phenomenon is seen as an object or a process, depending on the scale of time, the scale of distance, and human purposes. Consider nine different time-scales—astronomical time, geologic time, evolutionary time, historic time, human lifetime, daily time, firing-squad time, switching time for a microchip, unstable particle lifetime.

A smaller scale in space, or speeding up time, turns an object into a process. A larger scale in space, or slowing down time, turns a process into an object.

In brief, *an object is a slow process. A process is a speedy object.*

What about mathematics? In mathematics don't we have "abstract objects," not located in time or space? Like "the" equilateral triangle? or "the" number 9?

From Pythagoras to Frege, philosophy gave mathematical objects an idyllic existence, free of blemishes such as temporality, impermanence, and indefiniteness. They were thought to be absolutely still, free of any process properties. Plato's Ideas were pure objects. Frege's abstract objects are the nineteenth century version of Plato's Ideas. Instead of timeless, they're "tenseless."

The example of numbers and triangles is supposed to prove that "abstract objects" exist. On the other hand, the puzzle about the mode of existence of numbers and triangles is supposed to be solved by calling them "abstract objects."

What's the meaning of "abstract"? Abstract objects are *not* mental, *not* physical, *not* historical, *not* social or intersubjective.

How do I get acquainted with them? No answer.

We can see that mathematical objects aren't mental or physical. It would be an over-polysyllabic tongue-twister to call them "nonmental-nonphysical objects." So make an abbreviation: "abstract objects." A more honest name would be "transcendental objects."

Nowhere in heaven or on earth is there a pure object, totally free of process aspects—free of change.

Only in mathematics we think we have pure objects. There, it is thought, we find *nothing but* pure objects. Infinitely many of them!

Could this thinking come from seeing mathematics in too short a time scale? Wouldn't a view that encompassed centuries show mathematics evolving—a process?

This book argues that mathematics is a social-historical-cultural phenomenon, without need of anything abstract-Platonic-nonhuman, without need of formalist or intuitionist reduction, but this view implies that mathematical objects, like other objects, are also processes. They change, whether in plain sight or too slowly to be noticed.

The next section takes the number 2 as a case study of a changing mathematical object.

Change

From Pythagoras to us, 2 has changed, as surely as music has changed and religion has changed.

In some ways 2 has eroded; in many ways it has expanded.

$1 + 2 = 3$

is true for us. It was true for Pythagoras. Yet his meaning wasn't identical with ours. "1" was not just our 1. It also was God, unity. 2 was the female principle. 3 was the male principle.

$$1 + 2 = 3$$

was more than a rule of making change in the market. It was a religio-philosophical cosmic truth.

The Pythagoreans discovered: "$\sqrt{2}$ does not exist." That is to say, there's no pair of whole numbers, p and q, such that

$$(p/q)^2 = 2.^{**}$$

Today, the real numbers are available to help understand a natural number like 2. Your calculator tells you $\sqrt{2}$ is some close approximation to an irrational number

$$1.414213562\ldots$$

So now $\sqrt{2}$ exists, and consequently the meaning of 2 has changed.

Since Pythagoras mathematics has been regarded as unchanging and eternal.

$$2 + 2$$

always was and always must be 4. Euclid knew it, and so (hopefully) will our descendants millennia hence. How can I claim that

$$2 + 2 = 4$$

isn't timeless?

First a small point. Euclid never saw the formula $2 + 2 = 4$. The symbols "2," "4," "+," and "="—and the practice of putting symbols together to make formulas—were alien to his time.

But never mind the formula, let's stick to the facts. When Euclid went to market, he knew that two oboli plus two oboli was worth four oboli. If we're talking about "2" and "4" as *adjectives* modifying "oboli," and + and = as a commercial operation and a commercial relation, we and Euclid agree, just as we would agree that the sun rises in the East. But if we're talking about the *nouns* "2" and "4"—meaning some sort of autonomous objects—and operations and relations on *them*, there are differences between us and Euclid. (See the section on Certainty in Chapter 4.)

2 is no longer only a counting (natural) number. Now it has an additive inverse, -2. (It's an integer.)

2 is no longer isolated. Now it's a rational number. As such, an element in a dense ordered set. 2 is now a point on the continuous number line.

2 is even a point in the complex plane, participating in analytic functions and conformal maps.**

2 has matured into a complex creature, with resonances, possibilities, and connections of unfathomed subtlety.

And most remarkable—it has a *pair* of square roots—nonexistent for Pythagoras and Euclid!

As we add new structures, we embed the old ones into them. It's more efficient to make the natural numbers, the integers, the rational numbers, the real numbers subsets of the complex numbers, rather than separate objects isomorphic to subsets of the complex numbers.

To appease any fussy formalists who join our conversation, we might try more explicit notations: 2_N for the natural number 2, 2_I for the integer 2, 2_Q for the rational number 2, 2_R for the real number 2, 2_C for the complex number 2. But no one working with 2 would bother with this. It would just slow you down. What's the Platonist's alternative? Are there uncountably many undiscovered twos still waiting to be discovered? Or is there and was there, already in Pythagoras's time, one majestic, unique, eternal 2, already an integer, already a rational number, already a real number, already a complex number, *and* who knows what else? These are fables you can believe if you want to.

As mathematics grows and changes, the numbers change.

Euclid had line segments and we have line segments, but they're not the same. Euclid's is a simple thing. It has two endpoints and "lies evenly on itself." Ours is a grand mystery, a set with an uncountable infinity of elements (points) and subsets of undecidable sizes. It participates in the operations of algebra and analysis, and is linked to an unfamiliar non-Euclidean cousin. Euclid knew what he was talking about, but it wasn't quite what we talk about.

As mathematics grows and changes, geometry changes.

A more current example. Until the midtwentieth century, the "derivative" or "slope" of a function at a point existed only if at that point the graph of the function was smooth—had a definite direction, and no jumps. Now mathematicians have adopted Laurent Schwartz's generalized functions.** *Every* function, no matter how rough, has a derivative.

The Heaviside function H(t) consists of two pieces.
On the left, when t is less than 0, H(t) = 0 identically.
On the right, when t is greater than 0, H(t) = 1 identically.
When t = 0, H(t) jumps from 0 to 1.

Classically, H'(t), the slope of H(t), exists only for t greater than 0 and for t less than 0. At those points the graph of H(t) is flat, and its slope H'(t) is 0.

At t = 0, where H(t) jumps, H'(t) is classically undefined. The derivative doesn't exist there.

Nowadays there's a derivative for H(t) as a whole, *including at the jump*. It's Dirac's "delta function,"** a gadget introduced by the physicist Paul Dirac. By the classical definition of function, the "delta function" isn't a function. We ought to write, not "delta function," but delta "function."

Dirac's delta function is zero for all t except $t = 0$. When $t = 0$, it's infinite. Its "graph" is an infinitely thin, infinitely high spike at $t = 0$. And the area under this weird graph is 1!

Classically, this is nonsense. "Infinite" isn't a number. Dirac's spike would be a "rectangle" with infinite height and thickness 0, so the "area" would be infinity times 0. But infinity times 0 can be anything at all.

Why bother with such peculiar stuff? For Dirac, it came in handy in quantum mechanical calculation. But it's also a model for some ordinary physics. Slam a ping-pong ball with your paddle. Its speed goes instantly from 0 to 1. As a function of time, its speed is a Heaviside function. Its acceleration is Dirac's delta function!

The meaning of differentiation has changed. Newton and Leibniz's differentiation operator has become something more general. Our generalized differentiation includes the old differentiation, and it's much more powerful.

As mathematics grows and changes, functions and operators change.

A familiar cliché says that while other sciences throw away old theories, mathematics throws away nothing. But the old mathematics isn't preserved intact. Mathematics is intensely interconnected and self-interactive. The new is vitally linked to the old. The old is revitalized, enriched, and complexified by interaction with the new.

An excerpt from Mary Tiles (p. 151):

> The introduction of ideal elements does not leave the pre-existing domain fixed. The conception of number as the measure of a magnitude is what militates against the admission of either 0 or negative numbers as numbers. If they are admitted as ideal elements, to complete the system, then they have no interpretation as proper numbers (magnitudes.) But as they come to form a single representation system with the positive numbers, they come to have applications. The result is that the concept of number itself is no longer tied to that of magnitude. In this sense even the finitary significance of numbers has not, historically, been without change. Peano's axiomatization of the natural number system, making it essentially a system of entities generated from 0 by repeated application of the successor operation, again opens up a pathway to the field of recursive function theory, the theory of algorithms and the whole modern computational life of the number system. Computation as understood by this route is not what it was before those developments. . . . What emerges is not a static picture of a set of entities on the one hand with a determinate and fixed set of reasoning procedures which can, once and for all, be characterized and certified as reliable.

Nonexist/Exist

There are different kinds of existence. There are also different kinds of nonexistence. Speaking of Pegasus, the beautiful winged horse of Greek legend, I can say

There's no such myth (false).
There's no such physical object (true).
There's no such entity in biology (true).
There's no such thought in my mind (false).

In mathematics, nonexistence usually is a matter of impossibility.

"A solution to this problem does not exist" means "It's impossible to solve this problem."

Often nonexistence of one thing is equivalent to existence of another thing. Euclid proves there's no largest prime number—no prime number greater than all other primes. Nonexistence! Today the usual statement of this fact is: "There exist infinitely many primes." Infinite existence!

Back in the early Stone Age when we couldn't count past 20, 20 was an upper bound of our number system. Some easy questions were hard. Like,

$$15 + 15 = ?$$

We overcame this difficulty and enlarged the number system. We reached our system of natural numbers, which has no upper bound. First-graders today know you can go on counting forever. And we can write

$$15 + 15 = 30.$$

In imagination we can add any two numbers, no matter how big. For this enlargement we pay a penalty. Problems arise that couldn't be imagined in a system bounded by 20. We have a number system which is ultimately unknowable.

Two old conjectures we can't prove or disprove are: "Every even number is the sum of two primes" (Goldbach); and "There are infinitely many prime pairs"

(like $\{11,13\}$, $\{29,31\}$, $\{41,43\}$).

We enlarge a mathematical system to make it simpler in some sense, and we thereby create worse complications in some other senses. The new complications generate new problems, which drive us to enlarge the system still further.

Mathematics is the only science in which "exist" is a technical term. Contention about mathematics between philosophical schools is mainly about existence of mathematical objects. An advanced mathematics graduate student can tell whether, in her specialty, existence has been proved. But what is ***meant*** by existence? ***In what sense*** does something exist?

Two kinds of mathematical existence are sometimes distinguished—constructive and indirect. Constructive existence means that the object is gotten from known objects by a finite number of steps, or within arbitrarily small error by a finite number of steps.

To "construct" the rationals, I say, "Consider the set of ordered pairs (a b), etc." Voilà! The rationals have been constructed!**

Step back from the math for a moment. Isn't this amazing? We've actually constructed infinitely many distinct entities, in a few minutes of pencil-pushing, at little effort, and no expense! Isn't there something strange about that?

On the other hand, most people think all these numbers existed way back in time, long before we started this conversation. If that's so, have we constructed anything?

Yet this kind of "construction" is considered straightforward. There is controversy about indirect proof, in which you prove "A must be true" by proving "not-A is impossible." Indirect proofs accepted by classical mathematicians are rejected by intuitionists and constructivists. (See Brouwer and Bishop in Chapter 9, and "Finite/Infinite" above.)

How Mathematics Grows

Predicting the growth of mathematics would require knowing what mathematicians are trying to do, not just what they have done so far. A judgment on whether their goals are attainable would help prediction. In 1900 David Hilbert laid 23 problems before the mathematical world. One could have tried to guess which problems would be attractive and approachable, and thereby made a short-term prediction about future mathematics.

A deeper attack on the prediction problem would look, not only at what mathematicians have done recently, and what they're now attempting, but at what the present state of mathematical knowledge naturally calls for—what we *should* be trying to do. That's what Hilbert did to make his list of problems. Because he presented it in 1900 at an International Congress, and because he was David Hilbert, it was a self-fulfilling prophecy.

Some mathematicians say that if history started all over, mathematics would evolve into much the same thing, in much the same order. The opportunities and questions arising from what we know decide what advances we make, what we'll know next year. Igor Shafarevich, a Russian mystic and anti-Semitic algebraic geometer, says mathematicians don't make mathematics, they're instruments for mathematics to make itself. This strange-sounding theory is supported by many instances of repeated or simultaneous discovery.

Desargues discovered projective geometry in the seventeenth century. In the shadow of the analytic geometry of Fermat and Descartes it was overlooked and forgotten. In the nineteenth century Monge and Poncelet rediscovered it.

Gauss discovered Abelian integrals before Abel and the method of least squares before Legendre.

The "Argand diagram" of complex numbers was found by Caspar Wessel before Argand, by Gauss after Argand, and by Euler before any of them.

The Ascoli theorem of the 1920s became the Arzela-Ascoli theorem when it was found in a paper Arzela had published 39 years before Ascoli.

Polya's counting formula was published by Redfield 30 years before Polya.

Jesse Douglas and Tibor Radó had a priority quarrel about the Plateau problem.

And so on.

These examples suggest that mathematical discoveries force themselves on mathematicians. Mathematics unfolds, like a flower opening, or a tree plunging roots deeper and crown higher.

Michael Dummett said that mathematics grows as if it moved on rails projecting a little way into the future. This means short-run prediction can work, long-run prediction can't. I distinguish two kinds of "rails" into the future of mathematics. One kind is solving problems already recognized and stated. (One kind of "solution" would be a proof that the problem is unsolvable.) Since these problems are under attack today, their solutions, though yet unknown, are already determined. Once a problem is solved, we see how the shape of the solution was predetermined by the statement of the problem. This kind of rail is rigid but not long—few problems stay unsolved more than a few dozen years.

Dummett's idea of rails that go only a little way forward is an appealing metaphor. How seriously can we take it? What are the rails? How far do they go? Some people talk as if math is already determined forever by the axioms of set theory—because all theorems are in principle already determined by the axioms and rules of inference. But the mathematics that will ever be known and used is a tiny fraction of all the "in principle" consequences. Real-life mathematics can only include consequences that somebody thought useful or interesting. To what extent do notions of useful or interesting endure? (v. Polanyi). Some theories lose interest. New ideas appear unexpectedly.

A different kind of "rail" is created by unrecognized potentialities of what we know. The whole theory of analytic functions was predetermined when sixteenth-century algebraists introduced the square root of -1. Some of today's familiar concepts are pregnant with beautiful developments, which will seem inevitable once we notice them.

New theories allow more individual leeway than solutions of explicitly stated problems. They're less rigid but stick further into the future.

Mathematics is made of theories in the first place, of objects only secondarily. We study the theory of numbers, not numbers per se. The theory of distributions was created by Laurent Schwartz in the 1930s. A Platonist would say Schwartz's distributions always existed in the set theoretic universe. It's more interesting to see how aspects of his theory were foreshadowed, for example, by Hadamard's "finite part" of a divergent integral. Saying distributions already existed in an abstract set universe "somewhere" is of little interest to mathematicians.

We have to distinguish two senses of "exist"—potential and actual. The statue of David already existed potentially. Michelangelo gave it actual existence. The passage from potential to actual is the interesting event.

The distinction between potential and actual existence isn't limited to mathematics. It arises whenever you consider the future. Possible futures have potential existence. Some will be realized, many will not. Whatever is ultimately actualized must already have been possible (potential). Related notions are common in physics, biology, and political science.

In ***Mathematical Notes and Comments***, in the section on "Imaginary Becomes Reality," we give examples of one way mathematics grows—by creating new entities or theories, to make unsolvable problems solvable.

Is the philosophy of mathematics concerned with potential existence or actual existence? Frege and Gödel, and set-theoretic Platonists in general, seem to think of the potential as if it were already actual.

Before Socrates, Parmenides taught that whatever isn't contradictory already exists. Brouwer's intuitionism, on the other hand, defines mathematics as the thinking of the Creative Mathematician. It focuses on the actual. One could class many formalists with intuitionists in this respect. The present work emphasizes both the distinction and the connection between potential and actual. Mathematical discovery or creation is transformation of potential to actual.

Part Two

six

Mainstream Before the Crisis

The next four chapters tell the mainstream philosophies of mathematics. They try to be entertaining and relevant, not definitive or exhaustive. Most of the facts are well known. The arrangement and some interpretation are novel.

The name "foundationism" was invented by a prolific name-giver, Imre Lakatos. It refers to Gottlob Frege in his prime, Bertrand Russell in his full logicist phase, Luitjens Brouwer, guru of intuitionism, and David Hilbert, prime advocate of formalism. Lakatos saw that despite their disagreements, they all were hooked on the same delusion: ***Mathematics must have a firm foundation.*** They differ on what the foundation should be.

Foundationism has ancient roots. Behind Frege, Hilbert, and Brouwer stands Immanuel Kant. Behind Kant, Gottfried Leibniz. Behind Leibniz, Baruch Spinoza, and René Descartes. Behind all of them, Thomas Aquinas, Augustine of Hippo, Plato, and the great grandfather of foundationism—Pythagoras.

We will find that the roots of foundationism are tangled with religion and theology. In Pythagoras and Plato, this intimacy is public. In Kant, it's half covered. In Frege, it's out of sight. Then in Georg Cantor, Bertrand Russell, David Hilbert, and Luitjens Brouwer, it pops up like a jack-in-the-box.

In the twentieth century, we look at Russell, Brouwer, Hilbert, Edmund Husserl, Ludwig Wittgenstein, Kurt Gödel, Rudolph Carnap, Willard V. O. Quine, and a small sample of today's authors. Philip Kitcher said the philosophy of mathematics is generally supposed to begin with Frege—before Frege there was only "prehistory." Frege transformed the issues constituting philosophy of mathematics. In that sense earlier philosophy can be called prehistoric. But to understand Frege you must see him as a Kantian. To understand Kant you must see his response to Newton, Leibniz, and Hume. Those three go back to Descartes, and through him to Plato. Plato was a Pythagorean. The thread from Pythagoras to Hilbert and Gödel is unbroken. I aim to tell a connected story from Pythagoras to the present—where foundationism came from, where it left us.

Instead of going straight through from Pythagoras, I've split the story into two parallel streams—the first section is about the "Mainstream." The second is about the "humanists and mavericks."

For the Mainstream, mathematics is superhuman—abstract, ideal, infallible, eternal. So many great names: Pythagoras, Plato, Descartes, Spinoza, Leibniz, Kant, Frege, Russell, Carnap. (For Kant, membership in this group is partial.)

Humanists see mathematics as a human activity, a human creation. Aristotle was a humanist in that sense, as were Locke, Hume, and Mill. Modern philosophers outside the Russell tradition—mavericks—include Peirce, Dewey, Roy Sellars, Wittgenstein, Popper, Lakatos, Wang, Tymoczko, and Kitcher (a self-styled maverick). There are some interesting authors who aren't labeled philosophers: psychologist Jean Piaget; anthropologist Leslie White; sociologist David Bloor; chemist Michael Polányi; physicist Mario Bunge; educationists Paul Ernest, Gila Hanna, Anna Sfard; mathematicians Henri Poincaré, Alfréd Rényi, George Pólya, Raymond Wilder, Phil Davis, and Brian Rotman.

First we honor the great grandfather of the mainstream.

Pythagoras (fl. 540–510 B.C.)

Religion Is Mathematics—Mathematics Is Religion

No one knows when mathematics started. Koehler reports on studies by comparative psychologists of counting by jackdaws and pigeons. Human arithmetic goes back at least as far as possession of property—clam shells, fish heads, whatever. Geometry in the most elementary sense goes back even further, to the animal that must run straight toward a prey or straight away from a predator.

The philosophy of mathematics, on the other hand, seems to start with Pythagoras (about 572–479 B.C. or a little later; Heath, 1981, p. 67).

Schoolchildren know his namesake, the indispensable relation between the sides of a right triangle. Whatever Pythagoras had to do with that formula, it was known centuries before him, in China and Babylon. More than about Pythagoras, we know about the Pythagoreans, the secret society he supposedly founded. They combined mysticism and superstition with geometry and arithmetic in a way incomprehensible today. According to Heath (1981, p. 66):

> It is difficult to disentangle the portions of the Pythagorean philosophy which can safely be attributed to the founder of the school. Aristotle evidently felt this difficulty; . . . when he speaks of the Pythagorean system, he always refers it to "the Pythagoreans," sometimes even to "the so-called Pythagoreans."

But

> It is certain that the Theory of Numbers originated in the school of Pythagoras; and with regard to Pythagoras himself, we are told by Aristoxenus that "he seems to have attached supreme importance to the study of arithmetic, which he advanced and took out of the region of commercial utility."

Boyer and Merzbach (1991, p. 53) tell us that "Many early civilizations shared various aspects of numerology, but the Pythagoreans carried number worship to its extreme, basing their philosophy and their way of life upon it. The number one, they argued, is the generator of numbers and the number of reason; the number two is the first even or female number, the number of opinion; three is the first true male number, the number of harmony, being composed of unity and diversity; four is the number of justice or retribution, indicating the squaring of accounts; five is the number of marriage, the union of the first true male and female numbers; and six is the number of creation. Each number had its peculiar attributes. The holiest of all was the number ten, or the *tetractys*, for it represented the number of the universe, including the sum of all possible dimensions. [See also Heath, 1981, p. 75.] A single point is the generator of dimensions, two points determine a line of dimension one, three points S (not on a line) determine a triangle with area of dimension two, and four points (not in a plane) determine a tetrahedron with volume of dimension three; the sum of the numbers representing all dimensions, therefore, is . . . ten. It is a tribute to the abstraction of Pythagorean mathematics that the veneration of the number ten evidently was not dictated by anatomy of the human hand or foot."

R. Tarnas writes, in *The Passion of the Western Mind* p. 46, "For Pythagoreans, as later for Platonists, the mathematical patterns discoverable in the natural world secreted, as it were, a deeper meaning that led the philosopher beyond the material level of reality. To uncover the regulative mathematical forms in nature was to reveal the divine intelligence itself, governing its creation with transcendent perfection and order. The Pythagorean discovery that the harmonics of music were mathematical, that harmonious tones were produced by strings whose measurements were determined by simple numerical ratios, was regarded as a religious revelation. Those mathematical harmonies maintained a timeless existence as spiritual exemplars, from which all audible musical tones derived. The Pythagoreans believed that the universe in its entirety, especially the heavens, was ordered according to esoteric principles of harmony, mathematical configurations that expressed a celestial music. To understand mathematics was to have found the key to the divine creative wisdom. . . . Through intellectual and moral discipline, the human mind can arrive at the existence and properties of the mathematical Forms, and then begin to unravel the mysteries of nature and the human soul."

Heath (1981, pp. 68–69) thinks that in Pythagorean doctrine "the number in the heavens" means the number of visible stars. He asks, "may this not be the origin of the theory that all things are numbers, a theory which of course would be confirmed when the further capital discovery was made that musical harmonies depend on numerical ratios, the octave representing the ratio 2:1 in length of string, the fifth 3:2, and the fourth 4:3?"

Plutarch (40 A.D.–120 A.D.) connects the Pythagoreans to the Isis cult of Egypt, blending sacred history and mathematical theorems: "The Egyptians relate that the death of Osiris occurred on the seventeenth (of the month), when the full moon is most obviously waning. Therefore the Pythagoreans call this day the 'barricading' and they entirely abominate this number. For the number seventeen, intervening between the square number sixteen and the rectangular number eighteen, two numbers which alone of plane numbers have their perimeters equal to the areas enclosed by them (***proving this makes a nice problem for the reader with pencil and paper handy***), bars, discretes, and separates them from one another, being divided into unequal parts in the ratio of nine to eight. The number of twenty-eight years is said by some to have been the extent of the life of Osiris, by others of his reign; for such is the number of the moon's illuminations and in so many days does it revolve through its own cycle. When they cut the wood in the so-called burials of Osiris, they prepare a crescent-shaped chest because the moon, whenever it approaches the sun, becomes crescent-shaped and suffers eclipse. The dismemberment of Osiris into fourteen parts is interpreted in relation to the days in which the planet wanes after the full moon until a new moon occurs."

Nicomachus of Gerasa (c. 100 A.D.) was a later Pythagorean. (Gerasa is in Judaea east of the Jordan; Heath, 1981, p. 97). His *Introductio Arithmetica* came out in several versions. After a religio-philosophical introduction, it's a straightforward account of number theory as known to the Pythagoreans and to Euclid. He writes (p.187) "arithmetic . . . existed before all the others [of the quadrivium: music, geometry, and astronomy] in the mind of the creating God like some universal and exemplary plan, relying upon which as a design and archetypal example the creator of the universe sets in order his material creations and makes them attain to their proper ends" (p. 189). "All that has by nature with systematic method been arranged in the universe seems both in part and as a whole to have been determined and ordered in accordance with number, by the forethought and the mind of him that created all things; for the pattern was fixed, like a preliminary sketch, by the domination of number preexistent in the mind of the world-creating God, number conceptual only and immaterial in every way, but at the same time the true and the eternal essence, so that with reference to it, as to an artistic plan, should be created all these things time, motion, the heavens, the stars. . . ."

Heath thinks (1981, pp. 97–99) that the success of Nicomachus's book is "difficult to explain except on the hypothesis that it was at first read by philosophers rather than mathematicians, and afterward became generally popular at a time when there were no mathematicians left, but only philosophers who incidentally took an interest in mathematics." Van der Waerden, on the other hand, (p. 97) finds Nicomachus entertaining.

The Pythagoreans are not talked about in philosophy courses. Their thinking seems unworthy to be called philosophy. But Cornford says (p. 194): "Parmenides, the discoverer of logic, was an offshoot of Pythagoreanism, and Plato himself [found] in Pythagoreanism the chief source of his inspiration."

Bertrand Russell (1945) says, "Pythagoras was intellectually one of the most important men that ever lived. . . . The influence of mathematics on philosophy, partly owing to him, has, ever since his time, been both profound and unfortunate (p. 29). . . . He may be described, briefly, as a combination of Einstein and Mary Baker Eddy. He founded a religion, of which the main tenets were the transmigration of souls and the sinfulness of eating beans" (p. 31).

"The combination of mathematics and theology, which began with Pythagoras, characterized religious philosophy in Greece, in the Middle Ages, and in modern times down to Kant. [My addition: and also down to Frege and Russell] . . . in Plato, Saint Augustine, Thomas Aquinas, Descartes, Spinoza, and Kant there is an intimate blending of religion and reasoning, of moral aspiration with logical admiration of what is timeless, which comes from Pythagoras, and distinguishes the intellectualized theology of Europe from the more straightforward mysticism of Asia. . . . I do not know of any other man who has been as influential as he was in the sphere of thought . . . what appears as Platonism is . . . in essence Pythagoreanism. The whole conception of an eternal world, revealed to the intellect but not to the senses, is derived from him" (p. 37).

Plato (428-7–348-7 B.C.)
Pythagoras Is Refined—Numbers Live in Heaven

R. Tarnas in *Passion of the Western Mind*, p. 10 ff., writes, "The paradigmatic example of Ideas for Plato was mathematics. Following the Pythagoreans, with whose philosophy he seems to have been especially intimate, Plato understood the physical universe to be organized in accordance with the mathematical Ideas of number and geometry. These Ideas are invisible, apprehensible by intelligence only, and yet can be discovered to be the formative causes and regulators of all empirical visible objects and processes. But again, the Platonic and Pythagorean conception of mathematical ordering principles in nature was essentially different from the conventional modern view. In Plato's understanding, circles, triangles, and numbers are not merely formal or quantitative structures imposed by the human mind on natural phenomena, nor are they only mechanically present in phenomena as a brute fact of their concrete being. Rather, they are numinous and transcendent entities, existing independently of both the phenomena they order and the human mind that perceives them. While the concrete phenomena are transient and imperfect, the mathematical Ideas ordering those phenomena are perfect, eternal, and changeless. Hence the basic Platonic belief—that there exists a deeper, timeless order of absolutes behind the surface confusion and randomness of the temporal world—found in mathematics, it was thought, a particularly graphic demonstration. The training of the mind in mathematics was therefore deemed by Plato essential to the philosophical enterprise, and according to tradition, above the door to his Academy were placed the words, 'Let no one unacquainted with geometry enter here.'

"It had become evident that several celestial bodies did not move with the same eternal regularity as did the rest, but instead they "wandered" (the Greek root for the word "planet," *planetes*, means "wanderer" and signified the Sun and Moon as well as the other five visible planets—Mercury, Venus, Mars, Jupiter, and Saturn.) Not only did the Sun (in the course of a year) and the Moon (in a month) move gradually eastward across the starry sphere in an opposite direction from the westward diurnal movement of the entire heavens. More puzzling, the other five planets had glaringly inconsistent cycles in which they complete those eastward orbits, periodically appearing to speed up or slow down relative to the fixed stars, and sometimes to stop altogether and reverse direction while emitting varying degrees of brightness. The planets were inexplicably defying the perfect symmetry and circular uniformity of the heavenly motions.

"Because of his equation of divinity with order, or intelligence and soul with perfect mathematical regularity, the paradox of the planetary movements seems to have been felt most acutely by Plato, who first articulated the problem and gave directions for its solution.

"To Plato the proof of divinity in the universe was of the utmost importance for only with such certainty could human ethical and political activity have a firm foundation. In the *Laws*, he cited two reasons for belief in divinity—his theory of the soul (that all being and motion is caused by soul, which is immortal and superior to the physical things it animates) and his conception of the heavens as divine bodies governed by a supreme intelligence and world soul. The planetary irregularities and multiple wanderings seemingly contradicted that perfect divine order, thereby endangering human faith in the divinity of the universe. Therein lay the significance of the problem. Part of the religious bulwark of Platonic philosophy was at stake. Indeed, Plato considered it blasphemous to call any celestial bodies 'wanderers.'

"But Plato not only isolated the problem and defined its significance. He also advanced, with remarkable confidence, a specific—and in the long run extremely fruitful—hypothesis: namely that the planets, in apparent contradiction to the empirical evidence, actually move in single uniform orbits of perfect regularity. Although there would seem to have been little but Plato's faith in mathematics and the heavens' divinity that could have supported such a belief, he enjoined future philosophers to grapple with the planetary data and find 'what are the uniform and ordered movements by the assumption of which the apparent movements of the planets can be accounted for'—to discover the ideal mathematical forms that would resolve the empirical discrepancies and reveal the true motions" (Heath, 1913; and Plato's *Laws,* 1961, pp. 821–22).

Unlike Pythagoras, Plato left us plenty of written records. As a youth, he studied with Socrates, who was convicted in King Archon's court for impiety and corrupting young men. He was executed by drinking hemlock in 399 B.C., when Plato was about 31 (Ryle). We know little of Socrates's life. Plato made him the central figure in most of his *Dialogues.* We can only guess which of Socrates's speeches are Socrates's thoughts, which the thoughts of Plato.

The *Dialogues* aren't systematic philosophical expositions. Nowhere is there a coherent statement of Plato's philosophy of mathematics, which we could compare with "Platonism" today. They achieve trenchant philosophical analysis by lively conversation among living, breathing Athenians. Plato's views changed over the years when the *Dialogues* were written. In midlife, on a trip to Syracuse, he became a Pythagorean. It's said that unknown to the young Plato Socrates too had belonged to a Pythagorean cell. The Pythagoreans were a political as well as a religio-philosophical group, and there was reason for secrecy.

Plato's *Republic* is a manifesto for dictatorship by philosopher-kings, trained under Plato's principles.

It contains the famous metaphor of the cave. Here is the end of Socrates's concluding speech about the cave:

"What appears to me is, that in the world of the known last of all is the idea of the good, and with what toil to be seen! And seen, this must be inferred to be the cause of all right and beautiful things for all, which gives birth to light and the king of light in the world of sight, and, in the world of mind, herself the queen produces truth and reason; and she must be seen by one who is to act with reason publicly or privately."

Here are rules for educating "guardians" or philosopher-kings, from Book 7 of *The Republic* ("Great Dialogues of Plato," pp. 315–16, 323–31). Plato gave Socrates the first person, "I."

> My dear Glaucon, what study could draw the soul from the world of becoming to the world of being? . . . this, which they all have in common, which is used in addition by all arts and all sciences and ways of thinking, which is one of the first things every man must learn of necessity."
>
> "What's that?" he asked again.
>
> "Just this trifle,' I said—to distinguish between one and two and three: I mean, in short, number and calculation . . ."

"Number, then, appears to lead towards the truth?"

"That is abundantly clear."

"Then, as it seems, this would be one of the studies we seek; for this is necessary for the soldier to learn because of arranging his troops, and for the philosopher, because he must rise up out of the world of becoming and lay hold of real being or he will never become a reckoner."

"That is true," said he.

"Again, our guardian is really both soldier and philosopher."

"Certainly."

"Then, my dear Glaucon, it is proper to lay down that study by law, and to persuade those who are to share in the highest things in the city to go for and tackle the art of calculation, and not as amateurs; they must keep hold of it until they are led to contemplate the very nature of numbers by thought alone, practicing it not for the purpose of buying and selling like merchants or hucksters, but for war, and for the soul itself, to make easier the change from the world of becoming to real being and truth."

"Excellently said," he answered.

"And besides," I said, "it comes into my mind, now the study of calculations has been mentioned, how refined that is and useful to us in many ways for what we want, if it is followed for the sake of knowledge and not for chaffering."

"How so?" he asked.

"In this way, as we said just now; how it leads the soul forcibly into some upper region and compels it to debate about numbers in themselves; it nowhere accepts any account of numbers as having tacked onto them bodies which can be seen or touched. . . . I think they are speaking of what can only be conceived in the mind, which it is impossible to deal with in any other way."

"You see then, my friend, said I, that really this seems to be the study we need, since it clearly compels the soul to use pure reason in order to find out the truth."

"So it most certainly does. . . .

"For all these reasons, the best natures must be trained in it."

After arithmetic, Glaucon and Socrates consider geometry.

Says Socrates, "The knowledge the geometricians seek is not knowledge of something which comes into being and passes, but knowledge of what always is."

"Agreed with all my heart, said he, for geometrical knowledge is of that which always is."

> "A generous admission! Then it would attract the soul toward truth, and work out the philosopher's mind so as to direct upwards what we now improperly keep downwards."
>
> After arithmetic and plane geometry, Glaucon proposes astronomy as the third subject in the curriculum of the Guardians. Socrates objects; solid geometry is more appropriate, he says.
>
> "Quite so," says Glaucon, "but it seems that those problems have not yet been solved."
>
> "For two reasons," I said, "because no city holds them in honour, they are weakly pursued, being difficult. Again, the seekers lack a guide, without whom they could not discover; it is hard to find one in the first place, and if they could, as things now are, the seekers in these matters would be too conceited to obey him. But if any whole city should hold these things honourable and take a united lead and supervise, they would obey, and solutions sought constantly and earnestly would become clear. Indeed even now, although dishonoured by the multitude, and held back by the seekers themselves having no conception of the objects for which they are useful, these things do nevertheless force on and grow against all this by their own charm, and I should not be too surprised if they should really come to light. . . .
>
> "Let us put astronomy as the fourth study, assuming that solid geometry, which we leave aside now, is there for us if only the city would support it."

Plato didn't have a "philosophy of mathematics" as we understand that phrase today. Mathematics is central in his philosophy. His believes the physical world of visible, changeable entities is illusion. What's real is invisible, immaterial, eternal. Mathematics is real *because* it's immaterial and eternal. It's tied to religion, as a stepping stone in one's ascent toward "the good," the loftiest aspect of invisible reality. A challenge to his notion of mathematics would be a challenge to his religion.

In a famous mathematical episode in Plato's *Meno*, Socrates is working, as usual, on the problem of virtue.

What is virtue?

How can we know it?

Examples show it's not learned by observation or from teachers. Socrates makes an analogy between virtue and mathematics. To make his point about the nature of mathematics, a "slave boy" is brought in. (His name is never mentioned.) Socrates makes the slave boy draw a small square inscribed at the midpoints of the sides of a larger square. What's the area of the small square? Socrates wants to make the slave boy *remember* the answer.

Socrates (to Meno): I shall do nothing more than ask questions and not teach him. Watch whether you find me teaching and explaining things to him instead of asking for his opinion.

(to the slave boy): You tell me, is this not a four-foot figure? You understand?

Slave boy: I do.

Socrates: We add to it this figure which is equal to it?

Slave boy: Yes.

Socrates: And we add this third figure equal to each of them?

Slave boy: Yes.

Socrates: Could we then fill in the space in the corner?

Slave boy: Certainly.

Socrates: So we have these four equal figures?

Slave boy: Yes.

Socrates: Well then, how many times is the whole figure larger than this one?

Slave boy: Four times.

By more questioning Socrates gets the slave boy to correct his mistake, and to say that the small square has area, not one fourth, but one half that of the big square.

Socrates: What do you think, Meno? Has he, in his answers, expressed any opinion that was not his own?

Meno: No, they were all his own . . .

Socrates: And he will know it without having been taught but only questioned, and find the knowledge within himself.

Meno: Yes

Socrates: And is not finding knowledge within oneself recollection?

Meno: Certainly.

Socrates: Must he not either have at some time acquired the knowledge he now possesses, or else have always possessed it?

Meno: Yes.

Socrates: If he always had it, he would always have known. If he acquired it, he cannot have done so in his present life. Or has someone taught him geometry? . . . You should know, especially as he has been born and brought up in your house.

Meno: But I know that no one has taught him . . .

Socrates: If he has not acquired them in his present life, is it not clear that he had them and had learned them at some other time?

Meno: It seems so.

Socrates: Then that was the time he was not a human being?

Meno: Yes.

You expect Socrates to argue next that knowledge of virtue, like knowledge of geometry, is remembered from Heaven. But he says something a little different:

"Virtue appears to be present in those of us who may possess it as a gift from the gods."

Bertrand Russell (1945) wrote, "When . . . in the *Meno*, [Socrates] applies his method to geometrical problems, he has to ask leading questions which any judge would disallow" (*A History of Western Philosophy*, p. 92).

The *Meno* shows that Plato understood geometrical reasoning. As proof that we remember geometry from before birth, it's a hoax. It is one of the most popular of Plato's *Dialogs*. But its popularity is from literary merits, not philosophical ones.

"Tradition is the forgetting of the origins"—Edmund Husserl, quoted by Philip J. Davis.

Once, long ago, men and women first captured fire, from a grass fire started by lightning. Lightning was the thunderbolt of the gods, so it seemed they had stolen fire from the gods.

Millennia passed. The true origin of fire was forgotten. In its place, an origin myth was created.

"The demigod Prometheus loved mankind. He stole fire from the gods, and gave it to us. In revenge, Jupiter chained him to a rock, where an eagle eternally plucks at his liver."

Number, like fire, can seem a divine gift to mankind. We forgot how we captured fire, and we forgot how we invented number. The *Meno* is an origin myth for numbers, a variant on the Prometheus myth for the origin of fire.

Modern Platonism is a descendant of Plato's Platonism. It too is a kind of origin myth.

An even more influential dialogue of Plato was the *Timaeus*. It's a creation story, like Genesis in the *Old Testament*—a myth of how a Demiurge or Divine Architect created the universe. In Cornford's book, *Plato's Cosmology*, he explains how in this dialogue Plato makes the "series" 1, 2, 4, 8 and 1, 3, 9, 27 "the basis for the harmony of the world-soul" ("harmony" in both musical and spiritual senses). This is Pythagorean thinking. In the Middle Ages *Timaeus* was the most popular Platonic dialogue, judging by copies found in old abbeys and monasteries. It transmitted Pythagorean numerology to the Middle Ages, and thence to modern times.

There's a well-known puzzle about Plato's teaching. The reports by Aristotle and other followers of Plato's teaching don't match what we read in Plato's

Dialogues. Yet "Aristotle, who was 37 when Plato died, had belonged to Plato's Academy for the last twenty years of Plato's life" (Ryle). It is believed that there was an Unwritten Doctrine, presented by Plato only in speech, perhaps because it never attained a satisfactory final form, perhaps because of cultic secrecy. In the Unwritten Doctrine, numbers are not merely examples or samples of the Ideas, the unmaterial eternal realities. *All* the Ideas are numbers! Perhaps just the first four numbers (1, 2, 3, 4), or perhaps even just the first two numbers (1, 2). If this seems inconceivable, remember that in twentieth-century set-theoretic Platonism, the set-theoretic superstructure, which contains all present, past and future mathematics, is generated *from the empty set.***

The Unwritten Doctrine has two kinds of numbers, Ideals and Mathematicals. The Ideals are what we understand from *The Republic*—there's an Ideal 5, an Ideal 6, an Ideal 7, 8, and 9. The Mathematicals are "lower" than the Ideals. In the equation,

$$5 + 5 = 10,$$

we see on the left of the equal sign *two* 5s. But there is only *one* Ideal 5, so what can these *two* 5s be?

When we notice that there's only one Ideal Horse but many material horses, we say the material horses are physical manifestations or exemplifications of the Ideal Horse. But the 5s in our equation are not material quintuples, they are ideal or abstract 5s, yet there are two of them! To escape this dilemma, Plato invented "Mathematicals." They are immaterial entities below the Ideal. This is an early example of the Philosophers' Rule: "When stopped by a contradiction, invent a distinction."

These matters are discussed by Wedberg, Findlay and Dillon.

Neoplatonists—Still in Heaven

The Neoplatonists, mystics in the Hellenistic world of the late Roman Empire, are interesting mathematically as a link between Plato-Pythagoras and Saint Augustine. I quote a few secondary sources, to give the flavor. From Saunders, 218 ff., Philo Judaeus of Alexandria (20? B.C.–54? A.D.):

"I doubt whether anyone could adequately celebrate the properties of the number 7, for they are beyond all words. . . . So August is the dignity inherent by nature in the number 7, that it has a unique relation distinguishing it from all the other numbers, for of these some beget without being begotten, some are begotten but do not beget, some do both these, both beget and are begotten: 7 alone is found in no such category. We must establish this assertion by giving proof of it."

(He means, if n is between 1 and 10, and neither n/2 nor 2n is between 1 and 10, then n = 7. *What about n = 9??* He must mean, if no multiple of n is between 1 and 10, and no factor of n is between 1 and 10, then n = 7. Which is correct.)

From Pistorius, p. 160, Plotinus:

"The only practical direction to the philosopher [given by Plotinus (204–270)] is a course in mathematics to train him in abstract thought, and in a faith in the unembodied."

From Cornford, p. 204 ff.: "The whole nature of things, all the essential properties of physis, were believed by the Pythagoreans to be contained in the tetractys of the decad . . . the Pythagorean One, or Monad, splits into two principles, male and female, the Even and the Odd, which are the elements of all numbers and so of the universe. The analogy reminds us that the One is not simply a numerical unit, which gives rise to other numbers by a process of addition. . . . We must think of the One (which is not itself a number at all) as the primary, undifferentiated group-soul, or physis, of the universe, and numbers must arise from it by a process of differentiation or separating out. . . . Similarly, each of these numbers is not a collection of units, built up by addition, but itself a sort of minor group-soul—a distinct 'nature', with various mystical properties."

Aurelius Augustinus, Bishop of Hippo Regius (354–430)
Pythagoras Is Christianized

In St. Augustine, neo-Platonist philosophy is wedded to Catholic theology. His life is told in his eloquent *Confessions.* From the age of 17 until 32 he lived with a dearly beloved concubine. Then he was converted to Christianity by his saintly mother. He wrote: "My concubine was torn from my side as a hindrance to my marriage; my heart which clave to her was torn and bleeding. And she returned to Africa, vowing unto Thee never to know any other man, leaving with me my son by her."

Thereafter he devoted himself to God. His obsession with human depravity, and his insistence that salvation comes only by Divine Grace, make him the forerunner of Calvinists and Jansenists a thousand years later.

One of his sentences is alarming: "Those impostors then, whom they style Mathematicians, I consulted without scruple; because they seemed to use no sacrifice, nor to pray to any spirit for their divinations; which art, however, Christian and true piety consistently rejects and condemns" (*Confessions,* p. 50). This quote is sometimes misused by failing to explain that by "mathematicians" Augustine meant "astrologers." In fact, he was deeply interested in arithmetic. Chapter VIII of *On Free Choice of the Will* (Mourant, p. 89) is headed: "The reason of numbers is perceived by no bodily sense; by anyone who understands it, it is perceived as one and immutable." In following paragraphs we find these arguments:

"Seven and three are ten, not only now but always; nor was there ever a time when seven and three were not ten, nor will ever be a time when seven and three will not be ten. I say, therefore, that this incorruptible truth of number is common to me and to any reasoning person whatsoever. . . .

"We cannot perceive the number one by our bodily senses. For whatever reaches us through such a sense is clearly seen to be not one but many, for it is body, and therefore has innumerable parts. . . . Moreover, if we do not perceive one by a sense of the body, we do not perceive any number by that sense . . . for there is no one of them that does not get its name from the number of times it contains one; and one is not perceived by the bodily sense. . . . The ***number*** that we call two because it contains twice that which is simply one, its half part, is something which is itself simply one, and cannot have a half or a third or any part whatsoever, because it is simply and truly one. . . .

"After one we get two, which comparing with one we find to be its double; the double of two does not follow in immediate succession, but four, which is the double of two, follows after the interposition of three. And this rule extends to all the other numbers by a most certain and immutable law . . . that the double of any number is just as many numbers after it as that number itself is from the very beginning. [*Today we say, for all x,* $2x - x = x - 0$.]

"But now, when we perceive this thing to be for all numbers fixed and inviolate, whence comes this perception? For no one has touched all numbers by any sense of the body, for they are innumerable. Whence then do we know this for all . . . this truth of number throughout things innumerable, if we do not perceive it by that inner light which the sense of the body knows not? By this and many like proofs, those to whom God has given an inquiring mind and who are not blinded by obstinacy are compelled to acknowledge that the reason and truth of numbers is not related to the bodily senses. . . . For not for nothing is number joined to wisdom in the Sacred Books, where it is said: "I and my heart have gone round, that I might know, and consider, and inquire the wisdom and the number" (Ecclesiastes, 7:25).

From Chapter 11, pp. 97–99: "I would much like to know whether these two, wisdom and number, are contained in some one genus. . . . For I should not have ventured to say that wisdom arises from number or consists in number, for wisdom strikes me as far more venerable than number, because I have known many calculators or computers, or whatever else you may call them, who reckon superbly and marvelously—but of wise men only a very few or possibly none. . . .

"When I reflect upon the immutable truth of numbers and upon its lair, as it were, or shrine or region or whatever else we may appropriately call the dwelling place and seat as it were of numbers, I am far removed from the body; and finding maybe something which I can think, but not finding anything which I can put into words, I return as if wearied to this world of ours. . . .

"Even if it cannot be clear to us whether number is in wisdom or from wisdom, or wisdom itself is in number or from number, or whether it can be shown that they are names for one thing; it is certainly manifest that both are true and true immutably."

Instead of this elevated Platonism, his *On the Trinity* contains naked Pythagorean number worship. Chapter IV is headed, "The ratio of the single to the double comes from the perfection of the senary number. The perfection of the senary number is commended in the scriptures. The year abounds in senary numbers."

The "senary number" is 6. In Pythagorean arithmetic, "perfect" means equal to the sum of the factors. Six is perfect, because $1 + 2 + 3 = 6$. Perfect numbers are scarce, and they are still a topic in recreational number theory. To Augustine, "perfect" has religious as well as arithmetical meaning.

Chapter V continues his praise of 6. It is headed, "The number six is also commended in the building up of the body of Christ and of the temple in Jerusalem."

It seems Solomon's temple was forty-six years in building. "And six times forty-six makes two hundred and seventy-six. And this number of days completes nine months and six days . . . the perfection of the body of the Lord is found to have been brought in so many days to the birth. . . . For He is believed to have been conceived on the 25th of March. . . . But he was born, according to tradition, upon December the 25th. If then, you reckon from that day to this you find two hundred and seventy-six days, which is forty-six times six. And in this number of years the temple was built, because in that number of sixes the body of the Lord was perfect, which being destroyed by the suffering of death, He raised again on the third day . . . 'the Son of man be three days and three nights in the heart of the earth.' From the evening of the burial to the dawn of the resurrection are thirty-six hours which is six squared . . . there is no one surely so foolish or so absurd as to contend that [these numbers] are so put in the Scriptures for no purpose at all, and that there are no mystical reasons why those numbers are there mentioned."

Augustine and Plato share the vision of mathematics as an aspect of the Divine. Said Augustine, "Among the disciples of Socrates, Plato was the one who shone with a glory which far excelled that of the others. . . . Why discuss with the other philosophers? It is evident that none come nearer to us than the Platonists" (1950, p. 23).

An impressive chapter in *On the Trinity* (Chapter XII, Book XV, pp. 849–50) uses a recursion argument that is unmistakably mathematical. It could have inspired Descartes's immortal one-liner, "I think, therefore I am."

"The knowledge . . . that we live is the most inward of all knowledge, of which even the Academic cannot insinuate: perhaps you are asleep, and do not know it, and you see things in your sleep . . . [Where Augustine says 'Academic' we say 'skeptic.'] he who is certain of the knowledge of his own life, does not say, I know I am awake, but, I know I am alive; therefore, whether he be asleep or awake, he is alive. . . . Nor can the Academic again say, in confutation of knowledge: perhaps you are mad, and do not know it: for what madmen see is precisely like what they also see who are sane; but he who is mad is alive. Nor does he answer the Academic by saying, I know I am not mad, but, I know I am alive. Therefore he who says he knows he is alive, can neither be deceived nor lie. . . . For he who says, I know I am alive, says that he knows one single thing. Further, if he says, I know that I know I am alive, now there are two; but that he knows these two is a third thing to know. And so he can add a fourth, and a fifth, and innumerable others, if he holds out. But since he cannot either comprehend an innumerable number by additions of units, or say a thing innumerable times, he comprehends this at least, and with perfect certainty, viz. that this is both true and so innumerable that he cannot truly comprehend and say its infinite number."

Emperor, Pope, and Magic Number

Arithmetic Is Sanctified

To the medieval mind, a sacred number was part of the divine spiritual order.

In 1240, western Europe was disturbed by rumors of a great king in the Far East who was making his way relentlessly westward. One Islamic kingdom after another had fallen to him. Some thought the news meant the coming of the legendary Prester John, who would join the Christian kings and destroy Islam. European Jews thought the Eastern monarch was David's son King Messiah, and prepared to meet him joyously. Actually, the eastern king was Batu, son of Genghis Khan.

Why was it thought that the Messiah was arriving? The year 1240 A.D. corresponded to the Jewish year 5000. Some thought the Messiah should come at year 5000.

One type of number mysticism is "gematria." It associates a number with a word by adding up the numbers of its letters. The name "Innocentius Papa" (Pope Innocent IV) has the number 666. This is the Number of the Beast in Revelations 13:18. Hence Pope Innocent is Antichrist!

Thomas Aquinas (1225?—1274)

Aristotle Is Christianized

St. Thomas is the principal author of Christian theology and philosophy. He led in incorporating Aristotle into Church doctrine. Mathematics is one of the few major topics on which he didn't write at length. It's hard to find any but passing reference to mathematics in his voluminous works or those of his commentators. In *The Pocket Aquinas* I found some tantalizing crumbs.

The Exposition of Boethius on the Trinity, Book 5, Chapter 1, has a passage classifying the sciences. To distinguish mathematics from physics and natural science, which deal with material objects, and theology or metaphysics, which deal with immaterial objects, Aquinas explains, "There are some things which, though they depend on matter according to their act of being, do not do so when they are understood, because sensible matter is not included within their definitions: for instance, lines and numbers. Now, *mathematics* deals with these objects."

The same work, Book 6, Chapter 1: "In mathematical science, the process [of demonstrative reasoning] works through those items that pertain merely to the essence of a thing, since these sciences demonstrate only by means of the formal cause. So, in them something is not demonstrated about one thing by means of another thing, but by the proper definition of that thing. For, although some demonstrations are given concerning the circle from the triangle, or conversely, this is done only because the triangle is potentially in the circle, or the converse . . . to proceed according to a learning method is characteristic of mathematical science; not that it alone progresses by learning but because this is especially appropriate to it. Indeed, since to learn simply means to get scientific knowledge from another, we are said to use the method of learning when our procedure leads to such knowledge which is called science. Now this happens chiefly in the mathematical sciences.

"Since mathematics is intermediate between natural and divine science, it is more certain than either of them; more so than the natural because its thinking is cut off from motion and matter; while the thinking of the natural scientist is directed to matter and motion. . . .

"Moreover, the procedure of mathematics is more certain than that of divine science because the things that divine science studies are more removed from the objects of sensation from which our knowledge takes its start . . . the objects of mathematics do fall with sense experience and are subjects for the imagination; for instance, figures, lines, numbers, and the like. Thus, human understanding grasps the knowledge of these items from the phantasms more easily and more certainly than the knowledge about an intelligence, or even than about a substantial essence, act, potency, and other similar items. So, it is clear that mathematical thinking is easier and more certain than physical or theological, and much more so than that of the other operative sciences."

From the *Exposition of the Book of Causes*, Lecture 1: "[The philosophers] put the science of the first causes last in the order of learning, and assigned its consideration to the last period in a person's life. They began first with logic, for it teaches the method of the sciences. Then they went to mathematics, which even boys can grasp." Here Aquinas is misrepresenting Plato, who explicitly included women in his school for guardians.

From the *Exposition of Aristotle's Physics*, Book 1: "A curve, though it cannot exist except in sensible matter, does not, however, include sensible matter in its definition. This is the way that all mathematicals are, for instance, number, size, and figure. . . . Mathematics is concerned with these things which depend on sensible matter for their actual being but not for their rational meaning. The natural science that is called physics is concerned with those things which depend on matter not only for actual being but also for their rational meaning."

Thinking of Aquinas as a Catholic philosopher, and seeing him place mathematics between physics and theology, it may seem he is a Platonist. But his analysis of mathematics is Aristotelian, not Platonic. Mathematics studies real objects, defined to exclude their materiality. This is a restatement of Aristotle's analysis of mathematical objects as abstractions from real objects. Aristotle says: to get a mathematical object, ignore all qualities of the real object except mathematical qualities. Aquinas says: define the mathematical object to be free of the material properties of the real object.

I'm tempted to move Aquinas over to the Humanists and Mavericks, but if anybody is Mainstream, he is. The classification between mainstream philosophers and humanist or maverick philosophers (like any classification) is not always à propos.

Cardinal Nicholas Krebs of Cusa (1401–1464)
Infinity in Theology

In the *Encyclopedia of Philosophy* Nicholas Krebs is called "theologian, philosopher, mathematician." In addition to many religious-political and theological-philosophical works, Maurer lists four mathematical-scientific books: *De Staticis Experimentis* (1450), *De Transmutationibus Geometricis* (1450), *De Mathematicis Complementis* (1453), and *De Mathematica Perfectione* (1458). Jasper says Cusa produced eleven mathematical writings from 1445 to 1459. Among his subjects: "the quadrature of the circle, the reform of the calendar, the improvement of the Alfonsine Tables (planetary tables which improved those left by Ptolemy), the heliocentric theory of the universe (he looked on it as a paradox, not a scientific probability) and the theory of numbers. Wallis said Nicholas was the first writer known to have worked on the cycloid, but this is not supported by the evidence" (Davis).

The son of a fisherman, Nicholas rose to become a diplomat and counselor for the Church. "He was a member of the commission sent to Constantinople to negotiate with the Eastern church for reunion with Rome, which was temporarily effected at the Council of Florence (1439)." In 1448 he became cardinal and governor of Rome.

Cusa was not a philosopher of mathematics. He was a philosopher whose thinking was imbued with mathematical images, so that he used mathematics to teach theology. He knew that there are different degrees of infinity. He said, amazingly, that the physical universe is finite but unbounded. He showed that a geometric figure can be both a maximum and a minimum, depending on how it's parametrized.

Again from the *Encyclopedia*, "According to Cusa, a man is wise only if he is aware of the limits of the mind in knowing the truth. . . . Knowledge is learned ignorance (*docta ignorantia*). Endowed with a natural desire for truth, man seeks it through rational inquiry, which is a movement of the reason from something presupposed as certain to a conclusion that is still in doubt. . . . As a polygon inscribed in a circle increases in number of sides but never becomes a circle, so the mind approximates to truth but never coincides with it. . . . Thus knowledge at best is conjecture (*coniectura*)."

Cusa was a Platonist at a time when Aristotelians were dominant. "He constantly criticized the Aristotelians for insisting on the principle of noncontradiction and stubbornly refusing to admit the compatibility of contradictories in reality. It takes almost a miracle, he complained, to get them to admit this; and yet without this admission the ascent of mystical theology is impossible. . . . He constantly strove to see unity and simplicity where the Aristotelians could see only plurality and contradiction.

"Cusa was most concerned with showing the coincidence of opposites in God. God is the absolute maximum or infinite being, in the sense that he has the fullness of perfection. There is nothing outside him to oppose him or to limit him. He is the all. He is also the maximum, but not in the sense of the supreme degree in a series. As infinite being he does not enter into relation or proportion with finite beings. As the absolute, he excludes all degrees. If we say he is the maximum, we can also say he is the minimum. He is at once all extremes. . . . The coincidence of the maximum and minimum in infinity is illustrated by mathematical figures. For example, imagine a circle with a finite diameter. As the size of the circle is increased, the curvature of the circumference decreases. When the diameter is infinite, the circumference is an absolutely straight line. Thus, in infinity the maximum of straightness is identical with the minimum of curvature. . . .

"Cusa denied that the universe is positively infinite; only God, in his view, could be described in these terms. But he asserted that the universe has no circumference, and consequently that it is boundless or undetermined—a revolutionary notion in cosmology. . . . Just as the universe has no circumference, said Cusa, so it has no fixed center. The earth is not at the center of the universe, nor is it absolutely at rest. Like everything else, it moves in space with a motion that is not absolute but is relative to the observer. . . .

"Beneath the oppositions and contradictions of Christianity and other religions, he believed there is a fundamental unity and harmony, which, when it is recognized by all men, will be the basis of universal peace."

To these quotes from the *Encyclopedia of Philosophy*, I adjoin an excerpt from Cusa's *De Mente*. This is a dialogue between Layman and Philosopher. Layman is speaking, p. 59: "Just as from God's viewpoint the plurality of things comes from the divine mind, so from our perspective the plurality of things proceeds from our mind. Mind alone counts; if mind is removed, distinct numbers do not exist. Because mind grasps one and the same thing individually and in signs and we consider this itself (for we say something is one from the fact that mind understands a single thing once and individually), mind is truly the equality of unity. But when mind grasps a single thing individually and by multiplying, we judge that there are many things by speaking of a two because the mind grasps singly one and the same thing twice or by doubling it; and so on for the rest. . . . The plurality of things arises from this, that the mind of God understands one thing in a certain way and a second in another way. If you attend sharply you will discover that the plurality of things is no more than the way the divine mind understands. I conjecture that one can say without blame that the first exemplar of things in the mind of their maker is number. The delight and beauty inherent in all things demonstrates this, for it consists in proportion, and proportion in number. So number is the principal clue which leads to wisdom."

Philosopher. First the Pythagoreans, then the Platonists said that, and Severinus Boethius followed them.

Layman. In like manner I say number is the exemplar of our mental concepts. Mind can do nothing without number.

In the *Monadology* Leibniz mentions Cusa as one of his sources.

René Descartes (1596–1650)

Skepticism Is Refuted—Geometry Is Born Again

Philosophy students are supposed to read Descartes's *Discourse on Method* (1637). They don't realize that the complete *Discourse* includes Descartes' mathematical masterpiece, the *Geometry*. The full title is *Discourse on Method, Optics, Geometry, and Meteorology.*

On the other hand, mathematics students also are miseducated. They're supposed to know that Descartes was a founder of analytic geometry, but not that his *Geometry* was part of a great work on philosophy.

Descartes's Method consists of four simple rules:

> The first of these was to accept nothing as true which I did not clearly recognize to be so: that is to say, carefully to avoid precipitation and prejudice in judgments, and to accept in them nothing more than what was presented to my mind so clearly and distinctly that I could have no occasion to doubt it.
>
> The second was to divide up each of the difficulties which I examined into as many parts as possible, and as seemed requisite in order that it might be resolved in the best manner possible.
>
> The third was to carry on my reflections in due order, commencing with objects that were the most simple and easy to understand, in order to rise little by

> little, or by degrees, to knowledge of the most complex, assuming an order, even if a fictitious one, among those which do not follow a natural sequence relatively to one another.
>
> The last was in all cases to make enumerations so complete and reviews so general that I should be certain of having omitted nothing" (Vol. I, p. 92, of Haldane and Ross).

Leibniz sneered "famously" that these "rules" amount to little more than "Take what you need, do what you should, and you will get what you want." But what's left out of the rules is more important than what is stated. By omission, Descartes advises the investigator to respect no authorities, neither Aristotle nor Church Fathers nor Holy Scripture! His Method proclaims individual autonomy in the search for truth.

D'Alembert wrote that it was Descartes who first "dared . . . to show intelligent minds how to throw off the yoke of scholasticism, of opinion, of authority—in a word, of prejudices and barbarism. . . . He can be thought of as a leader of conspirators who, before anyone else, had the courage to arise against a despotic and arbitrary power, and who, in preparing a resounding revolution, laid the foundations of a more just and happier government which he himself was not able to see established."

Philosophers of the scholastic persuasion pointed to the dangerous parallel between Descartes's scientific individualism and the outlawed Protestant heresy. Descartes said individual thinkers could find scientific truth; Protestants said individual souls could find direct communion with the Almighty. But the Holy Roman Catholic Church knew that individual souls and thinkers could be deceived. It took the experience and wisdom of the Church to prevent the seeker from wandering astray. Despite such scholastic criticism, Descartes quickly came to dominate West European intellectual life.

For Descartes, mathematics was the central subject. "I was delighted with Mathematics because of the certainty of its demonstrations and the evidence of its reasoning. . . . I was astonished that, seeing how firm and solid was its basis, no loftier edifice had been reared thereupon" (Vol. I, p. 85).

He wrote (p. 7), "Arithmetic and geometry alone deal with an object so pure and uncomplicated, that they need make no assumptions at all which experience renders uncertain, but wholly consist in the rational deduction of consequences. They are on that account much the easiest and clearest of all . . . in them it is scarce humanly possible for anyone to err except by inadvertence . . . in our search for the direct road towards truth we should busy ourselves with no object about which we cannot attain a certitude equal to that of the demonstrations of Arithmetic and Geometry" (Vol. I, p. 3). Isaac Beeckman visited Descartes in 1628, and wrote: "He told me that insofar as arithmetic and geometry were concerned, he had nothing more to discover, for in these branches during the past nine years he had made as much progress as was possible for the human mind. He gave me decisive proofs of this affirmation and promised to send me shortly his Algebra, which he said was finished and by which not only had he arrived at a perfect knowledge of geometry but also he claimed to embrace the whole of human thought" (Vrooman, p. 78).

This lofty goal is hardly to be attained by one mind! Descartes's mathematical tools were certainly insufficient for this grandiose purpose. Like Galileo, Descartes recognized mathematics as the principal tool for revealing truths of nature. He was more explicit than Galileo about how to do it. In every scientific problem, said Descartes, find an algebraic equation relating an unknown variable to a known one. Then solve the algebraic equation! With the development of calculus, Descartes's doctrine was essentially justified. Today we don't say "find an algebraic equation." We say "construct a mathematical model." This is only a technical generalization of Descartes's idea. Our scientific technology is an inheritance from Descartes.

In *Rules for the Direction of the Mind*, Descartes wrote: "The first principles themselves are given by intuition alone, while, on the contrary, the remote conclusions are furnished only by deduction. . . . These two methods are the most certain routes to knowledge, and the mind should admit no others. All the rest should be rejected as suspect of error and dangerous."

Descartes was embracing the Euclidean ideal: Start from self-evident axioms, proceed by infallible deductions. But in his own research, Descartes forgot the Euclidean ideal. Nowhere in the *Geometry* do we find the label Axiom, Theorem, or Proof.

In classical Greece, and again in the Renaissance and after, mathematicians distinguished two ways of proceeding—the "synthetic" and the "analytic." The synthetic way was Euclid's: from axioms through deductions to theorems. In the analytic mode, you start with a problem and "analyze" it to find a solution. Today we might call this a "heuristic" or "problem-solving" approach.

In formal presentation of academic mathematics, the synthetic was and still is the norm. Foundationist schools of the nineteenth and twentieth centuries identify mathematics with its synthetic mode—true axioms followed by correct deductions to yield guaranteed true conclusions.

In his *Rules for the Direction of the Mind*, Descartes insists on the synthetic method. But his own research, in the *Geometry*, uses *only* the analytic mode. He solves problems. He finds efficient methods for solving problems. Never does he bother with axioms.

Descartes's conviction of the certainty of mathematics might lead readers to expect that at least Descartes's own mathematics is error-free. But of course, as we will see, the *Geometry*, like every other math book, has mistakes. Certitude is only a goal.

In the *Geometry* Descartes makes no pretense of unquestionable axioms or irrefutable reasoning. He follows the heuristic, pragmatic style normal in mathematical research. This starkly contradicts his *Method*.

Replying to criticism Descartes wrote, "Analysis shows the true way by which a thing was methodically discovered and derived, as it were effect from cause. . . . Synthesis contrariwise employs an opposite procedure. . . . It was this synthesis alone that the ancient Geometers employed in their writings, not because they were wholly ignorant of the analytic method, but, in my opinion, because they set so high a value on it that they wished to keep it to themselves as an important secret" (Vol. II, pp. 48–49).

He admits that some of his arguments are deficient: "I have not yet said on the basis of what reasons I venture to assure you that a thing is or is not possible. But if you take note that, with the method I use, everything falling under the geometer's consideration can be reduced to a single class of problem—namely, that of looking for the value of the roots of a certain equation—you will clearly see that it is not difficult to enumerate all the ways through which they can be found, which is sufficient to prove that the simplest and the most general one has been chosen" (Vol. I, Book 3, p. 251).

It is strange that in the vast body of writing about Descartes accumulated in three centuries, almost no one seems to have called attention to this bizarre misfit—Euclidean certainty boldly advertised in the *Method* and shamelessly ditched in the *Geometry*. The Dutch mathematician Willem Kuyk is the only author I know who has noticed this remarkable fact. Maybe philosophers don't read the *Geometry*, and mathematicians don't read the *Method*.

The *Method* is an essential link in the chain from Pythagoras to modern foundationists. Descartes's ignoring it in his own research casts doubt on it as a viable, realistic methodology.

You won't find in the *Geometry* the method we teach nowadays as Cartesian or "analytic" geometry. Our analytic geometry is based on rectangular coordinates (which we call "Cartesian"). To every point in the plane we associate a pair of real numbers, the "x" and "y" coordinates of the point. To an equation relating x and y corresponds a "graph"—the set of points whose x and y coordinates satisfy the equation. For an equation of first degree, the graph is a straight line. For an equation of second degree, it's a circle or other conic section. Our idea is to solve geometric problems by reducing them to algebra. Nowhere in Descartes's book do we see these familiar horizontal and vertical axes! Boyer says it was Newton who first used orthogonal coordinate axes in analytic geometry.

The conceptual essence of analytic geometry, the "isomorphism" or exact translation between algebra and geometry, was understood more clearly by Fermat than by Descartes. Fermat's analytic geometry predated Descartes's, but it wasn't published until 1679. The modern formulation comes from a long development. Fermat and Descartes were the first steps. Instead of systematically developing the technique of orthogonal coordinate axes, the *Geometry* studies a group of problems centering around a problem of Pappus of Alexandria (third century A.D.). To solve Pappus's problem Descartes develops an algebraic-geometric procedure. First he derives an algebraic equation relating known and unknown lengths in the problem. But he doesn't then look for an algebraic or numerical solution, as we would do. He is faithful to the Greek conception, that by a solution to a geometric problem is meant a construction with specified instruments. When possible Descartes uses the Euclidean straight edge and compass. When necessary, he brings in his own instrument, an apparatus of hinged rulers. Algebra is an intermediate device, in going from geometric problem to geometric solution. Its role is to reduce a complicated curve to a simpler one whose construction is known. He solves third and fourth-degree equations by reducing them to second degree—to conic sections. He solves certain fifth- and sixth-degree equations by reducing them to third degree. A modern reader knows that the general equation of fifth degree can't be solved by extraction of roots. So he's skeptical about Descartes's claim that his hinged rulers can solve equations of degree six and higher. Descartes was mistaken on several points. In themselves, these are of little interest today. But they discredit his claim of absolute certainty. ***Descartes's mathematics refutes his epistemology.***

Emily Grosholz and Carl Boyer point out errors in the *Geometry.* "When he turns his attention to the locus of five lines, he considers only a few cases, not bothering to complete the task, because, as he says, his method furnishes a way to describe them. But Descartes could not have completed the task, which amounted to giving a catalogue of the cubics. . . . Newton, because he was able to move with confidence between graph and equation, first attempted a catalogue of the cubics; he distinguished seventy-two species of cubics, and even then omitted six" (Grosholz, referring to Whiteside).

"Descartes stratified his hierarchy into levels of pairs of degrees, since (so he thought) from curves of degree n and n + 1 his apparatus of hinged rulers produced curves of degree n + 2 and n + 3. Fermat gave a counter-example to this generalization.

"Descartes classification into orders of two degrees each was based on the fact that the algebraic solution of the quartic leads to a resolvent cubic from which Descartes rashly concluded—incorrectly, as Hudde later showed—that an equation of degree 2n would in all cases lead to a resolvent of degree 2n − 1. . . . The method Descartes proposed for the study of the properties of a space curve is to project it upon two mutually perpendicular planes and to consider the two curves of projection. Unfortunately, the only illustrative property given here is erroneous for one reads that the normal to a curve in three-space at a point P on the curve is the line of intersection of the two planes through P, determined by the normal lines to the curves of projection at the points corresponding to P. This would be true of the tangent line, but does not in general hold for a normal. . . . Descartes, in these casual remarks, seems not to have been aware of the fact that for space of more than two dimensions a normal is not uniquely determined for a point on a curve."

From these descriptions by Boyer and Grosholz, it's clear that Descartes's mistakes weren't merely inadvertent. We can hardly imagine Descartes nodding asleep as he composes the *Geometry*! These are *conceptual* errors. Descartes failed to understand certain aspects of his subject, and consequently made false statements while in full possession of his senses. If mathematics were indubitable axioms followed by infallible reasoning, this would be impossible. Even when mathematics is presented in the synthetic mode, mistakes happen. But the analytic mode used by Descartes is the usual mode of mathematical thinking. It isn't infallible, in Descartes's hands, yours, or mine. Therefore any philosophical claim based on the infallibility of mathematics is discredited.

Back to Boyer: "One gets the impression that Descartes wrote *La géometrie* not to explain, but to boast about the power of his method. He built it about a difficult problem and the most important part of his method is presented, all too concisely, in the middle of the treatise for the reason that it was necesary for the solution of this problem."

Do you think Boyer is unfair? In a letter to Florimond De Beaune, Descartes wrote, on February 20, 1639 (Adam and Tannery, Vol. 2, p. 511): "In the case of the tangents, I have only given a simple example of analysis, taken indeed from a rather difficult aspect and I have left out many of the things which could have been added so as to make the practice of the analysis more easy. I can assure you, nevertheless, that I have omitted all that quite deliberately, except in the case of the asymptote, which I forgot. But I felt sure that certain people, who boast that they know everything, would not miss the chance of saying that they knew already what I had written, if I had made myself easily intelligible to them. I should not then have had the pleasure, which I have since enjoyed, of noting the irrelevance of their objections."

Boyer goes on, "In concluding the work, Descartes justifies the inadequacy of exposition by the incongruous remark that he had left much unsaid in order not to rob the reader of the joy of discovery" (Boyer, p. 104). Here is the quote to which Boyer refers: "Finally, I have not demonstrated here most of what I have said, because the demonstrations seem to me so simple that, provided you take the pains to see methodically whether I have been mistaken, they will present themselves to you; and it will be of much more value to you to learn them this way than by reading them" (Vol. I, p. 244).

Says Boyer: "Either this was sarcasm or else the author grossly misjudged the abilities of his readers to profit by what he had written. It is no wonder that the number of editions of his *Geometrie* was relatively small during the seventeenth century and has been still smaller since then" (Boyer, p. 104).

Descartes claimed his Method was infallible in science and mathematics. He was more cautious with religion. He didn't derive Holy Scripture or divine revelation by self-evident axioms and infallible deductions. When he heard that Galileo's *Dialogue on the Two Chief Systems* was condemned by the Holy Church, he suppressed his first book, *Le Monde,* even though he was living in Holland, safe from the Church. (Galileo was kept under house arrest at first. For three years he had to recite the seven penitential psalms every week.) Descartes wrote to Father Mersenne, "I would not want for anything in the world to be the author of a work where there was the slightest word of which the Church might disapprove."

Following the four rules of the *Method* is an addendum: "But this does not prevent us from believing matters that have been divinely revealed as being more certain than our surest knowledge, since belief in these things, as all faith in obscure matters, is an action not of our intelligence but of our will. They should be heeded also since, if they have any basis in our understanding, they can and ought to be, more than all things else, discovered by one or other of the ways above-mentioned, as we hope perhaps to show at greater length on some future opportunity" (Vol I., p. 11).

Like Pascal, Newton, and Leibniz, Descartes may have valued his contributions to theology above his mathematics. His struggle against skeptics and heretics is the major half of his philosophy, more explicit than his battles with scholastics. "In Descartes' reply to the objections of Father Bourdin, he announced that he was the first of all men to overthrow the doubts of the Sceptics . . . he discovered how the best minds of the day either spent their time advocating scepticism, or accepted only probable and possibly uncertain views, instead of seeking absolute truth. . . . It was in the light of this awakening to the sceptical menace, that when he was in Paris Descartes set in motion his philosophical revolution by discovering something so certain and so assured that all the most extravagant suppositions brought forward by the sceptics were incapable of shaking . . . in the tradition of the greatest medieval minds, (he) sought to secure man's natural knowledge to the strongest possible foundation, the all-powerful eternal God" (Popkin, p. 72).

The essence of the *Meditations* is a proof that the world exists by first proving Descartes exists, and then, by contemplating Descartes's thoughts, proving that God exists ***and is not a deceiver.*** Once a nondeceiving God exists, everything else is easy.

In the fifth meditation Descartes gives a "mathematical" proof that God exists, based on the geometry of the triangle. In essence it's exactly the argument that Plato originated, and passed on to Locke, Berkeley, and Leibniz: The certainty of mathematics implies the certainty of religion.

Here it is: "When I imagine a triangle, although there may nowhere in the world be such a figure outside my thought, or ever have been, there is nevertheless in this figure a certain determinate nature, form, or essence, which is immutable and eternal, which I have not invented, and which in no wise depends on my mind, as appears from the fact that diverse properties of that triangle can be demonstrated, viz. that its three angles are equal to two right angles, that the greatest side is subtended by the greatest angle, and the like, which now, whether I wish it or do not wish it, I recognize very clearly as pertaining to it, although I never thought of the matter at all when I imagined a triangle for the first time, and which therefore cannot be said to have been invented by me . . . but now, if just because I can draw the idea of something from my thought, it follows that all which I know clearly and distinctly as pertaining to this object does really belong to it, may I not derive from this an argument demonstrating the existence of God? It is certain that I no less find the idea of God, that is to say, the idea of a supremely perfect Being, in me, than that of any figure or number whatever it is; and I do not know any less clearly and distinctly that an actual and eternal existence pertains to this nature than I know that all which I am able to demonstrate of some figure or number truly pertains to the nature of this figure or number, and therefore although all that I concluded in the preceding Meditations were found to be false, the existence of God would pass with me as at least as certain as I have ever held the truths of mathematics to be. . . . I clearly see that existence can no more be separated from the essence of God than can its having its three angles equal to two right angles be separated from the essence of a rectilinear triangle, or the idea of a mountain from the idea of a valley; and so there is not any less repugnance to our conceiving a God (that is, a Being supremely perfect) to whom existence is lacking (that is to say, to whom a certain perfection is lacking) than to conceive of a mountain without a valley. . . . But after I have recognized that there is a God—because at the same time I have also recognized that all things depend upon Him, and that He is not a deceiver, and from that have inferred that what I perceive clearly and distinctly cannot fail to be true. . . . And this same knowledge extends likewise to all other things which I recollect having formerly demonstrated, such as the truths of geometry and the like. . . . And now that I know him I have the means of acquiring a perfect knowledge of an infinitude of things, not only of those which relate to God Himself, and other intellectual matters, but also of those which pertain to corporeal nature in so far as it is the object of pure mathematics" ("Meditation V, Vol. 1, pp. 180 ff.).

There is an obvious difficulty with this and the other antiskeptical arguments Descartes invented. Skeptics like Pierre Gassendi immediately pointed out that Descartes argues from what he sees in *his own* mind. Others find things otherwise in their minds. To Descartes it may have seemed that his use of the Method in geometry and in theology was all of a piece. Today, we see two different men: Descartes the mathematician, inventive and ingenious; and Descartes the theologian, self-duped by trivially fallacious arguments.

The two aspects of Descartes split apart in his intellectual heritage. His *Geometry* instructed Newton and Leibniz, and is forever integrated into the living body of mathematics. Mathematicians remember him with the name "Cartesian product" for ordered pairs in set theory and geometry.

In theology, the story is more complex. It would be unfair to suppose Descartes' bows to the Church were insincere. He really was a devout Catholic. His commitment to philosophy and science originated in a dream-vision of the Blessed Virgin.

Descartes was very considerate of the Church's worries. Still some denounced him as a skeptic or crypto-skeptic. His First Meditation raises profound doubt. Does the Third Meditation really dispel it? Will his "clear and distinct idea" really revive faith, once his doubt has shaken it? In 1663 he was put on the Index (Vrooman, p. 252). In 1679 Leibniz wrote of Descartes's philosophy, "I do not hesitate to say absolutely that it leads to atheism" (Leibniz, p. 1).

In the following century, nevertheless, Cartesianism became popular among Church apologists. But Descartes's follower, the "God-intoxicated" Spinoza, was denounced as an atheist, and "Spinozism" became a synonym for atheism. Cartesians were prominent denouncers of Spinoza (Balz, pp. 218–41).

seven

Mainstream Philosophy at Its Peak

Baruch Spinoza (1632–1677)

The Higher Criticism. Ethics à l'Euclide.

Benedictus de Spinoza, born Baruch Spinoza, was excommunicated by the Jews of Amsterdam, but never baptized Christian. His native language was Portuguese; he spoke Dutch with difficulty.

The Amsterdam Jews had fled to Holland from the Holy Inquisition. Like similar communities today, they were grateful for refuge, and eager not to disturb their Protestant hosts. The bill of excommunication, proclaimed in Portuguese on July 27, 1656, reads: "The chiefs of the council, having long known the evil opinions and works of Baruch de Espinoza . . . have had every day more knowledge of the abominable heresies practiced and taught by him. . . . With the judgment of the angels and of the saints we excommunicate, cut off, curse, and anathematize Baruch de Espinoza. . . . Cursed be he by day and cursed be he by night. Cursed be he in sleeping and cursed be he in waking, cursed in going out and cursed in coming in . . . none may speak with him by word of mouth nor by writing, nor show any favour to him, nor be under one roof with him, nor come within four cubits of him, nor read any paper composed or written by him." Before the excommunication he had been attacked with a dagger (Pollock, pp. 17–18).

Spinoza learned of the anathema while visiting a friend in "the small dissenting community of Remonstrants or Collegiants, heretics anathematized by the Synod of Dort . . . men who without priests or set forms of worship carried out the precepts of simple piety" (Pollock, p. 17). It seems Spinoza was something like a Quaker! (See Fell, 1987.)

His first book, *Tractatus Theologico-Politicus*, scandalized Amsterdam's Orthodox. Like Descartes, Spinoza believed that reason should guide him in understanding the world. Unlike Descartes, who tried not to tread on the Church's toes, Spinoza followed reason without regard to consequences. He made no exception of Scripture and revealed religion. His *Tractatus* is the first example of what later became known as "higher criticism" of the Bible. Spinoza saw the *Old Testament* as an incomplete, corrupted collection of documents about the history of the Jews.

He wrote: "It is thus clearer than the sun at noonday that the Pentateuch was not written by Moses, but by someone who lived long after Moses. . . . The history relates not only the manner of Moses death and burial and the thirty days mourning of the Hebrews, but further compares him with all the prophets who came after him, and states that he surpassed them all. . . . Such testimony cannot have been given of Moses by himself" (VII, p. 124).

"Moses . . . conceived God as a ruler, a legislator, a king, as merciful, just, &c., whereas such qualities are simply attributes of human nature and utterly alien from the nature of the Deity" (IV, p. 63).

"What pretension will not people in their folly advance! They have no single sound idea concerning either God or human nature, they confound God's decrees with human decrees, they conceive nature as so limited that they believe man to be its chief part!" (VI, p. 81).

(Richard Popkin has called attention to a book by the Quaker Samuel Fisher. He was part of the Quaker missionary effort in Holland, and acquainted with Spinoza. Fisher's book, published ten years before Spinoza's, has remarkably similar arguments and examples.)

"The States of Holland and West Friesland, being satisfied that the book . . . entitled 'B.D.S. Opera Posthuma' 'labefactated' various essential articles of the faith and 'vilipended the authority of miracles', expressed 'the highest indignation' at the disseminating thereof, declared it profane, atheistic, and blasphemous, and forbad printing, selling, and dealing in it, on pain of their high displeasure."

"The stupid Cartesians," Spinoza wrote to Oldenburg, "being suspected of favoring me, endeavored to remove the aspersion by abusing everywhere my opinions and writings, a course which they still pursue" (Ratner, 1927).

Spinoza took up lens-grinding. The rest of his life he maintained himself in poverty by that trade. He was offered a philosophy chair at Heidelberg, a pension from Louis XIV of France, and annuities by friends in the Netherlands. He always declined, valuing intellectual independence above physical comfort.

After meeting with a Frenchman during the war between the Netherlands and France, he was accused of spying. His host was in fear of an attack on his home. He told his host: "So soon as the crowd makes the least noise at your door I will go out and make straight for them though they should serve me as they have done the unhappy De Witts. I am a good republican, and never had any aim but the honour and welfare of the State." The De Witts were two brothers, one a friend of Spinoza, who were lynched by a mob at the Hague (Pollock, pp. 36–37). The feared attack did not take place.

Spinoza died in 1677 of consumption, aggravated by inhaling glass particles while grinding lenses. "The funeral took place on the 25th February, being attended by many illustrious persons, and followed by six coaches" (Spinoza, 1895, p. xx).

Spinoza didn't write explicitly about philosophy of mathematics, but he was a major influence in general philosophy, and it's interesting to surmise what he thought about mathematics.

When Spinoza wanted to say something was indubitable, he said it was as certain as that the angle sum of a triangle equals two right angles. This familiar theorem of Euclid was his paradigm of the indubitable. (*Ethics*, p. 55, for example.) But this "indubitable fact" is negated in non-Euclidean geometry. (See the section on Certainty in Chapter 4.) What to Spinoza was indubitable, today is dubitable for every mathematics student.

The Ethics is the only work of Spinoza read in philosophy class today. It has five parts: "Concerning God," "Of the Nature and Origin of the Mind," "On the Origin and Nature of the Emotions," "Of Human Bondage or the Strength of the Emotions," and "On the Power of the Understanding, or of Human Freedom." Spinoza treats these matters "geometrically," that is, in the style of Euclid's *Elements*. Each part starts with Axioms and Definitions and produces Propositions.

In Part I, Definition I is: "By that which is self-caused, I mean that of which the essence involves existence, or that of which the nature is only conceivable as existent." Axiom I of Part I is "Everything which exists, exists either in itself or in something else."

From such ingredients, Spinoza obtains Proposition 33: "Things could not have been brought into being by God in any manner or in any order different from that which has in fact obtained."

Here's the proof: "All things necessarily follow from the nature of God (Proposition xvi), and by the nature of God are conditioned to exist and act in a particular way (Proposition 29). If things, therefore, could have been of a different nature, or have been conditioned to act in a different way, so that the order of nature would have been different, God's nature would also have been able to be different from what it now is; and therefore (by Proposition 11) that different nature also would have perforce existed, and consequently there would have been able to be two or more Gods. This (by Proposition 14, Corollary 1) is absurd. Therefore things could not have been brought into being by God in any other manner, etc. Q.E.D."

To a modern reader this argumentation is embarrassing. Spinoza's proof is not what we call a proof. Roth (41) thinks that "It is . . . not a method of proof, but an order of presentation, as may be proved . . . by the fact that Spinoza proposed to deal in precisely the same way with the intricacies of Hebrew Grammar. In the *Ethics* itself the geometric form, even as an order, is dropped at convenience. . . . He adopted it for a definite reason, and that was its impersonality. Mathematics recognizes and has no place for personal prejudice. It neither laughs nor weeps at the objects of its study, because its aim is to understand them. . . . The mathematical method, therefore, meant to Spinoza the free unprejudiced inquiry of the human mind, uncramped by the veto of theology and theological philosophy" (Roth, p. 43).

Spinoza wrote, in the preface to Part III of the *Ethics*: "Persons who would rather abuse or deride human emotions than understand them . . . will doubtless think it strange that I should attempt to treat of human vice and folly geometrically . . . the passions of hatred, anger, envy, and so on, considered in themselves, follow from [the] necessity and efficacy of nature. . . . I shall therefore treat of the nature and strength of the emotions . . . in exactly the same manner, as though I were concerned with lines, planes, and solids."

Another glimpse at Spinoza's idea of mathematics, as cited by Roth, is in the appendix to the *Ethics*, pp. 72–73. "[Superstitious people] laid down as an axiom, that God's judgments far transcend human understanding. Such a doctrine might well have sufficed to conceal the truth from the human race for all eternity, if mathematics had not furnished another standard of verity in considering solely the essences and properties of figures without regard to their final causes."

In the century after Spinoza's death, he suffered universal condemnation and near oblivion. Protestants and Catholics attacked him for rejecting their dogmas. Free thinkers and liberals repudiated him for fear of guilt by association. Since Spinoza's only openly published work was an exposition of Descartes, the Cartesian sect was particularly active in denouncing Spinozism.

Toward the end of the eighteenth century the German playwright Lessing privately confessed to being a Spinozist. Then Goethe openly praised and studied Spinoza. By the time of Hegel, you didn't know philosophy if you didn't know Spinoza. Samuel Coleridge, the poet, critic, and importer of advanced German culture, awoke Spinozism in England.

Although Spinoza didn't directly enter into the philosophy of mathematics, his indirect influence is important. Until the late eighteenth century, religion was rarely questioned in scholarly writing. The philosophy of mathematics and theology were twin trees holding each other up. Plato's *Meno* and Descartes's fifth meditation show mathematics shoring up religion. Writings of Berkeley and Leibniz show religion supporting the philosophy of mathematics. The notion that nonphysical reality or even all reality, physical and nonphysical, is located in the mind of God was a complete answer to the question of the nature of mathematics. Recent troubles in philosophy of mathematics are ultimately a consequence of the banishing of religion from science.

Religion in general is not decaying. Fundamentalist religion is spreading like a noxious weed. In intellectual circles you meet disciples of Buber, Kierkegaard,

Maharishi, and Nagarjuna. But religion is no longer granted a role in science. Newton said his discoveries redounded to the glory of God, but Laplace told Napoleon he had found God to be an unnecessary hypothesis. Today's scientist may attend church or *shul*, but he wouldn't dream of using the will of God to explain a puzzle in microbiology or particle physics.

Kant is the dividing line. As we shall see, he still has God in his philosophy. But his principal work, *The Critique of Pure Reason* is secular. God appears in the *Critique of Practical Reason*, in second place, so to speak.

Following Kant, mainstream philosophy became secular. Perhaps theology had reached the stage of self-sufficient independence, and, like psychology, broke off from philosophy. Perhaps philosophers, yearning for respect from scientists, followed the example of scientists in ghettoizing religion.

The important role of Spinoza in the philosophy of mathematics is his contribution to discrediting Scripture and Revelation, and ultimately secularizing science. By so doing he helped create the modern dilemma of the philosophy of mathematics.

Gottfried Wilhelm Freiherr von Leibniz (1646–1716)

Monads;Infinitesimals.The Differential Calculus.

Leibniz was a giant in both mathematics and philosophy. He's the founder of German idealism, the father of Hegel and the Hegelians. Christian von Wolff produced a popularization of Leibnizism, which became virtually official religion in Germany in the late eighteenth century.

(In his *Preliminary Discourse on Philosophy in General*, Wolff writes: "mathematical knowledge must be joined to philosophy if you desire the highest possible certitude. For this reason, we also grant a place in philosophy to mathematical knowledge, even though we have distinguished it from philosophy. For we hold that nothing is more important than certitude.")

Leibniz was the great polymath, expert and active in every field of learning and human affairs. He was rich, lived where and how he pleased. Yet by his own ambitions he was a failure.

At age 22 he produced a Latin tractate, *ordine geometrico demonstrata* (a l'Euclide), supporting the aspiration of the Count Palatine, Philip William of Neuburg, for the vacant throne of Poland. Philip William lost the election.

In 1672 Louis XIV's wars of conquest were turning Europe upside-down. Leibniz "urged the Christian powers of Europe instead of fighting one another to combine against the infidel, and suggested that in such a united war against the Turks, Egypt, one of the best situated lands in the world, would fall to France . . . the same idea commended itself more than a century later to Napoleon . . . [who discovered] Leibniz's memorial when he took possession of Hanover in 1803" (Carr, p. 12). Louis did not follow Leibniz's urging. But in

Paris Leibniz met Christian Huygens (1629–1695) who became his mathematics tutor, and he also met leading thinkers Antoine Arnauld, Jacques Bossuet (1627–1704), and Nicolas Malebranche (1638–1715). He visited London and was elected to the Royal Society.

At age 27 he accepted the post of librarian to John Frederick, of the House of Brunswick-Luneburg, Duke of Hanover. "In his correspondence he always refers to its head as 'my prince'" (Carr, p. 13). He continued to circulate in Paris, Vienna, and Rome, with ideas and projects, notably the reunion of Protestant and Catholic churches (he was Protestant).

Duke George (successor to Ernest Augustus, who was successor to John Frederick) grew vexed with Leibniz's slowness on the *History of the Guelph Family.* In 1714 when the duke was called to London to become George I, he commanded 68-year-old Leibniz to stay in Hannover until he finished the *History.* Leibniz died two years later, having got to the year 1005. The *History of the Guelph Family* is still incomplete.

Leibniz was a compulsive writer. He published only a few short works, but left a large secret cabinet in Hannover with "a great mass of hardly legible drafts" (Meyer, pp. 1–4). Johann Edward Erdmann spent years in the Leibniz archive. He wrote, "I leave to those who come after me not honeycombs, but pure wax." The Berlin Academy, which Leibniz founded, plans a 40-volume *Collected Works.* A dozen or so have appeared.

Leibniz was co-inventor, with Isaac Newton, of the infinitesimal calculus. Their priority fight is the most famous of its kind. As a consequence, English mathematicians ignored continental mathematics for a century, to their own great detriment.

Unlike Newton, Leibniz worked with actual infinitesimals,** though he couldn't explain coherently what they were. Cavalieri and others had calculated areas by dividing regions into infinitely many strips, each having infinitesimal positive area. unfortunately, as Huygens showed, this method could give wrong answers.

Leibniz had intimations of a formal language, but not in connection with infinitesimals. He dreamed of a language in which all correct reasoning would be straightforward calculation. He wrote (Nidditch, pp. 19–23) that "the invention of an ABC of man's thoughts was needed, and by putting together the letters of this ABC and by taking to bits the words made up from them, we would have an instrument for the discovery and testing of everything. . . . If we had a body of signs that were right for the purpose of our talking about all our ideas as clearly and in as true and as detailed a way as numbers are talked about in Arithmetic or lines are talked about in the Geometry of Analysis, we would be able to do for every question, in so far as it is under the control of reasoning, all that one is able to do in Arithmetic and Geometry. . . . It would not be necessary for our heads to be broken in hard work as much as they are now and we would certainly be able to get all the knowledge possible from the material given. In addition, we would have everyone in agreement about whatever would have been worked out, because it would be simple to have the working gone into by doing it again or by attempting tests like that of 'putting out nines' in Arithmetic. And if anyone had doubts of one of my statements I would say to him: let us do the question by using numbers, and in this way, taking pen and ink we would quickly come to an answer."

What a naive dream! But calculating machines were known to Leibniz. Pascal invented an adder while he was a teenager, and Leibniz taught it to multiply. Leibniz's unpublished thoughts were a forerunner of modern logic.

Leibniz's metaphysics is a most fascinating fantasy. At the request of Prince Eugene of Savoy (Carr, p. 3), Leibniz wrote his ***Monadology***. Leibniz's idealistic atomism makes him think there must be "simple" parts or "monads" out of which the world is made. Therefore, apart from these monads there can't be anything real. It follows that their relations can't be real. They can't see each other. They're "windowless"!

Leibniz was aware that things do seem to interact. He explained this in a most beautiful and incredible way. Imagine two clocks that always show the same time. It might seem that they are communicating with each other. But they aren't. They're just separately keeping correct time. There's no connection, only a "preestablished harmony." So, said Leibniz, body and soul are two separate windowless monads, keeping the same time. When you and I converse, we each speak independently, as predetermined by God. He sees our seeming conversation. His knowledge of it is its only reality.

Leibniz's world is beautifully described by Gottfried Martin (p. 1 ff.). "All monads are first defined as living. By a bold leap the whole knowable universe is then filled with a dense sea of living things, that is to say with a dense sea of monads. Everything is alive; everything unfruitful, everything sterile, everything dead in the universe is only outward illusion. . . . In this great ocean of living things there are no empty places. Wherever one looks there surges an infinite world of creatures, of living beings, of animals, of entelechies, of souls. Every single particle of matter, however small, is a garden full of plants, a pond full of fishes, and every twig of these plants in this garden and every drop in the blood of these fishes is another garden full of plants, another pond full of fishes, and so forth to infinity. Everywhere, in the infinitely great and in the infinitely small, there is life, everywhere there are monads. Every monad perceives and every monad wills . . . [My comment: Infinitely small monads are reminiscent of Leibniz's mathematical infinitesimal. There's a reverberation between Leibniz's mathematics and his metaphysics.]"

Martin continues, "Outside and between the individual existence of monads there is nonreality. Since only monads and modifications of monads have real existence in this sense, relations cannot have real existence . . . to use an alternative expression which Leibniz often uses, they only have mental existence . . . [relations] include number, time, duration, space, the extension of bodies. . . . But the understanding that thinks relations is the understanding of God. Relations have their substantial reality taken away from them by being referred to an understanding, but because the understanding that carries them is the divine understanding, they . . . receive back again a new reality. . . ."

"Matter," Leibniz keeps repeating, "is a mere phenomenon, and it is even clearer for him that space is a mere phenomenon . . . but a *phaenomenon Dei.*" [Reminiscent of Berkeley, the "British empiricist" who conventionally is classed in opposition to Leibniz the rationalist. For Berkeley all seeming existence, including physical existence, is illusory except as a thought in the mind of God.]

Martin goes on, "The divine thinking also provides the ground of the possibility of mathematics, because in his continuous thinking of possible worlds God also thinks continuously the eternal truths that hold for these worlds, in particular all numerical and spatial relations. This holds especially of Euclidean space which, as the only possible space, is the same for all possible worlds. The being of Euclidean space thus consists primarily in being thought without contradiction by God. . . .

"The propositions of three-dimensional Euclidean geometry are primarily and continuously thought by God, and when the mathematician discovers, understands, and proves an adequately formulated proposition in geometry, this knowing is a repetition of the primary divine knowing. . . ."

[I can't help repeating a remark in the last section: What a beautifully simple solution to the ontology of mathematics! We needn't "break our heads" figuring out how numbers have existence independent of human thought. The thought is God's! The present trouble with the ontology of mathematics is an after-effect of the spread of atheism. But mysticism still hasn't vanished from mathematics. See Paul Erdös and "The Book" in Chapter 1.]

"The profound consequences of Leibniz's connections with Plato . . . begin to unfold at this point. Plato pronounced repeatedly that thinking the truth means becoming like God. Leibniz may have met the ancient Platonic thesis . . . in Malebranche, in the proposition that we see all things in God . . . Leibniz was carrying on a great tradition, following Plato, Plotinus, Augustine, and Malebranche. . ." (Martin).

In his *Discourse on Metaphysics*, section 26, Leibniz summarizes Plato's dialogue, the *Meno*. (See the section above on Plato.) In this dialogue Socrates coaches a "slave boy" to discover that a small square inscribed at the midpoints of the sides of a large square has half the area of the large square. Socrates claims that in so doing he proved that the "slave boy" remembers this bit of geometry from before his birth, in Heaven. Leibniz calls the *Meno* "a beautiful experiment." He thinks Plato actually makes a strong case for his doctrine of "reminiscence." A fine example of intertwining between religion and philosophy of mathematics!

Some of Leibniz's pieties are worthy of his parody, Voltaire's Dr. Pangloss. "There will be no good action unrewarded and no evil action unpunished; everything must turn out for the well-being of the good; that is to say of those who are not disaffected . . . if we were able to understand sufficiently well the order of the universe, we should find that it surpasses all the desires of the wisest of us, and that it is impossible to render it better than it is, not only for all in general, but also for each one of us in particular, provided that we have the proper attachment for the author of all . . . who alone can make us happy" (1992, Vol. IV; 1965, p. 90).

As Kant stands behind Frege, Hilbert, and Brouwer, Leibniz and Hume stand behind Kant. Kant's metaphysics is sometimes described as an effort to wed Leibniz's rationalism to Hume's empiricism.

The "pre-critical" Kant (before the *Critique of Pure Reason*) had been soaked in Christian von Wolff's popular Leibnizism. Reading Hume "woke him from dogmatic slumber." He responded by trying to rehabilitate Leibniz's rationalism from Hume's attack.

"Kant can therefore justly say, looking back on his philosophical life-work, 'The *Critique of Pure Reason* is, after all, the real apologia for Leibniz, even against his followers who exalt him with praises which do him no honor' " (Martin).

Bishop George Berkeley (1685—1753)

Ghosts of Departed Quantities

George Berkeley said something interesting about calculus. His book *The Analyst* (1734) is the only philosophical critique that challenges mathematicians as mathematicians. *The Analyst* is never taught in Philosophy 101, perhaps because philosophy professors don't think it's philosophy, perhaps because calculus isn't a prerequisite for the students or the professor.

Students learn that Berkeley was the empiricist between Locke and Hume. This is misleading. Berkeley's empiricism proved that nothing exists, except in the mind of God—a conclusion far from those of Locke or Hume.

The Analyst is an attack on the differential calculus of Newton and Leibniz.** Its subtitle is, "A Discourse Addressed to an Infidel Mathematician, Wherein it is examined whether the object, principles, and inferences of the modern Analysis are more distinctly conceived, or more evidently deduced, than religious Mysteries and points of Faith." The Infidel Mathematician was the Astronomer Royal, Edmund Halley, who gave his name to a comet, and paid for publishing Newton's *Principia*.

According to Berkeley's brother, Berkeley wrote *The Analyst* in response to the circumstances of the death of the King's physician, Dr. Garth. Joseph Addison, the essayist, visited Garth to remind him to "prepare for his approaching dissolution." Said the dying man, "Surely, Addison, I have good reason not to believe those trifles, since my friend Dr. Halley who has dealt so much in demonstration has assured me that the doctrines of Christianity are incomprehensible, and the religion itself an imposture" (Berkeley, *Works*, Vol. 4, Introduction).

Halley deprived Garth of eternal bliss!

Berkeley counter-attacked, not by defending religion, but by showing that the mathematics of Newton and Leibniz is more obscure than the Church's deepest mystery.

"I shall claim the privilege of a Free-thinker," he wrote, "and take the liberty to inquire into the object, principles, and method of demonstration admitted by the mathematicians of the present date, with the same freedom that you presume to treat the principles and mysteries of Religion."

Bishop Gibson of London wrote gratefully to Bishop Berkeley: "the men of science (a conceited generation) are the greatest sticklers against revealed religion . . . we are much obliged to your Lordship for retorting their arguments upon them."

Berkeley's attack on mathematics went beyond calculus. He also denounced the mathematicians' claim that a line segment can be divided into arbitrarily short subsegments. For then the line segment would have infinitely many parts. But that is inconceivable to any "man of sense." What any man of sense cannot conceive is impossible.

Berkeley didn't say what was the shortest possible length, or even claim it could be known. His was a purely existential proof by contradiction. Infinite divisibility is inconceivable, so, by the law of the excluded middle, divisibility must be finite, so there must be "atoms" of length. That would mean that finding instantaneous velocities is impossible in principle. Instead of instantaneous velocity, we would have minimal-distance velocity—the minimal distance divided by the time elapsed in traversing it. Berkeley didn't go into these ramifications. But he did reject Euclid's construction for bisecting a line segment. For if the segment happens to be made of an odd number of indivisible length-atoms, it obviously can't be bisected.

Neither did he accept the diagonal of the unit square. All line segments are made of length-atoms, so they're all commensurable. Since no rational number equals $\sqrt{2}$, there's no line segment of length $\sqrt{2}$. So there's no line segment connecting opposite corners of a unit square. The supposed diagonal misses one corner or the other by a little bit.

I don't think Berkeley considered whether time is also discrete. Hume later proved that both time and space are discrete. This idea has been revived in our time, in an effort to make sense of quantum mechanics.

Berkeley also had an original idea about arithmetic: Numbers don't exist. Numerals are meaningless symbols. This was part of his doctrine that abstractions and general concepts don't exist. In today's philosophical lingo, he was a strict nominalist. More than that, he said that the sun, the moon, the teeth in your mouth, all are mere appearances. They, we, and all else are only thoughts in the mind of God.

He was a remarkable anomaly, a "British empiricist" who attacked a whole tribe of scientists—the mathematicians. Plato, Descartes, and Spinoza wanted to use the supposed certainty of mathematics to advance religion. Berkeley used the *deficiency* of mathematics to advance religion. His attack on mathematicians is unique since St. Augustine.

Immanuel Kant (1724–1804)

Synthetic a priori. Non-Euclidean Geometry.

Classical philosophy reached its peak at the end of the eighteenth century in Kant. Kant's metaphysics is a continuation of the Platonic search for certainty and timelessness in human knowledge. He wanted to rebut Hume's denial of certainty. To do so, he made a sharp distinction between noumena, things in themselves, which we can never know, and phenomena, appearances, which our senses tell us. His goal was knowledge a priori—knowledge timeless and independent of experience.

He distinguished two kinds of a priori knowledge. The "analytic a priori" is the kind we know by logical analysis, by the meanings of the terms being used. Like the rationalists, Kant believed we also possess a priori knowledge that is not logical truism. This is his "synthetic a priori." Our intuitions of time and space are such knowledge, he believed. He explained their a priori nature by saying they're intuitions—inherent properties of the human mind. Our intuition of time is systematized in arithmetic, based on the intuition of *succession*. Our intuition of space is systematized in geometry. For Kant, as for all earlier thinkers, there's only one geometry—the one we call Euclidean. The truths of geometry and arithmetic are forced on us by the way our minds work; this explains why they are (supposedly) true for everyone, independent of experience. The intuitions of time and space, on which arithmetic and geometry are based, are objective in the sense that they're valid for every human mind. No claim is made for existence outside the human mind. Yet the Euclid myth (see below) remains central in Kantian philosophy.

Indeed, mathematics is central for Kant. His *Prolegomena to any Future Metaphysics Which Will Be Able to Come Forth as a Science*, has three parts. Part One is, "How Is Pure Mathematics Possible?" (a question I discussed in Chapter 1).

Kant's fundamental presupposition is that "contentful knowledge independent of experience (the 'synthetic a priori') can be established on the basis of universal human intuition." In *The Critique of Pure Reason*, he gives the two examples already mentioned: (1) space intuition, the foundation of geometry, and (2) time intuition, the foundation of arithmetic. In *The Critique of Practical Reason*, without using the term "synthetic a priori," he gives a third intuition: (3) moral intuition, the foundation of religion.

I will discuss intuitions (1) and (3), although (2) is still interesting today. It deals with the question, "Are primitive counting notions universal and invariable?" Most writers think so. Wittgenstein disagreed. Frege also disagreed; he thought arithmetic was analytic a priori—based on logic—rather than synthetic a priori—based on time intuition.

I discuss three topics on Kant:

1. Synthetic a priori. Intuition of space and time.
2. Effect of non-Euclidean geometry on Kant's theory of space intuition.
3. Intuition of duty, God, and the parallel with intuitions of space and time.

1. The *Prolegomena*, p. 21: "Weary therefore of dogmatism [Leibniz], which teaches us nothing, and of skepticism [Hume], which does not even promise us anything—even the quiet state of a contented ignorance—disquieted by the importance of knowledge so much needed, and rendered suspicious by long experience of all knowledge which we believe we possess or which offers itself in the name of pure reason, there remains but one critical question, on the answer to which our future procedure depends, namely, "Is metaphysics at all possible?". . . The *Prolegomena* must therefore rest upon something already known as trustworthy, from which we can set out with confidence and ascend to sources as yet unknown, the discovery of which will not only explain to us what we knew but exhibit a sphere of many cognitions which all spring from the same sources. The method of prolegomena, especially of those designed as a preparation for future metaphysics, is consequently analytical.

"But it happens, fortunately, that though we cannot assume metaphysics to be an actual science, we can say with confidence that there is actually given certain pure *a priori* synthetical cognitions, pure mathematics and pure physics; for both contain propositions which are unanimously recognized, partly apodictically certain by mere reason, partly by general consent arising from experience and yet as independent of experience. We have therefore at least some uncontested synthetical knowledge *a priori*, and need not ask *whether* it be possible, for, it is actual, but *how* it is possible, in order that we deduce from the principle which makes the given knowledge possible the possibility of the rest."

Against synthetic (contentful) knowledge he contrasted analytic knowledge, which is derived from logic and the meaning of words.

Again from the *Prolegomena*:

"Analytical judgments express nothing in the predicate but what has been already actually thought in the concept of the subject, though not so distinctly or with the same (full) consciousness. When I say, 'All bodies are extended,' I have not amplified in the least my concept of body, but have only analyzed it, as extension was really thought to belong to that concept before the judgment was made, though it was not expressed. This judgment is therefore analytical. On the contrary, this judgment, "All bodies have weight," contains in its predicate something not actually thought in the universal concept of body. It amplifies my knowledge by adding something to my concept, and must therefore be called synthetical.

"First of all, we must observe that all strictly mathematical judgments are a priori, and not empirical, because they carry with them necessity, which cannot be obtained from experience. . . . It must at first be thought that the proposition $7 + 5 = 12$ is a mere analytical judgment, following from the concept of the sum, of seven and five, according to the law of contradiction. But on close examination it appears that the concept of the sum of $7 + 5$ contains merely their union in a single number . . . Just as little is any principle of geometry analytical." (This is the point at which Frege turned away from Kant.)

Richard Tarnas writes (p. 342): "The clarity and strict necessity of mathematical truth had long provided the rationalists—above all Descartes, Spinoza and Leibniz—with the assurance that, in the world of modern doubt the human mind had at least one solid basis for attaining certain knowledge. Kant himself had long been convinced that natural science was scientific to the precise extent that it approximated to the ideal of mathematics. . . . By Hume's reasoning, with which Kant had to agree, the certain laws of Euclidean geometry could not have been derived from empirical observation. Yet Newtonian science was explicitly based upon Euclidean geometry. . . . Kant began by noting that if all content that could be derived from experience was withdrawn from mathematical judgments, the ideas of space and time will remain. From this he inferred that any event experienced by the senses is located automatically in a framework of spatial and temporal relations. Space and time are 'a priori forms of human sensibility': They condition whatever is apprehended through the senses. Mathematics could accurately describe the empirical world because mathematical principles necessarily involve a context of space and time, and space and time lay at the basis of all sensory experience: they condition and structure any empirical observation. . . . Because [geometrical] propositions are based on direct intuitions of spatial relations, they are '*a priori*'—constructed by the mind and not derived from experience—and yet they are also valid for experience, which will by necessity conform to the *a priori* form of space."

Kant's intuitions are supposed to explain, not how we might or could, but how we *actually do* conceive of time and space. There's no claim that they correspond to an objective reality. They're properties of Mind.

For Kant and his predecessors, mathematics and Mind are unchanging, eternal, and universal. Kant's intuitions are supposed to be eternal, universal features of Mind. But the Mind Kant knows is the mind of eighteenth-century Europe, plus the books in his library. He assumes this constitutes all human thinking.

2. Kant's views came to dominate West European philosophy, in spite of a development in geometry that made Kant's account of space untenable. That development was non-Euclidean geometry.**

The fifth axiom of Euclid's *Elements*, the parallel postulate, for centuries was considered a blot on the fair cheek of geometry. This postulate says: "If a line A crossing two lines B and C makes the sum of the interior angles on one side of A less than two right angles, then B and C meet on that side."

An equivalent axiom, the usual one in geometry books, is Playfair's: "Through a point not on a given line passes one parallel to the line."

This parallel axiom, everybody agreed, is intuitively true. Yet it isn't as "self-evident" as the other axioms. It says something happens at a point that possibly is very remote, where our intuition isn't as firm as nearby. Mathematicians wanted it proved, not assumed as Euclid did. Many tried, no one succeeded.

Then, as I mentioned in Chapter 4, Gauss, Bolyai, and Lobachevsky had the same brilliant idea: Suppose the fifth postulate *is* false, and then see what happens! Each of them got a new geometry! A possibility never before conceived.

Later Beltrami, Klein, and Poincaré showed that Euclidean and non-Euclidean geometry are "equiconsistent." If either is consistent, so is the other. Since no one doubts that Euclidean geometry is consistent, non-Euclidean also is believed to be consistent.

Kant's theory of spatial intuition meant Euclidean geometry was inescapable. But the establishment of non-Euclidean geometry gives us choices. Which geometry works best in physics? The question becomes empirical, to be settled by observation.

In 1915, Einstein published his theory of general relativity. The cosmos is a non-Euclidean curved space-time, more general than the hyperbolic space of Gauss, Lobatchevsky, and Bolyai. So non-Euclidean geometry is not just consistent, it governs the universe! Non-Euclidean geometry is used to represent relativistic velocity vectors. Physics doesn't prefer Euclid to non-Euclid.

Our intuitive notion of space is learned on a small scale, compared to the universe as a whole. Locally, "in the small," the difference between Euclidean and non-Euclidean geometries is too tiny to notice. The belief that the Euclidean angle sum theorem is "indubitable" or "absolute" is based on belief in an infallible spatial intuition. That belief is discredited by non-Euclidean geometry and general relativity.

Decades before non-Euclidean geometry was discovered by Kant's countryman Karl Friedrich Gauss, it was "almost known" to Johann Heinrich Lambert (1728–1777), a German mathematician who was actually an acquaintance or friend of Kant! Lambert came to a crucial recognition—that *if* the "postulate of the acute angle" were true it would lead to a strange new geometry. This already would have refuted Kant's theory that Euclidean geometry is an unavoidable innate intuition of the human mind.

Did Kant know Lambert's work? Martin thinks he did, but disregarded it as a "mere abstraction."

Körner says Kant didn't deny the abstract conceivability of non-Euclidean geometries; he thought they could never be realized in real time and space. This idea was wiped out by the advance of science.

Even though Kant's philosophy of space had already been exploded by non-Euclidean geometry, Philip Kitcher shows that all three foundationist gurus—Frege, Hilbert, and Brouwer—were Kantians. That was a consequence of the dominance of Kantianism in their early milieus, and the usual tendency of research mathematicians toward an idealist viewpoint. When they became disturbed by the "crisis in foundations" they couldn't help thinking in Kant's categories, in particular, his analytic and synthetic a priori. But instead of talking about the synthetic a priori, they talked about restoring the indubitability of mathematics—building or finding a solid foundation.

Non-Euclidean geometry makes Kant's philosophy of space untenable. But mathematicians avoid philosophical disputation by not mentioning the issue. To this day, texts on non-Euclidean geometry ignore its revolutionary philosophical implications. The first direct statement of the contradiction seems to be by Hermann Helmholtz, in *Mind* in 1877 (the birth year of that august journal.) In the next volume of *Mind* a Dutch philosopher, H. K. Land, replied that, by the nature of things, nothing in mathematics could be relevant to Kant's theory. Modern philosophy texts and lecturers on Kant seem to follow Land's principle. They don't mention non-Euclidean geometry.

3. Kant may have been the last philosopher or mathematician in the chain from Pythagoras to the present who explicitly made theology part of his philosophy. There's a half-hidden connection between Kant's a prioristic philosophy of mathematics and his moral-intuition version of Christianity.

In the *Critique of Practical Reason* he demolishes the three standard proofs of the existence of God. "Ontological": By definition, God is Perfect. Nonexistence would be an imperfection. "Cosmological": Every event has a cause. To avoid infinite regress, there had to have been a First Cause (God). "Teleological": A watch has a watch-maker. The World is more intricate than a watch, so it has a World-Maker (God).

Kant tears these proofs to shreds. He says they're the only proofs "speculative reason" (Leibnizian rationalism) could ever give. Kant isn't doubting God's existence. He's showing the superiority of his own proof, based on intuition. Not so different from his intuitions of time and space. Everyone has an intuition of duty, Kant thinks, of right and wrong. He doesn't say this *proves* God exists. He says it *justifies the postulate* "God exists."

"The moral law leads us to postulate not only the immortality of the soul, but the existence of God. . . . This second postulate of the existence of God rests upon the necessity of presupposing the existence of a cause adequate to the effect which has to be explained . . . a being who is a part of the world and is dependent upon it . . . ought to seek to promote the highest good, and therefore the highest good must be possible. . . . There is therefore implied, in the idea of the highest good, a being who is the supreme cause of nature, and who is the cause or author of nature through his intelligence and will, that is, God . . . or, in other words, it is morally necessary to hold the existence of God."

And in the *Prolegomena*, paras. 354–55, p. 103: "We must therefore think an immaterial being, a world of understanding, and a Supreme Being (all mere noumena) because in them only, as things in themselves, reason finds that completion and satisfaction which it can never hope for in the derivation of appearances from their homogeneous grounds, and because these actually have reference to something distinct from them (and totally heterogeneous), as appearances always presuppose an object in itself, and therefore suggest its existence whether we can know more of it or not."

Tarnas again (p. 350): "It is clear that at heart Kant believed that the laws moving the planets and stars ultimately stood in some fundamental harmonious relation to the moral imperatives he experienced within himself. 'Two things fill the heart with ever new and always increasing awe and admiration: the starry heavens above me and the moral law within me.' But Kant also knew he could not prove that relation, and in his delimitation of human knowledge to appearances, the Cartesian schism between the human mind and the material cosmos continued in a new and deepened form.

"In the subsequent course of Western thought, it was to be Kant's fate that, as regards both religion and science, the power of his epistemological critique tended to outweigh his positive affirmations. On the one hand, the room he made for religious belief began to resemble a vacuum, since religious faith had now lost any external support from either the empirical world or pure reason, and increasingly seemed to lack internal plausibility and appropriateness for secular modern man's psychological character. On the other hand, the certainty of scientific knowledge, already unsupported by any external mind-independent necessity after Hume and Kant, became unsupported as well by any internal cognitive necessity with the dramatic controversion by twentieth century physics of the Newtonian and Euclidean categories which Kant had assumed were absolute" (Tarnas, p. 350).

As the universal intuition of space is refuted by non-Euclidean geometry, the universal intuition of duty is refuted by history. For Winston Churchill and Harry Truman, fire-bombing German and Japanese civilians was duty. In the police stations of the world, torturing prisoners is duty. In Nazi Germany, genocide was duty.

What's the connection between Kant's philosophy of mathematics and his moral-intuition version of religion? Unlike Descartes and Leibniz, Kant does not use the certainty of mathematics (time and space) to support the certainty of God's existence. He considers the intuition of duty independently of the intuitions of time or space. He keeps his theory of God separate from his theory of mathematics. But they both have the same logic. Both rely on intuition: knowledge coming, not from the senses, study, or learning, but from the nature of Mind. Right and wrong, like time and space, are universal intuitions. Our space intuition leads to geometry, our time intuition leads to arithmetic, our duty intuition leads to Divinity.

In God's mind, the difficulties and puzzles in philosophy of mathematics disappear. How do numbers exist? Why do mathematical facts seem certain and timeless? Why does mathematics work in the "real world"?

In the mind of God, it's no problem.

The trouble with today's Platonism is that it gives up God, but wants to keep mathematics a thought in the mind of God.

Euclid as a Myth. Nobody's Perfect.

The myth of Euclid is the belief that Euclid's *Elements* contain indubitable truths about the universe. Even today, most educated people still believe the Euclid myth. Up to the middle or late nineteenth century, the myth was unquestioned. It has been the major support for metaphysical philosophy—philosophy that sought a priori certainty about the nature of reality.

The roots of our philosophy of mathematics are in classical Greece. For the Greeks, mathematics was geometry. In Plato and Aristotle, philosophy of mathematics is philosophy of geometry.

Rationalism served science by denying the intellectual supremacy of religious authority, while defending the truth of religion. This equivocation gave science room to grow without being strangled as a rebel. It claimed for science the right to independence from the Church. Yet this independence didn't threaten the Church, since science was the study of God's handiwork. "The heavens proclaim the glory of God and the firmament showeth His handiwork."

The existence of mathematical objects as ideas independent of human minds was no problem for Newton or Leibniz; they took for granted the existence of a Divine Mind. In that belief, the problem is rather to account for the existence of nonideal, material objects.

After rationalism displaced medieval scholasticism, it was challenged by materialism and empiricism; by Locke and Hobbes in Britain, by the encyclopedists in France. The advance of science on the basis of the experimental method gave the victory to empiricism. The conventional wisdom became: "The material universe is the fundamental reality. Experiment and observation are the only legitimate means of studying it."

The empiricists held that all knowledge *except mathematical* comes from observation. They usually didn't try to explain how mathematical knowledge originates. In the controversies, first between rationalism and scholasticism, later between rationalism and empiricism, the sanctity of geometry was unchallenged.

Philosophers disputed whether we proceed from Reason (a gift from the Divine) to discover the properties of the world, or whether only our bodily senses can do so. Both sides took it for granted that geometrical knowledge is not problematical, even if all other knowledge is. Hume exempted books of mathematics and of natural science from his outcry, "Commit it to the flames."

For rationalists, mathematics was the main example to confirm their view of the world. For empiricists, it was an embarrassing counter-example, which had to be ignored or explained away. If, as seemed obvious, mathematics contains knowledge independent of sense perception, then empiricism is inadequate as an explanation of all human knowledge. This embarrassment is still with us; it's a reason for the difficulties of philosophy of mathematics.

Mathematics always had a special place in the battle between rationalism and empiricism. The mathematician-in-the-street, with his common-sense belief in mathematics as knowledge, is the last vestige of rationalism.

The modern scientific outlook took ascendancy in the nineteenth century. By the time of Russell and Whitehead, only logic and mathematics could still claim to be nonempirical knowledge, obtained directly by Reason.

From the customary viewpoint among scientists now, the Platonism of most mathematicians is an anomaly. For many years the accepted assumptions in science have been materialism in ontology, empiricism in epistemology. The world is all one stuff, "matter," which physics studies. If matter gets into certain complicated configurations, it falls under a special science with its own methodology—chemistry, geology, and biology. We learn about the world by looking at it and thinking about what we see. Until we look, we have nothing to think about.

Yet in mathematics we have knowledge of things we can never observe. At least, this is the natural point of view when we aren't trying to be philosophical.

Until well into the nineteenth century, the Euclid myth was universal among mathematicians as well as philosophers. Geometry was the firmest, most reliable branch of knowledge. Mathematical analysis—calculus and its extensions and ramifications—derived legitimacy from its link with geometry. We needn't say "Euclidean geometry." The qualifier became necessary only after non-Euclidean geometry had been recognized. Before that, geometry was simply geometry—the study of the properties of space. These were exact, eternal, and knowable with certainty by the human mind.

eight

Mainstream Since the Crisis

How Did We Get Here? Can We Get Out?

Vacillation between two unacceptable philosophies wasn't always the prevalent mode. Where did it come from?

Until the nineteenth century, geometry was regarded by everybody, ***including mathematicians***, as the most reliable branch of knowledge. Analysis got its meaning and its legitimacy from its link with geometry.

In the nineteenth century, two disasters befell. One was the recognition that there's more than one thinkable geometry. This was a consequence of the discovery of non-Euclidean geometries.

A second disaster was the overtaking of geometrical intuition by analysis. Space-filling curves** and continuous nowhere-differentiable curves** were shocking surprises. They exposed the fallibility of the geometric intuition on which mathematics rested.

The situation was intolerable. Geometry served from the time of Plato as proof that certainty is possible in human knowledge—including religious certainty. Descartes and Spinoza followed the geometrical style in establishing the existence of God. Loss of certainty in geometry threatened loss of all certainty.

Mathematicians of the nineteenth century rose to the challenge. Led by Dedekind and Weierstrass, they replaced geometry with arithmetic as a foundation for mathematics. This required constructing the continuum—the unbroken line segment—from the natural numbers. Dedekind,** Cantor, and Weierstrass found ways to do this. It turned out that no matter how it was done, building the continuum out of the natural numbers required new mathematical entities—infinite sets.

Foundationism—Our Inheritance

The textbook picture of the philosophy of mathematics is strangely fragmentary. You get the impression that the subject popped up in the late nineteenth century because of difficulties in Cantor's set theory. There was talk of a "crisis in the foundations." To repair the foundations, three schools appeared. They spent thirty or forty years quarreling. But none of the three could fix the foundations. The story ends some sixty years ago. Whitehead and Russell abandoned logicism. Gödel's incompleteness theorem checkmated Hilbert's formalism. Brouwer remained in Amsterdam, preaching constructivism, ignored by most of the mathematical world.

This episode was a critical period in the philosophy of mathematics. By a striking shift in meaning of words, the domination of philosophy of mathematics by foundationism became the *identification* of philosophy of mathematics with foundations. We're left with a peculiar impression: The philosophy of mathematics was awakened by contradictions in set theory. It was active for forty or fifty years. Then it went back to sleep.

Of course there has always been a philosophical background to mathematical thinking. In the foundationist period, leading mathematicians engaged in public controversy about philosophical issues. To make sense of that period, look at what went before and after. Two strands of history have to be followed, philosophy of mathematics and mathematics itself. The "crisis" manifested a long-standing discrepancy between the Euclid myth, and the reality, the actual practice of mathematicians.

In discussions of foundations three dogmas are presented: Platonism, formalism, and intuitionism. Platonism was described in Chapter 1. I remind the reader what it says: "Mathematical objects are real. Their existence is an objective fact, independent of our knowledge of them. Infinite sets, uncountably infinite sets, infinite-dimensional manifolds, space-filling curves—all the denizens of the mathematical zoo—are definite objects, with definite properties. Some of their properties are known, some are unknown. These objects aren't physical or material. They're outside space and time. They're immutable. They're uncreated. A meaningful statement about one of these objects is true or false, whether we know it or not. Mathematicians are empirical scientists, like botanists. We can't invent anything; it's there already. We try to discover."

In recent times Platonism has sometimes been identified with logicism. *If* a Platonist makes an effort to explain the nature of his nonhuman mathematical objects, it's usually in terms of logic and/or set theory.

According to formalism, on the other hand, there are *no* mathematical objects. Mathematics is *axioms, definitions, and theorems*—in brief, formulas. A strong version of formalism says that there are rules to derive one formula from another, but the formulas aren't *about* anything. They're strings of meaningless symbols. Of course the formalist knows that mathematical formulas are being applied to physics. When a formula gets a physical interpretation, ***then*** it acquires meaning. ***Then*** it can be true or false. But the truth or falsity refers only to the physical interpretation. As a mathematical formula apart from any interpretation, it has no meaning and can be neither true nor false.

The difference between formalist and Platonist is clear in their attitudes to Cantor's continuum hypothesis. Cantor conjectured that there's no infinite cardinal number greater than $\aleph_0$ (the cardinality of the integers) and smaller than c (the cardinality of the real numbers). Kurt Gödel and Paul J. Cohen showed that on the basis of the Zermelo—Fraenkel axioms of set theory, the continuum hypothesis can neither be disproved (Gödel, 1937) nor proved (Cohen, 1964). To the Platonist, this means our axioms of sets are incomplete. The continuum hypothesis *is* either true or false. We just don't understand the real numbers well enough to tell which is the case.

To the formalist, the Platonist interpretation makes no sense, because there *is* no real number system, except as we "create" it by laying down axioms to describe it. We're free to change these axioms, for convenience, usefulness, or any criterion that appeals to us. But the criterion can't be better correspondence with reality, because there's no reality to correspond with.

Formalists and Platonists take opposite sides on existence and reality. On the principles of mathematical proof, they have no quarrel. Opposed to both of them are the constructivists. Constructivists accept only mathematics that's obtained from the natural numbers by a finite construction. The set of real numbers, and any other infinite set, cannot be so obtained. Consequently, the constructivist accepts neither the Platonist not the formalist view of Cantor's hypothesis. Cantor's hypothesis is meaningless. Any answer is a waste of breath.

Today some mathematicians still call themselves formalists or constructivists, but in philosophical circles one speaks more often of Platonists versus fictionalists. Fictionalists reject Platonism. They can be formalists, constructivists, or something else (see Chapter 10).

Philosophers like to call their arguments and counter-arguments "moves." It's a standard move to finesse a dispute by declaring it meaningless. This was favored by the logical positivists in days of yore. They decreed that the meaning of any statement is no more or less than its truth conditions. It followed, where logical positivists were in charge, that metaphysics, ethics, and much else was thrown out of philosophy. In time, by the same rule, logical positivism was thrown out too.

Two large facts about mathematics are hardly doubted.

Fact one: Mathematics is a human product.

It may seem unfair to expect a Platonist to admit this; it may seem like asking him to give the game away. Nevertheless, contributions to mathematics are made every day, by specific, particular human beings. Many contributions are signed by an author or authors. No one questions the claim of these authors for their results.

The quibble between discovery and invention or creation was discussed in Chapter 5. But no one doubts that the mathematics we know comes from the work of human beings. In fact, it is sometimes possible to account for features of a mathematical discovery by the interests, tastes, and attitudes of the discoverer and sometimes also by the needs or traditions of his country. This fact is the bulwark of "fictionalism." In its way of coming into being, in the way in which it's thought of by its creators, mathematics is like an art such as fiction or sculpture.

Fact two: We can choose a problem to work on, but we can't choose what the answer should be.

When you resolve a mathematical difficulty, you sense that the answer was already "there," waiting to be found. Even if the answer is, "There's no answer," as in Cantor's continuum problem, *that* is the answer, like it or not.

The number 6,785,123,080,772,901,001 is either prime or composite. I don't know which. But I know that any method I use will give the same answer—prime or composite. 6,785,123,080,772,901,001 is what it is, regardless of what I think or know. In this respect numbers are independent of their creators. Did I just bring 6,785,123,080,772,901 into existence? Or was it waiting and ready, some*how* if not some*where*, along with billions, quadrillions, and quintillions of cousins?

The philosophers more impressed by the *objectivity* of mathematics are Platonists. They say numbers exist apart from human consciousness. Those more impressed with the *human role* in creating mathematics are anti-Platonists. Depending on what they offer to replace Platonism, they may be fictionalists, formalists, constructivists, intuitionists, conventionalists.

What Is logic? What Should It Be?

Is it the rules of correct thinking?

Everyday experience, and ample study by psychologists, show that most of our thinking doesn't follow logic.

This might mean most human thinking is wrong. Or it might mean the scope of logic is too narrow.

Computing machines do almost always obey logic.

That's the answer! Logic is the rules of computing machinery! Logic also applies to people when they try to be computing machines.

Once upon a time logic and mathematics were separate. Then George Boole figured out how to make logic part of mathematics.

Russell claimed the opposite—that mathematics is nothing but logic. But the paradoxes made that idea unpalatable. Far from a solid foundation for mathematics, set theory/logic is now a branch of mathematics, and the least trustworthy branch at that.

Like other branches, logic has expanded greatly in scope and power. It offers problems and challenges, techniques and tools to other parts of mathematics. And it renounces any desire or duty to check up on other branches of mathematics, or to tell people how to think. For today's mathematical logician, logic is just another branch of mathematics like geometry or number theory. He disowns philosophical responsibilities.

"This book does not propose to teach the reader how to think. The word 'logic' is sometimes used to refer to remedial thinking, but not by us" (Enderton).

In U.S. philosophy departments, on the other hand, "analytic philosophy," a kind of left-over from logicism and logical positivism, lingers on. Kitcher gave it the fitting sobriquet "neo-Fregeanism" (see Carnap and Quine in Chapter 9).

Analytic philosophers mustn't be confused with mathematical logicians. A few outstanding logicians do encompass both mathematical and philosophical logic; that is, they are competent by both mathematical and philosophical standards.

Of course logical blunders aren't acceptable in mathematical reasoning. In that sense, mathematicians (and other scientists) are subject to logic. This isn't the business of logicians. It's the business of the mathematician and her referees.

Gottlob Frege (1848–1925)

Grandpa of the Mainstream

Frege is the first *full-time* philosopher of mathematics. According to Baum, "Although Frege is sometimes spoken of as being the first philosopher of mathematics, he was at most the initiator of the recent period of intensive concentration on this area by specialists using the tools of mathematical logic. Frege considered himself to be working entirely within the tradition of Plato, Descartes, Leibniz, etc. with regard to his work on the philosophy of mathematics" (Baum, p. 263). Frege's greatest contribution to learning is the ***Begriffsschrift*** (***Idea Script***), where he introduced quantifiers—symbols for "there exists" (now written backward E) and for "for all" (now written upside-down A). Quantifiers were independently invented by O. H. Mitchell, a student of Charles Sanders Peirce (Lewis, 1918; Putnam, 1990). Frege's introduction of quantifiers is considered the birth of modern logic. His technical logic is a means to a philosophical end. He wants to establish arithmetic as a part of logic.

It's believed that logic with quantifiers (usually called "predicate calculus") can express any reasoning mathematicians use in strict, formal proof. (We also do heuristic, intuitive, informal reasoning.) In principle, using Frege's notation or others developed later, it seems possible to write any complete mathematical proof in a form that a computer can check.

Kant thought geometry is based on space intuition, and arithmetic on time intuition. That made both geometry and arithmetic "synthetic a priori." About geometry, Frege agreed with Kant that it is a synthetic intuition. About arithmetic, he agreed with Leibniz: It is not *synthetic* but *analytic*. That is, it doesn't depend on an intuition of time. It comes from logic. For Frege and Leibniz, "logic" means the intuitively obvious rules of correct reasoning. These are supposed to be certain and indubitable, independent of anybody's thought or experience. Deriving arithmetic from logic would make arithmetic equally certain and indubitable.

One really cannot speak of Frege's philosophy of mathematics. He had a philosophy of arithmetic, and a different philosophy of geometry. Arithmetic is logic; geometry is space intuition. Fitting them together is as awkward as yoking an ape and an alligator. If arithmetic is part of logic, why not geometry as well (since we construct spaces from numbers by using coordinates)? On the other hand, if geometry is space intuition, why may not arithmetic be time intuition, as Kant had it?

Frege's Grundlagen. Logicism's Koran

In his *Grundlagen der Arithmetik* (Foundations of Arithmetic) Frege constructed the natural numbers out of logic. This achievement was ignored for 16 years, until Bertrand Russell took up the same project and made Frege known to the world. "Today, Frege's *Grundlagen* is widely appreciated as a philosophical masterpiece. In retrospect the mathematicians who ignored it appear as men who failed to recognize a pioneering work" (Kitcher, "Frege, Dedekind . . ."). Since all classical mathematics can be built from the natural numbers, Russell claimed that *all mathematics* is logic. This is called logicism.

Before giving his definition of number in the *Grundlagen*, Frege tries to demolish all previous definitions. He carries out a merciless, hilarious campaign against psychologism (numbers are ideas in someone's head—Berkeley, Schloemilch); historicism (numbers evolve); and empiricism (numbers are things in the physical world—Mill). To this day, philosophers of mathematics hardly dare contemplate psychologism or historicism.

First Frege trounces Mill, who based arithmetic on empirical experience (Frege, 1980, pp. 9–10): "The number 3 . . . consists, according to him, in this, that collections of objects exist, which while they impress the senses thus,

0 0 ,
0

they may be separated into two parts, thus,

00 0.

What a mercy, then, that not everything in the world is nailed down; for if it were, we should not be able to bring off this separation, and 2 + 1 would not be 3!

What a pity that Mill did not also illustrate the physical facts underlying the number 0 and 1! . . . From this we can see that it is really incorrect to speak of three strokes when the clock strikes three, or to call sweet, sour and bitter three sensations of taste, and equally unwarrantable is the expression 'three methods of solving an equation'. For none of these is a parcel which ever impresses the senses thus,

0 0."
 0

(A bit unfair! Mill actually mentions strokes of the clock as an example of counting. See the article on Mill in the next chapter.)

Mill's lack of precision makes him an easy mark for Frege. Nevertheless, Mill is right to say number has something to do with physical reality. Every child learns arithmetic from the pebbles and ginger snaps Frege laughs at. If our ancestors didn't need to keep track of coconuts or fish heads, they wouldn't have invented arithmetic. Much deeper is the discovery, many times repeated, that mathematics is the language of nature. Mill grapples with the relation between numbers and physical reality. Frege brushes it aside. In that respect, Mill did a service to human understanding, Frege a disservice.

Next is Heinrich Hankel. Frege writes, "the first question to be faced is whether number is definable." Hankel thought not, and expressed himself in this unfortunate manner: "What we mean by thinking or putting a thing once, twice, three times, and so on, cannot be defined, because of the simplicity in principle of the concept of putting." Replies Frege, "But the point is surely not so much the putting as the once, twice, and three times. If this could be defined, the indefinability of putting would scarcely worry us" (p. 26).

Next Frege turns on George Berkeley (p. 33), whom he quotes: "Number . . . is nothing fixed and settled, really existing in things themselves. It is entirely the creature of the mind. . . . We call a window one, a chimney one, and yet a house in which there are many windows, and many chimneys, hath an equal right to be called one, and many houses go to the making of one city." In fact, as mentioned in an earlier chapter, Berkeley thought numbers don't exist. To him, numerals were meaningless symbols.

Frege's answer: "This line of thought may easily lead us to regard number as something subjective . . . number is no whit more an object of psychology or a product of mental processes than, let us say, the North Sea is. The objectivity of the North Sea is not affected by the fact that it is a matter of our arbitrary choice which part of all the water on the earth's surface we mark off and elect to call the North Sea. This is no reason for deciding to investigate the North Sea by psychological methods. In the same way, number too, is something objective. If we say 'The North Sea is 10,000 square miles in extent' then neither by the 'North Sea' nor by '10,000' do we refer to any state of or process in our minds: on the contrary, we assert something quite objective, which is independent of our ideas and everything of the sort."

His attack on psychologism means to prove that numbers aren't ideas. To do so, he assumes surreptitiously that ideas are property only of individuals, uncorrelated with other people's ideas or with physical reality. An indefensible assumption!

Here is his next assault, against Schloemilch: "I cannot agree with Schloemilch either (p. 36), when he calls number the idea of the position of an item in a series. If number were an idea, then arithmetic would be psychology. But arithmetic is no more psychology than, say astronomy is. Astronomy is concerned, not with ideas of the planets, but with the planets themselves, and by the same token the objects of arithmetic are not ideas either. If the number two were an idea, then it would have straight away to be private to me only. [No! No!] Another man's idea is, *ex vi termini*, another idea. We should then have it might be many millions of twos on our hands. We should have to speak of my two and your two, of one two and all twos. If we accept latent or unconscious ideas, we would have unconscious twos among them, which would then return subsequently to consciousness. As new generations of children grew up new generations of twos would continually be being born and in the course of millennia these might evolve, for all we could tell, to such a pitch that two of them would make five. Yet, in spite of all this, it would still be doubtful whether there existed infinitely many numbers, as we ordinarily suppose. 10^{10}, perhaps, might be only an empty symbol, and there might exist no idea at all, in any being whatever, to answer to the name.

"Weird and wonderful, as we see, are the results of taking seriously the suggestion that number is an idea. And we are driven to the conclusion that number is neither spatial and physical, like Mill's piles of pebbles and gingersnaps, nor yet subjective, like ideas, but non-sensible and objective. Now objectivity cannot, of course, be based on any sense-impression, which as an affection of our mind is entirely subjective, but only, so far as I can see, on the reason. It would be strange if the most exact of all the sciences had to seek support from psychology, which is still feeling its way none too surely."

It's too late to defend Mill or Hankel or Schloemilch. But we must reject Frege's argument against "psychologism"—the belief that mathematical objects are ideas. He says 2 cannot be an idea, because different people have different ideas, and there is only one 2.

Frege is confounding private and public senses of "idea." It's not unusual to say "We have the same idea." Frege assumes an idea resides only in one person's head (private ideas). But ideas can be shared by several people, even millions of people (public ideas). Cheap and dear, legal and illegal, sacred and profane, patriotic and treasonous—all ideas, but not ideas of a particular person. Public ideas, part of society, history, and culture. (Philosophers say "intersubjective" to avoid "society" and "culture.") The existence of language, society, and all social institutions prove that people sometimes do have the same idea. Not in the sense of subjective inner consciousnesses; in the sense of verbal and practical understanding and agreement. A piece of green paper called a dollar is worth a quart of milk because many people agree that it is. All these people have the same idea—the equal value of a quart of milk and a dollar bill.

Whoever told Frege, "2 is an idea" intended the public meaning of "idea." Frege replaced that with the private meaning of "idea," and then had fun throwing stones at Schloemilch. Could Frege prove that number is ***not*** an intersubjective, social-cultural object? No. His sarcasm about psychologism has no bearing on my proposal that mathematical objects are ideas on the social level.

After Frege makes mincemeat of empiricism, historicism, psychologism, and (in another book, the *Grundgesetze*) formalism, you are ready for his solution: *NUMBERS ARE ABSTRACT OBJECTS.* Objects which are real, but not physically, not psychologically, real in an ***abstract*** sense.

What is "abstract"?

Evidently, ***not*** mental or physical. What qualities are possessed by abstract objects? They're timeless or tenseless. Aren't born, do not die.

What an astonishing kinship to Plato's Ideas! They were neither mental nor physical, but eternal and changeless. Frege is a Platonist as well as a Kantian.

Frege's abstract objects include numbers. Everything else has been proved wrong, so this must be right. But the elimination argument isn't valid, because Frege hasn't considered the alternative we offer: mathematics as part of the social-cultural-historical side of human knowledge. (He did attack the notion of numbers as historical entities.)

Frege's argument against formalism, psychologism, and empiricism comes down to a declaration: "Anyone can see that $7 + 3 = 10$. There's no possible doubt of it. Clearly it's true a priori, now and forever, certainly and indubitably." The same argument was given 1,300 years earlier by Augustine, Bishop of Hippo: "Seven and three are ten, not only now but always; nor was there ever a time when seven and three were not ten, nor will ever be a time when seven and three will not be ten. I say, therefore, that this incorruptible truth of number is common to me and to any reasoning person whatsoever."

Augustine's arithmetical Platonism went with his theology. The certainty of mathematics supported the certainty of religion. By Frege's time, the association between Christian theology and mathematical Platonism had gone underground. The success of secular science made it bad form to bring religion into logic or mathematics. Even so, David Hilbert and Bertrand Russell, unlike Frege, were frank about their religious motives.

The important thing is Frege's analysis of number. *Numbers are classes.* More precisely, *they're equivalence classes of classes,* under the equivalence relation of one-to-one mappings. For example, *two is the class of all pairs.* This class of all pairs exists objectively, timelessly, independently of us. It's an abstract object.

Some readers haven't seen arithmetic built up from logic. It's easy. Frege's logic uses the concept of "class." This is almost what we mean today by set. A set is defined by who are its ***members*** ("extension"). A class is defined by the ***property*** that decides whether or not you're a member ("intension").

Frege says "Two is the class of all pairs. Three is the class of all triplets. And so on." The point of this construction is to prove the "a priority" of the numbers, their independence of experience.

One slight problem: These definitions are circular. Knowing what's a "pair" is already knowing what's 2. A better statement is: 2 is by definition the class of all classes equivalent to {A,B}. 2 is called the cardinality of any such class (commonly known as a pair). 3 is the class of all classes equivalent to {A,B,C}. 3 is called the "cardinality" of any such class (commonly known as a triplet).

What's "equivalent"? Classes are equivalent if their members can be matched with nothing left over. Any two pairs are equivalent. Any two triples are equivalent. Any natural number, including 0 and 1, is a class of equivalent classes. The reader is encouraged to think through these two special cases.

A little explanation will help you sympathize with Frege. When he defines 2 as the equivalence class of all pairs, he's assuming that the notion, "equivalence class of all pairs" is free of ambiguity. To ***justify*** his definition of "2," we have to see if there ***is*** a "class of all pairs." If it doesn't exist, we needn't bother about it! That such a thing exists may have been crystal clear to Frege. It's not so clear today. Today, with caution learned from Frege's burnt fingers, "the class of all pairs" or "the set of all sets equivalent to {0,1}" would not go down so easily. "Pairs of what?" the students would rightfully demand.

Not pairs of numbers; we're trying to ***define*** numbers. Not shoes or socks; they're too earthy for a transcendental theory. Probably pairs of ***abstract*** objects. But where are they and what are they? To create the mathematical universe from scratch, I have no ingredients available. If I want to create numbers as collections, I need something to collect! In today's mathematics, one doesn't take for granted that any specification written in English or German is meaningful to define a set. Nowadays we want a set to be ***located***, a subset of a given universal set. You may not simply "define" some infinite set. You must show that the definition isn't self-contradictory. Our caution is due partly to the disaster that befell Frege: the Russell paradox.**

This proposal opened a new direction of thinking in foundations. It was the basis of Frege's plan to make arithmetic part of logic/set theory.

Frege's influence is not so much his semi-Kantian philosophy as his statement of the issue—***establish mathematics on a solid, indubitable foundation.***

To an unprepared mind, Frege's definition of number is bizarre. It explains the clear and simple, number, by the complicated and obscure: infinite equivalence classes. The bizarrerie is mitigated if you remember the point—to reduce arithmetic to logic. Mathematicians know how to build analysis and geometry on arithmetic. Frege and Russell believe logic is rock-solid. If they could have built arithmetic on logic, that would have made *all* mathematics as solid as logic itself. It didn't work out that way.

0 is particularly nice. It's the class of sets equivalent to the set of all objects unequal to themselves! *No* object is unequal to itself, so 0 is the class of all empty sets. But all empty sets have the same members—none! So they're not merely *equivalent* to each other—they're all *the same* set. There's only one empty set! (A set is characterized by its membership list. There's no way to tell one empty membership list from another. Therefore all empty sets are the same thing!)

Once I have *the* empty set, I can use a trick of von Neumann as an alternative way to construct the number 1. Consider the class of *all* empty sets. This class has exactly one member: the unique empty set. It's a singleton. "Out of nothing" I have made a *singleton* set—a "canonical representative" for the cardinal number 1. 1 is the class of all singletons—all sets with but a single element. To avoid circularity: "1 is the class of all sets equivalent to the set [{ }]." In words, 1 is the class of all sets equivalent to the set whose only element is the empty set. Continuing, you get pairs, triplets, and so on. Von Neumann recursively constructs the whole set of natural numbers out of sets of sets of sets of nothing.

Set theory was introduced by Georg Cantor as a fundamental new branch of mathematics. The idea of set—any collection of distinct objects—was so simple and fundamental, it looked like a brick out of which all mathematics could be constructed. Even arithmetic could be downgraded (or upgraded) from primary to secondary rank, for the natural numbers could be constructed, as we have just seen, from nothing—i.e., the empty set—by operations of set theory.

At first set theory seemed to be the same as logic. The set-theoretic relation of inclusion, "A is a subset of B," is the same as the logical relation of implication, "If A, then B." "Logic" here means the fundamental laws of reason, of contradiction and implication—the objective, indubitable bedrock of the universe. To show mathematics is part of logic would show it's objective and indubitable. It would justify Platonism, passing to the rest of mathematics the indubitability of logic.

This was the "logicist program" of Russell and Whitehead's *Principia Mathematica*. The logicist school was philosophically (not technically) similar to Hilbert's formalist school. For the logicists it was logic that was indubitable a priori; for the formalists, it was finite combinatorics. The difference between the logicists and Kant is that they give up his claim that mathematics is synthetic a priori. They settle for analytic a priori. According to Bertrand Russell, mathematics is a vast tautology.

The logicists proposed to redeem all mathematics by injecting it with the soundness of logic. First of all, to reduce arithmetic to rock-solid logic. Is the notion of class rock-solid? Even an infinite class? No. If we include infinite sets, logic isn't rock-solid any more. But the class of singletons is already infinite. Without infinite sets, there's no mathematics.

Frege regarded "set" or "class" as equivalent to "property." To any property corresponds the set of things having that property. To any set corresponds the property of membership in it.

Frege's Fifth Basic Law says that to any properly specified property corresponds a set (Furth, 1964). Definition by properties gives a "concept." To Frege, defining numbers as sets is automatically defining them as concepts—not notions in someone's head, but abstract objects.

Frege was about to publish a monumental work in which arithmetic was reconstructed on the foundation of set theory. His hope was shattered in one of the most poignant episodes in the history of philosophy. Russell found a contradiction in the notion of set as he and Frege used it! After struggling for weeks to escape, he sent Frege a letter (van Heijenoort).

Frege added this postscript to his treatise: "A scientist can hardly meet with anything more undesirable than to have the foundations give way just as the work is finished. In this position I was put by a letter from Mr. Bertrand Russell, as the work was nearly through the press."

The axioms from which Russell and Frege attempted to construct mathematics are contradictory!

"My Basic Law concerning courses-of-values (V) . . . the (unrestricted) Axiom of Set Abstraction states that there exists, for any property we describe via an open formula, a set of things which possess the property. From this Axiom we can easily derive Russell's Paradox" (Musgrave, 1964, p. 101). Russell's paradox is catastrophic because it exhibits a legitimate property that is self-contradictory—a property to which no set can correspond.

The Russell paradox and the other "antinomies" showed that intuitive logic is riskier than classical mathematics, for it led to contradictions in a way that never happens in arithmetic or geometry. This was the "crisis in foundations," the central issue in the famous controversies of the first quarter of this century. Three remedies were proposed—logicism, intuitionism, and formalism. As we have already mentioned, all failed.

The response of "logicism," the school of Frege and Russell, was to reformulate set theory to avoid the Russell paradox, and thereby save the Frege-Russell-Whitehead project of establishing mathematics on logic as a foundation.

Work on this program played a role in the development of logic. But in terms of foundationism, it was a failure. To exclude the paradoxes, set theory had to be patched up into a complicated structure. It acquired new axioms such as the axiom of replacement (a complex recreation of Frege's Axiom 5) and the axiom of infinity (there exists an infinite set).

"There is something profoundly unsatisfactory about the axiom of infinity. It cannot be described as a truth of logic in any reasonable use of that term and so the introduction of it as a primitive proposition amounts in effect to the abandonment of Frege's project of exhibiting arithmetic as a development of logic" (Kneale and Kneale, p. 699).

This patched up set theory could not be identified with logic in the philosophical sense of "rules for correct reasoning." You can build mathematics out of this reformed set theory, but it no longer passes as a foundation, in the sense of justifying the indubitability of mathematics. Mathematics was not shown to be part of logic in the classical sense, as Russell and Whitehead dreamed. It became untenable to claim, as Russell had done, that mathematics is one vast tautology.

"Among all mathematical theories it is just the theory of sets that requires clarification more than any other" (Mostowski).

After Frege's first shock, he continued his foundationalist labors. Russell searched for a way out for a long time. He came up with a modified form of set theory, the theory of types. Zermelo introduced the axiom of foundation, which says any chain of set membership terminates in finitely many steps. This outlaws Russell's paradox by outlawing "Russell sets"—sets that belong to themselves—since the membership relation for a Russell set cycles round ad infinitum. (Recently a British computer scientist, Peter Aczel, published a version of set theory in which Zermelo's axiom of foundation is negated. This theory permits self-membership, and has applications in computer science. It has been proved to be relatively consistent!)

But set theory doctored up with the axiom of infinity, and Zermelo's axiom of foundation was no longer the perspicuous elementary set theory that had aroused foundationist hopes. Russell's paradox was unexpected. Are other paradoxes lurking?

The Russell paradox doomed that hope. Despite this philosophical failure, logico-set theoreticism dominates the philosophy of mathematics today. Philip Kitcher writes that "mathematical philosophy in the last 30 years is a series of footnotes to Frege." This suggests that mathematical philosophy is ready for new ideas and problems. Perhaps ideas and problems rising from today's mathematical practice.

Logicism never recovered from the Russell paradox. Eventually both Frege and Russell gave it up. Set theory had become, not clear and indubitable like elementary logic, but unclear and dubitable. To define a set by a property, I must show the property isn't self-contradictory. Such a demonstration can be harder than the problem it was supposed to clarify. "Reducing" arithmetic to logic was a disappointment. Instead of being anchored to the rock of logic, it was suspended from the balloon of set theory.

Frege's construction of number is defensible. But it's not sufficient to convince doubters that arithmetic is a priori. Its long-range importance wasn't Frege's philosophical goal but the stimulation it gave to logic and foundations. The Frege-Russell definition or the equivalent von Neumann definition let us derive arithmetic from facts about sets. Mainstream mathematicians ignored it for a long time. Frege at Jena was an unknown outsider, but even when the respected Richard Dedekind wrote on the foundation of the natural numbers, he too aroused little interest among mathematicians.

Mathematicians don't regard the natural numbers as a problem. With millennia of experience behind us, and deep, complex problems before us, we're not worrying about elementary arithmetic. Dedekind and Frege may object that we have only vague notions of what's meant by 0 or 1 or 2. Nevertheless, we have no qualms about 0, 1, 2.

Frege and Russell weren't mainly concerned with the opinion of the ordinary, unphilosophical mathematician. They were concerned with establishing mathematics on a solid foundation.

Frege always allowed geometry to rest on space intuition. In his old age he decided arithmetic too was based on geometry and space intuition (1979, pp. 267–81).

Hilbert published his *Grundlagen der Geometrie*, an epoch-making book that led to universal acceptance of the axiomatic method as the right way to present mathematics—in principle. In this book Hilbert (following Pasch and Peano) filled in the gaps in Euclid, making Euclidean geometry for the first time the rigorously logical subject it had always claimed to be. He did more. He showed that the axioms are *independent* (can't be deduced from each other) by giving examples in which all the axioms were satisfied except one.

Frege's Kantian views on geometry led him to attack Hilbert. He told Hilbert that Hilbert didn't know the difference between a definition and an axiom. Hilbert answered Frege's first letter or two (1979, pp. 167–73). Thereafter he ignored him. But Frege continued to crow. He even insinuated that Hilbert's failure to keep up the controversy was because Hilbert was afraid his results might be false!

Musgrave: "By 1924 Frege had come to the conclusion that 'the paradoxes of set theory have destroyed set theory.' He continued: 'The more I thought about it the more convinced I became that arithmetic and geometry grew from the same foundation, indeed from the geometrical one; so that the whole of mathematics is actually geometry.'" (These two remarks are quoted by Bynum in his Introduction to Frege [1972], cf. pp 53–54.)

Bertrand Russell (1872–1970)

A Loss of Faith

It wouldn't be too wrong to say philosophy of science in the twentieth century is mostly Bertrand Russell. Two other leading thinkers—Frege and Wittgenstein—are both Russell proteges. He didn't create them as philosophers, of course. But his enthusiasm for them is in part their compatibility with his logical atomism. In helping them become influential, he indirectly advances his own point of view.

Russell is frank about his motives, so far as he understands them. In philosophy of science, his leading motive is to establish certainty. In this, he confesses, he's seeking to replace the Christian faith he has rejected. He is also continuing an old tradition: Plato, Descartes, Leibniz, Kant. From "Reflections on My Eightieth Birthday" in *Portraits from Memory*:

"I wanted certainty in the kind of way in which people want religious faith. I thought that certainty is more likely to be found in mathematics than elsewhere. But I discovered that many mathematical demonstrations, which my teachers expected me to accept, were full of fallacies, and that, if certainty were indeed discoverable in mathematics, it would be in a new field of mathematics, with more solid foundations than those that had hitherto been thought secure. But as the work proceeded, I was continually reminded of the fable about the elephant and the tortoise. Having constructed an elephant upon which the mathematical world could rest, I found the elephant tottering, and proceeded to construct a tortoise to keep the elephant from falling. But the tortoise was no more secure then the elephant, and after some twenty years of very arduous toil, I came to the conclusion that there was nothing more that I could do in the way of making mathematical knowledge indubitable."

"Mathematics is, I believe," says Russell, "the chief source of the belief in eternal and exact truth, as well as in a super-sensible intelligible world. Geometry deals with exact circles, but no sensible object is *exactly* circular; however carefully we may use our compasses, there will be some imperfections and irregularities. This suggests the view that all exact reasoning applies to ideal as opposed to sensible objects; it is natural to go further, and to argue that thought is nobler than sense, and the objects of thought more real than those of sense-perception. Mystical doctrines as to the relation of time to eternity are also reinforced by pure mathematics, for mathematical objects, such as number, if real at all, are eternal and not in time. Such eternal objects can be conceived as God's thoughts. Hence Plato's doctrine that God is a geometer, and Sir James Jeans' belief that He is addicted to arithmetic. Rationalistic as opposed to apocalyptic religion has been, ever since Pythagoras, and notably ever since Plato, very completely dominated by mathematics and mathematical method.

"So it compels the soul to contemplate being, it is proper; if to contemplate becoming, it is not proper" (*Republic*, p. 326). For Plato, the "becoming" or the "unreal" is anything visible, ponderable, changeable. The "being," the "real," is invisible, immaterial, unchangeable. That means mathematics.

Russell calls himself a "logical atomist," in opposition to both the classical and evolutionist trends in early twentieth-century philosophy.

"Philosophy is to be rendered scientific" (p. 28).

"The philosophy which is to be genuinely inspired by the scientific spirit . . . brings with it—as a new and powerful method of investigation always does—a sense of power and a hope of progress more reliable and better grounded than any that rests on hasty and fallacious generalization as to the nature of the universe at large. . . . Many hopes which inspired philosophers in the past it cannot claim to fulfil; but other hopes, more purely intellectual, it can satisfy more fully than former ages could have deemed possible for human minds" (p. 20).

He's good at understated sarcasm. "The classical tradition in philosophy is the last surviving child of two very diverse parents: the Greek belief in reason, and the medieval belief in the tidiness of the universe. To the schoolmen, who lived amid wars, massacres, and pestilences, nothing appeared so delightful as safety and order . . . the universe of Thomas Aquinas or Dante is as small and neat as a Dutch interior. . . . To us, to whom safety has become monotony . . . the world of dreams is very different . . . the barbaric substratum of human nature, unsatisfied in action, finds an outlet in imagination (Written before August, 1914)."

Alan Musgrave (1977) quotes Russell, ***An Essay on the Foundations of Geometry***, 1897, p. 1: "Geometry, throughout the 17th and 18th centuries, remained, in the war against empiricism, an impregnable fortress of the idealists. Those who held—as was generally held on the Continent—that certain knowledge, independent of experience, was possible about the real world, had only to point to Geometry: none but a madman, they said, would throw doubt on its validity, and none but a fool would deny its objective reference. The English Empiricists, in this matter, had, therefore, a somewhat difficult task; either they had to ignore the problem, or, if, like Hume and Mill, they ventured on the assault, they were driven into the apparently paradoxical assertion that Geometry at bottom, had no certainty of a different ***kind*** from that of Mechanics."

P. H. Nidditch
Frank Talk on Logicism

The logician and historian P. H. Nidditch gave a fair summing up of the logicist struggle to save the foundations of mathematics. "The effect of these discoveries (Russell & Burali-Forti antinomies) on the development of Mathematical Logic has been very great. The fear that the current systems of mathematics might not have consistency has been chiefly responsible for the change in the direction of Mathematical Logic towards metamathematics, for the purpose of becoming free from the disease of doubting if mathematics is resting on a solid base. A special reason for being troubled is that the theory of classes is used in all parts of mathematics; so if it is wrong in some way, they are possibly in error. Further, quite separately from the theory of classes, might not discoveries of opposite theorems in algebra, geometry or Mathematical Analysis suddenly come into view, as the discoveries of Burali-Forti and Russell had done? It has been seen that common sense is not good enough as a lighthouse for keeping one safe from being broken against the overhanging slope of sharp logic. To become certain with good reason that the systems of mathematics are all right it is necessary for the details of these structures to be looked at with care and for demonstrations to be given that, with those structures, consistency is present.

"This last point and the fears and troubled mind that we have been talking about in these lines have been and are common among workers in what is named 'the foundations of mathematics,' that is, axiom systems of logic-classes-and-arithmetic. However, some persons, with whom the present writer is in agreement, have a different opinion. They would say the well being of mathematics is not dependent on its 'foundations.' The value of mathematics is in the fruits of its branches more than in its 'roots'; in the great number of surprising and interesting theorems of algebra, analysis, geometry, topology, theory of numbers and theory of chances more than in attempts to get a bit of arithmetic or topology as simply a development of logic itself. They would say that the name 'foundations of mathematics' is a bad one in so far as it sends a wrong picture into one's mind of the relations between logic and higher mathematics. Higher mathematics is not resting on logic or formed from logic. They would say that the troubles in the theory of classes came from most special examples of classes, and such classes are not used in higher mathematics. They would say further that though to be certain of consistency is to be desired if such certain knowledge is possible to us, a knowledge of the consistency of the theories of mathematics that is probable is generally enough and the only sort of knowledge of consistency that one does in fact generally have. And they would say that such probable knowledge is well supported if the theories of mathematics have been worked out much and opposite theorems in them have not come to light. There is no suggestion in all this that Mathematical Logic is not an important part of mathematics; the view put forward is that there is much more to mathematics than Mathematical Logic is and might ever become. And there is no suggestion that the questions of consistency and like questions, and discovery of ways of answering them are not important; to no small degree Mathematical Logic now is as interesting and important as it is because of its interest in such questions and answers."

Luitjens E. J. Brouwer (1882–1966)

An Angry Topologist

After logicism came intuitionism, the doctrine of the great Dutch topologist L. E. J. Brouwer. The name intuitionism displays its descent from Kant's intuitionist theory of mathematical knowledge. Brouwer followed Kant in saying that mathematics is founded on intuitive truths. Brouwer's impact came not only from the force of his philosophy, but also from his wonderful discoveries in topology, and from his dominating presence, which led some people to see him as a leader.

Here is the manifesto Brouwer called the FIRST ACT OF INTUITIONISM: "Completely separating mathematics from mathematical language and hence from the phenomena of language described by theoretical logic, recognizing that intuitionistic mathematics is an essentially languageless activity of the mind having its origin in the perception of a move of time. This perception of a move of time may be described as the falling apart of a life moment into two distinct things, one of which gives way to the other, but is retained by memory. If the twoity thus born is divested of all quality, [echo of Aritotles's 'abstraction'] it passes into the empty form of the common substratum of all twoities. And it is this common substratum, this empty form, which is the basic intuition of mathematics" (Brouwer, 1981).

According to Brouwer, the natural numbers are given by the fundamental intuition of "a move in time." This intuition is the starting point for all mathematics, and all mathematics must be based *constructively* on the natural numbers. But the notion "constructive" cannot and need not be explained. Supposed mathematical objects not constructively based on the natural numbers are not meaningful. Their existence would not be established, even if it were shown that assuming their nonexistence leads to a contradiction.

From the First Act flows the main dogma that separates intuitionistic mathematics from ordinary "classical" mathematics: "The belief in the universal validity of the principle of the excluded third [excluded middle] in mathematics is considered by the intuitionists as a phenomenon of the history of civilization of the same kind as the former belief in the rationality of π, or in the rotation of the firmament about the earth." Misapplication of the "Law of the Excluded Middle" is the great evil in mathematics. Classical "true" and "false" should be replaced by "constructively true," "constructively false," and "neither."

Before Brouwer, the French analysts Henri Poincaré, Émile Borel, and Henri Lebesgue had misgivings and disagreements with nonconstructive methods and free use of infinite sets. (Brouwer called them pre-intuitionists.) But Brouwer's demand to restructure analysis from the ground up went much further. To most mathematicians it seemed excessive.

There is also the SECOND ACT OF INTUITIONISM: "Admitting two ways of creating new mathematical entities: firstly in the shape of more or less freely proceeding infinite sequences of mathematical entities previously acquired (so that, for example, infinite decimal fractions having neither exact values nor any guarantee of ever getting exact values are admitted; secondly, in the shape of mathematical species, i.e. properties supposable for mathematical entities previously acquired, satisfying the condition that if they hold for a certain mathematical entity, they also hold for all mathematical entities which have been defined to be 'equal' to it, definitions of equality having to satisfy the conditions of symmetry, reflexivity and transitivity."

Brouwer's most famous contribution to topology was his "fixed point theorem."** It's a powerful tool in classical and applied branches such as differential equations. But it isn't constructive. Nor is the rest of his great work in topology. Ultimately he decided his own best work was wrong.

His interests went beyond mathematics. In youth he wrote a strange book called *Life, Art and Mysticism*. His biographer, van Stigt, calls it "the manifesto of an 'angry young man' rejecting and attacking all he sees at the surface of human society." Some quotes from the book:

"Intellect has done mankind a devil's service by linking the two phantasies of means and end . . . there are others (scientists) who do not know when to stop, who keep on and on until they go mad. They grow bald, short-sighted and fat, their stomachs stop working, and moaning with asthma and gastric trouble they fancy that in this way equilibrium is within reach and almost reached. . . . So much for science, the last flower and ossification of culture."

We present a famous example of Brouwer's intuitionism. It's about the law of trichotomy, which says: "Every real number is either positive, negative, or zero." Brouwer says the Law is false, and gives a counter-example—a real number that is neither positive, negative, nor zero! Most mathematicians vehemently reject this claim. His number *is* either zero, negative or positive, we say! We simply don't know how to determine which it is.

Since there's hardly any mathematics that doesn't depend on the real numbers, the example shows that Platonism is intimately associated with the practice of mathematics today.

To give the example, start with π, the ratio of the circumference of a circle to its diameter. From its decimal expansion we will shortly define a second number, π^ (read: "pi-hat"). The use of π in this construction is arbitrary. We could start with almost any other irrational number. We need two properties: (1) as with π, the capacity to (in principle!) compute its decimal expansion as far out as we wish; (2) some property of the expansion—for instance, appearance somewhere in it of a sequence of 100 successive zeros—that is "accidental" in the sense that we know no reason why this property is excluded or required by the definition of π. To determine whether somewhere in the decimal expansion of π there is a row of 100 successive zeros, we have no recourse but to generate the decimal expansion of π. If there is such a row, eventually (!!!) we'll find it. If there isn't, we won't know that's the case until we look at the complete infinite expansion—that is, never!

Let P be the statement, "In the decimal expansion of π, somewhere occurs a sequence of 100 successive zeros." Let –P be the negation, "In the decimal expansion of π, there's no sequence of 100 successive zeros."

What about the statement "Either P or –P" ? True or false? Is it true or false that P is either true or false? Most folks say "True." The law of the excluded middle (L.E.M.) says "True."

Brouwer says no! The law of the excluded middle doesn't apply. "The expansion of π" is a mythical beast. There's no such thing. The mistaken belief that either P or –P must be true comes from the delusion that the expansion of π exists as a completed object. All that exists, however, all we know how to construct, is a finite piece of this expansion.

The argument may seem a bit theological. Why does it matter?

In fact, if we give up our belief in the expansion of π, in the truth of either P or –P, we must restructure all analysis.

We show how Brouwer disproved the law of trichotomy, one of the fundamental properties of the real numbers.

We define the new number π^, by a rule that successively determines the first thousand, million, or hundred billion digits of the decimal expansion of π^. That's all that's meant by "defining" a real number.

π^ looks a lot like π. In fact, it's the same as π in the first hundred, thousand, even the first hundred thousand places. The rule is: expand π until you find a row of 100 successive zeros (or until you reach the desired precision for π^, whichever comes first). Up to the first run of 100 successive zeros, the expansion of π^ is identical to that of π. Suppose the first run of 100 successive zeros starts in the 93d place. Then π^ terminates with its 93d digit. This makes π^ less than π. If the first run of 100 successive zeroes starts in the 94th place, put a 1 in the next, the 95th place, and then terminate. This makes π^ greater than π. If the first 100-zero run starts in any other place, it has to be either an even-numbered or an odd-numbered place. If in an odd-numbered place, π^ terminates there. If in an even-numbered place, π^ stops with a 1 in the next, odd-numbered place.

We don't know, and quite possibly never will know, if there is a place where a 100-zero row starts. There may not be any. Nevertheless, our recipe for constructing π^ is perfectly definite; we know it to as many decimal places as we know π. If π doesn't include 100 successive zeros anywhere, π^ = π. If it does include such a sequence, and that sequence starts at an even-numbered place, π^ > π. If it starts at an odd-numbered place, π^ < π.

Now consider the difference, π^ –π. Call it Q. Is Q positive, negative, or zero? Try to find out by calculating the expansion of π on a computer. You won't get an answer until the computer finds 100 successive zeros in π. If the computer runs a million years and doesn't find 100 successive zeroes, you still don't know if Q is positive, negative, or zero. If there actually is no hundred-zero sequence, you'll never know.

π has been expanded to billions of places by the brothers Gregory and David Chudnovsky in New York and Tanaka in Tokyo (Preston). No sequence of 100 zeroes has occurred so far. We know nothing about the next hundred billion digits.

Even if 100 successive zeros turn up tomorrow, we can ask instead about 1,000 successive 9s (for example) and again have an open question. There are plenty of questions like this that we'll never answer.

So what about the "law of trichotomy"? It says Q is either positive, negative, or zero, regardless of the fact that we may never know which.

The constructivist says, "None of the three is true! Q *will* be zero, positive or negative *when* someone determines which of the three is the case. Until then, it's none of the three. Any conclusion based on the compound statement

'Either $Q > O$, $Q = O$, or $Q < O$'

is unjustified. Any conclusion about an infinite set is defective if it relies on the law of the excluded middle. As the example shows, a statement may be (in the constructive sense) neither true nor false."

The standard mathematician finds the argument annoying. He has no intention of giving up classical mathematics for a more restricted version. Neither does he admit that his mathematical practice depends on a Platonist ontology. He neither defends Platonism nor reconsiders it. He just pretends nothing happened.

Some aspects of the intuitionist viewpoint are attractive to mathematicians who want to escape Platonism and formalism. The intuitionists insist that mathematics is meaningful, that it's a kind of human mental activity. You can accept these ideas, without saying that classical mathematics is lacking in meaning.

Errett Bishop
À bas with L.E.M.

The U.S. analyst Errett Bishop revised intuitionism, and created a cleaned up, streamlined version he called "constructivism." Constructivism is concerned above all with throwing out the law of the excluded middle for infinite sets. It's closer to normal mathematical practice than Brouwer. It's not tainted with mysticism. Bishop's book *Constructive Analysis* goes a long way toward reconstructing analysis constructively. Here and there in the mathematical community, a few cells of constructivists are still active. But their dream of converting the rest of us is dead. The overwhelming majority long since rejected intuitionism and constructivism, or never even heard of them.

Like Brouwer, Bishop said a lot of standard mathematics is meaningless. He went far beyond Brouwer in remaking it constructively. Some quotes:

"One gets the impression that some of the model-builders are no longer interested in reality. Their models have become autonomous. This has clearly happened in mathematical philosophy: the models (formal systems) are accepted as the preferred tools for investigating the nature of mathematics, and even as the fount of meaning" (p. 2).

"One of the hardest concepts to communicate to the undergraduate is the concept of a proof. With good reason, the concept *is* esoteric. Most mathematicians, when pressed to say what they mean by a proof, will have recourse to formal criteria. The constructive notion of proof by contrast is very simple, as we shall see in due course. Equally esoteric, and perhaps more troublesome, is the concept of existence. Some of the problems associated with this concept have already been mentioned, and we shall return to the subject again. Finally, I wish to point to the esoteric nature of the classical concept of truth. As we shall see later, truth is not a source of trouble to the contructivist, because of his emphasis on meaning."

"One could probably make a long list of schizophrenic attributes of contemporary mathematics, but I think the following short list covers most of the ground: rejection of common sense in favor of formalism, debasement of meaning by the willful refusal to accommodate certain aspects of reality; inappropriateness of means to ends; the esoteric quality of the communication; and fragmentation" (p. 1).

"The codification of insight is commendable only to the extent that the resulting methodology is not elevated to dogma and thereby allowed to impede the formation of new insight. Contemporary mathematics has witnessed the triumph of formalist dogma, which had its inception in the important insight that most arguments of modern mathematics can be broken down and presented as successive applications of a few basic schemes. The experts now routinely equate the panorama of mathematics with productions of this or that formal system. Proofs are thought of as manipulations of strings of symbols. Mathematical philosophy consists of the creation, comparison and investigation of formal systems. Consistency is the goal. In consequence meaning is debased, and even ceases to exist at a primary level.

"The debasement of meaning has yet another source, the wilful refusal of the contemporary mathematician to examine the content of certain of his terms, such as the phrase, 'there exists.' He refuses to distinguish among the different meanings that might be ascribed to this phrase. Moreover he is vague about what meaning it has for him. When pressed he is apt to take refuge in formalism, declaring that the meaning of the phrase and the statement of which it forms a part can only be understood in the context of the entire set of assumptions and techniques at his command. Thus he inverts the natural order, which would be to develop meaning first, and then to base his assumptions and techniques on the rock of meaning. Concern about this debasement of meaning is a principal force behind constructivism."

While other mathematicians say "there exists" as if existence were a clear, unproblematical notion, Bishop says "meaning" and "meaningful" as if those were clear, unproblematical notions. Has he simply shifted the fundamental ambiguity from one place to another?

Most mathematicians responded to Bishop's work with indifference or hostility. We nonconstructivists should do better. We should try to state our philosophies as clearly as the constructivists state theirs. We have a right to our viewpoint, but we ought to be able to say what it is.

The account of constructivism here is given from the viewpoint of classical mathematics. That means it's unacceptable from the constructivist viewpoint.

From that point of view, classical mathematics is the aberration, a jumble of myth and reality. Constructivism is just refusing to accept a myth.

Hilbert, Formalism, Gödel
Beautiful Idea; Didn't Work

Formalism is credited to David Hilbert, the outstanding mathematician of the first half of the twentieth century. It's said that his dive into philosophy of mathematics was a response to the flirtation of his favorite pupil, Hermann Weyl, with Brouwer's intuitionism. Hilbert was alarmed. He said that depriving the mathematician of proof by contradiction was like tying a boxer's hands behind his back.

"What Weyl and Brouwer do comes to the same thing as to follow in the footsteps of Kronecker! They seek to save mathematics by throwing overboard all that which is troublesome. . . . They would chop up and mangle the science. If we would follow such a reform as the one they suggest, we would run the risk of losing a great part of our most valuable treasure!" (C. Reid, p. 155).

Hilbert met the crisis in foundations by inventing proof theory. He proposed to prove that mathematics is *consistent*. To do this, he had a brilliant idea: work with formulas, not content. He intended to do so by purely finitistic, combinatorial arguments—arguments Brouwer couldn't reject! His program had three steps:

1. Introduce a formal language and formal rules of inference, so every classical proof could be replaced by a formal derivation from formal axioms by mechanically checkable steps. This had already been accomplished in large part by Frege, Russell, and Whitehead. Once this was done, the axioms of mathematics could be treated as strings of meaningless symbols. The theorems would be other meaningless strings. The transformation from axioms to theorems—the *proof*—could be treated as a rearrangement of symbols.

2. Develop a combinatorial theory of these "proof" rearrangements. The rules of inference now will be regarded as rules for rearranging formulas. This theory was called "meta-mathematics."

3. Permutations of symbols are finite mathematical objects, studied by "combinatorics" or "combinatorial analysis." To prove mathematics is consistent, Hilbert had to use finite combinatorial arguments to prove that the permutations allowed in mathematical proof, starting with the axioms, could never yield a falsehood, such as

$$1 = 0.$$

That is, to prove by purely finite arguments that a contradiction, for example, $1 = 0$, cannot be derived within the system.

In this way, mathematics would be given a guarantee of consistency. As a foundation this would have been weaker than one known to be *true* (as geometry was once believed to be true) or impossible to doubt (like, possibly, the laws of elementary logic.)

Hilbert's formalism, like logicism, offered certainty and reliability for a price. The logicist would save mathematics by turning it into a tautology. The formalist would save it by turning it into a meaningless game. After mathematics is coded in a formal language and its proofs written in a way checkable by machine, the meaning of the symbols becomes extramathematical.

It's very instructive that Hilbert's writing and conversation displayed full conviction that mathematical problems are about real objects, and have answers that are true in the same sense that any statement about reality is true. He advocated a formalist interpretation of mathematics only as the price of obtaining certainty.

"The goal of my theory is to establish once and for all the certitude of mathematical methods. . . . The present state of affairs where we run up against the paradoxes is intolerable. Just think, the definitions and deductive methods which everyone learns, teaches and uses in mathematics, the paragon of truth and certitude, lead to absurdities! If mathematical thinking is defective, where are we to find truth and certitude?" (D. Hilbert, "On the Infinite," in *Philosophy of Mathematics* by Benacerraf and Putnam).

As it happened, certainty could not be had, even at this price. A few years later, Kurt Gödel proved consistency could never be proved by the methods of proof Hilbert allowed. Gödel's incompleteness theorems showed that Hilbert's program was hopeless. Any formal system strong enough to contain arithmetic could never prove its own consistency. This theorem of Gödel's is usually cited as the death blow to Hilbert's program, and to formalism as a philosophy of mathematics.

(A simple new proof of Gödel's theorem by George Boolos is given in the mathematical Notes and Comments.)

The search for secure foundations has never recovered from this defeat.

John von Neumann tells how "working mathematicians" responded to Brouwer, Hilbert, and Gödel:

1. "Only very few mathematicians were willing to accept the new, exigent standards for their own daily use. Very many, however, admitted that Weyl and Brouwer were *prima facie* right, but they themselves continued to trespass, that is, to do their own mathematics in the old, 'easy' fashion-probably in the hope that somebody else, at some other time, might find the answer to the intuitionistic critique and thereby justify them *a posteriori*.

2. "Hilbert came forward with the following ingenious idea to justify 'classical' (i.e., pre-intuitionist) mathematics: Even in the intuitionistic system it is possible to give a rigorous account of how classical mathematics operates, that is, one can describe how the classical system works, although one cannot justify its workings. It might therefore be possible to demonstrate intuitionistically that classical procedures can never lead into contradiction—into conflicts with each

other. It was clear that such a proof would be very difficult, but there were certain indications how it might be attempted. Had this scheme worked, it would have provided a most remarkable justification of classical mathematics on the basis of the opposing intuitionistic system itself! At least, this interpretation would have been legitimate in a system of the philosophy of mathematics which most mathematicians were willing to accept.

3. "After about a decade of attempts to carry out this program, Gödel produced a most remarkable result. This result cannot be stated absolutely precisely without several clauses and caveats which are too technical to be formulated here. Its essential import, however, was this: If a system of mathematics does not lead into contradiction, then this fact cannot be demonstrated with the procedures of that system. Gödel's proof satisfied the strictest criterion of mathematical rigor—the intuitionistic one. Its influence on Hilbert's program is somewhat controversial, for reasons which again are too technical for this occasion. My personal opinion, which is shared by many others, is, that Gödel has shown that Hilbert's program is essentially hopeless.

4. "The main hope of justification of classical mathematics—in the sense of Hilbert or of Brouwer and Weyl—being gone, most mathematicians decided to use that system anyway. After all, classical mathematics was producing results which were both elegant and useful, and, even though one could never again be absolutely certain of its reliability, it stood on at least as sound a foundation as, for example, the existence of the electron. Hence, as one was willing to accept the sciences, one might as well accept the classical system of mathematics. Such views turned out to be acceptable even to some of the original protagonists of the intuitionistic system. At present the controversy about the 'foundations' is certainly not closed, but it seems most unlikely that the classical system should be abandoned by any but a small minority.

"I have told this story of this controversy in such detail, because I think that it constitutes the best caution against taking the immovable rigor of mathematics too much for granted. This happened in our own lifetime, and I know myself how humiliatingly easily my own views regarding the absolute mathematical truth changed during the episode, and how they changed three times in succession!" (pp. 2058–59).

Instead of providing foundations for mathematics, Russell's logic and Hilbert's proof theory became starting points for new mathematics. Model theory and proof theory became integral parts of contemporary mathematics. They need foundations as much or little as the rest of mathematics.

Hilbert's program rested on two unexamined premises. First, the Kantian premise: *Something* in mathematics—at least the purely "finitary part"—is a solid foundation, is indubitable. Second, the formalist premise: A solid theory of formal sentences could validate the mathematical activity of real life, where the possibility of formalization is in the remote background, if present at all.

The first premise was shared by the constructivists; the second, of course, they rejected. Formalization amounts to mapping set theory and analysis into a part of itself—finite combinatorics. Even if Hilbert had been able to carry out his program, the best he could have claimed would have been that all mathematics is consistent *if* the "finitistic" principle allowed in "metamathematics" is reliable. Still looking for the last tortoise under the last elephant!

The bottom tortoise is the Kantian synthetic a priori, the intuition. Although Hilbert doesn't explicitly refer to Kant, his conviction that mathematics must provide truth and certainty is in the Platonic heritage transmitted through the rationalists to Kant, and thereby to intellectual nineteenth-century western Europe. In this respect, Hilbert is as much a Kantian as Brouwer, whose label of intuitionism avows his Kantian descent.

To Brouwer, the Hilbert program was misconceived at Step 1, because it rested on identifying mathematics itself with formulas used to represent or express it. But it was only by this transition to languages and formulas that Hilbert was able to envision the possibility of a *mathematical* justification of mathematics.

Like Hilbert, Brouwer was sure that mathematics had to be established on a sound and firm foundation. He took the other road, insisting that mathematics must start from the intuitively given, the finite, and must contain only what is obtained in a constructive way from this intuitively given starting point. Intuition here means the intuition of *counting* and that alone. For both Brouwer and Hilbert, the acceptance of geometric intuition as a basic or fundamental "given" on a par with arithmetic would have seemed utterly retrograde and unacceptable *within the context of foundational discussions.* Like Brouwer, Hilbert, the formalist, regarded the finitistic part of mathematics as indubitable. His way of securing mathematics, making it free of doubt, was to reduce the infinitistic part—analysis and set theory—to the finite part by use of the finite *formulas*, which described these nonfinite structures.

In the mid-twentieth century, formalism became the predominant philosophical attitude in textbooks and other official writing on mathematics. Constructivism remained a heresy with a few adherents. Platonism was and is believed by nearly all mathematicians. Like an underground religion, it's observed in private, rarely mentioned in public.

Contemporary formalism is descended from Hilbert's formalism, but it's not the same thing. Hilbert believed in the reality of finite mathematics. He invented metamathematics to justify the mathematics of the infinite. Today's formalist doesn't bother with this distinction. For him, all mathematics, from arithmetic on up, is a game of logical deduction.

He defines mathematics as the science of rigorous proof. In other fields some theory may be advocated on the basis of experience or plausibility, but in mathematics, he says, we have a proof or we have nothing.

Any proof has a starting point. So a mathematician must start with some undefined terms, and some unproved statements. These are "assumptions" or "axioms." In geometry we have undefined terms "point" and "line" and the axiom "Through any two distinct points passes exactly one straight line." The formalist points out that the logical import of this statement doesn't depend on the mental picture we associate with it. Nothing keeps us from using other words—"Any two distinct bleeps ook exactly one bloop." If we give interpretations to the terms bleep, ook, and bloop, or the terms point, pass, and line, the axioms may become true or false. To pure mathematics, any such interpretation is irrelevant. It's concerned only with logical deductions from them.

Results deduced in this way are called theorems. You can't say a theorem is true, any more than you can say an axiom is true. As a statement in pure mathematics, it's neither true nor false, since it talks about undefined terms. All mathematics can say is whether the theorem follows logically from the axioms. Mathematical theorems have no content; they're not *about* anything. On the other hand, they're absolutely free of doubt or error, because a rigorous proof has no gaps or loopholes.

In some textbooks the formalist viewpoint is stated as simple matter of fact. The unwary student may swallow it as the official view. It's no simple matter of fact, but a matter of controversial interpretation. The reader has the right to be skeptical, and to demand evidence to justify this view.

Indeed, formalism contradicts ordinary mathematical experience. Every school teacher talks about "facts of arithmetic" or "facts of geometry." In high school the Pythagorean theorem and the prime factorization theorem are learned as true statements about right triangles or about natural numbers. Yet the formalist says any talk of facts or truth is incorrect.

One argument for formalism comes from the dethronement of Euclidean geometry.

For Euclid, the axioms of geometry were not assumptions but self-evident truths. The formalist view results, in part, from rejecting the idea of self-evident truths.

In Chapter 6 we saw how the attempt to prove Euclid's fifth postulate led to discovery of non-Euclidean geometries, in which Euclid's parallel postulate is assumed to be false.

Can we claim that Euclid's parallel postulate and its negation are *both* true? The formalist concludes that to keep our freedom to study both Euclidean and non-Euclidean geometries, we must give up the idea that either is true. They need only be consistent.

But Euclidean and non-Euclidean geometry conflict only if we believe in an objective physical space, which obeys a single set of laws, which both theories attempt to describe. If we give up this belief, Euclidean and non-Euclidean geometry are no longer rival candidates for solving the same problem, but two different mathematical theories. The parallel postulate is true for the Euclidean straight line, false for the non-Euclidean.

Are the theorems of geometry meaningful apart from physical interpretation? May we still use the words "true" and "false" about statements in geometry? The formalist says no, the statements aren't true or false, they aren't about anything and don't mean anything. The Platonist says yes, since mathematical objects exist in their own world, apart from the physical world. The humanist says yes, they exist in the shared conceptual world of mathematical ideas and practices.

The formalist makes a distinction between geometry as a deductive structure and geometry as a descriptive science. Only the first is mathematical. The use of pictures or diagrams or mental imagery is nonmathematical. In principle, they are unnecessary. He may even regard them as inappropriate in a mathematics text or a mathematics class.

Why give *this* definition and not another?

Why *these* axioms and not others?

To the formalist, such questions are premathematical. If they're in his text or his course, they'll be in parentheses, and in brief.

What examples or applications come from the general theory he develops? This is not really relevant. It may be a parenthetical remark, or left as a problem.

For the formalist, you don't get started doing mathematics until you state hypotheses and begin a proof. Once you reach your conclusion, the mathematics is over. Any more is superfluous. You measure the progress of your class by how much you prove in your lectures. What was understood and retained is a nonmathematical question.

One reason for the past dominance of formalism was its link with logical positivism. This was the dominant trend in philosophy of science during the 1940s and 1950s. Its aftereffects linger on, for nothing definitive has replaced it. (See the section on the "Vienna circle" in Chapter 9.) Logical positivists advocated a unified science coded in a formal logical calculus with a single deductive method. Formalization was the goal for all science. Formalization meant choosing a basic vocabulary of terms, stating fundamental laws, and logically developing a theory from fundamental laws. Classical and quantum mechanics were the models.

The most influential formalists in mathematical exposition was the group "Nicolas Bourbaki." Under this pseudonym they produced graduate texts that had worldwide influence in the 1950s and 1960s. The formalist style dripped down into undergraduate teaching and even reached kindergarten, with preschool texts on set theory. A game called "WFF and Proof" was used to help grade-school children learn about "well-formed formulas" (WFF's) according to formal logic.

In recent years, a reaction against formalism has grown. There's a turn toward the concrete and the applicable. There's more respect for examples, less strictness in formal exposition. The formalist philosophy of mathematics is the source of the formalist style of mathematical work. The signs say that the formalist philosophy is losing its privileged status.

nine

Foundationism Dies/ Mainstream Lives

This chapter attempts to bring the story of the Mainstream up to date, which largely means surveying recent analytic philosophy of mathematics. Husserl isn't in the analytic mainstream, but his international stature and his mathematical qualifications justify including him. Carnap and Quine are the two unavoidable analytic philosophers. Both are "icons" among academic philosophers of science and mathematics.

Among "younger" philosophers who seem to me to be mainstream, I have benefited from reading Charles Castonguay, Charles Chihara, Hartry Field, Juliet Floyd, Penelope Maddy, Michael Resnik, Stuart Shapiro, David Sherry, and Mark Steiner.

I report briefly on structuralism and fictionalism. These alternatives to Platonism are attracting attention as I write. I regret the limitations that keep me from describing the work of other interesting authors. For this I recommend Aspray and Kitcher. I reserve for Chapter 12 the philosophers writing today whom I regard as fellow travelers in the humanist direction. There also I've been unable to pay due attention to most of them.

Edmund Husserl (1859–1938)

Phenomenologist/Weierstrass Student

Husserl is the creator of phenomenology, intellectual father of Karl Heidegger and grandfather of Jean-Paul Sartre. He wrote his doctoral dissertation on the calculus of variations under Karl Weierstrass, one of the greatest nineteenth-century mathematicians, and took lifelong pride in being Weierstrass's pupil. Husserl's first philosophical work was in philosophy of mathematics. Yet he's never mentioned today in talk about the philosophy of mathematics, which has been monopolized by analytic philosophers, descendants of Bertrand Russell.

In a posthumous essay, Gödel expressed high hopes for Husserl's phenomenology. ". . . the certainty of mathematics is to be secured . . . by cultivating (deepening) knowledge of the abstract concepts themselves. . . . Now in fact, there exists today the beginning of a science which claims to possess a systematic method for such a clarification of meaning, and that is the phenomenology founded by Husserl. Here clarification of meaning consists in focusing more sharply on the concepts concerned by directing our attention in a certain way, namely onto our own acts in the use of these concepts, onto our powers in carrying out our acts, etc. . . . I believe there is no reason at all to reject such a procedure at the outset as hopeless . . . quite divergent directions have developed out of Kant's thought—none of which, however, really did justice to the core of Kant's thought. This requirement seems to me to be met for the first time by phenomenology, which, entirely as intended by Kant, avoids both the death-defying leaps of idealism into a new metaphysics as well as the positivistic rejection of all metaphysics." (Gödel, 1995, p. 383, 387)

In a review of Husserl's first book, Frege charged him with psychologism. Husserl faithfully avoided psychologism ever after.

Husserl's early works on philosophy of mathematics came before he developed his major ideas on phenomenology. They are influenced by logicism and formalism. I present some of his mature thinking about mathematics, the well-known essay, "The Origin of Geometry."

There he argues that since geometry has a historic origin, someone "must have" made the first geometric discovery. For that primal geometer, geometric terms and concepts "must have" had clear, unmistakable meaning. Centuries passed. New generations enlarged geometry. We inherit it as a technology and a logical structure, but we've lost the meaning of the subject.

We must recover this meaning.

"Our interest shall be the inquiry back into the most original sense in which geometry once arose, was present as the tradition of millennia. . . . The progress of deduction follows formal-logical self-evidence, but without the actually developed capacity for reactivating the original activities contained within its fundamental concepts, i.e. without the "what" and the "how" of its prescientific materials, geometry would be a tradition empty of meaning; and if we ourselves did not have this capacity, we could never even know whether geometry had or ever did have a genuine meaning, one that could really be 'cashed in.' This is our situation, and that of the whole modern age.

"By exhibiting the essential presuppositions upon which rests the historical possibility of a genuine tradition, true to its origins, of sciences like geometry, we can understand how such sciences can vitally develop through the centuries and still not be genuine. The inheritance of propositions and of the method of logically constructing new propositions and idealities can continue without interruption from one period to the next, while the capacity for reactivating the primal beginnings, i.e. the sources of meaning for everything that comes later, has not been handed down with it. What is lacking is this, precisely what had given and had to give meaning to all propositions and theories, a meaning arising from the primal sources which can be made self-evident again and again. . . ."

Husserl isn't asking for the usual fact-obsessed historical research. Nor for the usual theorem-obsessed geometrical research. Unfortunately, he doesn't give an example of what he is asking for.

Yet he ends on a transcendent note:

"Do we not stand here before the great and profound problem-horizon of reason, the same reason that functions in every man, the ***animal rationale***, no matter how primitive he is?"

Rota presents a remarkably readable expert account of phenomenology and mathematics.

Vienna Circle. Carnap, etc.

In the late 1920s the logicist tradition was picked up by the Vienna Circle of logical positivists. They tacked together philosophy of language from Wittgenstein's *Tractatus* and logicism from Frege and Russell to fabricate what they considered "scientific philosophy." For them, thinking about mathematics meant thinking about logic and axiomatic set theory. The proper model for all science was mechanics. Ernst Mach had arranged classical mechanics in a deductive system. Mass, length, and time are his undefined terms. His axioms are Newton's laws. Classical mechanics has rules to interpret measurements as values of mass, length, or time.

From the axioms and interpretation rules, everything else must be deduced.

The Vienna Circle ordered all science to conform to that model. To each science, its own axioms and undefined terms. (The undefined terms are [informally] empirical measurements.) To each science, its own interpretation rules to connect theory and data. To do science, you should:

1. Choose basic observables.
2. Find formulas for their relationships (called "axioms").
3. Express all other observables as functions of the basic observables.
4. From (1,2,3) derive the rest of the subject by mathematics. For this school of philosophy, mathematics is a language, and a tool for formulating and developing physical theory. The fundamental laws of science are equations and inequalities. (In mechanics, they're differential equations.) The scientist does mathematical calculations to derive consequences of the fundamental laws. But mathematics has no content of its own. Indeed, mathematics has no empirical observations to which to apply interpretation rules! For logical positivism, mathematics is ***nothing but*** a language for science, a ***contentless*** formal structure.

Rudolf Carnap wrote, "Thus we arrived at the conception that all valid statements of mathematics are analytic in the specific sense, that they hold in all possible cases and therefore do not have any factual content" (*Autobiography*).

So logical positivism in philosophy of science matches formalism in philosophy of mathematics. (This is so, even though Carnap's philosophy of mathematics was logicist, not formalist.) As an account of the nature of mathematics, formalism is incompatible with the thinking of working mathematicians. But this was no problem for the positivists! Entirely oriented on theoretical physics, they saw mathematics only as a tool, not a living, growing subject. For a physicist or other user it may be convenient to identify mathematics itself with a particular axiomatic presentation of it. For the producer of mathematics, quite the contrary. Axiomatics is an embellishment added after the main work is done. But this was irrelevant to philosophers whose idea of mathematics came from logic and foundationist philosophy.

With this philosophy came a test of meaningfulness. If a statement isn't "in principle" refutable by the senses, it's meaningless. That kind of statement is no more than a grunt or a groan. In particular, esthetic and ethical judgments have no *factual* content. I say "The *Emperor Concerto* is beautiful. Hitler is evil." I'm just saying, "I like the *Emperor Concerto*. I don't like Hitler."

In retribution, an embarrassment dogged logical positivism. Its own philosophical edicts can't be empirically refuted—not even in principle So by its own test, its own edicts were—mere grunts and groans!

Despite this glitch, logical positivism reigned over American philosophy of science in the 1930s and 1940s, under a group of brilliant refugees from Hitler. Foremost was Rudolph Carnap. W. V. O. Quine wrote, "Carnap more than anyone else was the embodiment of logical positivism, logical empiricism, the Vienna Circle" (*The Ways of Paradox, and Other Essays*, 1966–76, pp. 40–41).

Look at his influential *Introduction to Symbolic Logic*.

Part One describes three formal languages, A, B, and C. Some simple properties of these languages are proved. No nonobvious property or nontrivial problem is stated.

Part Two, "Applications of Symbolic Logic," has a chapter on theory languages and a chapter on coordinate languages. It presents axiom systems for geometry, physics, biology, and set theory/arithmetic. The "applications" are specializing one of his three languages to one of his four subjects.

He doesn't discuss whether these formalizations should interest mathematicians, physicists, or biologists. Such formalizations are rarely seen in biology, physics, or even mainstream mathematics (analysis, algebra, number theory, geometry). Carnap does cite Woodger and some of his own papers.

On p. 21 he says, "if certain scientific elements—concepts, theories, assertions, derivations, and the like—are to be analyzed logically, often the best procedure is to translate them in symbolic language." He provides no support for this claim. In 1957 it evidently was possible to believe in ever-increasing interest among mathematicians, philosophers, and "those working in quite specialized fields who give attention to the analysis of the concepts of their discipline." There's no longer such a belief.

"Symbolic logic" (now called "formal logic") did turn out to be vitally useful in designing and programming digital computers. Carnap doesn't mention any such application. Contrary to his statement, we rarely use formal languages to analyze scientific concepts. We use natural language and mathematics. Carnap's identification of philosophy of mathematics with formalization of mathematics was a dead end. Language A, Language B, and Language C are dead.

On page 49 of his *Autobiography* he wrote: "According to my principle of tolerance, I emphasized that, whereas it is important to make distinctions between constructivist and non-constructivist definitions and proofs, it seems advisable not to prohibit certain forms of procedure but to investigate all practically useful forms. It is true that certain procedures, e.g., those admitted by constructivism or intuitionism, are safer than others. Therefore it is advisable to apply these procedures as far as possible. However, there are other forms and methods which, though less safe because we do not have a proof of their consistency, appear to be practically indispensable for physics. In such a case there seems to be no good reason for prohibiting these procedures so long as no contradictions have been found."

As if Carnap ever had the authority to "prohibit these procedures"!

By the 1950s the sway of logical positivism was shaky. Physicists never accepted its description of their work. Few physicists are interested in axiomatics. Physicists take more pleasure in tearing down axioms. They look for provocative conjectures, and experiments to disprove them.

Hao Wang quotes Carnap's *Intellectual Autobiography*. "From 1952 to 1954 he was at the Princeton Institute and had separate talks 'with John von Neumann, Wolfgang Pauli and some specialists in statistical mechanics on some questions of theoretical physics with which I was concerned.' He had expected that 'we would reach, if not agreement, then at least mutual understanding.' But they failed despite their serious efforts. One physicist said, 'Physics is not like geometry; in physics there are no definitions and no axioms'" (*Intellectual Autobiography*, in Schilpp, pp. 36–37). Wang provides a meticulous criticism of Carnap.

Russell, Frege, and Wittgenstein brought philosophy of mathematics under the sway of analytic philosophy: The central problem is meaning, the essential tool is logic. Since mathematics is the branch of knowledge whose logical structure is best understood, some philosophers think philosophy of mathematics is a model for all philosophy. As the dominant style of Anglo-American philosophy, analytic philosophy perpetuates identification of philosophy of mathematics with logic and the study of formal systems. Central problems for the mathematician become invisible—the development of pre-formal mathematics, and how pre-formal mathematics relates to formalization.

Willard Van Ormond Quine

Most Influential Living Philosopher

Quine is "the most distinguished and influential of living philosophers" says the eminent English philosopher P. F. Strawson (on the jacket of *Quiddities*, Quine's latest book as of 1994).

Quine proved that the real numbers exist—exist philosophically, not just mathematically. He proved that you're guilty of bad faith if you say the real numbers are fictions. We will present and refute Quine's argument. First we sketch a few of his other contributions.

Quine makes no separation between philosophy and logic. Formalization—presentation in a formal language—makes a philosophical theory legitimate. Apart from what can be said in a formal language, it makes no sense to talk philosophy.

Quine granted an interview to *Harvard Magazine*. "'Someone who was a student here many years ago recently sent me a copy of *Methods of Logic* and asked me to inscribe it for him and to write something about my philosophy of life.' (The last three words spoken in gravelly disbelief.)

'And what did you write?'

'Life is agid. Life is fulgid. Life is what the least of us make most of us feel the least of us make the most of. Life is a burgeoning, a quickening of the dim primordial urge in the murky wastes of time.'

'Agid?'

'Yes, it's a made-up word.'

'What you're saying is it's not a serious question.'

'That's right, it's not a serious question. Not a question you can make adequate sense of.'

"For Quine it is important for philosophy to be a technical, specialized discipline (with subdisciplines) and give up contact with people" (Hao Wang, p. 205. Wang provides an infinitely detailed and complete report on Quine's publications, with fascinating critiques.).

Quine's most famous bon mot is his definition of existence: "To be is to be the value of a variable."

This has the merit of shock value.

In the *Old Testament*, Yahweh roars "I am that I am." Must we construe this as: "I, the value of a variable, am the value of a variable!"

Or Hamlet's "To be the value of a variable or not to be the value of a variable?"

Or Descartes's *Meditations*: "I think, therefore I am the value of a variable."

To all this, Quine would have a quick reply. Yahweh, Shakespeare, and Descartes are like the ex-student who asked about his philosophy of life. They all talk nonsense.

The only "existence" of philosophical interest is the existence associated with the existential quantifier of formal logic.

Quine's definition loses its charm when you see he has simply "conflated" the domain of formal logic with the whole material and spiritual universe. His definition could be paraphrased: "To someone interested in existence only as a term in formal logic, to be is. . . ."

"To be" is "to be visible through W. V. O. Quine's personal filter, which is formal logic."

Like a monomaniac photographer saying, "To be is to be recorded on my film," or Geraldo Rivera saying, "To be is to be seen on the Geraldo Rivera show."

Professor Quine also "famously" discovered that translation doesn't exist. (They say he's fluent in six languages.) The insult to common sense is what gets attention. Someone not seeking to shock would say, "*Perfect* or *precise* translation is impossible." That would be banal. Better say something shocking and false.

A real question is being overlooked. Why does the impossibility of perfect translation make no difference in practice? Such an investigation would be empirical, particular, detailed—not Quine's cup of tea.

Our concern with Quine is his new, original argument for mathematical Platonism—for actual existence of real numbers and the set structure logicists erect under them.

Quine calls his idea "ontological commitment." Physics, he tells us, is inextricably interwoven with the real numbers, to such a pitch that it's impossible to make sense of physics without believing real numbers exist. Anyone who turns on a VCR or tests a "nuclear device" believes in physics. It's "bad faith" to drive a car or switch off an electric light without accepting the reality of the real numbers.

In "The Scope and Language of Science" Quine writes: "Certain things we want to say in science may compel us to admit into the range of values of the variables of quantification not only physical objects but also classes and relations of them; also numbers, functions, and other objects of pure mathematics. For, mathematics—not uninterpreted mathematics, but genuine set theory, logic, number theory, algebra of real and complex numbers, differential and integral calculus, and so on—is best looked upon as an integral part of science, on a par with the physics, economics, etc., in which mathematics is said to receive its applications.

"Researches in the foundations of mathematics have made it clear that *all* of mathematics in the above sense can be got down to logic and set theory, and that the objects needed for mathematics in this sense can be got down to a single category; that of *classes*—including classes of classes, classes of classes of classes, and so on." (His "class" is virtually our "set." A real number** is a set of sets of rational numbers, each of which is a set of pairs of natural numbers, each of which is a set of sets.**)

"Our tentative ontology for science, our tentative range of values for the variables of quantification, comes therefore to this: physical objects, classes of them, classes in turn of the elements of this combined domain, and so on up."

This argument of Professsor Quine's is taken seriously. In *Science Without Numbers* Hartry Field says it's the only proof of existence of the real numbers worthy of attention. Field is a nominalist. He denies that numbers exist, so he has to knock down Professor Quine. He writes, "This objection to fictionalism about mathematics can be undercut by showing that there is an alternative formulation of science that does not require the use of any part of mathematics that refers to quantifiers over abstract entities. I believe that such a formulation is possible; consequently, without intellectual doublethink, I can deny that there are abstract entities."

Field says the best exposition of Quine's existence argument is in Putnam's *Philosophy of Logic* (Chapter 5). But after publishing *Philosophy of Logic* Putnam reconsidered Quinism. In *Realism with a Human Face*, in a chapter titled "The Greatest Logical Positivist," Putnam wrote: [In Quine's theory] "mathematical statements, for example, are only justified insofar as they help to make successful predictions in physics, engineering, and so forth." He finds "this claim almost totally unsupported by actual mathematical practice." I agree.

Because we use phone and TV, Quine says we have an "ontological commitment" to the reality of the real numbers (pun intended) and therefore to uncountable sets. In view of his definition of "is," is he saying that the *set* of real numbers is the value of some variable? Which variable?

To Quine it's irrelevant that almost all mathematicians would say they're working on something real, with or without any connection with physics. As far as I can see, he has three options. All are unpleasant and unacceptable. (A) All mathematics is used in physics. (B) The part of mathematics not used in physics doesn't matter. (C) The part of mathematics used in physics not only exists, but somehow causes the unphysical part also to exist. I don't know if Professor Quine upheld any of these absurdities, or if he still thinks about the matter.

What's important is that Professor Quine's leading "insight"—the reality of physics implies the reality of math—is wrong.

The following paragraphs expound a simple remark that shows Quine's argument is without merit (which is what Field wants to do).

Think about digital computers. They're ubiquitous in physics. Physical calculations are either short enough to do by hand, or too long. The ones too long are done on a calculator or a computer. The short ones could also be done on a calculator or computer, if one wished to do so.

To go into a digital calculator or computer, information must be discretized and finitized. Digital computers only accept finite amounts of discrete information.

(A Turing machine is defined to have an infinite tape. But a Turing machine isn't a real machine, it's a mathematical construct. In the whole world, now and to the end of humanity, there's only going to be finitely many miles of tape.)

"Discretized" means there's a smallest increment that the machine and the program read. Anything smaller is read as zero. If a machine and program have a smallest increment of 2^{-100}, and the largest number they accept is 2^{100}, then they work in a number system of 2^{200} numbers. (The number of steps of size 2^{-100} to climb from 2^{-100} to 2^{100}.) 2^{200} is large, but finite.

Real numbers are written as infinite decimals. Nearly all of them need infinitely many digits for their complete description. A computer can't accept such a real number. The biggest computer ever built doesn't have space for even *one* infinite nonterminating nonrepeating decimal number. How does it cope? It *truncates*—keeps the first 100 or the first 1,000 digits, drops the rest.

So physics is dependent on machines that accept only finite decimals. Physics dispenses with real numbers!

Here's an objection. The calculations are done after a theory is formulated. Formation of theory is done by humans, not machines. In formulating a theory, physicists use classical mathematics with real numbers. Doesn't that save Quine's argument?

No.

Quantum mechanics is set in an infinite-dimensional blow-up of Euclidean space known as "Hilbert space." Any coordinate system for Hilbert space has infinitely many "basis vectors." A point in Hilbert space represents a "state" of a quantum-mechanical system. A typical "state" has infinitely many coordinates, and each coordinate is a real number, that is, an infinite decimal. However, not every infinite sequence of real numbers defines a vector in the Hilbert space. The vector must have "finite norm." That means the sum of the squares of the coordinates must be finite.

$$(1, 1, 1, \ldots)$$

isn't a Hilbert space vector, but

$$(1, .5, .25, .125, .0625 \ldots)$$

is a Hilbert space vector. The condition of finite norm means that the "tail," the last part of the coordinate expansion, is negligibly small. For some large finite number N we need only look at the first N terms in the expansion of our vector. In geometrical language, the N-dimensional projection of our infinite-dimensional vector is so close to the vector itself that the distance between them is negligible. Each infinite-dimensional Hilbert-space vector is approximated, as closely as we wish, by N-dimensional vectors, finite-dimensional vectors.

But aren't the component-coordinates of this finite-dimensional projection each separately a real number—an infinite decimal? Yes, but we can choose any N we like, and truncate that infinite decimal after N digits, making an error of only 10^{-N}. If N is large, this is physically undetectable. So the state vector is represented, to arbitrarily high accuracy, by an N-tuple of finite decimals, each containing N digits.

We are interested not only in vectors ***in*** Hilbert space—(creatures with infinitely many coordinates that I just introduced)—but especially in linear transformations or operators ***on*** the space. If we choose convenient coordinates, such an operator is represented by an infinite-by-infinite matrix, all of whose infinitely many entries or elements are real numbers. To be a legitimate operator, the rows and columns of the matrix must satisfy a requirement similar to the one satisfied by vectors. If you go far out along a row or column of this infinite matrix, eventually the elements there will be so small they are physically undetectable. There you can truncate the matrix, make it finite. Then you can store the truncated matrix, whose elements are truncated real numbers, in your computer.

The sophisticated reader may ask if this ***finite*** mathematical system is "isometrically isomorphic" to Quine's (equivalent in a precise mathematical sense.) Isomorphic structures differ only in the names of their elements. But my proposed alternative is certainly ***not*** isomorphic to the reals. The reals are uncountably infinite; my substitute is finite.

Another objection may come from the Quinist. "You say the computer is a finite-state machine. But the machine is made of silicon, copper, and plastic. It's a physical object and it obeys the laws of physics. It's subject to infinitely many different states of temperature, electrostatic field, kinetic and dynamical variables. There's no such thing as a finite-state machine."

Agreed, the real computer in the computer room is a physical object. We think of it as a finite state machine for simplicity and convenience. But to claim that we can't describe this piece of metal and plastic without using the full system of real numbers is just repeating Quine's original claim that physics requires the real number system. It doesn't. That's true of the physics of digital computers. To consider a Cray or a Sun as a physical system, not an ideal computing machine, we need a much more detailed description of it. The much more detailed description will still be finite. Any description we give of anything is finite.

Another defender of Quine might say, "The real number system developed out of necessity. Mathematical analysis and mathematical physics are impossible without the completeness property—the ability to define or construct a number as the limit of a convergent sequence. How can you do anything without π or $\sqrt{2}$?"

Answer: π and $\sqrt{2}$ exist conceptually, not physically or computationally. Computationally,

$$3.141592653589793238462643383279502884197169399375105820$$

is the circumference of the unit circle, and 1.414213562 is the length of the diagonal of the unit square (the square root of 2.) These finite decimals have errors smaller than we can detect by any physical measurement. Such error is of mathematical interest, not physical interest.

Mathematicians defined an infinite decimal as a "real number" to get a theory in which these negligible errors actually vanish. To compute, we go back to finite decimals or rational numbers.

The same discretization/finitization works for infinite-dimensional manifolds, Lie groups, Lie algebras, and so forth. If some infinite mathematical structure couldn't be approximated by a finite structure, then in general and in principle it would be impossible to carry out physical computations in it. Such a structure might fascinate mathematicians or logicians, but it wouldn't interest physicists. The uncountability of the reals fascinates mathematicians and philosophers. Physically it's meaningless. Physicists don't need infinite sets, and they don't need to compare infinities.

We use real numbers in physical theory out of convenience, tradition, and habit. For physical purposes we could start and end with finite, discrete models. Physical measurements are discrete, and finite in size and accuracy. To compute with them, we have discretized, finitized models physically indistinguishable from the real number model. The mesh size (increment size) must be small enough, the upper bound (maximum admitted number) must be big enough, and our computing algorithm must be stable. Real numbers make calculus convenient. Mathematics is smoother and more pleasant in the garden of real numbers. But they aren't essential for theoretical physics, and they aren't used for real calculations.

It's strange that Quine offers this argument about the real numbers, for it makes the same error he attacked in philosophy of science. In Quine's famous contribution to philosophy of science, "Two Dogmas of Empiricism," he showed that physical theory isn't completely determined by data. Physical theory is a loose-hanging network connected to data along its boundary. For any experimental finding, several explanations are possible. No experiment by itself can establish or refute a physical theory. The data do not determine the physical theory uniquely. If a prediction of a theory is refuted, it may not thereby determine which axiom in the theory needs to be revised.

The traditional view of Bacon, Mill, and Popper said each statement in a physical theory can be tested by experiment or measurement. If the measurement obeys the claim, the statement is confirmed (Mill) or at least not disconfirmed (Popper) and can be retained in our world picture. If measurement contradicts claim, we revise the claim.

However, Poincaré "famously" pointed out that no observation could compel us to consider physical space as Euclidean or non-Euclidean. Anomalous behavior by light rays could instead be explained by anomalous light-transmitting properties of the medium. Rejected theories like Ptolemy's theory of planetary motions, the phlogiston theory of fire, and the ether theory of light propagation all could continue to explain the phenomena by ad-hoc adjustments ("finagling"). Their successors—Copernicus's heliocentric solar system, Dalton's oxidation,

Einstein's relativistic light propagation—weren't the only possible explanation of observations. They were the simplest, most convenient, and therefore most credible.

Yet Quine, having shown the nonuniqueness of physical theory, takes for granted the uniqueness of mathematical theory. The mathematical part of the theory is determined uniquely, he thinks, and it must be the real number system and the set theory that logicians prop underneath the real number system. This is just as wrong as the idea that data determine a physical theory.

Hilary Putnam
Somewhat Influential Living Philosopher

In "What is Mathematical Truth?" Putnam parted company from Quine. Mathematical statements are true, not about objects, but about possibilities. Mathematics has conditional objects, not absolute objects. Putnam refers to Kreisel's remark that mathematics needs objectivity, not objects.

Putnam cites the use in mathematics of nondemonstrative, heuristic reasoning and the difficulty of believing in unspecified, unearthly abstract objects. Implicitly, he excludes physical or mental objects as mathematical objects.

Kreisel was right. In the first instance, mathematics needs objectivity rather than objects. Mathematical truths are objective, in the sense that they're accepted by all qualified persons, regardless of race, color, age, gender, or political or religious belief. What's correct in Seoul is correct in Winnipeg. This "invariance" of mathematics is its very essence. Since Pythagoras and Plato, philosophers have used it to support religion. Putnam's objectivity without objects, like standard Platonism, can be regarded as another form of mathematical spiritualism.

But need we really settle for objectivity without objects? "Not only are the 'objects' of pure mathematics conditional upon material objects," writes Putnam; "they are in a sense merely abstract possibilities. Studying how mathematical objects behave might be better described as studying what structures are abstractly possible and what structures are not abstractly possible. The important thing is that the mathematician is studying something objective, even if he is not studying an unconditional 'reality' of nonmaterial things, . . . mathematical knowledge resembles *empirical* knowledge—that is, the criterion of truth in mathematics just as much as in physics it is success of our ideas in practice and that mathematical knowledge is corrigible and not absolute. . . . What he asserts is that certain things are *possible* and certain things are *impossible*—in a strong and uniquely mathematical sense of 'possible' and 'impossible.' In short, mathematics is essentially *modal* rather than existential."

Possible in what sense? Perhaps logically possible—noncontradictory. If so, he's agreeing with Poincaré from a hundred years ago, and Parmenides from two millennia ago. The mathematician is studying something "real"—the consistency or inconsistency of his ideas. This is close to Frege.

On the other hand, maybe Putnam doesn't mean *logically* possible. Maybe he means *physically* possible. Is an infinite-dimensional infinitely smooth manifold of infinite connectivity "physically possible"? Does he mean there "could be" physical objects modeled by such manifolds? He seems to be running up against Lesson 1 in Applied Mathematics, cited above against Professor Quine: *No real phenomenon (physical, biological, or social) is perfectly described by any mathematical model. There's usually a choice among several incompatible models, each more or less suitable.*

Putnam doesn't think a possibility is an object. He doesn't explain what he means by "object," so it's hard to know if he's right or wrong. Is an object something that can affect human life or consciousness? If so, some probabilities are objects.

"The main burden of this paper is that one does not have to buy Platonist epistemology to be a realist in the philosophy of mathematics. The theory of mathematics as the study of special objects has a certain implausibility which, in my view, the theory of mathematics as the study of ordinary objects with the aid of a special concept does not. . . ."

The argument claims that the consistency and fertility of classical mathematics is evidence that it or most of it is true under *some interpretation.* The interpretation might not be a realist one.

The doctrine of objectivity without objects is not easy to understand or believe. It's proposed because of inability to find appropriate objects to correspond to numbers and spaces. Abstract objects are vacuous. Mental or physical objects are ruled out. So Kreisel and Putnam think no kind of object can be a mathematical object. They overlook the kind of object that works. Social-historic objects.

Patterns/Structuralism

Shades of Bourbaki

The dichotomy between neo-Fregean and humanist maverick must be applied with a light touch. There are neo-Fregeans, there are humanist mavericks, and there are others. In this and the following section, I present two influential recent trends in the philosophy of mathematics. I don't classify them one way or the other.

Structuralism—defining mathematics as "the science of patterns"—may be new to some philosophers, but not to mathematicians. Bourbaki said as much, and called it structuralism. Before Bourbaki there was Hardy: "A mathematician, like a painter or a poet, is a maker of patterns . . . the mathematician's patterns, like the painter's or the poet's, must be beautiful. . . . There is no permanent place in the world for ugly mathematics" (Hardy, 1940). Structuralism is the core of Saunders MacLane's (1986).

It has been adopted by philosophers Michael Resnik and Stuart Shapiro. The definition, "science of patterns," is appealing. It's closer to the mark than "the science that draws necessary conclusions" (Benjamin Peirce) or "the study of form and quantity" (*Webster's Unabridged Dictionary*). Unlike formalism, structuralism allows mathematics a subject matter. Unlike Platonism, is doesn't rely on a transcendental abstract reality. Structuralism grants mathematics unlimited generality and applicability. Watch a mathematician working, and you indeed see her studying patterns.

Structuralism is valid as a partial description of mathematics—an illuminating comment. As a complete description, it's unsatisfactory. Saunders MacLane in *Philosophia Mathematica* has pointed out that elementary and analytic number theory, for example, are understood much more plausibly as about objects than about patterns.

Not everyone who studies patterns is a mathematician. What about a dressmaker's patterns? What about "pattern makers" in machine factories? Resnik and Shapiro don't mean to call machinists and dressmakers mathematicians. By "pattern" they mean, not a piece of paper or sheet metal, but a nonmaterial pattern. (Though mathematicians do use physical models on occasion!)

Then can we say "Mathematics is the science of nonmaterial patterns"? No.

There are physicists, astronomers, chemists, biologists, geologists, historians, ethnographers, sociologists, psychologists, literary critics, journalists of the better class, novelists, and poets who also study "nonmaterial patterns."

The cure of this over-inclusiveness is simple. Resnik and Shapiro mean to define mathematics as the study of *mathematical* patterns. But it's no easier to explain "mathematical pattern" than to explain "mathematics."

And that would still leave a difficulty. "Mathematics is the study of mathematical patterns" would no longer be over-inclusive in subject matter, but still would be over-inclusive in methodology. Some people are studying mathematical patterns—geometric patterns, for example—by computer graphics or by physical models or by statistical sampling. By any of various empirical models, without major use of demonstrative reasoning.

People who do this *are* studying mathematical patterns. They aren't mathematicians. The definition of mathematics has to involve methodology, the "mathematical way of thinking." I believe it would be accurate finally to say "Mathematics is the mathematical study of mathematical patterns." Accurate, but not exciting.

In *Realism in Mathematics* Penelope Maddy said that structuralism differs only verbally from her set-theoretic realism. I have difficulty seeing structuralism and set-theoretic realism as essentially the same. We have a much more definite idea of "set" than of "pattern."

It's easy to give examples of pattern. Less easy to give a coherent, inclusive definition. Maclane quotes and rejects Bourbaki's "precise" definition of "structure."

Maybe Resnik and Shapiro mean that a pattern is a structure; they do call their pattern doctrine "structuralism." "Patternsism" is a bit awkward.

"Pattern" is like "game," Wittgenstein's example whose referents can't be isolated by any explicit definition, only by "family resemblance." Solitaire can be connected by a chain of intermediate games to soccer. In the same way, very likely, the pattern of continuity-discontinuity in mathematical analysis could be connected by a chain of intermediate patterns to the pattern of quotient rings and ideals in algebra.

The structuralist definition fits mathematical practice, because it's all-inclusive. All mathematics easily falls under its scope. Almost everything else falls under it too. The set-theoretic picture of mathematics is unconvincing because set theory is irrelevant to the bulk of mainstream mathematics. Structuralism suffers the opposite shortcoming. It recognizes something present not only in mathematics, but in all analytical thinking. The set-theoretic picture is restrictive; the structuralist picture, over-inclusive.

Fictionalism
Hamlet/Hypotenuse

Recall the three familiar number systems: natural, rational, and real.

Natural numbers are for counting. Most everybody seems to think the natural numbers exist in some sense, though they are infinitely many, and some people gag at anything infinite.

Rational numbers are fractions, positive and negative. They're just pairs of natural numbers. (The rational number 1/2 is the pair [1, 2]). No special difficulty.

The irrational real numbers are the headache. They can be defined as non-repeating infinite decimals. But no one has ever seen an infinite decimal written out all the way to infinity. Does the irrational number π exist? Its first ***billion*** decimal places were computed by the Chudnovskys and Tanaka, as I mentioned above. But even after two and three billion decimal places, the Chudnovsky's will only see a finite piece of π. The unseen piece will still be infinite. Wherever we quit, we'll be in the dark about infinitely many digits in the decimal expansion of π.

Cantor proved there are uncountably many real numbers.** That's too many to grasp. Any list of real numbers leaves out nearly all of them! Do so many real numbers really exist, or are they a fairy tale?

Ordinary mathematicians say "They exist!" Fictionalists say "They don't!" They try to show that science doesn't require (as W. V. O. Quine claimed) actual existence of mathematical entities. You can do science, they say, while regarding mathematical entities as fictional—not actually existing. Among philosophers of mathematics who can be called fictionalists are Charles Chihara, Hartry Field, and Charles Castonguay.

What does either side mean by "exist"? They don't say! In today's Platonist-fictionalist argument, as in yesterday's foundationist controversies, this question is ducked. But what you mean by "exist" determines what you believe exists. If only physical entities exist, then numbers don't exist. If *relations* between physical entities exist, then small positive whole numbers exist. If exist means "having its own properties independent of what anybody thinks," then real numbers exist.

Could this difference of opinion affect mathematical practice? What evidence could settle it? I know no practical consequence of this dispute, nor any way to settle it. Argument about it usually comes down to, "To me, this opinion is more palatable than that one."

Some fictionalists are materialists. They notice that mathematics is imponderable, without location or size. Since, as they think, only material objects, ponderable and volume-occupying, are real, mathematics isn't real. They state this unreality by saying mathematics is a fiction. What it means to be a "fiction" isn't further explained.

"Fiction" means a "made-up story." And mathematics *is* a kind of made-up story. But its uniqueness, its difference from other made-up stories, is what we care about. Can the literary notion of fiction assist the philosophy of mathematics?

Aristotle wrote: "The distinction between non-fiction author and fiction author consists really in this, that the one describes the thing that has been, and the other a kind of thing that might be. Hence fiction is something more philosophic and of graver import than non-fiction, since its statements are of the nature rather of universals, whereas those of non-fiction are singular" (*Poetics*, p. 1451). (Modernized translation, replacing "poetry" with "fiction," "history" with "non-fiction.")

Charles Chihara makes an analogy between mathematics and Shakespeare's Prince Hamlet. Hamlet is a fiction, yet we know a lot about him. A stage director putting on *Hamlet* knows more about Hamlet than Shakespeare wrote down. Hamlet probably ate grapes. Certainly didn't eat fried armadillo. In mathematics we also reach conclusions about things that don't exist. Numbers are fictions like Hamlet.

Nonfiction corresponds to empirical science; fiction corresponds to mathematics!

Fictionalism rejects Platonism. In that sense, I'm a fictionalist. But Chihara, Field and I aren't in the same boat.

What do you mean by real? By fictional? Only if that's made clear can we say mathematics is a fiction. How is it distinguished from other fictions? Is Huck Finn the same kind of thing as the hypotenuse of a right triangle?

It's good to reject Platonism. To replace it, we need to understand its powerful hold. Then it becomes possible to let it go.

Our conviction when we work with mathematics that we're working with something real isn't a mass delusion. To each of us, mathematics is an external reality. Working with it demands we submit to its objective character. It's what it is, not what we want it to be. It's ineffective to deny the reality of mathematics without confronting this objectivity.

Fictionalism is refreshingly disrespectful to that holy of holies—mathematical truth. It puts human creativity at center stage. But it's a metaphor, not a theory.

The difficulty is failing to recognize ***different levels*** of existence. Numbers aren't physical objects. Yet they exist outside our individual consciousness. We encounter them as external entities. They're as real as homework grades or speeding tickets. They're real in the sense of social-cultural constructs. Their existence is as palpable as that of other social constructs that we must recognize or get our heads banged. That's why it's wrong to call numbers fiction, even though they possess neither physical nor transcendental reality, even though they are, like Hamlet, creations of human mind/brains.

We need to start by recognizing nonmaterial realities—mental reality and social-cultural-historical reality. Then it becomes apparent that mathematics is a social-cultural-historical reality with mental and physical aspects. Does that make it a fiction? Sure, if the U.S. Supreme Court and the prime interest rate and the baseball pennant race are all fiction. But they're not. Mathematics *is* at once a fictional reality and a realistic fiction. The interesting question is the intertwining of real and fictitious that make it what it is.

ten

Humanists and Mavericks of Old

About the Humanist Trend in Philosophy of Mathematics

The idea of mathematics as a human creation has been advocated many times, by Aristotle, by the empiricists John Locke, David Hume, and John Stuart Mill, and by many others.

I use "humanist and maverick" to include all these writers. I call my own slant on humanist philosophy of mathematics "social-cultural-historic" or just "social-historic."

Some humanist mavericks weren't primarily philosophers: Jean Piaget, psychologist; Leslie White, anthropologist; Michael Polányi, chemist; Paul Ernest, educationist; and Alfréd Rényi, George Pólya, Raymond Wilder, mathematicians.

Those who were mainly philosophers were nonstandard or off-beat: the pragmatist, Charles S. Peirce; the Hegelian mystical historicist, Oswald Spengler; the critical realist, Roy Sellars; the quasi-empiricist or fallibilist, Imre Lakatos; the scientific materialist, Mario Bunge; the objectivist, Karl Popper; the naturalist, Philip Kitcher; and the quasi-empiricist, Thomas Tymoczko.

Aristotle (384–322 B.C.)

The First Scientist

The first humanist in the philosophy of mathematics is modern compared to Pythagoras or Plato. Aristotle's first concern is careful logical analysis of terms and concepts. His next concern is whether speculative theories conform to known facts. There's no mysticism in his dry reports of the mystics Pythagoras, Plato, and Plato's successors Speusippus and Xenocrates.

H. G. Apostle collected Aristotle's writings on mathematics. I made it my main source on Aristotle. Apostle writes, "Of Aristotle's extant works no one treats of mathematics systematically. . . . However, numerous passages on mathematics are distributed throughout the works we possess and indicate a definite philosophy of mathematics."

In Aristotle's philosophy of mathematics, the key concept is abstraction. Numbers and geometrical figures are abstracted from physical objects by setting aside irrelevant properties—color, location, price, etc.—until nothing's left but size and shape (in the case of geometric figures) or "numerosity" (in the case of finite sets). As an account of elementary mathematics, this is not bad. Today it's inadequate, because mathematics includes much more than circles, triangles, and the counting numbers. His account of abstraction is clear and reasonable. But by twentieth-century standards it's not precise. It would be difficult to give a formal definition of abstraction.

Aristotle gets bad press in the survey course on Western Civilization where many people meet him. We learn there that modern science was born in the struggle of Galileo, Copernicus, and Descartes against the followers of Aquinas and Aristotle. But history is more complicated than that. Much of European philosophical thought developed as a contest between Platonists and Aristotelians. From the time of Augustine (fifth century), Plato was Church dogma. For centuries, Aristotle's writings were lost from Western Europe. They were retrieved in the twelfth century, thanks to Arab scholars of North Africa and Spain.

The recovery of Aristotle's writings led to a turn toward scientific realism under Church control. Thomas Aquinas was a leader in bringing Aristotle's philosophy into the Church.

Later there was a revival of Platonism (Vico), as part of a humanist opposition to Descartes's scientific rationalism. Descartes put observation and experiment above authority. By then, Aristotelian scholasticism really had become antiscientific. But in the longer perspective of the centuries-old competition between Aristotle and Plato, Aristotle favored scientific rationalism, Plato, transcendental mysticism.

For a glimpse of a remarkable mind, some samples from Apostle's anthology of Aristotle:

"The infinite cannot be a number or something having a number, for a number is numerable and hence exhaustible. Moreover, if the infinite were an odd number, then by the removal of a unit the resulting number would be even and still infinite; for as finite it could have only finite numbers as parts—and likewise if it were an even number. But it cannot be both odd and even. Further, if the infinite, after the removal of a unit is considered as odd, were divided into two equal parts, then two infinite numbers would result; and, if this were continued, an infinite number could be divided into as many infinite numbers as one pleased . . . (p. 69).

If we bisect the straight line AZ at B, and again the line BZ on the right at C, and continue the bisection of the part which remains on the right, that part exists from the beginning; and the parts which are taken away from it and added to the left still remain. Here, it is also evident that there is both an infinite by division and an infinite by addition at the same time and in the same straight line AZ; and this is true for any finite magnitude. Along with the division there is a corresponding addition taking place, for corresponding to a given division, say at E, there is a magnitude DE added to the magnitudes AD already taken; and just as there is more division (indeed an endless division) to be made in the remaining magnitude EZ, so there are more and more parts in EZ to be taken away and added to AE without the possibility of an end. Yet, as the bisection continues on and on, the sum of the parts taken tend more and more to a certain limit, AZ, which is never reached (p. 73).

"Antiphon, in attempting to square the circle makes the fallacy in thinking that the parts outside to be taken will finally come to an end. He inscribes a square within the circle, erects isosceles triangles on the sides to the square as bases and with the vertices on the circle, continues this process, and concludes that the increasing side of the circumscribed polygon will finally coincide with the points on the circle. Thus, since a square can be erected equal in area to each set of isosceles triangles added to the previous inscribed regular polygon, the circle itself will ultimately be squared. But this is impossible, for the diminishing sides of the inscribed regular polygon will never become points, and there will always be isosceles triangles outside yet to be taken" (p. 76).

This critique of Antiphon is followed by an analysis of Zeno's paradoxes against motion. This is so clear, one wonders why anyone after Aristotle ever bothered with those paradoxes.

"A difficulty in connection with the infinite concerns the mathematician. If the infinite is not actual and the magnitude of the universe is finite, his theorems concerning numbers will not be true for all numbers but only for a finite number of them; and he will not be able to extend his straight lines and planes indefinitely to demonstrate certain theorems in geometry. . . . If at a certain time the number one million does not exist, then it is false to say that the theorem is true for the number one million at that time; and the theorem is false, not simply, but in a qualified way, at such-and-such a time and for such-and-such a number. The theorem is stated in universal terms and has a potential nature; it is true for any number, not at this or that time or place but whenever and wherever a number exists. The fact that numbers exist or can exist shows that arithmetic does not deal with not-being" (p. 78).

He produced pages criticizing Plato's "Ideas" on grounds of vagueness and inconsistency. From p. 182: "There is also a difficulty in defining an Idea, if we are to predicate definitions of them as we predicate definitions of the corresponding species and genera of things. A definition is a predicate of many individuals and not of only one, but an Idea is one individual. If a man is defined as a rational animal, will the definition of the Idea of Man be 'Absolute Rational

Absolute Animal' or 'Absolute Rational Animal' or 'Absolute Rationality and Absolute Animal'? Moreover, can we truly say that Absolute Man is Rational or Absolute Rational if Absolute Man and Absolute Rationality are two different individual Ideas? This will be equivalent to saying that Plato is Socrates. If Absolute Triangle is defined as "three-sided figure," and "three-sided figure" is a predicate of Absolute Triangle, then "three-sided figure" will also be a predicate of Absolute Isosceles Triangle; and Absolute Triangle will be a part of Absolute Isosceles Triangle and will not exist separately. Further, Absolute Triangle is numerically one, and, since one of two contradictories is true, either it is isosceles (or Isosceles or else Absolute Isosceles) or it is not. If it is isosceles, then it does not differ from Absolute Isosceles Triangle, and the two Ideas will be one Idea; besides, it has just as much reason to be isosceles as to be equilateral or scalene, and it cannot be all of them. If it is not isosceles, then for the same reason it is not equilateral or scalene; but it must be one of them, for the three sides must be related in one of the three ways. Also, whatever is a predicate of an isosceles is also a predicate of an isosceles triangle and conversely. Hence, Absolute Isosceles and Absolute Isosceles Triangle turn out to be one and the same Idea. Again, if the One is ultimately the formal cause of an Idea and "one" is a predicate of the Idea, perhaps the One should be in the Idea. But the One is unique, and so is each Idea of its form. Hence, it would be absurd to try to show that all things have One as form."

In reading this passage, I am reminded of Frege. The direction of their arguments is opposite, for Aristotle is tearing up the Platonic Idea of number, which Frege upholds. But the tone—the cat-like delight in chewing up a philosophical mouse! Across 2,000 years, the two could be cousins.

Euclid

Axioms and Diagrams

The philosophy of mathematics of the Greeks ought to include not only the Greek philosophers but the great mathematicians Archimedes, Eudoxus, Euclid, and Apollonius. Unfortunately, the fragments left to us are insufficient to draw firm conclusions. We may suspect that the philosophy of mathematics accepted by the Greek mathematicians may not have been identical with the teachings of the philosophers.

We are told that the Greeks despised applications. Plutarch says Archimedes didn't think his great military engineering in defense of Syracuse was worth being preserved in writing. Astronomy, of which Ptolemy was the preeminent practitioner, must have been put on a higher plane than earthly calculations.

It's said that Euclid's axiomatics was "material axiomatics"—statements of true facts about real objects—unlike modern "formal axiomatics," which isn't about anything. It's risky to say much more, with so little evidence.

In Chapter 3 I wrote, "In the middle half of the twentieth century formalism was the philosophy of mathematics most advocated in public. In that period, the style in mathematical journals, texts, and treatises was: Insist on details of definitions and proofs; tell little or nothing of why a problem is interesting or why a method of proof is used."

Does that sound like Euclid? Was Euclid a formalist?

Yes, it sounds like Euclid if we look at Euclid with a formalist eye—if we see his text as essential, and his diagrams as unfortunate breaches of rigor.

Without the diagrams, the *Elements* could pass for a formalist text. But Euclid comes with diagrams, not without!

A recent book called *Proof without Words* is instructive. This is a charming collection of proofs using only pictures and diagrams, not a single word. There are theorems on plane geometry, finite and infinite sums and integrals, algebraic inequalities, and more.

The introduction contains a depressing disclaimer. The editor warns us that the proofs in his book aren't really proofs. Only the usual verbal or symbolic proof is really a proof. A sad testimony to the grip of formalism.

A proof can be words only, of course. It can be, as in Euclid, words and diagrams. Or it can be, as in *Proofs without Words*, diagram only. There's no textual or historical evidence that Euclid's diagrams were thought to be unimportant or unnecessary. They supply the motivation and insight that are lacking in the text. They free Euclid from suspicion of formalism.

John Locke (1632–1704)

Tabula Rasa

Locke doesn't have an inclusive, comprehensive philosophy of mathematics. But he understands that mathematics is a creation and an activity of the human mind. Unlike Berkeley and Hume, he has no criticism to make of mathematicians.

In freshman philosophy "Locke-Berkeley-Hume" are presented as opposites of "Descartes-Spinoza-Leibniz"—English empiricists versus continental rationalists. But among the empiricists, Berkeley's goal was utterly different from that of Locke or Hume—closer to that of the rationalist Leibniz, which was not to uphold science against scholasticism, but to prove that matter exists only in the Mind of God.

What did Locke-Berkeley-Hume think about mathematics? No two were alike. Take them in chronological order. Locke insisted everything in common knowledge and in natural science comes from observation. There's no innate or nonempirical knowledge of the exterior world. One might think this position would have led him to attempt an empiricist explanation of mathematics. But he didn't try to reduce mathematical knowledge to the empirical, as Mill did 200 years later. Instead, he saw it virtually as introspective.

"The knowledge we have of mathematical truths is not only certain but real knowledge; and not the bare empty vision of vain, insignificant chimeras of the brain: and yet, if we will consider, we shall find that it is only of our own ideas. The mathematician considers the truth and properties belonging to a rectangle or circle only as they are an idea in his own mind. For it is possible he never found either of them existing mathematically, i.e., precisely true, in his life. But yet the knowledge he has of any truths or properties belonging to a circle, or any other mathematical figure, are never the less true and certain even of real things existing; because real things are no further concerned, nor intended to be meant, by any such propositions, than as things really agree to those archetypes in his mind. Is it true of the idea of a triangle, that its three angles are equal to two right ones? It is true also of a triangle wherever it really exists" *(Essay*, IV, 6).

"All the discourses of the mathematicians about the squaring of a circle, conic sections, or any other part of mathematics, concern not the existence of any of those figures, but their demonstrations, which depend on their ideas, are the same, whether there be any square or circle existing in the world or no" (para. 8).

This sounds like rationalism; triangle and circle as archetypes in our minds. The next sentence makes an analogy between moral knowledge and mathematical knowledge, and says that both types of knowledge "abstract" from observation. "In the same manner, the truth and certainty of moral discourses abstracts from the lives of men, and the existence of those virtues in the world whereof they treat: nor are Tully's *Offices* less true, because there is nobody in the world that exactly practises his rules." In this sentence "abstract" means "idealized." Moral truths, like mathematical truths, refer to ideal, not to empirical reality. Locke equivocates between internal and external as the source of mathematical and moral conceptions. He doesn't, in the manner of the rationalists, try to use mathematical certainty to justify religious certainty. He does use mathematical knowledge as an analogue to moral knowledge.

From Baum, p. 126: "When we nicely reflect upon them, we shall find that general ideas are fictions and contrivances of the mind, that carry difficulty with them, and do not so easily offer themselves as we are apt to imagine. For example, does it not require some pains and skill to form the general idea of a triangle (which is yet none of the most abstract, comprehensive, and difficult), for it must be neither oblique nor rectangle, neither equilateral, equicrural, nor scalenon; but all and one of these at once. (Echo of Aristotle!) In effect, it is something imperfect, that cannot exist; an idea wherein some parts of several different and inconsistent ideas are put together.

"Amongst all the ideas we have, there is none more simple, than that of unity, or one; . . . by repeating this idea in our minds, and adding the repetitions together, we come by the complex ideas of the modes of it. Thus, by adding one to one, we have the complex idea of a couple; by putting twelve units together, we have the complex idea of a dozen, and of a score, or a million, or any other number. . . . Because the ideas of numbers are more precise and distinguishable than in extension, where every equality and excess are not so easy to be observed or measured, because our thoughts cannot in space carrie at any determined smallness beyond which it cannot go, as an unit; and therefore the quantity or proportion of any the least excess cannot be discovered. . . . This I think to be the reason why some American [Indians] I have spoken with (who were otherwise of quick and rational parts enough), could not, as we do, by any means count to 1000, nor had any distinct idea of that number, though they could reckon very well to 20. Because their language being scanty, and accommodated only to the few necessaries of a needy, simple life, unacquainted either with trade or mathematics, had no words in it to stand for 1000. . . . Let a man collect into one sum as great a number as he please, this multitude, how great soever, lessens not one jot the power of adding to it, or brings him any nearer the end of the inexhaustible stock of number, where still there remains as much to be added as if none were taken out. And this ***endless addition*** or ***addibility*** of numbers, so apparent to the mind, is that, I think, which gives us the clearest and most distinct idea of infinity." (Anticipating Poincaré by two centuries!)

Isaiah Berlin thinks that "About mathematical knowledge Locke shows great acumen. He sees that, for example, geometrical propositions are true of certain ideal constructions of the human mind and not of, e.g., chalk marks or surveyor's chains in the real world" (Berlin, p. 108). The circle and the line are ideas in the mathematician's head, which is why the mathematician can discover facts about them by mere contemplation. Locke doesn't seem to mind letting mathematics be an exception to his empiricist doctrine. Nor is he troubled by the objection Frege would make later: if a circle is in the mathematician's head, then there must be many different circles, one in each mathematician's head.

Locke also has his proof of the existence of God, based on "intuitively clear" ideas, such as "I exist," "I came from somewhere," "there had to be a first cause," and so forth.

Berkeley is conventionally presented in chronological order, between Locke and Hume. But in the history of the philosophy of mathematics, he cannot be counted as a forerunner of the social-cultural tendency. We treated Berkeley in our previous historical section, the history of Mainstream formalism and mysticism.

David Hume (1711–1776)

Commit It to the Flames!

Locke and Berkeley's successor David Hume was the only well-known philosopher before the twentieth century to recognize that mathematics, like the other sciences, gives only probable knowledge. He explicitly describes the role of the community of mathematicians in the process of mathematical growth and discovery. "In all demonstrative sciences the rules are certain and infallible, but when we apply them, our fallible and incertain faculties are very apt to depart from them, and fall to error. . . . By this means all knowledge degenerates into probability; . . . There is no Algebraist nor Mathematician so expert in his science, as to place entire confidence in any truth immediately upon his discovery of it, or regard it as any thing, but a mere probability. Every time he runs over his proofs, his confidence encreases; but still more by the approbation of his friends; and is rais'd to its utmost perfection by the universal assent and applauses of the learned world. Now 'tis evident that this gradual encrease of assurance is nothing but the addition of new probabilities" (*Treatise*, p. 231).

I respond, "How true!" One of the rare statements about mathematics in philosophy texts to which I so respond.

Hume came under Berkeley's influence early in life. "Perhaps the most important formative influence in Hume's formative years was his membership in the Rankenian Club of Edinburgh during the early 1720's. . . . The club carried on a correspondence with Berkeley, whose philosophy was apparently one of the central topics for discussion. . . . Nowhere, according to [Berkeley], was he better understood" (Turbayne).

Nearly all of Part 2, Book 1, of Hume's *Treatise of Human Nature* is devoted to refuting the infinite divisibility of the line. This doctrine has been universally accepted in geometry at least since Euclid. Euclid's simple construction to bisect a line segment, when repeated enough times, produces a line segment as short as you please. But Hume argues that time and space must consist of indivisible atoms. "For the same reason that the year 1737 cannot concur with the present year 1738, every moment must be distinct from and posterior or antecedent to another. 'Tis certain then, that time, as it exists, must be composed of indivisible moments. For if in time we could never arrive at an end of division, and if each moment, as it succeeds another, were not perfectly single and indivisible, there would be an infinite number of co-existent moments, or parts of time; which I believe will be allowed to be an arrant contradiction. The infinite divisibility of space implies that of time, as is evident from the nature of motion. If the latter, therefor be impossible, the former must be equally so. . . . Tis an establish'd maxim in metaphysics, That whatever the mind clearly conceives includes the idea of possible existence, or in other words, that nothing we imagine is absolutely impossible."

Hume and Berkeley are sure there must be shortest, indivisible units of length, because, they think, "the Mind" cannot conceive of a finite interval being composed of infinitely many parts. By saying so, they imply an attack on the differential calculus of Newton and Leibniz, where arbitrarily short intervals play a central role, and which at this very time in the hands of Huygens, the Bernoullis, and Euler, was making wonderful progress, establishing the dynamics of particles, rigid bodies, fluids, and solids. Did Hume realize he was challenging the best mathematics and science of his time? To Hume it's axiomatic that an event is possible if and only if "the Mind" can conceive it clearly. This thinking goes back to Descartes.

Hume likes to say an infinitely divisible line segment is as inconceivable (and therefore as impossible) as a square circle.** More than geometry, it was modern physics that wiped out the doctrine that an event is possible if and only if the Mind finds it conceivable. The quantum-mechanical uncertainty principle, the complementarity of particle and wave, the relativization of simultaneity by Einstein, the cosmos as ***curved four-dimensional*** space-time—all were ***inconceivable***! They're incompatible with our deepest convictions about the world. Nevertheless, they are effective explanatory theories. Grandpa's rule of possibility says, "If it happens, it's possible!" To understand the world, we have to stretch the limits of what our minds can conceive.

"Thus it appears that the definitions of mathematics destroy the pretended demonstrations, and that if we have the idea of indivisible points, lines and surfaces conformable to the definition, their existence is certainly possible; but if we have no such idea, 'tis impossible we can ever conceive the termination of any figure, without which conception there can be no geometrical demonstration. The first principles are founded on the imagination and senses: The conclusion, therefor, can never go beyond, much less contradict these faculties. No geometrical demonstration for the infinite divisibility of extension can have so much force as what we naturally attribute to every argument, which is supported by such magnificent pretensions. . . . For tis evident, that as no idea of quantity is infinitely divisible, there cannot be imagin'd a more glaring absurdity than to endeavor to prove that quantity itself admits of such a division."

One fallacy here is "conflation" of mathematical space and physical space. He uses an intuitive ***physical*** argument to prove that infinite divisibility is ***mathematically*** impossible. We cannot fault him for this. The distinction between physical space and mathematical space was yet unborn (until the discovery of non-Euclidean geometry).

Even if ***physical*** space were granular, that wouldn't stop us from using infinite divisibility in a ***mathematical*** theory, just as we use such exotica as infinite-dimensional spaces, the set of all subsets of an uncountable set, or a well-ordering of the real numbers.

As a matter of fact, Hume is wrong both physically and mathematically. Physically, he claims to prove that there's a minimum possible size of a material particle—without benefit of any physical measurement, based only on his inability to conceive. But you can't limit physical reality by the limits of your imagination.

Mathematically, he's saying the notion of an infinitely divisible interval is nonsense because an interval can't be divided into infinitely many pieces of ***equal*** length. But division into an arbitrarily large finite number of pieces is enough to refute his idea of a shortest possible length. And can he not have noticed the possibility of infinitely many pieces of decreasing length? The success of differential calculus refutes his claim that infinite divisibility is incoherent.

He thinks that points are indivisible, and line segments are made of finitely many points, so the minimum line segment must have positive length. This puzzle comes up in modern measure theory. How can a segment of positive length be composed of points of zero length? Hume is tripping over a consequential matter—the uncountability of the continuum, a path-breaking discovery of George Cantor, in the late nineteenth century.

When Hume says that instants of time are linearly ordered, few would disagree. But then he says that since they're linearly ordered, they're discrete. This is a mistake. We mustn't blame him for that. Unlike Berkeley, whose arguments on this matter are insubstantial, ***Hume is grappling with a real mathematical problem.*** The example of an ordered set that is dense, not discrete, was there for all to see, in the set of rational numbers. But it wasn't noticed until Cantor did so, a century later. Hume has not received credit for raising the problem, though of course he couldn't solve it.

Hume's major contribution to epistemology was to show that the "law of cause and effect" is not deductively valid. Our faith in it rests only on habit, so we can't have certain knowledge about anything external or material.

This corrosive skepticism attacks science as well as religion. Science doesn't hope for absolute certainty, but it assumes that knowledge is possible, if only partial knowledge or tentative knowledge. Hume destructively discounts the possibility of fundamental scientific advance. Elasticity, gravity, and other observed physical phenomena should be taken at face value, he says; to seek for deeper explanation would be in vain.

"As to the causes of these general causes, we should in vain attempt their discovery, nor shall we ever be able to satisfy ourselves by any particular explication of them. These ultimate springs and principles are totally shut up from human curiosity and inquiry. Elasticity, gravity, cohesion of parts, communication of motion by impulse—these are probably the ultimate causes and principles which we shall ever discover in nature; and we may esteem ourselves sufficiently happy if, by accurate inquiry and reasoning, we can trace up the particular phenomena to, or near to, these general principles . . . the observation of human blindness and weakness is the result of all philosophy, and meets us at every turn. Nor is geometry . . . able to remedy this defect . . . by all the accuracy of reasoning for which it is so justly celebrated" (*Inquiry*, p. 45).

Still, he does acknowledge that mathematical knowledge is different from empirical knowledge. "From (cause and effect reasoning) is derived all philosophy excepting only geometry and arithmetic" *(Abstract*, p. 187).

So, in his famous peroration, he spares us from the bonfire:

"If we take in our hand any volume—of divinity or school metaphysics, for instance—let us ask, Does it contain any abstract reasoning concerning quantity or number? No. Does it contain any experimental reasoning concerning matter of fact and existence? No. Commit it then to the flames, for it can contain nothing but sophistry and illusion" (*Inquiry*, p. 173).

Jean Le Rond D'Alembert (1717–1783)
At Last, Enlightenment

D'Alembert, Diderot, Rousseau, and Condillac were the leading "Encyclopedists" who played a major role in forming the modern mind.

"D'Alembert was the natural son of a soldier aristocrat, the chevalier Destouches, and Madame de Tencin, one of the most notorious and fascinating aristocratic women of the century. A renegade nun, she acquired a fortune as mistress to the powerful minister, Cardinal Dubois, and after a successful career of political scheming, she rounded out her life by establishing a salon which attracted the most brilliant writers and philosophers of France. It is reported that d'Alembert was not the first of the inconvenient offspring she abandoned. In any case, he was found shortly after his birth on the steps of the Parisian church of Saint-Jean Lerond [note the name]. He was raised by a humble nurse, Madame Rousseau, whom he treated as his mother and with whom he lived until long after he achieved international fame."

A self-taught physicist and mathematician, he published his *Treatise on Dynamics* in 1743, at age 26. His formula for the one-dimensional linear wave equation with constant density is beloved by every student of applied mathematics. "A combination of virtuosity, ambition, aggressiveness, and personal charm eventually won him a most honored position in the intellectual community of Europe, including the lasting friendship of both Voltaire and Frederick the Great. . . . His entry in the Académie Française in 1754 marked a major victory for the encyclopedic party. Eventually he became the perpetual secretary of that academy. . . ."

"He never married. However, after 1754 he became the intimate friend of an aristocratic lady, likewise of illegitimate birth, the famous Julie de Lespinasse, with whom he was ever more closely bound until her death left him desolate in 1776. Their strictly spiritual and intellectual relationship was something exceptional among the *philosophes*" (Schwab in D'Alembert, 1963, p. 15 ff.).

D'Alembert is on the short list of notable contributors to both mathematics and philosophy. His word of encouragement to a beginner in calculus is often quoted: "Continue. Eventually, faith will come."

In his *Preliminary Discourse to the Encyclopedia of Diderot,* he stated his opinions on the nature of mathematics and logic:

"We will note two limits within which almost all of the certain knowledge that is accorded to our natural intelligence is concentrated, so to speak. One of those limits, our point of departure, is the idea of ourselves, which leads to that of the Omnipotent Being, and of our principal duties. [Echo of Descartes!] The other is that part of mathematics whose object is the general properties of bodies, of extension and magnitude. Between these two boundaries is an immense gap where the Supreme Intelligence seems to have tried to tantalize the human curiosity, as much by the innumerable clouds it has spread there as by the rays of light that seem to break out at intervals to attract us. . . ."

"With respect to the mathematical sciences, which constitute the second of the limits of which we have spoken, their nature and their number should not overawe us. It is principally to the simplicity of their object that they owe their certitude. Indeed, one must confess that, since all the parts of mathematics do not have an equally simple aim, so also certainty, which is founded, properly speaking, on necessarily true and self-evident principles, does not belong equally or in the same way to all these parts. . . . Only those that deal with the calculation of magnitudes and with the general properties of extension, that is, Algebra, Geometry, and Mechanics, can be regarded as stamped by the seal of evidence. . . . The broader the object they embrace and the more it is considered in a general and abstract manner, the more also their principles are exempt from obscurities. It is for this reason that geometry is simpler than mechanics, and both are less simple than Algebra. . . . Thus one can hardly avoid admitting that the mind is not satisfied to the same degree by all the parts of mathematical knowledge."

[Quite different from the earlier views of Plato, and the later views of foundationism, that all mathematics must and should be absolutely certain.]

"Let us go further and examine without bias the essentials to which this knowledge may be reduced. Viewed at first glance, the information of mathematics is very considerable, and even in a way inexhaustible. But when, after having gathered it together, we make a philosophical enumeration of it, we perceive that we are far less rich than we believed ourselves to be. I am not speaking here of the meager application and usage to which a number of these mathematical truths lend themselves; that would perhaps be a rather feeble argument against them. I speak of these truths considered in themselves. What indeed are most of those axioms of which Geometry is so proud, if not the expression of a single simple idea by means of two different signs or words? Does he who says two and two equals four have more knowledge than the person who would be content to say two and two equals two and two? Are not the ideas of 'all,' of 'part,' of 'larger,' and of 'smaller,' strictly speaking, the same simple and individual idea, since we cannot have the one without all the others presenting themselves at the same time? As some philosophers have observed, we owe many errors to the abuse of words. It is perhaps to this same abuse that we owe axioms. My intention is not, however, to condemn their use; I wish only to point out that their true purpose is merely to render simple ideas more familiar to us by usage, and more suitable for the different uses to which we can apply them. I say virtually the same thing of the use of mathematical theorems, although with the appropriate qualifications. Viewed without prejudice, they are reducible to a rather small number of primary truths. If one examines a succession of geometrical propositions, deduced one from the other so that two neighboring propositions are immediately contiguous without any interval between them, it will be observed that they are all only the first proposition which is successively and gradually reshaped, so to speak, as it passes from one consequence to the next, but which, nevertheless, has not really been multiplied by this chain of connections; it has merely received different forms. It is almost as if one were trying to express this proposition by means of a language whose nature was being imperceptibly altered, so that the proposition was successively expressed in different ways representing the different states through which the language had passed. Each of these states would be recognized in the one immediately neighboring it; but in a more remote state we would no longer make it out, although it would still be dependent upon those states which preceded it, and designed to transmit the same ideas. Thus, the chain of connection of several geometrical truths can be regarded as more or less different and more or less complicated translations of the same proposition and often of the same hypothesis. These translations are, to be sure, highly advantageous in that they put us in a position to make various uses of the theorem they express—uses more estimable or less, in proportion to their import and consequence. But, while conceding the substantial merit of the mathematical translation of a proposition, we must recognize also that this merit resides originally in the proposition itself. This should make us realize how much we owe to the inventive geniuses who have substantially enriched Geometry and extended its domain by discovering one of these fundamental truths which are the source, and, so to speak, the original of a large number of others . . . (p. 25 ff.).

"The advantage men found in enlarging the sphere of their ideas, whether by their own efforts or by the aid of their fellows, made them think that it would be useful to reduce to an art the very manner of acquiring information and of reciprocally communicating their own ideas. This art was found and named Logic. . . . It teaches how to arrange ideas in the most natural order, how to link them together in the most direct sequence, how to break up those which include too large a number of simple ideas, how to view ideas in all their facets, and finally how to present them to others in a form that makes them easy to grasp. . . . This is what constitutes this science of reasoning, which is rightly considered the key to all our knowledge. [This looks like a veiled reference to the great *Port-Royal Logic* of the Jansenist Antoine Arnauld.] However, it should not be thought that it [the formal discipline of Logic] belongs among the first in the order of discovery. . . . The art of reasoning is a gift which Nature bestows of her own accord upon men of intelligence, and it can be said that the books which treat this subject are hardly useful except to those who can get along without them. People reasoned validly long before Logic, reduced to principles, taught them how to recognize false reasoning, and sometimes even how to cloak them in a subtle and deceiving form" (p. 30).

John Stuart Mill (1808–1873)
Classic Liberal

Mill was a father-created child prodigy, like William Rowen Hamilton and Norbert Wiener. He performed wonders as an infantile multilingual classical scholar. He's remembered today for courageous writings in defense of liberty and against the subjection of women. He and Harriet Taylor achieved the first famous collaboration between male and female authors.

His well-known *Utilitarianism* has this provocative remark: "Confusion and uncertainty exist respecting the principles of all the sciences, not excepting that which is deemed the most certain of them—mathematics, without much impairing, generally indeed without impairing at all, the trustworthiness of the conclusions of those sciences. An apparent anomaly, the explanation of which is that the detailed doctrines of a science are not usually deduced from, nor depend for their evidence upon, what are called its first principles. Were it not so, there would be no science more precarious, or whose conclusions were more insufficiently made out, than algebra, which derives none of its certainty from what are commonly taught to learners as its elements, since these, as laid down by some of its most eminent teachers, are as full of fictions as English law, and of mysteries as theology. The truths which are ultimately accepted as the first principles of a science are really the last results of metaphysical analysis practiced on the elementary notions with which the science is conversant; and their relation to the science is not that of foundations to an edifice, but of roots to a tree, which perform their office equally well though they be never dug down to and exposed to light."

A rejection of foundationism before foundationism came into flower!

Mill's major contribution to philosophy of mathematics is the *System of logic* . . . , finished in 1841. Today's student knows of this book only by Frege's merciless attack in the *Grundlagen*. But one can't rely on Frege for a fair account. Frege wrote, claiming to paraphrase Mill: "From this we can see that it is really incorrect to speak of three strokes when the clock strikes three." But on page 189 Mill writes, "*Ten* must mean ten bodies, or ten sounds, or ten beatings of the pulse." Today's revival of empiricism in philosophy of mathematics calls for a new look at Mill.

Mill's main thesis was that laws of mathematics are objective truths about physical reality. They are derived from elementary principles or laws (*not* arbitrary axioms) that we learn by observing the world. To Mill, the number 3 is defined independently of 1 and 2, by generalization from observed triples.

"$2 + 1 = 3$" is a truth of observation, not essentially different from "all swans are white." It turned out that in Australia there are black swans. It could turn out that $2 + 1$ sometimes isn't three. But since we have a tremendous amount of confirmation of this law, we rightly have tremendous confidence in it. Mill's idea is close to Aristotle's. But in Mill it seems radical or paradoxical, because by Mill's time rationalist-idealist philosophy of mathematics had become dominant.

Mill fought against two different theories of arithmetic. The nominalists (Hobbes, Condillac, and J. S. Mill's father James Mill) said $2 + 1$ is the ***definition*** of 3; therefore (foreshadowing Bertrand Russell) the equation $2 + 1 = 3$ is an empty tautology. To Mill the formula $2 + 1 = 3$ is not tautologous but informative; it says a triple can be separated into a pair and a singleton, or a pair and a singleton united to become a triple.

The other philosophy that Mill opposed was called intuitionist (no connection with Brouwer!). Its leading advocate was William Whewell, a famous philosopher of science who was then tremendously influential in university mathematics. Whewell said anything inconceivable must be false. (Echoes of Berkeley and Hume!) Therefore, the negation of anything inconceivable must be true! It's inconceivable that

> $2 + 1$ isn't equal to 3;
> therefore, $2 + 1 = 3$.

In reply, Mill recalled "inconceivables" of previous times: a spherical earth, the earth revolving round the sun, instantaneous action at a distance by gravitation. "There was a time when men of the most cultivated intellects, and the most emancipated from the domain of early prejudice, could not credit the existence of antipodes; were unable to conceive, in opposition to old association, the force of gravity acting upward instead of downward. The Cartesians long rejected the Newtonian doctrine of the gravitation of all bodies toward one another, on the faith of a general proposition, the reverse of which seemed to them to be inconceivable—the proposition that a body can not act where it is not. . . . And they no doubt found it as impossible to conceive that a body should act upon the earth from the distance of the sun or moon, as we find it to conceive an end to space or time, or two straight lines enclosing a space" (Mill, pp. 178–79).

What's inconceivable last century is common sense in this.

Mill even quoted Whewell against himself. In Whewell's ***Philosophy of the Inductive Sciences***, Mill found these lines: "We now despise those who, in the Copernican controversy, could not conceive the apparent motion of the sun on the heliocentric hypothesis. . . . The very essence of these triumphs is that they lead us to regard the views we reject as not only false but inconceivable."

Mill didn't confront the difficulties of the empiricism he advocated. It's believed that there are finitely many physical objects, but there are infinitely many numbers. Already in Mill's time mathematics included non-Euclidean geometry, which had (as yet) no physical interpretation.

From Kubitz, pp. 267–68: "The axioms which the *Logic* examines are the laws of causation and the axioms of mathematics. Mill gave an empirical explanation of these because "***the notion that truths external to the mind may be known by***

intuition or consciousness, independently of observation and experience,*"** was the "great intellectual support of false doctrines and bad institutions.***" [My emphasis.] By the aid of this theory, every inveterate belief and every intense feeling of which the origin is not remembered, is enabled to dispense with the obligation of justifying itself by reason, and is erected into its own all-sufficient voucher and justification. There never was such an instrument devised for consecrating all deep-seated prejudices. And the chief strength of this false philosophy in morals, politics, and religion, lies in the appeal which it is accustomed to make to the evidence of mathematics and of the cognate branches of physical science."

The *System of Logic* explained the character of necessary truths on the basis of experience and education. The existence of "necessary truths, which is adduced as proof that their evidence must come from a deeper source than experience," was explained in such a way as to help justify the need for political reform. More on this point in Chapter 13.

eleven

Modern Humanists and Mavericks

Charles Sanders Peirce (1839–1914)

American Tragedy

Charles Sanders Peirce was a great logician, a great philosopher, and a respectable mathematician.[1]

He was a life-long freelancer in logic, mathematics, physics, and philosophy. Independently of Frege, Peirce's student, O. H. Mitchell, introduced the famous quantifiers "there exists" and "for all." Peirce was the founder of semiotics—the abstract, general study of signs and meaning.

I quote Christopher Hookway in the *Companion to Epistemology.* "From his earliest writings Peirce was critical of Cartesian approaches to epistemology. He charged that the method of doubt encouraged people to pretend to doubt what they did not doubt in their hearts, and criticized its individualist insistence that 'the ultimate test of certainty is to be found in the individual consciousness.' We should rather begin from what we cannot in fact doubt, progressing towards the truth as part of a community of inquirers trusting to the multitude and variety of our reasoning rather than to the strength of any one. He claimed to be a contrite Fallibilist and urged that our reasoning should not form a chain that is not stronger than its weakest link, but a cable whose fibres may be ever so slender, provided they are sufficiently numerous and intimately connected."

Peirce's view of mathematics is radically different from the foundationist sects. His surprising separation of mathematics and logic, and his full acceptance of the role of the mathematics community and of intuition in mathematics, put him on the humanist side. I quote from his essay "The Essence Of Mathematics." His remarks on the nature of mathematics could have been written yesterday.

[1] The first chapter of Corrington's book tells the tragic and embarrassing story of Peirce's life.

"Mathematics is distinguished from all other sciences, except only ethics, in standing in no need of ethics" (p. 267).

"Mathematics, along with ethics and logic alone of the sciences, stands in no need of logic. Make of logic what the majority of treatises in the past have made of it—that is to say, mainly formal logic, and the formal logic represented as an art of reasoning—and in my opinion, this objection is more than sound, for such logic is a great hindrance to right reasoning. True mathematical reasoning is so much more evident than it is possible to render any doctrine of logic proper—without just such reasoning—that an appeal in mathematics to logic could only embroil a situation. On the contrary, such difficulties as may arise concerning necessary reasoning have to be solved by the logician by reducing them to questions of mathematics."

"One singular consequence of the notion, which prevailed during the greater part of the history of philosophy, that metaphysical reasoning ought to be similar to that of mathematics, only more so, has been that sundry mathematicians have thought themselves, as mathematicians, qualified to discuss philosophy; and no worse metaphysics than theirs is to be found."

"In the major theorems it will not do to confine oneself to general terms. It is necessary to set down some individual and definite schema, or diagram—in geometry, a figure composed of lines with letters attached; in algebra an array of letters of which some are repeated. After the schema has been constructed according to the precept virtually contained in these, the assertion of the theorem is evidently true. Thinking in general terms is not enough. It is necessary that something should be *done*" (pp. 260–61).

From p. 267: "It is a remarkable historical fact that there is a branch of science in which there has never been a prolonged dispute concerning the proper objects of that science. It is the mathematics. Mistakes in mathematics occur not infrequently, and not being detected give rise to false doctrine, which may continue a long time. Thus, a mistake in the evaluation of a definite integral by Laplace in his *Mécanique céleste*, led to an erroneous doctrine about the motion of the moon which remained undetected for nearly half a century. But after the question had once been raised, all dispute was brought to a close within a year" (1960, para. 3,426; reprinted in the *American Mathematical Monthly*, 1978, p. 275).

Besides his path-breaking contributions to semiotics and logic, Peirce was the founder of pragmatism. The great American pragmatists William James and John Dewey were his followers. All three of them thought of mathematics as a human activity rather than a transcendental hyper-reality. Martin Gardner writes in his book *The Meaning of Truth*, that "James argues for a mind-dependent view of mathematics very close to that of Davis, Hersh, and Kline." Gardner thinks that "James makes a good case for a cultural approach to mathematics that was shared by F. C. S. Schiller and (he thinks) by John Dewey." Gardner himself, I should say, thinks otherwise.

Henri Poincaré (1854–1912)
Mozart of Mathematics

Poincaré was one of the supreme mathematicians, rivaled or surpassed in his time only by David Hilbert. Like other French masters of the turn of the century, he was a virtuoso at complex function theory. His qualitative study of the three-body problem was a fountainhead of algebraic topology, one of the great achievements of twentieth-century mathematics.

Poincaré's work on the Lorentz transformation and Maxwell's equations could have led him to special relativity. But in physics Poincaré was a conventionalist. He thought it a matter of convenience which mathematical model one uses to describe a physical situation. For example, he said, it makes no sense to ask if physical space is Euclidean or non-Euclidean. Whichever is convenient is best. His conventionalism hid from him the deep physical meaning of his mathematical results on relativity.

Einstein was not a conventionalist but a realist. He opposed Ostwald and Mach, who thought atoms and molecules were only convenient fictions. His 1905 paper on Brownian motion proved that molecules actually have definite volumes—they are real! As a philosophical realist, Einstein could more readily appreciate the physical consequences of mathematical relativity.

Poincaré disliked Peano's work on a formal language for mathematics, then called "logistic." He wrote of Russell's paradox, with evident satisfaction, "Logistic has finally proved that it is not completely sterile. At last it has given birth—to a contradiction."

With his colleagues Emile Borel and Henri Lebesgue, he opposed uninhibited proliferation of infinite sets. Arithmetic should be the common starting point of all mathematics. Mathematical induction is the fundamental source of novelty in mathematics. In this he was a forerunner of Brouwer's intuitionism. But he never joined Brouwer in condemning indirect proof (the law of the excluded middle.)

His brilliant, penetrating articles on philosophy of science and mathematics still find admiring readers.

Ludwig Wittgenstein (1889–1951)
Vienna in Cambridge

Ludwig Wittgenstein is one of the remarkable personalities of the century. His father was a top capitalist in Austrian steel. He had two brothers who committed suicide, and he spent a lifetime with self-hatred and religious guilt. In philosophy he created two revolutions.

He went to England to work on aeronautics. Bertrand Russell's *Philosophy of Mathematics* drew him to philosophy. Russell told Wittgenstein's sister that Wittgenstein was the philosopher of the future.

In 1921 he published the *Tractatus Logico-Philosophicus.* Russell wrote an admiring introduction to it. Wittgenstein said Russell didn't understand it. Later it became a bible for the Vienna Circle.

The message of the *Tractatus* is: exact correspondence between language, logic, and the world. Language and logic match perfectly everything of which it's possible to speak. The rest must pass in silence.

After the *Tractatus,* no more philosophy was needed. He went to teach school in the Austrian mountains. After a few difficult years, he gave up school teaching and returned to Cambridge as a professor. His opinions had changed. He repudiated the *Tractatus.* The task of philosophy is to show that there are no philosophical problems, "to show the fly the way out of the fly bottle."

In his last years, he gave courses and wrote notes on the philosophy of mathematics. Alan Turing, already famous for his theory of computation, was in the class of 1939. After Wittgenstein died, some of his notes were published as *Remarks on the Foundations of Mathematics.* Then class notes by Bosanquet, Malcolm, Rhees, and Smythies were published as *Wittgenstein's Lectures on the Foundations of Mathematics.*

Wittgenstein's *Remarks* did not meet universal enthusiasm. Logician Georg Kreisel called them "the surprisingly insignificant product of a sparkling mind." Logician Paul Bernays wrote, "He sees himself in the part of the free-thinker combating superstition. The latter's goal, however, is freedom of the mind; whereas it is this very mind which Wittgenstein in many ways restricts, through a mental asceticism for the benefit of an irrationality whose goal is quite undetermined."

Ernest, Klenk, Kielkopf, Shanker, and Wright offer more favorable interpretations of Wittgenstein's philosophy of mathematics. Of course Ernest, Klenk, Kielkopf, Shanker, and Wright disagree with each other. Kielkopf says Wittgenstein was a "strict finitist." Klenk says he wasn't a finitist at all.

Wittgenstein's bête noir was the idea that mathematics exists apart from human activity—Platonism. He rejected the logicist Platonism of his philosophical parents Russell and Frege. In fact, he denied any connection of mathematics with concepts or thought. Mathematics is nothing but calculation, and the rules of calculation are arbitrary. The rules of counting and adding are only custom and habit, "the way we do it." If others do it otherwise, there's no way to say who's right and who's wrong.

The *Remarks* and the *Lectures* are collections of provocative fragments. Many of the *Remarks* are about adding. Not adding big numbers, just adding 1. He doesn't get to subtracting or multiplying.

He doesn't wish to justify adding—quite the contrary! Going to the opposite extreme of Frege and Russell, he denies the logical necessity of the addition table. Even counting is debunked. Wittgenstein says that, given the question

1, 2, 3, 4, ???

"someone" ***might*** answer

1, 2, 3, 4, 100.

Then "we" would say "100" is wrong.

(I put quotes on "we," because I don't know if "we" means his auditors, or educated twentieth-century Europeans, or all sane adult humans.)

"We" might say, "That person doesn't understand that for all n, the n'th number is n."

But suppose "someone" understands it ***differently***, says Wittgenstein.

Then there is an impasse. Explicit formalization will work only if "someone" catches on to it.

Wittgenstein says

$$3 + 5 = 8$$

only because, "That's how we do it."

"We" could say

$$3 + 5 = 9$$

if "we" chose.

"Someone" might get 9 when he adds 3 and 5. He might insist that he's following the rules he learned in school. In that case, according to Wittgenstein, there's nothing more to say. "Someone" does it his way; "we" do it ours.

An alert student might interrupt: "Excuse me, Professor Wittgenstein. The axioms of arithmetic may be arbitrary for abstract logic, but not from a practical viewpoint. And once we fix the axioms, 3 + 5 equals 8. I can show you a proof."

Wittgenstein would not be troubled. He could reply, "Your proof is a proof only if we accept it as a proof. If someone says, 'I don't see that you've proved it,' there's no way to compel him to agree that you have proved it."

He admits that the rules of arithmetic may have been chosen for practical reasons—measuring wood, perhaps. But that wouldn't make them necessary. Only convenient.

Wittgenstein's position can be summarized: "Rules aren't self-enforcing." They're enforced by agreement of the people concerned.

He's not questioning just the axioms of arithmetic. He's questioning whether *any* axioms compel *any* theorem—except for the reason, "That's how we do it."

Some do a subtler reading of Wittgenstein, based on two plausible premises:

Premise 1: Any fool knows 3 + 5 = 8.
Premise 2: Wittgenstein was no fool.

Then his skepticism about

$$3 + 5 = 8$$

must have been a trick, perhaps to show that we don't know what numbers are, perhaps to teach by example how to be a philosopher. This is sometimes called "the Harvard interpretation." It has the merit of opening the door to a semi-infinite sequence of philosophy dissertations. But, it seems to me, elementary courtesy requires us to accept that what Wittgenstein says is what Wittgenstein means.

The *Remarks* is a list of numbered paragraphs. Here are my favorite Wittgensteinisms, mostly from the *Remarks,* numbered by page and paragraph. I comment in brackets.

"Mathematics consists entirely of calculations. In mathematics everything is algorithm and nothing is meaning, even when it doesn't look like that because we seem to be using words to talk about mathematical things. Even these words are used to construct an algorithm" (*Remarks,* p. 468).

"'Mathematical logic' has completely deformed the thinking of mathematicians and of philosophers, by setting up a superficial interpretation of the forms of our everyday language as an analysis of the structures of facts. Of course, in this it has only continued to build on the Aristotelian logic" (*Remarks,* p. 156, para. 48).

"If in the midst of life we are in death, so in sanity we are surrounded by madness." (157, 53)

"Nothing is hidden, everything is visible."

[But science is the struggle to reveal what's hidden!]

"Imagine someone bewitched so that he calculated

{1, 2, 3; 3, 4, 5; 5, 6, 7; 7, 8, 9; 9, 10}.

Now he is to apply this calculation. He takes 3 nuts four times over, and then 2 more, and he divides them among 10 people and each one gets *one* nut; for he shares them out in a way corresponding to the loops of the calculation, and as often as he gives someone a second nut it disappears" (42, 136). [Classical Wittgenstein. Imagine a fantastic "computation" to prove that computation is arbitrary. Here's an analogy. I insist on crawling instead of walking. I insist that crawling is a natural, proper locomotion. Then, for Wittgenstein, there's no way to prove I'm wrong. Wittgenstein's conclusion: We walk rather than crawl only because "That's the way we do it."

[Another of his "examples" is about people who sell wood. They spread it on the ground, and charge according to the *area* covered, regardless of whether the wood is piled high or low. "What's wrong with that?" asks Wittgenstein.

Easy to answer! Say that in the north woodlot, wood is stacked 2 feet high. In the south woodlot, it's stacked 4 feet high. Customers buy all the high-stacked wood from the south lot and none from the north. A competing wood seller sets price by volume, and Wittgenstein's silly wood seller goes broke. Despite Wittgenstein's off-the-wall "examples" of imaginary "tribes," the universal requirements of buying and selling compel

3 + 5 = 8.]

"A proof convinces you that there is a root of an equation (without giving you any idea *where*). How do you know that you understand the proposition that there is a root? How do you know that you are really convinced of anything? You may be convinced that the application of the proved proposition will turn up. But you do not understand the proposition so long as you have not found the application" (146, 25) [A quibble on "understand." I'd be interested to know if there's a hungry lion in the house with me, even if I don't know his exact location.]

"If you know a mathematical proposition, that's not to say you yet know *anything*" (160, 2).

"If mathematics teaches us to count, then why doesn't it also teach us to compare colors?" (187, 38).

"The idea of a (Dedekind) 'cut'** is such a *dangerous* illusion" (148, 29).

"Fractions cannot be arranged in an order of magnitude. At first this sounds extremely interesting and remarkable. One would like to say of it, e.g. 'It introduces us to the mysteries of the mathematical world.' This is the aspect against which I want to give a warning. When . . . I form the picture of an unending row of things, and between each thing and its neighbour new things appear, and more new ones again between each of these things and its neighbour, and so on without end, then certainly there is something here to make one dizzy. But once we see that this picture, though very exciting, is all the same not appropriate, that we ought not to let ourselves be trapped by the words 'series,' 'order,' 'exist,' and others, we shall fall back on the *technique* of calculating fractions, about which there is no longer anything *queer*" (60.11, *Philosophical Grammar*). [What trap? Why dizzy? Fall back from what?]

"Someone makes an addition to mathematics, gives new definitions and discovers new theorems—and in a *certain* respect he can be said to not know what he is doing.—He has a vague imagination of having *discovered* something like a space (at which point he thinks of a room), of having opened up a kingdom, and when asked about it he would talk a great deal of nonsense." [Wittgenstein is mocking his friend G. H. Hardy.] "We are always being told that a mathematician works by instinct (or that he doesn't proceed mechanically like a chess player or the like), but we aren't told what that's supposed to have to do with the nature of mathematics. If such a psychological phenomenon does play a part in mathematics we need to know how far we can speak about mathematics with complete exactitude, and how far we can only speak with the indeterminacy we must use in speaking of instincts etc" (p. 295). [Wittgenstein is unwittingly revealing that to hold onto his desiccated notion of mathematics he refuses to listen to better information.]

"How a proposition is verified is what it says" (p. 458).

[The Pythagorean theorem can be verified in many ways. But what it says is always: If a, b, and c are sides of a right triangle, $a^2 + b^2 = c^2$. According to Wittgenstein there are hundreds of different Pythagorean theorems.

You can arrive in London by land, sea or air. That doesn't mean there are three different Londons.]

"The only point there can be to elegance in a mathematical proof is to reveal certain analogies in a particularly striking manner when that is what is wanted; otherwise it is a product of stupidity and its only effect is to obscure what ought to be clear and manifest. The stupid pursuit of elegance is a principal cause of the mathematicians' failure to understand their own operations, or perhaps the lack of understanding and the pursuit of elegance have a common origin" (top, p. 462). [Dislike of mathematical elegance disqualifies anyone from talking about mathematics. In his preface to the *Tractatus*, Wittgenstein says its propositions are obviously true, he doesn't care if they're original. In fact, most of them are blatantly false. Their charm is originality and *elegance*—qualities he despises.]

Wittgenstein confuses what's logically undetermined and what's arbitrary. A mathematical rule may be determined by *convenience*. Such a determination isn't arbitrary, even though it's not compelled logically.

Besides their usual rules, arithmetic and mathematics have an unstated metarule:

Preserve the old rules!

Hermann Hankel called it "the law of preservation of forms." It's a principle of maximal convenience.

In the natural numbers, there's a rule that everything except 1 is greater than 1. When we introduce 0 and negative numbers, we give up that rule. Giving it up leads to a useful extension, the negative numbers and the integers—*if we preserve the other old rules.*

When we introduce negative numbers, we have to decide, what is

$$-1 \times -1\ ?$$

-1 is new, not previously defined or regulated.

So logically, we could choose any value we please, for example,

$$-1 \times -1 = -95.$$

But then -1 would violate the associative and distributive laws. If we want to keep the associative and distributive laws for negative numbers as well as positive, we find**

$$-1 \times -1 = 1.$$

In a sense this formula is just a convention. In another sense it's a theorem. It's *not* arbitrary.

Wittgenstein detected conventions in mathematics. He didn't know or didn't care that important convenience dictated those conventions. He acknowledged kitchen convenience—measuring wood, counting potatoes. He didn't recognize ***mathematical*** convenience.

He says:

1. Mathematics is just something we do.
2. We could just as well do it any other way.

1. Is correct.
2. Is nonsense.

The absurd (2) obscures his important insight (1).

(1) Can be restated: "Mathematics is an activity of the community. It doesn't exist apart from people." That's right. It's a courageous corrective to Frege's Platonism.

(2) Can be restated: "We can do mathematics any way we please." That's so wrong that it obscures the merit of the first statement.

Wittgenstein may have been misled by an analogy between mathematics and language. Language and mathematics both have rules. In language, the surface forms of grammar are indeed conventional, or in a sense arbitrary. They're not determined by intrinsic necessity. Mathematics is different. It's more than a semitransparent transmission medium. It has content. Its rules are not arbitrary. They're determined by mathematical convenience and mathematical necessity.

If Wittgenstein were right, why do we never hear of anyone who thinks

$$3 + 5 = 9?$$

It's true, school and society tell us

$$3 + 5 = 8.$$

But in politics, in music, or in sexual orientation, some people reject the dictate of school and society. Some people dare question the Holy Trinity, the American flag, whether God should save Our Gracious Queen, and so on. But ***nobody*** questions elementary arithmetic. A few poor souls trisect angles by compass and straight-edge, despite a famous proof that it can't be done. But in that problem the discoverer easily confuses himself. We ***never*** get letters claiming that

$$3 + 5 = 9.$$

If arithmetic can be whatever you like, why has no one in recorded history written

$$3 + 5 = 9?$$

Anthropologists in the Sepik Valley of New Guinea find surprising practices and beliefs about medicine, about rain, about gods and devils. Not about arithmetic.

"Mathematics is a practice," Wittgenstein said. Right.

He says that since it's a practice, it's not a body of knowledge.

Non sequitur!

The practice of carpentry has a body of knowledge. The practice of swimming has a body of knowledge. The practice of flamenco dancing has a body of knowledge. Mathematics is the example par excellence of a practice inseparable from its theory, its body of knowledge. Wittgenstein's claim that theory is just an adjunct secondary to calculation betrays embarrassing unfamiliarity with mathematics.

I will present my own example of Wittgenstein arithmetic.[2]

It's cousin to an example of Saul Kripke, with the merit of coming from real life. In this example,

$$12 + 1 = 14$$

and

$$12 + 2 = 15.$$

I explain.

I live on the twelfth floor. There's no thirteenth floor in my building. (13 is unlucky, so it's hard to rent apartments on 13.) Carlos is one flight above me, on 14.

Veronka took the elevator to Carlos on 14. Then, looking for me on 12, she walked down two flights, to arrive—at 11!

Some tenants say the floors from 14 up are misnumbered. "What's marked 14 is really 13."

Others say, "The name of a floor is whatever the people who live there call it." If it's marked 14, it is 14.

Veronka's experience shows that in this building,

$$12 + 1 = 14.$$

The management provides an addition table for the convenience of delivery boys. The "eights" row reads:

$$8 + 1 = 9 \qquad 8 + 2 = 10 \qquad 8 + 3 = 11$$
$$8 + 4 = 12 \qquad 8 + 5 = 14 \qquad 8 + 6 = 15 \text{ etc.}$$

This example proves Wittgenstein's theory. The standard addition table isn't the only one possible!

Some say this is a weird addition table.

Some say it isn't real addition. It's some other funny operation.

To a mathematician, there's no argument. We have many functions of two variables at our disposal. The one called "addition" or "+" is useful. At times

[2] After the fact, I found a connected example on page 83, Lecture VIII.

another may be more appropriate. It doesn't matter if we call the other function "alternative addition" or just a different function. The two peacefully coexist.

Imre Lakatos (1922–1973)

Hegel Comes In

If we want to separate the history of humanistic philosophy of mathematics from its prehistory, we could say that Aristotle, Locke, Hume, Mill, Peirce, and Wittgenstein are our prehistory. Our history starts not with Frege, but with Lakatos.

From the *London Times*, Wednesday, February 6, 1974:

> Professor Imre Lakatos died suddenly on February 3, at the age of 51. He was the foremost philosopher of mathematics in his generation, and a gifted and original philosopher of empirical science, and a forceful and colorful personality.
>
> He was born in Hungary on November 9, 1922. After a brilliant school and university career he graduated from Debrecen in Mathematics, Physics and Philosophy in 1944. Under the Nazi occupation he joined the underground resistance. He avoided capture, but his mother and grandmother, who had brought him up, were deported and perished in Auschwitz.
>
> After the war he became a research student at Budapest University. He was briefly associated with Lukacs. At this period he was a convinced communist. In 1947 he had the post of "Secretary" in the Ministry of Education, and was virtually in charge of the democratic reform of higher education in Hungary. He spent 1949 at Moscow University.
>
> His political prominence soon got him into trouble. He was arrested in the spring of 1950. He used to say afterward that two factors helped him to survive: his unwavering communist faith and his resolve not to fabricate evidence. (He also said, and one believes it, that the strain of interrogation proved too much—for one of his interrogators!)
>
> He was released late in 1952. He had no job and had been deprived of every material possession (with the exception of his watch, which was returned to him and which he wore until his death.) In 1954, the mathematician Rényi got him a job in the Mathematical Research Institute of the Hungarian Academy of Science translating mathematical works. One of them was Pólya's *How to Solve It*, which introduced him to the subject in which he later became preeminent, the logic of mathematical discovery. He now

> had access to a library containing books, not publicly available, by western thinkers, including Hayek and Popper. This opened his eyes to the possibility of an approach to social and political questions that was non-Marxist yet scientific. His communist certainties began to dissolve.
>
> After the Hungarian uprising he escaped to Vienna. On Victor Kraft's advice, and with the help of a Rockefeller fellowship, he went to Cambridge to study under Braithwaite and Smiley. . . .
>
> In 1958 he met Pólya, who put him on to the history of the "Descartes-Euler conjecture" for his doctorate. This grew into his *Proofs and Refutations* (1963–64), a brilliant imaginary dialogue that recapitulates the historical development. It is full of originality, wit, and scholarship. It founded a new, quasi-empiricist philosophy of mathematics.
>
> In England the man whose ideas came to attract him most was Professor (now Sir Karl) Popper, whom he joined at LSE in 1960. (There he rose rapidly, becoming Professor of Logic in 1969.). . .
>
> When he lectured, the room would be crowded, the atmosphere electric, and from time to time there would be a gale of laughter. He inspired a group of young scholars to do original research; he would often spend days with them on their manuscripts before publication. With his sharp tongue and strong opinions he sometimes seemed authoritarian; but he was "Imre" to everyone; and he invited searching criticism of his ideas and of his writings over which he took endless trouble before they were finally allowed to appear in print. . . .
>
> He was not without enemies; for he was a fighter and went for the things he believed in fearlessly and tirelessly. But he had friends all over the world who will be deeply shocked by his untimely death.

Foundationism, the attempt to establish a basis for mathematical indubitability, dominated the philosophy of mathematics in the twentieth century. Imre Lakatos offered a radically different alternative. It grew out of new trends in philosophy of science.

In science, the search for foundations leads to the classical problem of inductive logic. How can you derive general laws from a few experiments and observations? In 1934 Karl Popper revolutionized philosophy of science when he said that justifying inductive reasoning is neither possible nor necessary. Popper said scientific theories aren't derived inductively from facts. They're invented as hypotheses, speculations, guesses, then subjected to experimental test in which critics try to refute them. A theory is scientific, said Popper, if it's in principle capable of being tested, risking refutation. If a theory survives such testing, it gains credibility, and may be tentatively established; but it's never proved. Even if a scientific theory is objectively true, we can never know it to be so with certainty.

Popper's ideas are sometimes considered one-sided or incomplete, but his criticism of the inductivist dogma made a fundamental change in how people think about scientific knowledge.

While Popper and others transformed philosophy of science, philosophy of mathematics stagnated. This is the aftermath of the foundationist controversies of the early twentieth century. Formalism, intuitionism, and logicism each left its trace as a mathematical research program that made its contribution to mathematics. As *philosophical* programs, as attempts at a secure foundation for mathematical knowledge, they all ran their course, petered out, and dried up. Yet there remains a residue, an unstated consensus that philosophy of mathematics *is* foundations. If I find foundations uninteresting, I conclude I'm not interested in philosophy—and lose the chance to confront my uncertainties about the meaning or significance of mathematical work.

The introduction to *Proofs and Refutations* is a blistering attack on formalism, which Lakatos defines as the school "which tends to identify mathematics with its formal axiomatic abstraction and the philosophy of mathematics with metamathematics. Formalism disconnects the history of mathematics from the philosophy of mathematics. Formalism denies the status of mathematics to most of what has been commonly understood to be mathematics, and can say nothing about its growth.

"Under the present dominance of formalism, one is tempted to paraphrase Kant: the history of mathematics, lacking the guidance of philosophy, has become blind, while the philosophy of mathematics, turning its back on the most intriguing phenomena in the history of mathematics, has become empty. . . . The formalist philosophy of mathematics has very deep roots. It is the latest link in the long chain of dogmatist philosophies of mathematics. For more than 2,000 years there has been an argument between dogmatists and sceptics. In this great debate, mathematics has been the proud fortress of dogmatism. . . . A challenge is now overdue."

Lakatos doesn't make that overdue challenge. He writes,

"The core of this case-study will challenge mathematical formalism, but will not challenge directly the ultimate positions of mathematical dogmatism. Its modest aim is to elaborate the point that informal, quasi-empirical, mathematics does not grow through a monotonous increase of the number of indubitably established theorems, but through the incessant improvement of guesses by speculation and criticism, by the logic of proofs and refutations."

Instead of symbols and rules of combination, Lakatos presents human beings, a teacher and students. *Proofs and Refutations* is a classroom dialogue, continuing one in Pólya's *Induction and Analogy in Mathematics.*

Pólya considers various polyhedra: prisms, pyramids, double pyramids, roofed prisms, and so on. If the number of faces of a polyhedron is F, the number of vertices is V, and the number of edges is E, he shows that in "all" cases

$$V - E + F = 2 \text{ (the Descartes-Euler formula).}$$

Lakatos continues from this "inductive " introduction by Pólya. The teacher presents Cauchy's proof of Euler's formula:

Stretch the edges of your polyhedron onto a plane. Then simplify the resulting network, removing one edge or two edges at a time. At each removal, check that although V, E, and F are reduced, the expression $V - E + F$ is unchanged. After enough reductions, the network becomes a single triangle. In this simple case $V = 3$, $E = 3$, and $F = 2$. (One face inside the triangle, one face outside.) And

$$3 - 3 + 2 = 2.$$

This completes the "proof."

No sooner is the proof finished than the class produces a whole menagerie of counter-examples. The battle is on. What did the proof prove? What do we know in mathematics, and how do we know it? The discussion reaches ever greater sophistication, mathematical and logical. There are always several viewpoints in contest, and occasional about-faces when one student takes up a position abandoned by his antagonist.

Instead of a general system starting from first principles, Lakatos presents clashing views, arguments, and counter-arguments; instead of fossilized mathematics, mathematics growing unpredictably out of a problem and a conjecture. In the heat of debate and disagreement, a theory takes shape. Doubt gives way to certainty, then to new doubt.

In counterpoint with these dialectical fireworks, footnotes tell the genuine history of the Euler-Descartes conjecture in amazing complexity. The main text is in part a "rational reconstruction" of the actual history. Lakatos once said that the actual history is a parody of its rational reconstruction. *Proofs and Refutations* is overwhelming in its historical learning, complex argument, and self-conscious intellectual sophistication. Its polemical brilliance dazzles.

For fifteen years *Proofs and Refutations* was an underground classic, read by the venturesome few who browse in the *British Journal for Philosophy of Science*. In 1976, three years after Lakatos's death at age 51, it was published by Cambridge University Press.

Proofs and Refutations takes history as the text for its sermon: Mathematics, like natural science, is fallible. It too grows by criticism and correction of theories, which may always be subject to ambiguity, error, or oversight. Starting from a problem or a conjecture, there's a search for *both* proof and counter-examples.

Proof explains counter-examples, counter examples undermine proof. Proof isn't a mechanical procedure that carries an unbreakable chain of truth from assumption to conclusion. It's explanation, justification, and elaboration, which make the conjecture convincing, while the counter-examples make it detailed and accurate. Each step of the proof is subject to criticism, which may be mere skepticism or may be a counter-example to a particular argument. Lakatos calls a counter-example that challenges one step in the argument a "local counter-example"; one that violates the conclusion itself, he calls a "global counter-example."

Lakatos does epistemological analysis of *informal* mathematics, mathematics in process of growth and discovery, the mathematics known to mathematicians and mathematics students. Formalized mathematics, to which philosophy has been devoted, is hardly found on earth outside texts of symbolic logic.

Lakatos *argues* that dogmatic philosophies of mathematics (logicist or formalist) are unacceptable. He does not argue, but *shows* that a Popperian philosophy of mathematics is possible. But he doesn't discover an ontology to go with his fallibilist epistemology. In the main text of *Proofs and Refutations* we hear the author's puppets, not the author himself. He shows us mathematics as it's lived. But he doesn't tell the import of what he is showing us. Or rather, he states its import in the critical sense, in all-out tooth-and-nail attack on formalism. But what is its import in the positive sense? We need to know what mathematics is *about*. The Platonist says it's about objectively existing ideal entities, which a certain intellectual faculty lets us perceive, as eyesight lets us perceive physical objects. But few modern readers, certainly not Lakatos, are prepared to contemplate seriously the existence, objectively, timelessly, and spacelessly, of all the entities in set theory, let alone in future theories yet to be revealed.

The formalist says mathematics isn't about anything, it just *is*. A mathematical formula is just a formula. Our belief that it has content is an illusion. This position is tenable if you forget that informal mathematics *is* mathematics. Formalization is merely an abstract possibility, which one rarely wants or is able to carry out.

Lakatos thinks informal mathematics is science in the sense of Popper. It grows by successive criticism and refinement of theories and the introduction of new, competing theories—*not* the deductive pattern of formalized mathematics. But in natural science Popper's doctrine depends on the objective existence of the world of nature. Singular spatio-temporal statements such as "The voltmeter showed a reading of 2.6" provide the test whereby scientific theories are criticized and sometimes refuted. Popper calls these "basic statements" the "potential falsifiers." If informal mathematics is like natural science, it needs its objects. What are the data, the "basic statements," which provide potential falsifiers to theories in informal mathematics? This question is not even posed in *Proofs and Refutations*. Yet it's the main question if you want a fallibilist philosophy of mathematics.

After *Proofs and Refutations* Lakatos fought about philosophy of science with Rudolf Carnap, Karl Popper, Thomas Kuhn, Michael Polányi, Stephen Toulmin, and Paul Feyerabend. He never returned to philosophy of mathematics. His posthumous papers contain an article, "A renaissance of empiricism in the philosophy of mathematics?" with quotations from eminent mathematicians and logicians, both logicist and formalist, all agreeing that the search for foundations has been given up. The only reason to believe in mathematics is—it works! John von Neumann said modern mathematics is no worse than modern physics, which many people believe in. Having cut the ground under his opponents by showing that his "heretical" view isn't opposed to the mathematical establishment, Lakatos contrasts "Euclidean" theories like the traditional foundationist philosophies of mathematics, with "quasi-empiricist" theories that regard mathematics as conjectural and fallible. His own theory is quasi-empiricist, not empiricist *tout court*, because the potential falsifiers or basic statements of mathematics, unlike those of natural science, are not singular spatio-temporal statements (e.g., "the reading on the volt-meter was 6.2"). For formalized mathematical theories, he said, the potential falsifiers are informal theories. To decide whether to accept some axiom for set theory, for example, we would investigate how it conforms to the informal set theory we know. We may decide, as Lakatos acknowledges, not to fit the formal theory to the informal one, but instead to modify the informal theory. The choice may be complex and controversial. At this point, he's facing the main problem. What are the "objects" of *informal* mathematical theories?

When we talk about numbers or triangles apart from axioms or definitions, what kinds of entities are we talking about? There are many answers. Some go back to Aristotle and Plato. All have difficulties, and long attempts to evade the difficulties. The fallibilist position should lead to a critique of the old answers, perhaps to a new answer that would bring the philosophy of mathematics into the mainstream of contemporary philosophy of science. Lakatos doesn't commit himself. "The answer will scarcely be a monolithic one. Careful historico-critical case-studies will probably lead to a sophisticated and composite solution." A reasonable answer. But a disappointing one.

Except for a summary by Paul Ernest in *Mathematical Reviews*, I know no public response to *Proofs and Refutations* before its publication in 1976 by Cambridge University Press.

The response is in the book itself, in notes by editors Elie Zahar and John Worrall. On p. 138 they correct Lakatos's statement that to revise the infallibilist philosophy of mathematics "one had to give up the idea that our deductive inferential intuition is infallible." They write, "This passage seems to us mistaken and we have no doubt that Lakatos, who came to have the highest regard for formal deductive logic, would himself have changed it. First-order logic has arrived at a characterization of the validity of an inference which (relative to a characterization of the "logical" terms of a language) does make valid inference essentially infallible." The claim is repeated elsewhere. Lakatos "underplays a little the achievements of the mathematical 'rigorists'." "There is no serious sense in which such proofs are fallible." They correct him wherever he questions the achievement of a complete solution of the problem of mathematical rigor. A naive reader might conclude that today there's no doubt whether a rigorous mathematical proof is valid. A modern formal deductive proof is said to be infallible; doubt can come only from doubting the premises. If you think of a theorem as a conditional statement: "If the hypotheses are true, the conclusion is true," then in this conditional form the achievements of logic make it indubitable. Thus Lakatos's fallibilism is incorrect.

But Lakatos is right, Zahar and Worral wrong. It is remarkable that their objections repeat the very same error Lakatos attacked in his introduction—the error of identifying mathematics itself (what real mathematicians do in real life) with its model or representation, metamathematics, first-order logic.

Worrall and Zahar say a formal derivation in first-order logic isn't fallible in any serious sense. But such formal derivations don't exist, except for toy problems—homework exercises in logic courses.

On one side we have mathematics, with proofs established by "consensus of the qualified." Real proofs aren't checkable by machine, or by live mathematicians not privy to the mode of thinking of the appropriate field of mathematics. Even qualified readers may differ whether a real proof (one that's actually spoken or written down) is complete and correct. Such doubts are resolved by communication and explanation.

Once a well-known analyst at lunch told us how as a graduate student he was caught reading *Logic for Mathematicians*, by Paul Rosenbloom. His major professor ordered him to get rid of the book, saying, "You'll have time for that stuff when you're too old and tired for real mathematics." The group at lunch laughed. We weren't shocked. In fact, studying logic at that earlier time wouldn't have helped an analyst, and might have interfered with him. Today this is no longer true. Mathematical logic has produced tools that have been used by analyst or algebraist. But that has nothing to do with justifying proofs by rewriting them in first-order logic. Once a proof is accepted, its conclusion is regarded as true with high probability. To detect an error in a proof can take generations. If a theorem is widely used, if alternate proofs are found, if it is applied and generalized, if it's analogous to other accepted results, it can become "rock bottom." Arithmetic and Euclidean geometry are rock bottom.

In contrast to the role of proof in mainstream mathematics, there's "metamathematics" or "first-order logic." It's about mathematics in a certain sense, and at the same time it's part of mathematics. It lets us study mathematically the consequences of an imagined ability to construct infallible proofs. For example, the consequences of constructivist variations on classical deduction.

How does this metamathematical picture of mathematics affect our understanding and practice of real mathematics? Worrall and Zahar think the problem of fallibility in real proofs, which Lakatos is talking about, has been settled by the notion of infallible proof in metamathematics. (This term of Hilbert's is old-fashioned, but convenient.) How would they justify such a claim?

I guess they would say a real proof is merely an abbreviated or incomplete formal proof. This raises several difficulties. In mathematical practice we distinguish between a complete informal proof and an incomplete informal proof. In a complete informal proof, every step is convincing to the intended reader. In an incomplete informal proof, some step fails to convince. As *formal* proofs, *both* are incomplete. We considered this question in Chapter 4.

So what does it mean to say a real mathematical proof is an abbreviation of a formal one, since the same can be said whether an informal "proof" is correct or incorrect?

"Formalists," as Lakatos called people like Zahar and Worrall who "conflate" mathematics with its metathematical model, don't explain *in what sense* formal systems are a model of mathematics. Normatively, mathematics *should* be like a formal system? Or descriptively, mathematics *is* like a formal system?

If descriptive, then it has to be judged by how faithful it is to what it purports to describe. If normative, it has to explain why mathematics is prospering so well while paying so little heed to its norms.

Logicians today claim that their work is descriptive. Sol Feferman writes that the aim of a logical theory is "to model the reasoning of an idealized Platonistic or an idealized constructivistic mathematician." Comparison is sometimes made between logicians' use of formal systems to study mathematical reasoning and physicists' use of differential equations to study physical problems. But he points out that the analogy breaks down, since there's no analogue of the physicists' experimental method for the logicians to use in testing their model against experience.

"We have no such tests of logical theories. Rather, it is primarily a matter of individual judgment how well these square with ordinary experience. The accumulation of favorable judgment by many individuals is of course significant."

There is no infallibility.

Proofs and Refutations presents a picture of mathematics utterly at variance with the one presented by logic and metamathematics. It's clear which is truer to life. Feferman writes, "Clearly logic as it stands fails to give a direct account of either the historical growth of mathematics or the day-to-day experience of its practitioners. It is also clear that the search for ultimate foundations via formal systems has failed to arrive at any convincing conclusion."

Feferman has reservations about Lakatos's work. Lakatos's scheme of proofs and refutations is not adequate to explain the growth of all branches of mathematics. Other principles, such as the drive toward the unification of diverse topics, seem to provide the best explanation for the development of abstract group theory or of point-set topology. But Feferman acknowledges Lakatos's achievement. "Many of those who are interested in the practice, teaching and/or history of mathematics will respond with larger sympathy to Lakatos' program. It fits well with the increasingly critical and anti-authoritarian temper of these times. Personally, I have found much to agree with both in his general approach and in his detailed analysis."

Hungarian Style

In addition to Imre Lakatos, four other brilliant Hungarian Jews belong in our story. George Pólya and John von Neumann became Americans; Michael Polányi escaped to England. Only Alfréd Rényi lived and died in Budapest.

Pólya, like Wilder, had a second career after years of mathematical research. Pólya's late-life specialty was heuristics. How do we solve problems? How do we teach others to solve problems? (See the 4-cube in Chapter 1.)

In expounding the heuristic side of mathematics—what I called "the back" in Chapter 3—he was implicitly exposing the incompleteness of the formalist and logicist pictures of mathematics. Like Wilder and Gauss, Pólya disliked philosophical controversy. Perhaps, like Gauss, he was impatient with "Boeotians." (See non-Euclidean geometry.**) Pólya once said he became a mathematician because he wasn't good enough for physics, but too good for philosophy.

In the preface to ***Mathematics and Plausible Reasoning***, Pólya wrote: "Finished mathematics presented in a finished form appears as purely demonstrative, consisting of proofs only. Yet mathematics in the making resembles any other human knowledge in the making. You have to guess a mathematical theorem before you prove it; you have to have the idea of the proof before you carry through the details. You have to combine observations and follow analogies; you have to try and try again."

But a few pages later, "I do not know whether the contents of these four chapters deserve to be called philosophy. If this is philosophy, it is certainly a pretty low-brow kind of philosophy, more concerned with understanding concrete examples and the concrete behavior of people than with expounding generalities."

Polányi was a Hungarian-English chemist who in the fullness of his fame became philosopher of science. His opinions came out of the laboratory and lecture hall, out of collaboration and clash with other chemists. Like Lakatos and Pólya, he paid little mind to the disputations of philosophers of science. His books were read by scientists and the public, widely ignored by philosophers.

He contributed the notion of "tacit knowledge." We know more than we can say, and this tacit knowledge is essential to scientific discovery. This doctrine is indigestible to people who equate knowledge with written text. It clashes with the logicist and formalist pictures of mathematical knowledge as explicit formulas and sentences. It also contradicts Wittgenstein's much-quoted mot: "What we cannot speak about, we must pass over in silence."

I must quote these paragraphs of Polányi's (1962, pp. 187–88): "Even supposing mathematics were wholly consistent, the criterion of consistency, which the tautology doctrine is intended to support, would still be ludicrously inadequate for defining mathematics. One might as well regard a machine which goes on printing letters and typographical signs at random as producing the text of all future scientific discoveries, poems, laws, speeches, editorials, etc. for just as only a tiny fraction of true statements about matters of fact constitute science and only a tiny fraction of conceivable operational principles constitute technology, so also only a tiny fraction of statements believed to be consistent constitute mathematics. Mathematics cannot be properly defined without appeal to the principle which distinguishes this tiny fraction from the overwhelmingly predominant aggregate of other non-self-contradictory statements.

"We may try to supply this criterion by defining mathematics as the totality of theorems derived from certain axioms according to certain operations which will assure their self-consistency, provided the axioms themselves are mutually consistent. But this is still inadequate. First, because it leaves completely unaccounted for the choice of axioms, which hence must appear arbitrary—which it is not; second, because not all mathematics considered to be well established has ever been completely formalized according to strict procedure; and third—as K. R. Popper has pointed out—among the propositions that can be derived from some accepted set of axioms there are still, for every single one that represents a significant mathematical theorem, an infinite number that are trivial.

"All these difficulties are but consequences of our refusal to see that mathematics cannot be defined without acknowledging its most obvious feature: namely, that it is interesting. Nowhere is intellectual beauty so deeply felt and fastidiously appreciated in its various grades and qualities, as in mathematics, and only the informal appreciation of mathematical value can distinguish what is mathematics from a welter of formally similar, yet altogether trivial statements and operations."

Mathematics is created and pursued by people. It must have meaning or value to them or else they would not create or pursue it. Whatever mathematics we look at, out of the huge pile that humanity has created, one thing can always be said. It was interesting to someone, some time, and somewhere.

The next Hungarian on our list is Alfréd Rényi.

Rényi was the son of an engineer of wide learning and grandson of Bernat Alexander, a most influential professor of philosophy and aesthetics at Budapest. His uncle was the psychoanalyst, Franz Alexander. He went to a humanistic gymnasium, and maintained life-long interest in classical Greece. He had the rare ability to be equally at home in pure and applied mathematics.

His *Dialogues on Mathematics* are beautiful examinations of mathematical truth and meaning. They deal in profound and original ways with fundamental philosophical issues, yet their light touch and dramatic flair make them readable by anyone. There are dialogues with Socrates, Archimedes, and Galileo. "For Zeus's sake," asks Socrates, "is it not mysterious that one can know more about things which do not exist than about things which do exist?" Socrates not only asks this penetrating question, he answers it. It's astonishing that to this day (to my knowledge) no philosopher of mathematics has responded to Rényi's *Dialogues.*

Finally, among the great Hungarian Jewish mathematicians, we remember Neumann Janos—in the United States, John von Neumann (1903–1957). In breadth and power, the supreme mathematician of the 1930s, 1940s, and 1950s—along with perhaps Andrei Kolmogorov and Israel Moiseyevich Gel'fand. His contributions to games, economics, quantum mechanics, sets, lattices, linear operators, nuclear weapons, computers, numerical analysis, and fluid dynamics are legendary. His philosophical writings are sparse. In youth he worked on Hilbert's formalism. Later he was philosophically uncommitted. I quote his essay "The Mathematician" in the section on Hilbert in Chapter 8.

One can ask, what did these Hungarians have in common, besides Budapest? Widely as their specialties and their politics differed, there is a "Hungarian style." It's hard to describe, but not hard to recognize. Lively affection for the concrete, specific, and human. Quiet avoidance of blown-up pretension, vapid generality, words for words' sake. Intellectual and cultural breadth, encompassing history, literature and philosophy. Understanding that mathematics flowers as part of human culture.

Leslie White/Raymond Wilder

Anthropology Comes In

Wilder was a topologist at Ann Arbor. White was an anthropologist there. They became friends. In his later years, retired at Santa Barbara, Wilder had a second career as philosopher/historian of mathematics. He credited White for the insight that mathematics is a cultural phenomenon, which can be studied with the methods of anthropology. Wilder knew this viewpoint was opposed to Platonism, formalism, and intuitionism. But in his writings he tried to avoid philosophical controversy. Like Gauss who suppressed his discovery of non-Euclidean geometry, he usually preferred not to stir up the "Boeotians" (Aristophanes's name for ignorant hicks).

White's essay is a beautiful statement of the locus of mathematical reality. Its weakness is failure to confront the uniqueness of mathematics—what makes mathematics different from other social phenomena. He wrote: "We can see how the belief that mathematical truths and realities lie outside the human mind arose and flourished. They *do* lie outside the mind of each individual organism. They enter the individual mind as Durkheim says from the outside. . . . Mathematics is not something that is secreted, like bile; it is something drunk, like wine. . . . Heinrich Hertz, the discoverer of wireless waves, once said: 'One cannot escape the feeling that these mathematical formulas have an independent existence and an intelligence of their own, that they are wiser than we are, wiser even than their discoverers, that we get more out of them than was originally put into them.'. . . The concept of culture clarifies the entire situation. Mathematical formulas, like other aspects of culture, do have in a sense 'independent existence and intelligence of their own.' The English language has, in a sense, 'an independent existence of its own.' Not independent of the human species, of course, but independent of any individual or group of individuals, race or nation. It has, in a sense, an 'intelligence of its own.' That is, it behaves, grows and changes in accordance with principles with are inherent in the language itself, not in the human mind. . . . Mathematical concepts are independent of the individual mind but lie wholly within the mind of the species, i.e., culture."

twelve

Contemporary Humanists and Mavericks

Sellars, Popper, Medawar, Leavis, Bunge

The autonomy of the social/cultural/historic isn't a new idea. A school of "critical realism" advocated "levels of reality" and "emergent evolution" 70 and 80 years ago. (The father of the analytic philosopher Wilfred Sellars was the critical realist Roy Sellars.) In the 1930s emergent evolution was buried under analytic philosophy and logical empiricism, and almost forgotten. But it is re-emerging. In 1992 David Blitz wrote, "Despite its temporary eclipse by reductionist and physicalist philosophies in the period from the mid-1930s to the mid-1950's, emergent evolution is an active trend of thought at the interface between philosophy and science." I sketch the ideas of some recent authors.

In Chapter 1 I mentioned Karl Popper's role in dethroning the doctrine of inductive reasoning in empirical science. More recently he introduced a notion of "World 3"—a world of scientific and artistic knowledge, distinct from the physical world (World 1) and the world of mind or thought (World 2). In "Epistemology without a knowing subject" and "On the theory of the Objective Mind" he uses "World 3" to mean a world of intelligibles: ideas in the objective sense, possible objects of thought, theories and their logical relations, arguments, and problem situations (1974, p. 154). It seems Popper wants to put a Platonic world of Ideal Forms alongside the mental and physical worlds.

Peter Medawar was impressed by Popper's World 3. "Popper's new ontology does away with subjectivism in the world of the mind. Human beings, he says, inhabit or interact with three quite distinct worlds; World 1 is the ordinary physical world, or world of physical states; World 2 is the mental world; World 3 is the world of actual or possible objects of thought—the world of concepts, ideas, theories, theorems, arguments, and explanations—the world, let us say, of all artifacts of the mind. The elements of this world interact with each other much like the ordinary objects of the material world; two theories interact and lead to the formulation of a third; Wagner's music influences Strauss's and his in turn all music written since. . . . The existence of World 3, inseparably bound up with human language, is the most distinctly human of all our possessions. The third world is not a fiction, Popper insists, but exists 'in reality.' It is a product of the human mind but yet is in large measure autonomous. This was the conception I had been looking for: The third world is the greater and more important part of human inheritance. It's handing on from generation to generation is what above all else distinguishes man from beast."

But the difficulty of explaining the interaction between these worlds is fatal for Popper, as for his predecessors.

F. R. Leavis in his book *Nor Shall My Sword* calls the physical world "public," the mental world "private." Then he talks about "objects of the third kind," which are neither wholly public not wholly private. This is close to my meaning.

Mario Bunge, a prolific philosopher of physics, attacked Popper's World 3 in *The Mind-Body Problem* (pp. 169–73)—but then developed his own reformulation of it.

Bunge is a materialist. With respect to mathematics, he's a fictionalist. He thinks that claims for autonomy of mind interfere with the scientific study of mind as brain activity.

He writes: Popper's "autonomy of the 'world' of the creations of the human mind is supposed to be relative to the latter as well as to the physical world. Thus the laws of logic are neither psychological nor physiological (granted)—and, moreover, they are supposed to be 'objective' in some unspecified sense. The idea is probably that the formulas of logic (and mathematics), once guessed and proved, hold, come what may. However, this does not prove that conceptual (e.g. mathematical) objects, or any other members of 'world 3,' lead autonomous existences. It only shows that, since they do not represent the real world, their truth does not depend upon it, and, therefore, we can *feign* that they are autonomous objects. However, this is not what Popper claims: he assigns his 'world 3' reality and causal efficacy. . . . We pretend that there are infinitely many integers even though we can think of only finitely many of them—and this because we assign the infinite set of all integers definite properties, such as that of being included in the set of rational numbers. Likewise we make believe that every deductive system contains infinitely many theorems—and this because, if pressed, we can prove any of them. All such fictions are mental creations and, far from being idle, or merely entertaining, are an essential part of modern culture. But it would be sheer animism to endow such fictions with real autonomous existence and causal efficacy. . . . In short, ideas in themselves are fictions and as such have no physical existence; only the brain processes of thinking them up are real. . . ."

"But what about the force of ideas? Do not ideas move mountains? Has it not been in the names of ideas, and particularly ideals, that entire societies have been built or destroyed, nations subjugated or liberated, families formed or wiped out, individuals exalted or crushed? Surely even the crassest of materialists must recognize the power of ideas, particularly those our grandparents used to call idées-forces, such as those of fatherland and freedom. Answer: Certainly, ideation—a brain process—can be powerful. Moreover, ideation is powerful when it consists in imagining and planning courses of action engaging the cooperation of vast masses of individuals. But ideas in themselves, being fictions, are impotent. The power of ideation stems from its materiality, not from its ideality."

This isn't right. The power of ideas stems from their meaning, not their materiality. Someone shouts "Liberté, égalité, fraternité!" That shout inspires someone else to pull a cobblestone from the Rue de la Paix and throw it at a *gendarme*. It wasn't the air vibrating from the shout that budged the cobblestone. It was the *meaning* of the slogan that moved the stone. For that meaning activated a mind-brain that ordered hands to pull up a stone and hurl it. We can't explain the Revolution with neurons and hormones. We explain it with economics, politics, and history.

Yes, a slogan has effect only by actions of people. Yes, people have an effect only by physical action, whether hurling a cobblestone or whispering a word. But the power of the word is its meaning—not its decibels.

Later (p. 214) Bunge has second thoughts: "We have been preaching the reduction of psychology to neurophysiology, but have also warned that such reduction can only be partial or weak, and this for two reasons. One reason is that psychology contains certain concepts and statements that are not to be found in today's neuroscience. Consequently neuroscience must be enriched with some such constructs if it is to yield the known psychological regularities and, a fortiori, the new ones we would like to know. The second reason for the incomplete reducibility of psychology to neurophysiology is that neuroscience does not handle sociological variables, which are essential to account for the behavior and mentation of the social higher vertebrates. For these reasons the reductionistic effort should be supplemented by an integrative one. Let me explain. Behavior and mentation are activities of systems that cross a number of real (not just cognitive) levels, from the physical level to the societal one. Hence they cannot be handled by any one-level science. Whenever the object of study is a multilevel system, only a multi disciplinary approach—one covering all of the intervening levels—holds promise. In such cases pig-headed reductionism is bound to fail for insisting on ab initio procedures that cannot be implemented for want of the necessary cross-level assumptions. (Take into account that it has not been possible to write down, let alone solve, the Schrodinger equation for a biomolecule, let alone for a neuron, even less for a neuronal system.) In such cases, pressing for reduction is quixotic; it is not a fruitful research strategy. In such cases only the opportunistic (or catch-as-catch-can) strategy suggested by systemism and a multilevel world view can bring success, for it is the one that integrates the physical, the chemical, the biological and the sociological approaches, and the one that builds bridges among them. . . (p. 216). Our rejecting psychophysical dualism does not force us to adopt eliminative or vulgar materialism in either of its versions—i.e. the theses that mind and brain are identical, that there is no mind, or that the capacities for perception, imagination and reasoning are inherent in all animals or even in all things. Psychobiology suggests not just psychoneural monism but also emergentism, i.e., the thesis that mentality is an emergent property possessed only by animals endowed with an extremely complex and plastic nervous system. . . . In short, minds do not constitute a supra-organic level, because they form no level at all. But psychosystems do. To repeat the same idea in different words: one can hold that the mental is emergent relative to the merely physical without reifying the former. . . . And so emergentist (or systemic) materialism—unlike eliminative materialism—is seen to be compatible with overall pluralism, or the world view that proclaims the qualitative variety."

What Bunge taketh with one hand, he restoreth with the other. Minds aren't real, but "psychophysical systems" are. Out of habit, we will probably go on calling psychophysical systems "minds."

Still, there is a difference between Bunge's emergent monism and the mind-body dualisms he refutes.

Bunge and others had an easy time attacking Popper. Such attacks would be off the mark with reference to my level 3. I don't claim it's independent of mind and body. On the contrary, a socio-cultural-historical object exists only in some representation, whether physical (books, computer "memories," musical scores and recordings, photographs, drawings) or mental (knowledge or consciousness of people) or both.

There's no thinking without a brain. But the mental is autonomous in the sense that the evolution and interaction of minds has to be understood in terms of thoughts, emotions, habits, desires, and so forth, not just chemistry and electricity. In the same way that the mental exists through the physical, so the social-cultural exists through both the mental and physical.

N.B. In response, Professor Bunge explains that he does not say minds don't exist, only that they don't exist autonomously.

Philip Kitcher

Two important contributions to the humanist philosophy of mathematics are Kitcher's *Nature of Mathematical Knowledge* (NMK) and his edited book with William Aspray, *History and Philosophy of Modern Mathematics* (HP).

The introduction to HP explains that in both the philosophy of mathematics and the history of mathematics, an old tradition persists while a new trend challenges it. The philosophy of mathematics, we're told, was created by Gottlob Frege. (Earlier thinkers about the nature of mathematics were prehistoric.) After Frege came Whitehead and Russell, Hilbert, Brouwer, Wittgenstein, Gödel, the

Vienna Circle, Quine, and so on. In the 1950s philosophy of mathematics arrived at "neo-Fregeanism." This is an orthodoxy that decrees that mathematics is all about sets. For a mathematical statement to be true means it corresponds to the state of affairs in some set. Aspray and Kitcher call their own challenge to neo-Fregean philosophy "maverick."

Kitcher's book, *The Nature of Mathematical Knowledge,* starts with a painstaking critique of "a priorist" epistemologies. Traditional philosophy of mathematics wants to justify mathematical knowledge. To formalism, the justification for a theory is that the theorems follow from the axioms. Neither axioms nor theorems have truth value beyond their logical connections.

To intuitionism, elementary arithmetic is given directly to the intuition; other mathematics gets legitimacy from its connection to arithmetic. Kitcher makes an unsparing critique of both formalism and intuitionism.

His own viewpoint blends empiricism and evolutionism. All mathematical knowledge comes by a rationally explicable process of growth, starting from a basic core, the arithmetic of small numbers and the directly visual properties of simple plane figures. The basic core is experienced in physical acts of collecting, ordering, matching, and counting. The formula

$$2 + 2 = 4$$

can be proved as a theorem in a formal axiomatic system, but it derives its force and conviction from its physical model of collecting coins or pebbles.

Arithmetic is an idealized theory of matching and counting; set theory is an idealized theory of forming collections. An upper bound can be given for all the collections that will ever actually be formed by humans—but in mathematics we allow arbitrarily large sets and numbers, even infinite sets and numbers. This doesn't destroy the empirical nature of arithmetic, any more than the ideal gas hypothesis destroys the empirical character of gas dynamics.

To the question, what is mathematics about? Kitcher answers, it's about collecting. This is a constructivized re-interpretation of the idea that mathematics is about sets—what Kitcher calls neo-Fregeanism. But Kitcher is more generous than the constructivists in what his idealized mathematician can do. For instance, he can collect uncountably many elements into his collections.

Mathematics is a lawful, comprehensible evolution from a basic core. It develops in response to internal strain (here a definition would help) and external pressure. The mathematical knowledge of one generation is rooted in that of its parent's generation. The mathematics of the research journals is validated by the mathematical community through criticism and refereeing. Because most mathematical papers use reasoning too long to survey at a glance, acceptance is tentative. We reconsider our claim if it's disputed by a competent skeptic. Thus, our belief is *not* prior to all experience: it's conditioned on our *social* experience!

This is one of Kitcher's major points. The mathematician described by philosophy of mathematics must be, like the flesh-and-blood mathematician, a social being, not a self-sustaining isolate. This insight suffices to discredit the claim that mathematical knowledge is a priori, for communication with other humans is a precondition to mathematical knowledge.

The classical philosophical account of advanced mathematics relies on the possibility of reducing it all to set theory. Everything in partial differential equations or ergodic theory can be thought of as a set, so the only existence questions one need consider are about existence of sets. This reduction belies the perception of the working mathematician, who sees his own subject as central and autonomous, and set theory as peripheral or irrelevant. Kitcher is with the mathematicians, not the foundationists (philosophical logicians and set theorists). To the question, "How do we acquire knowledge of mathematics?" Kitcher gives the truthful answer: "We learn it at school."

This leads to the next question, "How does mathematical knowledge increase?" Answering this is the historian's job, and Kitcher takes on the historian's responsibility as well as the philosopher's. In a long account of mathematical analysis in the eighteenth and nineteenth centuries, he gives historical reasons for the establishment of real numbers as a foundation for calculus. This account is history at the service of philosophy, concretely answering the question, "What is mathematics?"

Kitcher adopts an important principle of Thomas Kuhn: scientific change means change in practice, not just in theory. He identifies *five components of mathematical practice*: language, metamathematical views, accepted questions, accepted statements, accepted reasonings. The five must be compatible. If one changes, the others must change accordingly.

He identifies *five rational principles* according to which mathematics develops: rigorization, generalization, question-generalization, question-answering, and systematization. These principles yield "rational interpractice transitions." "When these occur in sequence, the mathematical practice may be dramatically changed through a series of rational steps." The validity of today's mathematics comes from its connection, rational step by rational step, to the empirically valid basic mathematics of counting and collecting.

This explanatory scheme is powerful and convincing. It should be a lasting contribution to the historical analysis of mathematics.

Jean Piaget (1896–1982) & Lev Vygotsky (1896–1934)

Piaget was a psychologist, not a philosopher, but he has a respectful entry in the *Encyclopedia of Philosophy.* His painstaking observation of the growth of abstract thinking in children revolutionized cognitive psychology in the decades after the Second World War. In place of statistics on running rats, he used understanding, insight, and open-mindedness in observing children. His books report children's thinking about physics, logic, number, geometry, space, time, and chance, among other things.

Piaget's most popular idea was "stages": The brain can't absorb certain concepts until it attains the right stage of maturation. This idea was harmful to education. "Children in a classroom must be at the same maturational stage. It's useless to try teaching something to a child who hasn't reached the right stage." This idea wasn't supported by later research.

With a Dutch logician, Beth, Piaget wrote a book on the logico-psychological foundations of mathematics. Beth took it for granted that the foundation of mathematics is logic and set theory, so the book is now outdated.

Piaget and the Bourbakiste Jean Dieudonné had an intellectual encounter that gratified them both. Bourbaki had decided that mathematics is built from three fundamental structures: (1) order (in the sense of putting things one after another); (2) algebra (sets with operations, such as addition, multiplication, reflection, etc.); and (3) topology (neighborhoods, closeness). Piaget had independently decided that children's mathematical ideas are built from the same three elements. Naturally, they were happy to encounter each other. This Bourbaki-Piaget philosophy was called "structuralism." Mathematics was a collection of structures built from the three basic structures.

Bourbakisme and structuralism went out of fashion with change of styles. Later Piaget took up category theory. His book *Morphisms and Categories* has a critical introduction by the computer scientist Seymour Papert, inventor of LOGO, who was Piaget's pupil. The "structuralism" now being advanced by Shapiro and Resnik is not the same thing.

From my point of view, what's interesting in Piaget isn't his theory of stages or his encounters with Beth and Dieudonné. It's his epistemology. This has largely been ignored by mainstream philosophy. He presented his epistemology in several books. We look at *Genetic Epistemology*.

Piaget observed that children learn actively: picking up, putting down, moving, using. Presumably the bodily movements associated with two-ness (for example) leave a trace in the mind/brain. That trace would *be* the child's concept of two. Generalizing this idea, Piaget proposed that the concept of counting and natural number are acquired through activity. Real physical activity—not just observation or talk. The child picks up and puts down, manipulates, handles buttons, coins, and pebbles. By these activities the fundamental properties of discrete, permanent objects attain a representation in the child's mind/brain.

This may sound like Aristotle's explanation of number and shape as abstractions. But Aristotle's abstraction was passive observation. We see three apples, three pebbles, three coins, and we abstract "three." We notice that two beans added to two beans makes four beans. But we can't make these observations until we have learned that number is an interesting property. A cat sees two beans added to two beans, but it doesn't discover that 2 + 2 = 4. A cat doesn't handle, play with, pick up, and put down the beans or the beads. A child does.

From counting, go to motion. As the concept of discreteness comes from playing with beans and beads, the concepts of three-dimensional motion and space come from moving our three-dimensional bodies in three-dimensional space. Raise your arm, turn your head, lower your arm. By doing so, you learn the structure of the three-dimensional continuum. Without that knowledge, you couldn't find your way to breakfast.

This viewpoint is an escape from the Platonist conception of number and space as abstract objects, independent of both mental and physical reality.

If our concepts of number and space are the mental effects of childhood activity, what about more developed mathematics? P-adic analysis, measurable cardinals, square-integrable martingales, the well-ordering theorem, the continuum hypothesis? We can escape from the formalist conception, that each of these entities is nothing more than a formal definition in the context of a certain formal theory. We can recognize the crucial fact that the concept comes before the formal definition. The concept of an abstract group, for example, is the mental effect of calculations and reasoning and mental struggles with concrete groups. The trace or the effect of this activity and struggle on our mind-brain is experienced as an intuitive concept. With effort the mathematician may formalize it, interacting with the history of the subject and the thinking and writing of colleagues and competitors.

Piaget's epistemology is close to Marx's. Marx said our knowledge of natural science comes from interaction with Nature in productive labor. This was part of the doctrine that all classes but the Proletariat could be dispensed with. This Marxism is as dated as Bourbakisme. In rooting mathematical concepts in physical activity, Piaget unlike Marx had no ideological ax to grind.

Kitcher's *The Nature of Mathematical Knowledge*, Ernest's *The Philosophy of Mathematics Education* and Castonguay are among the few books where Piaget receives his due as a philosopher of mathematics.

Among cognitive psychologists today there's increasing interest in Lev Vygotsky, a Russian psychologist active in the 1920s and early 1930s. Vygotsky didn't write about mathematics, but his theory of learning is relevant to mathematics, and several psychologists are developing Vygotskian theories of mathematics education.

Vygotsky insisted that learning and intellectual activity are fundamentally social, not individual. The content of learning and thinking comes from social structures, and is assimilated by the individual learner.

This opinion wouldn't surprise an anthropologist or a sociologist, but psychologists, including Piaget, usually take it for granted that their task is to study the individual mind, detached from other people or from society. Vygotsky was a Marxist, but his intellectual daring and his deep love for western literature made him suspect in Stalin's Russia. (He wrote a book about Shakespeare's *Hamlet*, with a startling new interpretation.) He died of tuberculosis in his 30s, before the worst of the purges and persecutions. His books were banned in the Soviet Union for 30 years.

Paul Ernest

In the young movement to rebuild philosophy of mathematics as part of social reality, Ernest's *Philosophy of Mathematics Education* is one of the most comprehensive and comprehensible. It consists of an introduction and two parts—philosophy of mathematics is Part 1, philosophy of mathematics education is Part 2. Ernest starts with a critique of absolutist philosophies of mathematics. Absolutist here means philosophies that say mathematical truth is or should be absolute (free of all possible manner of doubt). This includes the three standard philosophies Imre Lakatos collected under the label "foundationalists": logicists, formalists, intuitionists. Ernest's critique is in the name of "fallibilism," a term sometimes joined to the name of Lakatos. It simply means the denial of absolutism. Next comes a chapter "reconceptualizing" the philosophy of mathematics, and then three chapters expounding "social constructivism."

This phrase is popular in "behavioral science" and "humanities." Ernest is familiar with and respects social constructivism in sociology and psychology. He provides generous summaries of Piaget and Vygotsky in constructivist psychology, Bloor and Restivo in constructivist sociology. Ernest seems to be the first to speak of social constructivism in philosophy of mathematics.

With each theory he describes he presents both the positive assertions of the theory and the objections that could be made against it. This includes the "conventionalism" of Wittgenstein and the "quasi-empiricism" of Lakatos, which he names as his foundation stones. Of course, it's one thing to state objections, another to deal with them. Once in a while Ernest dismisses with a sentence or two difficulties that deserve many chapters.

From his introductory summary:

"Social constructivism views mathematics as a social construction. It draws on conventionalism [read "Wittgenstein"], in accepting that human language, rules and agreement play a key role in establishing and justifying the truths of mathematics. It takes from quasi-empiricism [read "Lakatos"] its fallibilist epistemology, including the view that mathematical knowledge and concepts develop and change. It also adopts Lakatos' philosophical thesis that mathematical knowledge grows through conjectures and refutations, utilizing a logic of mathematical discovery. . . . A central focus of social constructivism is the genesis of mathematical knowledge, rather than just its justification . . . subjective and objective knowledge of mathematics each contributes to the creation and recreation of the other."

On p. 83, "In summary, the social constructivist thesis is that objective knowledge of mathematics exists in and through the social world of human action, interactions and rules, supported by individuals' subjective knowledge of mathematics (and language and social life), which need constant re-creation. Thus subjective knowledge recreates objective knowledge, without the latter being reducible to the former." Parts of the second half of the book will seem exotic to readers not acquainted with the English school system. Here Ernest classifies philosophies of education by their practice and practical effect as well as their theory. A remarkable table displays five educational ideologies: "industrial trainer," "technological pragmatist," "old humanist," "progressive educator," and "public educator" (in order from worst to best). Listed under each is its political ideology, view of mathematics, moral values, mathematical aims, theory of teaching mathematics, and theories of learning, ability, society, the child, resources, assessment in mathematics, and social diversity.

Of course the construction of the table is partly impressionistic and anecdotal, but it sounds right. Just for a taste, here are the views of maths and of society, listed below each teaching ideology.

Industrial Trainer	*Technological Pragmatist*	*Old Humanist*
Set of truths and rules	Unquestioned body of useful knowledge	Body of structured pure knowledge
Rigid hierarchy Market place	Meritocratic Hierarchy	Elitist class-stratified

Progressive Educator	*Public Educator*
Process view Personalized maths	Social constructivism
Soft hierarchy Welfare state	Inequitable hierarchy Needing reform

The second part of the book advocates the ideology of the public educator. This is a descendant of Deweyan progressivism, politicized and radicalized.

Ernest's forthcoming *Social Constructivism as a Philosophy of Mathematics* (Albany, SUNY) is close to the present book in point of view.

Ethnomathematics

Marcia Ascher has written an instructive book with this title.

Mathematical ideas, like artistic ideas or religious ideas, are a universal part of human culture. This forthright claim isn't made by Ascher, but her book compels me to that conclusion. Mathematics as we know it was invented by the

Greeks. But mathematical ideas involving number and space, probability and logic, even graph theory and group theory—these are present in preliterate societies in North and South America, Africa, the South Pacific, and doubtless many other places if anyone bothers to look.

This is not to say that everybody can do mathematics, any more than everybody can play an instrument or succeed in politics. Many people do not have mathematical or musical or political ability. But every society has its music and its politics; so too, it seems, every society has its mathematics.

Some people count by tens, others by twenties. "There is an oft-repeated idea that numerals involving cycles based on ten are somehow more logical because of human fingers. The Yuki of California are said to believe that their cycles based on eight are most appropriate for *exactly* the same reason. The Yuki, however, are referring to the interfinger spaces." And how about Toba, a language of western South America, in which "the word with value five implies (two plus three), six implies (two times three), and seven implies (two times three) plus one. Then eight implies (two times four), nine implies (two times four) plus one, and ten is (two times four) plus two."

Professor Ascher knows of three cultures that trace patterns in sand—the Bushoong in Zaire, the Tshokwe in Zaire and Angola, and the Malekula in Vanuatu (islands between Fiji and Australia formerly called the New Hebrides). Sand drawings play a different role in each culture. "Among the Malekula, passage to the Land of the Dead is dependent on figures traced in the sand. Generally the entrance is guarded by a ghost or spider-related ogre who is seated on a rock and challenges those trying to enter. There is a figure in the sand in front of the guardian and, as the ghost of the newly dead person approaches, the guardian erases half the figure. The challenge is to complete the figure which should have been learned during life, and failure results in being eaten. . . . The tales emphasize the need to know one's figures *properly* and demonstrate their cultural importance by involving them in the most fundamental of questions—mortality and (survival) beyond death. The figures vary in complexity from simple closed curves to having more than one hundred vertices, some with degrees of 10 or 12."

In all three cultures, there's special concern for Eulerian paths—paths that can be traced through every vertex without tracing any edge more than once. (The seven bridges of Königsberg!) All three seem to know that an Eulerian path is possible if and only if there are zero or two vertices of odd degree.

The Maori of New Zealand play a game of skill called mu torere. The game is played by two players; the "board" is an eight-pointed star. Each player has four markers—pebbles, or bits of broken china. Prof. Ascher shows that with any number of points except eight, the game would be uninteresting. "Mu torere, with four markers per player on an eight-pointed star, is the most enjoyable version of the game."

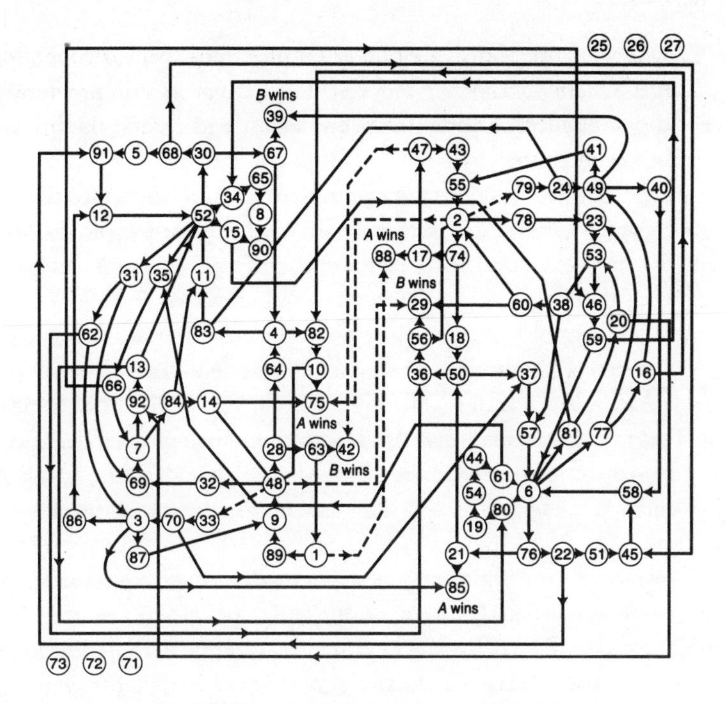

Figure 1. The flow of the game of mu torere.

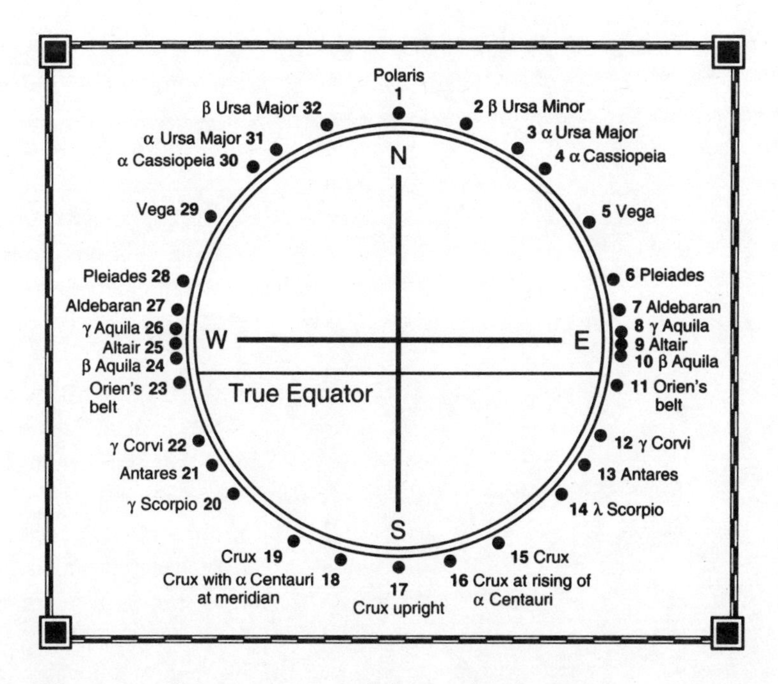

Figure 2. The start compass of the Caroline navigators.

The Caroline Islanders north of New Guinea cross hundreds of miles of empty ocean to Guam or Saipan. "The Caroline navigators do not use any navigational equipment such as our rulers, compasses, and charts; they travel only with what they carry in their minds."

Professor Ascher reminds us that leading anthropologists once taught that preliterate peoples were at an early stage of evolution. (Western society was advanced.) Later it was said that preliterate peoples ("savages") had an utterly different way of thinking from us. They were prelogical. We were logical.

Nowadays anthropologists say there's no objective way to rank societies as more or less advanced, higher or lower. Each is uniquely itself.

Professor Ascher's research is related to ethnomathematics as an educational program. This movement asks schools to respect and use the mathematical skills pupils bring with them—even if they differ from what's taught in school. By increasing understanding and respect for ethnomathematics, this work may benefit education.

There's a lesson for the philosophy of mathematics. Mathematics as an abstract deductive system is associated with our culture. But people created mathematical ideas long before there were abstract deductive systems. Perhaps mathematical ideas will be here after abstract deductive systems have had their day and passed on.

Summary and Recapitulation

thirteen

Mathematics Is a Form of Life

Self-graded Report Card Could Be Worse

Chapter 2 considered a list of criteria for a philosophy of mathematics. How do we look according to our own tests? I reprint the list:

1. Breadth
2. Connected with epistemology and philosophy of science
3. Valid against practice: research, applications, teaching, history, computing, intuition
4. Elegance
5. Economy
6. Comprehensibility
7. Precision
8. Simplicity
9. Consistency
10. Originality/novelty
11. Certitude/indubitability
12. Acceptability

Take them in order.

1. *Breadth.* Philosophy of mathematics should try to take into account all major parts and aspects of mathematics. The neo-Fregean dogma that set theory alone has philosophical interest, is an unacceptable excuse for ignorance. This book reflects an inside view of mathematical life. It's based on 20 years doing research on partial differential equations, stochastic processes, linear operators, and nonstandard analysis, 35 years teaching graduates and undergraduates, and many long hours listening, talking to, and reading philosophers.

2. *Epistemology/philosophy of Science.* Your philosophy of mathematics must fit your theory of knowledge and your philosophy of science. It just doesn't work, to be a Platonist in mathematics and a materialist empiricist in physical science.

Today Platonism is disconnected from *any* epistemology or philosophy of science. In Berkeley and Leibniz it was connected by their over-arching idealism. For them, mathematical knowledge was an aspect of spiritual knowledge, knowledge in the mind of God. Contemporary Platonists just ignore the question of how philosophy of mathematics connects with the rest of philosophy.

Empiricists like Mill and pragmatists like Quine do want to relate philosophy of mathematics to philosophy of science. But they do so without understanding the special character of mathematics. By recognizing mathematics as the study of certain social-cultural-historic objects, humanism connects philosophy of mathematics to the rest of philosophy.

3. *Valid against Practice*

a. *Research.* Taking seriously the actual experience of mathematical research is the main distinguishing feature of this book.
b. *Applications.* From Pythagoras to Russell and beyond, philosophy of mathematics rarely paid serious attention to applied mathematics. (Stephen Körner is one exception.) The present book recognizes the interlinking of the pure and applied viewpoints. The picture of mathematics as a social-cultural-historical phenomenon naturally includes applied mathematics.
c. *Teaching.* The philosophy of mathematics must let it be comprehensible that mathematics is actually taught. This issue is discussed by philosophically concerned educationists, not by philosophers. Tymoczko was an exception. Logicism, formalism, and intuitionism, each in a different way, make the possibility of teaching mathematics a deep mystery. Seeing mathematics as a cultural-social-historical entity makes it obvious that it's taught and learned. I elaborate on this in the section below titled "Teaching."
d. *History.* Like Lakatos and Kitcher, I explicitly incorporate history into the philosophy of mathematics, by identifying mathematics as the study of certain social-historic-cultural objects.
e. *Computing.* Logicist and formalist writers base their theories on an idealized, infallible, nonhuman, nonmaterial notion of computing. Intuitionists see computing as a purely mental activity. I discuss how mathematics is affected by real computing, an activity of real people and real machines. The section on "Proof" in Chapter 4 has more detail on this.
f. *Intuition.* See the section "Intuition" in Chapter 4.

4. *Elegance.* My theory is neither clean, complete, nor self-contained. I have put aside elegance in favor of other criteria.

5. *Economy.* Related to elegance. It means using the smallest number of basic concepts. I fare poorly, for the same reasons as in regard to point 4, elegance.

6. *Comprehensibility.* Some would say this criterion is not philosophical but merely literary. I strive to be clear both conceptually and verbally.

7. *Precision. Mathematics* is precise. *Philosophy* isn't. Mathematicians often mistakenly expect philosophy of mathematics to be part of mathematics. It isn't. It's part of philosophy. If it were part of mathematics, it could be precise. But it's no more necessary or possible for philosophy of mathematics to be precise than for any other branch of philosophy to be precise. (Gian-Carlo Rota brilliantly scores this point in *Indiscreet Thoughts.*) Frege, Russell, Hilbert, Brouwer, and Bishop, in trying to create mathematical foundations for mathematics, actually did contribute to mathematics and logic. In this book I have no intention to do mathematics.

8. *Simplicity.* Related to economy and elegance. Monism, whether materialist or idealist, is simple. Humanism, recognizing three kinds of existence, isn't so simple. If simplicity conflicts with truthfulness, adequacy, or accuracy, it should take second place.

9. *Consistency.* Most philosophers make consistency the chief desideratum, but in mathematics it's a secondary issue. Usually we can patch things up to be consistent. We can't so easily patch them up to be comprehensive or true to life. It's believed that if a set of statements is true, it must be consistent. In that spirit, I claim that what I offer here is consistent.

10. *Originality.* Pólya and Lakatos and White had a big influence on me. Professor Hao Wang of Rockefeller University corrected blunders and gave me courage to persist. Kitcher, Tymoczko, and Ernest wrote related ideas. Everyone in the Acknowledgments made an impact on this book. Nevertheless, this is the only book of its kind. There's no other quite like it.

11. *Certitude or Indubitability.* This book does nothing to establish mathematical certitude or indubitability. On the contrary, it says to forget about them. Move past foundationism.

12. *Acceptability.* Are my proposals acceptable, worthy of criticism by experts and authorities? That remains to be seen.

If you don't like these criteria, throw them out. Put others in. Apply *your* criteria to humanism, Platonism, logicism, intuitionism, constructivism, conventionalism, structuralism, and fictionalism. Find out which one looks good to you.

For Teaching, Philosophy Makes a Difference

What's the connection between philosophy of mathematics and teaching of mathematics? Each influences the other. The teaching of mathematics *should* affect the philosophy of mathematics, in the sense that philosophy of mathematics must be compatible with the fact that mathematics can be taught. A philosophy that obscures the teachability of mathematics is unacceptable. Platonists and formalists ignore this question. If mathematical objects were an other-worldly, nonhuman reality (Platonism), or symbols and formulas whose meaning is irrelevant (formalism), it would be a mystery how we can teach it or learn it. Its teachability is the heart of the humanist conception of mathematics.

In the other direction, the philosophy of mathematics held by the teacher can't help but affect her teaching. The student takes in the teacher's philosophy through her ears and the textbook's philosophy through her eyes. The devastating effect of formalism on teaching has been described by others. (See Khinchin or Ernest.) I haven't seen the effect of Platonism on teaching described in print. But at a teachers' meeting I heard this:

"Teacher thinks she perceives other-worldly mathematics. Student is convinced teacher really does perceive other-worldly mathematics. No way does student believe *he's* about to perceive other-worldly mathematics."

Platonism can justify a student's certainty that it's impossible for her/him to understand mathematics. Platonism can justify the belief that some people can't learn math. Elitism in education and Platonism in philosophy naturally fit together. Humanist philosophy, on the other hand, links mathematics with people, with society, with history. It can't do damage the way formalism and Platonism can. It could even do good. It could narrow the gap between pupil and subject matter.

Such a result would depend on many other factors. But if other factors are compatible, adoption by teachers of a humanist philosophy of mathematics could benefit mathematics education.

This possible educational value is not a warrant for *correctness* of humanist philosophy. In earlier chapters I argued the correctness of humanism. But it's not unexpected that a philosophy epistemologically superior is educationally superior.

Philosophy and Ideology
Politics Makes a Difference

In our half of the twentieth century, it's unacceptable to import ideology into scholarship. We remember the destruction of Russian linguistics and genetics by Stalin's political correctness. Philosophy was mutilated too. Only dialectical materialism was allowed in Moscow.

Nearly forgotten is Hitler's ideology-philosophy. (Think with your blood! When I hear "culture," I reach for my revolver!) Martin Heidegger, deemed by some to be the supreme philosopher of our time, quickly and easily accommodated to Nazism.

Ideological philosophy is twice shameful. Shameful intellectually, by fostering meretricious hacks while crushing genuine intellect. Shameful politically, by complicity in the regime's crimes.

Therefore, one doesn't ask whether a philosophical view is socially harmful or beneficial, or whether it's favored by aristocrats, generals, or coal miners. For intellectual value, none of that should matter. Is it true? Is it interesting? Is it beautiful?

That said, the fact remains, philosophers are human beings. We should care how a philosophy connects with other realms of thought, how it connects with society at large.

In historical Part 2, I described philosophers' religious beliefs in relation to their philosophies of mathematics. Philosophers also hold political beliefs. Can there also be a connection between political position and philosophy of mathematics?

To compare philosophers of different periods and places, we want a uniform terminology for comparative political positions. In classical Athens, democrat/oligarch; during the Enlightenment, clerical/anticlerical, royalist/republican; in the 1930s, Fascist/anti-Fascist. Rather than "progressive/conservative" or "popular/aristocratic," I use "left-wing/right-wing." These terms belong to the French Chamber of Deputies in the nineteenth century, but for convenience I call any politics that restricts popular political rights "right wing"; politics that increases them, "left wing." (Gödel said simply "left" and "right," as quoted in the Preface.)

I start with the Mainstream, as defined in Part 2, with ***Pythagoras and the Pythagoreans***. They were progressive in admitting women to their school. Their maxims, such as (Wheelwright) "Abstain from beans" and "Do not urinate in the direction of the sun" seem apolitical. But "it was to the young men of well-to-do families that Pythagoras made his appeal. Pretending to have the power of divination, given at all times to mysticism, and possessed in a remarkable degree of personal magnetism, he gathered about him some 300 of the noble and wealthy young men of Magna Graecia and established a brotherhood that has ever since served as a model for all the secret societies in Europe and America" (Smith, p. 72).

Moreover, "The doctrines of the Pythagoreans and Eleatics may be understood, partly at least, in the light of social patterns which were congenial to the philosophers; these thinkers were not unaffected in their theorizing about eternal values by the actual political structure of which they were a part. The Pythagoreans interpreted the world in terms of order and symmetry, based on fixed mathematical ratios and found similar satisfactory order and symmetry in existing aristocratic schemes of government. . . . As the body must be held in subjection by the soul, so in every society there must be wise and benevolent masters over obedient and grateful inferiors; and of course they had no doubt as to who were qualified to be the masters. Their religious brotherhoods became powerful political influences in Italiot Greece, a training school for aristocratic leadership. . . . Zeno defended the same thesis by a clever series of paradoxes (Agard, p. 42). . . . The philosophy that appealed most to the best families in

Athens was that of the Pythagorean brotherhoods, whose chief intellectual concern was mathematics but whose practical interest lay in a determined defense of aristocratic regimes against the inroads of democracy. They approved especially of the Pythagorean loyalty to ancient laws and customs even if they might be in certain respects inferior to new ones, on the principle that change is in itself a dangerous thing, the greatest sin is anarchy, and in the nature of things some are fitted to rule, others to obey" (Agard, p. 180). I count the Pythagoreans as *right wing.*

Plato is known for his totalitarian politics even more than for his philosophy of mathematics. Popper called the *Republic* a blueprint for fascism. Stone documented Plato's elitism and authoritarianism. Some are offended by Popper and Stone, but nobody claims Plato as a liberal. *Right wing.*

The Platonist Catholic philosophers *Augustine of Hippo* and *Nicolas Cusanus* thought of mathematics mystically. The Aristotelian Catholic *Thomas Aquinas* thought of mathematics more scientifically. But all three took for granted the Church's right and duty to guide society and its morality. *Right wing.*

Descartes's Method was radical in its rejection of authority in scientific work. Nevertheless, he was an obedient son of the Church. "His political views were also extremely orthodox, and closely linked with his religious ones . . . his deep respect for nobility and particularly sovereigns verged on the passionate if not the religious" (Vrooman, p. 42). Descartes's deep respect for royalty proved fatal. He spent his fifty-fourth winter in frigid Stockholm, rising to give philosophy lessons three times a week to Queen Christina at 5 A.M. He caught pneumonia, and died. A *right-winger.*

Spinoza was a subverter of Scripture, cursed by Protestant, Catholic, and Jew. "The *Tractatus Theologico-Politicus* is an eloquent plea for religious liberty. True religion is shown to consist in the practice of simple piety, and to be quite independent of philosophical speculations. The elaborate systems of dogmas framed by theologians are based on superstition, resulting from fear" (Pollock, *Trac.* intro., p. 31).

The *Tractatus* made Spinoza famous in Europe as a dangerous atheist. "Spinozism" was a top-ranking evil. To sell the book to the market created by banning it, booksellers printed it *with a false title page*!

In his *Tractatus Theologico-Politicus* (Chapter xvi), Spinoza wrote: "A Democracy may be defined as a society which wields all its power as a whole. . . . In a democracy, irrational commands are less to be feared: For it is almost impossible that the majority of a people, especially if it be a large one, should agree in an irrational design; and, moreover, the basis and aim of a democracy is to avoid the desires as irrational, and to bring men as far as possible under the control of reason, so that they may live in peace and harmony. . . . I think I have now shown sufficiently clearly the basis of a democracy. I have especially desired to do so, for I believe it to be of all forms of government the most natural, and the most consonant with individual liberty. In it no one transfers his natural right so absolutely that he has no further voice in affairs; he only hands it over to the majority of a society, where he is a unit. Thus all men remain, as they are in the state of Nature, equals.

"This is the only form of government which I have treated at length, for it is the one most akin to my purpose of showing the benefits of freedom in a state" (Ratner, pp. 304–7).

Feuer writes (p. 5), "Spinoza is the early prototype of the European Jewish radical. He was a pioneer in forging methods of scientific study in history and politics. He was cosmopolitan, with scorn for the notion of a privileged people. Above all, Spinoza was attracted to radical political ideas. From his teacher Van den Ende he had learned more than Latin. He had evidently imbibed something of the spirit of that revolutionist whose life was to end on the gallows. . . . In his youth, furthermore, Spinoza's closest friends were Mennonites, members of a sect around which there still hovered the suggestion of an Anabaptist, communistic heritage." A *lefty.*

An Anglican Bishop, *Berkeley* had the right-wing politics expected of his position. Yet when Ireland was wracked by famine, he worked strenuously to get food for the hungry. Still, a *right-winger.*

Leibniz thought he was the man able to save Europe from war and revolution. He wanted to be chief counselor to some principal monarch, to the emperor, or to the Pope (Meyer, pp. 2–4). He wrote, "Those who are not satisfied with what God does seem to me like dissatisfied subjects whose attitude is not very different from that of rebels . . . to act conformably to the love of God it is not sufficient to force oneself to be patient, we must really be satisfied with all that comes to us according to his will" (1992, p. 4). This was the Leibniz caricatured in Voltaire's *Candide. Right wing.*

Kant was a moderate. His life and writings upheld the status quo. Yet in religion he leaned toward free thought, and politically he wasn't out of sympathy with the French Revolution. He was ordered by the King of Prussia not to write about religion. Popper calls him an ardent liberal. *A leftish.*

Frege actually died a Nazi. Sluga reports: "Frege confided in his diary in 1924 that he had once thought of himself as a liberal and was an admirer of Bismarck, but his heroes now were General Ludendorff and Adolf Hitler. This was after the two had tried to topple the elected democratic government in a coup in November 1923. In his diary Frege also used all his analytic skills to devise plans for expelling the Jews from Germany and for suppressing the Social Democrats." Michael Dummett tells of his shock to discover, while reading Frege's diary, that his hero was an outspoken anti-Semite (1973). *Right wing.*

Russell was the "cream" of English society. His grandfather John Russell was a Whig foreign minister. Yet Bertie became a socialist. In World War I he went to jail as a conscientious objector. Then in the 1920s and 1930s he alienated left-wingers by hostility to the Soviet Union. After the explosion of the U.S.-British atom bombs, and before the Soviet atom bomb, he favored a preventive atom-bomb attack on the Soviet Union. But after the Soviet atomic explosion, he became a fervent opponent of atom bombs and an advocate of peace with the Soviet Union. He was on an international tribunal that condemned U.S. war crimes in Vietnam. *Left wing.*

Brouwer was a fanatically antifemale, pro-German eccentric. "There is less difference between a woman in her innermost nature and an animal such as a lioness than between two twin brothers. . . . The usurpation of any work by women will automatically and inexorably debase that work and make it ignoble. . . . When all productive labor has been made dull and ignoble by socialism it will be done exclusively by women. In the meantime men will occupy their time according to their ability and aptitude in sport, gymnastics, fighting, studying philosophy, gardening, wood-carving, traveling, training animals and anything that at the time is regarded as noble work, even gambling away what their wives have earned. For this really is much nobler than building bridges or digging mines" (*Life, Art and Mysticism*). His biographer writes, "At the time, a few months after his wedding, he was supported by his wife, who was herself working for a degree while running her pharmacy. . . . There is a touch of insincerity about most of Brouwer's strong condemnations. He ridicules fashions and many of the human weaknesses which mark his own life, such as ambition, lust for power, jealousy and hypochondria. His condemnation of those seeking security by amassing capital rings rather hollow in a man whose life was so obsessed with money. . . . *Life, Art and Mysticism* cannot be written off as a rash, 'teen-age effort.' . . . He wrote it in 1905, after his doctoral examination. Far from disowning his 'booklet' as a youthful aberration, Brouwer backed it all his life. He discussed the possibility of an English translation as late as 1964. It proudly features in every one of his entries for various biographical dictionaries as the first of his two books. Most important, it is the clearest expression of his philosophy of life, which inspired his intuitionism" (Van Stift).

After World War II his university convicted him of collaboration with the Nazi occupation. Van Stift explains that Brouwer was not so much anti-Semitic as anti-French. "The German occupation seriously affected academic life in Holland during the war years. Brouwer's pragmatic attitude to politics, his concern for the continuance of academic life and his fear of 'becoming involved' laid him open to petty accusations of the many enemies he had made in local government and at the University. In the postwar hysteria these were blown up into serious crimes before a kangaroo court of Amsterdam University. He was reprimanded and suspended from his duties for nine months." *Right wing.*

W. V. O. Quine is a self-identified Republican. *Right wing.*

So among the Mainstream we have Pythagoras, Plato, Augustine, Cusa, Descartes, Berkeley, Leibniz, Frege, Brouwer, and Quine on the right; Spinoza, Kant and Russell on the left. *10 righties, 3 lefties.*

Now the humanists.

Aristotle criticized all forms of government, especially the two competing in Athens—democracy and oligarchy. In the end, he comes out as a cautious, critical democrat. From Barker's translation of *The Politics*, published in 1962 by Oxford:

"It is possible, however, to defend the alternative that the people should be sovereign. The people, when they are assembled, have a combination of qualities which enables them to deliberate wisely and to judge soundly. This suggests that they have a claim to be the sovereign body. . . . It may be argued that experts are better judges than the non-expert, but this objection may be met by reference to (a) the combination of qualities in the assembled people (which makes them collectively better than the expert) and (b) their 'knowing how the shoe pinches' (which enables them to pass judgment on the behavior of magistrates). . . (p. 123).

"Each individual may indeed be a worse judge than the experts; but all, when they meet together, are either better than experts or at any rate no worse. . . . There are a number of arts in which the creative artist is not the only, or even the best judge. . . (p. 126). [This seems to be a cautious suggestion that the art of government is this kind.]

"It is therefore just and proper that the people, from whom the assembly, the council, and the court are constituted, should be sovereign on issues more important than those assigned to the better sort of citizens" p. 127). *A lefty*

John *Locke* was a father of modern democracy. The French Philosophes took him as their teacher. So did Thomas Jefferson, in writing the U.S. Declaration of Independence. A *lefty*.

Hume also belonged to the eighteenth-century enlightenment, which gave intellectual nourishment to the eighteenth-century revolutions. *A lefty*.

d'Alembert was a leader of the Philosophes, who laid the ideological groundwork for the French Revolution. On the *left* side.

Mill is better known for liberal politics than for philosophy of mathematics (Kubitz, p. 277). There's a surprising linkage between his philosophy of mathematics and his politics. This is revealed in two book reviews by an anonymous writer who may have been Mill.

"In October, 1830, the *Westminster Review*, the radical periodical with which both Mills were associated, brought a review of *The First Book of Euclid's Elements with Alterations and Familiar Notes. Being an Attempt to get rid of "Axioms" altogether; and to establish the Theory of Parallel Lines, without the introduction of any principle not common to other parts of the Elements.* By a member of the Univ. of Cambridge. 3rd Ed. R. Heward, 1830. The reviewer hails this work with an interesting passage, which, if coming from the hand of Mill, would give us a significant expression about axioms from the time when he was occupied with the question left him by Whately, 'how can the truths of deductive science be all wrapt up in its axioms?' The reviewer, whoever he is, rejoices,

"This is an attempt to carry radicalism into Geometry; always meaning by radicalism, the application of sound reason to tracing consequences to their *roots.* To those who do not happen to be familiar with the facts, it may be useful to be told, that after all the boast of geometricians of possessing an *exact science*, their science has really been founded on taking for granted a number of propositions under the title of Axioms, some of which were only specimens of slovenly acquiescence in assertion where demonstration might easily have been had, but others were in reality the begging of questions which had quite as much need of demonstration, as the generality of those to which demonstration was applied." In July 1833, the reviewer of Whewell's *First Principles of Mechanics* in the *Westminster Review* remarks in the same vein, "Axiom is a word in bad odour, as having been used to signify a lazy sort of *petitio principi* introduced to save the trouble of inquiry into cause. . . ." "These passages make it all the more evident that the group with which Mill was associated were not oppposed to axioms on speculative, but on the practical grounds of political reform" (Kubitz). (Today again, some writers are making a connection between political philosophy and the epistemology and pedagogy of mathematics.) Mill counts on the *left.*

Peirce kept away from politics. He was a notorious elitist and snob. He took great pride in his family connections with Boston "aristocracy." During the Civil War he managed to evade service in the Northern army. Yet his lifelong friend and supporter was the great liberal, William James. For decades he reviewed science and mathematics for the liberal magazine *The Nation*. He wrote a denunciation of "social Darwinism," an ideology supporting the right to rule of the moneyed classes. We class him as borderline.

Rényi, Lakatos, Polányi, Pólya, and Von Neumann were Hungarian Jews. In 1944 Rényi was dragged to a Fascist labor camp, but escaped when his company was sent west. (In the late 1930s the Horthy regime set up labor camps for men "unfit" for military service—Communists, Jews, gypsies. Sometimes they suffered extreme danger and hardship.) For half a year he hid with false papers.

His parents were in the Budapest ghetto. Jews there were being rounded up for annihilation. Rényi stole a soldier's uniform at a Turkish bath, walked into the ghetto, and marched his parents out. One must be familiar with the circumstances to appreciate his courage and skill.

Starting in 1950 he directed the Mathematical Institute of the Hungarian Academy of Science. Under his leadership it became an international research center, and the heart of Hungarian mathematical life. A *lefty.*

Pólya and Polányi were liberal exiles from Horthy's clerical fascism. As a student at Budapest University, Pólya belonged to the liberal Galileo Society. *Two lefties.*

Lakatos was a Communist before his arrest and imprisonment in 1950. By the time he left Hungary after the 1956 uprising, he was an ardent anti-Communist. I classify him as a *right-winger.*

Von Neumann was a hawk in the Cold War with the Soviet Union. He became science adviser to the United States and member of the Atomic Energy Commission. Since we haven't been able to classify him as Mainstream or humanist, he doesn't affect our tabulation.

Hilbert was first and last a mathematician, neither a right nor a left. Still less can *Wittgenstein* be put on the left or the right. *Poincaré* may have had political views, but I haven't found them. (His cousin Raymond, with whom he shared a flat in the Latin Quarter while the two were students, became President of the Republic during World War I, and Premier from 1924 to 1929. Raymond was a moderate bourgeois, working between the royalists and the socialists.) Three insufficient informations.

Omitting the inconclusive Peirce, Poincaré, Hilbert, and Wittgenstein, we have among the humanists Aristotle, Locke, Hume, d'Alembert, Mill, Rényi, Polányi, and Pólya, on the left; Aquinas and Lakatos on the right (*8 lefties, 2 righties*).

It appears the Mainstream are mostly rightish, humanists mostly leftish.

Following the example of Paul Ernest in *Philosophy of Mathematics*, think of the four entries as proportional to conditional probabilities. There are

$$10 + 2 + 3 + 8 = 23$$

philosophers in the matrix. Pick one at random. If he happens to be a Mainstream, he's "rightish" with probability 10/13, or 77 percent; leftish with probability 3/13, or 23 percent.

If you picked a humanist, the probability he's rightish is much less. Only 2/10, or 20 percent. The probability he's leftish is 8/10, or 80 percent.

Look at it the other way. If the philosopher you pick happens to be rightish, the probability that he's philosophically Mainstream is 10/12, or 83 percent. The probability that he's a humanist is only 2/12, or 17 percent.

If you pick a lefty, the probability he's Mainstream is much less. Only 3/11, or 27 percent. The probability he's a humanist is 8/11, or 73 percent.

Why so?

In Chapter 1 and in a previous section of this chapter I argue that philosophy of mathematics makes a difference for mathematics education. I claim that Platonist philosophy is anti-educational, while humanist philosophy can be pro-educational. I emphasize that the social consequence of a philosophy is ***not*** the same as its validity. But doesn't make the social consequence unimportant.

Conservative politics and Platonist philosophy of mathematics don't ***imply*** each other. This is proved by exceptions like Russell and Lakatos. The numbers do suggest a *correlation*.

What can we say that's neither dogmatic nor mere guesswork?

I simply ask: doesn't it make sense?

Political conservatism opposes change. Mathematical Platonism says the world of mathematics never changes.

Political conservatism favors an elite over the lower orders. In mathematics teaching, Platonism suggests that the student either can "see" mathematical reality or she/he can't.

A humanist/social constructivist/social conceptualist/quasi-empiricist/naturalist/maverick philosophy of mathematics pulls mathematics out of the sky and sets it on earth. This fits with left-wing anti-elitism—its historic striving for universal literacy, universal higher education, universal access to knowledge, and culture.

If the Platonist view of number is associated with political conservatism, and the humanist view of number with democratic politics, is that a big surprise?

The Blind Men and the Elephant
(Six Men of Indostan)

J. GODFREY SAXE (1816–1887)

It was six men of Indostan
To learning much inclined,
Who went to see the Elephant
(Though all of them were blind).
That each by observation
Might satisfy his mind.

The First approached the Elephant,
And happening to fall
Against his broad and sturdy side,
At once began to bawl:
"God bless me! but the Elephant
Is very like a wall!"

The Second, feeling of the tusk,
Cried, "Ho! what have we here
So very round and smooth and sharp?
To me 'tis mighty clear
This wonder of an Elephant
Is very like a spear!"

The Third approached the animal,
And happening to take
The squirming trunk within his hands,
Thus boldly up and spake:
"I see," quoth he, "the Elephant
Is very like a snake!"

The Fourth reached out an eager hand,
And felt about the knee.
"What most this wondrous beast is like
Is mighty plain," quoth he;
" 'Tis clear enough the Elephant
Is very like a tree!"

The Fifth who chanced to touch the ear,
    Said: "E'en the blindest man
Can tell what this resembles most:
    Deny the fact who can,
This marvel of an Elephant
    Is very like a fan!"

The Sixth no sooner had begun
    About the beast to grope,
Than, seizing on the swinging tail
    That fell within his scope,
"I see," quoth he, "the Elephant
    Is very like a rope!"

And so these men of Indostan
    Disputed loud and long,
Each in his own opinion
    Exceeding stiff and strong,
Though each was partly in the right
    And all were in the wrong!

This doggerel is a metaphor for the philosophy of mathematics, with its Wise Men groping at the wondrous beast, Mathematics.

What do the six men of Indostan have in common? (We'll call them 6I, to give the conversation a mathematical tinge.)

They obey a common axiom: *Axiom 6I: Cling to an incomprehensible partial truth to avoid a larger, more inclusive truth.*

Let FP denote a Famous Philosopher. (Continuing with the pseudomathematical terminology.) We're ready to formulate our problem in *precise language*:

Problem: Given an FP, does he satisfy axiom 6I?

Poetically speaking, is the Famous Philosopher one of the Six Men of Indostan?

I'll probe this question only with respect to the formalists. I leave the Platonists and intuitionists for the reader's pleasure.

With respect to the formalists, the answer to the *Problem* is YES. David Hilbert proposed that, for philosophical purposes, statements about infinitary objects (including all of calculus and analysis) be regarded as mere formulas, devoid of reference, meaning, or interpretation. This was forgotten whenever Hilbert took off his philosopher hat and resumed his true identity as mathematician. It wasn't taken seriously in real mathematical life. It was just a tactic in Hilbert's campaign against Brouwer's intuitionism.

Because of Hilbert's pre-eminence in mathematics, the formalist philosophy was identified with his name. Formalism was recognized long before; it was one of the bugbears Frege tried to squash. Decades after Hilbert, the Bourbakistes of Paris went deeper into formalism than Hilbert, regarding *all* mathematical statements, finitary or infinitary, as meaning-free formulas. More precisely, any meaning such a formula carries is irrelevant in mathematics, which is concerned with the formulas themselves.

(But Dieudonné shamelessly blurted out, as I quoted in Chapter 1, that actually Bourbaki didn't believe any of this. It was all a trick, to fend off philosophers.)

Whichever formalist represents the tribe, whether David Hilbert, "Nicolas Bourbaki," Haskell Curry, or David Henle, when we ask, "Does he satisfy Axiom 6I?" we must answer, "Yes." Everyone agrees that formalization is ***an aspect*** of mathematics. On the other hand, formalism doesn't have the whole beast in its view-finder. We need only point to the vital part of mathematics that formalism negates—the intuitive, or informal (see Chapter 4).

To save repetition, I make two declarations:

Declaration 1: The intuitive is an essential aspect of mathematics. Without the intuitive, mathematics wouldn't be mathematics as we have always conceived it.

Declaration 2: The intuitive, by its very conception and definition, is not formalizable.

The formalist can't reject Declaration 1. To do so would be simply incredible. He can't reject Declaration 2. To do so would be a contradiction in terms.

He can give up any claim to describe mathematics in its totality, and aspire only to describe the formalizable part—***the part that formalism can describe.*** This would be correct—but vacuous. The man of Indostan feels the part of the elephant that is like a snake—and calls out, "It's very like a snake!"

It would be good to repeat this discussion, replacing formalists by intuitionists (who see the mental side of the elephant mathematics, but are blind to its social-historic parts) and by logicists and Platonists (who see the dynamically evolving, socially interacting beast of mathematics as a frozen abstraction in the sky). I leave these exercises to the interested reader.

Summary

Mathematics is like money, war, or religion—not physical, not mental, but social. Dealing with mathematics (or money or religion) is impossible in purely physical terms—inches and pounds—or in purely mental terms— thoughts and emotions, habits and reflexes. It can only be done in social-cultural-historic terms. This isn't controversial. It's a fact of life.

Saying that mathematics, like money, war, or religion, is a social-historic phenomenon, is not saying it's the ***same*** as money or war or religion. Money is different from war, money is different from religion, religion is different from war. But all four are social-historic phenomena. Mathematics is another particular, special social-historical phenomenon. Its most salient special feature is the uniquely high consensus it attains.

War or money don't exist apart from human minds and bodies. Without bodies, no minds; without people's minds and bodies, no society, culture, or history. The emergent social level does not emerge from a vacuum; it emerges from the mental and physical levels. Yet it has qualities and phenomena that can't be understood in terms of the previous levels.

Recognizing that mathematics is a social-cultural-historical entity doesn't automatically solve the big puzzles in the philosophy of mathematics. ***It puts those puzzles in the right context, with a new possibility of solving them.***

This is like a standard move in mathematics—***widen the context.*** Consider the equation:

$$x^2 + 1 = 0$$

Among the real numbers, it has no solution. If we enlarge the context to the complex numbers, we find two solutions, $+i$ and $-i$. This is the first step to a beautiful and powerful theory.**

Yet from the viewpoint of the original problem, these solutions don't really exist. They aren't fair. They "don't count."

So it is here. Problems intractable from the foundationist or neo-Fregean viewpoint are approachable from the humanist viewpoint. But from the foundationist viewpoint, humanist solutions are no solutions. They're unfair. It's not allowed to give up certainty, indubitability, timelessness, or tenselessness. These restrictions in philosophy of mathematics act like the restriction to the real line in algebra. Dropping the insistence on certainty and indubitability is like moving off the line into the complex plane.

When mathematicians move into the complex plane, we don't throw away all sound sense. We keep the rules of algebra. Dropping indubitability from philosophy of mathematics doesn't mean throwing away all sound sense. The guiding principles remain: intelligibility, consistency with experience, compatibility with philosophy of science and general philosophy. A humanist philosophy of mathematics respects these principles.

Epilogue

The line between humanism and Mainstream is more than a philosophical preference. It's tied to religion and politics.

The humanist philosophy of mathematics has a pedigree as venerable as that of Mainstream. Its advocates are respected thinkers.

The Mainstream continues to dominate. That doesn't sanctify it. Religious obscurantism and propertied self-interest still dominate society. They're not sanctified.

Who's interested in escaping from neo-Fregeanism? Philosophically concerned mathematicians. And people interested in the foundations of mathematics education. Putnam's disconnection from Quinism may have been a straw in the wind. The blossoming of humanism among mathematics educators is another.

Mathematical audiences are impatient with neo-Fregeanism. They show lively interest in alternatives. Yet the typical journal article on philosophy of mathematics still plods after Carnap and Quine. I suspect that game is played out.

Neo-Fregeans will disagree. But neo-Fregeanism no longer reigns by inherited right. If it cares to be taken seriously much longer, it has to face the challenge of humanism.

Mathematical Notes/Comments

Most of these notes keep promises made in previous chapters. "Square circles" is in a light-hearted vein. No one need be offended. David Hume was my favorite philosopher in college.

Except for their use of algebra, the articles on "How Imaginary Becomes Reality" could have come earlier. They show natural, inevitable ways that mathematics grows, with no mystery about invention vs. discovery.

"Calculus refresher" is included because it isn't possible to talk sense about mathematics without an acquaintance with or recollection of calculus. By omitting exercises and formal computations, I present a semester of calculus in a few easy pages.[1]

The last article is a wonderful piece of mathematical artistry. The late George Boolos proves Gödel's great incompleteness theorem in three simple pages![2]

Arithmetic

[1] It's taken, with minor improvements, from the Teacher's Guide which accompanies the study edition of *The Mathematical Experience*, co-authored with Philip J. Davis and Elena Anne Marchisotto, Birkhauser Boston, 1995.

[2] It's reproduced from the *Notices* of the American Mathematical Society to make it available to a larger readership.

Logic

Sets

Geometry

How Imaginary Becomes Reality

Calculus

More Logic

ARITHMETIC

What Are the Dedekind-Peano Axioms?

Instead of constructing the natural numbers out of sets à la Frege-Russell, we can take them as basic, and describe them by axioms from which their other properties can be derived.

The axioms should be consistent, of course; chaos could ensue if they were contradictory (see below). We would like them to be minimal—not include any redundant axioms.

The standard axioms for the natural numbers were given by Richard Dedekind, inventor of the Dedekind cut. Following the usual rule of misattribution in mathematical nomenclature, they are called "Peano's postulates." The undefined terms are "1" and "successor of."

1. 1 is a number.
2. 1 isn't the successor of a number.
3. The successor of any number is a number.
4. No two numbers have the same successor.
5. (Postulate of mathematical induction) If a set contains 1, and if the successor of any number in the set also belongs to the set, then every number belongs to the set.

How to Add 1's

We show that Dedekind's axioms imply

$$2 + 2 = 4.$$

None of the symbols in this equation appears in the axioms, so we must define all four of them.

"=" is defined by the rule that for any x and y, if $x = y$, then in any formula y may be replaced by x and vice versa. This rule is called "substitution."

For present purposes, we need only define addition by 1 and 2, for all n. Let S stand for the successor operation.

Define 2 as S(1), 3 as S(2), 4 as S(3).
Define "n + 1" as S(n) and "n + 2" as S(S(n)).

Then by substitution

(A) 2 + 2 = S(S(2)).

Again by substitution,

(B) 4 = S(S(2)).
Voilà! 2 + 2 = 4.

To define n + k for all n and k would take more work, using recursion on both n and k.

Is 2231 prime?

I know how to find out. 2231 is prime if it's not divisible by any number between 1 and 2231. I could just divide 2231 by all the numbers from 2 to 2230.

This labor can be cut down a lot. If 2231 is factorable, it factors into two numbers, one larger, one smaller or both equal to each other. It's sufficient to find the smaller. Since

$47^2 = 2249,$

the smaller factor has to be less than 47. Moreover, it's not necessary to divide by any composite number, because if 2231 has a composite factor, that composite factor has prime factors that also factor 2231.

So we only have to check the prime numbers less than 47—2, 3, 5, 7, 11, 13, 17, 19, 23, 29, 31, 37, 41, 43.

Now 2231 is odd, so 2 isn't a factor. The sum of the digits is 8, which is not divisible by 3, so 3 isn't a factor. It doesn't end in 5 or 0, so 5 isn't a factor. The alternating sum

+ 2 − 2 + 3 − 1 isn't 0,

so 11 isn't a factor. Get your calculator and divide 2231 by 7, 13, 17, 19, 23, 29, 31, 37, 41, and 43. If none of them divides 2231 without remainder, 2231 is prime.

Logic

Zermelo-Fraenkel, Axiom of Choice, and the Unbelievable Banach-Tarski Theorem

"Given any collection of nonempty sets, it is possible to form a set that contains exactly one element from each set in the collection."

Surely a harmless-sounding assumption to make about finite collections of finite sets, and even countably infinite collections of countably infinite sets. But when it's applied to collections and sets of arbitrarily great uncountable cardinality, trouble comes! Consequences follow, which many mathematicians would rather not believe. Zermelo proved, using the axiom of choice, that any set—for instance, the uncountable set of real numbers—can be rearranged to be well-ordered. (But no one can actually do it, and no one expects anyone to be able to do it.) Stefan Banach and Alfred Tarski proved, using the axiom of choice, that it's possible to divide a pea (or a grape or a marshmallow) into 5 pieces such that the pieces can be moved around (translated and rotated) to have volume greater than the sun (see Wagon). As mentioned in Chapter 4 on proof, a transitory movement to avoid the axiom of choice has long been given up.

1/0 Doesn't Work (0 into 1 Doesn't Go)

Division by 0 is not allowed. Why not? If it's allowed to introduce a symbol i and say it's the square root of -1 ***which doesn't have a square root,*** why not introduce some symbol, say Q, for $1/0$?

We introduce new numbers, whether negative, fractional, irrational, or complex, to preserve and extend our calculating power. We relax one rule, but preserve the others. After we bring in i, for example, we still add, subtract, multiply, and divide as before. I now show that there's no way to define $(1/0) \times 0$ that preserves the rules of arithmetic.

One basic rule is,

$0 \times (\text{any number}) = 0.$

(Formula I) So $0 \times (1/0) = 0.$

Another basic rule is

$(x) \times (1/x) = 1$, provided x isn't zero. (But if we want $1/0$ to be a number, this proviso becomes obsolete.)

(Formula II) So $0 \times (1/0) = 1$

Putting Formulas I and II together,

$1 = 0.$

Addition gives

$2 = 0$, $3 = 0$, and so on, $n = 0$

for every integer n.

Since all numbers equal zero, all numbers equal each other.

There's only one number—0.

The supposition that $1/0$ exists and satisfies the laws of arithmetic leads to collapse of the number system. Nothing is left, except—nothing.

What Is Modus Ponens?

In scholastic (medieval Aristotelian) logic, Latin names were given to the different permutations of Aristotle's syllogisms. Modus ponens is the simple argument: "If

A implies B, and A is true, then B is true." In modern formal logic the other syllogistic arguments can be eliminated. Modus ponens turns out to be sufficient.

Formalizable

A statement is "formalized" when it's translated into a formal language. Computer languages like Basic, Pascal, Lisp, C, and others—are formal languages. The notion of a formal language goes back to Peano, Frege, Russell, and Leibniz. A formal language has a vocabulary specified in advance—$x_1, x_2, +, \times, =$, etc. It has an explicit grammar, which prescribes the admissible permutations of the vocabulary. Whether a sentence in natural language is formalizable depends on the formal language under consideration. To be formalizable a sentence is supposed to be unambiguous, and to mention only objects that have names in the formal language.

How One Contradiction Makes Total Chaos

Suppose some sentence A and its negation "not-A" are both true.

We claim that (A and not-A) together implies B, no matter what B says. First, notice that:

(I) not-(A and not-a) means the same thing as (not-A or A), which is a "tautology"—it's true, no matter what A says. Also, notice that a tautology is implied by any sentence at all, because an implication is false only when the antecedent is true and the conclusion is false; if the conclusion is a tautology, it can't be false. Therefore

(II) not-B implies the tautology (not-A or A)

Now, by the definition of "implies," if a sentence P implies a sentence Q, then not-Q implies not-P. This deduction rule is called "contrapositive."

So, applying contrapositive to II,

(III) not-(not-A or A) implies not-(not-B)

(IV) But not-(not-B) is the same as B (double negative.) So, substituting (IV) into (III),

(V) not-(not-A or A) implies B.

Now applying negation ("not") to both sides of (I), we get

(VI) (A and not-a) means the same thing as not-(not-A or A)

Combining (V) and (VI), we have, as claimed,

(A and not-a) implies B, for any B.

The way (A and not-A) makes the whole logical universe collapse is rather like the way 1/0 makes the whole number system collapse. Is there a connection?

Sets

The Natural Numbers Come Out of the Empty Set

I will describe the sequence of "constructions" by which we "create" the real number system out of "nothing." It has philosophical interest, and it's ingenious.

In practice, however, we think of numbers in terms of their behavior in calculation, not in terms of this "construction."

Start with the empty set. We define it as "the set of all objects not equal to themselves," since there are no such objects. All empty sets have the same members—no members at all! Therefore, as sets they're identical, by definition of identity of sets. In other words, there's only one empty set. This unique empty set is our building block. Next comes the set whose only member is—the empty set. This set is *not* empty. Think of a hat sitting on a table—an empty hat. An example of an empty set. Then put the hat into a box. The hat is still empty. The box containing one thing—an empty hat—is an example of a set whose single member is an empty set. We say the contents of the box has cardinality 1. We have so far two entities, the empty hat and the box containing the empty hat. Now put box and another hat together into a bigger box. The contents of the bigger box has cardinality 2. The interested reader can now construct sets with cardinality three, four, and so on. From an empty set we construct the natural number system!

How the Rational Numbers Are Dense but Countable

The natural numbers are discrete—each is separated from its two nearest neighbors by steps of size 1. On the other hand, the rational numbers (fractions) are "dense." Between any two you can find a third—the average of the two. Repeat the argument, and you see that between any two rationals there are infinitely many. (A fact intensely irritating to Ludwig Wittgenstein. He called it "a dangerous illusion." See Chapter 11.)

This seems to mean there are many more rationals than naturals. But that's not true. There are just as many!

Georg Cantor thought of a simple way to associate the rationals to the naturals. To each natural a rational, to each rational a natural.

Arrange the rational numbers in rows according to denominators. In the first row, all the fractions with denominator 1, numerators in increasing order:

$$\frac{0}{1}, \quad \frac{1}{1}, \quad \frac{2}{1}, \quad \text{and so on.}$$

In the second row, the fractions with denominator 2, numerators in increasing order:

$$\frac{0}{2}, \quad \frac{1}{2}, \quad \frac{2}{2}, \quad \text{and so on.}$$

Each row is endless, and the succession of rows is also endless.

Starting in the upper left corner at $0/1$, draw a zigzag line: go down one step, then go diagonally up and to the right to the top row (with ones in the denominator). Go one step to the right, then go diagonally down and to the left to the first column (the fractions with 0 in the numerator). Go another step down, and diagonally up and right again, and so on and on. This jagged line passes exactly once through every fraction in the doubly infinite array. That means you've arranged the fractions in linear order. There's now a first, a second, a third, and so on. Every rational number appears many times in this array (only once in lowest terms), so we have a mapping of the rational numbers onto a subset of the natural numbers. We describe this relationship by saying the rationals are countable.

Yet the real numbers, obtained by filling in the gaps in the rationals, are uncountable!

How the Real Numbers Are Uncountable

The basic infinite set is N, the natural or counting numbers. Many other sets can be matched one to one with N—for example, the even or odd numbers, the squares, cubes, or any other power, the positive and negative integers, and even the rationals, as explained in the previous article. Therefore it comes as a shock that the ***real*** numbers can't be put in one-to-one correspondence with the naturals. Any attempt to make a list of the real numbers is bound to leave some out!

The proof is simple.

Any real number can be written as an infinite decimal, like

3.14159 . . .

From any list of real numbers written as infinite decimals, Cantor found a way to produce another number ***not on the list***. It doesn't matter how the list was constructed. So all the real numbers can never be written in a list.

How does Cantor produce his unlisted number? Step by step. It is an infinite decimal, constructed one digit at a time.

Look at the *first* real number on the list. Look at its *first* digit. Choose some other number from 0 to 9—any other number. That's the first digit in your new, unlisted number. Now go to the ***second*** real number on the list. Look at its *second* digit. Choose any other number from 0 to 9. That's the second digit of your new, unlisted number. And so on. The n'th digit of your new unlisted number is obtained by looking at the n'th digit of the nth real number on the list, and picking some ***other*** number for the nth digit in your new, unlisted number.

This construction doesn't terminate. But in calculus a number is well-defined if you can approximate it with ***arbitrarily high*** accuracy. By going out far enough in its decimal expansion, you approximate the unlisted number as accurately as you wish.

How do you know the new number isn't on the original list? It can't be the first number on the list, because they differ in the first digit. It can't be the second number on the list, because they differ in the second digit. No matter what n you choose, your new number isn't the n'th number on the list, because they differ in the n'th digit. The new number can't be the same as any number on the list! It's not on the list!

Geometry

What's "Between"? What's "Straight"?

What is the "straightness" of the straight line? There's more in this notion than we know, more than we can state in words or formulas. Here's an instance of this "more."

a, b, c, d are points on a line. b is between a and c. c is between b and d.

What about a, b, and d? How are they arranged?

It won't take you long to see that *b has to be between a and d.*

This simple conclusion, amazingly, can't be proved from Euclid's axioms! It needs to be added, an additional axiom in Euclidean plane geometry. This oversight by Euclid wasn't noticed until 1882 (by Moritz Pasch). A gap in Euclid's proof was overlooked for 2000 years!*

Some theorems in Euclid require Pasch's axiom. Without it, the proof is incomplete. The intuitive notion of the line segment wasn't completely described by the axioms meant to describe it.

More recently, the Norwegian logician Thorolf Skolem discovered mathematical structures that satisfy the axioms of arithmetic, but are much larger and more complicated than the system of natural numbers. These nonstandard arithmetics include *infinitely large integers.* In reasoning about the natural numbers, we rely on our mental picture to exclude infinities. Skolem's discovery shows that there's more in that picture than is stated in the Dedekind-Peano axioms. In the same way, in reasoning about plane geometry, mathematicians used intuitions that were not fully captured by Euclid's axioms.

The conclusion that b is between a and d is trivial. You see it must be so by just drawing a little picture. Arrange the dots according to directions, and you see b has to be between a and d. You're using a pencil line on paper to find a property of the ideal line, the mathematical line. What could be simpler?

But there are difficulties. The mathematical line isn't quite the same as your pencil line. Your pencil line has thickness, color, weight not shared by the mathematical line. In using the pencil line to reason about the mathematical line, how can you be sure you're using *only* those properties of the pencil line that the mathematical line shares?

In the figure for Pasch's axiom, we put a, b, c, and d *somewhere* and get our picture. What if we put the dots in other positions? How can we be sure the answer would be the same, "b is between a and d"? We draw one picture, and we believe it represents all possible pictures. What makes us think so?

The answer has to do with our sharing a definite intuitive notion, about which we have reliable knowledge. But our knowledge of this intuitive notion

* H. Guggenheimer showed that another version of Pasch's axiom can be derived as a theorem using Euclid's fifth postulate.

isn't complete—not even implicitly, in the sense of a base from which we could derive complete information.

A few simple questions to ponder while shaving or when stuck in traffic:

Is "straight line" a mathematical concept?

When you walk a straight line are you doing math?

When you *think* about a straight line, are you doing math?

Appletown, Beantown, and Crabtown are situated on a north-south straight line.

Must one be between the other two? Can more than one of the three be between two others? How do you know? Can you prove it?

Dogtown, Eggtown, and Flytown are on a *circle*, center at Grubtown. On that circle, must one be between the other two? Can more than one be between two others? How do you know? Could it be proved?

Is a straight line something you know from observation? From a definition in a book? Or how?

Is it something in your head?

Is the straight line in your head the same as the one in my head? Could we find out?

Is Euclid's straight line the same as Einstein's?

Is the straight line of a great-grandma in the interior of New Guinea the same as Hillary Rodham Clinton's? If Hillary Clinton visited her and they had a common language, could she find out?

Euclid's Alternate Angle Theorem

This is the first part of theorem 29, Book 1 of Euclid. "A straight line falling on parallel straight lines makes the alternate angles equal to one another." *Proof*: Let AB and CD be parallel. Let EF cross them, intersecting AB at G and CD at H. We claim the alternate angles AGH and GHD are equal, for they are both supplementary to angle CHG (adding to two right angles). For by construction

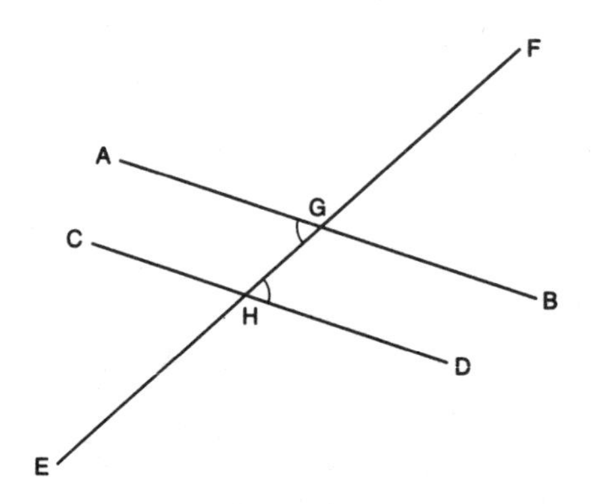

Figure 3. Alternate angles.

CHG and DHG add up to a straight angle, or two right angles. And by Euclid's fifth postulate (his definition of parallel lines) AB and CD parallel means the interior angles AGH and CHG add to two right angles.

$$\text{CHG} + \text{DHG} = 2\text{R} \qquad\qquad \text{CHG} + \text{AGH} = 2\text{R}$$
$$\text{DHG} = 2\text{R} - \text{CHG} = \text{AGH}$$

The proof is complete.

Euclid's Angle Sum Theorem

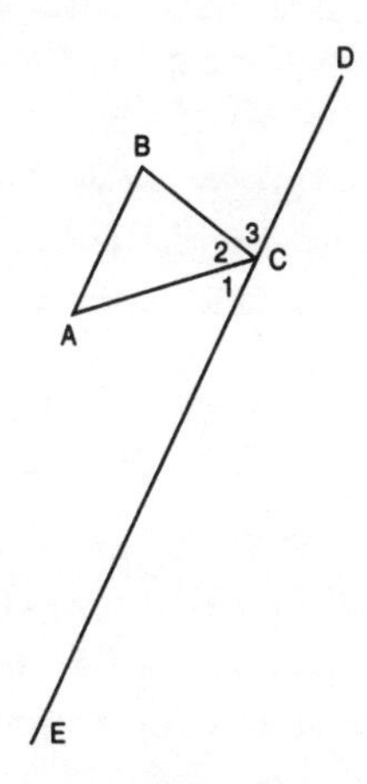

Figure 4.

In an arbitrary triangle ABC, choose a vertex, say C. Through C draw a line DCE parallel to AB. At C the three angles 1, 2, 3 add up to the sum of two right angles (180 degrees). Angle 2 is the same as angle C in triangle ABC. Angle 1 equals angle A, since they are alternate angles between two parallel lines (using Euclid's alternate angle theorem proved above). Similarly, angle 3 equals angle B. Adding,

Angle A + Angle B + Angle C =
Angle 1 + Angle 2 + Angle 3
= two right angles, q.e.d.

The Triangle Inequality

Here's an inequality valid for any six real numbers a, b, c, d, e, f:

$$\sqrt{(a-c)^2 + (b-d)^2} \le \sqrt{(a-e)^2 + (b-f)^2} + \sqrt{(c-e)^2 + (d-f)^2}.$$

This algebraic inequality has a geometric name—the "triangle inequality."

Why?

Let the three pairs (a,b), (c,d), (e,f) be rectangular coordinates of three points P, Q, and R in the plane. Then this inequality says the distance from P to Q is less or equal the distance from P to R plus the distance from R to Q.

If P, Q, and R are vertices of a triangle, the last statement says any side of a triangle is shorter or equal to the sum of the other sides. This is the triangle inequality—than which nothing could be visually more obvious. "A straight line is the shortest distance between two points."

The complicated-looking formula is the translation into algebra of this simple geometric fact. The geometric fact "motivates" the algebraic formula. One can, with effort, give an algebraic proof of the algebraic formula, and thereby give a complicated proof of a very simple geometric fact.

But you could just as well turn the procedure around. The triangle inequality is geometrically evident. Therefore its complicated-looking algebraic statement is also true. To prove the messy algebraic inequality, use its geometric interpretation with its simple visual proof.

What's Non-Euclidean Geometry?

(See the sections on Certainty, Chapter 4, and Kant, Chapter 7.) The fifth axiom of Euclid's *Elements*, the parallel postulate, was long considered a stain on the fair cheek of geometry. This postulate says: "If a line A crossing two lines B and C makes the sum of the interior angles on one side of A less than two right angles, then B and C meet on that side."

The usual version in geometry textbooks is credited to an English mathematician named Playfair: "Through any point P not on a given line L there passes exactly one line parallel to L." This is equivalent, and easier to understand.

This parallel postulate was true, everybody agreed. Yet it wasn't as self-evident as the other axioms. Euclid's version says that something happens, but perhaps very far away, where our intuition isn't as clear as nearby. From Ptolemy to Legendre, mathematicians tried to prove the parallel postulate. No one succeeded.

Many so-called "proofs" were found. But each "proof" depended on some "obvious" principle which was only a disguised version of the parallel postulate. Posidonius and Geminus assumed there is a pair of coplanar lines everywhere equally distant from each other. Lambert and Clairaut assumed that if in a quadrilateral three angles are right angles, the fourth angle is also a right angle. Gauss assumed that there are triangles of arbitrarily large area. Each of these different-sounding hypotheses is equivalent to the fifth postulate

In the early nineteenth century Gauss, Lobachevsky, and Bolyai all had the same idea: Suppose the fifth postulate is *false*!

Euclid's axiom can be replaced in two different ways. Either "Through P pass *more than one* line parallel to L" or "Through P pass *no* lines parallel to L." The first is called "the postulate of the acute angle." The second is "the postulate of the obtuse angle." These postulates generate two different non-Euclidean geometries, called "hyperbolic" and "elliptic." The hyperbolic was studied first, and is often referred to as just "non-Euclidean geometry."

An elegant contrast between the three geometries mentioned already in Chapter 4 is the sum of the angles in a triangle. In Euclidean geometry, as we proved above, the sum equals two right angles. In elliptic geometry, the sum of the angles of every triangle is ***more*** than two right angles. And in hyperbolic geometry it's *less.*

Gauss was the earliest of the three discoverers. As I mentioned earlier, in the section on Kant, he didn't publish his work, to avoid "howls from the Boeotians." In classical Athens, "Boeotions" meant "ignorant hicks." To Gauss, it meant perhaps "followers of Kant." They would say non-Euclidean geometry is nonsense, since Kant proved there can be no geometry but Euclid's.

Gauss's fear was justified. When non-Euclidean geometry became public, Kantian philosophers did say it wasn't really geometry. One of them was Gottlob Frege, the founder of modern logic.

Before Gauss, deep penetrations into the problem had been made by the Italian Jesuit priest Saccheri, by Lagrange, and by Johann Heinrich Lambert (1728–1777), a leading German mathematician who was an acquaintance or friend of Kant. Decades before Gauss, Lambert wrote:

> Under the (hypothesis of the acute angle) we would have an absolute measure of length for every line, of area for every surface and of volume for every physical space. . . . There is something exquisite about this consequence, something that makes one wish that the third hypothesis be true! In spite of this gain I would not want it to be so, for this would result in countless inconveniences. Trigonometric tables would be infinitely large, similarity and proportionality of figures would be entirely absent, no figure could be imagined in any but its absolute magnitude, astronomers would have a hard time, and so on. But all these are arguments dictated by love and hate, which must have no place either in geometry or in science as a whole. . . . I should almost conclude that the third hypothesis holds on some imaginary sphere. At least there must be something that accounts for the fact that, unlike the second hypothesis (of the obtuse angle), it has for so long resisted refutation on planes.

It's astonishing that Lambert actually gives the acute angle hypothesis a fair chance. The issue is to be decided by mathematical reasoning, not by "universal intuition." He honestly contemplates the possibility that a non-Euclidean geometry may be valid. His very ability to do so refutes Kant's universal innate Euclidean intuition.

In the end, Lambert slips into the same ditch as Legendre and Saccheri. He "proves" the Euclidean postulate by getting a "contradiction" out of the acute angle postulate. Laptev exposes Lambert's fallacy.

Beltrami, Klein, and Poincaré constructed models that showed that Euclidean and non-Euclidean geometry are "equiconsistent." If either one is consistent, so is the other. Since no one doubts that Euclidean geometry is consistent, non-Euclidean is also believed to be consistent.

Kant said that only one geometry is thinkable (see Chapter 7). But the establishment of non-Euclidean geometry offers a choice between several geometries. Which works best in physics? The choice must be empirical, to be settled by observation.

It's tempting to simply declare that "obviously" or "intuitively" Euclid is correct. This was not believed by Gauss. There's a legend that he tried to settle the question by measuring angles of a gigantic triangle whose vertices were three mountain tops. (The larger the triangle, the likelier that there would be a measurable deviation from Euclideanness.) Supposedly the measurement was inconclusive. Perhaps the triangle wasn't big enough.

What's a Rotation Group?

A "group" is a closed collection of reversible actions. For instance, multiplication by the positive real numbers is a group, since the product or quotient of two positive real numbers is a positive real number. The set of rotations in 3 dimensions is a group. Motions of your arm can be thought of as rotations around your shoulder and your elbow. So awareness of how your arm moves is an intuitive acquaintance with the 3-dimensional rotation group.

The Four-Color Theorem

In political maps, countries that share a border of positive length (not just some isolated points) are required to have different colors. It turns out that four colors always suffice to meet this requirement. This was stated as a mathematical conjecture in 1852. It was first proved by Haken and Appel in 1976. They broke the problem into a great many arduous calculations, which were performed on a computer. There followed discussion and dispute on whether this way of proving was new and different in mathematics. (See article on proof, Chapter 4.)

Two Bizarre Curves

A function has a curve as its graph; and a curve (subject to mild restrictions) is the graph of a function. Today we teach the function as primary. The graph is derived from the function. Until a hundred years ago or so, it was the other way around. As a geometric object, the curve was part of the best understood branch of mathematics. Functions leaned on geometry. Mathematicians were upset when, late in the nineteenth century, they learned of functions with wild graphs impossible to visualize. Example 1 is the Riemann-Weierstrass curve. It's continuous, but at every point it has no direction! Example 2 is the Peano-Hilbert curve. It fills a two-dimensional region—actually passes through *every* point of a square.

I'll give brief sketches of these monsters.

For example 1, van der Waerden's construction is simpler than Riemann or Weierstrass. Start with two connected line segments in the x-y plane. The piece on the left has slope 1 and rises from the x-axis to height 1. There it meets the second piece, which descends back down to the x-axis with slope −1. At the corner where the two segments meet, the slope is undefined.

From this first step, define a second with two peaks, having slope twice that of the first, but height or "amplitude" only half as great. It oscillates twice as fast as the first, but rises only half as high.

In this manner define a sequence of graphs made of connected line segments, each half as high and twice as steep as the previous one, with corners half as far apart, or twice as frequent.

Then—add them up!

The sum converges, because the terms are getting smaller in a ratio of 1:2. As you add more and more terms, the corners get closer and closer, and the slope in between gets bigger and bigger. In the limit, the corners are dense, and the slope in between is infinite. There is no direction.

In example two, start with a square. Cut it into four subsquares, then 16 subsquares, then $256 = 16^2$ subsquares, and so on. At each stage, draw a broken line (polygonal curve) connecting the centers of all the subsquares. You obtain a polygonal line through the centers of many small squares that cover the whole original square. This sequence of polygonal curves converges to a limit curve, which actually passes through every point of the original square.

Square Circles

> Mad Mathesis alone was unconfin'd,
> Too mad for mere material chains to bind,
> Now to pure Space lifts her ecstatic stare,
> Now running round the circle finds it square.
>
> —Alexander Pope, *The Dunciad, Book IV*

This article is an imaginary conversation with David Hume, who rashly presumed there could be no such thing as a square circle (see Chapter 10).

We are not concerned here with the classic Greek problem, proved in modern times to be unsolvable, of constructing with ruler and compass a square with area equal to that of a given circle. By "circle" we mean, as usual, a plane figure in which every point has a fixed distance (the radius) from a fixed point (the center.)

Suppose I live in a flattened, building-less war zone. Transportation is by taxi. Taxis charge a dollar a mile. There are no buildings, so they can run anywhere, but for safety, they're required to stick to the four principal directions: east, west, north, and south.

People measure distance by taxi fare. If two points are on the same east-west or north-south line, the fare in dollars equals the straight-line distance in miles. Otherwise, the fare in dollars equals the shortest distance in miles, traveling only east-west and north-south.

The taxi company has a map showing the points where you can go for $1. These points form a square, with corners a mile north, south, east, and west of the taxi office. In the taxicab metric, *this square is a circle*—it's the set of points $1 from the center.

Yes, a square circle! Inconceivable, yet here it is!

But Hume says, "You can't just change the meaning of 'distance' that way. You know I mean the regular Euclidean distance."

"Very well, David. What's a square?"

"A quadrilateral with equal sides and equal angles."

"Fine. Take your regular Euclidean circle, and inscribe four equally spaced points on it. Then the circumference is divided into four equal sides, and they all meet at the same angle, 180 degrees. Another square circle!"

"No, no!" cries Hume in exasperation. "I mean a quadrilateral with straight sides, not curved sides!"

"O.K. Let the regular Euclidean circle be the equator of the earth. That's a great circle, as straight as a line can be, here on earth. Doesn't *that* have four equal *straight* sides and four equal angles?"

"No, no, no! The four equal angles have to be *right* angles!" shouts Hume.

"That wasn't part of your definition."

"Any way," says he, "call the equator straight if you like, but it isn't! It's a circle! No line on the surface of the earth is straight."

"What's this? You say that what the Mind can conceive is possible, and what the Mind can't conceive is impossible. Now you tell me there's no straight line on the surface of the earth! I grant your mind conceives that, but most minds can't conceive it. Either give up geometry, or give up your notion that what you can't conceive is impossible."

To carry the argument a step further, I leave David Hume behind, and introduce the equation

$$(x^p) + (y^p) = 1.$$

Here x and y are the usual rectangular "Cartesian" coordinates. To avoid irrelevant complications, take x and y in their absolute values, so the graphs are symmetric in the x- and y-axes. p is an arbitrary positive real number, a parameter.

For each different value of p we have a different equation and a different graph in the x - y plane.

If $p = 2$, $x^2 + y^2 = 1$

The graph is the familiar Euclidean circle. For any p bigger than 1 the graph is a smooth convex curve passing through four special points:

$$(1, 1), (-1, 1), (1, -1), (-1, -1).$$

This curve is the unit circle in a new metric, where distance from the origin to the point (x,y) is defined as

the p'th root of $(x^p + y^p)$.

Two cases are especially simple:

$p = 1$ and p infinite.
For $p = 1$,

the graph is exactly the unit square of the "taxicab metric" defined above!

For p infinite, the graph is a larger square, with horizontal and vertical sides: the four lines

$$x = +1, \quad x = -1, \quad y = +1, \quad y = -1.$$

The square for $p = 1$ is inscribed in the square for p infinite. Its corners are at the midpoints of the sides of the larger square.

Call the small square inner, and the large one outer. If you let p increase, starting with $p = 1$, the graph on your computer's monitor will expand smoothly from the inner square through a family of smooth convex curves to become the outer square. A student who watched this transformation could inform Hume, "We have infinitely many unit circles of various shapes. The first and the last are square!"

Embedded Minimal Surfaces

Soap bubbles and soap films are constrained by a physical force called "surface tension." This force makes the surface area as small as possible, subject to appropriate side conditions. In a bubble, the side condition is the volume occupied by the air inside. In a soap film, the side condition is where the film is attached to something—a bubble pipe, another bubble, or the fingers of a child.

It's easy to guess or observe that for a soap bubble the minimal surface is a sphere. For a soap film, with endless different possible boundaries, the problem is more complicated.

One basic fact is clear. In order to have minimal area "in the large" (globally) the film must have minimal area "in the small" (locally). That is, if you mentally mark out any small simple closed curve in the soap film, the area of film enclosed by that curve must be the smallest area that can be enclosed by that curve.

Because of this "local" property, the soap film surface satisfies a certain complicated-looking partial differential equation discovered by Joseph-Louis Lagrange in 1760.

This suggests the famous "Plateau's problem": Given a curve in 3-space, construct a soap film (a minimal surface) having that curve as boundary. J. A. F. Plateau was a blind Belgian physicist who made the problem known to the world in 1873. Jesse Douglas, a New York mathematician, won a Fields Medal in 1936 for his solution.

In the late nineteenth-century Karl Weierstrass, Bernhard Riemann, Hermann Amandus Schwarz, and A. Enneper discovered a number of interesting new minimal surfaces. For years the corridors of university mathematics departments were lined with glass-fronted cabinets displaying plaster models of these surfaces.

It turned out that a mathematical minimal surface need not be a physical one. The mathematical conditions permit the surface to intersect itself, which soap film doesn't ordinarily do. A surface is called "embedded" if it has no self-intersection.

The use of computer graphics, starting some fifteen years ago, has revitalized the subject by making possible the visualization of complicated surfaces formerly described only by equations. The computer pictures often reveal instantly whether a surface has self-intersections, and may show other properties of the surface that can be used to provide rigorous proofs. An infinite number of new complete embedded minimal surfaces have been found in this way. David Hoffman and Jim Hoffman, whose computer-generated video is the basis of our dust-jacket picture, have been leaders in this work.

How Imaginary Becomes Reality

In these notes we have already "constructed" the natural numbers from the empty set. Now we go the rest of the way. Step by step, we construct the integers (positive and negative whole numbers); then the fractions or rational numbers; then the real numbers, rational and irrational; and at last the complex numbers. I show five different ways to construct the complex numbers! And then we go even further, to exotic creatures called quaternions. These extensions of number systems show where the axiom-theorem model is misleading. We don't just obey axioms, we modify them.

Creating the Integers

From the natural numbers we wish to construct the integers—the natural numbers *plus* zero *plus* the negative whole numbers. We can subtract one natural number from another—for instance, 3 from 7—if the former isn't bigger than the latter. But with the natural numbers, we can't subtract a larger from a smaller. "Seven from three you can't take," say the first-graders.

Mathematicians have two ways to extend a mathematical system. One is brute force—create a needed object by fiat. For instance, there's no natural number x such that

$$x + 1 = 0.$$

But we might need such a number. (For instance, to keep track of money we owe.) No problem. Just make up a symbol: -1 and state ***as a definition*** that

$$-1 + 1 = 0.$$

That's how it's done in school.

But can we really create what doesn't exist, just by definition? Who gave us a license for such presumption?

The fact is, it's not necessary to "create" anything. We can use a more sophisticated approach known as "equivalence classes." Since we want -1 to equal $(0 - 1)$ and $(1 - 2)$ and $(2 - 3)$ and so on, we just collect all these ordered pairs into a great big "equivalence class":

$$\{ (0 , 1), (1, 2), (2, 3) \ldots \}$$

It includes infinitely many elements; each element is an ordered pair of natural numbers. Yet it deserves and has the name you expect: -1.

My definition of equivalence class was vague. I just wrote down three members of this infinite class "to give you the idea," and then wrote three dots. . . . I can't possibly write the complete list. We need a membership rule for the class.

We find this by elementary-school arithmetic. When does

$$a - b = c - d$$

without using minus signs? Of course, when

$$a + d = b + c.$$

So we make a precise definition of equivalence:

$(a - b)$ is equivalent to $(c - d)$ if and only if
$a + d = b + c$.

To make a fresh start, not depending on a fiat, we temporarily give up the expression $a - b$, and instead write $\{a,b\}$—an ordered pair in curly brackets.

$$\{a, b\} = \{c, d\} \text{ if and only if } a + d = b + c.$$

We have to figure out how to add, subtract, and multiply the equivalence class of $\{a,b\}$ and the equivalence class of $\{c,d\}$. (Division isn't generally possible, since we don't have fractions yet.) Now, from junior high school we know the rules to add, subtract, and multiply positive and negative whole numbers. It's simple to rewrite those rules in terms of ordered pairs in curly brackets. You can do it, if you wish.

One class plays a special role:

$$\{ (0, 0), (1, 1) \dots \}.$$

It's the additive identity, and has a special name—zero. But that name is already in use, for the whole number before 1. So we allow the same word to have two meanings.

We say two classes are "negatives" of each other, or "additive inverses," if their sum is zero. You can easily check that for any natural numbers *a* and *b*, the class of (a, b) is the negative of the class of (b, a). That is to say,

$$\{a, b\} + \{b, a\} = \{0, 0\}.$$

This means, in particular, that each equivalence class has one and only one negative or additive inverse.

In every equivalence class, there's exactly one ordered pair that includes the natural number 0. If that pair is {a, 0}, we call that equivalence class "a." (It's in a sense the "same" as the natural number a.) And if that pair is {0, a}, we call it −a.

We Have Constructed the Integers, Including the Negative Numbers!

"Constructed" out of some infinite sets, to be sure. Rather than ordered into existence "by fiat."

Why $-1 \times -1 = 1$

In extending from positive whole numbers to integers we preserve ***all*** the rules of arithmetic except ***one***. The rule we give up is:

No number comes before 0.

But we still have:

Rule A $a \times 0 = 0 \times a = 0$, for all a.

So in particular

Rule A′ $-1 \times 0 = 0.$

And we still have

Rule B $a \times 1 = 1 \times a = a$, for all a.

So in particular

Rule B′ $-1 \times 1 = 1 \times -1 = -1$

And we still have the distributive law:

Rule C $x \times (y + z) = (x \times y) + (x \times z)$.

Consider

$$(-1) \times (-1 + 1).$$

This is -1×0, which is zero, by Rule A'. On the other hand, by Rule C,

$$(-1) \times (-1 + 1) = (-1 \times -1) + (-1 \times 1).$$

The last term on the right equals -1, by Rule B'. So

$$0 = (-1 \times -1) + (-1).$$

That is to say,

$$(-1 \times -1)$$

is the additive inverse of -1. But -1 has a unique additive inverse: 1.

So -1×-1 is 1, as we claimed.[3]

Creating the Rationals

In the course of human progress people acquired property and money. So they needed fractions. When a baker had a whole loaf and a half, he had to know he had three halves to sell. Anybody can add

$$\frac{1}{2} + \frac{1}{2} = \frac{2}{2} = 1.$$

Confusion arises with improper fractions:

$$\frac{4}{5} + \frac{4}{5} = \frac{8}{5},$$

and still worse with unequal divisors:

$$\frac{1}{3} + \frac{1}{8} = ??.$$

But people did manage to extend addition and multiplication to fractions (both positive and negative.) This enlarged system is called "the rational numbers." ("Rational" meaning, not "reasonable" or "logical," but just *ratio* of whole

[3] Thanks to Howard Gruber for suggesting this example.

numbers.) With these numbers, any problem of addition, subtraction, multiplication, or *division* (except by zero) has a solution.

We pay a penalty for this enlargement. The natural numbers are ordered "discretely"—every one has a unique follower, and all but 1 have a unique predecessor. This beautiful property makes possible a powerful method—"proof by induction." It's no longer true for the rationals.

Ordinarily we write fractions as a/b, but I will temporarily write them as ordered pairs (in *square* brackets, [a,b]).

Since

$$\frac{1}{2} = \frac{2}{4} = \frac{3}{6},$$

the rule now is,

$$[a,b] = [c,d] \text{ (or } a/b = c/d) \text{ if } ad = bc.$$

This defines "equivalence classes of pairs of integers"—what we call "equal fractions" in the fourth grade. The ones we used when we practiced reducing fractions to lowest terms.

With fractions, as with negatives, we need rules for calculating with these new ordered pairs. And again, we just take the known rules for fractions and rewrite them in terms of ordered pairs in square brackets. We have constructed the rational numbers! Jacob Klein shows that to the Greeks, number meant "positive whole number greater or equal to 2." Number 1 wasn't like other numbers. Fractions were a commercial and practical necessity, but they weren't *numbers*. Klein writes that the broadening of "number" to include positive fractions took place only in the late middle ages and early Renaissance, and with difficulty.

Why $\sqrt{2}$ Is irrational

The most famous theorem in Euclid is the "Pythagorean": "In any right triangle, the sum of the squares of the lengths of the two shorter sides equals the square of the length of the long side" (the "hypotenuse"). You can construct a pair of right triangles by drawing a diagonal in a square of side 1. Then Pythagoras's theorem says the diagonal has length $\sqrt{2}$. On the other hand, the Pythagoreans also discovered that *there is no ratio equal to* $\sqrt{2}$!

Since it doesn't exist, there's nothing to exhibit or construct. All the proof can do is show that the presumption such a ratio exists is absurd. This is called "indirect proof." Suppose that for some pair of numbers p and q,

$$(p/q)^2 = 2.$$

If so, p/q can be put in "lowest terms"—p and q should have no common factor. In particular, they don't both have 2 as a factor—they aren't both even. Multiplying both sides by q^2 gives

$$p^2 = 2q^2.$$

A factor 2 is visible on the right side, so the right side of the equation is an even number. Therefore p^2, the left side of the equation, is also even. It's easy to check that the square of an even number is always even, the square of an odd number always odd. Since p^2 is even, p is even. That means p is twice some other whole number. Let's call it r, so $p = 2r$. Then

$$p^2 = 4r^2.$$

We replace p^2 by $4r^2$ in the previous equation, and get

$$4r^2 = 2q^2,$$

which simplifies to

$$2r^2 = q^2.$$

This is just like the equation $p^2 = 2q^2$ we started with, but p is replaced by q and q by r. So the same argument as before proves that q is even, as p was proved to be even. But p and q aren't allowed to both be even. CONTRADICTION!

The contradiction shows that the presumption that such a fraction p/q exists is impossible, or absurd.

If we only have whole numbers and ratios, we're stuck with the conclusion that $\sqrt{2}$ doesn't exist. It exists as a line segment, the diagonal of the unit square, but not as a number. The diagonal of the unit square does not have a length! Yet, using operations of Euclidean geometry, we can add it to other line segments, and also subtract, multiply, and divide. Line segments constitute an arithmetical system richer than the system of arithmetical numbers! This impasse suggests we go beyond the rational numbers. We need a theory of irrational numbers.

Creating the Real Numbers—Dedekind's Cut

So we want the "real numbers"—rationals and irrationals together. (The name "real" is in contrast to the imaginary and complex numbers, which we will meet shortly.) We use $\sqrt{2}$ to motivate our construction of the irrationals. No rational number when squared can equal 2 (proof is above). Yet we can approximate $\sqrt{2}$:

1
1.4
1.41
1.414

and so on, as far as our computing budget permits. This sequence converges, but what does it converge *to*? $\sqrt{2}$, naturally. But what *is* $\sqrt{2}$, if it can't be a rational number?

We want mathematics to include $\sqrt{2}$ —and many other irrational numbers, of course. We have to somehow take such "convergent" sequences of rationals, which don't have rational limits, and make them into numbers—"real numbers."

Georg Cantor, Karl Weierstrass and Richard Dedekind each found a way to do this. Dedekind's is especially easy.

Arrange the rational numbers in a row or a line in the usual way, increasing from negative to positive as you go from left to right. By a "cut" Dedekind means a separation of this row into two pieces, one on the left, one on the right. The row can be cut in infinitely many different places. Dedekind regards such a split or "cut" in the rationals as being a new kind of number! He shows in a natural way how to add, subtract, multiply, or divide any two cuts (not dividing by zero, of course). In an equally natural way, he defines the relation "less than" for cuts, and the limit of a sequence of cuts. Once these rules of calculation are laid out, the cuts are established as a number system.

Every rational number x defines an associated cut. The left piece is simply the set of rational numbers less or equal x, and the right piece is the set of rationals greater than x. By this association between cuts and rational numbers, we make the rational numbers a subsystem of the system of cuts. To identify Dedekind cuts as the sought-for "real number system," we must show that they include *all* the rationals and irrationals—all the numbers that can be approximated with arbitrary accuracy by rationals.

I'll be satisfied to show that one particular irrational is included as a Dedekind cut— $\sqrt{2}$. To do so, I must identify a left half-line and right half-line associated with $\sqrt{2}$. What rationals are less than $\sqrt{2}$? Certainly all the negative ones, and also all those whose squares are less than 2. All numbers x such that either $x < 0$ or $x^2 < 2$. That specifies the left piece of the cut, the left half-line associated to $\sqrt{2}$. Its complement is the corresponding right half-line. It's easily verified that when this cut is multiplied by itself, it produces the cut identified with the rational number 2. Among Dedekind cuts 2 does have a square root!

All that's left to prove is that no numbers are missing. Dedekind's cuts provide a limit for every convergent sequence of rationals, but we need more. We need a limit for every convergent sequence of *real* numbers—every convergent sequence of cuts. This property is called completeness. The proof is in every text on real analysis and many texts on advanced calculus. I give the essence of it. Let a_n be a convergent sequence of Dedekind cuts (real numbers.) We want to produce a cut a which is the limit of this sequence. We know that every cut a_n is the limit (in many ways) of a convergent sequence of rational numbers. So we replace each a_n by an approximating rational number, choosing the rational approximation more and more accurately as we go out in the a_n sequence. This is easily shown to be a convergent sequence of rationals, and it's easily shown that its limit cut is the limit of the original sequence of cuts.

These constructions are "existence proofs." If you believe Dedekind cuts exist, you have proved that the real numbers exist.

What's the Square Root of -1?

Does $\sqrt{-1}$ exist? There's no real number that yields -1 when squared. That's the reason we say $\sqrt{-1}$ doesn't exist.

Yet in our next breath we bring it into existence!

I'll show you five different ways to do it.

The simplest way is the high-school way. Just ***define*** i as a "quantity" that obeys the laws of arithmetic and algebra in all respects, except that

$$i^2 = -1.$$

If you wish, instead of a "quantity" you may call it a "symbol," which ***by definition*** satisfies

$$i^2 = -1.$$

This approach is direct. It is clear cut. i is treated algebraically like any "letter" or "indeterminate." It can be added and multiplied. These operations and their inverses obey the same commutative, distributive, and associative laws as the real numbers do. The only difference is

$$i^2 = -1.$$

Real multiples of i, like $2i$ or $-3i$, are called "imaginary" or "pure imaginary." Numbers of the form $z = x + iy$, where x and y are real numbers, are called "complex." x is called "the real part" of z, and y is "the imaginary part."

Either x or y or both can be 0, so the imaginary numbers and the real numbers are among the complex numbers! (0 is the only complex number that is both real and imaginary.) This shouldn't be a shock. The positive and negative whole numbers (the integers) are among the rational numbers (fractions.) When we enlarge a number system, we want the numbers we start with to be included among the numbers we "construct."

But since no real number satisfies $x^2 = -1$, is it legitimate to simply "introduce" the square root of -1? Isn't this cheating?

We've seen that pretending some number equals $1/0$ leads to disaster. If $1/0$ is fatal, how can we be sure $\sqrt{-1}$ is O.K.?

One answer might be that analysis with complex numbers is a powerful theory that has never led to a contradiction. That would be saying, "We never had trouble so far, so we never will have trouble." A dubious defense. To resolve such worries, we renounce "introducing" or "creating" the square root of -1. Instead, we'll ***find*** it, already there! As promised, we'll do it in five different ways.

1. A point in the x-y coordinate plane.
2. An ordered pair of real numbers.
3. A 2-by-2 matrix of real numbers.

4. An equivalence class of real polynomials.
5. In the Grand Universal Super-Structure of Sets.

1. After centuries of skepticism, mathematicians accepted complex numbers when they found them "already there," as points in the x-y plane. The complex number $3 + 4i$, for example, is associated to the point with coordinates $x = 3$, $y = 4$. In this way, every complex number gets a point in the coordinate plane, and every point in the plane gets a complex number.

Addition and multiplication of complex numbers turn out to be elementary geometric operations! Addition is just shifting. Adding $3 + 4i$, for example, shifts any complex number 3 units to the right and 4 units up.

Multiplying is stretching and turning. To see this, use polar coordinates. The "polar distance r" of a point $x + iy$ is its distance from the origin. For

$$3 + 4i,$$

the Pythagorean theorem gives $r = 5$.

The "polar angle Q" of a point is the angle between the positive x-axis and the ray from the origin to that point. Multiplying by $3 + 4i$ then turns out to be simply *multiplying* distance by $r = 5$ and *increasing* polar angle by Q.

For $i = 0 + 1i$, evidently $x = 0$ and $y = 1$. The point corresponding to i is on the (vertical) y-axis. So we call the y-axis the "imaginary axis." The "imaginary unit" i is there, one unit above the origin. The (horizontal) x-axis is the "real axis."

For the point i, polar distance r is 1, and polar angle Q is a right angle, 90 degrees.

Multiplying $i \times i$ results in *squaring* r and *doubling* Q.

Since $r = 1$, $r^2 = 1$.

Since Q is a right angle, 90 degrees, its double is two right angles—180 degrees.

This means that i^2 is on the x-axis (the real axis) one unit *left* of the origin. It has coordinates $(-1,0)$. Its complex number is $-1 + 0i$, or simply -1. We have demonstrated geometrically that

$$i^2 = -1.$$

That is, the point i or $0 + i$ is a square root of -1!

Since classical times geometry was the most venerated part of mathematics. Identifying the complex numbers with plane geometry made them respectable.

2. From a more critical viewpoint, something is still missing. The complex numbers are defined by laws of arithmetical operations. They're an independent *algebraic* system, defined prior to their geometric interpretation. We should give an *algebraic* proof of consistency. This was done by Ireland's greatest mathematician, William Rowan Hamilton (remembered also for quaternions, the Hamilton-Jacobi equations, and Hamiltonian systems of differential equations).

To construct the complex numbers, Hamilton creates from the real numbers a simple new kind of thing: an ***ordered pair*** of real numbers. This will look a lot like how we constructed the integers and the rationals—but historically Hamilton's construction of the complex numbers came first!

He defines equality of his ordered pairs:

$$(a, b) = (c, d) \text{ if and only if both } a = c \text{ and } b = d.$$

He defines addition in a very natural way:

$$(a, b) + (c, d) = (a + c, b + d)$$

Multiplication is more complicated:

$$(a, b) \times (c, d) = (ac - bd, ad + bc).$$

Hamilton didn't pull this multiplication rule out of thin air. He just translated the known multiplication of $(a + bi)$ times $(c + di)$ into his notation of ordered pairs. Seen this way, the whole performance looks trivial. But it gets rid of the suspicious i, and replaces it by the innocent $(0, 1)$. Please check that its square is $(-1, 0)$, which is $-1 + 0i$, which is -1.

One should verify the arithmetical laws that complex numbers share with real numbers: commutative laws of addition and multiplication, associative laws of addition and multiplication, and distributive law of multiplication over addition. These verifications are straightforward calculations that the interested reader can carry out.

Notice that ordered pairs whose second component is 0 behave just like real numbers. The zero in the second place never "gets in the way." The multiplicative identity is $(1,0)$; it's algebraically "the same" as 1, the multiplicative identity of the reals.

The pair $(-1, 0)$ is algebraically "the same" as the real number -1. The additive identity is $(0, 0)$; it's algebraically "the same" as the real number 0.

It's straightforward to define subtraction:

$$-(3, 4) = (-3, -4) \qquad -(a,b) = (-a, -b)$$

and to check that

$$(a, b) + (-(a, b)) = (0, 0).$$

Division is trickier. I'll save time by just telling you how to do it—you can check that it works. First, for the special example (3,4),

$$\frac{1}{(3, 4)} = \frac{(3, -4)}{(3^2 + 4^2)} = \left(\frac{3}{25}, \frac{-4}{25}\right).$$

And in general,

$$\frac{1}{(a, b)} = \left(\frac{a}{[a^2 + b^2]}, \frac{-b}{[a^2 + b^2]}\right).$$

which you can check by multiplying

$$(a,b) \times \frac{1}{(a, b)}$$

using the multiplication rule given above. The answer is (1 , 0), or simply 1.

If the definitions of multiplication and division seem baffling, go back to the geometric interpretation of complex numbers to make them intuitively clear.

From a strict formal point of view, one oughtn't to write

$$a + 0i = a.$$

That's "equating apples and oranges." A single real number a just isn't the same as the pair of real numbers (a, 0). Instead of "=" one could say "is isomorphic to."

3. Another way to construct complex numbers uses 2×2 matrices of real numbers instead of ordered pairs. The complex number $a + bi$ corresponds to the matrix

$$\begin{pmatrix} a & b \\ -b & a \end{pmatrix}$$

If you know how to multiply 2×2 matrices, you can check that the usual rules of matrix algebra correspond to the usual rules of addition and multiplication of complex numbers. The number -1 corresponds to the matrix

$$\begin{pmatrix} -1 & 0 \\ 0 & -1 \end{pmatrix}$$

The matrix

$$\begin{pmatrix} 0 & 1 \\ -1 & 0 \end{pmatrix}$$

gives -1 when squared. We are entitled to call this matrix "i"!

We found a square root of -1 by interpreting -1 as a 2×2 matrix. What does this say about existence of $\sqrt{-1}$? It exists if you interpret -1 the right way!

4. A fourth way of finding $\sqrt{-1}$ is inspired by a branch of modern algebra called Galois theory, after Evariste Galois. He was a student, killed in a duel in 1838 at age 21, before being recognized as a precocious genius.

Instead of matrices or ordered pairs we use "polynomials with real coefficients." For instance,

$$5x^4 + 3x^3 + 7x^2 - x^1 + 5.$$

We divide all our polynomials by $x^2 + 1$. Why? Because the thing we're after, i, is a root of $x^2 + 1$!

As in division of numbers, so in division of these polynomials, we get a quotient and a remainder. And the remainder has ***degree*** less than the ***degree*** of the divisor. We're dividing by $x^2 + 1$—a second degree polynomial—so the remainder has degree 1 or 0. There might be ***zero*** remainder, or a constant remainder different from zero (a zero-degree polynomial), or a first-degree remainder—a polynomial of the simple form $ax + b$.

Two polynomials, whatever their degree, are ***equivalent*** if they have the ***same remainder*** on division by $x^2 + 1$. This equivalence splits the polynomials into equivalence classes—sets of polynomials having the same remainder on division by $x^2 + 1$.

An equivalence class is a sack. We're putting polynomials into sacks. All the polynomials in any sack have the same remainder, which is some polynomial of degree 1 or 0. The polynomials that are multiples of $x^2 + 1$, including the number 0, all have zero remainder, so that sack, or if you will that equivalence class, is the zero class, the zero of this algebra.

It's straightforward to define operations between classes or sacks—multiplication, addition, subtraction, division, additive inverse, multiplicative inverse. Everything is done

"mod $(x^2 + 1)$."

Meaning: "Whenever $x^2 + 1$ shows up, throw it away." It's equivalent to zero, because the remainder of $(x^2 + 1)$ on division by $(x^2 + 1)$ is 0.

The multiplicative inverse of the sack of polynomials with remainder $(ax + b)$ is the sack of polynomials with remainder

$$\frac{(-ax + b)}{(a^2 + b^2)}$$

Why? When multiplied together, they yield a polynomial whose remainder is 1, which, naturally, is the multiplicative unit in this algebraic structure.

And of course its additive inverse, which we denote by -1, is the sack of polynomials that leaves the remainder -1.

Now the big question. What about a square root of -1? The answer is so easy, it feels like swindle.

Since $x^2 + 1$ is equivalent to 0, or as an equation, $x^2 + 1 = 0$, then subtracting 1 from both sides,

$$x^2 = -1.$$

Hey, that's it! We've found the thing that when squared equals minus one! It's the equivalence class containing the simple special polynomial x. If you prefer, its the class with remainder $ax + b$, where $a = 1, b = 0$.

In a fussier notation, $x^2 = -1$ modulo $(x^2 + 1)$. So x^2 is "congruent" (equivalent) to -1, and x is equivalent to $\sqrt{-1}$.

I'll say it once more. x^2 is equivalent to -1 because both give the same remainder on division by $x^2 + 1$. (Or, put even more simply, adding 1 to either gives 0.) If x^2 is equivalent to -1, that means x is equivalent to the square root of -1. So x in our algebra of polynomial equivalence classes is "the same" as the complex number i! Polynomials with remainder 1 are equivalent to the real number 1. A combination of x and 1, say, $ax + b$, corresponds to the complex number $ai + b$. All our equivalence classes correspond to remainders of the form $ax + b$, so the equivalence classes and the complex numbers are in a one-to-one correspondence. They're "isomorphic." These sacks correspond precisely to the complex numbers!

Why go through all this when we can just adjoin *i*? Because adjoining something new and prescribing rules for it to follow is a leap in the dark. In using equivalence classes, on the other hand, we add nothing and risk nothing. We just notice what's there. The step from real numbers to real polynomials involves bringing in *x*, but we don't require x to satisfy any weird conditions (like $x^2 = -1$.) We just divide by the polynomial

$$x^2 + 1$$

and look at the remainder. Given two polynomials, we can find their remainders on division by

$$x^2 + 1$$

and see if they're the same or different. This relation automatically sorts the polynomials into classes. Then behold! These equivalence classes are the complex numbers!

Let's compare the three constructions—by ordered pairs, by 2×2 matrices, and by polynomials mod $(x^2 + 1)$.

The construction by ordered pairs uses an algebraic structure created specifically for constructing the complex numbers. Conceptually and computationally, it's the simplest.

The construction by matrices uses something already available—the algebra of 2×2 matrices. It isolates a special subset of them—those whose diagonal elements are equal, and whose off-diagonal elements are equal in absolute value but opposite in sign. One checks that this matrix algebra is closed—sums, products, and inverses of matrices of this type are again of this type. Then, since we know that the identity element is

$$\begin{pmatrix} 1 & 0 \\ 0 & 1 \end{pmatrix}$$

we know that -1 corresponds to

$$\begin{pmatrix} -1 & 0 \\ 0 & -1 \end{pmatrix}$$

and simply check that

$$\begin{pmatrix} 0 & 1 \\ -1 & 0 \end{pmatrix}$$

squared is

$$\begin{pmatrix} -1 & 0 \\ 0 & -1 \end{pmatrix}$$

Just call

$$\begin{pmatrix} 0 & 1 \\ -1 & 0 \end{pmatrix}$$

"i," and you have

$$i^2 = -1$$

You could say Hamilton "constructed" the complex numbers with his algebra of ordered pairs. In the matrix approach, you can't say anything has been *constructed*—the matrices are here already. You might say we "isolated" or "discovered" the complex numbers embedded in the algebra of 2×2 matrices.

What about the method of polynomials mod $(x^2 + 1)$? Here the objects that correspond to Hamilton's ordered pairs or to 2×2 matrices are equivalence classes of polynomials mod $(x^2 + 1)$. (See section below on "Equivalence Classes.") We take all the polynomials that have the same remainder, say $2x + 1$, and throw them into the same sack. We think of the sackful—the whole class of mutually equivalent polynomials—as a single object, which can be added to or multiplied by any other equivalence class. The multiplicative identity is the class of all polynomials with remainder 1. -1 is the class of all polynomials with remainder -1. x is a square root of -1, because the remainder when x^2 is divided by $(x^2 + 1)$ is -1.

Proof: $x^2 = [1 \times (x^2 + 1)] - 1.$

5. Finally, let's see how the complex numbers might be regarded by some anonymous set theorist.

There are two approaches to set theory. One is axiomatic. If something satisfies the 12 axioms of Zermelo and Frankel, it exists. The other way is constructive. Start with the empty set, and step by step, using axiomatically authorized set-theoretic operations, construct ever bigger uncountable sets. Everything you can get by iterating uncountably infinitely often the set-theoretic operations of enlargement is thought to have *already* existed, in advance. Modern set theory is a fascinating and difficult study.

The most famous example of constructing by means of equivalence classes was Frege's "construction" of the natural numbers as equivalence classes of sets. He ended his career resigned to the failure of his set-theoretic foundation. Yet those ideas continue to permeate philosophical logic and set theory.

Where does this put the complex numbers? In the number system they're at the top of the heap, but in the grand set-theoretic structure they're near the bottom. Whether there's a number whose square is -1 is of little set-theoretic interest. But if by chance you want a square root of -1, you have to look in the set-theoretical structure. There isn't any place else!

Recall Hamilton's ordered pairs. Forming ordered pairs is licensed by one of Zermelo's axioms, so all ordered pairs always existed, whether Hamilton knew it or not. Hamilton gave his ordered pairs an algebraic structure. How is that algebraic structure understood set-theoretically? To explain, let's go back to multiplication of natural numbers. We get a natural number as product. The formula $a \times b = c$ describes a ***function***, the "times" function, which operates on the pair a, b and yields the value c. So the formula is "really" a set of ordered triples, a, b, c. Therefore, it's a subset of the set of ***all*** ordered triples. Since it's a subset of a set, it's a set—it exists, by another of Zermelo's axions. This "proves," in a certain strange sense, that multiplication of natural numbers exists. If we go from the natural numbers to ordered pairs (rational numbers) we get the set of all ordered triples of pairs—sextuples. A certain subset of that set represents multiplication of rational numbers. It wouldn't be essentially different to treat a ***pair*** of operations, like "plus" and "times." Proceeding further, we would find that the real numbers, the complex numbers, and all their operations, are already there in the grand set-theoretic super-universe. The problem is to find them. That means showing that certain sets have certain required properties. To do that requires the same checking we've been doing.

Are more representations of the complex numbers waiting to be discovered? If you look in the right math book, you'll find a theorem, "There's only one system of complex numbers." If we line up our representations carefully, they look like merely verbal variants of each other. In Hamilton`s ordered pairs, (1,0) is the multiplicative unit, and (0,1) is the imaginary unit. In the matrix representation of complex numbers,

$$\begin{pmatrix} 1 & 0 \\ 0 & 1 \end{pmatrix}$$

is the multiplicative unit, and

$$\begin{pmatrix} 0 & 1 \\ -1 & 0 \end{pmatrix}$$

is the imaginary unit i. The correspondence is obvious.

In the polynomials mod $(x^2 + 1)$, the class of polynomials having remainder $1 + 0x$ is the multiplicative unit, the class having remainder $0 + 1x$ is the imaginary unit. So the standard complex number $a + b\textit{i}$, the ordered pair (a, b), the matrix

$$\begin{pmatrix} a & b \\ -b & a \end{pmatrix}$$

and the equivalence class of a + bx are four names for the same thing.

These correspondences between algebraic systems are called "isomorphisms." They are one-to-one invertible mappings, which preserve algebraic structure. In mathematics teaching an impression is often given that isomorphic systems should be regarded as the same.

The difference between (a, b), the equivalence class of a + bx and

$$\begin{pmatrix} a & b \\ -b & a \end{pmatrix}$$

is regarded as trivial or meaningless. Or their difference is mere notation, like the difference between x^2 as a function of x and t^2 as a function of t. This would be an error. The mathematician describes this situation by saying that the same structure has several representations. The structure is the abstract thing that each representation represents, in a particular language and from a particular viewpoint. An investigation often is possible only by means of some concrete representation. It can be advantageous to have several representations. Several famous theorems are representation theorems—the Riesz theorems, the Radon-Nikodym theorem, the spectral theorems. An attitude that structure is all, representations are trivial, is a serious misrepresentation.

This question comes up in use of coordinates in geometry. Geometric results should be independent of coordinates; therefore, the story goes, they should be proved without coordinates. Yet the coordinate proof may be more accessible to find and to teach.

Anything that claims to be a new representation must be ***substantially new***. Somebody could report a new representation for complex numbers by using 3×3 matrices instead of 2×2. He could simply augment the 2×2 representation by one more row and one more column, all zeroes. This would be new formally but not substantially. Such a change from 2×2 to 3×3 would be uninteresting and obvious. What we think interesting today isn't always what Euler thought interesting, nor what geniuses in 2997 will consider interesting. This question is esthetic. Esthetic questions play a small part in deciding what's correct, a major part in deciding what's interesting. Esthetic considerations are spared little space in the journals, but they're crucial for understanding the development of mathematics.

At present our number system looks stable, although Abraham Robinson's nonstandard real numbers have proved their worth, and John Conway's "surreal numbers" may have a future.

What Are Quaternions?

Hamilton's passion was to find a number system to do for 3 dimensions what the complex numbers do for 2. To define an algebraic structure, each element of which could be identified with a point of x-y-z-space, with addition and multiplication corresponding to translation and rotation in 3-space.

This proved impossible. Hamilton came as near as anyone could.

His quaternions include three independent "imaginaries," i, j, and k. Each of them squared yields -1!

A general quaternion has the form

$$a + bi + cj + dk$$

where a, b, c, d are real numbers.

To multiply quaternions, you have to multiply i, j, k by each other. This was the hard part. Hamilton discovered the system worked if

$$ij = k \qquad jk = i \qquad ki = j$$
$$ji = -k \qquad kj = -i \qquad ik = -j$$

The commutative law has vanished! Instead, an "anticommutative law." This was the first time anyone imagined an algebraic structure without commutativity. Hamilton was so delighted that he carved

$$ij = -ji$$

on a bridge he crossed going to church.

Hamilton and his disciples tried hard to make quaternions useful in mathematical physics. But Gibbs's vector analysis accomplished similar things more conveniently.

Quaternions are hyper-complex numbers. They add, subtract, multiply, and divide. Gibbs's vectors, on the contrary, have two different multiplications, but no division.

Quaternions are four-dimensional—a 3-vector linked to a number. They don't fit in higher dimensions. Gibbs vectors generalize to any dimensions.

Crowe reports the competition between quaternions and Gibbs-Heaviside vectors for modeling electromagnetism. Both formalisms can describe electromagnetic fields. But physicists preferred the one they found more convenient for calculation—Gibbs vectors.

Do and did quaternions exist? They existed as mathematical concepts from the day Hamilton discovered them. But they weren't sitting on that Irish bridge from the beginning of time, patiently waiting to be discovered. And they didn't start to exist on the day Hamilton started trying to fit them to physics.

They're a permanent piece of algebra, and they continue to be proposed for use in physics and engineering. But from the viewpoint of Platonist set theory, the quaternions were always ready and waiting in the grand abstract universal set structure, their anticommutative multiplication merely a certain subset of the set of all sets of sets of sets of sets of empty sets.

Extension of Structures and Equivalence Classes

The extensions of number systems we have just presented are in a sense optional, but in a stronger sense not optional. Nothing in the natural numbers ***logically*** forces us to introduce negatives. The enlargements to integers, rational numbers, real numbers, and complex numbers were all compelled, slowly and reluctantly.

For another example of how these optional enlargements are in a deep sense compulsory, look at the Fibonacci numbers. These are the sequence

$$1, 1, 2, 3, 5, 8, 13, 21 \ldots$$

Each number after the first two is the sum of the two previous ones.

It's obvious that all the Fibonacci numbers are positive integers.

A little analysis shows that they're all combinations of the solutions of this quadratic equation:

$$x^2 = x + 1.$$

You can solve this equation with your high-school quadratic formula. You find two roots, both involving the square root of 5 (in a combination known to fame as the "golden ratio"). Both roots are irrational. But if you combine them, with the right irrational coefficients, you get the Fibonacci numbers! This is a sequence of natural numbers, yet to write a formula for them, we're forced to use an irrational number, $\sqrt{5}$!

It's enough to make you think $\sqrt{5}$ was already there when we learned to count—or even before, since, as Martin Gardner tells us,

$$2 + 2 = 4$$

was already true with the dinosaurs.

No wonder the mathematician in the street thinks $\sqrt{5}$ existed even before Fibonacci.

Another example. The infinite series

$$1 + x^2 + x^4 + x^6 + \ldots$$

converges if the absolute value of x is less than 1. It diverges if absolute x is greater than 1. To see why, notice that if absolute x is greater than 1, then the terms farther and farther out in the series get bigger and bigger. But for convergence they must get smaller and smaller.

For absolute value of x less than 1, this series sums to

$$\frac{1}{(1 - x^2)}$$

This fraction blows up when $x = 1$, because it becomes $1/0$. This is a good reason why the series can't converge for $x = 1$.

On the other hand, there's the series

$$1 - x^2 + x^4 - x^6 + \ldots$$

Like the previous one, this converges if absolute x is less than 1, and diverges for absolute x greater or equal 1. It sums to

$$\frac{1}{(1 + x^2)}$$

This denominator is always greater than or equal to 1, so this fraction doesn't blow up for any real x. Then why should the series diverge, if it's equal to a fractional algebraic expression that is well behaved for all real x?

Try replacing x by $z = x + iy$. That means, let the independent variable run around the complex x-y plane, not just the real x-axis. If you choose $z = i$ (the square root of −1) then the denominator *is* zero—the fraction blows up. There's a singularity on the ***imaginary*** axis at $z = i$, one unit away from the real x-axis. The singularity on the ***imaginary axis*** is responsible for divergence on the ***real*** axis! A phenomenon in real analysis, which, in a reasonable sense, can't be understood in terms of real numbers only. The complex numbers, whether or not we recognize them, are already controlling some of our real-number computations!

Finally, consider the trigonometric functions sin x and cos x and the exponential function e^{ax}, where a is some positive real number that we can choose at will. If the variable x is real, the behavior of this exponential function is completely different from that of the trigonometric functions. The exponential grows steadily from zero at $x = -\infty$, and its rate of growth is a. If a is any positive number, the exponential function grows faster and faster as x increases. The trigonometric functions, in contrast, remain bounded for all x, however large. They oscillate periodically between a minimum of -1 and a maximum of 1. By use of complex numbers, Euler made the astounding and brilliant discovery that these functions are "essentially" the same! If you do the unorthodox thing—choose a to be, not real, but imaginary—then you find that

$$e^{ix} = \text{cox } x + i \sin x.$$

Making the domain of the exponential function imaginary turns it into a combination of sine and cosine!

A gap is yawning. A unification and deeper understanding beckon, which demand going out of the given mathematical structure—allowing the existence of $\sqrt{-1}$ —*changing the axioms.*

How to change the axioms? How to change or enlarge our mathematical structure? These questions go beyond axioms and theorems. As well as working within given axiomatic structures, mathematicians tear structures down, to replace them with others more powerful.

Calculus

Newton, Leibniz, Berkeley

Berkeley's famous *Analyst* (famous in the history of mathematics, forgotten in the history of philosophy) is an attack on the differential calculus of Newton and

Leibniz. The fallacy Berkeley exposed is simple. To compute the speed of a moving body, you divide the distance traveled by the time elapsed. If the speed is variable, this fraction depends on how much time elapses. But we want the speed at one instant—a time interval of length *zero*. For a falling stone, for example, we want its speed when it hits the ground—its final or "ultimate" velocity. But that seems to require dividing by zero—which is impossible.

Newton explained: "By the ultimate velocity is meant that with which the body is moved, neither before it arrives at its last place, when the motion ceases, nor after, but at the very instant when it arrives. . . . And in like manner, by the ultimate ratio of evanescent quantities is to be understood the ratio of the quantities, not before they vanish, nor after, but that with which they vanish."

This gives us a physical intuition of ultimate velocity. But when Newton calculated he used a mathematical algorithm, not physical intuition. Starting with a time interval of positive duration (call it h), he got an average speed depending on h. He simplified the answer algebraically, and finally set $h = 0$. The resulting expression was the instantaneous speed. Newton called it the "fluxion," and the associated distance function the "fluent."

"But," wrote Berkeley, "It should seem that this reasoning is not fair or conclusive. . . . For when it is said, let the increments vanish, let the increments be nothing, or let there be no increments, the former supposition that the increments were something, or that there were increments, is destroyed, and yet a consequence of that supposition, i.e., an expression got by virtue thereof, is retained. Which is a false way of reasoning. . . . Nor will it avail to say that [the term neglected] is a quantity exceedingly small; since we are told that *in rebus mathematicis errores quan minimi non sunt contemnendi.*" ("In mathematics not even the smallest errors are ignored.")

Berkeley admitted that Newton got the right answer, and that his use of it in physics was correct. He merely showed that Newton's reasoning was obscure.

Leibniz was co-inventor, with Isaac Newton, of the infinitesimal calculus. Unlike Newton, Leibniz used "actual infinitesimals," though he couldn't explain coherently what they were. Cavalieri and others had calculated areas by dividing regions into infinitely many strips, each having infinitesimal positive area. Unfortunately, as Huygens showed, this method could give wrong answers. Problems of rates and velocities also led to infinitesimals. Think of a stone that in 2 seconds falls a distance of 4 feet, in 3 seconds a distance of 9 feet, and in general in t seconds a distance of t^2 feet. Leibniz got the stone's instantaneous speed by calculating its average speed over a time interval of infinitesimal duration. The calculation is so easy we do it right now.

Let dt be the duration of an infinitesimal time interval. At the beginning of the interval your watch reads t seconds, where t is some positive number. The distance fallen up to that time is t^2 feet. An infinitesimal time interval of duration dt elapses. Your watch then reads $(t + dt)$ seconds, and the stone has fallen $(t + dt)^2$ feet. So in the infinitesimal time dt, from instant t to instant $t + dt$, the distance

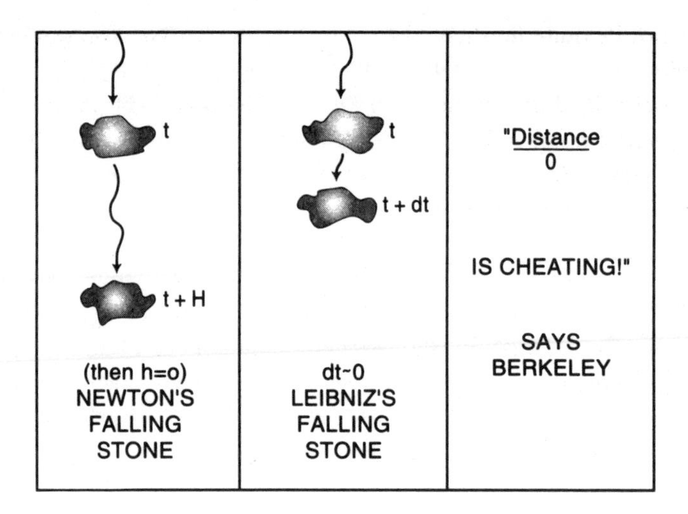

Figure 5. Falling body.

traveled is the distance from its starting point, which was t^2 feet below the stone's initial height, to its ending point, which is $(t+ dt)^2$ feet below the stone's initial height. The distance between the two points is the difference, $(t + dt)^2 - t^2$. "Average speed" is defined in general as a ratio: distance traveled divided by time elapsed. In the present case, that ratio is $[(t + dt)^2 - t^2]/(dt)$. A little algebra simplifies this expression to $(2t + dt)$. dt is infinitesimal, so we "neglect" it—throw it away—and find the instantaneous speed after t seconds: exactly 2t feet per second.

Leibniz's algebra was just like Newton's. He got the same answer as Newton after algebraic simplification, except that his formula had the infinitesimal dt where Newton had the small finite h. Then, instead of setting h equal to zero, as Newton did, Leibniz simply threw away the terms involving dt, *because they were infinitesimal*—negligible compared to the finite part of the answer.

This reasoning was also torn to shreds by Bishop Berkeley. He admitted that the answer, 2t, is right. Berkeley rightly objected to "throwing away" anything not equal to zero, no matter how small. He pointed out that, infinitesimal or not, dt has to be either zero or not zero. If it's not zero, then $2t + dt$ isn't the same as 2t. If it's zero, Leibniz had no right to divide by it. Either way, a fallacy.

Today Berkeley's objections don't disturb us. We show that the average speed *converges to a limit* as the time interval gets shorter. That limit is then *defined* as the instantaneous speed. This limit-and-continuity approach was developed by Cauchy and Weierstrass in the nineteenth century. It is adequate to demystify calculus.

Newton and Leibniz didn't have an explicit definition of limit. The careful use of limits requires explication of the real number system. This subtle task even now may not be quite finished. Still, we see today that Newton was essentially using limits.

Leibniz explained that his infinitesimal dt is "fictitious." This fiction is like an ordinary positive number, but smaller than any ordinary positive number. This is not easy to grasp. How do we decide which properties of ordinary positive numbers apply to dt because it's "just like an ordinary positive number," and which ones don't apply because "it's smaller than any ordinary positive number"? What's the square root of dt? It must be infinitesimal, yet bigger than dt. How many infinitesimals are we going to need? What about the cube root, the fourth root, the tenth root? These puzzles were solved by Abraham Robinson 200 years later, using the theory of formal languages—modern mathematical logic.

The infinitesimal has a fascinating history. At least as far back as Archimedes, it's been used by mathematicians who were perfectly aware that it didn't make sense. It surfaced from underground in the 1960s, when Robinson legitimized it with his "nonstandard analysis."

Nonstandard analysis is the fruit of a century's development of mathematical logic. The basis of it is to regard the language in which we talk about mathematics as itself a mathematical object, obeying explicit formal rules. This formal language then is subject to mathematical reasoning. (Which we carry on, as usual, in ordinary, everyday language, just as we use ordinary language to talk about Basic or C.) Then it makes mathematical sense to say that an infinitesimal is greater than zero and smaller than all the positive numbers ***expressible in the formal language***. When Robinson rehabilitated infinitesimals with his nonstandard analysis, he borrowed the word "monad" from Leibniz's metaphysics. In his nonstandard analysis, a monad is an infinitesimal neighborhood (the set of points infinitely close to some given point.)

So today we have two distinct rigorous formulations of calculus. The creators of the calculus were using tools whose theories were centuries in the future.

A Calculus Refresher

Calculus is the heart of "modern mathematics"—mathematics since Newton. It's the part of mathematics most important in science and technology, the part engineers must know.

It's built around two main problems. The central discovery of calculus is that these problems are related—in fact, as we will see, they're opposites.

The first problem is speed. How fast is something changing? The second main problem is area. How big is some curved region?

First, speed. The speed is simple if it's constant:

$$\text{Speed} = \frac{\text{Distance}}{\text{Time}}$$

Divide distance traveled by time elapsed.

Figure 6. Motion at constant speed.

But speed isn't constant. When you drive you start at speed zero, gradually go to the speed limit, then finally slow down to zero. Your speed varies from instant to instant. What is your speed at some particular instant?

Figure 7.

Example: a body falling in vacuum near the surface of the earth travels 16 t^2 feet in t seconds. How fast is it falling after 2 seconds?

In the time interval between 2 seconds and 2.1 seconds—time lapse of .1 second—it falls

$$16\,(2.1)^2 - 16\,(2^2) \text{ feet} = 6.56 \text{ feet}.$$

Dividing distance by time (.1 second), its average speed was 65.6 ft/sec.

Exercise. Repeat the calculation with a time lapse of .01 second. (You'll get an average speed of 64.16 ft/sec, between time 2 seconds and time 2.01 seconds.) Do it again, with a *very small* time lapse, .001 seconds. (Its average velocity over this time period is 64.016 ft/sec.) ### (### means "end of exercise.")

But I don't want an *average* speed. I want the *exact* speed after 2 seconds! That means a time lapse of *zero.* Division by zero is impossible. The formula

$$\text{Speed} = \frac{\text{Distance}}{\text{Time}}$$

becomes meaningless.

However, without setting time lapse to 0, you've crept closer and closer to 0. You used lapses of .1, .01, .001. and found speeds of 65.6, 64.16, and 64.016.

NOW! A giant conceptual leap! If the average speeds approach a limit as the time lapse approaches zero, we declare, ***as a definition***, the instantaneous speed *is* that limit! In this example, the limit is 64 ft/sec when $t = 2$. It makes sense! We agree, that's what we'll mean by instantaneous velocity.

The notion of speed as a limit took centuries to formulate. Medieval and Renaissance mathematicians calculated rates of change without defining mathematically what they wanted. The founders of the calculus, Isaac Newton and Gottlob Leibniz, fought bitterly about who had priority in the fundamental theorem of calculus (explained below.)

Exercise. Make a graph of this falling body function: distance = time squared, or $d = t^2$.

(I dropped the 16 to simplify your graphing and my calculating.) This is a quadratic function. Its graph is a parabola. Mark the points (2, 4) and (2.1, 4.41) on the parabola. The second is above and right of the first. Draw a straight line (*secant*) between the two. What's the slope of this line? ("Rise over run.")

$\text{Rise} = 2.1^2 - 2.0^2 = .41$
$\text{Run} = 2.1 - 2.0 = .1$

Slope = .41/.1 = 4.1, which we just found is the average velocity (allowing for the factor of 16 which we took out). ***The average rate of change of a function***

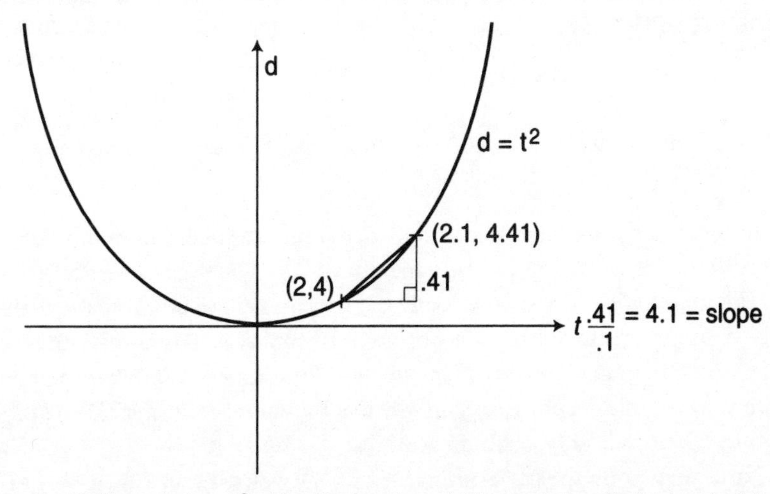

Figure 8. Differentiating x^2 = finding its slope (this graph is called a parabola).

of time is identical to the slope of its secant! Again replace .1 by .01 and .001. The corresponding marks on the graph are creeping closer and closer to (2, 4). The slopes of the secants are exactly the numbers you found to approximate the instantaneous rate of change. As the two points approach closer and closer, and the denominator approaches zero, the secant becomes a tangent, and its slope becomes the instantaneous speed, the ***derivative.*** ###

Calculating the derivative (the speed) is called ***differentiation.*** Simple functions often have simple derivatives. The derivative of t^n is n t^{n-1}. (n is any number, integer or fraction, positive or negative.) The derivative of the natural logarithm of t is 1/t. The derivative of e^t is e^t. The derivative of sine t is cosine t; of cosine t, −sine t.

Exercise. In a way similar to how you found the rate of change of $f(t) = t^2$ at $t = 2$, find the rate of change of that function at an arbitrary time t. Do the same for the cubic $f(t) = t^3$. Check with the formula in the previous paragraph for t^n. ###

Now the second main problem of calculus, area. First, a different-sounding problem. Given the velocity of a moving body, calculate the total distance traveled, at every instant of the trip. This is the opposite of the problem above, where we had the distance and found the velocity.

Start with the simplest case—constant velocity. From 2 P.M. to 3 P.M. you drive a steady 50 miles an hour. How far do you go in that hour? in half an hour? at any time t between 2 P.M. and 3 P.M.?

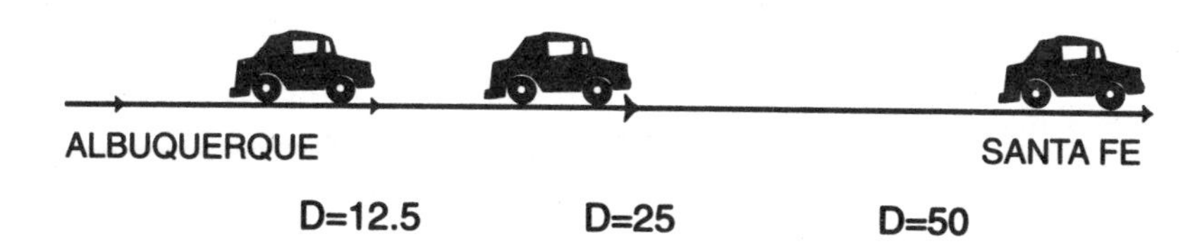

Figure 9. Driving 50 miles at a constant speed of 50 m.p.h.

In one hour you go 50 miles. In half an hour, 25 miles. 50 m.p.h. for t hours goes 50 t miles—where t can be a fraction.

The graph of the constant velocity is a horizontal line 50 units above the time axis. Time is measured on the horizontal time axis. Distance is speed times time—in the graph, height of the velocity line times length on the time axis from start to finish. ***The product of these two lengths is the area of the rectangle they enclose.*** Distance is graphically an area!

Now vary your speed. The graph of v(t), velocity as a function of time, becomes a curve, not a horizontal line. How can we find the distance traveled

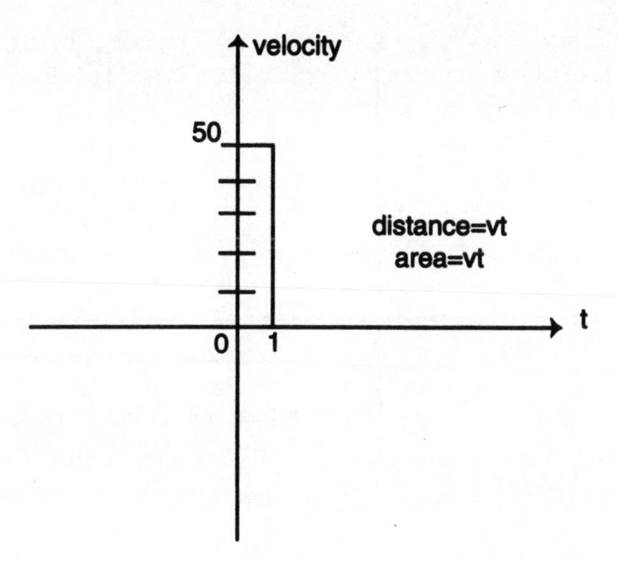

Figure 10. On a time-velocity graph, distance = area.

now? Since we know how to do it in the case of constant speed (horizontal graph), ***replace the curved graph by a piecewise horizontal graph.***

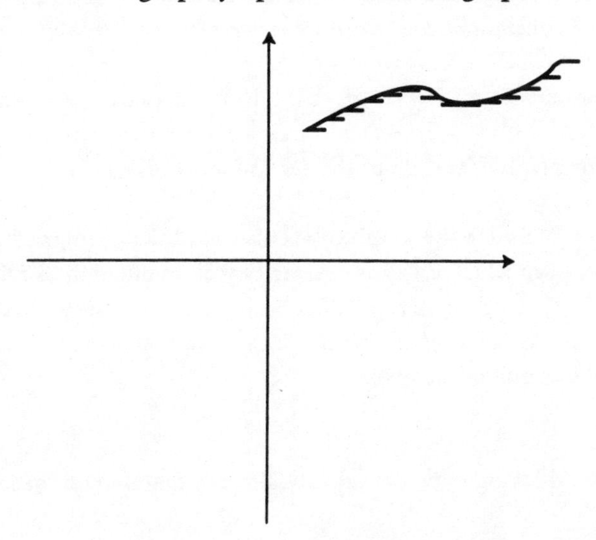

Figure 11. Variable velocity is approximated by piecewise-constant velocity.

Make the speed constant for a second, then a different constant for the next second. The sum of the distances traveled by this piecewise constant, rapidly changing velocity is close to the actual distance. The distance traveled in each second equals the speed in miles per second times the time, one second. In the graph it's

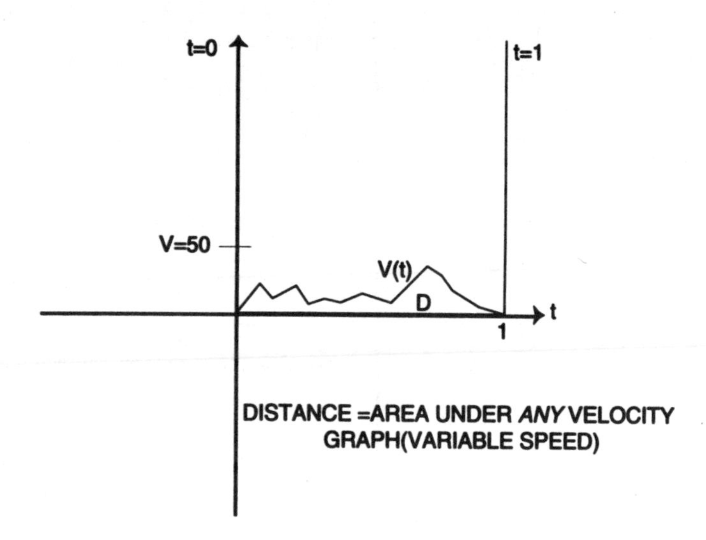

Figure 12.

the area of a skinny vertical rectangle one second wide. These skinny rectangles ***nearly*** fill the region under the velocity curve. The sum of their areas is very close to the total area. If we make their times shorter and shorter, we see that as in the case of constant speed, ***distance traveled equals area under the velocity curve.***

To summarize: To a ***distance*** function d(t) is associated a velocity function v(t), the derivative of d(t). To v(t) in turn is associated an ***area*** function A(t), the area under the graph of v(t) up to the vertical line t. ***The area A(t) is the same as the distance d(t).***

The area A(t) under the graph of v(t) is called the "integral" of v(t). The function d(t), from which v(t) was obtained by differentiation, is called the antiderivative of v(t). Finding A(t) is called "integrating" v(t). We have just found the "Fundamental Theorem of Calculus": the area function of v (the integral of v) is equal to the antiderivative of v:

$$A(t) = d(t).$$

We've been thinking of v(t) as velocity. But any function can be interpreted as a velocity!

So the Fundamental Theorem says: the integral of the derivative of any function is the function itself (except possibly for an additive constant.)

Computing the derivative directly from its definition often is easy; computing the integral directly from its definition often is hard. The Fundamental Theorem shows you how to do the hard part by doing the easy part. Make a collection of differentiation formulas. If in your collection you find a function w(t) whose derivative is v(t), then the integral of v(t) is w(t).

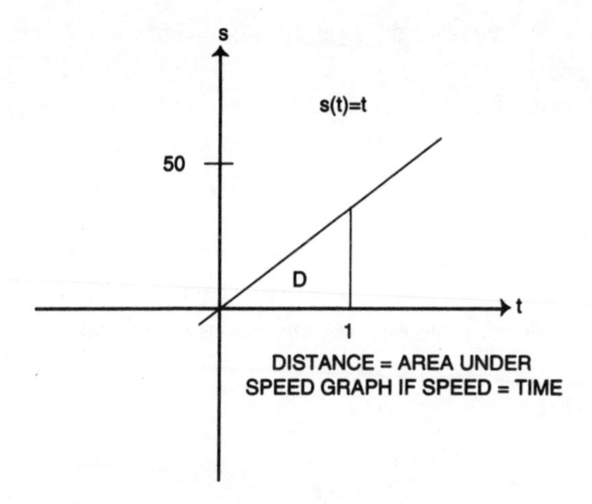

Figure 13.

Let's do a simple area problem—an isosceles right triangle with sides of length 1. By high school geometry the area is 1/2. To do it by calculus, take one side on the x axis, the other side on the vertical line x = 1.

The hypotenuse is the upper boundary of the triangle. It has slope 1 and passes through the origin, so it's a segment of the graph of y = x. The left boundary, x = 0, is a point. The right boundary at x = 1 is the vertical segment $0 < y < 1$.

Cut up the triangle with vertical lines .01 apart. Each piece is long and skinny, almost a rectangle, with a tiny triangle at the top of the rectangle. Each rectangle is .01 wide. How high? The upper boundary of each rectangle is part of hypotenuse, which is on the line y = x. The point on the graph above x = .23, for example, has y-coordinate .23. That's the height of the rectangle at the 23d

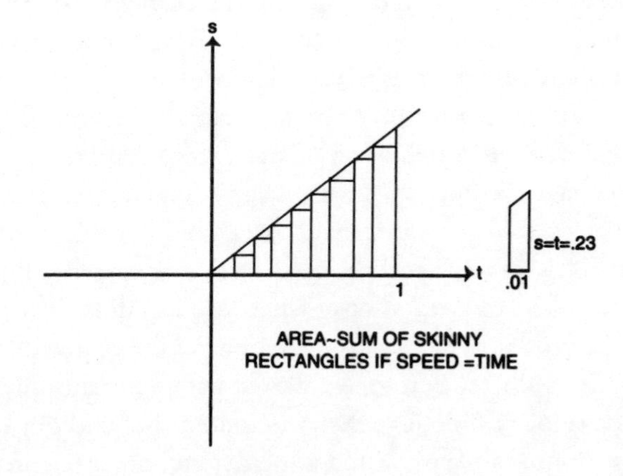

Figure 14.

piece. The area of any rectangle is height times width, so the area of this rectangle is .01 times .23 = .0023.

To approximate the whole area, add the areas of all the rectangles. As the 23d has area .0023, the 38th has area .0038, and so on. Add them all up and factor out .0001. For your approximation to the area of the whole triangle you get

.0001 times (1 + 2 + 3 + . . .).

How many rectangles are there? Your base of length 1 is in pieces .01 = 1/100 wide. So there are 100 rectangles. The last term of the sum in parentheses is 100.

This is a nice puzzle:

1 + 2 + 3 . . . + 100 = ?

A lovely trick does the job. It was discovered by the famous mathematician Karl Friedrich Gauss in school in the first grade. Karl noticed he could write the sum twice—once forward, once backward. The number in the first sum plus its neighbor below in the second sum always add to 101. He had 100 such pairs. So the two sums together equal 100 times 101. The single sum is half of that: 5,050.

In our area calculation we must multiply by .0001. We get for the approximate area .5050. Not too far off from the exact answer, .5. The error, .0050, comes from the little triangles on top of the skinny rectangles. Make the skinny rectangles skinnier and skinnier. The error gets smaller and smaller, as you see from the picture.

It wouldn't do to set the thickness of each little piece *equal* to zero. Then each little rectangle would have area *zero*, and they'd all add up to *zero*, which is wrong.

Calculating area by adding tiny rectangles is called "*integration.*" It's exactly what we did to calculate total distance from variable speed. The method works, it makes sense, so we *define* it to be correct! The area under a curve is *defined* to be the limit of the sum of areas of very skinny inscribed or circumscribed rectangles. There can't be a *proof* that the limit equals the area, because for curved regions we have no other definition of area except that limit!

What you have accomplished isn't just a roundabout way to measure triangles. It works for about any area that comes along. Problems on arc length, volume, probability, mass, electrical capacity, work, inertia, linear and angular momentum, all lead to integrations such as you just did.

What if we do our two calculus operations in the opposite order—first integrate, then differentiate? Before, we started with a distance function, differentiated to get a velocity function, then integrated that and got back our original distance function. Now start with a velocity function, integrate it to get an area and distance function, then differentiate that.

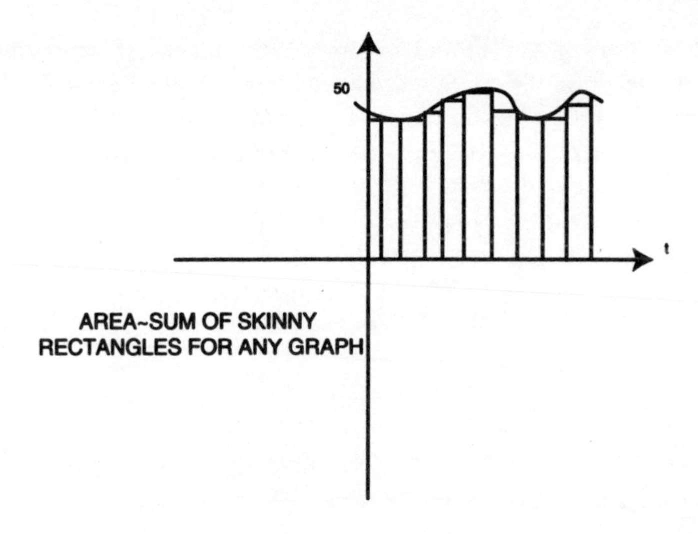

Figure 15.

We can use Figure 13, but now the right side of the triangle is movable. Let t be the distance from the origin. It is a variable. The right boundary could be a stick you slide to the right. As t increases, the stick moves to the right and the area A increases. A is a function of t, call it A(t).

You're going to calculate the rate of change of the area as the stick moves to the right. That is, you will differentiate the variable which we have named A(t).

Our argument does not require the region to be a triangle. As long as it's bounded below by the x-axis, on the left by the y-axis, and on the right by the moving vertical line $x = t$, its upper boundary can be the graph of any function v(x) you like. (Figure 15)

To differentiate—find a rate of change—you must increase the independent variable t by a little bit h, see how much your function increases from A(t) to A(t + h), and divide that increase in A(t) by h. This quotient is the average increase over the interval from t to t + h. Since h is small, the numerator and denominator are both close to zero. Their ratio is close to a limit, which limit you defined as the rate of change of A(t).

In applying the definition of rate of change to the area function A(t), you're working with *two different pictures.* The integration picture computes the area of D, the region under v(x), by cutting up D with many close vertical lines. The differentiation picture computes the rate of change of any function by drawing the secant through two nearby points on its graph. We're applying the differentiation picture to the integration picture, or, if you like, plugging the integration picture into the differentiation picture.

What happens to D and its area A if the right side moves a bit farther right? The region is enlarged by a little additional piece, which differs from a rectangle only in a very small bit at the top. Its width is h, the amount of increase of t. Its height is the height of the upper boundary of D, which is the graph of v(x). So the height of the little added rectangle is v(t), and its area is hv(t). This hv(t) is the increment of A(t). The derivative of A(t) is the increment hv(t) divided by h.

That's hv(t)/h = v(t)!

We have shown that the derivative of A is v. And A is the integral of v.

(The derivative of the integral of v) =
(the integral of the derivative of v) = v.

Symbolically,

D: A ⟶ v
I: v ⟶ A

Differentiation and integration reverse each other, in either order.

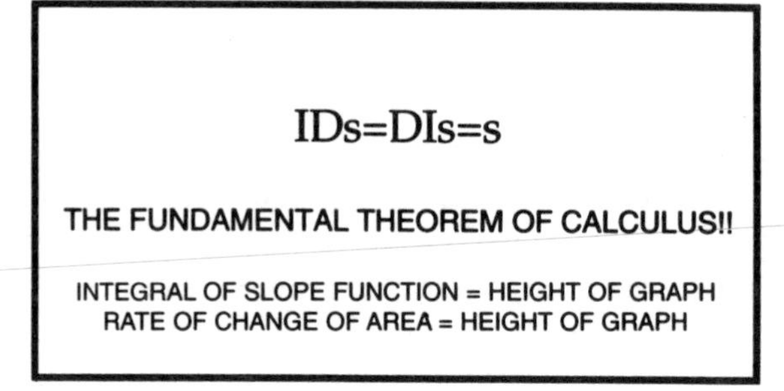

Figure 16.

The Fundamental Theorem is a powerful method of computing areas. Suppose we want to know the area A under the parabola $y = 3x^2$, between $x = 0$ and $x = 3$.

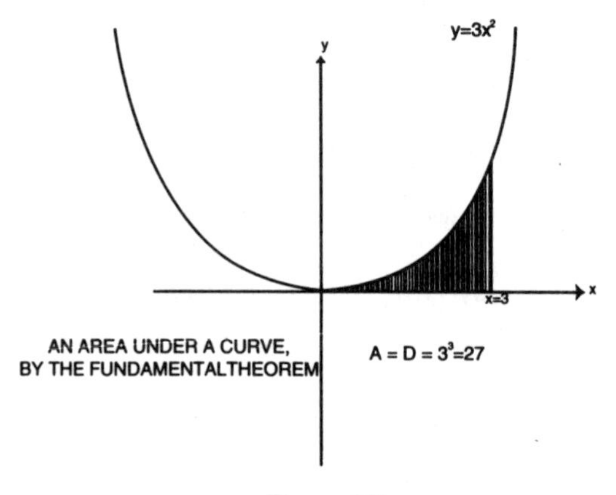

Figure 17.

According to your work a few paragraphs above, this function is the derivative of x^3. Therefore, by the Fundamental Theorem, its area function A(x) is x^3. Between 0 and 2, the area under $3x^2$ is therefore $2^3 = 8$.

Exercise. Use the Fundamental Theorem to compute the areas under the curves $y = x^2$, $y = x^3$, $y = x^4$, all for $0 < x < 1$. ###

The important scientific use of calculus is in solving differential equations. These involve an unknown function and its derivatives. In case the unknown function is a distance, the first derivative is the velocity, and the second derivative is the rate of change of the velocity, the "acceleration." The fundamental law of mechanics, Newton's third law, says

$$f = ma.$$

Force equals mass times acceleration. Often the force is given by some fundamental principle governing the motion under study. Since acceleration is a second derivative of position, Newton's law is a second-order differential equation. To find out how a body moves under the influence of a force, we try to solve this differential equation.

In the case of the planets and the sun, the force is gravity, which is directly proportional to the masses of the attracting bodies, and inversely proportional to the square of their distance. In the case of only two bodies, such as the earth and the sun, the differential equation can actually be solved. By doing so Newton proved that the three laws of Kepler (elliptic orbits; position vector covering equal areas in equal times; and length of year proportional to the 3/2 power of the radius) are equivalent to his law of gravity and his third law. This calculation requires more technique than we assume here.

This triumph of Newtonian calculus and physics ignored the mutual attractions of the planets. If we think of Mars, the earth, and the sun as a system of three bodies none of whose mutual interactions may be ignored, we have the stubbornly intractable three-body problem, which has been tempting and frustrating us for 300 years.

To glimpse how differential equations solve problems of motion, suppose I throw a ball up into the air in a room with a 16-foot ceiling. I want the ball to just barely touch the ceiling. This depends on the velocity V with which I toss the ball up. Determine the correct V.

As usual in elementary treatments, we ignore air resistance. The only force is gravity, which creates a downward acceleration of 32 feet per second per second. We measure distance in feet from ground level, so the initial height of the ball is zero. Now Newton's third law is just

$$\text{Mass} \times \text{Acceleration} = -32 \times \text{Mass}.$$

The minus sign is needed because the acceleration of gravity is ***downward***, ***decreasing*** the height h(t) as a function of time t. Newton's equation gives the

Figure 18.

acceleration, which is the second derivative of h(t). To find h(t) from a(t) we must do two integrations. (Integrating is the opposite of differentiating!)

First divide both sides by the mass. Acceleration a(t)—the second derivative of h(t)—is the derivative of velocity v(t), the first derivative of h(t). We already know that the constant -32 is the derivative of -32t + K, where K is any constant. So from Newton's law, with m divided out,

$$a(t) = -32,$$

integration gives

$$v(t) = -32t + K.$$

To find K, see what the equation says when t = 0. It reduces to

$$v(0) = 0 + K.$$

So the constant of integration K is V, the initial velocity, which we want to find.

(NOTATION! We use the upper-case, capital letter V for "initial velocity"—how many feet per second the ball moves when I release it from my hand. The lower-case v(t) means the velocity varying with time, starting with t = 0, and continuing until the ball returns to the ground. Capital V is just another name for little v(0).)

Since velocity v(t) is the derivative of height h(t), and −32t + V is the derivative of

$$-16t^2 + Vt + L,$$

where L is another arbitrary constant, integrating the above equation for v(t) gives

$$h(t) = -16t^2 + Vt + L.$$

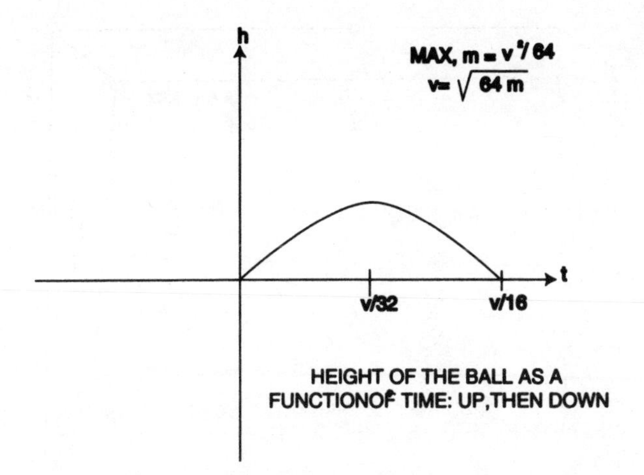

Figure 19.

Again set t = 0 to determine the arbitrary constant. Since h(0), the initial height, is 0, we get L = 0. So we have a formula for height as a function of time:

$$h(t) = -16t^2 + Vt.$$

When is h(t) = 0—when is the ball at the ground?

When h = 0 the equation for h(t) is easy to solve. We find that when h = 0, t = 0 or t = V/16. Why two answers? Because the ball is at the ground twice! First at time 0, before I throw it, and again at time V/16, when it returns to the ground.

Finally, what is M, the greatest height reached by the ball? The height is greatest when the ball stops rising and starts to fall. That is, when v(t) changes from positive to negative. That is, when v(t) = 0. We know v(t) = −32t + V. The maximum is when −32t + V = 0, or t = V/32. To find the maximum height then we need only calculate the height function h(t) when t = V/32. With a little arithmetic simplification, we get

$$h(t) = V^2/64.$$

This tells how high the ball goes, given V—how hard I threw it. But I wanted that height to be 16 feet. So

$$16 = V^2/64$$

and $V = \sqrt{(64 \times 16)} = 32$ feet per second. (In figure 19, instead of a ceiling height of 16 feet, we have allowed the ceiling height to be an arbitrary M.)

We computed this figure here on earth, where g, the acceleration due to gravity, is 32 feet per second per second. What if we visit the Moon? Or Mars? We

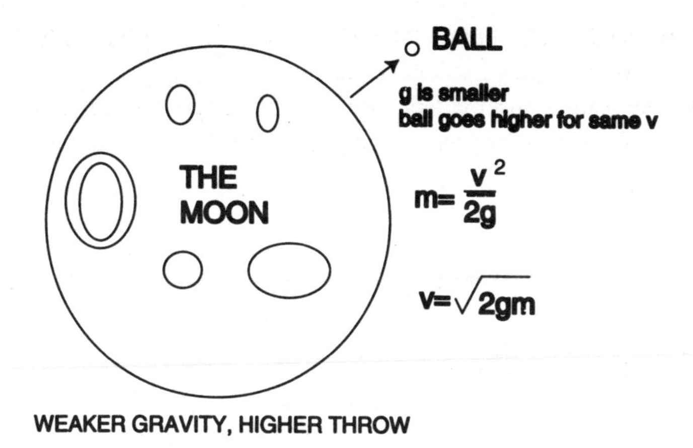

Figure 20.

must replace 32 by the gravitational constant there. On the moon g is much smaller, so we'd need a smaller initial velocity V for a given ceiling height M.

We could repeat the whole calculation, starting with an indeterminate g instead of 32. Or we can just look at our Earthly answer—

$$V = \sqrt{(64 \times 16)}$$

—and make the obvious guess for Mars or the moon. Replace 64 by 2g, 16 feet by M feet, and get $V = \sqrt{2gM}$.

Finally, let's go back to the falling body problem we started with. My friend Nicky fell off the First International Unpaid Debts Building in Miami. The Fire Department is there, life net ready to catch him. Is the net strong enough?

Figure 21. Will Nicky be saved?

Nicky weighs 150 pounds. The FIUDB is 1600 feet high. How fast will he reach the ground? How hard will he hit?

Like the planets and the ball, Nicky's body obeys Newton's third law, with gravitational force 32 feet per second per second. In the equation

$$\text{acceleration} = a(t) = 32$$

we again recognize that both sides are derivatives, or rates of change. Acceleration = $a(t)$ is the rate of change of

velocity = $v(t)$, and
32 is the rate of change (derivative) of
$32\,t$ + an arbitrary constant which we can call B. So
$v(t) = 32t + B$.

What's B? Set $t = 0$. The velocity equation now reads,

$$v(0) = \text{initial velocity} = B.$$

What was Nicky's initial velocity? When he slipped off the roof, he had just started to fall. His initial velocity was zero. So $B = 0$.

Our equation simplifies to

$$v(t) = 32\,t.$$

Again we recognize that both sides are rates of change. Velocity = $v(t)$ is the rate of change of distance fallen, and $32\,t$ is the rate of change (derivative) of

$16\,t^2$ + another arbitrary constant C.
So distance fallen = $d(t) = 16\,t^2 + C$

Nicky's distance fallen at time 0 is 0, so $d(0) = 0$ and therefore $C = 0$.

We have managed to find a formula for the distance Nicky falls as a function of time:

$$d(t) = 16\,t^2.$$

When does he hit the ground? He'll hit when he has fallen a distance equal to the height of the FIUDB, 1600 feet. That is, when

$$16\,t^2 = 1600.$$

Divide both sides by 16: $t^2 = 100$.
Take square roots of both sides: $t = 10$.
So he hits the ground after falling for ten seconds.

We already have his velocity as a function of time—

$$v(t) = 32t.$$

So his velocity at time $t = 10$ is

$$v(10) = 32 \times 10 = 320 \text{ feet/second.}$$

Multiplying by his weight, the life net must sustain a momentum of 150 times 320, or 48,000 foot-pounds per second!

Only calculus can provide this life-saving information.

Should You Believe the Intermediate Value Theorem?

Here's a theorem of elementary differential calculus, the "intermediate value theorem." It seems indubitable, yet it's denied by intuitionist and constructivist mathematicians.

Theorem: If f(x) is continuous, f(0) is negative, and f(1) is positive, then at one or more points c between 0 and 1, f(c) = 0.

Just picture in your mind the graph of f. It's below the x-axis at x = 0, above the axis at x = 1, and doesn't have any jumps. (It's continuous.) It's visually obvious that the graph can't get from below the axis at x = 0 to above the axis at x = 1 without crossing the axis at least once.

But for today's calculus teaching, this visual argument isn't rigorous enough. (A rigorous proof mustn't depend on a picture.)

Here's the rigorous proof:

To start, ask "What's f(.5)?" If it's zero, we're done—the proof is finished. If it isn't zero, it's positive or negative. (This is the Law of Trichotomy.)

Suppose it's positive. Then *f* has to change from negative to positive in the closed half-interval from 0 to .5.

What's f(.25)? Again, it's 0, + , or −. If +, *f* has to change from negative to positive in the closed quarter-interval from 0 to .25. If f(.25) is negative, *f* has to change from negative to positive in the closed quarter-interval from .25 to .5.

Continuing, we generate a nested sequence of closed subintervals of width 1/2, 1/8, 1/16, and so on. In each subinterval, *f* changes from negative to positive as x goes from left to right. Such a nested sequence of closed subintervals, with width converging to zero, contains exactly one common point. Call it c. The left sides of all the nested subintervals converge to that point, and so do the right sides. What's f(c)?

In all these nested subintervals, *f* is negative on the left side and positive on the right side. The left sides and right sides both converge to c. Since *f* is continuous, the values of *f* approach some limit there.

The value of *f* at all the left sides is negative. A sequence of negative numbers can converge to a negative number or to zero, but not to a positive number. On the other side, approaching c from the right, the values of *f* are positive, and their limit can't be negative. Since f is continuous, the two limits are the same, f(c).

So f(c) equals both a nonpositive limit from the left and a nonnegative limit from the right. There's only one number that's both nonpositive and nonnegative—zero. Therefore, f(c) = 0.

End of proof.

This is a fair sample of proof in modern (since 1870) analysis or calculus. Is it constructive? Teachers who teach it and pupils who learn it think it's constructive. The point c seems to be constructed—approximated with any degree of accuracy—by straightforward, elementary steps. At the first step we merely check whether f(x) is positive, negative, or zero at $x = 1/2$, $x = 1/4$, and $x = 3/4$. At the second step, we look at f(x) for $x = 1/8$ and $x = 3/8$; at the third step, for $x = 5/8$ and $x = 7/8$. And so on.

The visual proof is already enough for most people. The rigorous proof is perhaps a bit of overkill. What could be left to disagree about, after firing all those bullets into a dead horse?

The trouble is that in thinking of the continuous function *f*, you had in mind something like x^2, something easily computed. However, the theorem makes an assertion about *all* continuous functions, most of which are not elementary. The guru of intuitionism, L. E. J. Brouwer, showed a constructively continuous function f(x) such that it's impossible to constructively determine whether f(0) is positive, negative, or zero. Such determinations are the heart of the proof we just presented. There are constructively continuous functions that are negative at $x = 0$, positive at $x = 1$, but for which there is no constructively defined point c where $f = 0$.

Whether a proof is constructive depends not only on whether it uses the law of the excluded middle (proof by contradiction), but also on whether it needs other theorems whose proof is not constructive.

In this case, we needed the law of trichotomy:

"Every real number is either positive, negative, or zero." (See the section on Brouwer in Chapter 8.) This is a still more elementary theorem in beginning calculus. Its proof is indirect—uses the law of the excluded middle. Constructivists and intuitionists don't accept indirect proof or the law of the excluded middle. They don't accept the law of trichotomy. They don't accept the intermediate value theorem.

To be fair, I add that they don't stop with a purely negative position. Bishop presents *two* constructive intermediate value theorems. In one, he strengthens the hypotheses. In the other, he weakens the conclusion.

What Is a Fourier Series?

We want to calculate the sum of $1/n^2$ as n runs from 1 to infinity through the odd numbers.[4] We'll need a few tricks.

I have to remind you about "cosine," often shortened to "cos." Imagine a circle of radius 1. Through its center, the "origin," draw two perpendicular lines, one horizontal and one vertical. Now let P be any point on the circle, and draw a "ray" connecting it to the center.

We're interested in two numbers associated to P. The first number is the magnitude of its angle, measured between the rightward horizontal axis and the ray to P from the origin. Defying custom, I will call this angle-magnitude x.

[4] See Chapter 4, "Certainty."

The second number is the distance from P to the vertical axis.

If the angle x is given, then that distance, from P to the vertical axis, is thereby determined. Mathematicians refer to such a dependence of one number on another as a "function." This function—distance to the vertical axis as a function of the angle x—is called "the cosine of x," or "cos x" for short. It's a periodic function, because every time x makes a complete revolution around the circle, the cosine returns to its original value.

It's a deep fact of mathematics that every "decent" function f(x) (for example, every continuous function with a continuous derivative) has a Fourier cosine expansion. That is to say, there are numbers a_n such that

$f(x)$ = sum of $[a_n \cos(nx)]$, n running from 0 through all natural numbers.

This is not a triviality. You may accept it on authority, or read the proof in a text on Fourier analysis.

How do we find the coefficients a_n? This delightful trick is what makes Fourier series fun.

You multiply both sides of the equation by cos(mx), where m is an arbitrary integer.

Then integrate both sides from 0 to π.

According to a formula from first-year calculus, all the terms in the sum are zero, except one—the m'th! The infinite sum reduces to a single term!

So the left side of our formula is now the given function f(x), multiplied by cos(mx), and integrated from $x = 0$ to $x = \pi$. The right side is the unknown coefficient a_m, multiplied by the integral of $\cos^2(mx)$ from $x = 0$ to $x = \pi$.

Calculus tells us that if m isn't 0, the last mentioned integral equals $\pi/2$. If m is 0, the integral equals π.

So, dividing, we have (when m isn't 0)

$$a_m = \frac{2}{\pi} \int \cos(mx) f(x) dx$$

The limits of integration are still from 0 to π.

If $m = 0$, the coefficient in front of the integral is $1/\pi$.)

This formula for a_m is the fundamental formula of Fourier series.

Now back to our problem: to sum the squares $1/m^2$ for odd m. We can do it, using our formula for Fourier cosine expansions, if we know a function whose Fourier coefficients a_m are $1/m^2$ for odd m and zero for even m.

Forgive me if I now save time and trouble by doing the usual thing: pulling a rabbit out of a hat.

Here's the rabbit: consider the function

$$f(x) = \frac{\pi^2 - 2\pi x}{8} \qquad 0 < x < \pi$$

In its Fourier cosine expansion the even terms all integrate to zero. For odd m, we can look up the integral in standard tables, and find: $a_m = 1/m^2$.

Just right!

The function is very decent, so we know it's equal to its Fourier cosine expansion for all x between 0 and π. In particular, they're equal at $x = 0$. So we set $x = 0$ on both sides of the equation.

Voilà!

Sum of $1/m^2$ for odd m equals $\frac{(\pi^2)}{8}$.

Brouwer's Fixed Point

Suppose we want to solve an equation of the form

$$AX = X.$$

It is often possible to transform a different-looking equation into this form. X is an unknown function or vector. A is a given "operator." It operates by transforming a function or vector into another function or vector. Familiar operators are multiplication, translation, rotation, differentiation, integration. "AX = X" says the operator A leaves the function or vector X unchanged. We call X a fixed point of A. Solving the equation AX = X means finding the fixed points of A.

For instance, A could be the operation, "Rotate 3-space around the z-axis through 90 degrees," and X would be a vector in three-space. This operator leaves every vector parallel to the z axis fixed, so all those vectors are solutions of the equation. If we restrict A to operate on vectors in the x-y plane ($z = 0$), it becomes rotation around the origin ($x = 0, y = 0$). Now the zero vector is the only fixed point.

A second example is the simple differential equation,

$$\frac{d}{dt}\, b(t) \;=\; b(t).$$

The operator is now d/dt, differentiation with respect to t. The "fixed point" is any function that equals its own derivative. All such functions are of the form

$$f(t) = ce^t,$$

some multiple of the exponential function.

Some operators have ***no*** fixed points. An example is the operator of shifting the x-axis to the right. There's no fixed point. In our example above, rotating the plane through 90 degrees, suppose our plane had been "punctured"—the origin had been punched out. In such a punctured plane, rotation has no fixed point.

Brouwer proved that if A is continuous in a reasonable sense and it maps a ball (in any finite dimension) into itself, then it has at least one fixed point. That's "the Brouwer fixed point theorem." To use it in differential equations, think of the set of possible solutions as a "space." For the important special case of an "ordinary differential equation"—an equation in only one independent variable—the space of solutions is finite dimensional. We may be able to rewrite the equation in the form AX = X, where A maps some finite-dimensional space into itself. If we can show that A maps some "ball"—the set of solutions of "norm" less than or equal to 5, say—into itself, then we know a solution exists. ***The theorem doesn't say how to find it or construct it.***

What Is Dirac's Delta Function?

Enlarging number systems is no longer a popular sport of mathematicians, but we do often enlarge function spaces. Laurent Schwartz of Paris received the Field Medal for his "distributions" or "generalized functions." Important work in the same direction was also done by Solomon Bochner, J. L. Sobolev, and Kurt Otto Friedrichs.

We will construct a distribution by differentiating a nondifferentiable function. Recall that differentiating a function f(x) means finding the slope of its graph. (Slope is rise/run, y-increment divided by x-increment. See Calculus Refresher above.) At a smooth point on a curve the slope is a number, depending on the position x. It's called f'(x). If $f(x) = x^3$, for example, f'(x) is $3\,x^2$. I'll call this "the classical derivative," to distinguish it from the generalized derivative.

What if the graph has jumps or breaks? The simplest jump function is the Heaviside function, H(x). (Oliver Heaviside was a great telephone engineer and applied mathematician in Victorian England.) H(x) is 0 for all ***negative x***, and 1 for all ***positive x***. So the graph is horizontal everywhere, except at x = 0. There it jumps instantly, from 0 on the left to 1 on the right. H'(x), the derivative of H(x), is trivial to calculate where x is not zero. For x positive or negative, H is constant, its graph horizontal, and its slope or derivative H' is 0. But what happens at x = 0? If you connect the two pieces of the graph, you draw a vertical line segment rising ***instantly*** from

$$(x = 0, y = 0) \text{ to } (x = 0, y = 1).$$

The y-increment is 1 and the x increment is 0. The "slope" is 1/0. But there is no number equal to 1/0! (see above.) The "slope" of this vertical segment doesn't exist as a real number.

We're in the position of the Pythagoreans when they discovered that among the rational numbers, $\sqrt{2}$ doesn't exist. They couldn't find a number equal to $\sqrt{2}$. Before distributions, mathematicians couldn't find a function equal to the derivative of H(x). The key is a different way to think about functions.

The classical definition of function is: "a rule that associates to every point in its domain some point in its range." If domain and range are the real line, it's a real-valued function of a real variable, such as cos(x) and H(x). But there's a more sophisticated way to think of a function—as something we call a "functional."

First I'll explain functionals. Then I'll explain how a function can be regarded as a functional—the key insight in Schwartz's theory.

A functional is a function of functions. Its domain is some set of functions. To each function in that domain, it associates a number. A functional operates on a function to give rise to a number. In contrast, an ordinary function maps a number into a number. A functional maps a function into a number.

The simplest functional is "evaluate at $x = 0$." This applies to any function $w(x)$ that's defined at $x = 0$. The outcome is the number, $w(0)$.

Now we have to see how Schwartz interprets an ordinary function $f(x)$ as a functional. As a functional, it must operate on suitable "test functions" $w(x)$. We restrict the test functions to be identically zero for x very large. Say $w(x) = 0$ for all $x > a$ and all $x < b$, for some numbers a and b. Consider the product, $f(x)w(x)$ and its graph. Because $w(x)$ is zero if $x < a$ or $x > b$, the graph of $f(x)w(x)$, together with the x-axis, encloses a finite area, which is the integral of $f(x)w(x)$. (See "Calculus Refresher" above.)

This area or integral is *defined* to be the number that the functional f obtains by operating on the test function $w(x)$. It's written $f(w)$ or (f, w).

A functional by definition doesn't have a value at a number, because it operates on $w(x)$ as a whole, not on a point x. So the standard definition of derivative can't be applied directly to a functional. But there's a natural way to define a "generalized" derivative, using a formula from elementary calculus: integration by parts (explained in every calculus text).

By thinking of $H(x)$ as a functional, we can define its generalized derivative *without ignoring the crucial singularity, the discontinuity at $x = 0$.* The generalized derivative of $H(x)$ is a new entity, a functional. Although it's not a function, it's called the delta function, or *Dirac's delta function $\delta(x)$.*

The physicist Paul Dirac and his followers used this illegitimate creature to good effect without license from mathematicians. Dirac defined it as a "function" whose values are zero for all x different from zero, but infinite at $x = 0$, in such a way that its area or integral is 1! Mathematicians were amused. They knew that infinity is not a number, and the value of a function at a single point can't affect the value of its integral. Since Dirac's $\delta(x)$ equaled 0 at every point except $x = 0$, its integral must be 0, not 1!

But notice that the derivative of the Heaviside function $H(x)$, like $\delta(x)$, is zero everywhere except at $x = 0$. Where H' blows up is precisely where $\delta(x)$ is infinite! Could it be that if we look at things the right way, we'll see that $\delta(x) = H'(x)$? Yes! That's what distribution theory does for us.

If $\delta(x)$ doesn't make sense, how does it work in the physicists' calculations? Now that we understand it, there's no mystery. $\delta(x)$ can be *talked about* as a function, but it's *really* the very functional we started with—evaluation at 0!

As the meaning of "number" changes with enlargement of the number system, the meaning of "function" changes with enlargement of a function space. Generalized functions (distributions) are different from functions. But we can think of them as if they were functions, *because the usual operations on*

functions (addition, multiplication, differentiation, integration) extend to these functionals.

If f(x) is *smooth,* unlike the discontinuous H(x), we can prove that the functional associated with the classical derivative f'(x) is the same as the generalized derivative of the functional associated with f(x).

The result is that ***every function, no matter how rough, is infinitely differentiable in the generalized sense!*** Not only do we find $H'(x) = \delta(x)$, we are able to differentiate the highly singular pseudo function $\delta(x)$ as many times as we like.

Landau's Two Constants

Let f(z) be a function of a complex variable, analytic in a neighborhood of the origin. Then it has a power series expansion centered there, which has some radius of convergence, say R. If there's no point at which f = 0 and no point at which f = 1, then the radius of convergence R is not greater than a certain constant, which depends *only on f(0) and f'(0).* For details, see Epstein and Hahn.

More Logic

Russell's Paradox

Gottlob Frege's hope of making logic a solid, secure foundation for arithmetic was shattered in one of the most poignant episodes in the history of philosophy. Bertrand Russell found a contradiction in the notion of set as he and Frege used it!

Russell's paradox is a set-theoretic pun, a tongue-twister. To follow it, you must first see how it might be possible for a set to *belong to itself.*

Consider "The class of all sets that can be defined in less than 1,000 English words." I have just defined that class, and I used only 15 words. So it belongs to itself! (This is the Berry Paradox, which Boolos exploits in the following article.)

For a simpler example, how about, "The set of all sets." It's a set, so it's a member of itself.

Let's call those two, and any others that belong to themselves, "Russell sets." Let's call the sets that don't belong to themselves "non-Russell sets." Now, asks Russell, what about the class of *all* non-Russell sets? This class ought to be very big, so I'll call it "The Monster." The Monster contains all the non-Russell sets, and nothing else. Is The Monster a Russell set, or a non-Russell set?

According to the law of the excluded middle, every meaningful statement is true or false. According to Frege's Basic Law Five, the statement "*The Monster is a Russell set*" is meaningful. It must be true or false. However, Russell discovered, it can't be true, and it can't be false!

Suppose it's false. That is, *The Monster is a non-Russell set.* Then The Monster belongs to The Monster, by definition of The Monster! It belongs to itself! *The Monster is a Russell set*! But we came to this conclusion by supposing that The

Monster is a non-Russell set. Contradiction!! The supposition that The Monster is non-Russell is impossible.

The reader should repeat the above reasoning, starting with the presumption that The Monster is a Russell set. You will easily verify that, just as in the non-Russell case, the Russell case leads in two steps to a contradiction.

If The Monster is Russell, it must be non-Russell.

If The Monster is non-Russell, it must be Russell.

The law of the excluded middle says The Monster must be either Russell or non-Russell. Both alternatives are self-contradictory. The presumption that The Monster exists leads to either one contradiction or the other contradiction.

The Monster can't exist! There is no Monster!

A New Proof the Gödel Incompleteness Theorem
BY GEORGE BOOLOS[5]

> Many theorems have many proofs. After having given the fundamental theorem of algebra its first rigorous proof, Gauss gave it three more; a number of others have since been found. The Pythagorean theorem, older and easier than the FTA, has hundreds of proofs by now. Is there a great theorem with only one proof?
>
> In this note we shall give an easy new proof[6] of the Gödel Incompleteness Theorem in the following form: *There is no algorithm whose output contains all true statements of arithmetic and no false ones.* Our proof is quite different in character from the usual ones and presupposes only a slight acquaintance with formal mathematical logic. It is perfectly complete, except for a certain technical fact whose demonstration we will outline.
>
> Our proof exploits *Berry's paradox.* In a number of writings Bertrand Russell attributed to G. G. Berry, a librarian at Oxford University, the paradox of *the least integer not nameable in fewer than nineteen syllables.* The paradox, of course, is that that integer has just been named in eighteen syllables. Of Berry's paradox, Russell once said, "It has the merit of not going outside finite numbers."[7]
>
> Before we begin, we must say a word about algorithms and "statements of arithmetic," and about what "true" and "false" mean in the present context. Let's begin with "statements of arithmetic."
>
> The *language of arithmetic* contains signs + and × for addition and multiplication, a name *0* for zero, and a sign *s* for successor

[5] George Boolos was Professor of Philosophy at MIT.

[6] Saul Kripke has informed me that he noticed a proof somewhat similar to the present one in the early 1960s.

[7] Bertrand Russell, "On Insolubilia and Their Solution by Symbolic Logic," in *Essays in Analysis*, ed. Douglas Lackey, George Braziller, New York, 1973, p. 210.

(plus-one). It also contains the equals sign =, as well as the usual logical signs ¬ (not), ∧ (and), ∨ (or), → (if . . . then . . .), ↔ (. . . if and only if . . .), ∀ (for all), and ∃ (for some), and parentheses. The variables of the language of arithmetic are the expressions *x*, *x′*, *x″* . . . built up from the symbols *x* and ′: they are assumed to have the natural numbers (0,1,2, . . .) as their values. We'll abbreviate variables by single letters: *y*, *z*, etc.

We now understand sufficiently well what truth and falsity mean in the language of arithmetic; for example, $\forall \exists y x = sy$ is a *false* statement, because it's not the case that every natural number *x* is the successor of a natural number *y*. (Zero is a counter-example: it is not the successor of a ***natural*** number.) On the other hand, $\forall x \exists y (x = (y + y) \vee x = s(y + y))$ is a true statement: for every natural number *x* there is a natural number *y* such that either $x = 2y$ or $x = 2y + 1$. We also see that many notions can be expressed in the language of arithmetic, e.g., less than: $x < y$ can be defined: $\exists z (sz + x) = y$ (for some natural number *z*,the successor of *z* plus *x* equals *y*). And, you now see that $\forall x \forall y [ss0 \times (x \times x)) = (y \times y) \rightarrow x = 0]$ is—well, test yourself, is it true or false? (Big hint: $\sqrt{2}$ is irrational.)

For our purposes, it's not really necessary to be more formal than we have been about the syntax and semantics of the language of arithmetic.

By an ***algorithm***, we mean a computational (automatic, effective, mechanical) procedure or routine of the usual sort, e.g., a program in a computer language like C, Basic, Lisp, . . . , a Turing machine, register machine, Markov algorithm, . . . a formal system like Peano or Robinson arithmetic, . . . , or whatever. We assume that an algorithm has an ***output***, the set of things it "prints out" in the course of computation. (Of course an algorithm might have a ***null*** output.) If the algorithm is a formal system, then its output is just the set of statements that are provable in the system.

Although the language of arithmetic contains only the operation symbols *s*, +, and ×, it turns out that many statements of mathematics can be reformulated as statements in the language of arithmetic, including such famous propositions as Fermat's last theorem, Goldbach's conjecture, the Riemann hypothesis, and the widely held belief that $P \neq NP$. Thus if there were an algorithm that printed out all and only the true statements of arithmetic—as Gödel's theorem tells us there is not—we would have a way of finding out whether each of these as yet unproved propositions is true or not, and indeed a way of finding out whether or not any statement that can be formulated as a statement *S* of arithmetic is true: start the algorithm, and simply wait to see which of S and its negation $-S$ the algorithm prints out. (It must eventually print out exactly one of S and $-S$ if it prints out all truths and no falsehoods, for, certainly, exactly one of S and $-S$ is true.) But alas, there is no worry that the algorithm might take too long to come up with an answer to a question that interests us, for there is, as we shall now show, no algorithm to do the job, not even an infeasibly slow one.

To show that there is no algorithm whose output contains all true statements of arithmetic and no false ones, we suppose that M is an algorithm whose output contains no false statements of arithmetic. We shall show how to find a true statement of arithmetic that is not in M's output, which will prove the theorem.

For any natural number n, we let $[n]$ be the expression consisting of 0 preceded by n successor symbols s. For example, $[3]$ is $sss0$. Notice that the expression $[n]$ stands for the number n.

We need one further definition: We say that a formula $F(x)$ ***names*** the (natural) number x if the following statement is in the output of M: $\forall \times F(x) \leftrightarrow x = [n]$. (Observe that the definition of 'names' contains a reference to the algorithm M.) Thus, for example, if $\forall x(x + x = ssss0 \leftrightarrow x = ss0)$ is in the output of M, then the formula $x + x = ssss0$ names the number 2.

No formula can name two different numbers. For if both of $\forall x(F(x) \leftrightarrow x = [n])$ and $\forall x(F(x) \leftrightarrow x = [p])$ are true, then so are $\forall x(x = [n] \leftrightarrow x = [p])$ and $[n] = [p]$, and the number n must equal the number p. Moreover, for each number i, there are only finitely many different formulas that contain i symbols. (Since there are 16 primitive symbols of the language of arithmetic, there are at most 16^i formulas containing i symbols.) Thus for each i, there are only finitely many numbers named by formulas containing i symbols. For every m, then, only finitely many (indeed, $\leq 16^{m-1} + \ldots + 16^1 + 16^0$) numbers are named by formulas containing fewer than m symbols; some number is not named by any formula containing fewer than m symbols; and therefore there is a least number not named by any formula containing fewer than m symbols.

Let $C(x,z)$ be a formula of the language of arithmetic that says that x is a number that is named by some formula containing z symbols. The technical fact mentioned above that we need is that whatever sort of algorithm M may be, there is some such formula $C(x,z)$. We sketch the construction of $C(x,z)$ below, in 3).

Now let $B(x,y)$ be the formula $\ni z(z < y \wedge C(x,z))$. $B(x,y)$ says that x is named by some formula containing fewer than y symbols.

Let $A(x,y)$ be the formula $(\neg B(x,y) \wedge \forall a(a < x \rightarrow B(a,y)))$. $A(x,y)$ says that x is the least number not named by any formula containing fewer than y symbols.

Let k be the number of symbols in $A(x,y)$. $k > 3$.

Finally, let $F(x)$ be the formula $\exists y(y = ([10] \times [k] \wedge A(x,y))$. $F(x)$ says x is the least number not named by any formula containing fewer than $10k$ symbols.

How many symbols does F contain? Well, $[10]$ contains 11 symbols, $[k]$ contains $k + 1$, $A(x,y)$ contains k, and there are 12 others (since y is x^1): so $2k + 24$ in all. Since $k > 3$, $2k + 24 < 10k$, and $F(x)$ contains fewer than $10k$ symbols.

We saw above that for every m, there is a least number not named by any formula containing fewer than m symbols. Let n be the least such number for $m = 10k$. Then n is not named by $F(x)$; in other words, $\forall x(Fx) \leftrightarrow x = [n]$ is not in the output of M.

But $\forall x(F(x) \leftrightarrow x = [n])$ is a true statement, since n *is* the least number not named by any formula containing fewer than $10k$ symbols! Thus we have found a true statement that is not in the output of M, namely, $\forall x(F(x) \leftrightarrow x = [n])$. Q.E.D.

Some comments about the proof:

1. In our proof, the symbols are the "syllables," and just as "nineteen" contains $2 << 19$ syllables, so the term $([10]x[k])$ contains $k + 15 << 10k$ symbols.

2. In his memoir of Kurt Gödel,[8] Georg Kreisel reports that Gödel attributed his success not so much to mathematical invention as to attention to philosophical distinctions. Gregory Chaitin once commented that one of his own incompleteness proofs resembled Berry's paradox rather than Epimenides' paradox of the liar ("What I am now saying is not true").[9] Chaitin's proofs make use of the notion of the *complexity* of a natural number, i.e., the minimum number of instructions in the machine table of any Turing machine that prints out that number, and of various information-theoretic notions. None of these notions are found in our proof, for which the remarks of Kreisel and Chaitin, which the author read at more or less the same time, provided the impetus.

[8] Georg Kreisel, "Kurt Gödel, 28 April 1906–14 January 1978." *Biographical memoirs of Fellows of the Royal Society* 26 (1980), p. 150.

[9] Cf. Martin Davis, "What is a computation?" in *Mathematics Today*, ed. Lynn Arthur Steen, Vintage Books, New York, 1980, pp. 241–267, especially pp. 263–267, for an exposition of Chaitin's proof of incompleteness. Chaitin's observation is found in Chaitin, Gregory, "Computational complexity and Gödel's incompleteness theorem," (Abstract) *AMS Notices* 17 (1970), p. 672.

3. Let us now sketch the construction of a formula $C(x,z)$ that says that x is a number named by a formula containing z symbols. The main points are that algorithms like M can be regarded as operating on "expressions," i.e., finite sequences of symbols; that, in a matter reminiscent of ASCII codes, symbols can be assigned code numbers (logicians often call these code numbers *Gödel* numbers); that certain tricks of number theory enable one to code expressions as numbers and operations on expressions as operations on the numbers that code them; and that these numerical operations can all be defined in terms of addition, multiplication, and the notions of logic. Discussion of symbols, expressions (and finite sequences of expressions, etc.) can therefore be coded in the language of arithmetic as discussion of the natural numbers that code them. To construct a formula saying that n is named by some formula containing I symbols, one writes a formula saying that there is a sequence of operations of the algorithm M (which operates on expressions) that generates the expression consisting of $\forall$, x, (, the i symbols of some formula $F(x)$ of the language of arithmetic, $\leftrightarrow$, x, $=$, n consecutive successor symbols *s*, *0*, ***and***). Gödel numbering and tricks of number theory then allow all such talk of symbols, sequences, and the operations of M to be coded into formulas of arithmetic.

4. Both our proof and the standard one make use of Gödel numbering. Moreover, the unprovable truths in our proof and in the standard one can both be seen as obtained by the substitution of a name for a number in a certain crucial formula. There is, however, an important distinction between the two proofs. In the usual proof, the number whose name is substituted is the code for the formula into which it is substituted; in ours it is the true number of which the formula is ***true***. In view of this distinction, it seems justified to say that our proof, unlike the usual one, does not involve ***diagonalization***.

In a later issue of the journal where his proof appeared, Boolos made some interesting remarks about it. "Several readers of my 'New Proof of the Gödel Incompleteness Theorem,' (*Notices*, April 1989, pp. 388–90) have commented on its shortness, apparently supposing that the use it makes of Berry's paradox is responsible for that brevity. It would thus seem appropriate to remark that once syntax is arithmetized, an even briefer proof of incompleteness is at hand, essentially the one given by Gödel himself in the introduction to his famous "On Formally Undecidable Propositions. . . ."

Say that m applies to n if F[n] is in the output of M, where F(x) is the formula with Gödel number m. Let A(x,y) express "applies to" and let n be the Gödel number of −A(x,x). If n applies to n, the false statement −A([n],[n]) is in the output of M, impossible; thus n does not apply to n and −A([n],[n]) is a truth not in the output of M.

What is concealed in this argument is the large amount of work needed to construct a suitable formula A(x,y); proving the existence of the key formula C(x,z) in the "New Proof" via Berry's paradox requires at least as much effort. What strikes the author as of interest in the proof via Berry's paradox is not its brevity but that it provides ***a different sort of reason*** for the incompleteness of algorithms.

Bibliography

Aczel, P. *Non Well Founded Sets.* Stanford: Center for the Study of Language Information, 1988.

Agard, W. R. *What Democracy Meant to the Greeks.* Chapel Hill: University of North Carolina Press, 1942.

Angelelli, I. *Studies on Gottlob Frege and Traditional Philosophy.* Dordrecht: Reidel, 1967.

Anglin, W. S. *Mathematics, A Concise History and Philosophy.* New York: Springer-Verlag, 1994.

Anscombe, G. E. M. and P. Geach. *Three Philosophers.* Ithaca: Cornell University Press, 1961.

Apostle, H. G. *Aristotle's Philosophy of Mathematics.* Chicago: University of Chicago Press, 1952.

Appel, K. and W. Haken. "The Four-Color Problem." In L. A. Steen, ed., *Mathematics Today*, pp. 153–80. New York: Springer-Verlag, 1978.

Appel, K., W. Haken, and J. Koch. "Every Planar Map Is Four-Colorable." *Illinois Journal of Mathematics* 21: 429–567, 1977.

Aquinas, St. T. *The Pocket Aquinas.* Edited by Vernon J. Bourke. New York: Washington Square Press, 1960.

Archer-Hind, R. D. *The Timaeus of Plato.* London: Macmillan, 1888.

Aristotle. *Basic Works.* Edited by R. McKeon. New York: Random House, 1941.

———. *Physics, Book VI*, chapters 2 and 9. In R. M. Hutchins, ed., *Great Books of the Western World*, Vol. 8. Chicago: Encyclopedia Britannica, 1952.

———. *Poetics.* Chapel Hill: University of North Carolina Press, 1986, p. 1451

Ascher, M. *Ethnomathematics: A Multicultural View of Mathematical Ideas.* San Francisco: Brooks/Cole, 1991.

Aspray, W. and P. Kitcher, P., eds. *History and Philosophy of Modern Mathematics, Volume XI, Minnesota Studies in the Philosophy of Science.* Minneapolis: University of Minnesota Press, 1988.

Balz, A. G. A. *Cartesian Studies.* New York: Columbia University Press, 1951.

Bartley, W. W. *Wittgenstein.* Lasalle: Open Court, 1985.

Barwise, J. and J. Etchemendy. *The Liar.* New York: Oxford University Press, 1987.

Baum, R. J. *Philosophy and Mathematics from Plato to the Present.* San Francisco: Freeman, Cooper, 1973.

Beck, L. J. *The Method of Descartes.* New York: Oxford University Press, 1952.

Benacerraf, P. and H. Putnam. *Philosophy of Mathematics.* New York: Prentice-Hall, 1964.

Benacerraf, P. "Frege, The Last Logicist." In P. French, T. Ukehling, H. Wettstein, eds., *Midwest Studies in Philosophy VI.* Minneapolis: University of Minnesota Press, 1981.

Berkeley, G. *The Analyst.* London: J. Tonson, 1774.

Berlin, I., ed. *The Age of Enlightenment. The 18th Century Philosophers.* New York: New American Library, 1956.

Bernays, P. "Comments on Ludwig Wittgenstein's Remarks on the Foundations of Mathematics." In P. Benacerraf and H. Putnam.

Beth, E. W. and J. Piaget. *Mathematics, Epistemology and Psychology.* Translated by W. Mays. Dordrecht: Reidel, 1966.

Bishop, E. *Foundations of Constructive Analysis.* New York: McGraw Hill, 1967.

——. *Aspects of Constructivism.* Las Cruces, N.Mex.: Department of Mathematical Sciences, New Mexico State University, 1972.

——. "The Crisis in Contemporary Mathematics." *Historia Mathematica* 2: 507–17, 1975.

——. "Schizophrenia in Contemporary Mathematics." In Rosenblatt, M., ed., *Errett Bishop: Reflections on Him and His Research.* Providence: American Mathematical Society, 1985.

Bledsoe, W. W. and D. W. Loveland. *Automated Theorem Proving: After 25 Years.* Contemporary Mathematics 29. Providence: American Mathematical Society, 1984.

Blitz, D. *Emergent Evolution.* Boston: Kluwer, 1992.

Bloor, D. *Knowledge and Social Imagery.* Boston: Routledge and Kegan Paul, 1976.

——. *Wittgenstein: A Social Theory of Knowledge.* New York: Columbia University Press, 1983.

Bluck, R. S. *Plato's Meno.* New York: Cambridge University Press, 1964.

Boolos, G. "New Proof of the Gödel Incompleteness Theorem." *Notices of the American Mathematical Society* 36: 388–90, 1989.

——. "A Letter from George Boolos." *Notices of the American Mathematical Society* 36: 676, 1989.

Bourbaki, N. "Foundations of Mathematics for the Working Mathematician." *Journal of Symbolic Logic* 14: 1–8, 1949.

Boyer, C. "Galileo's Place in the History of Mathematics." In E. McMullin, ed., *Galileo Man of Science*. New York: Basic Books, 1968, p. 251.

——. *History of Analytic Geometry*. New York: Scripta Mathematica, 1956.

——. *The History of the Calculus and Its Conceptual Development*. New York: Dover, 1959.

——. *A History of Mathematics*, 2nd. ed., revised by Uta C. Merzbach. New York: Wiley, 1991.

Brouwer, L. E. J. *Life, Art and Mysticism*. In *Collected Works*. Amsterdam: North-Holland, 1975.

——. *Brouwer's Cambridge Lectures on Intuitionism*. Edited by D. Van Dalen. New York: Cambridge University Press, 1981.

Brown, M., ed. *Plato's Meno*. Indianapolis: Bobbs-Merrill, 1971.

Brumbaugh, R. S. *Plato's Mathematical Imagination*. Bloomington: Indiana University Press, 1954.

Bunge, M. *Intuition and Science*. Englewood Cliffs: Prentice Hall, 1962.

——. *The Mind-Body Problem*. New York: Pergamon, 1980.

Carnap, R. *Introduction to Symbolic Logic and Its Applications*. New York: Dover, 1958.

Carr, H. W. *Leibniz*. New York: Dover, 1960.

Castonguay, C. *Meaning and Existence in Mathematics*. New York: Springer-Verlag, 1972.

Chihara, C. *Ontology and the Vicious Circle Principle*. Ithaca: Cornell University Press, 1973.

——. "Wittgenstein's Discussion of the Paradoxes in his 1939 Lectures on the Foundations of Mathematics." *Philosophical Review* 86: 365–81, 1977.

Cohen, D. I. A. "The Superfluous Paradigm." In J. H. Johnson and M. J. Loomis, eds., *The Mathematical Revolution Inspired by Computing*. Oxford: Clarendon Press, 1991.

Cohen, P. "Comments on the Foundations of Set Theory." In Dana Scott, ed., *Axiomatic Set Theory*, pp. 9–15. Providence: American Mathematical Society, 1971.

——. *Set Theory and the Continuum Hypothesis*. New York: W. A. Benjamin, 1966.

Cohen, R. S. and L. Lauden. *Physics, Philosophy and Psychology: Essays in Honor of A. Grunbaum*. Dordrecht: Reidel, 1983.

Cornford, F. M. *Plato's Cosmology*. London: Routledge & Kegan Paul, 1937.

——. *From Religion to Philosophy*. New York: Harper and Row, 1957.

Corrington, R. S. *An Introduction to C. S. Peirce*. Lanham, Md.: Rowman & Littlefield, 1993.

Courant, R. and Robbins, H. *What Is Mathematics?* New York: Oxford University Press, 1948.

Crowe, M. J. *A History of Vector Analysis*. New York: Dover, 1967.

——. "Ten 'Laws' Concerning Patterns of Change in the History of Mathematics." *Historia Mathematica* 2: 161–66, 1975.

——. "Ten Misconceptions about Mathematics and Its History." In Aspray and Kitcher, 1988.

Curley, E. M. *Descartes Against the Skeptics.* Cambridge: Harvard University Press, 1978.

Currie, G. "Frege's Realism." *Inquiry* 21: 1978.

Curry, H. *Outline of a Formalist Philosophy of Mathematics.* Amsterdam: North-Holland, 1958.

D'Alembert, J. L. R. *Preliminary Discourse to the Encyclopedia of Diderot.* Translated by R. N. Schwab. Indianapolis: Bobbs-Merrill, 1963.

Damasio, A. *Descartes' Error: Emotion, Reason and the Human Mind.* New York: Putnam, 1994.

Dancy, J. and E. Sosa. *A Companion to Epistemology.* Oxford: Blackwell, 1992.

Dantzig, T. *Number, the Language of Science.* New York: Macmillan, 1959.

Davis, C. "Materialist Mathematics." *Boston Studies in the Philosophy of Science,* Vol. 15. Dordrecht: Reidel, 1974, pp. 37–66.

Davis, E. W. In Cajori, F., ed., *Teaching and History of Mathematics in the U.S.* Washington, D.C.: 1890.

Davis, M. and R. Hersh. "Nonstandard Analysis." *Scientific American* June 1972, pp. 768–84.

——. "Hilbert's Tenth Problem." *Scientific American* November 1973, pp. 84–91.

Davis, P. J. "Fidelity in Mathematical Discourse: Is 1 + 1 Really 2?" *American Mathematical Monthly* 78: 252–63, 1972.

——. "Proof, Completeness, Transcendentals, and Sampling." *Journal of the Association for Computing Machinery* 24: 298–310, 1977.

——. "Mathematics by Fiat?" *The Two-Year College Mathematics Journal* June 1980.

Davis P. J. and R. Hersh. *The Mathematical Experience.* Cambridge: Birkhauser, 1981.

Davydov, V. *Problemy Razvivayuschego Obucheniaya* (The problems of development-generated learning). Moscow: Pedagogika, 1986.

Dedekind, J. W. R. *Essays on the Theory of Numbers.* Lasalle: Open Court, 1901. Reprinted by Dover Books, 1963.

Dejnozka, J. "Zeno's Paradoxes and the Cosmological Argument." *Philosophy of Religion* 25: 65–81, 1989.

DeMillo, R. A., R. J. Lipton, and A. J. Perlis. "Social Processes and Proofs of Theorems and Programs." *Communications of the ACM* 22: 271–80, 1970.

Descartes, R. "Objectiones Septimae cum Notis Authoris sive Dissertatio de Prima Philosophia." In C. Adam and P. Tannery, eds., *Oeuvres,* 12 vols. Paris: L. Cert, 1897–1910.

——. *Correspondence.* Edited by C. Adam and G. Milhaud. Paris: Felix Alcan (Vols. 1–2). Presses Universitaires de France (Vols. 3–8), 1936–1963.

——. *Philosophical Works.* Translated by E. S. Haldane and G. R. T. Ross. New York: Dover, 1955.

——. *A Discourse on Method, Optics, Geometry and Meteorology.* Translated by Paul J. Olscamp. Indianapolis: Bobbs-Merrill, 1965.

de Villiers, M. "The Role and Function of Proof in Mathematics." *Pythagoras* 24: 17–24, 1990.

——. "Pupils' Needs for Conviction and Explanation within the Context of Geometry." *Pythagoras* 26: 18–27, 1991.

Dieudonné, J. "The Work of Nicholas Bourbaki." *American Mathematical Monthly* 77: 134–45, 1970.

Dillon, J. *The Middle Platonists.* Ithaca: Cornell University Press, 1977.

Dreben, B., P. Andrews, and S. Anderaa. "False Lemmas in Herbrand." *Proceedings of the American Mathematical Society* 69: 699–706, 1963.

Dummett, M. "Frege's Philosophy." In *Truth and Other Enigmas*, p. 89. Cambridge: Harvard University Press, 1978. Originally published as an article on Frege in P. Edwards, ed., *Encyclopedia of Philosophy.* New York: Macmillan, 1967.

——. *Frege: Philosophy of Language.* New York: Harper and Row, 1973.

——. "Frege as a Realist." *Inquiry* 19: 468, 1976.

——. *Elements of Intuitionism.* Oxford: Clarendon Press, 1977.

——. "Wittgenstein's Philosophy of Mathematics." In *Truth and Other Enigmas.* Cambridge: Harvard University Press, 1978.

——. "Frege and Wittgenstein." In Block, I., ed., *Perspectives on the Philosophy of Wittgenstein.* Cambridge: MIT Press, 1981.

Durkheim, E. *The Rules of Sociological Method.* Chicago: University of Chicago Press, 1938. Preface to 2nd edition, p. 61.

——. *Essays on Sociology and Philosophy.* New York: Harper and Row, 1964.

——. *The Elementary Forms of the Religious Life.* London: George Allen and Unwin, 1976.

Echeverria, J., A. Ibarra, and R. Mormann, eds. *The Space of Mathematics. Philosophical, Epistemological and Historical Explorations.* Berlin, New York: de Gruyter, 1992.

Eckstein, J. *The Platonic Method.* New York: Greenwood, 1968.

Edelman, G. M. *Neural Darwinism.* New York: Basic Books, 1987.

——. *Bright Air, Brilliant Fire.* New York: Basic Books, 1992.

Emerson, R. W. *Self-Reliance.* New York: Crowell, 1901.

Enderton, H. B. *A Mathematical Introduction to Logic.* New York: Academic Press, 1972.

Epstein, D. and S. Levy. "Experimentation and Proof in Mathematics." *Notices of the American Mathematical Society* 42: 670–74, 1995.

Ernest, P. *The Philosophy of Mathematics Education.* New York: Falmer, 1991.

——. *Social Constructivism in the Philosophy of Mathematics.* Albany: SUNY Press, 1997.

Euclid. *The Thirteen Books of Euclid's Elements.* Introduction and Commentary by T. L. Heath. New York: Dover, 1956.

Fann, K. T. *Wittgenstein's Conception of Philosophy.* Oxford: Blackwell, 1969.

Feferman, S. *The Logic of Mathematical Discovery vs. the Logical Structure of Mathematics.* Department of Mathematics, Stanford University, 1976.

——. "What Does Logic Have to Tell Us about Mathematical Proofs?" *Mathematical Intelligencer* 2: 20–24, 1979.

Fell, M. *Spinoza's Earliest Publication? A Loving Salutation*. In R. H. Popkin and M. A. Signer, eds. Assen, The Netherlands: Van Gorcum, 1987.

Fetisov, A. I. *Proof in Geometry*. Boston: D. C. Heath, 1963.

Feuer, L. S. *Spinoza and the Rise of Liberalism*. Boston: Beacon, 1966.

Field, H. *Science without Numbers*. Princeton: Princeton University Press, 1980.

Findlay, J. N. *Plato—The Written and Unwritten Doctrines*. London: Routledge & Kegan Paul, 1974.

Floyd, J., "Wittgenstein, Gödel and the Trisection of the Angle." To appear in J. Hintikkaa, ed., *The Foundations of Mathematics in the Early Twentieth Century*.

Floyd, J. "Wittgenstein on 2,2,2. . . : The Opening of *Remarks on the Foundations of Mathematics*." *Synthèse* 87: 143–80, 1991.

Fogelin, R. J. *Wittgenstein*. London: Routledge, 1976.

——. "Hume and Berkeley on the Proofs of Infinite Divisibility." *The Philosophical Review* 97: 47–69, 1988.

Fowler, D. H. *The Mathematics of Plato's Academy. A New Reconstruction*. Oxford: Clarendon Press, 1987.

Frascola, P. *Wittgenstein's Philosophy of Mathematics*. London: Routledge, 1994.

Frege, G. *The Thought: A Logical Inquiry*. A translation of part of Frege, *Der Gedanke*, 1919.

——. *Translations from the Philosophical Writings of Gottlob Frege*. In P. T. Geach and M. Black, eds. Oxford: Blackwell, 1950.

——. "Begriffsschrift." In J. K. van Heijenoort, ed., *From Frege to Godel: A Source Book in Mathematical Logic*. Cambridge: Harvard University Press, 1967.

——. *On the Foundations of Geometry and Formal Theories of Arithmetic*. Edited by E.-H. W. Kluge. New Haven: Yale University Press, 1972.

——. *Conceptual Notation and Related Articles*. Edited by T. W. Bynum. New York: Oxford University Press, 1972.

——. *Logical Investigations*. Translated by P. T. Geach and R. H. Stroothoff. Oxford: Blackwell, 1977.

——. In H. Hermes et al., eds., *Posthumous Writings*. Chicago: University of Chicago Press, 1979.

——. *The Foundations of Arithmetic*. Evanston: Northwestern University Press, 1980.

——. *Philosophical and Mathematical Correspondence*. Abridged, B. McGuinness. London: Blackwell. Chicago: University of Chicago Press, 1980.

——. *Collected Papers on Mathematics, Logic and Philosophy*. New York: Blackwell, 1984.

Freudenthal, H. *Mathematics as an Educational Task*. Dordrecht: Reidel, 1973.

Friedman, M. "Kant's Theory of Geometry." *The Philosophical Review* 94: 455–506, 1985.

Furth, M. *The Basic Laws of Arithmetic*. Berkeley: University of California Press, 1964.

Gale, D. "Proof as Explanation." *The Mathematical Intelligencer* 12: 4, 1991.

Gardner, M. *Order and Surprise.* Buffalo: Prometheus Books, 1983.

Geach P. T. "On Names of Expresssions." *Mind,* 1950.

Gerrard, S. "Wittgenstein's Philosophies of Mathematics." *Synthèse* 87: 125–42, 1991.

Gleick, J. *Chaos.* New York: Penguin, 1987.

Glimm, J., J. Impagliazzo, and I. Singer, eds. *The Legacy of John von Neumann. Proceedings of Symposia in Pure Mathematics,* Vol. 50. Providence: American Mathematical Society, 1990.

Gödel, K. *The Consistency of the Axiom of Choice and of the Generalized Continuum Hypothesis with the Axioms of Set Theory.* Princeton, 1940.

Gödel, K. "What Is Cantor's Continuum Problem?" In P. Benacerraf and H. Putnam, eds., 1988.

Gödel, K. *Collected Works,* Vol. III. New York: Oxford University Press, 1995.

Goffman, E. *The Presentation of Self in Everyday Life.* Garden City, N.Y.: Doubleday, 1959.

Goldmann, L. *Immanuel Kant.* London: NLB, 1971.

Goldstine, H. H. (1972). *The Computer from Pascal to von Neumann.* Princeton: Princeton University Press, 1972.

Goodman, N. "Mathematics as an Objective Science." *American Mathematics Monthly* 81: 354–65, 1974.

———. "Worlds of Individuals." In *Problems and Projects.* Indianapolis: Bobbs-Merrill, 1972.

Grabiner, J. "Descartes and Problem-Solving." *Mathematics Magazine* 68: 83–97, 1995.

Grosholz, E. R. "Descartes' Unification of Algebra and Geometry." In S. Gaukroger, ed., *Descartes Philosophy, Mathematics and Physics.* Sussex: Harvester, 1980, pp. 156–68.

Guggenheimer, H. "The Axioms of Betweenness in Euclid." *Dialectica* 31: 187–92, 1977.

Haaparanta, L. and J. Hintikka, eds. *Frege Synthesized,* pp. 299–343. Dordrecht: Reidel, 1986.

Hadamard, J. *The Psychology of Invention in the Mathematical Field.* New York: Dover, 1945.

Hahn, H. "The Crisis in Intuition." In J. R. Newman, ed., *The World of Mathematics,* 1957–1976. New York: Simon & Schuster, 1956.

Hahn, L. E. and P. A. Schilpp, ed. *The Philosophy of W. V. Quine.* La Salle, Illinois, Open Court, 1986.

Hahn, L. S. and B. Epstein. *Classical Complex Analysis.* Sudbury, Mass.: Jones and Bartlett, 1996.

Halmos, P. Address to 75th annual summer meeting of the Mathematical Association of America. Columbus, Ohio. Tape recording, 1990.

Hanna, G. *Rigorous Proof in Mathematics Education.* Toronto: OISE Press, 1983.

——. "Some Pedagogical Aspects of Proof." *Interchange* 21: 6–13, 1990.

Hardy, G. H: "Mathematical Proof." *Mind* 38: 1–25, 1929.

——. *A Mathematician's Apology*. New York: Cambridge University Press, 1940.

Hatfield, G. *The Natural and the Normative*. Cambridge: MIT Press, 1990.

Heath, Sir T. L. *Aristarchus of Samos; the Ancient Copernicus*. Oxford: Clarendon Press, 1913.

——. *Mathematics in Aristotle*. Oxford, 1949.

——. *A History of Greek Mathematics*. New York: Dover, 1981.

Heijenoort, J. van. *From Frege to Godel*. Cambridge: Harvard University Press, 1967.

Heims, S. J. *John von Neumann and Norbert Wiener*. Cambridge: MIT Press, 1980.

Helmholtz, H. *Epistemological Writings*. Boston: Reidel, 1977.

Henrici, P. "Reflections of a Teacher of Applied Mathematics." *Quarterly of Applied Mathematics* 30: 31–39, 1972.

Hersh, R. "Introducing Imre Lakatos." *Mathematical Intelligencer* 1: 148–51, 1978.

——. "Some Proposals for Reviving the Philosophy of Mathematics." *Advances in Mathematics* 31: 31–50, 1979.

——. "Inner Vision, Outer Truth." In R. Mickens, ed., *Mathematics and Science*. Singapore: World Scientific, 1990.

——. "Proving Is Convincing and Explaining." *Educational Studies in Mathematics* 24: 389–99, 1993.

Hersh, R. and V. John-Steiner. "A Visit to Hungarian Mathematics." *Mathematical Intelligencer* 15: 13–26, 1993.

Hesse, M. "Epistemology Socialized." In E. McMillen, ed., *Construction & Constraint*. Notre Dame University Press, 1988.

Hessen, B. *The Social and Economic Roots of Newton's Principia*. New York: H. Fertig, 1971.

Hilbert, D. "On the Infinite." In P. Benacerraf and H. Putnam, 1988.

Hintikka, J. "Kant on the Mathematical Method." In L. W. Beck, ed., *Kant Studies Today*. La Salle, Ill.: Open Court, 1969.

Holton, G. *Thematic Origins of Scientific Thought*. Cambridge: Harvard University Press, 1973.

Hone, J. N. and Rossi, M. M. *Bishop Berkeley, His Life, Writings, and Philosophy*. London: Faber, 1931.

Hopkins, J. *A Concise Introduction to the Philosophy of Nicholas of Cusa*. Minneapolis: University of Minnesota Press, 1978.

——. *Nicholas of Cusa's Metaphysic of Contraction*. Minneapolis: A. J. Banning, 1983.

Hume, D. *An Abstract of a Treatise of Human Nature*. Edited by J. M. Keynes and P. Sraffa. New York: Cambridge University Press, 1938.

——. *An Inquiry Concerning Human Understanding*. Indianapolis: Bobbs-Merrill, 1955.

——. *Philosophical Works*. London: T. H. Green and T. H. Grose, 1874–1975; Scientia Verlag Aalen, 1964.

——. *A Treatise of Human Nature*. New York: Penguin, 1969.

Husserl, E. "The Origin of Geometry." In *The Crisis of European Science*, Appendix VI. Evanston: Northwestern University Press, 1970.

Iliev, L. "Mathematics as the Science of Models." *Russian Mathematical Surveys* 27: 181–89, 1972.

Irvine, A. D., ed. *Physicalism in Mathematics*. Dordrecht: Kluwer, 1990

Isaacson, D. "Mathematical Intuition and Objectivity." In A. George, ed., *Mathematics and Mind*. New York: Oxford University Press, 1994.

Janik, A. and S. Toulmin. *Wittgenstein's Vienna*. New York: Simon & Schuster, 1973.

Jesseph, D. M. *Berkeley's Philosophy of Mathematics*. Dissertation, Princeton University, January 1987.

Johnson, M. *The Body in the Mind: The Bodily Basis of Meaning, Reason and Imagination*. Chicago: University of Chicago Press, 1987.

Kac, M., G.-C. Rota, and J. Schwartz. *Discrete Thoughts*. Boston: Birkhauser, 1986.

Kant, I. *Philosophy*. Translated by J. Watson. Glasgow: James Maclehose & Sons, 1901.

——. *Critique of Pure Reason*. Chicago: Encyclopedia Britannica, 1952.

——. *Critique of Practical Reason*. New York: Liberal Arts Press, 1956.

——. *Foundations of the Metaphysics of Morals*. Indianapolis: Bobbs-Merrill, 1959.

——. *Prolegomena to Any Future Metaphysics*. La Salle, Ill.: Open Court, 1967.

——. *Metaphysical Foundations of Natural Science*. Indianapolis: Bobbs-Merrill, 1970.

——. *Logic*. Indianapolis: Bobbs-Merrill, 1974.

Kielkopof, C. F. *Strict Finitism*. Paris, Mouton, 1970.

Kitchener, R. F. *Piaget's Theory of Knowledge*. New Haven: Yale University Press.

Kitcher, P. *Kant's Transcendental Psychology*. New York: Oxford University Press, 1990.

Kitcher, P. "Frege's Epistemology." *Philosophical Review* 88: 235–62, April 1979.

——. *The Nature of Mathematical Knowledge*. New York: Oxford University Press, 1983.

——. "Mathematical Naturalism." In Aspray and Kitcher, 1988.

Klein, F. Development of Mathematics in the 19th Century. Brookline, Mass.: Math Science Press, 1979. Translation by M. Ackerman of *Vorlesungen uber die Entwichlung der Mathematik in 19 Jahrhundert*. Teil I, Berlin: Springer-Verlag, 1929.

Klein, J. *A Commentary on Plato's Meno*. Chapel Hill: University of North Carolina Press, 1965.

——. *Greek Mathematical Thought and the Origin of Algebra*. Cambridge: MIT Press, 1968.

Klenk, V. H. *Wittgenstein's Philosophy of Mathematics*. The Hague: Nijhoff, 1976.

Kline, M. *Mathematical Thought from Ancient to Modern Times*. New York: Oxford University Press, 1972.

Kneale, W. and M. Kneale. *The Development of Logic.* New York: Oxford University Press, 1962.

Knuth, D. E. *The Art of Computer Programming.* Reading, Mass.: Addison Wesley, 1968–1973.

——. "Mathematics and Computer Science: Coping with Finiteness." *Science* 194: 1235–42, 1976.

Koehler, O. "The Ability of Birds to 'Count.'" In J. R. Newman, ed., *The World of Mathematics,* Vol. 1, pp. 489–96. New York: Simon & Schuster, 1956.

Kopell, N. and G. Stolzenberg. "Commentary on Bishop's Talk." *Historia Mathematica* 2: 519–21, 1975.

Körner, S. *Kant.* New York: Penguin, 1955.

——. *The Philosophy of Mathematics.* New York: Dover, 1968.

Koyré, A. *Discovering Plato.* New York: Columbia University Press, 1945.

Kozulin, A. *Vygotsky's Psychology, A Biography of Ideas.* Cambridge: Harvard University Press, 1990.

Kreisel, G. "Wittgenstein's Remarks on the Foundations of Mathematics." *British Journal for the Philosophy of Science* 9: 158, 1958.

——. "Critical Notice: 'Lectures on the Foundations of Mathematics'." In S. G. Shanker, ed., *Ludwig Wittgenstein: Critical Assessments.* London: Croom Helm, 1986, pp. 98–110.

Kripke, S. *Wittgenstein on Rules and Private Language.* New York: Oxford University Press, 1982.

Kubitz, O. A. *Development of John Stuart Mill's System of Logic.* Urbana: University of Illinois, 1932.

Kuyk, W. *Complementarity in the Philosophy of Mathematics.* Dordrecht: Reidel, 1977.

Lakatos, I. *Proofs and Refutations: The Logic of Mathematical Discovery.* New York: Cambridge University Press, 1976.

——. *Mathematics, Science and Epistemology. Philosophical Papers, Volume 2.* New York: Cambridge University Press, 1978.

Lakoff, G. and R. E. Nuñez. "The Metaphorical Structure of Mathematics: Sketching Out Cognitive Foundations For a Mind-Based Mathematics." To appear in L. English., ed., *Mathematical Reasoning: Analogies, Metaphors, and Images.* Hillsdale, N.J.: Erlbaum, 1996.

Lambert, J. H. *Theorie der Parallelinien.* Leipzig: 1786.

Land, J. P. "Kant's Space and Modern Mathematics." *Mind* 2: 38–46, 1877.

Lanford, O. E. III. "A Computer-assisted Proof of the Feigenbaum Conjectures." *Bulletin of the American Mathematical Society* (N.S.) 6: 427–34, 1982.

——. "Computer-assisted Proofs in Analysis." *Physica* 124A: 465–70, 1984.

——. "A Shorter Proof of the Existence of the Feigenbaum Fixed Point." *Communications in Mathematical Physics* 96: 521–38, 1984.

Laptev, B. L. *Lambert as a Geometer.* Istoriko-mat. issl. 25: 248–60 (Russian), 1980.

Lazerowitz, M. and A. Ambrose. *Essays in the Unknown Wittgenstein.* Buffalo: Prometheus, 1984.

Lear, J. "Aristotle's Philosophy of Mathematics." *Philosophical Review* 91: 161–92, 1982.

Leavis, F. R. *Nor Shall My Sword.* London: Chatto & Windus, 1972.

Lehmer, D. N. *List of Prime Numbers from 1 to 10,006,721.* Washington, D.C.: Carnegie Institution of Washington, Publication No. 163, 1914.

Leibniz, G. W. F. *Philosophical Works.* New Haven: Tuttle, Morehouse and Taylor, 1908.

——. *Monadology and Other Philosophical Essays.* Indianapolis: Bobbs-Merrill, 1965.

——. *Discourse on Metaphysics.* Buffalo: Prometheus, 1992.

Leron, U. "Structuring Mathematical Proofs." *American Mathematical Monthly* 90: 174–85, 1983.

Lewis, C. I. *A Survey of Symbolic Logic.* Berkeley: University of California Press, 1918.

Lichnerowicz, A. "Rémarques sur les mathématiques et la réalité." In *Logique at connaissance scientifique.* Dijon: Encyclopédie de la Pléaiade, 1967.

Lighthill, M. J. *Fourier Analysis and Generalized Functions.* New York: Cambridge University Press, 1964.

Locke, J. "An Essay Concerning Human Understanding (1690)." In *The Empiricists.* Garden City, N.Y.: Dolphin, 1961.

MacLane, S. *Mathematics: Form and Function.* New York: Springer-Verlag, 1986.

Macrae, N. *John von Neumann.* New York: Pantheon, 1992.

Maddy, P. *Realism in Mathematics.* New York: Oxford University Press, 1992.

Malcolm, N. *Ludwig Wittgenstein: A Memoir.* New York: Oxford University Press, 1984.

Manin, Yu. I. *A Course in Mathematical Logic* New York: Springer-Verlag, 1977.

Martin, G. *Kant's Metaphysics and Theory of Science.* New York: Barnes and Noble, 1955.

Maurer, A. A. "Nicholas of Cusa." In *The Encyclopedia of Philosophy.* New York: Macmillan, 1976, pp. 496–98.

Maziarz, E. A. and T. Greenwood. *Greek Mathematical Philosophy.* New York: Ungar, 1968.

McShea, R. J. *The Political Philosophy of Spinoza.* New York: Columbia University Press, 1968.

Medawar, P. *Pluto's Republic.* New York: Oxford University Press, 1982.

Mehrtens, H. T. "T. S. Kuhn's Theories and Mathematics." *Historia Mathematica* 3: 297–320, 1976.

Menger, K. *Selected Papers in Logic etc.* Dordrecht: Reidel, 1979, chapter 18, "Square Circles" (The Taxicab Geometry), p. 217 (ref. H. C. Curtis, *Am. Math. Monthly* 60: 1953); chapter 21, p. 237, "My Memories of L. E. J. Brouwer," 1978.

Meyer, A. R. "The Inherent Computational Complexity of Theories of Ordered Sets." *Proceedings of the International Congress of Mathematicians 1972* 2: 481, 1974.

Mill, J. S. *A System of Logic, Ratiocinative and Inductive, being a connected view of the principles of evidence and the methods of scientific investigation*, 8th ed. New York: Harper & Brothers, 1874.

——. *Utilitarianism*. Indianapolis: Hackett, 1979.

Miller, G. L. "Riemann's Hypothesis and Tests for Primality." *Journal Comp. Sys. Sci.* 13: 300–17, 1976.

Miller, J. P. *Number in Presence and Absence*. The Hague: Nijhoff, 1982.

Molland, A. G. "Shifting the Foundations: Descartes' Transformation of Ancient Geometry." *Historia Mathematica* 3: 21–79, 1976.

Monk, R. *Ludwig Wittgenstein*. New York: Free Press, 1992.

Mostowski, A. "Thirty Years of Foundational Studies." *Acta Philosophica Fennica* 17: 7, 1965 (quoted by Musgrave, p. 108).

Mueller, I. *Coping with Mathematics (The Greek Way)*. Chicago: Morris Fishbein Center for the Study of the History of Science and Medicine. Publication No. 2, 1980.

Musgrave, A. "Logicism Revisited." *British Journal for Philosophy of Science* 28: 99–127, 1977. Quotes Russell, *An Essay on the Foundations of Geometry*, 1897, p. 1.

Nelsen, R. B. *Proofs without Words. Exercises in Visual Thinking*. Washington, D.C.: Mathematical Association of America, 1993.

Nicholas de Cusa. *Idiota de Mente The Layman about Mind*. Translated by Clyde Lee Miller. New York: Abaris Books, 1979.

Nicomachus of Gerasa. *Introduction to Arithmetic*. Translated by Martin Luther D'Orge. Ann Arbor: University of Michigan Press, 1946.

Nidditch, P. H *The Development of Mathematical Logic*. Glencoe, Ill.: Free Press, 1962 (in basic English).

Orwell, G. *Down and Out in Paris and London*. New York: Harcourt Brace, 1961.

Parsons, C. "Quine on the Philosophy of Mathematics." In Hahn and Schilpp.

Passmore, J. *A Hundred Years of Philosophy*. Baltimore: Penguin, 1970, pp. 153–54.

Pears, D. F. *Ludwig Wittgenstein*. New York: Viking, 1970.

Peirce, C. S. "The Essence of Mathematics." In *Essays in the Philosophy of Science*. Indianapolis: Bobbs-Merrill, 1957.

——. *The New Elements of Mathematics*. The Hague: Mouton, 1976.

——. *Collected Papers*. Cambridge: Harvard University Press, 1960. Paragraph 3, p. 426. Reprinted in the *American Mathematical Monthly* 275: 1978.

Peppinghaus, B. "Some Aspects of Wittgenstein's Philosophy of Mathematics." In J. C. Bell, ed., *Proceedings of the Bertrand Russell Memorial Logic Conference*. Uldum, Denmark, 1971; Leeds, 1973.

Péter, R. *Playing with Infinity*. New York: Atheneum, 1964.

Piaget, J. *Growth of Logical Thinking from Childhood to Adolescence*. New York: Basic Books, 1958.

——. *The Child's Conception of Geometry*. With B. Inhelder and A. Szeminka. New York: Basic Books, 1960.

——. *The Child's Conception of Number*. New York: Norton, 1965.

——. *The Child's Conception of Physical Causality*. Totowa, N.J.: Littlefield, 1965.

——. *The Child's Conception of Space*. New York: Norton, 1967.

——. *Early Growth of Logic in the Child; Classification and Seriation*. New York: Norton, 1969.
——. *The Child's Conception of Movement and Speed*. New York: Basic Books, 1970.
——. *Genetic Epistemology*. New York: Columbia University Press, 1970.
——. *Insights and Illusions of Philosophy*. New York: World, 1971.
——. *Psychology and Epistemology*. New York: Grossman, 1971.
——. *Origin of the Idea of Chance in Children*. With B. Inhelder. New York: Norton, 1975.
——. *Epistemology and Psychology of Functions*. Dordrecht: Reidel, 1977.
——. *Morphisms and Categories*. With G. Henriques, E. Ascher, and T. Brown. Hillsdale, N.J.: Erlbaum, 1992.
Pistorius, P. V. *Plotinus and Neoplatonism*. Cambridge: Bowes, 1952.
Plato. *The Republic*. In *Great Dialogues*.
——. *Theaetetus*. Indianapolis: Bobbs-Merrill, 1949.
——. *Great Dialogues*. New York: New American Library, 1956.
——. *Timaeus*. Indianapolis: Bobbs-Merrill, 1959.
——. *Laws* 7: 821–22, *The Collected Dialogues*. Edited by E. Hamilton and H. Cairns. Princeton: Princeton University Press, 1961.
——. *Meno*. Indianapolis: Bobbs-Merrill, 1971.
Plutarch. *The Lives of the Noble Grecians and Romans*. Chicago: Encyclopedia Britannica, Vol. 14, 1982.
Poincaré, H. *The Foundations of Science*. New York: Science Press, 1913.
——. *Science and Method*. New York: Dover, 1952.
——. *Science and Hypothesis*. New York: Dover, 1952.
——. *Mathematics and Science; Last Essays*. New York: Dover, 1963.
——. *New Methods of Celestial Mechanics 1. Periodic and Asymptotic Solutions*. Introduction by D. L. Goroff. American Institute of Physics, 1993.
Polányi, M. *Personal Knowledge*. Chicago: University of Chicago Press, 1962.
——. *The Tacit Dimension*. New York: Doubleday, 1966.
Pole, D. *The Later Philosophy of Wittgenstein*. London: Athlone. 1958.
Pollock, F. *Spinoza, His Life and Philosophy*. London: C. Kegan Paul & Co., 1880.
Pólya, G. *How to Solve It*. Princeton: Princeton University Press, 1945.
——. *Mathematics and Plausible Reasoning*. Princeton: Princeton University Press, 1954.
Pont, J.-C. *L'Aventure des paralleles, Histoire de la géometrie non-euclidienne: précurseurs et attardés*. Berne: Lang, 1986, pp. 248–60 (Russian).
Pope, A. *Poetical Works*. New York: Thomas Y. Crowell, 1896.
Popkin, R. H. *The History of Scepticism from Erasmus to Spinoza*. Berkeley: University of California Press, 1979.
Popper, K. *The Open Society and Its Enemies*. Princeton: Princeton University Press, 1971.
——. "Epistemology without a Knowing Subject" and "On the Theory of the Objective Mind" in *Objective Knowledge*. Oxford: Clarendon Press, 1974.

Preston, R. "The Mountains of Pi." *New Yorker* 68: 36, 1992.

Pritchard, P. *Plato's Philosophy of Mathematics.* Sankt Augustin: Academia Verlag.

Putnam, H. "What Is Mathematical Truth?" In *Mathematics, Matter and Method*, Cambridge University Press; reprinted in T. Tymoczko, ed., *New Directions in the Philosophy of Mathematics.* Cambridge: Birkhauser, 1986.

———. *Representation and Reality.* Cambridge: MIT Press, 1988.

———. "Peirce the Logician." In *Realism with a Human Face.* Cambridge: Harvard University Press, 1990.

Quine, W. V. O. *Methods of Logic.* New York: Holt, 1959.

———. "On Frege's Way Out." In *Selected Logical Papers.* New York: Random House, 1966.

———. "The Scope and Language of Science." In *The Ways of Paradox.* Cambridge: Harvard University Press, 1976.

———. *Quiddities.* Cambridge: Harvard University Press, 1987.

Rabin, M. O. "Probabilistic Algorithms." In J. F. Traub, ed., *Algorithms and Complexity: New Directions and Recent Results.* New York: Academic Press, 1976.

Ratner, J., ed. *The Philosophy of Spinoza.* New York: Modern Library, 1927.

Regier, T. *The Human Semantic Potential.* Cambridge: MIT Press, 1996.

Reichenbach, H. *The Rise of Scientific Philosophy.* Berkeley: University of California Press, 1951.

Reid, C. *Hilbert.* New York: Springer-Verlag, 1970.

Rényi, A. *Dialogues on Mathematics.* San Francisco: Holden Day, 1967.

———. *Letters on Probability.* Detroit: Wayne State University Press, 1972.

———. *A Diary on Information Theory.* Budapest: Akadémiai Kiadó, 1984.

Renz, P. "Mathematical Proof: What It Is and What It Ought to Be." *The Two-Year College Mathematics Journal* 12: 83–103, 1981.

Resnik, M. D. *Frege and the Philosophy of Mathematics.* Ithaca: Cornell University Press, 1980.

Restivo, S. *The Social Relations of Physics, Mysticism, and Mathematics.* Dordrecht: Reidel, 1983.

———. *Mathematics in Society and History.* Dordrecht: Kluwer, 1992.

Restivo, S., J. P. Van Bendegem, and R. Fischer, eds. *Math Worlds.* Albany: State University of New York Press, 1993.

Robinson, A. "From a Formalist's Point of View." *Dialectica* 23: 45, 1969.

———. *Nonstandard Analysis.* Amsterdam: North-Holland, 1974.

Roche, W. J. "Measure, Number, and Weight in Saint Augustine." *New Scholasticism* XV: October 1941.

Rorty, R. *Objectivity, Relativism and Truth.* New York: Cambridge University Press, 1991.

Rosenfeld, B. A. *A History of Non-Euclidean Geometry.* New York: Springer-Verlag, 1988.

Rota, G.-C. *Indiscreet Thoughts.* Cambridge: Birkhauser, 1996.

Roth, L. *Spinoza Descartes & Maimonides.* New York: Russell, 1963.

Rotman, B. *Signifying Nothing: the semiotics of zero.* New York, St. Martin's Press, 1987.

Rotman, B. *Ad Infinitum—The Ghost in Turing's Machine.* Stanford, 1993.

Roxin, E. "A Living and Constructive View of Mathematics." Talk at Brown University, Department of Applied Mathematics, Seminar on Philosophy of Mathematics.

Russell, B. *The Principles of Mathematics.* London: Allen & Unwin, 1937.

———. *A History of Western Philosophy.* New York: Simon & Schuster, 1945.

———. "Reflections on My Eightieth Birthday." In *Portraits from Memory.* New York: Simon & Schuster, 1956.

Russell, B. and A. Whitehead. *Principia Mathematica.* Cambridge: University Press, 1925.

Ryle, G. "Plato." In *The Encyclopedia of Philosophy.* New York: Macmillan, 1967, Vol. 6, pp. 314–33.

Salmon, W. C., ed. *Zeno's Paradoxes.* Indianapolis: Bobbs-Merrill, 1970.

Saunders, J. L., ed. *Greek & Roman Philosophy after Aristotle.* New York: Free Press, 1966.

Schatz, J. A. *The Nature of Truth.* Unpublished manuscript.

Schilpp, P. A., ed. *The Philosophy of Rudoph Carnap.* La Salle, Ill.: Open Court, 1963.

Schirn, M., ed. *Studies on Frege.* Stuttgart: Bad Canstart, 1976.

Schwartz, J. T. "Fast Probabilistic Algorithms for Verification of Polynomial Identities." *Journal of the Association for Computing Machinery* 27: 701–17, 1980.

Scruton, R. *Spinoza.* New York: Oxford University Press, 1986.

Shanker, S. *Wittgenstein and the Turning Point in the Philosophy of Mathematics.* London: Croom Helm, 1987.

Shebar, W. "In Quest of Quine." *Harvard Magazine* 47–51, November–December 1987.

Shwayder, D. S. "Wittgenstein on Mathematics." In P. Winch, ed., *Studies in the Philosophy of Wittgenstein.* London: Routledge, 1969.

Skyrms, B. "Zeno's Paradox of Measure." In Cohen and Lauden.

Sluga, H. Review of *Nachgelassene Schrifte. Journal of Philosophy* 68: 265–272, 1971.

———. "Frege and the Rise of Analytic Philosophy." *Inquiry,* 18: 477, 1973.

———. *Gottlob Frege.* London: Routledge, 1980.

———. *Heidegger's Crisis.* Cambridge: Harvard University Press, 1993.

———, ed. *The Philosophy of Frege.* New York: Garland, 1993.

Smith, D. E. *History of Mathematics,* Vol. I. New York: Dover, 1958, p. 72.

Smorynski, C. "Mathematics as a Cultural System." *Mathematical Intelligencer* 5: (1), 1983.

Snapper, E. "What Is Mathematics?" *American Mathematical Monthly* 86: 551–57, 1979.

Spinoza, B. *Tractatus Theologico-Politicus* and *Tractatus Politicus.* London: Routledge, 1895.

———. *Improvement of the Understanding, Ethics and Correspondence.* Translated by R. H. M. Elwes, intro. by Frank Sewall. New York and London: M. W. Dunne, 1901.

St. Augustine of Hippo. *On the Trinity*, chapter IV. Translated by S. MacKenna. Washington, D. C.: Fathers of the Church Series No. 45, Catholic University of America.

———. *Basic Writings*. Edited by W. J. Oates. New York: Random House, 1948.

———. *The City of God*. New York: Modern Library, 1950.

———. *Introduction to the Philosophy of Saint Augustine Selected Readings and Commentaries*. Edited by J. A. Mourant. University Park, Pa.: Pennsylvania State University Press, 1964.

———. *The Confessions*. London: Collier, 1969.

———. *On Free Choice of the Will*. Indianapolis: Hackett, 1993, p. 89.

Steen, L. "The Science of Patterns." *Science* 240: 611–16, 1988.

Steiner, M. *Mathematical Knowledge*. Ithaca: Cornell University Press, 1975.

Stolzenberg, G. "Can an Inquiry into the Foundations of Mathematics Tell Us Anything Interesting about Mind?" In George Miller, ed. *Psychology and Biology of Language and Thought*. New York: Academic Press.

———. *Contemporary Mathematics*. Providence: American Mathematical Society, 1985, p. 39.

Stone, I. F. *The Trial of Socrates*. Boston: Little, Brown, 1988.

Swart, E. R. "The Philosophical Implications of the Four-color Theorem." *American Mathematical Monthly* 697–707, 1980.

Tarnas, R. *Passion of the Western Mind*. New York: Harmony, 1991.

Taylor, A. E. *Platonism and Its Influence*. New York: Cooper Square Publishers, 1963.

Thiel, C. *Sense and Reference in Frege's Logic*. Dordrecht: Reidel, 1968.

Thom, R. "Modern Mathematics: An Educational and Philosophical Error?" *American Scientist* 59: 695–99, 1971.

Thomas, J. *Musings on the Meno*. The Hague: Nijhoff, 1980.

Thomas, R. Private communication, 1996.

Tichy, P. *The Foundations of Frege's Logic*. New York: de Gruyter, 1988.

Tiles, M. *Mathematics and the Image of Reason*. London: Routledge, 1991.

Tragesser, R. S. *Husserl and Realism in Logic and Mathematics*. New York: Cambridge University Press, 1984.

Turbayne, C. M. *Introduction to Berkeley's Treatise Concerning the Principles of Human Knowledge*. Indianapolis: Bobbs Merrill, 1955–1979, pp. xviii–xix.

Tymoczko, T. "Finding a Place for the Mathematician in the Philosophy of Mathematics." *Mathematical Intelligencer*, 1981.

———. "The Four-Color Problem and Its Philosophical Significance." *Journal of Philosophy* 76: 57–83, 1979.

Ungar, P. Personal communication. 10 October 1989.

van Stift, W. P. *Brouwer's Intuitionism*. Amsterdam: North-Holland, 1990.

van der Waerden, B. L. *Science Awakening*. New York: Wiley, 1963.

van Bendegem, J. P. *Theory and Experiment*. Dordrecht: D. Reidel, 1980.

———"Zeno's Paradoxes and the Tile Argument." *Philosophy of Science* 54: 295–302, 1987.

Vartanian, A. *Diderot and Descartes.* Princeton: Princeton University Press, 1953.
Vlastos, G. "Zeno of Elea." In P. Edwards, ed., *The Encyclopedia of Philosophy*, Vol. vii, pp. 369–79. New York: Macmillan, 1967.
von Neumann, J. "The Mathematician." In R. B. Heywood, ed., *Works of the Mind.* Chicago: University of Chicago Press, 1947.
Vrooman, J. V. *René Descartes, A Biography.* New York: Putnam, 1970.
Walsh, W. H. "Immanuel Kant." In P. Edwards, ed., *The Encyclopedia of Philosophy*, pp. 305–24. New York: Macmillan, 1967.
Wang, H. "Proving Theorems by Pattern Recognition." *Communications of the Association for Computing Machinery* 3: April 1960.
———. *Popular Lectures on Mathematical Logic.* New York: van Nostrand Reinhold, 1971.
———. *From Mathematics to Philosophy.* London: Routledge & Kegan, 1972.
———. "Toward Mechanical Mathematics." International Business Machines Corporation, 1960, reprinted in J. Siekmann and G. Wrightson, eds., *Automation of Reasoning*, pp. 229–66. Berlin: Springer-Verlag, 1983.
———. *Beyond Analytic Philosophy . . . Doing Justice to What We Know.* Cambridge: MIT Press, 1988.
———. *Reflections on Kurt Gödel.* Cambridge: MIT Press, 1991.
———. "To and from Philosophy: Discussions with Gödel and Wittgenstein." *Synthèse* 88: 229–77, 1991.
———. "Computer Theorem Proving and Artificial Intelligence." In J.-L. Lassez and G. Plotkin, eds., *Computational Logic.* Cambridge: MIT Press, 1991.
———. "Imagined Discussions with Gödel and with Wittgenstein." In *Yearbook of the Kurt Gödel Society.* Vienna: Kurt Gödel Society, 1992.
———. "Quine's Logical Ideas in Historical Perspective." In Hahn and Schilpp.
Wedberg, A. *Plato's Philosophy of Mathematics.* Stockholm: Almsqvist & Wiksell, 1955.
Wheeler, J. *Magic without Magic.* San Francisco: Freeman, 1972.
Wheelwright, P. *The Presocratics.* Indianapolis: Bobbs-Merrill, 1981.
White, L. A. "The Locus of Mathematical Reality." *Philosophy of Science* 14: 289–303. Also chapter 10, *The Science of Culture: A Study of Man and Civilization.* New York: Farrar, Straus, 1949.
White, M. J. "Zeno's Arrow, Divisible Infinitesimals, and Chrysippus." *Phronesis* 27: 239–54, 1982.
———. *The Continuous and the Discrete.* New York: Oxford University Press, 1992.
Whiteside, D. T. *The Mathematical Papers of Isaac Newton*, Vol. 7. New York: Cambridge University Press, 1976.
Whitman, W. "Song of Myself," Section 51. In *Leaves of Grass.* Philadelphia: McKay, 1900.
Wiener, N. *Cybernetics.* New York: Wiley, 1948.
———. *Ex-Prodigy.* Cambridge: MIT Press, 1953.
———. *I Am a Mathematician.* Garden City, N.Y.: Doubleday, 1956.

Wilder, R. L. *Introduction to the Foundations of Mathematics.* New York: Wiley, 1968.

——. *Evolution of Mathematical Concepts An Elementary Study.* New York: Wiley, 1968.

——. *Mathematics as a Cultural System.* New York: Pergamon, 1981.

Wittgenstein, L. *Tractatus Logico-Philosophicus.* London: Routledge & Kegan Paul, 1922.

——. *Philosophical Investigations.* New York: Macmillan, 1953.

——. *The Blue and Brown Books.* London: Blackwell, 1958.

——. *Philosophical Grammar.* Berkeley: University of California Press, 1974.

——. *Lectures on the Foundations of Mathematics.* Ithaca: Cornell University Press, 1976.

——. *Remarks on Color.* Berkeley: University of California Press, 1977.

——. *Remarks on the Foundations of Mathematics.* Cambridge: MIT Press, 1983.

Wolff, C. *Preliminary Discourse on Philosophy in General.* Indianapolis: Bobbs-Merrill, 1963.

Woodger, J. *The Axiomatic Method in Biology.* New York: Cambridge University Press, 1937.

Wos, L. "The Impossibility of the Automation of Logical Reasoning." *Automated Deduction—CADE-11.* In D. Kapur, ed., pp. 1–3. New York: Springer-Verlag, 1992.

Wright, C. *Wittgenstein on the Foundations of Mathematics.* Cambridge: Harvard University Press, 1980.

Yates, F. *Giordano Bruno and the Hermetic Tradition.* Chicago: University of Chicago Press, 1964.

Zabeeh, F. "Hume's Scepticism with Regard to Deductive Reason." *Ratio* 2: 134–43, 1960.

Index

THE C

Divisions between Catholics and Protestants have been a feature of English history since the Reformation. Even into the industrial nineteenth century, age-old theological disagreements were the cause of religious and cultural conflicts. *The Old Enemies* asks why these ancient divisions were so deep, why they continued into the nineteenth century, and how novelists and poets, theologians and preachers, historians and essayists reinterpreted the religious debates. Michael Wheeler, a leading authority on the literature and theology of the period, explains how each side misunderstood the other's deeply held beliefs about history, authority, doctrine and spirituality, and, conversely, how these theological conflicts were a source of inspiration and creativity in the arts. This wide-ranging, well-illustrated study sheds much new light on nineteenth-century history, literature and religion.

MICHAEL WHEELER is a Visiting Professor at the Universities of Lancaster, Roehampton and Southampton, and a Lay Canon and member of Chapter of Winchester Cathedral. His other books include *Heaven, Hell and the Victorians* (Cambridge, 1994) and *Ruskin's God* (Cambridge, 1999).

THE OLD ENEMIES

Catholic and Protestant in Nineteenth-Century English Culture

MICHAEL WHEELER

CAMBRIDGE UNIVERSITY PRESS
Cambridge, New York, Melbourne, Madrid, Cape Town, Singapore, São Paulo

Cambridge University Press
The Edinburgh Building, Cambridge CB2 2RU, UK

Published in the United States of America by Cambridge University Press, New York

www.cambridge.org
Information on this title: www.cambridge.org/9780521828109

First published 2006

Printed in the United Kingdom at the University Press, Cambridge

A catalogue record for this book is available from the British Library

Library of Congress cataloguing in publication data
Wheeler, Michael, 1947–
The old enemies : Catholic and Protestant in nineteenth-century English culture / Michael Wheeler
p. cm.
Includes bibliographical references and index.
ISBN 0 521 82810 4 (hardback)
1. English literature – 19th century – History and criticism. 2. Christianity and literature – Great Britain – History – 19th century. 3. Protestants – Great Britain – History – 19th century. 4. Christian literature, English – History and criticism. 5. Catholics – Great Britain – History – 19th century. 6. Catholic Church – Relations – Protestant churches. 7. Protestant churches – Relations – Catholic Church. 8. Great Britain – Church history – 19th century. 9. Great Britain – Civilization – 19th century. I. Title.
PR468.R44W47 2005 820.9′3823 – dc22 2005018619

ISBN 13 978 0 521 82810 9 hardback
ISBN 10 0 521 82810 4 hardback

To the Warden and staff of
St Deiniol's Library, Hawarden,
the scholar's haven

Out of the same mouth proceedeth blessing and cursing.
My brethren, these things ought not so to be.

James 3.9–10

We live in a country which for three hundred years has been pervaded by a spirit of opposition to the Catholic Church. Everything round about us is full of antagonism to the Faith. The whole literature of this country is written by those who, sometimes unconsciously, sometimes consciously, assume an attitude of hostility to it.

Cardinal Manning

This is the supreme quarrel of all . . . This is not a dispute between sects and kingdoms; it is a conflict within a man's own nature – nay, between the noblest parts of man's nature arrayed against each other. On the one side obedience and faith, on the other, freedom and the reason.

Joseph Henry Shorthouse

Contents

Illustrations

Preface

Like so many literary projects, this one started life in a secondhand bookshop. During the 1990s, when scouring Carnforth Bookshop, near Lancaster, I came across a collection of twenty-four pamphlets entitled 'The Roman Catholic Question, 1850–1851'. Bound in with these pamphlets were a number of others, including a discourse by Nicholas Wiseman on the Gorham controversy, Sir Robert Peel's maiden speech on 'Papal Aggression', Gladstone's speech on the Ecclesiastical Titles Assumption Bill, Bishop Phillpott's pastoral letter to the clergy of the diocese of Exeter, and William Dodsworth's tract on Anglicanism. It struck me immediately that the huge subject of the conflict between Catholics and Protestants in nineteenth-century England would be worth exploring in depth, when time allowed.

Five years later, having left the University of Lancaster, and then Chawton House Library and the University of Southampton, I was able to write virtually full-time. On reviewing the 'Catholic question' in the mid-nineteenth century, it turned out that historians such as Arnstein, Chadwick, Norman and Paz had been active in the field, but that no literary critic had written a wide-ranging study on the subject and thus brought out its wider cultural implications. Three literary studies were later published during the writing of this book: Michael E. Schiefelbein's *The Lure of Babylon: Seven Protestant Novelists and Britain's Roman Catholic Renewal* (2001) and Ian Ker's *The Catholic Revival in English Literature: Newman, Hopkins, Belloc, Chesterton, Greene, Waugh, 1845–1936* (2003) both focus upon particular writers, and Susan M. Griffin's *Anti-Catholicism and Nineteenth-Century Fiction* (2004) examines the ways in which anti-Catholic themes are used by American and British novelists, many of them long forgotten, as a medium of general cultural critique.

The more I looked at the subject, the clearer it became that the divisions between the old enemies were both caused and exacerbated by a series of misreadings, and that the focus of my study should be upon the relationship between spirituality and language, between the doctrinal and devotional content of Catholic and Protestant traditions and the tropes and symbols which are open to different readings and misreadings by friends and foes of each tradition.

To most English people today, the phrase 'Catholic and Protestant' evokes the 'troubles' over there in Northern Ireland and football matches between Celtic and Rangers up there in Scotland. Yet 'Catholic and Protestant' has a long history in England over which each succeeding generation has fought. In 1569, during the Catholic 'Rising of the North', Thomas Percy and his companions celebrated the Mass and burned copies of the Bible and the official Protestant prayer book. In 1969, leading English Catholics were writing articles about 'dialogue' and 'convergence' with Anglicanism: old enemies were becoming new friends. In the intervening four hundred years, England's holy wars were repeated again and again, in the political sphere and in the domains of theology and ecclesiastical history, literature and criticism, painting and architecture. Yet the religion over which these battles raged is based upon a message of love, of peace and of hope.

The Old Enemies asks why these ancient divisions are so deep, why they continued into the industrial age, and how the writers of that age – novelists and poets, historians and essayists, theologians and pamphleteers – reinterpreted them. The book thus creates a Victorian viewing platform from which the reader can see the history of 'Catholic and Protestant' in England, from the distant past to the present, from a new perspective.

In many ways, anti-Catholicism can be described as the hatred of one patriarchal institution by another. This book also examines the contribution that English women – Catholic and Protestant – have made to these debates. It considers the ways in which Catholics and Protestants have fought over history – especially the history of the early Church and the Reformation – and over the vexed question of authority, for Catholics grounded upon the rock of St Peter, for Protestants upon the rock of the 'Scriptures'. It discusses the tension in Protestantism between an ingrained sense of repulsion from Catholicism, with its 'unnatural' celibate priests, monks and nuns, its 'superstitions' and its 'idolatrous' foreign forms of worship, and an attraction – often at an unconscious level – to Catholicism's apparent unity, its claim to be the true Church, its Marian theology, its doctrine of purgatory as an intermediate state between earthly and heavenly existence, and above all its access to divine 'mystery', made real in the Mass. In terms of ecclesiastical history, these conflicts and tensions can be explained as the result of mutual misunderstanding and prejudice, and of passionate belief in what each side regards as an exclusive saving truth. In terms of cultural history, the same conflicts and tensions are often to be found close to the source of the creative energy behind literature and the arts.

The Church of England is both Protestant and part of the 'Catholic and Apostolic Church'. Within Anglicanism I would describe myself as a Catholic, and within the Catholic wing of the Church of England a Modern Catholic – all very arcane to those outside Anglicanism's big tent. I write as an Anglican who is sympathetic towards the Roman Catholic Church. Today, as in the nineteenth century, 'Catholic' Anglicans do not like being called 'Protestants'. For the sake of concision, however, I use the term 'Protestant' in this book to mean all Anglicans and Protestant Dissenters taken together. Similarly, 'Catholic' means Roman Catholic, unless the context clearly indicates Anglo-Catholicism.

In exploring a wide field over a number of years, I have been supported by some remarkable people. My wife Viv was unfailingly supportive, especially when the going got rough. Three generous friends read all or part of the book, and special thanks are due to Michael Alexander, Olivia Thompson and Chris Walsh. Andrew Brown of Cambridge University Press encouraged me to write the book, and his colleague Linda Bree has been an excellent editor. Thanks, too, to all those who helped to get the book from computer to press, including Susan Beer, Alison Powell and Maartje Scheltens. I have also enjoyed and benefited from conversations with Graham Beck, Lida Kindersley, the late Linda Murray, Stephen Prickett and Trevor Robinson.

What started in a secondhand bookshop came to maturity in the British Library and at St Deiniol's Library, Hawarden. I am particularly grateful to The British Academy, which awarded me a travel grant, and to the staff of the Rare Books and Music Reading Room in the British Library. Also to the archival staff of Arundel Castle; St Deiniol's Library, Hawarden and its Warden, Peter Francis, and Librarian, Patricia Williams, to whom the book is dedicated; the Armstrong Browning Library, Baylor University, where I was a Visiting Fellow in February 2004, and its Director, Professor Stephen Prickett, and his staff; the University of Southampton Library; the Morley Library of Winchester Cathedral, and the curator, John Hardacre; and the Thorold and Lyttelton Library of the Diocese of Winchester.

Work in progress towards this book is reflected in the following publications: '"One of the larger lost continents": Religion in the Victorian novel', in *A Concise Companion to the Victorian Novel*, ed. Francis O' Gorman (Oxford and New York: Blackwell, 2004); 'The Variant and the Vatican: Catholic and Protestant Authority in Nineteenth-Century English Culture', in *Varianten – Variants – Variantes*, ed. Christa Jansohn and Bodo Plachta, Internationales Jahrbuch für Editionswissenschaften, 22 (Tübingen: Niemeyer, 2005); entries on 'Catholicism', 'Church of England', 'Keble', 'Manning', 'Newman', 'Religious literature', 'Wiseman', in *The Grolier Encyclopedia of the Victorian Era* (Danbury, CT: Scholastic, 2004).

I am grateful to owners for permission to reproduce the illustrations.

Note on referencing

Footnotes give short references to publications described more fully in the Bibliography. In the main text, when quoting from novels that are in print and readily available, chapter numbers only (with book or part numbers where appropriate) are given. The particular edition quoted is, however, listed in the Bibliography (Primary texts).

CHAPTER I

Introduction: 'Papal Aggression'

The enemy has come.

The Illustrated London News[1]

I

Queen Victoria sits on a raised throne at the opening of the Great Exhibition, listening to an address by her uniformed consort, Prince Albert, President of the Royal Commissioners (illustration 1). A less private communication between husband and wife is difficult to imagine. This couple have the highest profiles in the land and are here playing out their public roles. The speaker reads the address on behalf of a group. The listener attends to the address as the reigning monarch. The address is written with not only the royal listener in mind, but also the surrounding crowds of people, who themselves represent different constituencies, at home and abroad.

The Commissioners have encouraged exhibitors to send material from overseas and from all over Britain to London, capital of the world's first industrialized nation, where the latest products of modern manufacture can now be admired. The main political aim of the project is to display and celebrate Britain's own technological supremacy. Joseph Paxton's cathedral of glass, complete with 'nave' and 'transepts', is also a pleasure dome, a miracle of rare prefabrication. Large equestrian statues of the royal couple flank the dais. The galleries above are crowded with eager spectators. The tree that fills the space between the galleries has been cleared of birds by sparrow-hawks, on the advice of the aged Duke of Wellington, whose birthday it is today and who was cheered by the crowds on his arrival.

The opening of the Great Exhibition, on 1 May 1851, is a familiar enough scene. Two details are usually overlooked, however: the figures standing in

[1] Anon., 'The Papal Aggression', *Illustrated London News*, 9 November 1850, p. 358.

1. 'Inauguration of the Great Exhibition Building, by Her Majesty, May 1, 1851', *The Illustrated London News*, 18, 481 (3 May 1851), 350–1.

the foreground, facing Her Majesty, and the figure to her right, in front of the tree, who is dressed in a black gown, Geneva bands (like those of Swiss Calvinists) and a wig. The latter is the Archbishop of Canterbury, John Bird Sumner, whose role is to offer a special benediction, addressed to Almighty God on behalf of, and in the hearing of all who are present. This is to be followed by the 'Hallelujah Chorus' from Handel's *Messiah*.[2] Behind him stand other senior clerics, two of whom also wear wigs. The message is clear: the Church of England can give its blessing to modernity and to what the current Whig government at least regard as 'progress'. The figures in the foreground are foreign ambassadors, dressed in splendid ceremonial uniforms, and many of them are Catholics. They are witnessing a very English ceremony, mounted by Church and State. It is poised, polished, and Protestant.[3] For Continental Catholics, it must seem like another world – economically, politically, culturally and, above all, spiritually.

One of the presiding clergy on the great day was Charles James Blomfield, Bishop of London, who had preached at the Queen's coronation back in 1838, and in 1850 had appointed a committee 'for providing foreigners and other strangers with the means of attending divine worship during the period of the Exhibition'.[4] 'Let us not welcome them to this great emporium of the world's commerce', he said in his charge to the London clergy at St Paul's Cathedral, 'as though we looked only to the gratification of our national pride, or to mutual improvements in the arts, which minister to the enjoyment of this present life, and took no thought of the spiritual relation which subsists between all mankind as children of God, whom he desires to be saved through Jesus Christ'.[5] Although Blomfield's tone would have appealed particularly to the Evangelicals among the London clergy, their Broad Church brethren and the (often troublesome) High Church Tractarians would have broadly agreed with their tenor. Yet most of them, particularly in the Evangelical and Broad Church parties, would also have believed that the historic links between Church and State under the Crown were the key to the nation's material prosperity, and that politics and religion were inseparable.[6] Five years before the Duke of Wellington

2 Anon., 'The Great Exhibition', *Illustrated London News*, 3 May 1851, p. 349.

3 In her coronation oath, Queen Victoria had promised to 'maintain the Laws of God, the true Profession of the Gospel, and the Protestant Reformed Religion established by Law', and had received a copy of the Bible, the 'most valuable thing that this world affords': Anon., *Form and Order of the Service* (1838), pp. 27, 41.

4 Anon., 'Great Industrial Exhibition of 1851', *Illustrated London News*, 23 November 1850, p. 399.

5 Anon., 'The Bishop of London's Charge', in Anon., *Roman Catholic Question* (1851), 2nd series, p. 12.

6 A conservative, Anglican weekly newspaper entitled the *Church and State Gazette* was published between 1842 and 1856.

brought in unpopular Catholic emancipation legislation in 1829, Robert Southey had celebrated the Established Church of England on precisely these grounds, in *The Book of the Church*.[7]

Like other clerical and lay members of the British Establishment gathered in Hyde Park, however, Bishop Blomfield had witnessed a series of deeply unsettling events in England since the mid-1840s. Social and political tensions in the 'Hungry Forties' had been caused by increased Irish immigration during the railway boom and the potato famine (1845–9), the repeal of the Corn Laws (1846), a cholera epidemic (1847) and Chartist riots (1848). Fierce religious controversy had arisen on several occasions: when John Henry Newman, the spiritual leader of the Tractarians, converted to Rome and took others with him (1845); when Sir Robert Peel's Tory government subsidized the Catholic College of St Patrick, Maynooth (1845); when Lord John Russell – the next, Whig Prime Minister – offered the Bishopric of Hereford to Renn Dickson Hampden, whose liberal views were regarded as heterodox by the Tractarians (1847); when Bishop Henry Phillpotts of Exeter refused to institute George Cornelius Gorham – a case that turned on the doctrine of baptismal regeneration and the question of Church authority, and that deeply divided Evangelicals from Tractarians (1847–50); and, most recently, when the Pope restored the Roman Catholic episcopal hierarchy in England and Wales (1850).

Blomfield had felt an 'unwonted degree of anxiety and difficulty' in November 1850, when addressing his clergy, convened at St Paul's Cathedral for the sixth time since his appointment by the Duke of Wellington in 1828. In his charge, which was published and widely quoted as a public utterance, he first defended his position on 'The Gorham Controversy'.[8] Then he turned to the resulting 'Recent Recessions to Rome' and the vexed question of 'Romanising Innovations in Public Worship' – the subject of voluminous correspondence between himself and the more extreme ritualists among the Tractarians.[9] These 'Anglo-Catholic' tendencies caused widespread anxiety at mid-century. It was his next subject, however, that had convulsed the nation: 'Aggressive Movements of the Papacy'.[10] For on 29 September 1850, the feast of St Michael and All Angels, Pope Pius IX ('Pio Nono') had announced from Rome that the Catholic hierarchy of England was to be

[7] See pp. 144–5 below.

[8] The Bishop 'did not concur' with the judgment for Gorham: Anon., *Gorham* v. *Bishop of Exeter* (1850), p. 2.

[9] See A. Blomfield, *Memoir of Blomfield* (1863), II, 136–47.

[10] Anon., 'The Bishop of London's Charge', in Anon., *Roman Catholic Question* (1851), 2nd series, pp. 1–10.

2. 'Piux IX, Pontifex Maximus', *The Catholic Encyclopedia*, ed. Charles G. Herbermann *et al.*, 16 vols. (New York: Encyclopedia, 1913–14), XII, 134.

restored. Blomfield's response was representative: 'The assertion now first made of the Pope's right to erect Episcopal Sees in this country appears to me to be, not only an intentional insult to the Episcopate and clergy of England, but a daring though powerless invasion of the supremacy of the crown.'[11]

Pio Nono (illustration 2), a vilified 'foreign prince' in the autumn of 1850, had previously been regarded in England as a reforming, liberal

[11] For detailed accounts of the restoration of the episcopal hierarchy, see O. Chadwick, *Victorian Church* (1966–70), Pt I, pp. 271–309; E. R. Norman, *Anti-Catholicism* (1968) and *English Catholic Church* (1984); D. G. Paz, *Popular Anti-Catholicism* (1992).

Pope,[12] who had himself suffered at the hands of his own people, when he was forced to flee from Rome in disguise in November 1848, returning only in April 1850, with the aid of the French. Five months later, the long delayed bull was interpreted as 'Papal Aggression', through one of the strongest misreadings in modern British history. It was addressed to the universal Roman Church, but in particular the Catholic minority in England and Wales. It was also read, however, by the Protestant majority, of whom even the better educated were unfamiliar with its conventions – ecclesiastical, diplomatic, linguistic and spiritual. The 'Letters Apostolical' first addressed the most vehemently contested subject in the many battles between English Protestants and Catholics in the nineteenth century – English history.[13] The rich and complex story of the Reformation was remembered with pride by Dissenters and by Anglican Evangelicals and Broad Churchmen, notwithstanding Henry VIII's marital motives. The Papal bull simply referred to the 'Anglican schism of the sixteenth age',[14] and went on to explain that, in the seventeenth century, England had been divided into four administrative Districts, each overseen by a 'Vicar Apostolic'. In 1840, 'having taken into consideration the increase which the Catholic religion had received in that kingdom', Pope Gregory XVI had doubled their number.

Now, ten years later, having 'invoked the assistance of Mary the Virgin, Mother of God, and of those Saints who illustrated England by their virtues', Pio Nono decreed that, 'in the kingdom of England, according to the common rules of the Church, there be restored the Hierarchy of Ordinary Bishops, who shall be named from Sees, which we constitute in these our Letters, in the several districts of the Apostolic Vicariates'.[15] In order to further the 'fruitful and daily increasing extension' of Catholicity in England, the diocese of Westminster was elevated to 'the degree of the Metropolitan or Archiepiscopal dignity', the remaining twelve bishops to serve as his suffragans.

It was announced on the same day that Nicholas Wiseman, formerly Vicar Apostolic of the London District, was to be Archbishop of Westminster (illustration 3). To most English ears, this was to sound like a

[12] 'The Pope himself has turned reformer': W. Empson, 'The Papal States', *Edinburgh Review*, October 1847, p. 496.
[13] See chapters 3 and 4 below.
[14] 'Letters Apostolical, – Pius P. P. IX', in Anon., *Roman Catholic Question* (1851), 1st series, p. 1.
[15] Westminster, Southwark, Hexham, Beverley, Liverpool, Salford, Shrewsbury, St David's (Wales being regarded as part of England for these purposes), Clifton, Plymouth, Nottingham, Birmingham and Northampton. Ibid., p. 2.

3. 'Dr Wiseman, appointed by the Pope Cardinal Archbishop of Westminster', *The Illustrated London News*, 17, 453 (2 November 1850), 341.

contradiction in terms, as no such title existed and Westminster symbolized the nation itself. Parliament met in the Palace of Westminster. Westminster Abbey, England's church and a shrine to England's monarchs and fallen heroes, was drenched in national pride and a semi-mystical sense of antiquity.[16] The Abbey had no bishop's throne, let alone an archbishop's, as it was a 'royal peculiar', with a Dean appointed by Queen Victoria, who was the Supreme Governor of the Church of England and who had been crowned there, like almost all of her predecessors. Leading English commentators, including the historian Thomas Babington Macaulay, had formerly applied

[16] See, e. g., M. F. Tupper, 'The Abbey', in *Three Hundred Sonnets* (1860), p. 243.

territorial metaphors to the revival of Catholicism across Europe.[17] Now the Pope seemed to be invading the very citadel of England and its constitution.

Wiseman, still in Rome and, from 30 September, a cardinal with the title of St Pudentiana, made matters worse by sending a celebratory pastoral letter to English Catholics, 'From Without the Flaminian Gate of Rome', and couched in triumphalist terms, on 7 October. Again, he addressed his co-religionists but was also read, or rather misread by a Protestant nation. He and his fellow bishops were later to defend both the pastoral and the bull on the grounds that they referred to matters spiritual rather than temporal. This was not apparent at the time, however, and some of his phrasing offended Protestant readers when they saw a transcription of the pastoral in the newspapers. He wrote of continuing to 'govern' the counties under his administrative control, for example.[18] He believed that Catholic England had been 'restored to its orbit in the ecclesiastical firmament, from which its light had long vanished'.

When Victoria heard of all this she is said to have asked, 'Am I Queen of England or am I not?'[19] The story may be apocryphal, but it rings true. The Queen was in fact comparatively broad-minded on religious issues and disliked anti-Catholic propaganda directed against the Irish, who were her subjects. Her love of simple Bible-based worship, however, reflects a deeply Protestant spirituality, at a time when it was normal for the communion service to be celebrated only three or four times a year in Anglican churches. In the middle of the 'Papal Aggression' furore, she attended a church near Osborne in which the preaching of 'the Word' was literally elevated above all other aspects of worship: access to the chancel was gained by passing *under* the pulpit (illustration 4).[20] The contrast with the elaborate ceremonials favoured by the 'Ultramontane' Cardinal Wiseman, who emphasized Rome's jurisdiction beyond the Alps, could hardly be more striking (illustration 5).[21]

[17] 'During the eighteenth century, the influence of the Church of Rome was constantly on the decline. During the nineteenth century, this fallen Church has been gradually rising from her depressed state and reconquering her old dominion.' T. B. Macaulay, 'Von Ranke', in *Critical and Historical Essays* (1848), III, 253.

[18] N. P. S. Wiseman, 'Pastoral', in Anon., *Roman Catholic Question* (1851), 1st series, p. 5.

[19] E. R. Norman, *Anti-Catholicism* (1968), p. 56.

[20] 'Her Majesty and his Royal Highness, and the Princess Royal, and the ladies and gentlemen of the Royal household, attended divine service on Sunday morning, at Whippingham Church.' Anon., 'The Court at Osborne', *Illustrated London News*, 2 November 1850, p. 343.

[21] At his enthronement, 'a graceful canopy, fringed with silk and gold, was borne over his path by, as we were informed, the "converts" exclusively. . . . The mass was Haydn's, and performed by the choir in the most impressive style.' Anon., 'Papal Aggression', *Illustrated London News*, 14 December 1850, p. 457. On 'Word' and sacrament in Protestant and Catholic traditions, see G. Ebeling, *Word and Faith* (1963), p. 36.

4. 'Interior of Whippingham Church, Isle of Wight', *The Illustrated London News*, 17, 452 (26 October 1850), 337.

News of Pio Nono's bull reached London on 4 October 1850. Five days later, after a period of reflection, *The Times* initiated a campaign in defence of the constitution.[22] On 14 October, its first leader argued that the appointment of Wiseman was either 'a clumsy joke' or else 'one of the grossest acts of folly and impertinence which the Court of Rome [had] ventured to commit since the Crown and people of England threw off its yoke'. Referring to the Catholic revival within the Church of England (the 'Oxford Movement', or Tractarianism), it commented that the Pope had mistaken the 'renovated zeal of the Church in this country for a return towards Romish bondage'. Two words here – 'bondage' and 'impertinence' – provide clues to mainstream early-Victorian attitudes to Roman Catholicism. Their

[22] *Times*, 9, 14, 19 October 1850.

"ENTHRONIZATION" OF CARDINAL WISEMAN, IN ST. GEORGE'S CHURCH, LAMBETH.

5. '"Enthronization" of Cardinal Wiseman', *The Illustrated London News*, 17, 459 (14 December 1850), 457.

implication is that the Church of England is Protestant and independent, whereas the Tractarians emphasized that the Established Church was part of the 'Catholick' Church referred to in the creeds. (When referring to all Anglicans and Dissenters together, I will use the term 'Protestant', but only for convenience. 'Catholic' refers to Roman Catholicism, unless the context clearly indicates Anglo-Catholicism, but again, only for convenience.)

Protestantism, in all its varieties, emphasized the freedom of the enlightened individual conscience under God – a freedom that was constructed as 'natural' to humankind, God's special creation.[23] In the nineteenth century, 'liberty' was a key word in the lexicons not only of radicals and Dissenters, but also of conservative Anglicans, who used it as a weapon in defence of the British constitution of Church and State against supposed attacks from Rome.[24] Roman Catholicism seemed to be about the suppression of liberty, and about the tyranny of the priest over his flock and potentially of the Pope, a foreign prince, over the nation. Rome was therefore represented as a 'slave-master'.[25] Senior English Catholics were not slow, however, to

[23] On the concept of God's intervention in the defeat of the Armada on the side of England's 'liberty' – the title of James Thomson's poem of 1735–6 – see M. Dobson and N. J. Watson, *England's Elizabeth* (2002), p. 201. On 'liberty' and anti-Catholicism, see J. Wolffe, *Protestant Crusade* (1991), p. 127.

[24] Southey ends *The Book of the Church* with these words: 'From the time of the Revolution [of 1688] the Church of England has partaken of the stability and security of the State. . . . It has rescued us, first from heathenism, then from Papal idolatry and superstition; it has saved us from temporal as well as spiritual despotism . . . If the friends of the Constitution understand this as clearly as its enemies, and act upon it as consistently and as actively, then will the Church and State be safe, and with them the liberty and the prosperity of our country' (R. Southey, *Book of the Church* (1824), II, 528). Archbishop Sumner reminded the Secretaries of the Metropolitan Church that 'a liberty which has been enjoyed by all Churchmen from the Reformation to the present day, in the exposition of subjects of such deep mystery, should by all means be continued to them within the limits permitted by the revealed Word of God' (Anon., 'The Gorham Case', *Illustrated London News*, 17 August 1850, p. 135). On 8 November 1850, Benjamin Disraeli wrote to the Lord Lieutenant of Buckinghamshire: 'I have received numerous appeals from my constituents requesting that I would co-operate with them in addressing your Lordship to call a meeting of the county, in order that we may express our reprobation of the recent assault of the Court of Rome on the prerogatives of our Sovereign and the liberties of her subjects' (Anon., 'The Real Danger of the Church in England', in Anon., *Roman Catholic Question* (1850), 1st series, pp. 12–13). Queen Victoria spoke in the first person singular at the opening of parliament on 4 February 1851: 'The recent assumption of certain ecclesiastical titles conferred by a foreign Power has excited strong feelings in this country, and large bodies of my subjects have presented addresses to me, expressing attachment to the Throne, and praying that such assumptions should be resisted. I have assured them of my resolution to maintain the rights of my Crown, and the independence of the nation, against all encroachment, from whatever quarter it may proceed. I have, at the same time, expressed my earnest desire and firm determination, under God's blessing, to maintain unimpaired the religious liberty which is so justly prized by the people of this country' (Anon., 'The Queen's Speech', *Illustrated London News*, 8 February 1851, p. 107).

[25] 'When the light of the Reformation, in the sixteenth century, broke upon our world, and its sun-burst scattered the clouds and the darkness of preceding times – when the shadowy forms and the mysterious shapes of an enchaining superstition shrank abashed before the light, it was to the Church of Rome a day of "lamentation, and mourning, and woe." [Ezekiel 2.10] She had till then been the slave-master of the human mind.' M. H. Seymour, *Nature of Romanism* (1849), pp. 3–4.

throw words such as 'liberty' back at their antagonists. Bishop Ullathorne of Birmingham, who was personally involved in the process of restoring the episcopal hierarchy, wrote to *The Times* on 22 October 1850, arguing that it is 'unfair to confound this boon of liberty to the Catholic Church in England with ideas of aggression on the English Government and people as it is to confound the acts of Pius IX. as Pope with the notion of his temporal Sovereignty'.[26]

Three weeks after *The Times* referred to Rome's 'impertinence', the word was picked up by the more popular *Illustrated London News* in its lead story: 'The Roman Catholics . . . seem to have misunderstood the English people in this matter – if we may judge from the impertinent nomination just made by the Pope, and by the complacent glorification of Cardinal Wiseman (the *soi-disant* Archbishop of Westminster) in his pastoral letter, which was read in Roman Catholic churches of this metropolis on Sunday last'.[27] There is then some leaden play upon another negative 'im-' or 'in-'term – 'impudence': 'The nomination of Cardinal Wiseman – or St Pudentia, or Impudentia – is a direct insult to the Sovereign, the Parliament, and the people of his country.' Two days later, on 4 November 1850, the Prime Minister, Lord John Russell, quoted a variation on the theme in a letter to his old friend the Bishop of Durham, Dr Maltby, when he agreed with the bishop's statement that the recent 'Papal Aggression' was 'insolent and insidious'.[28] Again, *The Illustrated London News* echoed the word 'insolent', this time two weeks later, when commenting on the style of what it regarded as offensive Catholic documents.[29]

Insolence, impudence, impertinence: these terms of abuse, designed to denigrate the aspirations of the minority group and thus to confirm the powers that be, set the tone for the subsequent conflict. The war of words was fought out in hundreds of tracts, sermons, bishop's charges, letters (private and open) and newspaper articles. Some of these documents were written for the general reading public. Others, though addressed verbally or in writing to specific individuals or groups, were 'overheard' by a wider readership. A selection of them appeared in a series of twenty-four densely

[26] W. B. Ullathorne, 'To the Editor of the Times', in Anon., *Roman Catholic Question* (1850), 1st series, p. 7.

[27] Anon., 'The New Roman Catholic Bishops', *Illustrated London News*, 2 November 1850, pp. 341–2.

[28] Anon., 'Lord John Russell and the Pope: To the Right Rev. the Bishop of Durham', in Anon., *Roman Catholic Question* (1850), 1st series, p. 8.

[29] 'It was this kind of preaching and writing – it was this insolent and exulting, but foolish spirit, manifested somewhat too clumsily by the Roman Catholic priesthood – that aroused the Protestant feeling of England.' Anon., 'The Papal Vindication', *Illustrated London News*, 23 November 1850, p. 397.

printed pamphlets, published at least once a week, entitled *The Roman Catholic Question.* It demanded of its editor 'onerous labours',[30] for in the words of Bishop Ullathorne, 'Papal Aggression' had put the whole country 'in a boil'.[31]

II

There are a number of explanations for the strength of feeling associated with what Protestant England regarded as 'Papal Aggression'. In its suddenness, it came as a shock, although Wiseman argued that, since 1848, there had been 'no concealment, no attempt to take people by surprise'.[32] In its stated aims, it revived old fears of Papal ambitions, a year after the three hundredth anniversary of the first Act of Uniformity (1549). It seemed to confirm the hostile anti-Catholic views that had in fact gained ground in Britain after Catholic Emancipation in 1829, and that were reflected in the increased activity of anti-Catholic groups such as the Protestant Association and the Reformation Society.[33] After a series of other religious and political crises, and at a time of social unrest and a rising crime rate in the autumn of 1850, it seemed like the last straw. In reacting to 'Papal Aggression', all parties, political and religious, tried to turn it to their own advantage, thus intensifying the controversy.[34] Finally, existing anxieties concerning Tractarianism strengthened, so that cries of 'No Popery' were aimed at

[30] Anon., *Roman Catholic Question* (1851), p. iv.

[31] E. R. Norman, *English Catholic Church* (1984), p. 104. Bonamy Price wrote, 'Shame, fear, and anger convulsed the minds of the English nation': 'The Anglo-Catholic Theory', *Edinburgh Review*, October 1851, p. 528.

[32] N. P. S. Wiseman, *Appeal to the Reason* (1850), p. 6. See also W. B. Ullathorne, *History of the Restoration of the Catholic Hierarchy* (1871).

[33] 'There was a lull in . . . 1832 and 1833, but it revived in 1834, reached a peak in the summer of 1835, and remained at a high level until 1841. A new wave began with the anti-Maynooth agitation of 1845. This had by no means subsided when the creation of the Roman Catholic episcopal hierarchy in the autumn of 1850 was followed by widespread public indignation, raising "No Popery" frenzy to its highest pitch since the Gordon Riots of 1780.' J. Wolffe, *Protestant Crusade* (1991), pp.1–2.

[34] Disraeli, e.g., took a swipe at the Whigs in his open letter to the Lord Lieutenant of Buckingham: 'The fact is, that the whole question has been surrendered, and decided in favour of the Pope, by the present Government. . . . The policy of the present Government is, that there shall be no distinction between England and Ireland' (Anon., 'The Real Dangers', in Anon., *Roman Catholic Question* (1851), 1st series, p. 13). Protestant Dissenters uneasily positioned themselves in relation to the Roman Catholics, who also regarded themselves as Dissenters in England. But for 'papists', suggested a pseudonymous writer, 'where would Dissenters be now? Are we idiots as well as ingrates? It was the Catholic cause which redeemed opinion from the bondage of prelacy. The Irish members achieved for us the Reform Bill. It was Irish and Catholic majorities which, step by step, and inch by inch, against English Churchmen, obtained for us Municipal Reform, the repeal of Tests, the abolition of Negro slavery' (Anon. ('John Bull'), 'A Plain Appeal', in Anon., *Roman Catholic Question* (1851), 4th series, p. 5).

both Roman Catholics and Anglo-Catholics, sometimes in a confused and unfocused way, as we will now see.

The Prime Minister's letter to the Bishop of Durham, published by Maltby with Russell's permission on 7 November, was considered by moderates of all persuasions to be ill-judged, as it fanned the flames of the crisis even more than Wiseman's pastoral. First, Lord John Russell overstated Papal ambitions. 'There is an assumption of power', he wrote, 'in all the documents which have come from Rome – a pretension to supremacy over the realm of England, and a claim to sole and individual sway, which is inconsistent with the Queen's supremacy, with the rights of our bishops and clergy, and with the spiritual independence of the nation, as asserted even in Roman Catholic times.'[35] Secondly, while asserting that his 'alarm' was not equal to his 'indignation' concerning these documents, he went on to identify a danger which worried him 'much more than any aggression of a foreign Sovereign':

> Clergymen of our own Church . . . have been the most forward in leading their flocks, 'step by step, to the very verge of the precipice'. The honour paid to saints, the claim of infallibility for the Church, the superstitious use of the sign of the cross, the muttering of the Liturgy, so as to disguise the language in which it is written, the recommendation of auricular confession, and the administration of penance and absolution – all these things are pointed out by clergymen of the Church of England as worthy of adoption, and are now openly reprehended by the Bishop of London in his charge to the clergy of his diocese.
>
> What, then, is the danger to be apprehended from a foreign prince of no great power, compared to the danger within the gates from the unworthy sons of the Church of England herself?[36]

Russell uses the politician's trick of condemning one group through association with another that is currently in bad odour with the public. He inserts an attack upon the Tractarians between two blows against the Pope (that he has pretensions to power, but in fact has little power), thus channelling some of the energy in anti-Roman Catholic feeling towards the Anglo-Catholics – for him, and for most English people, the enemy within.

Since John Keble's Assize Sermon of 1833, the Tractarians, who were also known as 'Puseyites', had challenged the Church of England to rediscover her historical identity as 'one Catholick and Apostolick Church'.[37]

[35] 'Lord John Russell and the Pope', in Anon., *Roman Catholic Question* (1851), 1st series, p. 8.

[36] Ibid.

[37] Nicene Creed, The Order of Holy Communion, *Book of Common Prayer*. At her coronation, Queen Victoria had received a ring, the 'Ensign of Kingly Dignity, and of Defence of the Catholic Faith': Anon., *Form and Order of the Service*, p. 37.

Tractarians regarded themselves, not as Protestants – a term reserved for Dissenters, in their view – but as Catholics.[38] Moving in a quite different direction from that of the older and more influential Evangelical movement, with its emphasis upon personal conversion and the authority of 'Scripture', the Tractarians had taken great pains to raise the standard of Anglican public worship and private devotions, and had tried to revive monastic traditions. The more extreme manifestations of these aims smacked of 'Popery', or 'Papism', although those involved in the Anglo-Catholic revival saw it differently: their ideal was the beauty of holiness; their authority was grounded in tradition and the early Church; and their Catholicism was reformed.

Conspiracy theories were rife, however. It was widely believed that John Keble, author of *The Christian Year* (1827), John Henry Newman, Edward Bouverie Pusey, Canon of Christ Church and Professor of Hebrew, and their more extreme ritualist followers (of whom they disapproved), had a pact with the Pope and aimed to convert England to Roman Catholicism – a fear that seemed to be justified when Newman converted to Rome in 1845. Thus before the restoration of the Roman Catholic hierarchy in 1850, the Catholic wing of the Established Church of England had come to be regarded with even more suspicion than the 'Papists' themselves by Anglican Evangelicals and Broad Churchmen, as well as by Dissenters such as Methodists and Baptists, a development which greatly complicates the story of Roman Catholicism in nineteenth-century England. Lord John Russell was now playing upon existing fears and suspicions in the much more volatile atmosphere of 'Papal Aggression'.

Two days before the publication of Russell's letter had been Guy Fawkes day, when there were riots and demonstrations across the country. Dissenting fervour in Exeter was reflected in the annual pageant, which included guys in the form of the twelve Roman Catholic bishops *and* Puseyite clergymen of the Exeter diocese, 'flanked by officers of the Inquisition with instruments of torture for heretics'.[39] Later in the month there were 'No Popery' demonstrations in the City of London (illustration 6) and in Salisbury (illustration 7). *Punch* had a field day. Under 'Pontifical News', it reported that

[38] Faber, e.g., wrote in 1843, 'the Church of England is not Protestant': *Faber, Poet and Priest*, ed. R. Addington (1974), p. 100. Opponents of the Tractarians claimed that the 'assumption of the name "Catholic," [was] of very recent date': Anon., 'Tractarianism and Mr. Ward', *Dolman's Magazine*, March 1845, p. 73.

[39] R. Swift, 'Guy Fawkes Celebrations', *History Today*, November 1981, p. 7. Such celebrations in Lewes remain anti-Catholic to this day.

THE GREAT CITY MEETING AT GUILDHALL.—(EXTERIOR.)

6. 'The Great City Meeting at Guildhall – Exterior', *The Illustrated London News*, 17, 457 (30 November 1850), 420.

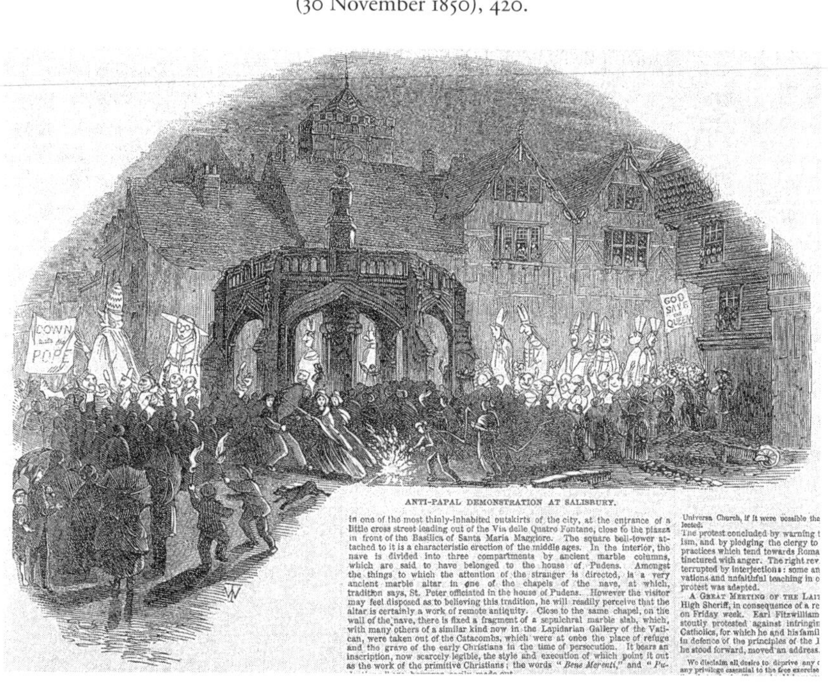

ANTI-PAPAL DEMONSTRATION AT SALISBURY.

in one of the most thinly-inhabited outskirts of the city, at the entrance of a little cross street leading out of the Via delle Quatro Fontane, close to the piazza in front of the Basilica of Santa Maria Maggiore. The square bell-tower attached to it is a characteristic erection of the middle ages. In the interior, the nave is divided into three compartments by ancient marble columns, which are said to have belonged to the house of Pudens. Amongst the things to which the attention of the stranger is directed, is a very ancient marble altar in one of the chapels of the nave, at which, tradition says, St. Peter officiated in the house of Pudens. However the visitor may feel disposed as to believing this tradition, he will readily perceive that the altar is certainly a work of remote antiquity. Close to the same chapel, on the wall of the nave, there is fixed a fragment of a sepulchral marble slab, which, with many others of a similar kind now in the Lapidarian Gallery of the Vatican, were taken out of the Catacombs, which were at once the place of refuge and the grave of the early Christians in the time of persecution. It bears an inscription, now scarcely legible, the style and execution of which point it out as the work of the primitive Christians; the words "*Bene Merenti*," and "*Pu-*

Universa Church, if it were possible the lected.
The protest concluded by warning t ism, and by pledging the clergy to practices which tend towards Roma tinctured with anger. The right rev terrupted by interjections: some an vations and unfaithful teaching in o protest was adopted.
A GREAT MEETING OF THE LAI High Sheriff, in consequence of a re on Friday week. Earl Fitzwilliam stoutly protested against infringin Catholics, for which he and his famil in defence of the principles of the I he stood forward, moved an address.
We disclaim all desire to deprive any c any privilege essential to the free exercise

7. 'Anti-Papal Demonstration at Salisbury', *The Illustrated London News*, 17, 457 (30 November 1850), 421.

His Eminence CARDINAL PANTALEONE, Legate of His Holiness, has arrived at the Golden Cross, Charing Cross: and is bearer of a message to the Chief of the British Government, demanding the usual acknowledgment on the part of the Sovereign of Great Britain, which has been always and from all time a fief of the Holy See.

In case of obstinate recusancy (which is not apprehended) his Eminence is commissioned to proclaim the PRINCE OF LUCCA as sovereign of these Islands, the prince being direct and undoubted descendant of those legitimate monarchs of England, who were driven by rebellion, the one to death, and the other to exile, from their neighbouring palace of Whitehall.[40]

In 'A Dream of Whitefriars', by Thackeray, the bigoted Mr Punch has a nightmare in which Roman Catholics take over the country. His dialogue with a friar ends with an argument about their respective martyrs in sixteenth-century London – one of the touchstones in disputes between Catholics and Protestants.[41] It was the cartoonists, however, who really seized the hour, producing a number of the most grotesque images associated with 'Papal Aggression' (illustrations 8–10).[42] *Punch*'s anti-Papal campaign was, however, hateful to one of their cartoonists, Richard Doyle, a devout Irish Catholic, who felt compelled to resign, suddenly, late in 1850, thus leaving the proprietors in the lurch and thereby, it has been suggested, making the campaign even more intemperate than before.[43] Meanwhile, *Punch* also lampooned the Tractarians mercilessly (illustration 11) and reinforced the popular misconception that Anglo-Catholicism and Roman Catholicism were virtually indistinguishable (illustration 12).[44]

Much of the agitation, physical and journalistic, was laid at the door of Lord John Russell, whose letter to the Bishop of Durham was widely regarded as its inspiration. (He was wrongly accused of stirring up the Guy Fawkes day riots, as some were misled by the date on the letter – 4 November.) John Arthur Roebuck, MP for Sheffield, wrote an open letter to Russell on 2 December, in which he drew parallels between the 'hate' and 'religious bigotry' aroused by his 'most unwise and unstatesmanlike letter' and the anti-Catholic fury aroused in the London mob by 'that madman Lord George Gordon' in 1780.[45] The following day, the *Morning*

[40] *Punch*, July–December 1850, p. 182. [41] Ibid, pp. 184–5. See chapter 3 below.

[42] Sir John Tenniel's first cartoon, 'representing Lord John Russell as David attacking Dr Wiseman, the Roman Goliath', was, with one exception, 'said to be the only cartoon founded on a strictly Biblical or Scriptural subject ever published in *Punch*'. M. H. Spielmann, *History of "Punch"* (1895), pp. 102–3.

[43] See M. H. Spielmann, *History of "Punch"* (1895), pp. 455–7.

[44] After the furore over 'Papal Aggression', '*Punch*'s religious war was directed chiefly against Puseyism and its "toys" – by which were designated the cross, candlesticks, and flowers.' 'Less fierce, but much more constant, was the ridicule meted out to the Jews.' Ibid., p. 103.

[45] Anon., 'Mr. Roebuck's Letter', in Anon., *Roman Catholic Question* (1851), 9th series, p. 2. Victorian responses to the Gordon Riots are discussed in chapter 4 below.

8. 'The Pope "Trying it on" Mr John Bull', *Punch*, 19 (July–December 1850), 192.

Chronicle published extracts from another letter to the Prime Minister, this time from William Bennett, Vicar of St Paul's Knightsbridge. Bennett complained that the 'No Popery' demonstrators who were making divine service impossible at St Barnabas's, the small sister church, and who made constant police protection essential there, had been directly influenced by Lord John Russell's letter.[46] Yet Russell, he pointed out, had himself worshipped at the Tractarian St Paul's for almost seven years, from its consecration by Bishop Blomfield in 1843, and had contributed generously to the building of St Barnabas.[47] Ending with a solemn warning to a premier who appeared to have sold his soul to Erastianism (the subordination of ecclesiastical to secular power), Bennett drew attention to the shibboleth of 'liberty': 'where is the justice of talking about civil and religious liberty,

[46] Anon., 'The Rev. Mr. Bennett's Letter', in Anon., *Roman Catholic Question* (1851), 9th series, p. 4.
[47] Ibid., pp. 8–9.

9. 'The Guy Fawkes of 1850, Preparing to Blow up all England!', *Punch*, 19 (July–December 1850), 197.

if the State is brought in to rule matters which do not belong, never have belonged, and never can belong to it?'[48] Bennett's resignation a fortnight later was widely discussed, being regarded as the first step in securing the Protestantism of the Church of England.[49]

[48] Ibid., p. 13.

[49] See O. Chadwick, *Victorian Church*, I, 303. Bennett moved to the living of Frome-Selwood, Somerset, where Christina Rossetti and her mother ran a school in 1853–4. See A. Grieve, 'The Pre-Raphaelite Brotherhood', *Burlington Magazine*, 1969, p. 294.

THE THIN END OF THE WEDGE.
DARING ATTEMPT TO BREAK INTO A CHURCH.

10. 'The Thin End of the Wedge: Daring Attempt to Break into a Church', *Punch*, 19 (July–December 1850), 207.

In Bishop Blomfield's charge to London's clergy of 2 November 1850, he compared the convert to Rome to a 'man who, having observed some instance of doubt or hesitation in his guide, in order to avoid mistaking the path on one side, rushes blindfold over a precipice on the other'.[50] While

[50] Anon., 'The Bishop of London's Charge', in Anon., *Roman Catholic Question* (1850), 2nd series, p. 8.

11. 'Fashions for 1850; or, A Page for the Puseyites', *Punch*, 19 (July–December 1850), 226–7.

What. Not come into my own house.
PETER DENS
VISITORS BELL
A PAGE FOR THE PUSEYITES.

11. (*cont.*)

12. 'Which is Popery, and Which is Puseyism?', *Punch*, 20 (January–June 1851), 15.

feeling 'bound to do justice' to the Tractarians' 'zeal and devotedness – their self-denial and charity', he believed that their current practices indicated that they felt separated from Rome by a 'faint and almost imperceptible line'.[51] Dr Gilbert Elliott, Dean of Bristol, felt that Blomfield was too gentle with the Tractarians. '*Remember*', he said in a speech to clergy on 6 November, '*it is not from dissent that Rome gains its victims, it is principally*

[51] Ibid, pp. 8–9.

from the Church of England'.[52] Elliott identified 'Tractarianism within our Church' as the 'principal and criminal cause of the increase of the Roman apostacy among us', and argued that its teaching is either identical or very close to Roman doctrines.

Whereas Blomfield and Elliott addressed meetings of Anglican clergy, Dr John Cumming, Minister of the National Scottish Church, Crown Court, Covent Garden, was renowned as a speaker who could stir up powerful emotions at large public meetings and lectures, and whose addresses were often published by the press.[53] When he lectured at the rooms in Hanover Square, on 7 November 1850, the Chairman who introduced him had some of the more lurid current catch-phrases to hand: 'The Pope of Rome – the man of sin – the head of the apostacy – the head of that system which was designated in the Scriptures as the ministry of iniquity, Babylon the Great, the mother of harlots, and the abomination of the earth [Revelation 17.5], had had the boldness and audacity to insult our Queen, our Church, our religion, and our laws; and they were called upon not to yield for one moment in submission to such an assumed authority as that. (Cheers.)'[54] He rejoiced that Bishop Blomfield had 'not only renounced Popery, but that also which was still worse – that great pest which was stalking through our country, and was now called "Puseyism." (Tremendous applause.)' Cumming himself attracted 'great cheering' when, at the beginning of his lecture, he quoted from 'that noble, Protestant, and faithful letter addressed by Lord John Russell to the Bishop of Durham', and then, towards the end, suggested that 'another result of the Cardinal's presence would be that Puseyism would disappear! They would have the real thing, and a sham one would not do.' Like nationalistic demagogues everywhere, Cumming wrapped himself in the flag: 'If we were to have Popery at all, let us have Italian Popery under an Italian flag, and not under the flag of old England. (Cheers.)'[55]

[52] Anon., 'Speech of the Dean of Bristol', in Anon., *Roman Catholic Question* (1851), 4th series, p. 14.

[53] 'The Victorian public meeting was a ceremony, planned in advance by a committee, in which participants elected a chairman, moved and seconded resolutions, fought among themselves, turned out dissenters, and adopted the petition to Parliament by acclamation.' The 'line of demarkation between the public meeting (to express in a constitutional way the opinion of "the people") and the public lecture (to inform and entertain) was blurred'. The public meeting was 'a form of entertainment', and, 'with its set speeches and resolutions, was perfectly suited to reporting'. D. G. Paz, *Popular Anti-Catholicism* (1992), pp. 23, 25, 33.

[54] Anon., 'Dr. Cumming on the Romish Aggression', in Anon., *Roman Catholic Question* (1851), 2nd series, p. 13. For a High Anglican reading of Babylon as Roman Catholicism, see C. Wordsworth, ed., *New Testament* (1877), II, 252–5.

[55] Anon., 'Dr. Cumming on the Romish Aggression', in Anon., *Roman Catholic Question* (1851), pp. 14, 16.

III

Today, many people in western Europe reject religion on the grounds that it has caused so much suffering. In the case of Christianity, Jesus preached a gospel of love and left his followers a peace that drives out fear (John 14.27). Yet his Church seems always to have been at war. The Church's civil wars, and most notably those between Protestants and Catholics, have been regarded by the combatants on both sides as just – wars that are necessary in defence of a saving truth so important that it is worth dying or killing for.[56] When that truth – in fact, arguably, a version or part of a larger truth – is challenged, love and peace are often driven out by fear.[57] What, then, were Protestants' specific fears regarding Roman Catholics in 1850?

As we have seen, from a Protestant perspective, whether it be from the position of the largely Anglican Establishment, or the church- and chapel-going middle classes, or the anti-'Papist' mobs, the restoration of the Catholic hierarchy was generally regarded as a threat to the constitution, and to religious and civil liberty. Significantly, for an island race less than four decades after Napoleon, the threat came from across the water – from Rome and from Ireland. First, there were the kind of fears associated with immigration.[58] In 1769, when James Watt had patented an improved version of the steam engine and Richard Arkwright had developed his spinning frame, there were seventy or eighty thousand Catholics in England and Wales out of a population of seven million (1%). By 1859, the year of Darwin's *Origin of Species* and the death of Isambard Kingdom Brunel, there were over a million out of a population of twenty million (5%), the influx of Irish Catholics having been enormous, especially in the 1840s.[59]

Secondly, unlike the 'old religion' of the English recusant Catholics, with its private chapels and its discreet clergy who dressed like lay gentlemen in public, the rapidly expanding Roman Church of nineteenth-century England looked, and indeed was 'Other' – Italianate in its ornament and

[56] George Townsend, Prebendary of Durham, wrote of the Reformation period: 'Our fathers believed that the chief importance of the disputes between the two churches consisted in this – that the salvation of the soul was endangered by the wilful errors of the church of Rome.' J. Foxe, *Acts and Monuments* (1843–9), I, 105. Cf. Prayer XII in the Anglican *Octave of Catholic Prayers*, ed. E. Phillips (1851), p. 23: 'We . . . especially thank thee for permitting our country still to be the faithful protester [sic] against the soul-destroying errors of Romanism.'

[57] 'There is no fear in love; but perfect love casteth out fear' (I John 4.18).

[58] In Kingsley's *Alton Locke* (1850), Crossthwaite talks about 'the competition of women, and children, and starving Irish' in the labour market (10).

[59] See p. 163 below. Today there are over five million Roman Catholics out of a population of fifty-three million (10%).

13. 'The "Chapelle Ardents," at Claremont', *The Illustrated London News*, 17, 445 (7 September 1850), 208.

liturgy, often Irish in its parochial clergy.[60] Moreover, new and cheaper printing methods – *The Illustrated London News* was founded by Herbert Ingram in 1842 – made it possible for people who lived far from towns and cities with a sizeable Roman Catholic presence actually to *see* what these 'Papist' priests looked like in their alien canonicals. They looked 'unnatural', confirming dark suspicions associated with their celibate state. Their ornate vestments smacked of idolatry. Furthermore, to an English Protestant eye, at a time when the influence of Evangelical 'Puritanism' on Victorian culture was strong, they looked un-English, unmanly and unwholesome.

Thirdly, the more visible Roman Catholics became, the more their activities, which were not seen, seemed secretive and therefore sinister. The darkness surrounding an illustration of the funeral of Louis Phillippe in *The Illustrated London News* in September 1850 signifies not only the sombre content of the Latin Mass for the dead, but also its inaccessibility (illustration 13).

[60] Thackeray presents a picture of an old Catholic family in the eighteenth century in *The History of Henry Esmond* (1852): see pp. 120–4 below.

ENTRANCE TO THE VAULT IN THE CHAPEL, AT WEYBRIDGE.—COUNT DE PARIS, DUC DE NEMOURS, DUC D'AUMALE, AND PRINCE DE JOINVILLE.

14. 'Entrance to the Vault in the Chapel, at Weybridge', *The Illustrated London News*, 17, 445 (7 September 1850), 208.

The illustrator can take the reader to the 'entrance to the vault', but not into the vault itself (illustration 14). In August 1850, when the Irish Roman Catholic hierarchy and many senior clergy from other countries gathered for the first synod in Ireland for two hundred years, it was possible to see images of them in procession and robing (illustrations 15–16). In its report, however, *The Illustrated London News* emphasized that their deliberations at Thules, in Ireland, would be 'conducted with the strictest secrecy'.[61] Worse still, some decisions might emerge only later, from that dark mystery of mysteries, the Vatican: 'it is not expected that the result of the deliberations will be published for some ten days after they shall have terminated, if, indeed, it be not necessary to wait until they shall have the approbation of the Pope'.

[61] Anon., 'Roman Catholic Synod', *Illustrated London News*, 24 August 1850, p. 167.

15. 'The Procession from the College to the Cathedral', *The Illustrated London News*, 17, 446 (14 September 1850), 224.

Fourthly, the loyalty of the Irish to the Crown was under question, and the majority of Roman Catholic clergy and laity in England in 1850 were Irish. At Thules, an official document was regarded by *The Illustrated London News* as 'chiefly remarkable for the unqualified condemnation which it pronounce[d] on the system of instruction adopted at the Queen's Colleges, and an announcement that the Synod have determined upon the establishment of a Roman Catholic University in Ireland'.[62] This was presented to the reader as an insult to 'our august, most gracious, and beloved Sovereign'.

All these fears are related to the threatening nature of alterity, or 'Otherness'.[63] That which seemed to be invasive, foreign, unnatural, secretive and insulting was demonized in language that revealed some of the profoundest anxieties, and specifically sexual anxieties of the age. What is more, that

[62] Anon., 'The Address of the Synod on the Queen's Colleges', *Illustrated London News,* 21 September 1850, p. 250.

[63] 'All is so novel, so strange, so unlike what is familiar to them, so unlike the Anglican prayer book . . . The mass is so difficult to follow, and we say prayers so very quickly, and we sit when we should stand, and we talk so freely when we should be reserved, and we keep Sunday so differently from them, and we have such notions of our own about marriage and celibacy, and we approve of vows, and we class virtues and sins on so unreasonable a standard; these and a thousand such details are decisive, in the case of numbers, that we deserve the hard names which are heaped on us by the world.' J. H. Newman, *Christ upon the Waters* (1850), p. 28.

16. 'The Bishops Robing', *The Illustrated London News*, 17, 446 (14 September 1850), 224.

language could be legitimized by judicious reference to scriptural sources. In the words of Cummings's chairman, quoted earlier, the threat to what 'we' Protestants count as holy – 'our Queen, our Church, our religion, and our laws' – comes from the head of a Church which, according to 'our' commentators, was described in the Revelation as 'Babylon the Great, the mother of harlots, and the abomination of the earth'. In February 1851, four months after the 'Papal Aggression' crisis began, an anonymous Catholic writer, quickly identified as Edward Bellasis, a leading parliamentary barrister and recent convert, supported his plea for a halt to hostilities by drawing attention to the language used by all the Anglican bishops during that period, in their 'communications with their clergy and people'.[64] His quotations

[64] E. Bellasis, *Anglican Bishops* (1851), p. 6. For examples of letters written in November 1850, see Anon., 'Letters and Replies', in Anon., *Roman Catholic Question* (1851), 8th series, pp. 2–11. Bellasis became a Serjeant-at-Law in 1844: see E. Bellasis, *Memorials* (1893), p. 10.

cover eight printed pages, and more than a third of the terms of abuse applied by the bishops to the Catholic Church, its doctrines and practices, its bishops and clergy, and its new hierarchy are those that were commonly applied to another threat to the purity of Victorian England – the 'harlot'.[65] The most frequently used term of all is 'corrupt', which recurs in various forms no fewer than thirteen times. 'Insulting' recurs nine times and 'audacious' seven times. Other significant terms are: false, subtle (both used four times); dangerous, dark, erring, insidious, offensive (three times); degraded, dishonest, monstrous, pagan, revolting, wanton (twice); base, blandishing, bold, cunning, defiling, disgraceful, disgusting, evil, indecent, infectious, insinuating, malignant, outrageous, pestilent, poisoning, polluting, profane, repugnant, satanic, scandalous, shameless, skilful, tainted, unclean, unholy, unlawful, violating (once).

In August 1850, only two months before the 'Papal Aggression' crisis began, the sixteenth number of Dickens's *David Copperfield* had opened with the chapter entitled 'Martha' (47), where he brought together two themes that were of deep public concern that summer: the pollution of the Thames in central London – since 1847 a stinking open sewer – and the problem of prostitution.[66] As David and Mr Peggotty follow Martha through the streets, hoping for news of Little Emily, Westminster Abbey is the point at which she passes 'from the lights and noise of the leading streets' and approaches the river, where she intends to follow the example of countless other fallen sisters by drowning herself. Character and setting merge in a vision of a hidden London that for Dickens was associated with the humiliations of his secret childhood, but here also bespeaks an adult sense of guilt. His terms are of the kind that the Anglican bishops will soon be applying to Catholicism: corruption (twice), dark, decay, decomposed, defiled, delapidated, disgrace, dust, grovelling, guilt, low-lying, plague, polluted, refuse, rotted, shadow, shame, wretchedness. It is as if Martha were a 'part of the refuse' that the river had 'cast out', Dickens writes.

Language that induces a feeling of repulsion from the corrupted body is also a commonplace of anti-Semitic propaganda, to which anti-Catholicism

[65] E. Bellasis, *Anglican Bishops* (1851), pp. 7–15.

[66] In 1847, the Metropolitan Commission of Sewers decided that all privy refuse was to be discharged directly into the sewers. The effluent was thus transported straight into the central reaches of the Thames. The smell was so bad that sheets soaked in chlorine had to be hung across the windows of the Palace of Westminster. In 1858, during the summer of the 'Great Stink', work finally began on a new sewage system. See P. Ackroyd, *London* (2001), p. 344; W. Acton, *Prostitution* (1857); M. Maison, *Making of Victorian Sexuality* (1994).

came only second in 'power and longevity'.[67] In nineteenth-century England, reforms relating to Catholic and Jewish political and religious 'disabilities' were often discussed at the same time. In the same month as the publication of the 'Martha' chapter, the 'Jewish Question' was in the news again, when Lord John Russell deferred a bill to admit Jews to the British legislature. *The Illustrated London News* showed Baron Rothschild taking the oath in the House of Commons, though not 'on the true faith of a Christian'.[68] *Punch* gave the story its customary anti-Semitic treatment (illustration 17).[69] Coincidentally, Halévy's *La Juive*, his only opera to become famous, had been presented for the second time at the Royal Italian Opera in London on 27 July 1850 (illustration 18). At the end of the opera, which is set in Constance at the time of the Council of 1414, Cardinal de Brogni, the former Chief Magistrate of Rome, asks Eleazor where his darling daughter may be found. The wealthy jeweller points to the scaffold, where Rachel, the girl he saved from de Brogni's burning palace in Rome, is to be thrown into a cauldron of boiling oil as a Jewish martyr, although Rachel is not in fact Jewish. For she and Eleazor have been condemned to death for working on a Catholic feast day. Act IV, Scene 2 opens in the public square, with the Chorus singing,

> Oh, what rapture! What delight!
> Now the heretics shall burn.
> Ere the dawn have we assembled,
> Eager now to watch them die.[70]

[67] R. D. Tumbleson argues that the 'sectarian dichotomy' of Protestant and Catholic is the 'obscured original of the one Edward Said sees as central to the colonialist and orientalist mentality, a "*positional* superiority" privileging the European subject over the foreign other that conceals itself in "the general liberal consensus that 'true' knowledge is fundamentally nonpolitical"'. For Tumbleson, anti-Catholicism is 'the ghost in the machine, the endless, neurotic repetition by self-consciously rational modernity of the primal scene in which it slew the premodern as embodied in the archetypal institution, arational and universal, of medieval Europe. Second only to anti-Semitism in power and longevity, it is a myth of iniquity whose pitiless prelatic villain has remained consistent for half a millennium, from John Foxe's martyrology to Dostoyevsky's Grand Inquisitor.' *Catholicism in the English Protestant Imagination* (1998), pp. 12–13.

[68] 'The House of Commons has twice affirmed its desire that the Baron Rothschild, twice elected by the city of London, should take his seat; and the House of Lords, acting in accordance with its constitutional right, has twice refused its assent to the measure introduced by the Ministry. Under these circumstances, it was clearly the duty of Lord John Russell to have forced the subject to an issue, and to have risked the existence of his administration upon it. At the last moment . . . his Lordship coolly announced his determination to postpone it until next session.' Anon., 'The Jewish Question', *Illustrated London News*, 3 August 1850, p. 93.

[69] 'Mark Lemon certainly did nothing to temper the flood of merciless derision which *Punch* for a while poured upon the whole house of Israel, and some of Brooks's verses are to this day quoted with keen relish in anti-Semitic circles.' M. H. Spielmann, *History of "Punch"* (1895), pp. 103–4.

[70] J. F. F. Halévy, *The Jewess* (1900), p. 19. On the play as a critique of the *ancien régime*, see D. Hallman, *Opera, Liberalism, and Anti-Semitism* (2002).

17. 'A Gentleman in Difficulties', *Punch*, 19 (July–December 1850), 75.

As Rachel goes to her death, the last words are the Chorus's:

> Thus shall all heretics and infidels at last
> Come to their doom.

This combination of Catholic oppression and Jewish martyrdom provided an uncannily prophetic historical backdrop to 'Papal Aggression' and the Rothschild affair.[71] Then Myerbeer's famous opera, *Les Huguenots*, which has parallels with *La Juive*, was presented for the eleventh time on

[71] Compare sermon VII, 'Popery Considered in Reference to the Jews', by the minister of the Jews' Episcopal Chapel, Bethnal Green: J. B. Cartwright, *No Popery!* (1850), pp. 160–84.

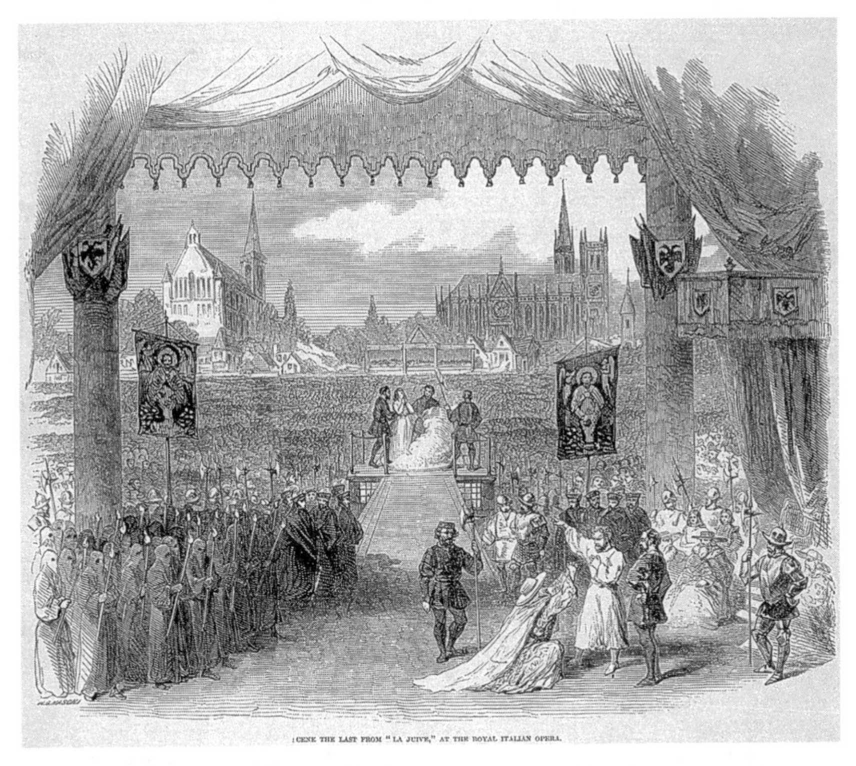

18. 'Royal Italian Opera', *The Illustrated London News*, 17, 439 (3 August 1850), 105.

17 August. This opera takes as its subject the struggle between Catholics and Protestants under Charles IX: the St Bartholomew Day massacre of 24 August 1572 provides the tragic climax.[72] The 'enthusiasm of the audience for this grand work', it was reported, appeared to be 'on the increase every season, inspiring the executants to fresh exertions'. The performance of the principal singers created a 'perfect *furore*'.[73]

As these operas were not only historical fictions, but were also set on the Continent, there was considerable imaginative distance between their plots and current events in England. Ancestral memory, however, associated with England's own bloody history, particularly in the sixteenth century, contributed more substantially to the climate of fear that tipped over into

[72] See D. Charlton, 'The Nineteenth Century: France', in *Oxford Illustrated History of Opera*, ed. R. Parker (1994), p. 151.

[73] Anon., 'Royal Italian Opera', *The Illustrated London News*, 24 August 1850, p. 175.

constitutional alarm the following October, with 'Papal Aggression'. Three hundred years earlier, for example, a forbear of the Prime Minister's, also named Lord John Russell, had suppressed a Catholic rebellion in the west country, resulting in four thousand rebel deaths. As a reward, he was made Earl of Bedford and received the abbey lands of Tavistock and Woburn in 1550. One of Wellington's ancestors, Walter Colley, received a warrant from Queen Elizabeth to execute martial law in County Kildare in Ireland, for which he was knighted.[74]

The three hundred years between 1550 and 1850 had provided inspiration for novelists and poets, painters and architects, as well as political and ecclesiastical historians, through a convulsive period of revolutions since the 1770s – agrarian, industrial and Romantic at home, American and French abroad – when British national identity was reinvented under pressure from within and without.[75] The last quarter of the eighteenth century witnessed the passing of Relief Acts which only slightly improved the lot of English Catholics, but were enough to trigger the Gordon Riots of June 1780, when a London mob torched Catholic houses to cries of 'No Popery' during six days and nights of drunken violence.[76] While Protestant and Catholic historians of England argued over the relative cruelty of the torments inflicted upon their own side's martyrs in the reigns of Queen Mary Tudor and Queen Elizabeth, the tragic romance associated with the Catholic Other was exploited by Ann Radcliffe in her novels and by English narrative painters in their depictions of the cult figure of Mary Queen of Scots.[77] For Protestant writers, artists and architects, the 'Gothic revival' offered an opportunity to reinterpret England's Catholic past – a project which continued in the novels of Sir Walter Scott and his Victorian successors.

Historical novels, operas and paintings brought past conflicts between Protestants and Catholics to bear upon present ones, and those who orchestrated the Protestant Establishment's response to 'Papal Aggression' were acutely conscious of the pressure of history upon them while they themselves made history. For example, in an age of spectacle, elaborate arrangements were made for a series of loyal addresses to be presented to Queen Victoria at Windsor Castle by the City of London and the Universities of Oxford and Cambridge. *The Times* reported that, on 10 December, the Great Western Railway Company 'placed at the disposal of the corporation the railway carriage hitherto used by her Majesty' and that the company

[74] See G. N. Wright, *Life of Wellington* (1841), I, 2.
[75] See L. Colley, *Britons* (1992). [76] See chapter 4 below.
[77] See R. Strong, *And when did you last See your Father?* (1978), pp. 128–35.

'took down seventy private carriages and two hundred horses for the accommodation of those who went by the special train'.[78] The procession from Slough station to Windsor was all that the spin doctors might have hoped, with the corporation, in the shape of the Lord Mayor and his large retinue of retainers, guarded front and rear by the representatives of law and order, in the shape of 'police-constables on foot, four abreast' and mounted police on the flanks. Although it was a cold, foggy day, almost all the inhabitants of Windsor turned out to 'gaze upon the variegated crowd of aldermen, town-councillors, doctors and bachelors of divinity, masters of arts, and graduates in various faculties, who filled the streets'.

After each of the five addresses that denounced 'Papal Aggression' and pledged loyalty to the Crown, Her Majesty responded personally. She referred to the 'great principles of civil and religious liberty' and rejoiced 'in the proofs which have been given of the zealous and undiminished attachment of the English people to the principles asserted at the Reformation'.[79] She was said to look 'very well, though somewhat flushed', while Lord John Russell's demeanour was more difficult to interpret. Finally, however, an irony of history intruded, when there was some 'good-humoured comment' that, in the hall in which lunch was served, portraits of a cardinal and a Pope 'seemed to smile benignantly on the royal deputations'. 'The portraits', the reporter added drily, 'had been placed there when the Waterloo-hall was first decorated'. On 14 December 1850, in the same issue as an article on Cardinal Wiseman's enthronement as Archbishop of Westminister, *The Illustrated London News* reported the 'Addresses to the Crown'.[80] Two weeks later, it printed a large engraving of the scene in St George's Hall, in which both the setting and the formal robes of those in attendance gave a reassuring impression of historical continuity and stability (illustration 19).

With the new year came a sense of being overwhelmed with reading matter on the crisis.[81] The blizzard of print was soon to increase in ferocity, however, when, on 7 February 1851, Lord John Russell opened a debate in the House of Commons on Ecclesiastical Titles legislation, designed to

[78] Anon, 'Addresses to the Crown (from the "Times")', in Anon., *Roman Catholic Question* (1851), 11th series, p. 1.

[79] Ibid., pp. 3, 6.

[80] Anon., 'Addresses to the Crown', *Illustrated London News*, 14 December 1850, p. 446.

[81] 'Our table is now covered with pamphlets of every size and colour, – the Bishop of London's "Charge" advancing in solemn array against Cardinal Wiseman's "Appeal," D'Israeli and Punch skirmishing against the Premier and the Papal Bull. Dr. Newman's sermon supplies eloquence, and a dozen Unitarian sermons honest logic and kindly good sense, to the heap. But at least half the fighting [e.g. Dr Cumming] is wide of the point destined to be lost and won.' Anon. ['C.'], 'The No-Popery Agitation', *Christian Reformer*, January–December 1851, p. 2.

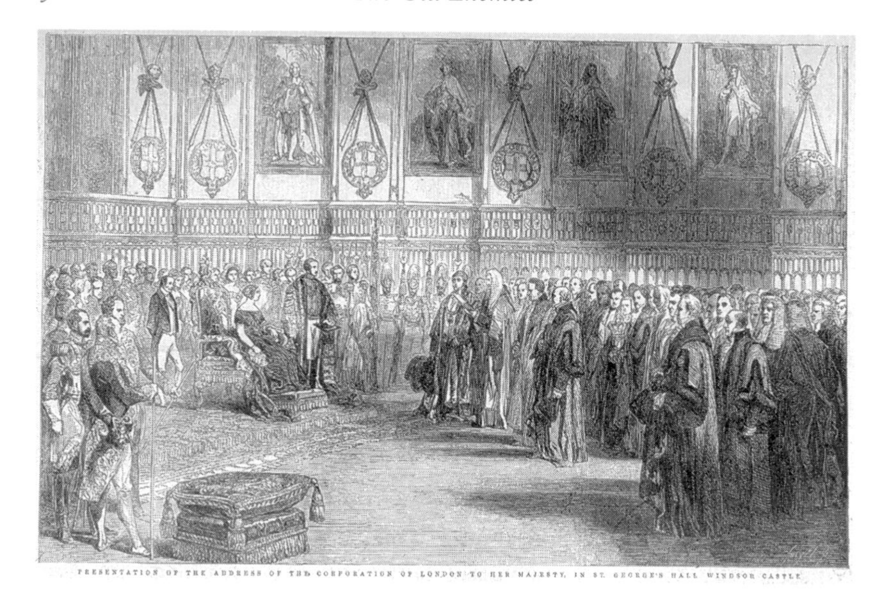

19. 'Presentation of the Address of the Corporation of London to Her Majesty, in St George's Hall, Windsor Castle', *The Illustrated London News*, 17, 462 (28 December 1850), 509.

make 'the assumption of any titles of archbishop, etc., of any place in the United Kingdom illegal, and to make any gift of property conveyed under such title null and void'.[82] The Bill was popular in the country, as it played to widespread anti-Catholic feeling. The Act, however, pleased nobody in parliament, was almost a dead-letter and little more than a nuisance in England, and was bitterly disliked in Ireland (illustration 20).[83] According to *The Illustrated London News*, it 'went too far for the enemies of the Church of England – not far enough for its friends – and affronted all Ireland, without being effectual as a remedy against the mischievous priestly domination it was intended to restrain'.[84] Meanwhile a number of cases against Roman

[82] Letter from Lord John Russell to Queen Victoria on 11 December 1850, quoted in E. R. Norman, *Anti-Catholicism* (1968), p. 72.

[83] Ibid., p. 79. Russell resigned on 22 February 1851, but his government resumed power on 3 March. Nobody was prosecuted under the Act, which Gladstone repealed twenty years later: see O. Chadwick, *Victorian Church* (1966–70), 1, 305.

[84] Anon., 'The Downfall of the Russell Administration', *Illustrated London News*, 1 March 1851, p. 170. Cf. 'The Peelites, Irish, and Radicals, upon whom the ministry depended for important support, found the measure repugnant, while Conservatives and some right-wing Whigs considered it too weak to meet the problem, for its penalties were relatively mild and its ban did not extend to Ireland.' D. G. Paz, *Popular Anti-Catholicism* (1992), p. 12.

THIS IS THE BOY WHO CHALKED UP "NO POPERY!"—AND THEN RAN AWAY!!

20. 'This is the Boy who Chalked up "No Popery" – and then Ran Away!', *Punch,* 20 (January–June 1851), 119.

Catholic priests, involving 'sex, money, and priestly domination', provided scandalous material for further anti-Catholic propaganda.[85]

IV

Naturally, Roman Catholic perspectives on these events and issues were different from those of the Protestant majority. Early in the nineteenth century, the battle had been for Catholic Emancipation, finally achieved in 1829.[86] Catholics could now vote, hold commissions in the army and navy, and qualify for ministerial office. Discriminatory laws remained, however,

[85] Anon., 'The Downfall of the Russell Administration', *Illustrated London News*, 1 March 1851, p. 170. See also the Achilli libel case against Newman, during which witnesses testified to rape in Italian sacristies by this former Catholic priest, who was now giving popular anti-Catholic lectures up and down the land: see O. Chadwick, *Victorian Church* (1966–70), 1, 306–8.

[86] See chapter 5 below.

and Catholics continued to be second-class citizens, unable to attend the universities, for example.[87]

Although this encouraged solidarity within the Catholic camp and the presentation of a united front to a hostile Protestant world, it was also natural for Catholics to disagree among themselves. The slowness of Catholic reforms since the 1770s had in fact been caused as much by internal divisions between Catholics as by Protestant opposition.[88] The bishops saw the restoration of the hierarchy in 1850 as a way of strengthening their control over their Church, while some Catholic laity and lower clergy remained discontented with a system that ignored their claims to a proper parochial system.[89] The old recusant English families were no more enamoured of Wiseman's Ultramontane Catholicism than they were of the Irish 'hordes'. Henry Charles Howard, 13th Duke of Norfolk and England's senior Catholic layman, declared himself to be strongly opposed to the new arrangements and worshipped at the Anglican parish church.[90]

The conversion of England ('Mary's Dowry'), so deeply feared by Protestants, was to remain a fervent though unfulfilled Catholic hope.[91] On the basis of increased numbers and a restored hierarchy, the Church at least built up its strength over the subsequent 150 years.[92] After 1850, Catholics felt that they were once again something of a power in the land[93] – and the story of Catholic and Protestant in the nineteenth century is partly

[87] 'Their churches could not have steeples; their priests could not wear clerical garb in public; their schools were denied state funding; they could not leave charitable bequests for purposes judged "superstitious" by Protestant standards. Roman Catholics also faced social discrimination. They were physically shunned; and the mass media of the day produced a torrent of tracts, books, magazines, and newspaper stories that reviled their beliefs, challenged their political loyalty, and depicted them as the deluded dupes of men who lusted for sex, money, and power.' D. G. Paz, *Popular Anti-Catholicism* (1992), pp. 1–2.

[88] See E. R. Norman, *English Catholic Church* (1984) and J. Bossy, *English Catholic Community* (1975).

[89] See E. R. Norman, *English Catholic Church* (1984), pp. 73, 106.

[90] See J. M. Robinson, *Dukes of Norfolk* (1995), p. 202, and p. 162 below.

[91] Newman preached movingly on 'The Second Spring' of English Catholicism at the first synod of the restored hierarchy, at Oscott in 1852, although he warned that it might turn out to be an English spring, of 'bright promise and budding hopes, yet withal, of keen blasts, and cold showers, and sudden storms'. (I. T. Ker, *John Henry Newman* (1988), p. 382.) In 1838, Ambrose Phillipps (later de Lisle) and Revd George Spencer founded the Association for Universal Prayer for the Conversion of England. (C. G. Herbermann, *et al.*, eds., *Catholic Encyclopedia* (1913–14), IV, 699.) On 2 May 1867, the Roman Catholic bishops resolved that the exposition and benediction of the Blessed Sacrament on the second Sunday of every month would be for the intention of the conversion of England. ('Prayers for the Conversion of England', in R. Challoner, *Garden of the Soul* (n.d.), p. 234.) Gerard Manley Hopkins looked for the conversion of England in 'The Wreck of the Deutschland', his unpublished poem of 1876.

[92] See G. A. Beck, ed., *English Catholics* (1950) and V. A. McClelland and M. Hodgetts, eds., *From Without the Flaminian Gate* (1999).

[93] By 1860, Wiseman thought that they had sufficient political power to determine the outcome of general elections. See W. L. Arnstein, *Protestant versus Catholic* (1982), p. 49.

about different kinds of power. As this study will show, however, it is also a story about different kinds of spirituality and of language. Indeed, stylistic contrasts between Catholic and Protestant writings of the period can sometimes be even more illuminating than contrasts between their doctrinal content, as they reveal the devotional hearts and minds of the different faith communities.

Take, as an extreme example, the following passage from an open letter by Ambrose Lisle Phillipps (later de Lisle), a leading lay convert, to his friend the Earl of Shrewsbury, who was due to convey the thanks of English Catholics to the Pope in November 1850: 'surely our hearts should expand with joy and gratitude, our voices should be lifted up to thank and to glorify the Successor of St. Peter, and to invoke upon His sacred head the choicest blessings of our Lord, whose chief vicar on earth He is'.[94] Thanksgiving and benediction are followed by some trenchant answers to anti-Catholic protests in the Protestant press. At the end of the letter, however, Phillipps again draws upon ancient traditions of Catholic writing. He comments that the choice of Westminster as the name of 'our Primatial See' is 'one of Catholic and happy omen'. He then recounts an anecdote concerning a vision of 'the Blessed Saint Edward the Confessor' shortly before his death in the royal palace of Westminster in January 1066, when he fell into a trance and saw 'two pious Benedictine Monks of Normandy whom he had loved in his youth when an exile in that country':

> These Monks foretold to the King, what was afterwards to happen in England, they declared that the wickedness of the English nation was exceeding great, and that it provoked the wrath of God, but that when it should be come to the full, He would send in His anger a Mission of wicked spirits into the land, who should grievously punish it, and sever the green tree thereof from its stock, for the space of three furlongs distance, but that at length God would have mercy upon England, when this same tree should return again to its own root without the help of any man's hand and bear fruit and flourish.[95]

'Our Catholic ancestors', Phillipps asserts, 'understood the Mission of wicked Spirits to signify that of the Protestant Innovators, who in the 16th century pretended to reform the English Church.' The three furlongs, he believes, represent three centuries. Whereas lay Anglicans tended to write on 'Papal Aggression' in the style of political pamphleteers, this romantic 'English Catholic gentleman' supports his defence of the Pope's actions by

[94] A. L. Phillipps, *Letter to the Earl of Shrewsbury* (1850), p. 3. (For extracts, with changes to capitalization, see Anon., *Roman Catholic Question* (1851), 1st series, pp. 14–15.)

[95] Ibid., p. 13.

invoking an ancient legend relating to a richly symbolic vision granted to a dying Saxon king. He inhabits a different spiritual and imaginative world.[96]

So, too, does Cardinal Wiseman, although his Ultramontane, Rome-centred Catholicism contrasts, and sometimes conflicts with Phillipps's adopted English tradition of Catholicism. Wiseman was born in Seville, of a Roman Catholic Irish family who had settled in Spain. After his father's death in 1805, the family moved to Waterford via London, before settling in England. He was educated at Ushaw from 1810, where he was influenced by the historian John Lingard and met George Errington. In 1818 he was sent to Rome, where he was first trained at the English College and then taught there. He also became an authority on Roman antiquities and developed his remarkable gift for languages. He was made a Doctor in Divinity in 1825 and ordained priest the following year. By 1828 he was Rector of the English College, which increased the frequency of his contacts with the Pope. In 1836 he founded the *Dublin Review*, to which he contributed many articles, and published eight lectures on *The Real Presence . . . in the Blessed Eucharist*, delivered in the English College. Having also lectured at St Mary Moorfields, in the City of London, that year – the first public presentation of Roman Catholic doctrines in industrial England – he moved to St Mary's College, Oscott, Sutton Coldfield, as President, in 1840, with the title of Bishop of Melipotamus. (Before 1850, Roman Catholic bishops in England were given honorary titles based on defunct foreign sees.) In February 1849, Wiseman was made Vicar Apostolic of the London District and in September 1850, Cardinal Archbishop of Westminster. Although an admirer of English manners and sense of 'fair play', he never really understood the old English Catholics and felt that Rome was his spiritual home. When he presided over grand ceremonial services in England, dressed in the full vestments of a cardinal, of which he was inordinately fond, he was said by Father Faber to look like 'some Japanese god'.[97] When he dined with the rich and powerful, he was regarded by some as rather vulgar.

Wiseman's greatest successes were his use of the restoration of the hierarchy to establish Ultramontanism in England and Wales[98] and his direct appeals to the fairness of the British public during the 'Papal Aggression' crisis. Three of his published statements that were widely circulated in 1850 are particularly revealing – politically, spiritually and stylistically. First, let us return to his pastoral letter, 'From Without the Flaminian Gate of Rome'

[96] It was a dream that made Phillipps take the final step to conversion at the age of sixteen. He searched for the miraculous throughout his life, regarding its cause as the suspension of nature's laws at privileged moments. See M. Pawley, *Faith and Family* (1993), pp. 19, 36.

[97] Quoted in E. R. Norman, *English Catholic Church* (1984), p. 118.

[98] Ibid., p. 127.

on 7 October, quoted earlier. This was written in Rome in a state of euphoria. Indeed, so transparent was his private satisfaction at receiving from the Pope, in separated ceremonies, his cardinal's hat and his archbishop's pallium, that Wiseman presented *Punch* with a broad target, both literally and metaphorically.[99] He was also joyful for the English Catholic Church, of course, and in this respect was naïve only in assuming that all his brethren would be equally ecstatic about his translation to the new archbishopric. Whereas pastoral letters by bishops of the Church of England tended to begin by underlining the writer's authority (sometimes nervously), Wiseman, as both Archbishop of Westminster and 'Administrator Apostolic of the Diocese of Southwark', spoke from the heart when explaining this dual ministry:

If this day we greet you under a new title, it is not, dearly beloved, with an altered affection. If in words we seem to divide those who till now have formed, under our rule, a single flock, our heart is as undivided as ever in your regard. For now truly do we feel closely bound to you by new and stronger ties of charity; now do we embrace you, in our Lord Christ Jesus, with more tender emotions of paternal love; now doth our soul yearn, and our mouth is open to you; though words must fail to express what we feel on being once again permitted to address you.[100]

Although some terms associated with authority are sprinkled through the text of the pastoral,[101] most of the key terms are associated with grace and glory, love and hope, light and joy.[102]

Roman Catholicism's claim to ultimate authority is based on its being universal, or 'catholic', firstly, in space, being global in its membership; secondly, in time, the Papal succession stretching back to St Peter himself; and thirdly, in truth, through the infallibility of the Church on matters of faith. It is with respect to time, however, that Roman Catholic writing is often most distinctive. In his pastoral letter, Wiseman writes of the English saints and martyrs, not as departed historical figures, but as present, living beings:

How must the saints of our country, whether Roman or British, Saxon or Norman, look down from their seats of bliss with beaming glance upon this new evidence of the Faith and Church which led them to glory, sympathising with those who

[99] N. P. S. Wiseman, 'Pastoral', in Anon., *Roman Catholic Question* (1851), 1st series, pp. 4–6. The pallium is a band of white wool worn on the shoulders. See 'The Cardinal's Hat', *Punch,* July–December 1850, p. 192 and illustration 8.

[100] N. P. S. Wiseman, 'Pastoral', in *Roman Catholic Question* (1851), 1st series, p. 4.

[101] E.g., authority, dignity, eminence, regulation, sanction, weight.

[102] E.g., admiration, affection, attachment, benevolence, blessing, charity, emotions, exaltation, fruit, glow, glory, gratitude, heart, hopes, hospitality, joy, light, love, prospects, thanksgiving, vigour.

have faithfully adhered to them through centuries of ill repute, for the truth's sake, and now reap the fruit of their patience and long-suffering. And all those blessed martyrs of these later ages, who have fought the battles of the Faith under such discouragement, who mourned, more than over their own fetters or their own pain, over the desolate ways of their own Sion and the departure of England's religious glory; oh! how must they bless God, who hath again visited His people, how take part in our joy, as they see the lamp of the temple again enkindled and re-brightening, as they behold the silver links of that chain which has connected their country with the See of Peter in its Vicarial Government changed into burnished gold; not stronger nor more closely knit, but more beautifully wrought and more brightly arrayed.[103]

Again, the passage moves towards a celebration of the beauty of the Church, this time described as the holy temple, in which the Pope represents on earth the Christ who is our great high priest in the heavenly Jerusalem.

Secondly, having written a pastoral letter which, together with the papal bull, ignited the 'Papal Aggression' furore, Wiseman sought to make amends by writing directly to his Protestant fellow countrymen in defence of Catholicism and the restoration of the hierarchy. 30,000 copies of his tract entitled *An Appeal to the Reason and Good Feeling of the English People, on the subject of the Catholic Hierarchy* were sold in the first three days, as he was pleased to tell everybody, and it was printed in full in five London daily newspapers on 20 November.[104] Much of it is devoted to parrying Protestant attacks, including what he regards as deceitful attempts to present the restoration of the hierarchy as 'something territorial'.[105] He quotes from a letter that he addressed to Lord John Russell on 3 November, *en route* from Rome to London, the tone of which, he suggests, indicates no 'insolent or insidious design'.[106] He also takes a debating point often made by statesmen in the Palace of Westminster, relating to the neglect of the nearby slums, and applies it to the Abbey, which owned some of the houses, in a manner reminiscent of Dickens, Disraeli[107] and early Pugin:[108]

[103] N. P. S. Wiseman, 'Pastoral', in Anon., *Roman Catholic Question* (1851), 1st series, p. 5.

[104] See E. R. Norman, *Anti-Catholicism* (1968), p. 63.

[105] 'Time will unmask the deceit, and show that not an inch of land, or a shilling of money, has been taken from Protestants and given to Catholics.' N. P. S. Wiseman, *Appeal to the Reason* (1850), p. 17.

[106] Ibid., p. 24.

[107] 'The Abbey of Westminster rises amid the strife of factions. Around its consecrated precinct some of the boldest and some of the worst deeds have been achieved or perpetrated: sacrilege, rapine, murder, and treason.' B. Disraeli, *Sybil* (1845), IV, 6.

[108] A. W. N. Pugin's *Contrasts* (1836) between modern architecture and that of the fourteenth and fifteenth centuries includes social commentary: an ugly, Utilitarian modern world is shown to neglect the poor, whereas a beautiful, pre-Reformation Catholic world is shown to care for them. Pugin, a convert to Catholicism, later recanted his own idealism in relation to the Middle Ages, in a pamphlet on the restoration of the Catholic hierarchy: see p. 78 below.

In ancient times, the existence of an Abbey on any spot, with a large staff of clergy, and ample revenues, would have sufficed to create around it a little paradise of comfort, cheerfulness, and ease. This, however, is not now the case. Close under the Abbey of Westminster there lie concealed labyrinths of lanes and courts, and alleys and slums, nests of ignorance, vice, depravity, and crime, as well as of squalor, wretchedness, and disease; whose atmosphere is typhus, whose ventilation is cholera; in which swarms a huge and almost countless population, in great measure, nominally at least, Catholic; haunts of filth which no sewage committee can reach – dark corners, which no lighting-board can brighten. This is the part of Westminster which alone I covet, and which I shall be glad to claim and to visit, as a blessed pasture in which sheep of holy Church are to be tended, in which a Bishop's godly work has to be done, of consoling, converting, and preserving.[109]

His argument that he had no designs upon the Abbey, but coveted only the surrounding slums, was dismissed by a hostile Protestant press as humbug.[110] *The Times* for 21 November sarcastically congratulated Wiseman on his 'recovery of the use of the English language': his pastoral letter's 'golden chain of St. Peter rings no longer in our ears'.[111] The broader stylistic analysis that follows reflects Protestant suspicions concerning Catholic language in general. Why, *The Times* asks, does Wiseman only now tell us the plain truth about the restoration of the hierarchy, after the public have protested loudly at his bombastic pastoral letter, 'From Without the Flaminian Gate of Rome'?

It is because the Roman Catholic Church has two languages, an esoteric and an exoteric – the first couched in the very terms of that more than mortal arrogance and insolence in which Hildebrand [Pope Gregory VII] and Innocent [III] thundered their decrees against trembling kings and prostrate emperors, the second artful, humble, and cajoling, seizing on every popular topic, enlisting in its behalf every claptrap argument, and systematically employing reasoning the validity of which the sophist himself would be the last to recognise.

Whether artful or not, Wiseman succeeded in associating the scandal of poverty in the slums of Westminster with Anglicanism. He also again appealed to the heart, as in the motto that Newman was to choose for his cardinal's coat of arms a quarter of a century later: '*Cor ad cor loquitur*' (heart speaks to heart).[112] In the fifth series of *The Roman Catholic Question*, 'John Bull' criticized statesmen and clergy for their aggressive partisanship

[109] N. P. S. Wiseman, *Appeal to the Reason* (1850), pp. 30–1. For a defence of the Dean and Chapter's attempts to improve conditions in the 1840s, see A. Milman, *Henry Hart Milman* (1900), pp. 138–42.

[110] See Anon., *Roman Catholic Question* (1851), 5th and 7th series *passim*. *The Examiner* of 23 November compared Wiseman to Uriah Heep: 7th series, p. 4. See also 'Papal Vindication', where Bishop Ullathorne's intervention is criticized alongside Wiseman's.

[111] Anon., *Roman Catholic Question* (1851), 5th series, pp. 12–13.

[112] See I. T. Ker, *John Henry Newman* (1988), p. 719.

and appealed to the reader, over the heads of civic and church leaders, for understanding between Catholics and Protestants: '*We* must be statesmen', he wrote, 'when statesmen become alarmists.'[113] Himself a 'Liberal as well as a Protestant', he celebrated Catholicism's appeal to the heart:

> Catholicism supplies a great want in the human heart, which no agency that is not virtually Catholic can supply. There are many who cannot be religious by the cold assent of the understanding, and cannot be led to their knees by the thorny path of the logical propositions which go by the name of Calvinism. The passions, the senses, the imagination, the lower affections of man, are incapable of a pious direction except by the service which will first catch the ear, and fill the eye, and help the flagging soul upward by the veiled idolatry, the secondary conceptions of the presence of the Deity, which fill the anxious benches of the Revival, faint, and sob, and cry aloud in the Irving Conventicle, or bow in awe before the Host, and thrill with the solemn harmonies of the pealing anthem. [114]

Although 'John Bull' can write, 'Go on, then, cardinal – toil, bishop – labour you, presbyter – there is room and need for you all', the assumption remains that Roman Catholicism, like Edward Irving's revivalist 'Catholic Apostolic Church', appeals to the lower rather than the higher 'affections of men'.[115]

The third of Wiseman's most significant documents of 1850 is the sermon that he delivered in a packed St George's Catholic Cathedral, Southwark, on Sunday 8 December, dressed, according to the *Morning Herald*, in the mitre that the Pope had presented to him, 'the ring, and the other gorgeous appointments of his dignity'.[116] On this day of devotion to the Conception of the Blessed Virgin Mary, he says, it is particularly appropriate for those who have 'crowded here' to celebrate this 'patron festival of our diocese' to have their thoughts filled with 'exultation and joy', and to show their 'gratitude to Almighty God for the benefits which He has bestowed upon us so lately'.[117] He then appeals to the heart once again, but this time with a specifically Marian emphasis: 'we wish . . . to renew in our hearts, and with the expression of our own lips, that earnest, deep, affectionate devotion towards the blessed mother of our Lord and Saviour Christ Jesus which now more than at any time deserves to be openly proclaimed and

[113] Anon., 'To Englishmen, Irishmen, and Scotchmen', in Anon., *Roman Catholic Question* (1851), 5th series, p. 2.

[114] Ibid., pp. 7, 11.

[115] Irvingite priests wore vestments and used incense.

[116] 'A Sermon preached at St. George's Catholic Church', in *Roman Catholic Question* (1851), 10th series, p. 4. The doctrine of the Immaculate Conception was to be defined by Pio Nono as a truth contained in the original teaching of the Apostles on 8 December 1854. St George's became a cathedral in 1851: see pp. 111–12 below.

[117] Ibid, p. 1.

professed by Catholic mouth, when outrage and blasphemy are most loud in her dishonour, and in that, consequently, of Him to whom she gave birth'.[118] Whereas Protestant opinion regarded 'Papal Aggression' as an assault upon Victoria, Our Dear Queen (to whom, incidentally, English Catholics frequently expressed their loyalty during the crisis[119]), Wiseman represents the Protestant backlash to the restoration of the hierarchy as an assault upon Mary, our Lady, Queen of Heaven – for Catholics, the object of profound love and affection, as well as of veneration.

The Cardinal Archbishop, who is also currently Administrator Apostolic of Southwark, then announces that the Pope has been pleased to agree to a special, compressed jubilee over the subsequent two weeks, when the diocese is called to repentance of sins, for the 'good of individuals' and the 'general benefit to the Church and to the world'.[120] With respect to individuals, he reminds his listeners that 'even the best-regulated heart, even the purest soul, will be conscious of that dropping of the impure dew of this world which accumulates within us'. Even for those who do not feel themselves to be 'immersed in vice or crime', the jubilee is a 'period of reconsideration of themselves, and of remodelling, if I may so speak, of their very heart, and shaping and forming it more truly in the mould of the Gospel'. All must therefore 'frequent the tribunal of confession', 'approach the holy table' and 'join in prayers and supplications'. With respect to the world, he continues, the Church – 'a kind, and merciful, and most loving mother' – bids us 'lift up our hands and our hearts together' in 'one solemn concert of pleading for mercy'.[121]

Protestant Evangelicalism has been called the 'religion of the heart'.[122] Whereas the Evangelical heart must be converted, however, through faith in Christ's atonement on the cross,[123] the Catholic heart must be remodelled, shaped and formed through works, under the guidance of Christ's

[118] The Revd Cartwright wrote, e.g.: 'The Virgin Mary is at this time the god of Romish idolatry.' J. B. Cartwright, *No Popery!* (1850), p. 45.

[119] A loyal 'Address of the Catholics of England to her Majesty', written by Wiseman, was placed in Catholic churches for signature in November 1850: Anon., *Roman Catholic Question* (1851), 4th series, p. 16.

[120] 'A Sermon preached at St. George's Catholic Church', in Anon., *Roman Catholic Question* (1851), 10th series, p. 2.

[121] Ibid., p. 3.

[122] E. Jay, *Religion of the Heart* (1979), p. 104. William Wilberforce argued that religion is seated in the heart, where its authority is recognized as supreme, whence by degrees it expels whatever is opposed to it, and where it gradually brings all the affections and desires under its complete controul and regulation': *Practical View of the Prevailing Religious* (1797), pp. 117–18.

[123] The Revd J. B. Cartwright, e.g., attacked the 'fables and idolatries of the mass' with its 'endless processions and elevations', and above all the Catholics' 'defective and unworthy views of the one all-sufficient atonement of our Lord Jesus Christ upon the cross, whereon by one offering He perfected for ever them that are sanctified'. J. B. Cartwright, *No Popery!* (1850), p. 37.

intermediaries, and especially his mother and his Church.[124] Wiseman ends his sermon with these words:

> Thus shall we all, dearly beloved, comply with the exhortation of our dear mother the Church; thus shall we fulfil her wishes; thus shall we respond to her desires for our eternal salvation; and thus will God hear our prayers and pour upon you and upon all, pour upon this glorious Church – pour upon this generous and powerful country, pour down upon the whole world, however oppressed and however afflicted it may be – the blessings which each may want, blessings here upon earth, but still more those blessings which alone are worthy of our desire, and which alone are to be found with God.

V

The chapters that follow are grouped in three parts. Part I, 'Bloody histories', begins with chapter 2, in which Catholic and Protestant treatments of the early Church, including Wiseman's, are discussed. Three historical novels, by Charles Kingsley, Wiseman and Newman, were published in the 1850s, and each had a female martyr of the early Church as its central character. The novel form becomes a vehicle for ecclesiastical debate between these clerical writers as they tackle the contentious issues of martyrdom and the heavenly reward, sexual abstinence and Christian vocation, idolatry and superstition. Behind these novels lies a Victorian obsession with the search for origins and for 'facts', as well as a rich tradition of early Church history.

In chapter 3 I turn to the writing of English Reformation history and associated historical fiction in the nineteenth century. The arguments elaborated by the historians are revisited, and often distorted, in novels by William Harrison Ainsworth, Emma Robinson and Kingsley. Protestant historical fiction is informed by works such as Foxe's *Book of Martyrs* – regarded as mere anti-Papist propaganda by Catholics – and the writings of Thomas Babington Macaulay and James Anthony Froude. Catholic historians include John Milner and John Lingard, whose interpretations of the Reformation are quite different, as is that of the radical patriot, William Cobbett.

Chapter 4 examines nineteenth-century responses to more recent history – the Gordon Riots of 1780 and an earlier threat to Protestant England's constitution, the Jacobite rebellion of 1745. Both were regarded by Catholic writers as examples of Protestant persecution, and by Protestants

[124] Cf. Newman's description, in a Chapter Address of February 1848, of St Philip Neri's wish to 'mould' his disciples into 'living laws, or, in the words of Scripture, to write the law on their hearts': see p. 171 below.

as examples of what Dickens in *Barnaby Rudge* (1841) called the 'veiling' of proceedings in 'mystery', which appealed to 'popular credulity' in their day. In 1845, English Protestants remembered the 'Forty-Five', which provided the background to novels by Scott and Thackeray.

Each of the three chapters in Part II, 'Creeds and crises', examines a turning-point in the troubled relationship between Catholics and Protestants during the first half of the nineteenth century at a time of crisis in the Established Church of England. In chapter 5 a symbolic comparison between the Revd George Croly and his fortress model of Protestant England and the Catholic Howards and their actual castle at Arundel reveals some of the complexities on both sides of the Protestant–Catholic divide on the issue of Catholic Emancipation (1829).

Chapter 6 focuses upon Newman's ideas and feelings relating to his conversion to and membership of the Roman Catholic Church. His *Essay on the Development of Christian Doctrine*, which is usually discussed in relation to Newman's thought, also draws upon his feelings about communities. His description of his conversion of 1845 as a call from God 'out of the war of tongues into a realm of peace and assurance' is compared with the experience of other converts, and finally illustrated in his novel, *Loss and Gain*.

The later 1840s witnessed a series of events which threw into relief the question of the authority of the Established Church of England, most dramatically in the Gorham case, the subject of chapter 7. In their addresses on the subject, Wiseman, Newman and Gladstone explored aspects of Church authority and of baptism which also engaged the attention of Ruskin and the Pre-Raphaelite Brotherhood.

The three chapters in Part III, 'Cultural spaces', consider the impact of Roman Catholicism upon English culture in the second half of the nineteenth century, after the restoration of the Catholic hierarchy. Chapter 8 compares masculine constructions of the feminine with the contribution that English women – both Catholic and Protestant – made to discussion on the role of women in society and the Church. Victorian representations of the Virgin Mary are related to ideas on women and 'Mother Church', particularly in relation to convents and 'mothers superior', including Charlotte Brontë's critique of Continental Catholicism in *Villette* (1853).

Chapter 9 discusses the debate between liberalism and dogma that unfolded over an extended period, before, during and after the Vatican Council of 1869–70, at which the doctrine of Papal Infallibility was defined. Matthew Arnold and Robert Browning, both liberal Protestants, though with different religious backgrounds, wrote during the build-up to the Council in Rome and the reform of the franchise at home. During the Council itself, Dr Cumming prophesied the end-time and Newman explored the nature of belief in his *Grammar of Assent*. In the aftermath of the Council, Newman, Manning, Gladstone and Arnold all published major statements, to which the obscure part-time novelist, John Henry Shorthouse, added an important coda, in *John Inglesant* (1880).

Finally, in chapter 10, I examine what has come to be known as the Decadence, a movement growing out of Symbolism which represented a critique of bourgeois Protestant Victorianism. Several of the key figures in the movement were converts to Catholicism, including John Gray, Lionel Johnson, Ernest Dowson, Aubrey Beardsley (shortly before his death), Oscar Wilde (on the very point of death) and Lord Alfred Douglas (in 1911). Having considered the special role of the London Oratory in relation to some of these writers, I focus upon some of the painful epiphanies that feature in the writings of the Decadence, in which shame and guilt are related to the pain and humiliation of Christ's Passion. I then show how aspects of Decadent aesthetics are informed by a sense of sin and a longing for purity, both of which are dramatized for some writers in Catholic ceremonial.

PART I

Bloody histories

CHAPTER 2

On the origin of churches

The blood of the martyrs is the seed of the Church.

Tertullian (197)

I

On the Origin of Species by means of Natural Selection (1859) made a greater immediate impact, and was subsequently more influential, than any other English book of the nineteenth century. Yet a collection of scholarly essays on biblical questions, produced the following year by six clergymen and one layman, sold as many copies in two years as Darwin's seminal text did in two decades.[1] Contemporary readers of the *Origin* and of *Essays and Reviews* (1860) would have been surprised to learn that, a century and a half later, in an unchurched, post-industrial Britain, Victorian religion is often treated by today's teachers of literature as merely the background to the more 'relevant' story of Victorian science.[2] For the Victorian obsession with origins was not limited to geology and natural history.[3] Crucially, in terms of the current discussion, it extended to ecclesiastical history. While Darwin was working on his theory of evolution, leading Catholic and Protestant scholars were feverishly researching the first centuries of the Christian era, in order to prove that their own traditions either originated in the early Church, through an unbroken line of Popes or bishops, or were most closely modelled upon the primitive Church described in the New

[1] See A. Desmond and J. Moore, *Darwin* (1992), p. 500. It was the wider implications of Darwin's theory, drawn out by his supporters, that shook his contemporaries, most of whom did not read the *Origin* itself.

[2] The implications of Darwin's theory for late Victorian religion were, of course, profound, and demanded a reassessment of the whole traditional scheme of redemption: if man had risen, what was the role of the fall? See J. H. Brooke, *Science and Religion* (1991), p. 313.

[3] 'In the Victorian period the Romantic search for the "One Life" had been set back in time and become a search for origins': G. Beer, *Darwin's Plots* (1985), p. 154.

Testament and in the writings of the Fathers, which needed to be translated from the *original* Greek and Latin, and then edited. Their subject was not the origin of species, but the origin of Churches.

Darwin's research began in the pre-Victorian period of social change and political reform described by George Eliot in *Middlemarch* (1871–2), where Dr Lydgate investigates 'certain primary webs or tissues' (15) in a town in which, 'during the agitation on the Catholic Question', many had given up the progressive newspaper, the *Pioneer* (37). In 1830, a year after Wellington and Peel brought in Catholic Emancipation, Charles Lyell published the first part of his *Principles of Geology* (1830–3), and in 1832, the year of Lord John Russell's Reform Act, Darwin embarked on the *Beagle*. Meanwhile, Newman was on a more sedentary voyage of discovery. In the Long Vacation of 1828 he had settled down to read the Fathers of the Church chronologically, beginning with St Ignatius and St Justin, and from around 1830 was writing a book on the Council of Nicæa. Thirty-five years later, he commented on this commission: 'It was to launch myself on an ocean with currents innumerable; and I was drifted back first to the ante-Nicene history, and then to the Church of Alexandria.'[4]

The product of Newman's investigation into the principles of the primitive Church was his first major work, *The Arians of the Fourth Century* (1832), published in the same year as *Origines Liturgicæ; or, Antiquities of the English Ritual, and a Dissertation on Primitive Liturgies*, by William Palmer of Worcester College, another Oxford High Churchman. Palmer's book became a standard work and was set reading for Anglican ordinands. Like Newman in *The Arians*, Palmer cites 'the learned Bingham', whose *Origines Ecclesiasticæ: The Antiquities of the Christian Church*, first published in 1708–22, was reprinted several times between 1821 and 1855, the period in which the debate between Catholics and Protestants over the origin of Churches was at its height, and in which Anglicans therefore needed ready access to historical source material.[5]

The argument over origins often focused upon the English Church, from its beginnings through to the Reformation. *The History and Antiquities of the Anglo-Saxon Church, containing an Account of its Origin, Government, Doctrines, Worship, Revenues, and Clerical and Monastic Institutions*, published by the great Catholic historian, John Lingard, in 1806, appeared in

[4] J. H. Newman, *Apologia Pro Vita Sua* (1967), p. 35. The first edition reads, 'It was launching myself . . .': *Apologia* (1864), p. 87.

[5] J. H. Newman, *Arians* (1871), p. 3, *et passim*; W. Palmer, *Origines Liturgicæ* (1845), I, 14. Multi-volume editions of Bingham appeared in 1821–9, 1838–40, 1840–5 and 1855, and Bohn published and reprinted a two-volume edition at mid-century: J. Bingham, *Origines Ecclesiasticæ* (1850).

Victorian bindings in 1845 and 1858. Mainstream Anglican perspectives were offered by Thomas Burgess, Bishop of St David's, for example, in *The Protestant's Catechism, in which it is clearly Proved, that the Ancient British Church existed several Centuries before Popery had any Footing in Great Britain; and many important Facts illustrative of the Fallacy of the Popish Doctrine of Infallibility, never before Published* (1817).[6] Other Protestant historians, writing in the years between Catholic Emancipation and the 'Papal Aggression' crisis, offered aggressive critiques of the origins of Papal supremacy, such as the anonymous *A History of Popery: Containing an Account of the Origin, Growth and Progress of the Papal Power* (1838) and Ingram Cobbin's *The Book of Popery: A Manual for Protestants, descriptive of the Origin, Progress, Doctrines, Rites, and Ceremonies of the Papal Church* (1840). The Jesuit, William Waterworth, who was prominent in the defence of English Catholic versions of history in the 1830s and 1840s, published his own response to Anglican claims in his *Origin and Developments of Anglicanism; or, A History of the Liturgies, Homilies, Articles, Bibles, Principles and Governmental System of the Church of England* (1854).

Behind all this impassioned scholarly activity there hovered the larger question of how the original Church of the first Apostles – the model for all that followed – was created and led. Among Anglican studies on the subject were the *History of the Christian Church, from the Ascension of Jesus Christ, to the Conversion of Constantine* (1836) by Edward Burton, Regius Professor of Divinity at Oxford, and two widely read books by Henry Hart Milman: *The History of Early Christianity* (1840), as it was popularly known, and the *History of Latin Christianity* (1854–5). In 1844 another study on the 'external history' of the Church during the first three centuries emerged from New College, Edinburgh, then recently established by the fledgling Free Church of Scotland, where David Welsh was Professor of Divinity and Church History. Initially, Welsh presses back to the foundation of the kingdom of Christ, and thus to the origins of mankind:

> The object of Church History is to give an account of the rise and progress, the vicissitudes and character of that spiritual kingdom, which the Almighty has established on the earth under the administration of his Son Jesus Christ.

[6] Cf. also William Hales, Rector of Killesandra, in *An Essay on the Origin and Purity of the Primitive Church of the British Isles, and its Independence upon the Church of Rome* (1819); Thomas Wood, whose book appeared the year after Newman's conversion to Rome and was elaborately entitled, *The Origin, Learning, Religion, and Customs of the Ancient Britons, with an Account of the Introduction of Christianity into Britain, and the Idolatry and Conversion of the Saxons; Remarks on the Errors and Progress of Popery; Considerations on the Christian Church, its Foundation, Superstructure, and Beauty, &c., &c.* (1846); and Edward Churton, Archdeacon of Cleveland, in his *Early English Church* (1840).

An account of the nature and design of this kingdom is to be found in the Scriptures, from which it appears, that it had its commencement with the parents of our race, being maintained throughout every age by an uninterrupted succession of members; who, under various forms, and in circumstances widely dissimilar, are distinguished from the world around them . . .[7]

For Dissenting Protestants like Welsh, the history of the Church is the history of its members, whereas for Roman Catholics, it is the history of the mystical body of Christ, founded and still ordered by those who were set apart – in modern times, the celibate clergy.

Like all modern historians since Gibbon, Welsh knew that he was working in an enormous lumber-room of inherited misinformation and half truths.[8] 'Party spirit', he complained, 'bigotry, a mistaken sense of duty, carelessness, credulity, prejudice, have often been active in those who have treated of the history of the Church, and they have led to the forgery of documents, to suppositious quotations, to garbled extracts, to false translations, to glosses that, while they seem to correspond with the letter, are foreign to the spirit of the original documents. Nor is this all. The original authors themselves must be perused with caution.'[9] In seeking to avoid bigotry and prejudice himself, he paid tribute to the sincerity and matchless industry of Claude Fleury, Louis Sébastien le Nain de Tillemont and Louis Ellies-Dupin, all Catholic ecclesiastical historians, born in the seventeenth century and familiar to English readers in the nineteenth.[10]

Welsh also acknowledged the contribution of the Germans, which had 'led to a more thorough sifting of the evidence for all the *facts favourable to the interests of Christianity*' (my emphasis).[11] Catholic and Protestant historians frequently accused each other of misconstruing, distorting or even inventing the facts of early Church history, and defended their own positions by claiming a close adherence to the facts.[12] Anti-Catholics were

[7] D. Welsh, *Elements of Church History* (1844), p. 1.

[8] 'The scanty and suspicious materials of ecclesiastical history seldom enable us to dispel the dark cloud that hangs over the first age of the church': E. Gibbon, *Decline and Fall* (1828–9), I, 264.

[9] D. Welsh, *Elements of Church History* (1844), pp. 19–20. [10] Ibid., p. 39. [11] Ibid., p. 44.

[12] On the Catholic side, Joseph Reeve, e.g., wrote, 'To remove the veil of misrepresentation, and to show, by facts, what the Roman Catholic Church for eighteen centuries has uniformly believed and taught, this historical epitome was first undertaken, and is now offered to the public' (*A Short View* (1802), I, xviii); Charles Butler claimed that, at the accession of Queen Elizabeth, the 'laity were divided, – but several facts indicate that a great majority must have been in favour of the catholic religion' (*Historical Memoirs* (1819–21), I, 272); and Newman reminded Protestant readers that it 'might do in books to talk of Antichrist, but that would not get rid of the plain fact of the saintliness of S. François de Sales, and S. Charles Borromeo' (*Apologia* (1967), p. 481). On the Protestant side, George Townsend attacked Butler, arguing that 'No apology can be necessary for any attempt to *elicit the truth of the facts*, upon which alone the decisions of the Romanist and the Protestant must be

particularly keen to explode Rome's claims that doctrines such as transubstantiation, the invocation of saints, Papal supremacy, clerical celibacy and auricular confession were doctrines of the early Church, as evidenced in the Bible and the writings of the Fathers. Catholic controversialists were therefore forced to present such evidence as clearly as possible, in a variety of published forms. The gathering of what Pugin called 'stern facts' was thus central not only to Victorian science, but also to Church history.[13] By comparing the historians' use of the term 'facts' with that of the scientists, however, we can distinguish between different usages, reflecting different kinds of evidence, in the religious writing of the period.

Darwin uses the word 'facts' in each of the first three sentences of the *Origin of Species*:

> When on board H.M.S. 'Beagle,' as naturalist, I was much struck with certain facts in the distribution of the organic beings inhabiting South America, and in the geological relations of the present to the past inhabitants of that continent. These facts, as will be seen in the latter chapters of this volume, seemed to throw some light on the origin of species . . . On my return home, it occurred to me, in 1837, that something might perhaps be made out on this question by patiently accumulating and reflecting on all sorts of facts which could possibly have any bearing on it.[14]

Having described how he observed the facts empirically and saw some kind of relationship between them, Darwin states that the data seemed to provide new evidence on the origin of species, and then recounts how the process of testing his hypothesis against this and other data began, over twenty years earlier.

Newman, writing as a Tractarian in the early 1840s, also regarded the patient accumulation of facts as an essential activity in the modern world. Indeed, he sounds not unlike Darwin, and quite Protestant and liberal, in his defence of the freedom of individual judgment, when arguing that Augustus Neander's *The History of the Christian Religion and Church during the Three First Centuries* (trans. 1831–41) is 'so full of theories, and those characteristic of his country, and facts are stated with so little attention to historical order and connection, that, valuable or rather necessary as his work is to the theological student, he does not come up to the demand of the present times, when men want to be put into possession of the plain

founded' (*The Accusations of History* (1825), p. 3), and Arthur Penrhyn Stanley welcomed the Gorham judgment, as it rested 'on a wider basis, on a more impregnable position, – the very foundation of the Church of England, as represented by the most indubitable testimony of historical facts' ('The Gorham Controversy', *Edinburgh Review*, July 1850, p. 266).

[13] A. W. N. Pugin, *Earnest Address* (1851), p. 13.

[14] C. Darwin, *Origin of Species* (1902), p. 1.

state of things, as it existed in ancient times, with the liberty of judging of them for themselves'.[15]

A few years later, as a recent convert to Roman Catholicism, he published his *Essay on the Development of Christian Doctrine* (1845), a work which has been described as the 'theological counterpart' of the *Origin of Species*.[16] Like Darwin, Newman peppers his opening paragraph with 'facts':

> Christianity has been long enough in the world to justify us in dealing with it as a fact in the world's history. . . . It may legitimately be made the subject-matter of theories; what is its moral and political excellence, what its due location in the range of ideas or of facts which we possess, whether it be divine or human, whether original or eclectic, or both at once, how far favourable to civilization or to literature, whether a religion for all ages or for a particular state of society, these are questions upon the fact, or professed solutions of the fact, and belong to the province of opinion; but to a fact do they relate, on an admitted fact do they turn, which must be ascertained as other facts, and surely has on the whole been so ascertained, unless the testimony of so many centuries is to go for nothing.[17]

Protestantism had ignored that testimony, in Newman's view, 'dispensing with historical Christianity altogether' and 'forming a Christianity from the Bible alone': 'It is shown by the long neglect of ecclesiastical history in England, which prevails even in the English Church. Our popular religion scarcely recognizes the fact of the twelve long ages which lie between the Councils of Nicaea and Trent.'[18]

Here we are far from Mr Gradgrind's demand for mechanically memorized facts at the beginning of Dickens's *Hard Times* (1854); but then 'facts' was a positive term in both revealed and natural theology.[19] In his widely read *Evidences of Christianity* (1794), the Anglican William Paley declared that the 'truth of Christianity depends upon its leading facts, and upon them alone':

> A Jewish peasant changed the religion of the world . . . The companions of this Person, after he himself had been put to death for his attempt, asserted his supernatural character . . . and, in testimony of the truth of their assertions . . . committed

[15] C. Fleury, *Ecclesiastical History*, ed. J. H. Newman (1842–44), 1, iv–v.

[16] I. Ker, *John Henry Newman* (1990), p. 300. For more detailed discussion of *Development*, see pp. 175–9 below.

[17] J. H. Newman, *Development of Christian Doctrine* (1974), p. 69. (Cf. also his *Lectures on certain Difficulties* (1850), p. 299.)

[18] Ibid., p. 72. Conyers Middleton, e.g., in his celebrated *A Free Inquiry into the Miraculous Powers* (1748), 'abandoning the entire Protestant tradition, handed over the primitive Church, lock, stock and barrel, to the Roman Catholic Church': O. Chadwick, *From Bossuet to Newman* (1987), p. 76.

[19] See M. Wheeler, *Ruskin's God* (1999), pp. 180–5.

themselves to the last extremities of persecution. . . . More particularly, a very few days after this Person had been publicly executed, . . . these his companions declared with one voice that his body was restored to life . . .[20]

Significantly, the second of Paley's three 'leading facts' relates to testimony through submission to persecution – a theme to which we will return. Thomas Scott, the Calvinist biblical commentator to whom the young Newman 'almost owed his soul', humanly speaking,[21] wrote on Matthew's account of a crucial moment of revelation in the baptism of Christ: 'Here, as in every part of the gospel, *facts* are simply related, without any studied remarks to awaken our attention: but what *facts* are they!'[22] Ruskin, who, like Newman, was brought up an Evangelical and much admired Thomas Scott, described the resurrection as 'the cardinal fact of Christianity'.[23] Catholics concurred, but also regarded certain aspects of tradition as cardinal facts. William Waterworth claimed that all Catholics, 'spread over the world, testify to Two Facts: 1° That the Church cannot err in matters of faith; and 2° That this belief has been handed down, as the declaration of our forefathers, from the days of the Apostles to these our times'.[24]

As a Catholic, Newman agreed with Waterworth. He also distanced himself from those Protestant writers who strove to harmonize theology and science. While Darwin continued to investigate the origin of species in the 1850s, Newman argued that 'induction is the instrument of Physics, and deduction only is the instrument of Theology. . . . Niebuhr may revolutionize history, Lavoisier chemistry, Newton astronomy; but God Himself is the author as well as the subject of theology'.[25] Whereas Darwin's findings were widely regarded as a challenge to Robert Chambers's assumption of the 'fact' that 'God created animated beings' and to Hugh Miller's grounding of his argument on the 'all-important fact of the Divine authorship of the universe', Newman looked, not to the 'testimony of the rocks', but to the testimony of the Church, in his search for truth.[26]

Newman also distanced himself from those Liberal churchmen who welcomed the new critical approaches to the Bible and ecclesiastical history,

[20] W. Paley, *Evidences of Christianity* (1807), II, 376.
[21] J. H. Newman, *Apologia* (1967), p. 18.
[22] *Holy Bible*, ed. T. Scott (1825), V. Scott's Bible was first published in 1788–92. Cf. W. Paley, *Evidences of Christianity* (1807), II, 376; G. Croly, *Divine Providence* (1834), p. xi.
[23] *Works of John Ruskin*, ed. E. T. Cook and A. Wedderburn (1903–12), XXIII, 368.
[24] W. Waterworth, S. J., *Origin of Anglicanism* (1854), p. 9.
[25] J. H. Newman, *Idea of a University* (1976), pp. 190–1. This passage is from a discourse which appeared in a second edition, entitled *The Scope and Nature of University Education; or, University Teaching considered in its abstract Scope and Nature* (1859).
[26] R. Chambers, *Vestiges of Creation* (1844), p. 152; H. Miller, *Testimony of the Rocks* (1862), p. 340 (1st edn published 1857).

irrespective of the damage that they might inflict upon the traditional touchstones of faith. Bishop Colenso of Natal chose as an epigraph to his controversial *The Pentateuch and Book of Joshua Critically Examined* (1862–79) a review of *Essays and Reviews* in the *Quarterly Review*: 'Not to exceed, and not to fall short of, facts, – not to add, and not to take away, – to state the truth, the whole truth, and nothing but the truth, – are the grand, the vital, maxims of Inductive Science, of English Law, and, let us add, of Christian faith . . . 1861.'[27] For Newman, however, the most important response to facts was not of the head, but of the heart, which is 'commonly reached, not through the reason, but through the imagination, by means of direct impressions, by the testimony of facts and events, by history, by description'. 'Persons influence us', he wrote, 'voices melt us, looks subdue us, deeds inflame us.'[28]

Newman could have been describing two popular kinds of writing here: the 'Lives of the Saints', best known through Alban Butler's frequently reprinted and revised volumes,[29] and full of facts – often miraculous facts - that are 'favourable to the interests of Christianity'; and the most successful literary form of the nineteenth century, the realist novel, where the reader engages with an imagined reality – a special kind of lying – based upon the known world, 'by the testimony of facts and events, by history, by description'. In the final section of this chapter, I will be discussing Newman's own novel, *Callista* (1856), as a response to two others: Charles Kingsley's *Hypatia* (1852–3) and Nicholas Wiseman's *Fabiola* (1855). Although all three novels appeal to the heart and describe the bloody history of a persecuted Church, drawing upon the 'Acts of the Martyrs',[30] their handling of the contested facts of ecclesiastical history, and their concepts of sacramentalism and sexuality, differ considerably, and not only across the divide between Catholic and Protestant theologies.

First, however, I want to look at three earlier treatments of the history of the primitive Church, which illustrate some of the major areas of contention between nineteenth-century Protestant and Catholic historians: Milman's

[27] J. W. Colenso, *Pentateuch* (1862–79), I, iii. In the Preface, Colenso stated that he 'knew for certain, on geological grounds, a fact', that 'a *Universal Deluge* . . . could not possibly have taken place in the way described in the Book of Genesis': ibid., I, vii–viii.

[28] J. H. Newman, *Discussions and Arguments* (1872), p. 293.

[29] A. Butler, *Lives of the Fathers, Martyrs, and other principal Saints, compiled from Original Monuments, and other authentick Records, illustrated with the Remarks of judicious modern Criticks and Historians*, 4 vols. (1756–9).

[30] The best of these Acts, many of which have been lost, were based on reports of the martyrs' trials. Of those that survive, the best known collections are the *Acta Sanctorum*, initiated by the Jesuit Bollandists in the seventeenth century and completed in 1940, and the *Acta Sanctorum Ordinis Sancti Benedicti*, published 1668–1701.

The History of Early Christianity, Berington and Kirk's *The Faith of Catholics*, and Newman's *The Arians of the Fourth Century*.

II

Henry Hart Milman is known to posterity as the 'Great Dean' of St Paul's. He was the son of George III's physician and was educated at Eton and Brasenose College, Oxford, where he won the Newdigate prize for poetry and was later a Fellow, Professor of Poetry and Bampton Lecturer. He became a famous poet, dramatist and liberal ecclesiastical historian. *The Fall of Jerusalem: a Dramatic Poem* (1820) and the controversial *History of the Jews* (1829), described as the 'first decisive inroad of German theology into England',[31] were written when he held the important living of St Mary's, Reading; his edition of Gibbon (1838–9) and the *History of Christianity to the Abolition of Paganism in the Roman Empire* (1840) when he was Canon of Westminster and Rector of St Margaret's, Westminster; and the *History of Latin Christianity to the Pontificate of Nicholas V* (1854–5) during his long tenure at St Paul's (1849–68).

Many of his fellow clergy did not trust him, for although he attained high office in the Church of England, his histories seemed to rinse the miraculous out of the grand narrative of the Bible. Some readers objected to the fact that Milman started his *History of Early Christianity* with the life of Christ, a decision that he defended in a later preface: 'I thought then, and still think, that life to be an integral and inseparable part of the History. It appeared to me necessary to the completeness of the history to trace it to its primal original.'[32] He was pleased to have anticipated many of the conclusions of Neander, and regarded his own 'clear, consistent, and probable narrative' of the gospel story as the 'best answer to Strauss'.[33] Instead of the usual subject-matter – the 'annals of the internal feuds and divisions in the Christian community, and the variations in doctrine and discipline' in the Church – Milman directed his attention mainly to the effects of Christianity on the 'social and even political condition of man, as it extended itself throughout the Roman world'. At a time when many Churchmen were looking for solidarity in their Church historians, especially on the subject of Rome, Milman's stated aim, 'entirely to discard all polemic views', would have been well received only by his more liberal colleagues.[34]

[31] Dean Stanley, cited in A. Milman, *Henry Hart Milman* (1900), p. 86.

[32] H. H. Milman, *History of Christianity* (1875), I, iii.

[33] Ibid., I, 114–15. David Friedrich Strauss's radical *Leben Jesu* (1835) was translated by George Eliot in 1846.

[34] Ibid., I, 46.

Milman's very liberalism, however, carried with it implicit criticism of positions held by Catholics, Evangelicals and Tractarians, and his handling of the origins of certain Catholic doctrines and practices reveals the fact that it was impossible to write the history of the early Church in the late 1830s without expressing, or at least suggesting, 'polemic views'. His interpretation of the key texts on the Petrine rock (John 1.42, Matthew 16.18), for example, is as far from Catholic tradition as it is possible to be: 'First among the twelve appears Simon, to whom Jesus, in allusion to the firmness of character which he was hereafter to exhibit, gave a name, or rather, perhaps, interpreted a name by which he was already known, Cephas, the Rock; and declared that his new religious community was to rest on a foundation as solid as that name seemed to signify.'[35]

Whereas St Paul, he writes, 'stands forth as the great central figure in the great unfolding Drama of the conversion of the world to Christianity', St Peter himself 'recedes from view'; and yet,

> The fame of St Peter, from whom she claims the supremacy of the Christian world, has eclipsed that of St Paul in the Eternal City. The most splendid temple which has been erected by Christian zeal, to rival or surpass the proudest edifices of heathen magnificence, bears the name of that Apostle, while that of St Paul rises in a remote and unwholesome suburb. Studious to avoid, if possible, the treacherous and slippery ground of polemic controversy, I must be permitted to express my surprise that in no part of the authentic Scripture occurs the slightest allusion to the personal history of St Peter, as connected with the Western Churches.[36]

Before the end of the third century, he adds later in the book, the 'lineal descent of her bishops from St Peter was unhesitatingly claimed, and obsequiously admitted by the Christian world'.[37]

On the vexed issue of celibacy, Milman first coolly links it to the Nativity, which was the 'consecration of sexual purity and maternal tenderness': 'No doubt by falling in, to a certain degree, with the ascetic spirit of Oriental enthusiasm, the former incidentally tended to confirm the sanctity of celibacy, which for so many years reigned paramount in the Church.'[38] Similarly, on monasticism (or 'monachism'), he initially states that the 'cloister or the religious foundation . . . became the place of refuge to all that remained of letters or of arts'.[39] Later, however, he is explicitly critical of monasticism: 'The deep and serious solicitude for the fate of that everlasting part of our being, the concentration of all its energies on its own individual welfare, withdrew it entirely within itself. A kind of sublime

[35] Ibid., I, 210. [36] Ibid., I, 386–7, 462.
[37] Ibid., III, 264. [38] Ibid., I, 96. [39] Ibid., III, 51.

selfishness excluded all subordinate considerations.'[40] In a note, Milman added:

It is remarkable how rarely, if ever (I cannot call to mind an instance), in the discussions on the comparative merits of marriage and celibacy, the social advantages appear to have occurred to the mind; the benefit to mankind of raising up a race born from Christian parents and brought up in Christian principles. It is always argued with relation to the interests and the perfection of the individual soul; and even with regard to that, the writers seem almost unconscious of the softening and humanising effect of the natural affections, the beauty of parental tenderness and filial love.[41]

In a reversal of the usual pattern, a Protestant defends communal values and criticizes Catholic individualism.

Milman's vain hope that he could avoid polemicism reflected the fact that the previous fifteen years had witnessed some bitter exchanges between Catholic and Protestant historians, as they fought over the origin of Churches. In the mid-1820s, when there was increased pressure for Catholic reforms, the ecclesiastical history of England had become a political football. As we will see in chapter 5, Lingard's famous *History of England to 1688* (1819–30), was attacked by Robert Southey in *The Book of the Church* (1824), which in turn drew a Catholic reply from the barrister and controversialist, Charles Butler, in *The Book of the Roman-Catholic Church* (1825).[42] All discussion on English Reformation history raised questions relating to the sources and authority of Anglicanism and Roman Catholicism, the answers to which lay in the history of the early Church and its 'original' documents; and those who wrote on the early Church were acutely aware of the relevance of their work to current debates about Catholicism.

So when William Keary, Rector of Nunnington, published his *Common-Place Book to the Fathers* in 1828, he referred to Butler in relation to Roman 'error and religious prejudice', reminding his readers of the 'wordy war' arising from the 'controversy with Rome', and presenting Bible-based Protestantism as the only true religion for the 'present enlightened era'.[43] Because the 'adversary' seeks to 'divert us, as much as possible, from the unerring testimony of Scripture', he argues, and to 'lead us into the wide field of christian antiquity, extending over the voluminous pages of the early fathers,

[40] Ibid., III, 195–6. [41] Ibid., III, 196. Cf. also III, 278–9, 282. [42] See pp. 145–6 below.

[43] W. Keary, *Common-Place Book to the Fathers* (1828), pp. 68, 1, 67. (Keary's *The Comments of the Early Fathers of the Church on Scripture, opposed to the Modern Interpretations of Rome* (1830) was intended as an appendix.)

the Protestant should therefore be prepared to meet him also upon this ground'.[44]

Having criticized Catholicism for diverting attention from the Bible – *the* original text for Protestants – Keary became involved in another, related debate, this time concerning the translation into English of the original texts of the Fathers of the Church in a Catholic publication entitled *The Faith of Catholics, confirmed by Scripture, and attested by the Fathers of the Five First Centuries of the Church* (1813), a substantial collection of passages from Christian antiquity on a range of important themes, compiled by Joseph Berington, a leading scholar, writer and priest, and John Kirk.[45] This book became the main source on the Fathers for English Catholic controversialists. Berington died in 1827, and when Kirk brought out a second edition in 1830 he related in the introduction how a friend had pointed out to him that 'free, instead of close versions' had been given of the most important passages, and that some words had been omitted, and other 'liberties' taken with the 'originals', without 'perverting or altering indeed the sense'.[46] He therefore now offered a 'more literal translation of some of the more important passages'.

In 1833, William Keary was back in print with a pamphlet describing a lengthy discussion on the invocation of saints, between himself, Richard Towers, the Prior of Ampleforth, and other disputants, during which *The Faith of the Catholics* had been cited. Keary claimed that, out of forty-two quotations 'adduced in support' of the doctrine by Berington and Kirk, twenty-two are 'palpable well-known forgeries; and the remainder either utterly irrelevant, or from writers of the fifth and sixth centuries, who are of no authority on the subject'.[47] To which James Waterworth, a Catholic priest in Newark, replied in a pamphlet that minced no words, accusing Keary of 'the most disgraceful shuffles', 'incorrect translation, alterations of the punctuation, and other similar artifices': 'The filth that the charitable Rector has taken so much trouble to rake together, will befoul no one but himself.'[48]

In 1840, however, the year of Milman's *History of Early Christianity*, another Protestant clergyman, the unaptly named Richard Pope, held the ostensibly improved translations in the second edition of *The Faith of Catholics* up against the 'originals', and again found them wanting. 'For

[44] Ibid., p. 2. [45] J. Berington and J. Kirk, *Faith of Catholics* (1830), p. iv. [46] Ibid., p. vi.

[47] W. Keary, *Continuation of the Ampleforth Discussion* (1833), cited in J. Waterworth, *Examination of the Evidence* (1834), p. 5

[48] J. Waterworth, *Examination of the Evidence* (1834), p. 7.

a long period', he comments, the 'attention of many has been directed to ancient Christian literature', but 'thoroughly to ascertain the minds of the Fathers is beset with no ordinary difficulties'.[49] His main charge, however, is that, '*whatever corrections may have been introduced into the edition of 1830, it presents the misquotations to which I am about to call attention, as they occur in the edition of 1813*'.[50]

It was presumably Pope's analysis of 'Roman misquotation', coming on top of Keary's derogatory comments seven years previously, that spurred James Waterworth to 'four years of severe study and reading' in order to revise and 'greatly enlarge' *The Faith of Catholics* in a third edition (1846).[51] Understandly, he states in the preface that he does not wish to go into the reasons, but it was thought necessary to 'read the entire works of the fathers, and ecclesiastical writers of the first five centuries; to give an entirely new translation of nearly all the extracts – especially those from the Greek writers; – and to use such aids as numerous authors have furnished towards distinguishing the genuine, from the spurious or doubtful, works of those early ages of the church'. It was the 'peculiar circumstances of the times' that forced him to 'enlarge the work very considerably'.[52]

Those 'peculiar circumstances' were created by the controversy surrounding Newman's conversion in 1845.[53] Waterworth's new edition of *The Faith of Catholics* was intended to defend his co-religionists from further attacks relating to translations of the Fathers, and to arm them for other, subsequent disputes. Meanwhile, the Anglicans also had new materials on the early Church readily to hand. Bingham's *Origines Ecclesiasticæ* was reprinted in 1838–40 and again in 1840–5. The Library of the Fathers, edited by Pusey and other Tractarians, initially in response to the Hampden crisis,[54] ran to forty-eight volumes (1838–85), nineteen of which appeared before 1845.[55] In the Prospectus, Pusey underlines the importance of 'exhibiting the real practical value of Catholic antiquity, which is disparaged by Romanists in

[49] R. T. P. Pope, *Roman Misquotation* (1840), pp. ix, 2.
[50] Ibid., p. vii. [51] J. Berington and J. Kirk, *Faith of Catholics* (1846), I, i.
[52] Cf., e. g., the section on 'The Invocation of Angels and Saints', to which Keary objected in 1833, in the three different editions: 1st edn (1813), pp. 430–52; 2nd edn (1830), pp. 427–51; 3rd edn (1846), III, 322–416. Although the second edition adds some sources, it broadly follows the first edition, whereas Waterworth's translations are quite new, and he adds numerous other sources and an elaborate scholarly apparatus.
[53] See pp. 166–85 below.
[54] In 1836, the Latitudinarian, Renn Dickson Hampden, was appointed Regius Professor of Divinity at Oxford, to the dismay of those, like Pusey, who regarded the university as a bastion of English orthodoxy.
[55] H. P. Liddon, *Life of Pusey* (1893), I, 409–47.

order to make way for the later Councils'.[56] It seems to him an 'act of special charity' to point out to dissatisfied Roman Catholics a 'body of ancient Catholic truth, free from the errors alike of modern Rome and of Ultra-Protestantism'.

Richard Pope commented in 1840 that, 'a few rare cases excepted, life itself, though possessed of all necessary qualifications, is utterly incompetent to wade through such an extensive mass of ecclesiastical lore'.[57] One of those rare cases was John Henry Newman. Having worked on the Fathers since 1828, he translated two volumes of St Athanasius for the Library of the Fathers (1842, 1844) and wrote short prefaces on St Cyril (1838), St Cyprian (1839), St Chrysostom on Galatians and Ephesians (1840) and St Athanasius's Historical Tracts (1843). It was from a well-informed and strongly dogmatic position that Newman, writing to Pusey from Littlemore in 1840, described rationalism as the 'great evil of the day', worried about future appointments to the Professorship of Divinity at Oxford, and regarded Milman's *History* as a 'sort of earnest' of a 'great battle' that was to come.[58] When he reviewed Milman the following year, he described him as one who wrote 'on the side of the world', regarded Christian history as a 'thing of earth', and attempted to 'touch the human element without handling also the divine'.[59] This 'external contemplation of Christianity', he argues, 'necessarily leads a man to write as a Socinian or Unitarian *would* write, whether he will or not'.[60] If we indulge Milman's 'speculations', he believes, Christianity will 'melt away in our hands like snow; we shall be unbelievers before we at all suspect where we are'.[61]

For Newman, the dangers associated with the speculations of a liberal like Milman were reminiscent of those associated with Arianism.[62] Nine years earlier, in *The Arians of the Fourth Century* (1832), Newman had described the heresy, in its 'original form', as a 'sceptical rather than a dogmatic teaching'.[63] His account of the Church of Alexandria, and its careful

[56] Ibid., I, 416. At this point Liddon cites Newman, who commented that Roman Catholics professed to 'appeal to primitive Christianity', but moved the dispute to another field when the controversy grew 'animated'.

[57] R. T. P. Pope, *Roman Misquotation* (1840), p. 3.

[58] J. H. Newman, *Apologia* (1967), p. 127.

[59] J. H. Newman, 'Milman's View of Christianity', in *Essays, Critical and Historical* (1872), II, 186–248 (pp. 196, 187–8).

[60] Ibid., p. 202.

[61] Ibid., p. 242.

[62] The belief that the Son was 'begotten of his Father before all worlds, ... Begotten, not made, Being of one substance with the Father, By whom all things were made', was defined by the Council of Nicæa in 325, in response to the Arian heresy, which denied the Son's consubstantiality with the Father, and thus virtually his Godhead. See 'Arius and Arianism', in W. E. Addis and T. Arnold, *Catholic Dictionary* (1897), pp. 52–6.

[63] J. H. Newman, *Arians* (1871), p. 28.

protection of true doctrine from such teaching, reads like a cautionary tale, addressed to his contemporaries during a period of reform, when the Established Church of England was rapidly losing both power and influence. The catechetical school in Alexandria, he points out, was noted for its 'diligent and systematic preparation of candidates for baptism'.[64] Its theologians wrote, 'not with the openness of Christian familiarity, but with the tenderness or the reserve with which we are accustomed to address those who do not sympathize with us, or whom we fear to mislead or to prejudice against the truth, by precipitate disclosures of its details'.[65] Writing only a year before Keble's Assize Sermon, Newman's praise for the 'economical method' of Athanasius, and his criticism of those who used the doctrine of the atonement to stimulate the 'affections' and thus win converts, prepared the ground for the coming conflict with English Evangelicalism.[66] Surely, he argues, in one of his most important early statements, the 'Sacred Volume was never intended, and is not adapted, to teach us our Creed': 'the Church should teach the truth, and then should appeal to Scripture in vindication of its own teaching'.[67] That divine truth is infinitely precious and mysterious. The 'secondary and distinct meaning' of prophecy, for example, is 'commonly hidden from view by the veil of the literal text, lest its immediate scope should be overlooked; when that is once fulfilled, the recesses of the sacred language seem to open, and give up the further truths deposited in them'.[68]

The contrast between Newman's reading of early Church history and Milman's is nowhere sharper than in their use of the word 'opinions' in relation to doctrine. Whereas Milman, in his *History of Latin Christianity*, was to refer to the 'Trinitarian opinions' that 'eventually triumphed through the whole of Christendom',[69] Newman argued in *The Arians* that the 'great doctrines of the faith' were 'facts, not opinions'.[70] They would 'scarcely occur to the primitive Christians', being the 'subject of an Apostolical Tradition'. They were the 'very truths which had been lately revealed to mankind'. They had been 'committed to the Church's keeping, and were dispensed by her to those who sought them, as a favour'.

Newman later claimed, in a famous passage of personal testimony in the *Apologia*, that his long road away from the Church of England and towards Rome began in the summer of 1839, when he was reading for the Library of the Fathers, and came face to face with what he regarded as a troubling but stubborn fact:

[64] Ibid., p. 42. [65] Ibid., p. 43. [66] Ibid., pp. 73, 47–8. [67] Ibid., p. 51–2.
[68] Ibid., p. 63. [69] H. H. Milman, *History of Latin Christianity* (1883), II, 441.
[70] J. H. Newman, *Arians* (1871), p. 138.

The Long Vacation of 1839 began early . . . About the middle of June I began to study and master the history of the Monophysites.[71] I was absorbed in the doctrinal question. This was from about June 13th to August 30th. It was during this course of reading that for the first time a doubt came upon me of the tenableness of Anglicanism . . .

. . . My stronghold was Antiquity; now here, in the middle of the fifth century, I found, as it seemed to me, Christendom of the sixteenth and the nineteenth centuries reflected. I saw my face in that mirror, and I was a Monophysite. The Church of the *Via Media* was in the position of the Oriental communion, Rome was, where she now is; and the Protestants were the Eutychians. Of all passages of history, since history has been, who would have thought of going to the sayings and doings of old Eutyches, that *delirius senex*, as (I think) Petavius calls him, and to the enormities of the unprincipled Dioscorus, in order to be converted to Rome![72]

Newman's self-mockery reflects the irony of a situation in which the very study of the Fathers which he and other Tractarians regarded as central to their search for the truth concerning the origin of Churches, led to the destruction of his own belief in the Catholicity of the Church of England, which he now increasingly regarded as a schismatic sect.

For Newman, the publication of Milman's *History of Early Christianity*, like the Hampden crisis, confirmed the fact that liberalism was making alarmingly rapid progress within the Anglican Establishment. He therefore defended Catholic orthodoxy within the Church of England by taking his stand upon the cardinal facts of Christianity, which for him, unlike for Milman, included miracles. And whereas Protestant historians were often sceptical of 'ecclesiastical miracles', performed after the Apostolic age that is recorded in the New Testament, and therefore possibly the later invention of 'Papists', Newman prefaced his edition of Fleury's *Ecclesiastical History* (1842–4) with an essay on the subject, which begins with this uncompromising statement:

Sacred History is distinguished from Profane by the nature of the facts which enter into its composition, and which are not always such as occur in the ordinary course of things, but are extraordinary and divine. Miracles are its characteristic, whether it be viewed as scriptural or ecclesiastical . . . It is a record of "the kingdom of heaven," a manifestation of the Hand of God . . . [73]

[71] The Monophysites, or Eutychians, believed that 'there was but one nature in Christ, and were condemned at the General Council of Chalcedon': W. E. Addis and T. Arnold, *Catholic Dictionary* (1897), p. 647.

[72] J. H. Newman, *Apologia* (1967), p. 108. On Newman's conversion, see also pp. 166–85 below. Dioscorus, Patriarch of Alexandria, supported Eutyches, c. 448, and was deposed and banished in 451.

[73] C. Fleury, *Ecclesiastical History*, ed. J. H. Newman (1842–44), I, xi. The essay was reprinted separately in 1843.

In 1845, the year of his conversion and the publication of his *Essay on the Development of Christian Doctrine*, Newman was accused by a reviewer of going 'instinctively to documents, not to life', and of thinking of the Church as a 'literature, and not a substance'.[74] Yet Newman was later to write a work that, perforce, treated of both the sacred and the profane, and that conveyed the life of the early Church in a work of literature – his novel *Callista,* Catholicism's second answer to Kingsley's *Hypatia.*

III

In the spring of 1872, John Samuel Bewley Monsell, Rector of St Nicholas, Guildford, sent a specimen copy of a new hymnal to several of his friends, inviting their evaluations. Charles Kingsley's marginal notes were brief, but telling, being pugnaciously hostile to both Roman Catholicism and Anglo-Catholicism. Next to 'O Paradise, O Paradise!', by the late Father Faber, he wrote: 'Whence did the author of this hymn learn that "the world is growing old?"'[75] He also objected to Faber's 'Hark! Hark, my soul, angelic songs are swelling', as people 'do *not* hear the angels singing over fields and seas', and to John Mason Neale's 'For thee, O dear, dear country', as congregations 'do *not* lie awake or weep, thinking of heaven'. Whereas Faber and Neale swooned for other worlds, Kingsley gloried in the teeming life of this one. His most interesting comment, on 'Sacred heart of Jesus, Heart of God in man', offers a glimpse of his theology: 'Should not this be "Heart of man in God?" Not by conversion of the God-head into flesh, but by taking of the manhood into God.'[76] Whereas Catholics adored the heart of Christ, 'not as mere flesh, but as united to the Divinity', Kingsley's Protestant theology started out from the 'heart of man', and the redeemable nature of humankind.

Kingsley lived and worked in the spirit of Monsell's own famous hymn:

> Fight the good fight with all thy might,
> Christ is thy Strength, and Christ thy Right;
> Lay hold on life, and it shall be
> Thy joy and crown eternally.[77]

[74] Anon., 'The Recent Schism', *Christian Remembrancer*, January 1846, p. 189.

[75] F. Kingsley, *Charles Kingsley* (1877), II, 382. See J. S. B. Monsell, *Parish Hymnal* (1873), p. 34.

[76] F. Kingsley, *Charles Kingsley* (1877), II, 383; J. S. B. Monsell, *Parish Hymnal* (1873), p. 122. The Feast of the Sacred Heart was extended to the whole Roman Catholic Church in 1856, but was not particularly popular in England, according to M. Heimann in *Catholic Devotion* (1995), pp. 43–4; but contrast W. E. Addis and T. Arnold, *Catholic Dictionary* (1897), p. 426.

[77] H. W. Baker, *et al.*, *Hymns Ancient and Modern* (1916), p. 637. This hymn is dated 1863.

The ferocity with which Kingsley took the fight to both the Tractarians and the Catholics has been related to anxieties about his own sexuality, and to memories of his early attraction to Rome and his 'rescue' of Frances, his future wife, from the clutches of Puseyism.[78] Having responded in his early fiction to the Chartist crisis of 1848 and associated issues of social and sanitary reform, his first historical novel, *Hypatia; or, New Foes with an Old Face* (1852–3), embodied his response to the 'Papal Aggression' crisis and to the threatening issues of monasticism and celibacy.[79] In *Hypatia*, as in *Westward Ho!* (1854), to be discussed in chapter 3, Kingsley ranges his own distinctive brand of Protestant militancy against what he regards as the slippery lies and secret homoeroticism of 'Papist' priests and religious.

One of his main targets in *Hypatia* is Catholic martyrology. When a monk named Ammonius hits Orestes, the Roman governor of Alexandria, with a stone, his punishment is crucifixion. The subsequent spiriting away of the body, and its honoured display in procession, is narrated from the viewpoint of Orestes, thus emptying the 'martyrdom' of spiritual meaning (20). Cyril the Patriarch, a saint who was highly regarded by Newman, is here suspiciously reminiscent of Cardinal Wiseman, as he processes in 'full pontificals' with 'monks from Nitria counted by the thousand', priest, deacons and archdeacons. When answering Arsenius's objection that this new canonization is not only scandalous, but also unrighteous, Cyril openly admits that he is driven by political expediency: he 'must have fuel', he says, 'wherewith to keep alight the flame of zeal', and thus ultimately get rid of the unpopular Orestes.

By choosing as his heroine Hypatia, the Platonist teacher and mathematician who was brutally murdered by a mob of unruly monks in 415, Kingsley can bring his novel to a climax with a well documented historical fact which reverses the pattern of Catholic martyrology, in which Christians are killed by pagans.[80] Whereas Ruskin's description of St Mark's, Venice, published a few months later in 1853, Protestantizes a Catholic interior,[81] Kingsley's description of the Cæsareum, to which the monks drag Hypatia, emphasizes its ambiguous Otherness:

[78] See S. Chitty, *Beast and the Monk* (1974), pp. 51–86; N. Vance, *Sinews of the Spirit* (1985), pp. 35–41; S. Prickett, *Origins of Narrative* (1996), pp. 222–47; V. A. Lankewish, 'Love Among the Ruins', *Victorian Literature and Culture*, 2000, pp. 239–73.

[79] One of Kingsley's first publications was an anonymous essay entitled 'Why should we Fear the Romish Priests?', *Fraser's Magazine*, April 1848.

[80] See J. Toland, *Hypatia* (1753), 1st edn (published 1721); T. Lewis, *History of Hypatia* (1721); E. Gibbon, *Decline and Fall* (1828–9), III, 272–3 (1st edn published 1776–88).

[81] See M. Wheeler, *Ruskin's God* (1999), pp. 91–2.

Into the cool dim shadow, with its fretted pillars, and lowering domes, and candles, and incense, and blazing altar, and great pictures looking from the walls athwart the gorgeous gloom; and right in front, above the altar, the colossal Christ watching unmoved from off the wall, His right hand raised to give a blessing – or a curse? (29)

Tennyson was shocked by Hypatia's nakedness in Kingsley's description of her murder.[82] Up at the altar, she

shook herself free from her tormentors, and springing back rose for one moment to her full height, naked, snow-white against the dusky mass around – shame and indignation in those wide clear eyes, but not a stain of fear. With one hand she clasped her golden locks around her; the other long white arm was stretched upward towards the great still Christ, appealing – and who dare say in vain? – from man to God. Her lips were opened to speak; but the words that should have come from them reached God's ear alone, for in an instant Peter struck her down, the dark mass closed over her again . . . and then wail on wail, long, wild, ear-piercing, rang along the vaulted roofs, and thrilled like the trumpet of avenging angels through Philammon's ears.

The 'dark mass' of monks closing over the lone naked victim, to tear her 'piecemeal', and 'worse than that', for perhaps an hour, is one of the most sinister moments in nineteenth-century fiction.

Kingsley's reference to the 'avenging angels' highlights the fact that his narrative drive is towards a fictional last judgment, as he, like other mid-Victorian novelists, plays God with his characters. The novel begins and ends with Philammon, the young monk whose spiritual journey is circular. Having left Laura, he averts his gaze from Pelagia, every 'swell' of whose 'bust and arms showed through the thin gauze robe' (3), and later discovers that she is his sister. He falls in love with Hypatia, whose death he witnesses, and finally returns with Pelagia to 'the desert and the hermit's cell, and thence into that fairy land of legend and miracle, wherein all saintly lives were destined to be enveloped for many a century thenceforth' (30). The reader's romantic expectations are fulfilled, however, when the hero, Raphael Aben-Ezra, mimics Kingsley's life by rescuing Victoria, the daughter of Majoricus, from a future in a nunnery, and marrying her. But again, irony lies in the fact that Raphael's repulsion from her father's plans for Victoria prevent him from converting to Christianity.

Kingsley's inversion of the conventions of martyrology extends beyond narrative structure to his treatment of the 'heart of man'. Rather than portray early Christians longing to emulate Christ in submitting to bloody

[82] See S. Chitty, *Beast and the Monk* (1974), p. 154.

persecution, and thereby achieving heavenly bliss for their souls, Kingsley sacramentalizes the 'honest flesh and blood' that he admired in his own wife.[83] Bishop Synesius, notorious in the early Church for refusing to give up his wife, loves blood sports, as Kingsley did (2). Only in Kingsley's fiction could an ostrich hunt turn into a battle with the Ausurians, as a 'yell of rage' rings from the hunting bishop's 'troop', who are blood-brothers: 'We eat together, work together, hunt together, fight together, jest together, and weep together' (21). So too are the Goths, whose blood-lust is as powerful as their lust for the women in their floating harem (3).

Kingsley's interest in race, 'blood' and heredity shapes his reading of history, and in *Hypatia* finds fullest expression in his admiration for the Goths and his interest in Raphael's Jewish identity. After the skirmish with the Ausurians, Raphael finds it strange to hear the 'grand old Hebrew psalms of his nation ring aloft' in a ruined church, as Augustine performs the evening service; and the blood of the man who wrote the words that are read aloud from Proverbs, in Latin, 'was flowing in Aben-Ezra's veins' (21). 'You felt the royal blood of Solomon within you!' says Miriam at the end of the novel, after revealing that she is his mother, who became pregnant during her time as a Christian nun, when she discovered that the 'Christian devils' were liars (30). She then dies, a 'dark stream of blood' flowing from her lips. Heredity is related to larger questions of origins, or what Kingsley was later to call the 'mystery of generation'.[84] Whereas Kingsley the theologian and scientist was committed to searching for the truth, Kingsley the novelist and historian was willing to distort the facts of ecclesiastical history, making Synesius a more attractive figure than Cyril, for example, as John Toland had in the eighteenth century,[85] in order to defend what he regarded as a higher truth – the truth of Protestantism, as opposed to the lies of Catholicism.[86]

That a Cardinal Archbishop of Westminster should have responded to Kingsley with a novel about Rome in the fourth century, written, he tells us in the preface, 'solely for recreation', in snatched moments on journeys and between appointments, seems extraordinary today. The Catholic Popular Library, however, which Wiseman inaugurated with *Fabiola: A Tale of the Catacombs* (1855), simply took the best medium then available for engaging people's imaginations, namely the novel, and used it to defend and

[83] Ibid. (1974), p. 55. [84] F. Kingsley, *Charles Kingsley* (1877), II, 171.

[85] See J. Toland, *Hypatia* (1753), pp. 29–30. Thomas Lewis accused Toland of dressing up the history of Hypatia, 'with Malice, Prejudice and Ignorance, on purpose to blast the Reputation of the Venerable St. *Cyril*, a strenuous Assertor of Orthodoxy and Church Discipline': *History of Hypatia* (1721), p. 3.

[86] Cf. J. H. Newman, *Apologia* (1967), p. xxiii.

explain Catholicism. In order to make his reader 'familiar with the usages, habits, conditions, ideas, feeling, and spirit of the early ages of Christianity', Wiseman drew upon writings such as the 'Acts of primitive Martyrs', with which he was familiar, rather than researching 'ecclesiastical antiquities' and producing a 'learned work'. The novel is 'not historical', he explains, and consists 'rather of a series of pictures than of a narrative of events'.

More significant, however, than Wiseman's tinkering with historical chronology, in order to focus upon a period of a few months, is his broader handling of time. Consecutive time (Gk *chronos*) is subordinated to the early Christians' sense of living in their own time or season (Gk *kairos*), as they anticipate their imminent end.[87] Whereas Kingsley offered a lively 'narrative of events', with some romantic interest, which appealed to a wide range of readers, including Queen Victoria and the Prince Consort,[88] Wiseman's 'pictures' are variations on the single theme of martyrdom, presented as the consummation of a believer's spiritual love affair with the Lord. Wiseman uses the word 'heart' over a hundred times in the novel, and the word 'blood' over fifty times; but hearts are led towards the love of neighbour (Gk *agape*, Lat. *caritas*) rather than romantic love (Gk *eros*), and blood becomes a sacramental sign of sacrifice.

Fabiola's journey towards conversion and baptism begins when her heart is touched, having seen 'disinterested love on earth between strangers' for the first time, in her servant, Syra, and the blind Cæcilia: 'as to charity, it was a word unknown to Greece or Rome' (I, 7). Later, as Syra explains the Christian faith to her, Fabiola's 'pagan heart' rises 'strong within her' in rebellion (I, 16). After a struggle, however, she becomes calm, and seems 'for the first time to feel the presence of One greater than herself, some one whom she feared, yet whom she would wish to love'. Her heart owns that it has a 'Master, and a Lord'. During a terrible period of persecution, described in the last third of the novel, Agnes's final wish before her martyrdom is for Fabiola's baptism: 'Waters of refreshment shall flow over your body, and oil of gladness shall embalm your flesh; and the soul shall be washed clean as driven snow, and the heart be softened as the babe's' (II, 29).

When Fabiola is finally baptized, deep underground, and makes her first communion, the 'rude baptistery' is described in guide-book style: 'The whole remains to this day, just as it was then, except that over the water is now to be seen a painting of St John baptising our Lord, added probably a century or two later.' The novel's many antiquarian facts, acquired by

[87] See M. Wheeler, *Death and the Future Life* (1990), p. 72.

[88] See S. Chitty, *Beast and the Monk* (1974), p. 154. Kingsley was soon to be appointed a Chaplain in Ordinary to the Queen (p. 201).

Wiseman during his time at the English College in Rome, and shared with the reader in long documentary asides, provide its realist underpinning and link the distant past with the Victorian present. (In one of his interpolations, the authorial narrator expresses the hope that the 'genuine Acts of SS. Perpetua and Felicitas' will be translated into English (II, 21).)

For Catholic readers, however, there is more, as Wiseman also emphasizes the continuity of tradition that such monuments represent. At times of persecution, he explains, the early Church's 'penitential code' for those who had confessed their sins was shortened, and clergy heard confessions throughout the night in order to prepare their flocks for their 'last public communion on earth' (II, 16):

> We need not remind our readers, that the office then performed was essentially, and in many details, the same as they daily witness at the Catholic altar. Not only was it considered, as now, to be the Sacrifice of Our Lord's Body and Blood, not only were the oblation, the consecration, the communion alike, but many of the prayers were identical; so that the Catholic hearing them recited, and still more the priest reciting them, in the same language as the Roman Church of the catacombs spoke, may feel himself in active and living communion with the martyrs who celebrated, and the martyrs who assisted, at those sublime mysteries. (II, 16)

Within the novel's historical time frame, Wiseman opens future vistas, thereby canonizing the martyred saints. When Sebastian is martyred, for example, his remains are collected and 'buried with honour where now stands his basilica' (II, 27). Similarly, when the young Pancratius is introduced, early in the novel, he has no 'vision of a venerable Basilica, eagerly visited 1600 years later by the sacred antiquary and the devout pilgrim, and giving his name, which it shall bear, to the neighbouring gate of Rome' (I, 3).

Wiseman uses similar techniques to suggest that living members of the Church militant, in both the fourth century and the nineteenth, are in communion with the Church triumphant, in heaven. In the first of many parallels between the sacrifice of martyrdom and the sacrifice of the Mass, Pancratius's mother, Lucina, shows him his father's blood, congealed in a sponge, and liquefies it with her tears, its colour glowing 'bright and warm' as if it has 'just left the martyr's heart'. In the novel's final episode, Fabiola's former slave, Jubala, is shot by a Numidian arrow, and is quickly baptized. As she dies, the 'water of regeneration' mingles with her 'blood of expiation' (III, 3), as in the crossing of the Red Sea, a type of baptism, the crucifixion and the eucharist. Fabiola's closing speech addresses the modern Catholic reader across the centuries:

The example of our Lord has made the martyrs; and the example of the martyrs leads us upwards to Him. Their blood softens our hearts; His alone cleanses our souls. Theirs pleads for mercy; His bestows it. May the Church, in her days of peace and of victories, never forget what she owes to the age of her martyrs.

Heaven is always close in *Fabiola*. As Pancratius stands with Sebastian, looking out across the city, he reflects that, if the underside of the firmament is so beautiful and bright, 'what must that upper-side be, down upon which the eye of boundless Glory deigns to glance', and concludes: 'I imagine it to be like a richly-embroidered veil, through the texture of which a few points of golden thread may be allowed to pass; and these only reach us' (I, 9). For the martyrs who die in the arena, that veil is the 'outstretched canvas which shades our spectators'. Had Fabiola's eyes been opened, after parting from Agnes in the previous chapter, and 'had she been able to look up above this world, she would have seen a soft cloud like incense, but tinged with a rich carnation, rising from the bed-side of a kneeling slave' (Syra), which, 'when it struck the crystal footstool of a mercy-seat in heaven, fell down again as a dew of gentlest grace upon her arid heart' (I, 8). For Wiseman, this vision is fact rather than fantasy: 'She could not indeed see this; yet it was no less true'. Similarly, as soldiers approach to take Agnes to her trial and martyrdom, she says to Fabiola, 'in an ecstasy of joy':

> *You* hear the measured tramp of the soldiers in the gallery. They are the bridesmen coming to summon me. But I see on high the white-robed bridesmaids borne on the bright clouds of morning, and beckoning me forward. Yes, my lamp is trimmed, and I go forth to meet the Bridegroom. Farewell, Fabiola, weep not for me. (II, 29)

Some of Wiseman's 'series of pictures' are reminiscent of the huge Baroque paintings and illusionist ceilings that he knew in Rome, in which supernatural beings break through the veil that separates earth and heaven, and communicate with the human figures below.[89]

Copies of Baroque paintings of this kind may well have hung in the Catholic church in which my own copy of a nineteenth-century American edition of *Fabiola* must have been stored for decades, judging by the lingering odour of incense. The Baroque was conducive to the Ultramontane tradition that Wiseman and Faber managed to impose upon English Catholicism after 1850, but not to Victorian culture as a whole. Similarly,

[89] On the 'Baroque preoccupation with the continuum of space' in the work of Barberini, Cortona, Baciccio and Pozzo, see J. R. Martin, *Baroque* (1977), pp. 164–74. On Ignation 'representation', and the 'glory' and limitations of such Baroque heavens, see H. U. von Balthasar, *Glory of the Lord* (1982–91), V, 102–14.

Wiseman's writing lacked the kind of general appeal that Newman's commanded, as the latter had a far deeper understanding of England and the English. Thus, in *Callista: A Sketch of the Third Century* (1856), Newman's plot is simpler than Wiseman's, focusing upon the unfolding spiritual biographies of the two leading characters, Callista and Agellius, and avoiding the excesses of Catholic euphoria and sentimentalism; his handling of persecution and martyrdom is more reserved; and he is relaxed about the questionable nature of some of the historical source material upon which he draws (32).

It is as if Newman's love plot were designed to disprove Kingsley's assertion in *Hypatia* that 'passionate friendships', 'fair and holy as they are when they link youth to youth, or girl to girl, reach their full perfection only between man and woman' (14). The epigraph on the title-page of *Callista* – an untitled poem from Aubrey de Vere's *The Waldenses* (1842) – announces the novel's central theme:

> Love thy God, and love Him only:
> And thy breast will ne'er be lonely.
> In that one great Spirit meet
> All things mighty, grave, and sweet.
> Vainly strives the soul to mingle
> With a being of our kind:
> Vainly hearts with hearts are twined;
> For the deepest still is single.
> An impalpable resistance
> Holds like natures still at distance.
> Mortal! love that Holy One!
> Or dwell for aye alone.[90]

Agellius, converted when he was serving as a guard of Christian martyrs, 'vainly strives the soul to mingle' with Callista, when confusing his desire to win her soul, for Christ, and her heart, for himself (11). Later, the pair share a spiritual advisor – the priest, Cæcilius, who turns out to be St Cyprian, Bishop of Carthage (20), one of the Fathers whom Newman particularly admired.[91]

Rather than focus upon ecstatic moments of vision, like Wiseman, Newman emphasizes the importance of confession – a form which suits the novel, offering access to the inner, spiritual life of characters through dialogue, over an extended period. 'Ah! but my father, my heart is below, not above', says Agellius: 'I want to tell you all' (13). Callista's journey

[90] A. de Vere, *The Waldenses* (1842), p. 122. [91] See p. 64 above.

towards conversion and martyrdom is traced largely through her dialogues with Cæcilius, whose counsel ventriloquizes some of Newman's most characteristic ideas and beliefs. Cæcilius's dialogues with Callista illustrate the motto that Newman was to adopt as a cardinal – '*Cor ad cor loquitur*'.[92] He recognizes that she has a 'heavy burden' at her heart, and assures her that she is in his heart (19). The good news that he offers her is of the heart: 'And in that human form He opens His arms and woos us to return to Him, our Maker. This is our Worship, this is our Love, Callista': 'It is an espousal for eternity.' Similarly, Cæcilius says to her, 'You cannot refuse to accept what is not an opinion, but a fact' (19), echoing Newman's comment in *The Arians,* cited earlier, that the 'great doctrines of the faith', were 'facts, not opinions'.[93]

What Cæcilius offers Callista is the peace that is to be found in the presence of the 'Eternal Word in human form' (3). He has a 'small golden pyx' around his neck in which he possesses the 'Holiest, his Lord and his God': 'That Everlasting Presence was his stay and guide amid his weary wanderings, his joy and consolation amid his overpowering anxieties' (18). Callista finds peace through self-surrender to God and through suffering as Christ suffered, becoming one of the 'Holy Martyrs'. When she is first arrested, she is not a Christian (21). Like Newman, who was rumoured to have become a Catholic before his conversion, she later becomes that for which she was earlier falsely accused. Again like Newman, who spent the whole of his adult life in various male religious communities, Callista's conversion is figured as an entry into community, rather than a purely personal event. Chione, Agellius and Cæcilius have told her that Christianity consists in the 'intimate Divine Presence in the heart': 'It was the friendship or mutual love of person with person.' (27) After her first trial, in the presence of the instruments of torture, she refuses to make a sacrifice to the 'genius' of the emperor, but yet denies that she is a Christian. She then reads Luke's gospel, which opens a 'view of a new state and community of beings, which only seemed too beautiful to be possible' (29). In her dream, the night before her martyrdom, the face of the Virgin Mary resolves itself into that of the crucified Christ, and on waking she can say, 'I am ready; I am going home' (33).

By excluding all sensationalism from his description of Callista's spiritual journey through her baptism, confirmation, and 'first and last communion', received from the hands of Bishop Cyprian (31), through her trials and periods of imprisonment, and finally on the rack of her martyrdom,

[92] See p. 43 above. [93] See p. 56 above.

Newman makes the reader receptive to the miracle that follows, which he describes, in the manner of the Acts of the Martyrs, as the kind of fact that is 'favourable to the interests of Christianity'. For the climax of the novel is not the martyrdom itself, but rather the description of the 'Corpo Santo' – the 'wonderful corpse' that lies, uncorrupted under the heat of an African sun and the chill of the moonlit night:

> A divine odour fills the air, issuing from that senseless, motionless, broken frame. A circle of light gleams round her brow, and, even when the daylight comes again, it there is faintly seen. Her features have reassumed their former majesty, but with an expression of childlike innocence and heavenly peace. The thongs have drawn blood at the wrists and ankles, which has run and soaked into the sand; but angels received the body from the soldiers when they took it off the rack, and it lies, sweetly and modestly composed, upon the ground. (35)

Word spreads through Sicca, and people return to see her again and again, 'for the mysterious and soothing effect she exerts upon them'. Again, it is the community that benefits from the miracle, and her funeral and burial are designed to benefit the community that is the Church, as Cyprian's sermon underlines: 'But yesterday one of a number, a grain of a vast heap, destined indiscriminately for the flame; to-day one of the elect souls, written from eternity in the book of life, and predestined to glory' (36). Many miracles follow the martyrdom, which 'may be said to have been the resurrection of the Church at Sicca'. And it is with a *fact* that Newman ends the novel:

> As to Agellius, if he be the bishop of that name who suffered at Sicca in his old age, in the persecution of Diocletian, we are possessed in this circumstance of a most interesting fact to terminate his history withal. What makes this more likely is, that this bishop is recorded to have removed the body of St. Callista from its original position, and placed it under the high altar, at which he said mass daily. After his own martyrdom, St. Agellius was placed under the high altar also.

As in the Acts of the Martyrs and Lives of the Saints, so in *Callista*: the heart is 'commonly reached, not through the reason, but through the imagination, by means of direct impressions, by the testimony of facts and events, by history, by description'.[94]

[94] See p. 58 above.

CHAPTER 3

England drawn and quartered

> The page of English history is covered with blood, where mention is made of Catholicity.
>
> William Waterworth S. J., *The Jesuits* (1852)[1]

I

Conflict between Protestants and Catholics in the nineteenth century often took the form of arguments over the history of the 'Reformation', or, as Catholics would say, 'English schism'. As James Anthony Froude put it at the end of the century, 'passion and prejudice . . . rule the judgment of all of us'.[2] At no time was this more obvious, however, than during the 'papal aggression' crisis. On 21 November 1850 the *Daily News* thundered: 'A Papal Bull, insulting to the nation, to its *history*, its *noble struggles*, and its as noble tolerance, is flung in the country's face' (my emphasis).[3] Lord John Russell, in his unfortunate letter to the Bishop of Durham, referred to the 'glorious principles and the immortal martyrs of the Reformation', whereas Ambrose Lisle Phillipps related the restoration of the hierarchy to Catholic martyrdom in his celebratory letter to the Earl of Shrewsbury: 'What our faithful ancestors for the last three centuries were constantly sighing for, what they prayed for in their dungeons and on the scaffold, we have at length been permitted to behold.'[4]

The popular versions of history that each side circulated in the nineteenth century are often reminiscent of the large murals which adorn end-terrace houses in militant areas of Belfast today: they are vivid, lurid and crude. In the heat of battle, the noble army of martyrs becomes the subject of a

[1] W. Waterworth, *Jesuits* (1852), p. 55. [2] J. A. Froude, *Council of Trent* (1896), p. 1.

[3] 'From the "Daily News"', in Anon., *Roman Catholic Question* (1851), 5th series, p. 14.

[4] A. L. Phillipps, *Letter to Earl of Shrewsbury* (1850), p. 3. (For extracts, see Anon., *Roman Catholic Question* (1851), 1st series, pp. 14–15.)

body count, with each side challenging the other's figures; the monarch, the Lord's anointed, is turned into an object of love or hate – 'The Virgin Queen'[5] or 'Bloody Mary'; and the Pope is either venerated as the 'Holy Father' or demonized as the 'Antichrist'.

On both sides of the divide, a handful of conscientious objectors to the war of words did manage to register their disapproval of the manipulation of historical evidence from the reigns of Henry VIII, Edward VI, Mary, Elizabeth I and James I in order to score political points during the 'Papal Aggression' crisis. The Catholic convert, Augustus Welby Pugin, for example, who recanted his idealism in relation to the Middle Ages in a pamphlet of 1851, called upon his co-religionists to acknowledge that senior Catholic clergy had been just as devious and cruel as their enemies during the English Reformation. He also appealed to both sides to stop using gruesome descriptions of torture and executions for propagandist purposes.[6] It was precisely these 'horrid acts perpetrated in the name of religion', however, that had always seized the public mind.

For generations, the Sunday reading available to most English readers, of high and low estate, had included three Protestant books: the Authorized Version of the Bible, Bunyan's *Pilgrim's Progress* and *Foxe's Book of Martyrs*, originally published as *Acts and Monuments of these latter Perilous Days, touching Matters of the Church* (Latin edition, Strasbourg, 1559; English edition, 1563).[7] Living on the Continent, in exile from Queen Mary's England, John Foxe laid particular emphasis in his martyrology upon those Protestants who suffered during her reign. On his return to England, he gained the respect of the Protestant Queen Elizabeth, who called him 'her *father Foxe*'.[8] A 'complete version' of *Foxe's Book of Martyrs*, edited by the Revd

[5] On the myth of Elizabeth, with her many names and titles – Eliza, Gloriana, the Fairy Queen, Cynthia, Good Queen Bess, Astraea, the Virgin Queen – see M. Dobson and N. J. Watson, *England's Elizabeth* (2002).

[6] 'To recapitulate horrid acts perpetrated in the name of religion on either side, to strike balances of burnings and bowellings, is only to add fuel to fire, to lead men to become cruel to each other from a very hatred of cruelty, and, in fine, to perpetuate those animosities and party feelings, which are alike unworthy of christians and injurious to the common weal.' (A. W. N. Pugin, *Earnest Address* (1851), p. 2.) Compare this comment earlier in the century: 'No religion can stand, if men, without regard to their God, and with regard only to controversy, shall rake out of the rubbish of antiquity the obsolete and quaint follies of the sectarians, and affront the majesty of the Almighty, with the impudent catalogue of their devices.' (H. Grattan, *et al.*, *Necessity for Universal Toleration* (1808), p. 33.) The great Catholic historian, John Lingard, wrote of the persecution of Protestants during Queen Mary's reign: 'To describe the sufferings of each individual would fatigue the patience, and torture the feelings of the reader': *History of England* (1819–30), v, 88.

[7] Many of the better educated also owned a copy of Milton's *Paradise Lost*.

[8] T. Fuller, *Church-History of Britain* (1655), quoted in H. Soames, *Elizabethan Religious History* (1839), p. 215.

George Townsend, Canon of Durham,[9] appeared in eight volumes in the 1840s – an important decade for the writing of English history.[10] Timely regional selections appeared in 1851, responding to 'Papal Aggression' and focusing upon martyrs from individual counties.[11] Later illustrated editions reinforced the impression that Mary Tudor was the real villain of the piece, as she sent brave Protestants to the stake (illustration 21). There were important historical lessons to be applied to the present. For example, three years after the Vatican Council's definition of Papal Infallibility (1870), William Bramley-Moore, incumbent of Gerrard's Cross, Buckinghamshire, alerted readers of his edition of *Foxe's Book of Martyrs* to the insidious effect upon the country of Popery and Puseyism by drawing parallels with the errors of the past.[12] In response, a new illustrated edition of Bishop Challoner's classic of English Catholic martyrology – his *Memoirs of Missionary Priests . . . 1577 to 1684* (1741) – appeared in 1878.[13]

From a Catholic perspective, the sixteenth-century Reformers celebrated by Foxe had effected England's schism from a united Christendom, when England itself was drawn and quartered. The Catholic barrister, Charles Butler, who had been the main driving force behind the agitation for Catholic reforms since 1782, commented that the flames in which Cranmer was consumed 'were those in which he himself had burned the anabaptists, and sought to burn the catholics'.[14] John Milner, an influential Catholic priest, and later bishop, wrote in 1800:

> . . . we find the lying *Acts and Monuments* of John Fox [sic], with large wooden prints of men and women encompassed with faggots and flames in every leaf of them, chained to the desks of many country churches, whilst abridgments of this inflammatory work are annually issued from the London presses, under the title of *The Book of Martyrs*. In the mean time it is carefully concealed from the public view, that Catholics have suffered persecution in this country to a much greater degree than they ever inflicted it, and that even the various sects of Protestants have persecuted each other, on account of their religious differences, to the extremity of death.[15]

Milner reminded his readers of the terrible manner in which most of Elizabeth's 204 Catholic victims were executed, mainly in the 1590s

[9] J. Foxe, *Acts and Monuments* (1843–9). In 1825, Townsend had published *The Accusations of History against the Church of Rome Examined* in response to Charles Butler's *Book of the Roman Catholic Church*, on which see p. 145 below.

[10] See pp. 95–8 below.

[11] See J. Foxe, *Sussex Martyrs* (1851) and *Popery as it Was and Is* (1851).

[12] See J. Foxe, *Book of Martyrs* (1873), pp. 383, 442–3, 600, 696.

[13] R. Challoner, *Martyrs to the Catholic Faith* (1878).

[14] C. Butler, *Book of the Roman-Catholic Church* (1825), p. 205.

[15] J. Milner, *Letters to a Prebendary* (1800), pp. 57–8. In fact around three hundred Protestants were executed under Mary and around two hundred Catholics under Elizabeth.

21. 'Be of good comfort, Master Ridley, and play the man', frontispiece, John Foxe, *The Book of Martyrs*, rev. William Bramley-Moore (London and New York: Cassell, 1873).

(illustration 22): 'After being hanged up, they were cut down, whilst they were alive, they were dismembered, ripped up, and their bowels burnt before their faces, after which they were beheaded and quartered. The time employed in this butchery was very considerable, and, in one instance, lasted above half an hour.'[16]

[16] Ibid., pp. 93–4. Cf. J. Reeve, *Short View* (1802), III, 187; J. Lingard, *History of England* (1819–30), V, 513–14. 'Dismembered' here means the cutting off of their 'privy parts': see R. Simpson, *Edmund Campion* (1867), pp. 308–9.

22. Nicolo Circignani and Giovanni Battista Cavalleriis, *Ecclesiae Anglicanae Trophaea* (Rome, 1584), 'A. Edmundus Campianus . . . conuersa sunt'.

Arguments over the 'horrid acts' perpetrated by both Catholics and Protestants during the English Reformation were revisited during periods of anxiety among Protestants throughout the nineteenth century. They were also taken up by historical novelists of the 1840s and 1850s who wrestled with the contradictory ideas and emotions associated with Catholicism in a period of Protestant cultural crisis. In the next section I will examine one of several lively debates over the English Reformation between historians writing in the early nineteenth century – that between the Catholic Milner and the Protestant Sturges – before turning to two novels by William Harrison Ainsworth, who pleaded for toleration: *The Tower of London* (1840) and *Guy Fawkes* (1840–1). In section III I move on to the historians of the 1840s and 1850s, including Macaulay, Carlyle and Froude, and two anti-Catholic novels: Emma Robinson's *Westminster Abbey* (1854) and Charles Kingsley's *Westward Ho!* (1855).

II

In the cathedral city of Winchester, on the cusp of the nineteenth century, an ecclesiastical battle took place which has since been forgotten, but which made a national impact at the time. John Milner, whose criticism of Foxe has been quoted, had been appointed as the Catholic priest at St Peter's, Winchester, in 1779.[17] He had witnessed the influx of émigré priests fleeing from Revolutionary France from 1791 onwards, and in Winchester numbering about a thousand at its peak. A learned man, he had been elected Fellow of the Society of Antiquaries in 1790. He was also an uncompromising controversialist in the Catholic cause, who, later in his career, frequently tried the patience not only of his brother bishops, but also that of the Pope himself.

Milner took great pains over his richly documented history of the city and its cathedral, the first volume of which appeared in 1798. While striving for accuracy, however, he could not but view the history of Winchester from his study at St Peter's and the history of England from the perspective of Rome.[18] With a degree of wish-fulfilment he writes that he 'imagines' that the people celebrated after the wedding of Queen Mary and Prince Philip of Spain in the cathedral.[19] He is on firmer ground, however, when describing the cruel treatment of the Catholic 'Winchester martyrs', whose sufferings between 1583 and 1593 are today memoralized in stained glass at St Peter's church.[20]

While Milner presided over his little Gothic chapel, built in 1792, John Sturges served as Chancellor of the diocese of Winchester and Prebendary (or Canon) of the ancient cathedral across the city – in the days of Catholic Christendom, a majestic church for the monks of St Swithun's Priory, and now a jewel in the crown of the Established Church of England. The

[17] See F. C. Husenbeth, *Life of Milner* (1862); B. Patten, *Catholicism and the Gothic Revival* (2001).

[18] See, e.g., his comments on the accession of Queen Mary: J. Milner, *History of Winchester* (1798–99), I, 350.

[19] Ibid., I, 353. [20] Ibid., I, 384.

peace of the close was disturbed when Sturges read Milner's history of Winchester, detected Catholic bias and decided to correct it. In the book that emerged, entitled *Reflections on the Principles and Institutions of Popery* (1799), he argued that there were lessons to be learned by both Catholics and Protestants from Milner's wrong-headed history.

Sturges's opening remarks imply that Milner, who had apparently been treated as a gentleman in Winchester (one of 'us'), had broken some kind of unwritten code of behaviour by writing in such an offensive manner.[21] What is more, the Winchester Establishment had been kind to the émigré Catholic priests, and yet here was Milner stirring up old antipathies. Milner's 'present performance', he writes, is 'an *aggression*, which seems to demand some animadversion' (my emphasis). Sturges's own reading of history reflects the kind of defensive conservatism that was characteristic of the period, when French republicanism posed a real threat to England and its constitution of Church and State under the Crown. Through the Reformation, he asserts, 'the Church of England was delivered from a foreign tyranny'. He disputes many of Milner's interpretations of history, arguing, for example, that Mary Tudor's 'inflexible cruelty' makes her reign 'more inauspicious, more melancholy, and more disgusting, than any which occurs in the English Annals'.

At a number of points, Sturges lays himself open to counter-attack, especially when his commentary broadens into anti-Catholic generalizations on the old bugbears, such as monasticism and celibacy.[22] When Milner published his counterblast against Sturges, entitled *Letters to a Prebendary* (1800), he drew attention to his antagonist's *locus standi* in Winchester.[23] Before the dissolution of the monastery, he implies, the cathedral was rich and the monks were poor, whereas today, the clergy are rich and the cathedral is poor. He had a point. In 1800, the Bishop of Winchester was the Hon. Brownlow North, a nepotist who married a wealthy and fashionable lady, spent four years in Italy during his episcopate and made his own son Warden of St Cross Hospital in Winchester. Milner thus anticipates Trollope, who was to caricature Bishop North as Dr Vesey Stanhope in *Barchester Towers* (1857) and make the financial scandal surrounding his son the subject of *The Warden* (1855), in which Mr Harding is ignorant of the source of his stipend. He also anticipates Cobbett, as we will see in chapter 5.

[21] J. Sturges, *Reflections on Popery* (1799), p. 3. Subsequent quotations in this paragraph are from pp. 4, 5, 22, 64.

[22] Ibid., p. 27.

[23] J. Milner, *Letters to a Prebendary* (1800), p. 42.

Milner answered Sturges point by point in his *Letters to a Prebendary*. He therefore had to address contemporary as well as historical issues, as when he turned Sturges's position on celibacy against him, for example. The benefit of having a celibate clergy was evident, he states, during Winchester's 'dreadful contagion', which 'raged amongst the prisoners of war, confined in the King's-house here, from eighteen to twenty years ago'.[24] Many hundreds died, together with those who looked after them. As Milner had mentioned in his history, 'a considerable number of the said prisoners' were French Protestants, who asked for Catholic priests to attend them, having seen how they ministered to Catholic prisoners. The Protestant clergy were no more afraid of death than the Catholic priests, but dreaded the thought of taking the contagion back to their families. So the Protestant prisoners were assisted by the priests.

Milner's annoyance, however, turns to anger when he describes Sturges's reading of post-Reformation history as an example of Anglican hypocrisy.[25] Having published, first, his history of Winchester and, secondly, his reply to Sturges's critique, Milner was now ready to 'follow up the success which the superiority of his cause gave him'.[26] He was persuaded, however, by the sympathetic Bishop Samuel Horsley, to hold back any further publication. Nevertheless, letters that he wrote in 1801–2 were finally published seventeen years later, in a volume of correspondence entitled *The End of Religious Controversy* (1818), a work which accelerated at least one conversion to Catholicism.[27] Here he explains, in a prefatory address to Thomas Burgess, the Bishop of St David's; that the reprinted (and slightly revised) letters in the book grew out of his disagreement with Sturges – a controversy that had 'made some noise in the public, and even in the Houses of Parliament, particularly in the Upper House'.[28] His comments to the bishop on Reformation history are as robust as ever. He argues, for example, that it is the duty of Catholics to 'repel these charges by proving that there was *reason*, and *religion*, and *loyalty*, and *good faith* among Christians, before Luther quarrelled with Leo X, and Henry VIII fell in love with Ann Bullen'.[29] A 'cogent reason' for the appearance of the work, however, was the recent publication of a catechism written by Burgess himself, who 'comes forward in his episcopal mitre, bearing in his hands a new *Protestant Catechism*, to be learnt by Protestants of every description, which teaches them to *hate* and *persecute* their elder brethren, the authors of their Christianity and

[24] Ibid., p. 55. [25] Ibid., p. 209. [26] J. Milner, *Vindication* (1822), p. 5.
[27] See T. H. Shaw, *McPhersons* (1879), p. 106.
[28] J. Milner, *End of Religious Controversy* (1819), p. iii. [29] Ibid., p. 5.

civilization!'[30] What distinguishes *The Protestant's Catechism* from other '*No-Popery* publications', Milner argues, is 'not so much the strength of its acrimony, as the boldness of its paradoxes'.[31]

Significantly, Milner also detects a general trend in 1819 towards the use among Protestants of abusive language in relation to Catholics:

> I trust your Lordship will not be the person to ask me, why the Letters, after having been so long suppressed, now appear? – You are witness, my Lord, of the increased and increasing virulence of the press against Catholics; and this, in many instances, directed by no ignoble or profane hands . . . . in short, we seldom find ourselves, or our religion mentioned, in modern sermons, or other theological works, unaccompanied with the epithets of *superstitious*, *idolatrous*, *impious*, *disloyal*, *perfidious*, and *sanguinary*.[32]

This kind of verbal abuse reflects the fact that, at a time of radical agitation for parliamentary reform, Catholicism was part of a wider threat to the established order. The subsequent eleven years were to witness increasing pressure for Catholic relief, culminating in Wellington and Peel's Catholic Emancipation legislation in 1829 – the subject of chapter 5. By the time of the 'Papal Aggression' crisis of 1850–1, the terms of abuse had become stronger.[33]

A year after the publication of Milner's *The End of Religious Controversy*, another Catholic priest, John Lingard, published the first three volumes of his *History of England to 1688* (8 vols, 1819–30). The book had initially been rejected by two Catholic publishers, but was then accepted by a Protestant one.[34] It was to go through seven English editions by 1883 and be translated into Italian, French and German. It was attacked by Milner for not being Catholic history at all, and thus confirming Protestant readers in their errors.[35] Lingard's ground-breaking use of original sources in the State Paper Office in the Tower of London, however, together with his moderation and 'fairness' – a term of his own that was often applied to his work – were what brought the *History* such widespread acclaim.[36] When describing Queen Mary's persecution of Protestants, for example, he comments that

[30] Ibid., p. 7. Milner must refer to the 1818 edition, 'by the Bishop of St David's': T. Burgess, *Protestant's Catechism* (1818). This includes the text of the first edition: T. Burgess, *Protestant's Catechism* (1817).

[31] J. Milner, *End of Religious Controversy* (1819), p. xii.

[32] Ibid., p. iv. [33] See pp. 29–30 above.

[34] E. R. Norman, *English Catholic Church* (1984), p. 291. John Lingard (1771–1851) was trained at Douay, ordained in 1795, taught at Ushaw College, and in 1811 moved to Hornby in Lancashire, where he devoted the next forty years to his historical writing.

[35] See J. P. Chinnici O. F. M., *English Catholic Enlightenment* (1980), pp. 119–26.

[36] Chinnici argues that Lingard 'carefully concealed his bias underneath a dispassionate narrative and a mass of documentary evidence': ibid., p. 112.

the mind is 'struck with horror' at the thought of almost 200 people being burnt alive over a period of four years, but later writes: 'It is, however, but fair to recollect . . . that the extirpation of erroneous doctrine was inculcated as a duty by the leaders of every religious party. Mary only practised what *they* taught. It was her misfortune, rather than her fault, that she was not more enlightened than the wisest of her contemporaries.'[37] Of her half-sister's reign, he writes: 'The historians, who celebrate the golden days of Elizabeth, have described with a glowing pencil, the happiness of the people under her sway. To them might be opposed the dismal picture of national misery, drawn by the catholic writers of the same period. But both have taken too contracted a view of the subject.'[38]

Interestingly, a Protestant historical novelist like William Harrison Ainsworth, writing in the early 1840s, could turn to Lingard's account of English Reformation history as his main source, even though antagonism towards Catholics was particularly strong at the time.[39] His *The Tower of London: A Historical Romance*, was published in monthly parts (January–December 1840), and *Guy Fawkes; or, The Gunpowder Treason: An Historical Romance*, was published serially in *Bentley's Miscellany* (January 1840–November 1841), which he also edited at the time. Ainsworth had made his name as a novelist with *Rookwood: A Romance* (1834), famous for its description of Dick Turpin's ride to York, and his controversial Newgate novel, *Jack Sheppard: A Romance* (1839), which outsold Dickens's *Oliver Twist*.[40] His romances are remarkable for their energy, having fast-moving plots that are full of exciting and often dangerous incident, and vivid evocations of the external appearance of people and places. They are not notable for their exploration of the inner life of the human heart and mind, or for their restraint. Tellingly, Ainsworth's 'first boyish passion' was for fireworks, but, as his biographer puts it, 'the rage for pyrotechnics soon gave way to dramatic and histrionic ambition'.[41]

With *The Tower of London*, which covers the period of Lady Jane Grey's rise and fall (10 July 1553–12 February 1554), Ainsworth reached the height of his popularity as an historical novelist, not least because he was partnered by the illustrator, George Cruikshank, who provided the idea for the book and made 40 plates and 58 woodcuts for it.[42] The woodcuts, interpolated in the

[37] J. Lingard, *History of England* (1819–30), v, 101, 134.

[38] Ibid., v, 625. See also Lingard's comparison between the numbers of executions following rebellions in the reigns of Mary and Elizabeth (v, 56–7).

[39] J. Wolffe, *Protestant Crusade* (1991), p. 2.

[40] W. E. A. Axon, *William Harrison Ainsworth* (1902), p. xxv.

[41] Ibid., p. xiv.

[42] Cruikshank claimed that Ainsworth 'wrote up to' the plates, which were sometimes printed before the text was even written: B. Jerrold, *Life of Cruikshank* (1971), pp. 167–9.

body of the text, illustrate every corner of the Tower, and reflect Ainsworth's fascination with the building itself, which he wished to see restored and made more fully accessible to the general public.[43] The vivid plates – full-page etchings on steel, many of them *tableaux vivants* – illustrate the plot in the novel's two sections: Book the First, 'Jane the Queen', and, almost three times longer, Book the Second, 'Mary the Queen'. Having explained his interest in the fabric of the Tower in the preface to the book edition, dated 28 November 1840, Ainsworth defended his treatment of the Queen who was usually reviled as 'Bloody Mary' by his contemporaries: 'To those, who conceive that the Author has treated the character of Queen Mary with too great leniency, he can only affirm that he has written according to his conviction of the truth. Mary's worst fault as a woman – her sole fault as a sovereign – was her bigotry: and it is time that the cloud, which prejudice has cast over her, should be dispersed.'[44]

The novel's opening chapter is entitled, 'Of the manner in which Queen Jane entered the Tower of London'. For a few days her palace, the Tower is soon to become the Protestant Queen's prison and eventually her place of execution, under Catholic Queen Mary. Cruikshank's frontispiece, entitled 'The Execution of Jane' (illustration 23), shows the axe about to descend on the swan-like neck on the block, the executioner's legs seeming to straddle the delicate head in an act of grotesque masculine aggression. From the very start, both Lady Jane Grey and the narrative are under the axe. We are reminded of this in a plate entitled 'Jane Imprisoned in the Brick Tower' (illustration 24), inserted early in Book the Second, where Jane is shown in the same posture as at her execution, her head lying on the seat of the 'curiously-carved chair' used by Queen Anne Boleyn (Princess Elizabeth's mother) during her imprisonment.[45] But we read for a further forty chapters before the execution, in the very last sentence of the novel: 'The axe then fell, and one of the fairest and wisest heads that ever sat on human shoulders, fell likewise.'[46]

Ainsworth exploited the cult associated with Lady Jane Grey, epitomized by Paul Delaroche's famous painting, *The Execution of Lady Jane Grey*, which seized the public imagination when it was first exhibited in the Paris Salon in 1834 and now hangs in the National Gallery, London.[47] He also reinforced the standard Protestant position, typified by Foxe, by presenting Lady Jane Grey as a martyr.[48] He was therefore freer to rehabilitate Queen

[43] W. H. Ainsworth, *Tower of London* (1903), p. vi.
[44] Ibid., p. vii. [45] Ibid., p. 135. [46] Ibid., p. 450.
[47] See R. C. Strong, *And when did you last See your Father* (1978), pp. 122–7.
[48] J. Foxe, *Book of Martyrs* (1873), pp. 302–13.

23. George Cruikshank, 'The Execution of Jane', in William Harrison Ainsworth, *The Tower of London: A Historical Romance*, new edn (London: Methuen, 1903), frontispiece.

Mary without being accused of pro-Catholic bias. Being Ainsworth, these larger themes are conveyed in a narrative which combines political intrigue with a romantic plot in which Mary and Elizabeth are rivals for the love of Edward Courtenay, a comic sub-plot associated with the Tower's warders, three of whom are giants and one a dwarf, and a highly detailed guide-book to the Tower itself.

Of these oddly assorted ingredients, the last is in a way the most justifiable, as Ainsworth's observation in the preface that he is 'desirous of exhibiting the Tower in its triple light of a palace, a prison, and a fortress' gives the clue to the dynamics of the historical narrative, which reflect the instabilities of the times. As I have said, for Lady Jane Grey the Tower is first a palace and then a prison. For Queen Mary it is first a palace and then,

24. George Cruikshank, 'Jane imprisoned in the Brick Tower', in William Harrison Ainsworth, *The Tower of London: A Historical Romance*, new edn (London: Methuen 1903), facing p. 135.

during attacks from outside, a fortress. For Princess Elizabeth, however, the Tower is first a palace and fortress ruled over by her sister and then a prison. So Mary's power is expressed through the Tower itself, within whose walls we witness terrible tortures, a prisoner condemned to be eaten alive by rats, beheadings and a particularly harrowing burning. Mary's first act upon entering the Tower precincts, however, is to release the Catholic prisoners, including Stephen Gardiner, Bishop of Winchester, whom she not only reinstates to his bishopric but also appoints Lord Chancellor of the realm.[49] So for Gardiner, the Tower is first a prison and then a palace and a fortress; he suffers at the hands of a Protestant regime before he attains highest office under Mary, and then wields power over his enemies. Indeed,

[49] W. H. Ainsworth, *Tower of London* (1903), p. 130.

it is Gardiner and other courtiers who order the cruel torture and execution of Protestants, rather than the Queen herself.

A less obvious, but more effective way in which Ainsworth gains the reader's sympathy for Mary, a devout, though 'bigoted' Catholic, is through his portrayal of her clergy engaging in various kinds of ministry in the novel, when the sound and fury of the plot are suddenly stilled, and a sense of spiritual peace is created. Ainsworth's love of pageantry is given full rein at the beginning of Book the Second, when he describes Mary's public entry into the City of London, on 3 August 1553. 'No one', he writes, 'has suffered more from misrepresentation than this queen. Not only have her failings been exaggerated, and ill qualities, which she did not possess, attributed to her, but the virtues that undoubtedly belonged to her, have been denied her . . . It was this "failure in religion" [Bishop Godwin] which has darkened her in the eyes of her Protestant posterity. With so many good qualities it is to be lamented that they were overshadowed by bigotry.'[50] Although he cites other sources at this point, his main debt is to Lingard's *History of England*, as comparison between his own and Lingard's descriptions of Mary and Elizabeth demonstrates.[51] More significantly, his handling of Mary's encounter with the Catholic clergy offers the reader some insight into Catholic spirituality. In contrast to Lady Jane Grey, at the beginning of Book the First, Mary is 'everywhere received with the loudest demonstrations of joy', as bells peal and cannons roar. Having admired a number of colourful pageants, mounted on stages along the route, Mary's attention is attracted to a stage which is

> filled with Romish priests in rich copes, with crosses and censers of silver, which they waved as the Queen approached, while an aged prelate advanced to pronounce a solemn benediction upon her. Mary immediately dismounted, and received it on her knees. This action was witnessed with some dislike by the multitude, and but few shouts were raised as she again mounted her palfry. But it was soon forgotten, and the same cheers that had hitherto attended her accompanied her to the Tower.[52]

The awkwardness of 'waved' reflects Ainsworth's Protestantism, while the Protestant reader of 1840 can empathize with the crowds who briefly express their disapproval of Mary's devotions, but are soon able to forget them, as

[50] Ibid., p. 125.
[51] Compare ibid., pp. 124, 126, with J. Lingard, *History of England* (1819–30), v, 14–15. Ainsworth embellishes Lingard, however, in his account of Lady Jane Grey encountering her husband's body on her way to the scaffold: contrast ibid., p. 449, with Lingard, v, 54.
[52] Ibid., pp. 127–8.

the procession, like the narrative, continues on its way. In a novel full of cursing, however, this act of blessing has real emotional and spiritual weight.

Following the execution of the Duke of Northumberland on Tower Hill, Mary continues to hold her court within the Tower, mainly because her 'plans for the restoration of the Catholic religion' can be 'more securely concerted within the walls of the fortress than elsewhere'.[53] Simon Renard, the ambassador of her cousin, Emperor Charles V, and now her confidential advisor, can visit her here 'unobserved', as can the Pope's envoy. Also within the fortress, however, are imprisoned some courageous Protestants who refuse to convert. Chief among them is Lady Jane Grey herself, who describes Catholicism as 'a pernicious and idolatrous religion, – a religion founded on the traditions of men, not on the word of God – a religion detracting from the merits of our Saviour – substituting mummery for the simple offices of prayer', and who refuses to convert, pointing to the Bible, the 'rock' on which her religion is built.[54] When Edward Underhill, the 'Hot-Gospeller', fails in his attempted assassination of Queen Mary, the narrator comments that his 'atrocious purpose' has been 'providentially defeated'.[55] Nevertheless, his cruel death at the stake is described as that of a Protestant martyr: 'his eyes starting from their sockets – his convulsed features – his erected hair, and writhing frame – proclaimed the extremity of his *agony*', and when his 'lower limbs were entirely consumed' he '*gave up the ghost*' (my emphases).[56]

Again, Ainsworth's judicious treatment of Protestantism frees him to deal sympathetically with Catholicism in another of his elaborate and picturesque set pieces. During the insurrection of Sir Thomas Wyatt, the Tower is not only proof against bombardment, but is also a fortress of Catholicism in liturgical terms. Throughout the siege, Mary maintains her 'accustomed firmness'. Having shared the physical dangers to which her troops are exposed on the battlements, she proceeds to St John's Chapel in the White Tower, where she is shown to have a heart:

> It was brilliantly illuminated, and high mass was being performed by Bonner and the whole of the priesthood assembled within the fortress. The transition from the roar and tumult without to this calm and sacred scene was singularly striking, and calculated to produce a strong effect on the feelings. There, all was strife and clamour; the air filled with smoke was almost stifling; and such places as were not lighted up by the blaze of the conflagration or the flashing of the ordnance and musquetry, were buried in profound gloom. Here, all was light, odour, serenity, sanctity. Without, fierce bands were engaged in deathly fight – nothing was heard

[53] Ibid., p.172. [54] Ibid., pp. 188, 194–5. [55] Ibid., p. 211. [56] Ibid., pp. 212–13, 274.

but the clash of arms, the thunder of cannon, the shouts of the victorious, the groans of the dying. Within, holy men were celebrating their religious rites, undisturbed by the terrible struggle around them, and apparently unconscious of it; tapers shone from every pillar; the atmosphere was heavy with incense; and the choral hymn mingled with the scarce-heard roar of cannon. Mary was so affected by the scene, that for the first time she appeared moved. Her bosom heaved, and a tear started to her eye.[57]

In case the reader has missed the point, Ainsworth makes Mary comment upon the contrast between 'the holy place' and 'the scene we have just quitted' to Bishop Gardiner. In evoking the sublimity of the old religion through chiarascuro descriptive effects, Ainsworth transforms the Tower into a metaphor for both the Queen and Catholicism – troubled and embattled without, peaceful and devout within. The nineteenth-century Protestant reader is manoeuvred into the unusual position of glimpsing something of the mystery of Catholicism and its appeal to the heart, as Mary advances to the altar, prostrates herself in fervent prayer, and leaves the chapel, as the 'holy rites' continue, 'with a heart strengthened and elated'.

Ainsworth uses a similar technique in *Guy Fawkes*, interrupting the violent action in order to describe Catholic worship. In that novel, however, the reader is presented with anti-Catholic oppression from the very beginning:

More than two hundred and thirty-five years ago, or, to speak with greater precision, in 1605, at the latter end of June, it was rumoured on a morning in Manchester that two seminary priests, condemned at the last assizes under the severe penal enactments then in forces against the Papists, were about to suffer death on that day. Attracted by the report, large crowds flocked towards the place of execution, which, in order to give great solemnity to the spectacle, had been fixed at the southern gate of the old Collegiate Church, where a scaffold was erected. Near it was a large blood-stained block, the use of which will be readily divined, and adjoining the block, upon a heap of blazing coals, smoked a caldron filled with boiling pitch, intended to receive the quarters of the miserable sufferers.[58]

The 'pursuivant', or official 'Papist'-hunter, scrutinizes his list of suspects, the executioner casually prepares his equipment, and a somewhat hostile crowd of Lancastrians – many of them Catholic – waits.[59] Then the 'mournful procession' approaches. Elizabeth Orton, a cave-dwelling prophetess, dressed like a Sister of Charity, kisses the hem of the garment of the nearest

[57] Ibid., p. 351. [58] W. H. Ainsworth, *Guy Fawkes* (1878), pp. 1–2.

[59] Manchester was to be a centre of Jacobite feeling during the rebellion of 1745: see D. Daiches, *Charles Edward Stuart* (1973), p. 163.

priest and has her name taken down for future punishment. The executioner then performs his 'horrible task', which is not described.

Ainsworth puts the reader in the front row of the waiting crowd, so that, like them, he or she is anxiously caught between revulsion from the 'revolting and sanguinary spectacle' about to be witnessed and a voyeuristic attraction to the pornography of violence.[60] In his preface, Ainsworth quotes a lengthy passage from Lingard on the 'tyrannical measures adopted against the Roman Catholics in the early part of the reign of James the First' – a 'fitting introduction to the present work'.[61] Ainsworth ends the preface by commenting that the 'one doctrine' that he has endeavoured to 'enforce throughout' is 'TOLERATION'.

The preface and the opening of the novel, then, unequivocally establish the lack of religious toleration in 1605. The hanging, drawing and quartering of the priests also reminds us of the fate awaiting Guy Fawkes himself. As in *The Tower of London*, the reader knows how the novel is going to end from the first page. So in place of suspense, Ainsworth provides exciting incident along the way and a build-up of tension as the anticipated ending approaches. There are sensational chases of 'Papists' by the pursuivant, and there are priest holes, sliding panels, disguises and conspirators aplenty, often described in a pedantically antiquarian style. (The place of execution at the beginning, for example, is 'guarded by a small band of soldiers, fully accoutred in corslets and morions, and armed with swords, half-pikes, and calivers'.)[62] Guy Fawkes foresees his own torture and execution several times during the narrative, through a variety of prophetic dreams and visions. The famous Dr Dee, who has foreknowledge of the Gunpowder Plot through his black art, takes Guy Fawkes to the chamber where the remains of the priests are collected, as in a butcher's shop: 'There were two heads, and, though the features were scarcely distinguishable, owing to the liquid in which they had been immersed, they still retained a terrific expression of agony.' Dee then raises the spirit of Elizabeth Orton from her recently buried corpse in this charnel house and she prophesies the failure of the plot and the horrible death of the conspirators. Later Guy Fawkes looks into Dr Dee's 'magic glass' and sees himself 'stretched upon the wheel, and writhing in the agonies of torture'.

In the midst of all this noisy and often noisome business, designed to keep the readership of a family magazine entertained, Ainsworth will suddenly adopt a quite different tone, capturing the devotional spirit of what he calls

[60] Ibid., p. 5. [61] Ibid., pp. v–vi.
[62] Ibid., p. 2. Other quotations in this paragraph are from pp. 61, 63, 71.

the 'Romish Church', and again replacing cursing with blessing. Consider, for example, the burial of Sir William Radcliffe inside the Collegiate Church at Manchester:

> Placing himself at the foot of the body, Garnet sprinkled it with holy water, which he had brought with him in a small silver consecrated vessel. He then recited the *De Profundis*, the *Miserere*, and other antiphons and prayers; placed incense in a burner, which he had likewise brought with him, and having lighted it, bowed reverently towards the altar, sprinkled the body thrice with holy water, at the sides, at the head, and the feet; and then walking round it with the incense-burner, dispersed its fragrant odour over it. This done, he recited another prayer, pronounced a solemn benediction over the place of sepulture, and the body was lowered into it.[63]

The 'loud knocking at the church door' that follows feels more like a rude interruption than the legitimate upholding of the penal laws that it actually is.

Gore must come to the fore, however, and at the end of the novel the vast London crowds who turn out to see the conspirators go to the scaffold over a couple of days relish the blood-letting. After the first four conspirators have suffered, the crowd separate to talk over the sight they have witnessed and to 'keep holiday during the remainder of the day; rejoicing that an equally exciting spectacle was in store for them on the morrow'.[64] We are not shown the executions, but the lieutenant of the Tower has reminded an unmoved Guy Fawkes, shattered by prolonged torture, precisely what his will involve. Yet when Fawkes's turn comes, his weakened state saves him from further agony. Indeed, he achieves an apotheosis in his dying. Seeing the woman he loved beckoning him 'to unfading happiness', he stretches out his arms and springs from the ladder: 'Before his frame was exposed to the executioner's knife, life was totally extinct.'[65] Ironically, the traditional celebrations on November the Fifth to which Ainsworth refers in the novel's final sentence, and which were to become increasingly elaborate and rowdy after 'Papal Aggression', mimic, not the hanging, drawing and quartering meted out to Catholics by Elizabeth and James I, but the burning of Protestants by Mary Tudor.[66]

[63] Ibid., p. 141. [64] Ibid., p. 400.

[65] Ibid., p. 403. Contrast Joseph Reeve's account: 'he placed his arms across, and was turned off the ladder. The executioner was about to cut him down immediately, as had been done to many others who suffered for religion, but the people would not permit it'. *General History of the Christian Church* (1851), p. 510. (Originally published as *A Short View of the History of the Christian Church* (1802).)

[66] See R. Swift, 'Guy Fawkes Celebrations in Victorian Exeter', *History Today* (1981), 5–9.

III

During the serial publication of both *The Tower of London* and *Guy Fawkes*, in 1840, two powerful and conflicting statements on Catholicism were laid before the reading public by leading literary figures of the day. First, in May, Thomas Carlyle delivered his lecture on Luther and Knox, 'The Hero as Priest', later published in *On Heroes and Hero-Worship*, in which he argued that the Reformation was inevitable because Catholicism had simply become incredible.[67] Secondly, in October, Thomas Babington Macaulay, then a Whig Cabinet Minister, published an important review essay in the *Edinburgh Review* on Leopold von Ranke's *The Ecclesiastical and Political History of the Popes of Rome during the Sixteenth and Seventeenth Centuries* (1840), translated by Sarah Austin. Whereas Carlyle's argument retains something of the Scottish Calvinist background from which he had moved away, Macaulay's suggests that he had retained nothing from his own Evangelical upbringing. Macaulay held liberal views on religious issues: he described Southey's *Book of the Church*, for example, as 'mere rubbish'.[68] In 1830, his maiden speech in the Commons had been in support of a bill for the Removal of Jewish Disabilities. Two years later, at an election meeting in Leeds, he had declared himself to be a Christian, but had strongly criticized the Methodist preacher who raised the question of the candidates' religious beliefs in public. On 14 October 1840, he told Macvey Napier, editor of the *Edinburgh*, that his article on 'Von Ranke' would attract 'plenty of abuse', but added that he had 'long ceased to care one straw'.[69]

'There is not', he writes, 'and there never was on this earth, a work of human policy so well deserving of examination as the Roman Catholic Church.'[70] He then opens up the widest of vistas, not only acknowledging the longevity and present vitality of the Catholic Church, but also, in a famous passage in which he gestures towards Piranesi's ruins of ancient Rome, imagining it outliving the Church of England in the future: 'The proudest royal houses are but of yesterday, when compared with the line of the Supreme Pontiffs . . . And she [the Catholic Church] may still exist in undiminished vigour when some traveller from New Zealand shall, in the

[67] 'Dante's sublime Catholicism, incredible now in theory, and defaced still worse by faithless, doubting and dishonest practice, has to be torn asunder by a Luther . . . The Reformation might bring what results it liked when it came, but the Reformation simply could not help coming. To all Popes and Popes' advocates, expostulating, lamenting and accusing, the answer of the world is: Once for all, your Popehood has become untrue. . . . Away with it.' T. Carlyle, *Works* (1884), I, 345, 361.

[68] T. B. Macaulay, *Essays* (1899), p. 101.

[69] G. Otto, *Life of Macaulay* (1899), pp. 115, 204, 394.

[70] T. B. Macaulay, 'Von Ranke', in *Essays* (1899), p. 548.

midst of a vast solitude, take his stand on a broken arch of London Bridge to sketch the ruins of St Paul's.'[71]

Macaulay's position on the English Reformation is that of the religiously uncommitted, and he treats what Southey called the 'prodigious' and 'astonishing' doctrine of transubstantiation with a degree of nonchalance.[72] At a time when Evangelicalism set the tone in polite English society, and when people were reading hagiographies of Reformers and pious histories of the Reformation on the Continent in translation,[73] Macaulay presents those involved in the Reformation as a mixed bunch.[74] Among the Protestant princes of the Reformation period, he argues, there was little or no 'hearty Protestant feeling': 'Elizabeth herself was a Protestant rather from policy than from firm conviction.'[75] In the south of Europe, meanwhile, the Counter-Reformation was achieving great things, largely through the Jesuits, and 'the Court of Rome itself was purified'.[76] Published during a period of heightened anxiety concerning Rome's ambitions towards England, Macaulay's conclusion concerning the contemporary situation did not make comfortable reading for John Bull: 'During the nineteenth century, this fallen Church has been gradually rising from her depressed state and reconquering her old dominion.'[77]

Macaulay's views on the Reformation were to be worked out more fully in the introductory chapter to his famous *History of England from the Accession of James II*, the first two volumes of which were published in 1849 and were said at the time to be 'the most popular historical work that ever issued from the English press'.[78] His maturation argument is a classic example of Whig history, which discerns progress in the march towards modernity: 'Those who hold that the influence of the Church of Rome in the dark ages was, on the whole, beneficial to mankind may yet with perfect consistency regard the Reformation as an inestimable blessing. The leading strings, which preserve and uphold the infant, would impede the full grown

[71] Four decades later, William Colenso was to complain that Macaulay's New Zealander was cited too often, particularly as the basic idea was not original: 'A few Remarks', in *Three Literary Papers* (1883).
[72] T. B. Macaulay, *Essays* (1899), p. 550.
[73] See, e.g., *Writings of Bradford* (1827); G. Stokes, *Lives of the British Reformers* (1834); L. Ranke, *Reformation in Germany* (1845–7); J. H. Merle d'Aubigné, *History of the Reformation* (1853). The fifth volume of d'Aubigné, on the English Reformation, 'took Protestant England by storm, and, of its kind, it must have been one of the best-sellers of the Victorian era': J. H. Merle d'Aubigné, *Reformation in England* (1962), p. 3.
[74] T. B. Macaulay, *Essays* (1899), p. 554.
[75] Ibid., p. 560. Cf. J. Lingard: 'It is pretty evident that the queen herself had formed no settled notions of religion': *History of England* (1819–30), v, 315.
[76] Ibid., pp. 558. [77] Ibid., pp. 569.
[78] J. Moncrieff, 'Macaulay's *History of England*', *Edinburgh Review*, July 1849, p. 249. The book sold more than 18,000 copies in the first six months.

man.'[79] Again, Macaulay is unusually relaxed in his judgments, commenting that 'it is difficult to say whether England owes more to the Roman Catholic religion or to the Reformation'.[80] He believes that Henry the Eighth 'attempted to constitute an Anglican Church differing from the Roman Catholic Church on the point of the supremacy, and on that point alone'.[81] Henry's success in this attempt was 'extraordinary', but his system 'died with him'.

Published only a year before the 'Papal Aggression' crisis, Macaulay's detailed account of the paranoid atmosphere in the court of Charles II, exploited by Titus Oates in his exposure of a non-existent 'Popish Plot', has a prophetic ring: 'the nation was in such a temper that the smallest spark might raise a flame. At this conjuncture fire was set in two places at once to the vast mass of combustible matter; and in a moment the whole was in a blaze'.[82] As we have seen, Bishop Ullathorne made a similar comment on 'Papal Aggression' in 1851, although he chose a different metaphor: the crisis had, he said, put the whole country 'in a boil'.[83] During the crisis, comic strip versions of English Reformation history featured in anti-Catholic sermons, speeches at public meetings, popular books and pamphlets, and newspapers and periodicals, most of which had specific sectarian affiliations. In such a heady atmosphere, the work of leading historians such as Lingard and Macaulay tended to be either forgotten or ignored. Preaching in the Episcopal Jews' Chapel, Bethnal Green, on 30 October 1850, James Cartwright, for example, took the unreconstructed anti-Catholic line on Queen Mary, appealing to 'our own murdered bishops and clergy, and people during the reign of that female sovereign whom Protestant posterity justly named Bloody, because she lent the authority of the British crown to the savage decrees of Popery'.[84] Several anti-Catholic manuals were adopted by the Reformation Society and the Scottish Protestant Association during the 1850s for use in their adult classes, which were taught using the catechistic method adopted in school textbooks, and satirized in the opening chapters of Dickens's *Hard Times* (1854). In one of them, Dr Blakeney deals with the 'reign of Mary, known as the Bloody Queen', by simply summarizing Foxe's descriptions of the executions of the leading Protestant martyrs, and asking for their names in the 'Questions and Answers' section at the end of the chapter.[85]

A much smaller and poorer Catholic propaganda machine responded as best it could. In response to the flood of Protestant material that poured

[79] T. B. Macaulay, *History of England* (1856), I, 46. [80] Ibid., I, 48. [81] Ibid., I, 49.
[82] Ibid., I, 183–4. [83] See p. 13 above. [84] J. B. Cartwright, *No Popery!* (1850), p. 71.
[85] R. P. Blakeney, *Popery in its Social Aspect* (n.d.), pp. 118–23.

from the presses, the Brotherhood of St Vincent of Paul published an extensive series of 'Clifton Tracts', for example, in which Catholic viewpoints were explained in simple terms. In one volume from its 'History Library', Henry VIII is described as 'a very wicked man' and his daughter, Elizabeth I, as a 'very artful and cunning woman'.[86] The religious beliefs of both monarchs are not the subject of discussion here – they are simply denied: 'it is quite clear that it was not from the love of God or from zeal for His truth, but for the sake of gratifying his own evil lusts and passions, that Henry rent this kingdom from the See of Rome, to which it had for centuries owed allegiance in spiritual things, and made himself supreme head of the Church of England'; 'I do not think that she [Elizabeth] had any real care for religion at all.'

Running alongside the battle over history were related battles over authority and ecclesiology, doctrine and liturgy. The Clifton Tracts also included a 'Library of Christian Doctrine' and a 'Library of Christian Devotion'. These tracts responded to anti-Catholic writings by explaining Catholicism and attacking Protestantism, as did works produced for more highly educated Catholics, such as a critical study on the *Origin and Developments of Anglicanism* by William Waterworth S. J.[87] Anti-Catholic tracts and handbooks of the early 1850s found a useful target in Pope Pius IV's 'profession of faith' (1564), defined at the Council of Trent.[88] Blakeney's *Manual of Romish Controversy* (1851) offered the reader a 'complete refutation' and provided questions and answers for class use. The *Manual* sold over 27,000 copies, yet there was still room in the market for Charles Stuart Stanford's *Handbook to the Romish Controversy* (1852), which also offered a 'refutation', and which matched the *Manual* in its sale figures.

Apart from the doctrine of transubstantiation, nothing disturbed the sensibilities of what George Eliot called 'English Puritanism' more than auricular confession – penance being one of the seven sacraments of the Catholic Church (illustration 25).[89] The dangers associated with the

[86] Anon., *Popish Persecution* (1851), pp. 4, 12.

[87] 'When we remember that Cranmer was the author of the *faith* of Protestantism, *the* individual specially selected to frame the *laws*, the *Articles*, the *Catechism*, the *books* out of which *the people were to be instructed, and the whole Church of England and Ireland was to pray*, we shrink with horror from the sight of such a man, and pity indeed the abettors and panegyrists of such a character. Even on the brink of the grave Cranmer was not sincere.' W. Waterworth S. J., *Origin and Developments* (1854), p. 75.

[88] The 'Profession of the Tridentine Faith' had to be sworn to by those in ecclesiastical office.

[89] George Eliot, *Middlemarch* (1986), p. 188. The other sacraments are Baptism, Confirmation, the Eucharist, Extreme Unction, Holy Orders and Matrimony. Only Baptism and the Eucharist are sacraments in the Church of England.

25. 'Expectations from Rome', *Punch*, 19 (July–December 1850), 193.

confessional seemed obvious to Blakeney. Mothers, daughters and wives, he points out, tell the priest 'their most secret thoughts and sins', and eighteenth-century treatises by Liguori, Dens and others, for the 'guidance of the confessional', are 'so polluted and filthy, that we do not use language too strong, when we say they are only fit for the abodes of hell'.[90] The depth of revulsion from auricular confession during the 'Papal Aggression' crisis is reflected in the fact that, in December 1850, the Revd Dr M'Neile of Liverpool, a respected Canon of Manchester, got carried away during an extempore sermon and called for all confessors to be executed.[91]

Whereas Victorian clergymen were reported in the newspapers when they made such wild and irresponsible suggestions, popular writers of historical romance were free to play out their own fantasies, and those of their readers, by describing confessors actually being executed. Prior 'Sancgraal' Bigod,

[90] R. P. Blakeney, *Manual of Romish Controversy* (n.d.), p. 87. Blakeney refers to St Alphonsus Liguori (1696–1787) and Peter Dens (1690–1775). Auricular confession had been a major theme in that popular farrago of anti-Catholic hysteria, *Maria Monk*, published in America in 1836: Anon., *Awful Disclosures* (n.d.), pp. 17, 25, 62, 89. On the subtle use of the confessional in Charlotte Brontë's anti-Catholic novel, *Villette* (1853), see pp. 225–8 below.

[91] M'Neile's statement that 'Death alone would prevent the evil' was reported in the press, along with his agonized retraction: Anon., *Roman Catholic Question* (1851), 13th series, p. 10.

Emma Robinson's villain in her anonymous *Westminster Abbey; or, The Days of the Reformation* (1854), is frequently described as the 'confessor', this being the pastoral role in which we see him most often in the novel, and which he most often abuses for his own dastardly ends.[92] At the end of the novel, Henry VIII condemns him to be burnt at the stake for murder and sacrilege, back to back with the novel's hero, Raphael Roodspere, a Cambridge Doctor of Divinity who dies a confessed heretic, having become a Lutheran 'Reformer' within the Church:

> 'Fire the pile who will – I cannot!' said Cromwell, throwing the lighted torch away. But Smiling Willie had no such scruples, and seizing the instrument, exclaimed, 'Let us burn the devil!' and set fire to the faggots beneath Sancgraal's feet. The pitch lighted in all directions, and in a moment a black cloud arose and enveloped both sufferers. A torrent of flames succeeded, rolling in an eddying and roaring whirl to the summit of the stake – and from the midst of the flames two voices came, uttering simultaneously, the words: 'God! destroy me utterly!' 'God! multiply thy mercies on all mankind, and LET THY KINGDOM COME!'[93]

This closing scene is a sensationalized version of the kind of narrative that Victorian readers recognized from John Foxe, whose account of John Lambert's trial in 1538 clearly formed the basis of Roodspere's in the penultimate chapter, where the strands of the novel's confession theme are drawn together.[94]

Westminster Abbey opens with a fine example of the kind of potted history that readers of popular railway novels – yesterday's airport novels – found acceptable: 'The doctrines of the German Reformers made but little progress in England until after the close of the first quarter of the sixteenth century.'[95] Unlike Ainsworth, who emphasized the need for toleration, Robinson is quick to show her party colours, when she comments in the third paragraph on the University of Cambridge in the year 1527: 'In this

[92] Emma Robinson (1814–90) was known as 'The Author of *Whitefriars*', a novel of 1845, set in the reign of Charles II. Little is known about her life. Her father disapproved of her writing and publicized the novels as being *not* by her. She was regarded as an eccentric in literary circles, wrote at least sixteen novels and some plays, never married, received a small Civil Service pension in 1862, and died insane in the London County Lunatic Asylum at Norwood.

[93] E. Robinson, *Westminster Abbey* (1859), p. 446.

[94] When Henry questions Roodspere on his 'hideous blasphemy in the matter of the Presence', the 'Reformer' says, 'I do deny the bread and wine to become the body and blood of Christ, in any but a spiritual sense!', to which the king replies, 'Then mark how thou art confuted even by Christ's own words, *Hoc est corpus meum!*' : ibid., p. 432. The closeness of this exchange to that recorded by Foxe would have been recognized by many Victorian readers: see Foxe, *Acts and Monuments* (1843–9), v, 230.

[95] E. Robinson, *Westminster Abbey* (1859), p. 1. George Routledge's 'yellowback' railway library of cheap single volume reprints included *Westminster Abbey* among around a thousand titles, the most popular of which were by Bulwer-Lytton. See J. Sutherland, *Longman Companion* (1988), p. 519.

university a theological party had gradually risen into notice, which excited the animadversion and alarm of the monks and bigots who formed by far the majority in it.'[96] Young Roodspere troubles Cranmer and Latimer by refusing to affirm his belief in transubstantiation. 'Only in spirituality and in essence', he says, can he 'admit of a superhuman presence in the cup of the Sacrament': 'There was a pause of dismayed silence among these founders of the Anglican Church, who were yet destined to place the doctrine thus timorously announced by the young apostle as the corner-stone in the great fabric of the Church they were to raise. It was in effect the denial of the Real Presence – the key of the arch of Protestantism!'[97] So Roodspere begins and ends the novel with a direct affirmation – or 'confession', in the sense that has always been acceptable to Protestants – of his rejection of the doctrine of transubstantiation. In between, this martyr in the making is associated with what Robinson clearly regarded as the light, purity and freedom of the Reformation.[98] What makes him interesting, however, is an inner struggle between an increasingly Lutheran head and a heart that cannot but be drawn to his natural father – Cardinal Wolsey.

His enemy, Sancgraal, is associated with the darkness, corruption and persecution which Robinson finds in unreformed Catholicism. At first sight, when a younger Sancgraal prepares to address the public before Marchant Hunne's corpse is burnt at the stake, he is presented as the physically repulsive other, embodying all that the innocent Roodspere, and the Victorian Protestant reader who empathizes with him, fear and detest: 'He was of a low stature, or else an habitual stoop in his figure produced the impression – lean and sinewy, and distinguished by a peculiar writhing movement in his gait, and the ominous addition of a club-foot. The popular imagination identified this deformity as an infallible characteristic of the Prince of Darkness himself.'[99] Sancgraal is an ambiguous figure, however, who exercises a mesmeric sexual power over women, like a stagey version of Milton's Satan.

[96] The future Queen Mary is also described as a bigot: 'The Princess Mary, a sallow, ill-looking girl of twelve years, whose expression already presaged her life of bigoted gloom and misery, walked beside her mother, crossing herself at frequent intervals from the moment when she caught sight of the church' (E. Robinson, *Westminster Abbey* (1859), p. 208).

[97] Ibid., p. 23. A Tractarian of the 1850s would not have agreed, for although the founders of Anglicanism were wary of the term 'real presence' because of its association with transubstantiation, they found circumlocutions for it, and Anglican tradition has in fact used the term widely since the time of Charles I.

[98] Ibid., pp. 26, 234, 286, 356, 435.

[99] Ibid., p. 36. Sancgraal later plays 'Sathanas' in one of the public spectacles for which he is responsible at the Abbey: ibid., p. 246.

Robinson uses the love plot as the vehicle for her religious themes, worked out through two different kinds of confession, as hero and villain vie for the hearts and souls of two women – the angelic novice, Lily-Virgin, who is loved by and later married to the hero, but is finally driven mad by the villain, and the cross-dressing precentor in the Abbey, 'Dan Gloria', who is in love with the hero, but is seduced and finally murdered by the villain. Roodspere's falling in love with the Lily-Virgin is presented as natural and liberating, and his marriage implicitly parallels that of Martin Luther to a nun. In stark contrast, Cardinal Wolsey's violent attempt to seduce Lily-Virgin, under the pretext of 'shriving' her in his oratory, is used to establish the novel's link between auricular confession and carnality. 'Save me – save me from worse than death!', cries Lily-Virgin, in true melodramatic style, before fainting in Roodspere's arms that are 'extended passionately and lovingly as its nest receives the throbbing fugitive back again', her clothes 'disordered' by the Cardinal.[100]

Four years after the Ultramontane Wiseman became Cardinal Archbishop of Westminster, looking like 'some Japanese god' in his full canonicals,[101] Robinson focused her readers' attention upon the Otherness of Cardinal Wolsey's oratory, where sexual abuse takes place behind its heavy, locked door:

> The chamber whence he had emerged rather resembled the cave of some eastern genie, blazing with a sunglow of jewels and gold, than the devote retirement of an ecclesiastic of the west . . . an altar there visibly was within, covered with ornaments of beaten gold, surrounded by blazing tapers that burned daylight; censers were still smoking with rich perfumes on its steps; magnificently-bound mass-books and other necessaries for the ceremonial worship – which, at all events, Wolsey never neglected – were profusely scattered about.[102]

Wolsey, magnificently dressed in the crimson of a cardinal, quails before Roodspere, his bastard son, an 'ill-clad, poverty-stricken student', whose beating against the door of the oratory hints at the impending dissolution of the monastery and the sweeping away of what the Reformers regarded as the idolatrous trappings of a corrupt, semi-pagan Church, in which the mass and the taking of virginities are sacriligiously conflated as 'sacrifices'.

Lily-Virgin, in Cromwell's eyes a 'snow-feathered pullet' among the Prioress of Clerkenwell's 'grey geese', is even more vulnerable when confessing to the scheming Sancgraal – one of the 'foxes' in the novel, reminiscent of *Punch* cartoons – in the crypt of the convent chapel.[103] Whereas Wolsey's

[100] Ibid., pp. 91–2, 97. [101] See p. 40 above.
[102] E. Robinson, *Westminster Abbey* (1859), p. 92. [103] Ibid., pp. 76, 85, 123, 223.

wanton kisses were rebuffed, Sancgraal's 'kiss of peace' on Lily-Virgin's brow is received with 'meek and tremulous humility', as an indignant Roodspere arrives in the 'gloomy chamber', too late to prevent the 'profanation'.[104] In the following chapter, entitled 'The Rival Confessors', Roodspere, who has officially replaced Sancgraal as her confessor, can make no progress in his attempt to persuade her that there is 'no salvation, no redemption, to be hoped from human works, or prayers, or sacrifice', as Sancgraal has skilfully turned her against him.[105]

In a novel that looks back to the Gothic sublime of Ann Radcliffe and forward to the sensationalism of Wilkie Collins, Sancgraal hears confessions in marginal locations, separated from the busy life of monastery and court, at extremes of depth and height. Driven out of the gloomy convent crypt, he retires to the dizzying height of his own lair under the roof of the cathedral, the 'Hermitage of St Wulfin', a vast loft in which instruments of mortification are prominently displayed, and from which, Quasimodo-like, he spies on all that passes far below.[106] Even Queen 'Katharine' (Catherine of Aragon) and the Duke of Norfolk are in Sancgraal's power as a confessor: 'The queen herself was said to have paid this lofty confessional a reverential visit; and here, at the moment we proceed in our record, no less a personage than that great duke of Norfolk who won Flodden Field, was kneeling in abject superstition, in confession, at the prior's feet!'[107] It is through Norfolk's confession that Sancgraal learns of Wolsey's plot to obtain a royal divorce. In contrast, Roodspere refuses to hear the confession of Dan Gloria, the precentor, on the grounds that 'Paternosters, Ave Marias, candles lighted to St Thomas and St Edward' are of no avail: 'The word of the Christ that pardoned, was not – go on a pilgrimage – whip yourself till the blood flows – give treasures to this shrine, or to that – but GO, AND SIN NO MORE!'[108] The colour drains from Dan Gloria's face, as Roodspere has discovered that she is in fact a woman. She was the wife of Marchant Hunne, who betrayed him to the monks and was seduced by Sancgraal, her confessor, with the promise of 'remission and pardon' for the 'sin passion sanctified'.[109]

The turning-point in the novel is Roodspere's extempore sermon in front of King Henry and his court. Whereas Sancgraal relishes officiating in the Cardinal's place at the service, and 'raising the gorgeous mass of plate from which he derived his name', Roodspere tears up his carefully prepared and self-protecting script, and chooses a startling text: 'Why seek ye the living

104 Ibid., p. 153. 105 Ibid., p. 158.
106 Victor Hugo's *Notre Dame de Paris* was published in 1831.
107 E. Robinson, *Westminster Abbey* (1859), pp. 169, 184.
108 Ibid., p. 178. 109 Ibid., p. 194.

among the dead?' (Luke 24.5). 'Nature' makes him eloquent, and he adopts 'a similitude of the decadence of the true Gospel Christianity to the betrayal, death, and burial of Christ himself, likening in every imaginable form the destruction effected in the practice and faith of the Church founded by that divine original – especially all that the monks had done to deprive it – to the death-swathings, spices, and massive stone used in the sepulchring of the crucified Saviour! He compares the new learning to the break of day over the tomb.'[110] He is now a marked man as a 'Reformer', and it is simply a matter of time before he becomes a martyr.

Sancgraal makes Roodspere's vivid analogy grotesquely literal when he drugs Dan Gloria and places her in the tomb of Edward the Confessor, the most sacred site in the cathedral, where she starves to death, having first gnawed away some of her own flesh. A 'dark ooze' later trickles from the middle recess of the tomb.[111] In contrast, the 'break of day' of the Reformation is marked by Roodspere's renunciation of the confessional and Anne Boleyn's smiling agreement to 'confess to one another . . . after old Christian fashion'.[112] Roodspere's subsequent trial, in the presence of the king, turns upon different kinds of confession.[113] Roodspere's is the most authoritative voice at the trial, as he confesses what the court regards as 'heresy': 'I have confessed, without blenching, what consigns men, in the least article thereof, to the flames.' He has thus both answered Wolsey's demand that he make full confession of his 'errors and offences', in the legal sense, and has made a testimony of faith. His denunciation of Sancgraal – 'this man – this priest – this confessor – rather this devilish incubus!' – links the monk's crimes with the third meaning of 'confession': the corrupted practice associated with the sacrament of penance. He explains to the court that the Lily-Virgin, his wife, has been 'maddened' by the 'superstitions' of the confessor, Sancgraal, and of Dame Barbara at the convent. Her death, smiling, in her husband's arms, is presented as 'love triumphant over death!'

Even in the final chapter, 'The Martyr', Robinson sustains her attack upon auricular confession.[114] When the hero and the villain are imprisoned together in the vertiginous Hermitage, the night before their execution, Roodspere manages to bring Sancgraal to repentance, and together they take the reformed sacrament 'of penance and conciliation by a spiritual instead of a physical oblation'. In a last desperate bid to avoid the horror of the flames that will soon 'drink' their 'living blood', however, Sancgraal appears to have a change of heart, and asks for confession and absolution

110 Ibid., p. 235. Compare Milman on Jesus as the 'primal original' of Christianity, p. 59 above.
111 Ibid., pp. 349, 369. 112 Ibid., p. 281. 113 Ibid., pp. 424–32. 114 Ibid., pp. 437–41.

from Father Giselbert, in a corner of the Hermitage. This is merely a ruse, however, to allow him to take poison, and it is thwarted. Struggling against his physical and spiritual extinction to the last, Sancgraal provides a ghastly contrast to the 'Reformer', as Roodspere is now called, who, in the style of Foxe's martyrs, refuses Cromwell's final invitation to recant before the faggots are lit: 'Nay, sir! for even now I behold the gates of paradise open to receive me, and one beckon me who I know will gladly welcome me! – I recommend my spirit to Him who gave it, and desire no further discourse with any in the flesh!'[115] He is already with Christ in glory, as Sancgraal is already in eternal torment.

The subject of auricular confession is doubly attractive to a novelist like Robinson. First, dialogue between penitent and priest in an intimate setting provides her with opportunities to relate the inner life of characters to the outer world. Secondly, confession and its abuse play a metonymic role in the novel, exemplifying the kind of intrigue in which both the Church authorities and the court of Henry VIII are engaged in the historical plot. King Henry presented nineteenth-century Protestant historians of the English Reformation with something of a challenge, in that, in their view, he did the right thing for the wrong reason – rejecting the authority of Rome, but doing so in order to divorce Catherine of Aragon. It was also difficult to present Anne Boleyn as a Protestant martyr. As a novelist, however, Robinson can distract the reader's attention from these historical problems by focusing mainly upon the corruption in the Church, which the novel's idealized Reformers seek to purge.

Charles Kingsley needed no such sleight of hand in *Westward Ho!*, published the following year, as there was an easily identifiable and widely acceptable association to be made in 1855 between a patriotic call to arms during the Crimean War and a positive treatment of England's glorious battles with Catholic Spain in the days of 'good Queen Bess'. Whereas *Westminster Abbey* is as claustrophobic as a confessional, *Westward Ho!* is a bracing, outdoor novel of imperial chivalry, which takes the reader from the quintessentially English countryside of rural Devon to the tropical islands of the Spanish Main. The boyhood books that the blinded hero, Amyas Leigh, finds in their old place on his return to Devon, at the end of the novel, are '"King Arthur," and "Foxe's Martyrs," and "The Cruelties of the Spaniards"' (33).

In contrast to the sensitive scholar, Roodspere, Kingsley's hero is a giant of a man, built to fight for queen and country, and trained in 'manhood,

[115] Ibid., p. 446.

virtue and godliness' (33). Before embarking on the voyage that dominates the second half of the novel, Amyas dines in Deptford Creek on board the *Pelican*, the ship in which Sir Francis Drake sailed round the world. He is surrounded by the adventurers of Elizabethan history, including Sir Walter Raleigh, Anthony Jenkinson, Christopher Carlile, Martin Frobisher, William Davis, John Winter and Sir Philip Sidney, together with their financial backers from the City of London (16). These are 'England's Forgotten Worthies', as James Anthony Froude called them, in an article published in the *Westminster Review* in 1852, and the inspiration for *Westward Ho!* (1). Shortly after the 'Papal Aggression' crisis, the time was ripe to draw out the heroic qualities of the Virgin Queen, Elizabeth, who, like Margaret Thatcher during the Falklands War, was represented by her supporters as the defender of manly John Bull qualities against the encroachments of dubious Catholic foreigners.[116]

Froude, himself a west countryman, regards the fight with the Spanish as a holy war with material benefits.[117] He writes of Elizabeth: 'She was able to paralyze the dying efforts with which, if a Stuart had been on the throne, the representatives of an effete system might have made the struggle a deadly one.'[118] Kingsley, too, denigrates the Stuarts – tainted with the effeminacy that he associates with Roman Catholicism – and praises Elizabeth as the defender of Protestant 'liberty' (7, 20, 29).[119] The 'manliness' of Protestantism was a commonplace of nineteenth-century Protestant historiography.[120] Kingsley's attacks upon Catholicism, however, and particularly his abhorrence of the rule of celibacy for priests and religious, reflect in their hysterical excess a complex psycho-sexual nature.[121] As we have seen, it has been argued that, in wooing his future wife, he was battling not only

[116] Compare eighteenth-century anti-Catholicism: 'the image of the cruel, dangerous Papist was projected, not on to Catholic neighbours, but on to faceless Catholics, Catholics abroad or at a distance'. C. Haydon, *Anti-Catholicism* (1993), p. 11.

[117] James Anthony Froude, 'England's Forgotten Worthies' (1852), in *Essays* (1906), p. 37. Froude goes on to suggest that 'these plain, massive tales' would be 'the most blessed antidote' to Ainsworth.

[118] Ibid., p. 43. Some editions have 'resolution' for 'inflexible will', 'appearance' for 'thunder birth', 'light' for 'flash', and 'shines' for 'roll and glitter'.

[119] On 26 January 1851, he wrote of the 'Puseyites', 'there is an element of foppery – even in dress and manner; a fastidious, maundering, die-away effeminacy', adding that 'the perfection of this aristocratic, exclusive method of soul-saving is only to be found in Romish Jesuitism': F. Kingsley, *Charles Kingsley* (1878), I, 249–50.

[120] Consider, for example, these closing words by Henry Soames, Rector of Shelley, in Essex: 'The masculine intellect of England is above a theatric worship, and superstitious toys; auricular confession, and sacramental absolution; mediators, whom neither Scripture warrants, nor reason says, can hear; a belief that Holy Writ is maimed of information, vital to the soul.' *Elizabethan Religious History* (1839), p. 594. Also cf. the final sentence in his *History of the Reformation* (1826–8), IV, 740.

[121] See p. 68 above.

against her Anglo-Catholicism, but also his own suppressed homosexuality.[122] Interestingly, however, Kingsley was clearly aware of the violence of his feelings against Catholicism, and adopted a number of defensive strategies in the novel when dealing with the subject.

Some of the more outrageous anti-Catholic statements, for example, are put into the mouths of his characters, such as Salvation Yeo, an Anabaptist seaman who bears the scars of the rack from the Inquisition, and who uses biblical texts like weapons of war (14). In the opening chapter, Captain Oxenham states that the Spanish 'pray to a woman, the idolatrous rascals! and no wonder they fight like women' (1). Amyas Leigh's brother, Frank, refers to 'Popish priests and friars, who are vowed not to be men' (18), and Lucy Passmore, a white witch and a 'virago', humiliates Father Campion – the most revered martyr of English Catholic tradition – by knocking his 'thin legs' from under him and demanding a higher price for the loan of a boat (6).

Like Emma Robinson, Kingsley pits his hero against a Catholic villain, Don Guzman Maria Magdalena de Soto, who first appears as Amyas's prisoner in Ireland, on Christmas Day 1580, when the English forces, including Sir Philip Sidney and Sir Walter Raleigh, are fighting a combined force of Spanish, Italian and Irish soldiers. Unlike the deformed Sancgraal, however, Don Guzman is an elegant and attractive figure, and Amyas's first reaction to him is similar to the mixture of conscious repulsion and unconscious attraction that characterized the response of many Victorian Protestants to Roman Catholicism:

> He was an exceedingly tall and graceful personage, of that *sangre azul* which marked high Visi-gothic descent; golden-haired and fair-skinned, with hands as small and white as a woman's; his lips were delicate, but thin, and compressed closely at the corners of the mouth; and his pale blue eyes had a glassy dulness. In spite of his beauty and his carriage, Amyas shrank from him instinctively . . . (9)

Amyas is thrown off balance by the Don's 'unexpected assurance and cool flattery', but the Spaniard's 'quiet mask' conceals 'terrible depths of fury and hatred'. Commenting on the terrible massacre that follows, the authorial narrator writes: 'The hint was severe, but it was sufficient. Many years passed before a Spaniard set foot again in Ireland.'

As in *Westminster Abbey*, several characters in *Westward Ho!*, including both Amyas and Frank Leigh, and their Catholic cousin, Eustace, are in love with the heroine, Rose Salterne, 'a thorough specimen of a west-coast

[122] See also K. L. Morris, 'John Bull and the Scarlet Woman', in *Recusant History*, (1996–7), p. 192.

maiden, full of passionate, impulsive affections, and wild, dreamy imaginations' (4). When Don Guzman woos this English Rose, Kingsley presents him as an Othello – the black Other – to Rose's Desdemona (10). Earlier, the 'chivalrous brotherhood of the Rose' toasted both 'Queen and Bible' and Rose Salterne – 'The Rose of Torridge' (8) – and when we finally set off, 'westward ho!', both to defeat the Spanish and liberate the heroine, now Dona Guzman at La Guayra in Caracas, where he is Governor, this 'Most Chivalrous Adventure' is in 'the Good Ship "Rose"' (16). The Revd John Brimblecombe has fiercely preached a 'crusade against the Spaniards' and has whipped up support so energetically that he is called 'a second Peter the Hermit' (16).

When they finally arrive at La Guayra, Amyas and Frank discover that their cousin is there before them, and is trying to convert Rose to Catholicism. For whereas her noble husband, Don Guzman, has in fact promised to 'respect and protect' her religion, Eustace insults her by talking of his love for her and attempting to complete her conversion 'to the bosom of that Church where a Virgin Mother stands stretching forth soft arms to embrace her wandering daughter, and cries to you all day long, "Come unto me, ye that are weary and heavy laden, and I will give you rest!"' (19). While somewhat crudely denigrating Catholicism by making Eustace put the words of Christ (Matthew 11.28) into the mouth of his mother, he becomes defensive when describing Rose remaining true to her Devonian and Protestant birthright, even under torture. At the beginning of chapter 22, 'The Inquisition in the Indies', it is as if he is answering Pugin's criticism of those who 'recapitulate horrid acts perpetrated in the name of religion': 'My next chapter is perhaps too sad; it shall be at least as short as I can make it; but it was needful to be written, that readers may judge fairly for themselves what sort of enemies the English nation had to face in those stern days.'[123] Eustace Leigh visits his tortured cousin, Frank, who is imprisoned with Rose and whose heart is 'still tender from the torture', as the Jesuits like to put it. He hears Rose's screams echo through the 'dreadful vaults' and hurries away to complete his novitiate and enter the Jesuit order.

There is then an extraordinary narrative turn of the kind that one might have expected in John Fowles, rather than in the Victorian fiction that he rewrites. Kingsley 'kills' the character of Eustace, not by constructing some suitable plotting that will finish him off, but by declaring him to be 'dead' in human terms and expunging him from the record:

[123] For Pugin's comment, see p. 78 above.

Eustace Leigh vanishes henceforth from these pages. He may have ended as General of his Order. He may have worn out his years in some tropic forest, 'conquering the souls' (including, of course, the bodies) of Indians; he may have gone back to his old work in England, and been the very Ballard who was hanged and quartered three years afterwards for his share in Babington's villainous conspiracy: I know not. This book is a history of men; of men's virtues and sins, victories and defeats: and Eustace is a man no longer; he is become a thing, a tool, a Jesuit; which goes only where it is sent, and does good or evil indifferently as it is bid; which, by an act of moral suicide, has lost its soul, in the hope of saving it; without a will, a conscience, a responsibility (as it fancies) to God or man, but only to 'The Society'. In a word, Eustace, as he says of himself is 'dead'. Twice dead, I fear. Let the dead bury their dead. We have no more concern with Eustace Leigh. (22)

Kingsley, an ordained clergyman of the Church of England, and soon to be Chaplain to the Queen, acts like his own distorted version of the Jesuits he attacks, turning a man into a thing: the Other has become an object. The deaths of Frank and Rose, on the other hand, are represented as Protestant martyrdoms. Lucy Passmore reports that they were burnt at the same stake, 'both very bold and steadfast', hand in hand: 'they did not feel it more than twenty minutes' (26). We are not invited to reflect that, from the viewpoint of the Spanish Inquisition, it is Frank and Rose who represent the Other, as impenitent heretics, along with 'a renegade Jew, and a negro who had been convicted of practising "Obi"', who are burnt with them, at another stake. No details are given of the Jew's and the negro's deaths, as they are also the Other to English Protestantism, and can be erased, like Eustace Leigh. We are close, here, to the sources of racism and religious hatred in every generation and in every culture.

Amyas Leigh must take his revenge, however, and the Armada offers him an opportunity to meet Don Guzman in battle once again. *Westward Ho!* is famous for the set-piece in chapter 31, 'The Great Armada': 'And now began that great sea-fight, which was to determine whether Popery and despotism, or Protestantism and freedom, were the law which God had appointed for the half of Europe, and the whole of future America.' On land, 'many a brave man', as he kneels beside his wife and daughters, feels his 'heart sink to the very pavement, at the thought of what those beloved ones might be enduring a few shorts days hence, from a profligate and fanatical soldiery, or from the more deliberate fiendishness of the Inquisition'. The Spanish invader is 'Antichrist himself'. When victory at sea comes, against all the odds, Amyas Leigh – now 'Sir Amyas' – pursues the Armada for sixteen days, right around Scotland and down to St David's Head, thirsting for revenge against Don Guzman. Thwarted when the Spanish galleon is smashed against rocks during a storm, Amyas throws his sword into the sea and is blinded by lightning (32). It is perhaps a penitent Kingsley, having enjoyed himself far too much, describing bloody battle scenes and 'Papist' horrors, who in 1855 speaks through Amyas Leigh when he repents of 'hating even the worst' of the Spaniards: 'But God has shown me my sin, and we have made up our quarrel for ever.' It was to take English Protestantism another hundred years to do the same.

CHAPTER 4

Jacobite claims and London mobs

> To surround anything, however monstrous or ridiculous, with an air of mystery, is to invest it with a secret charm, and power of attraction which to the crowd is irresistible. False priests, false prophets, false doctors, false patriots, false prodigies of every kind, veiling their proceedings in mystery, have always addressed themselves at an immense advantage to the popular credulity.
>
> Charles Dickens, *Barnaby Rudge*, 1841

I

The opening of chapter 37 in *Barnaby Rudge* reads like an introduction to a critique of Roman Catholicism, of a kind that was familiar enough in 1841. The impact of what follows – a devastating attack upon Lord George Gordon's anti-Catholic Protestant Association in 1780 – is thus all the greater. Yet Dickens had 'no sympathy with the Romish Church', as he reminded his readers in the preface. He would have been unimpressed by the spectacle presented to Londoners seven years later, on 4 July 1848, when St George's Catholic church – later cathedral – was dedicated, in the area of Southwark where Gordon had marshalled tens of thousands of anti-Catholic demonstrators on 2 June 1780, before marching to the Houses of Parliament. The official guide to the cathedral describes the 'long, stately, and heart-cheering procession, the like of which had not been seen in England for three centuries':

> Thurifer, cross-bearer, and acolytes, choir-boys and choristers, two-and-two. The secular clergy, about 230, two abreast. The religious orders in their respective costumes. The Institute of Charity, the Redemptorists, the Passionists, the Jesuits, the Franciscans, the Dominicans, the Cistercians, the Benedictines, &c. The Foreign Clergy and Canons. Then followed the Bishops, each attended by his chaplain and trainbearer, and vested in cope and mitre. The Right Revs., Bishops Davis, of Maitland, Morris, Gillis, Ullathorne, Sharples, Brown of Wales, Wareing, Brown

of Lancashire, Briggs. The Bishops of Elphin, Tournai, Liege, Treves and Luxembourg. . . . Mitre and Crosier Bearers. The Deacon and Subdeacon of the Mass. The Assistant Priest. The Bishop's Deacons. The Right Rev. Nicholas Wiseman, D. D., Pro. V. A. of the London District, (now Cardinal Archbishop of Westminster.) Trainbearers, &c., &c.[1]

'On the very spot where Lord George Gordon preached his wild crusade against Catholicity', the author exclaims in 1851, 'St George's Cathedral now stands!'[2]

This kind of Catholic triumphalism infuriated some of the more vocal Protestant observers during the 'Papal Aggression' crisis, when Wiseman preached exultantly at St George's, wearing his full regalia.[3] For their part, Protestants tended towards smugness rather than triumphalism when they compared their fellow countrymen's response to the crisis with that of the London mob in 1780. The Congregationalist, Henry Rogers, wrote in 1851:

We rejoice at another result of the recent movements. They afford palpable proof of the real progress which the mass of the people have made since the Lord George Gordon riots. That a great nation so deeply stirred throughout its length and breadth, should have spoken so decidedly and acted so moderately; should have uttered such vehement convictions, and yet maintained so much self-control, is a phenomenon equally novel and gratifying.[4]

In the 1890s, Lionel Johnson, a Catholic convert, was to describe the Gordon Riots as the 'last great persecution of the Catholic Church in England'.[5] From an ultra-Protestant perspective, they could be represented as the unfortunate result of a strong reaction to a Catholic threat, following the relaxation of laws designed to marginalize Catholics. In this chapter I will be discussing nineteenth-century Protestant and Catholic representations of both the Gordon Riots and an earlier threat to Protestant

[1] Anon., *Complete Description* (1851), pp. 7–8. [2] Ibid., pp. 5–6.

[3] See p. 44 above. Browning refers to St George's in the opening lines of 'Bishop Blougram's Apology' (1855):

No more wine? then we'll push back chairs and talk.
A final glass for me, though: cool, i'faith!
We ought to have our Abbey back, you see.
It's different, preaching in basilicas,
And doing duty in some masterpiece
Like this of brother Pugin's, bless his heart!
I doubt if they're half baked, those chalk rosettes,
Ciphers and stucco-twiddlings everywhere;
It's just like breathing in a lime-kiln: eh?

R. Browning, *Poems* (1981), I, 617.

[4] H. Rogers, 'Ultramontane Doubts', *Edinburgh Review*, April 1851, p. 573.

[5] L. Johnson, *Gordon Riots* (1893), p. I.

England's constitution – the Jacobite Rebellion of 1745–6. Both episodes were regarded by Catholic writers as examples of Protestant persecution, and by Protestants as examples of what Dickens called the 'veiling' of proceedings in 'mystery' – for Newman, two *positive* terms[6] – which appealed to 'popular credulity' in their day.

II

Following the 'Glorious Revolution' of 1688 and the exile in France of the Catholic James II, the most serious attempt to restore a Stuart to the English throne was the Jacobite rising of 1715. The ensuing punishments for treason revolted English public opinion, seeming like a throw-back to a less civilized age: two peers were beheaded in London and there were thirty-four hangings, drawings and quarterings in Lancashire, whereas the Scottish rebels escaped with their lives.[7] Looking back to the 'Fifteen' from the period of the Gordon Riots, the Catholic historian, Joseph Berington, made this significant observation: 'The Tories were, in principle at least, friends to Jacobitism, and so were the Papists; they should not therefore, it seemed, be great enemies to each other. From this time, and for many years to come, the words *Jacobite* and *Papist* remained inseparably united.'[8] In Berington's view, the identification of Jacobitism with Catholicism had serious repercussions thirty years later, at the time of the 'Forty-five', when

> a fresh impression was again made, which called up the former animosity of the nation, and it was said by many, that Papists would never peacably submit to a Protestant government. This was an ill-natured charge: For very few Catholics, I have observed, were engaged in the rebellion; and if the body must suffer for the follies of these few, surely the same should be the fate of Protestants.[9]

Catholics, he added, 'are now as sincere in their attachment to the Hanover family, as they ever were to the Stuarts'.[10]

Protestant accusations that Catholics would 'never peacably submit to a Protestant government' were often repeated in the nineteenth century, especially at times of crisis. It also suited anti-Catholic Protestants to perpetuate the myth that Jacobitism was synonymous with Catholicism. First, Catholicism could be associated with a misty Highlands romanticism which defied the steady march of progress, whereas the Protestantism that had defined

[6] See p. 65 above. [7] D. Daiches, *Charles Edward Stuart* (1973), p. 68.
[8] J. Berington, *State and Behaviour of English Catholics* (1781), p. 90. (Cf. J. Wolffe, *Protestant Crusade* (1991), pp. 9–10.)
[9] Ibid., p. 96. [10] Ibid., p. 133.

England's national identity after 1688 represented a cultural turn from mysticism and superstition to reason and common sense.[11] Secondly, after the accession of George III in 1760, Jacobitism was dead, which meant that the romantic elements of the 'Forty-five' could be relished by a Protestant culture without the risk of glamorizing a living tradition which threatened the state.[12] Thus, in the nineteenth century, Bonnie Prince Charlie took his place alongside Lady Jane Grey, Mary Queen of Scots, King Alfred, Harold, the Princes in the Tower, Charles I and Oliver Cromwell as one of the 'martyr' figures who were depicted in numerous English history paintings from the 1830s to the 1890s.[13]

A characteristically English admiration for the romantic failure, the figure who remains true to his or her beliefs in bravely losing a noble cause, is a feature of the sixteen-year-old Jane Austen's 'History of England', dated Saturday 26 November 1791. Having set out to 'prove the innocence of the Queen of Scotland' and to 'abuse Elizabeth', this 'partial, prejudiced, & ignorant Historian' ends her manuscript by explaining that she does not intend to give a particular account of the 'distresses' into which Charles I was 'involved through the misconduct & Cruelty of his Parliament':

> I shall satisfy myself with vindicating him from the Reproach of arbitrary & tyrannical Government with which he has often been charged. This, I feel, is not difficult to be done, for with one argument I am certain of satisfying every sensible & well disposed person whose opinions have been properly guided by a good Education – & this Argument is that he was a STUART.[14]

The mature Jane Austen's reputation as a novelist was boosted twenty-five years later, when Walter Scott reviewed *Emma* in the *Quarterly Review*, two years after his own first novel had been one of the publishing sensations of the age. A key to the success of *Waverley; or, 'Tis Sixty Years Since* (1814) is the close parallel that Scott creates between the young hero's inner conflict – English patriotism warring with the appeal of the Stuart cause, sense with sensibility – and the outward and visible struggles between the

[11] See R. D. Tumbleson, *Catholicism in the English Protestant Imagination* (1998), pp. 198–205. By the turn of the nineteenth century, John Sturges, Canon of Winchester, could write of Catholicism: 'All this combination of Enthusiasm and Superstition proceeds on the supposition, that to renounce the world is better, than to live well and usefully in it.' *Reflections on the Principles and Institutions of Popery* (1799), p. 32.

[12] 'It was not till the auspicious commencement of his present Majesty's reign, that these alarms from the pretensions of the Stewart Family, together with the fears of Popery adherent to them, subsided; and that Jacobitism, as a formidable party, expired.' J. Sturges, *Reflections* (1799), p. 93.

[13] See R. Strong *And when did you last See your Father?* (1978), pp. 40–5, 154, 168. Paul Delaroche's first scene from British history was Flora Macdonald succouring the Young Pretender (c. 1825, unlocated), exhibited at the Paris Salon in 1827.

[14] J. Austen, *History of England* (1993), n.p.

English redcoats and the Jacobites. In 'Chapter Fortieth', Edward Waverley commits himself to Bonnie Prince Charlie in a moment of passion which dissolves all sense of conflict:

To be thus personally solicited for assistance by a Prince, whose form and manners, as well as the spirit which he displayed in this singular enterprise, answered his ideas of a hero of romance; to be courted by him in the ancient halls of his paternal palace, recovered by the sword which he was already bending towards other conquests, gave Edward, in his eyes, the dignity and importance which he had ceased to consider as his attributes. Rejected, slandered, and threatened upon the one side, he was irresistibly attracted to the cause which the prejudices of education, and the political principles of his family, had already recommended as the most just. These thoughts rushed through his mind like a torrent, sweeping before them every consideration of an opposite tendency, – the time, besides, admitted of no deliberation, – and Waverley, kneeling to Charles Edward, devoted his heart and sword to the vindication of his rights!

Seven chapters later, it is the passion of the clan of Ivor that throws the orderly English battle lines into disarray at Prestonpans:

The English infantry, trained in the wars in Flanders, stood their ground with great courage. But their extended files were pierced and broken in many places by the close masses of the clans; and in the personal struggle which ensued, the nature of the Highlanders' weapons, and their extraordinary fierceness and activity, gave them a decided superiority over those who had been accustomed to trust much to their array and discipline, and felt that the one was broken and the other useless. (47)

Scott relates the depth of the Jacobites' passion to the height of their principles. In the final chapter, he presents the reader with a 'Postscript, which should have been a Preface', in which he reflects upon the rapid change in Scotland after the 'Forty-five':

There is no European nation which, within the course of half a century, or little more, has undergone so complete a change as this kingdom of Scotland. The effects of the insurrection of 1745 – the destruction of the patriarchal power of the Highland chiefs – the abolition of the heritable jurisdiction of the Lowland nobility and barons – the total eradication of the Jacobite party, which, averse to intermingle with the English, or adopt their customs, long continued to pride themselves upon maintaining ancient Scottish manners and customs – commenced this innovation. The gradual influx of wealth, and extension of commerce, have since united to render the present people of Scotland a class of beings as different from their grandfathers as the existing English are from those of Queen Elizabeth's time. (72)

He describes this change as 'steadily and rapidly progressive', but also 'gradual', and therefore difficult to detect at the time.

Scott celebrates progressive gradualism, implicitly contrasted with the violence and disruption witnessed in another 'European nation', revolutionary France, during the same period. He also recognizes, however, that something of value has been lost with the demise of Jacobitism, for all its 'absurd political prejudice': 'singular and disinterested attachment to the principles of loyalty which they received from their fathers, and of old Scottish faith, hospitality, worth, and honour'. This commentary on the recent history of Scotland also elucidates his own plot, and especially his hero's response to Fergus Mac-Ivor.

In 'Chapter Fifth' the authorial narrator apologizes for 'plaguing' the reader so long with 'old-fashioned politics, and Whig and Tory, and Hanoverians and Jacobites', but explains that this background is needed to make his story intelligible. He is not writing a romance for his 'fair readers' – 'a flying chariot drawn by hippogriffs, or moved by enchantment'. His is 'an humble English post-chaise, drawn upon four wheels, and keeping his Majesty's highway'. In the light of this realist, law-abiding and royalist agenda, young Waverley's defection to the Jacobite Highlanders led by Mac-Ivor can be read as a romantic diversion, a primrose path leading far from 'his Majesty's highway' and followed in the spirit of youthful idealism which flies in the face of historic inevitability. Scott lets his hero down gently, however. He also handles Mac-Ivor's trial and execution at the end of the novel with the tact of the moderate in matters of politics and religion.

Scott's successor, William Harrison Ainsworth, puts the reader right in front of the action at the beginning of *Guy Fawkes* (1840–1), when two seminary priests are brought to the scaffold to be hanged, drawn and quartered, being guilty of treason by practising as Catholic priests in England.[15] Mac-Ivor is to suffer a similar fate, as he too is guilty of treason, being a rebel leader. He is also a Catholic. Dawn breaks on the day of his execution to reveal a sleepless Waverly pacing up and down the 'esplanade in front of the old Gothic gate of Carlisle Castle' (69). Then the 'grating of the large old-fashioned bars and bolts, withdrawn for the purpose of admitting Edward, was answered by the clash of chains, as the unfortunate Chieftain, strongly and heavily fettered, shuffled along the stone floor of his prison to fling himself into his friend's arms'. Waverley leaves Mac-Ivor and Maccombrich, who is to suffer with him, for an hour, to allow their

[15] See p. 92 above.

confessor to administer 'the last rites of religion, in the mode which the Church of Rome prescribes'.

During the conversation that follows, Mac-Ivor says to Waverley:

'Nature has her tortures as well as art; and how happy should we think the man who escapes from the throes of a mortal and painful disorder, in the space of a short half hour? And this matter, spin it out as they will, cannot last longer. But what a dying man can suffer firmly, may kill a living friend to look upon. – This same law of high treason,' he continued, with astonishing firmness and composure, 'is one of the blessings, Edward, with which your free country has accommodated poor old Scotland: her own jurisprudence, as I have heard, was much milder. But I suppose one day or other – when there are no longer any wild Highlanders to benefit from its tender mercies – they will blot it from their records, as levelling them to a nation of cannibals.'

The power of Mac-Ivor's attack upon the English penal code lies in its inversion of the moral and political categories upon which both anti-Jacobitism and anti-Catholicism were based in the eighteenth century, as he challenges the assumption that England is civilized and advanced, whereas the Scottish Highlands are outlandish and backward.

The reader, like Edward Waverley, the vulnerable 'living friend', is spared the butchery. The horse-drawn sledge, or hurdle, on which the condemned are to be conveyed to the scaffold a 'mile distant from Carlisle', is painted black, and at one end of the vehicle sits 'the Executioner, a horrid-looking fellow, as beseemed his trade, with the broad axe in his hand'. The military procession sets off on its ritualized journey, leaving a distraught Waverley alone in the 'totally empty' court-yard. He hurries back to his inn where, about an hour and a half later, he hears the 'sound of drums and fifes, performing a lively air, and the confused murmur of the crowd which now filled the streets, so lately deserted', which tells him that all is over. Like Dickens, Scott had no sympathy for the Roman Catholic Church.[16] The Catholic Highlander, Mac-Ivor, however, epitomizes not only the 'singular and disinterested attachment to the principles of loyalty' which he received from his fathers, but also 'old Scottish faith, hospitality, worth, and honour'.

Robert Chambers took up the theme of Jacobite principles in his own historical account of the 'Forty-five':

In considering the various merits of the parties, it must be allowed, that, whatever the demerits of the Jacobites, they were personally disinterested – whatever the merits of the Whigs, they were ungenerous and self-seeking. The temperament

[16] When Catholic Emancipation legislation was being debated in 1829, Scott wrote in his diary: 'I hold Popery to be such a mean and depraving superstition, that I am not sure I could have found myself liberal enough for voting the repeal of the penal laws as they existed before 1780.' J. G. Lockhart, *Life of Scott* (1902–3), IX, 270.

of mind required for the formation of a Jacobite, seems to be that inconsiderate and poetical sort, which finds gratification in the joy of others, and is disposed to forego all earthly good for the sake of a visionary idol. The Whig, on the other hand, appears to have been characterised only by that vulgar good sense which keeps shops and makes money.[17]

Scott took a similar line in an Introduction that he added to *Redgauntlet: A Tale of the Eighteenth Century* in 1832:

Most Scottish readers who can count the number of sixty years, must recollect many respected acquaintances of their youth, who, as the established phrase gently worded it, had been *out in the Forty-five*. It may be said, that their political principles and plans no longer either gained proselytes or attracted terror, – those who held them had ceased to be the subjects either of fear or opposition. Jacobites were looked upon in society as men who had proved their sincerity by sacrificing their interest to their principles.

Yet in *Redgauntlet*, which received a lukewarm reception when it was published in 1824, Scott establishes an imaginary scenario in which an older and weaker Charles Edward Stuart returns to Scotland in the fond hope of reviving his cause, which, like the Pretender himself, now seems pathetic and irrelevant.

The novel's central characters, Alan Fairford and Darsie Latimer (whose father's head was displayed on a gate in Carlisle in 1746), represent not only different positions in relation to Jacobitism, but also two sides of Scott's own personality.[18] (In this, as in other ways, *Redgauntlet* anticipates Stevenson's masterpiece, *Kidnapped* (1886).) Scott is careful, however, to put the strong anti-Catholic sentiments that are expressed in the novel in the mouths of a wide range of minor figures – the characters who have received most critical acclaim. Joshua Geddes, the Quaker, speaks of 'the blinded times of Papistry' before the Reformation (letter 7), while Alan's barrister father, Alexander Fairford, warns Darsie of the danger of mixing with both 'Papists and Quakers' (letter 9). Thomas Trumbull, the 'rigid old Covenanter' and 'smuggler's ally', whose pornographic book is disguised as a Psalter, comments that the 'whore that sitteth on the seven hills ceaseth not yet to pour forth the cup of her abomination on these parts', and claims not to 'consort with Jacobites and mass-mongers' (ch. 13). Nanty Ewart talks of 'your jacobitical, old-fashioned Popish riff-raff' (ch. 15).

The association of Jacobitism with the 'Papists' is brought out most strongly in Darsie's journal, where he describes the activities of his fanatical

[17] R. Chambers, *History of the Rebellion* (1827), II, 283.
[18] E. Johnson, *Sir Walter Scott* (1970), p. 921.

uncle, 'Mr Herries', the Laird of Redgauntlet, who refuses to give up the good old cause. Darsie records that there are still 'partisans of the Stewart family' who, 'furnished with gold from Rome, moved, secretly and in disguise, through the various classes of society, and endeavoured to keep alive the expiring zeal of their party' (ch. 9). The Laird's critique of the English Protestant concept of 'liberty' reflects both his hidebound attachment to Scotland's past and his Catholicism: 'The privilege of free action belongs to no mortal – we are tied down by the fetters of duty – our moral path is limited by the regulation of honour – our most indifferent actions are but meshes of the web of destiny by which we are all surrounded.'

Any residual sense of mystery attached to Catholicism in the novel is, however, diffused by facetiousness, and the faded glamour of Jacobitism finally descends into bathos. Fairladies, the Catholic stronghold where Charles Edward is staying, disguised as 'Father Buonoventure', is described by Nanty Ewart as 'some sort of nun-shop', and is presided over by the Miss Arthurets, ludicrously paired as Angelica and Seraphina (ch. 16). Redgauntlet and Charles Edward are denied a heroic last stand, instead suffering the ultimate humiliation of not being taken seriously by the King, who orders General Campbell to arrest nobody, as the Jacobites pose no threat (ch. 24). Yet in the novel's final paragraph Charles Edward achieves a measure of dignity, akin to the respect paid to the dead by friend and enemy alike. When he has taken leave of his followers and his boat pulls away from the land, Dr Grumball – the divine from Oxford, 'the mother of learning and loyalty', who describes himself as 'an unworthy son of the Church of England' (ch. 23) – offers a blessing to which all find themselves giving their assent: he 'broke out into a loud benediction, in terms which General Campbell was too generous to criticise at the time, or to remember afterwards; – nay, it is said that, Whig and Campbell as he was, he could not help joining in the universal Amen! which resounded from the shore' (ch. 24).

Scott's enormous popularity with Victorian readers meant that their understanding of Jacobitism was shaped by his historical novels. The centenary of the 'Forty-five' was marked by the publication of Katharine Thomson's *Memoirs of the Jacobites of 1715 and 1745* (1845), based on research which also underpinned one of the novels that she published with Bentley for the circulating library market, *The Chevalier: A Romance of the Rebellion of 1745* (1844).[19] The most important treatment of Jacobitism in Victorian fiction, however, was to be published shortly after the 'Papal Aggression'

[19] She and her husband, an eminent physician, were prominent in London literary society for a time, and were friendly with Thackeray and Bulwer-Lytton. See J. Sutherland, *Companion to Victorian Fiction* (1988), p. 627.

crisis, when the Catholic element in Stuart history was sharply relevant. Thackeray, who knew Katharine Thomson, set *The History of Henry Esmond* (1852) in the reigns of William and Mary, Queen Anne and George I: Colonel Esmond finally settles in Virginia in 1718.[20] Like other heroes in Thackeray, Scott and Dickens, the younger Harry is portrayed as a product of both his difficult environment and his genetic inheritance.

The story of the Castlewoods reflects the instability of England's religio-political history and anticipates Harry's exposure to different religious traditions during his upbringing. Sir Francis was made the first Viscount Castlewood by a grateful James I, and his son, Sir George, remained in the Stuart court under Charles I, Charles II – in whose service he ruined himself – and James II (I, 2). Sir George's daughter, Isabella, was maid of honour to Queen Henrietta Maria. When she converted to Roman Catholicism, he, a 'weak man', followed her. His nephew and eventual successor, Sir Thomas, was the son of a Parliamentarian who was 'estranged from the chief of his house' during the Civil War. In order to secure the inheritance, Thomas gave up his life of 'duelling, brawling, vice, and play' on the Continent, converted to Catholicism, and married Isabella. Their only son died young, and when Thomas became the third Viscount and retired to Castlewood House, they were greeted with contempt during a period of 'No-Popery fervour'. As they were childless, Sir Thomas sent his Jesuit chaplain, Father Holt, to collect his bastard son, Harry, from the French Huguenot refugees who were bringing him up, like Dickens's Pip in *Great Expectations* (1860–1), by hand, in Ealing.

Father Holt displays the energy and commitment in which Jesuits of Thackeray's generation took great pride.[21] His task, as the 'director' of the family, is to wean Harry off Lutheran hymns and to convert him to Catholicism. He achieves these aims by surrounding his order with what Dickens called an 'air of mystery', and thus investing it with a 'secret charm, and power of attraction': 'Father Holt bade him keep his views secret, and to hide them as a great treasure which would escape him if it was revealed; and, proud of this confidence and secret vested in him, the lad became fondly attached to the master who initiated him into a mystery so wonderful and Awful' (I, 3). Holt's private meetings with visiting Catholic priests and his

[20] In its first book edition the novel is entitled *The History of Henry Esmond, Esq., a Colonel in the Service of Her Majesty Q. Anne, Written by Himself*, 3 vols. (1852).

[21] 'In spite of oceans and deserts, of hunger and pestilence, of spies and penal laws, of dungeons and racks, of gibbets and quartering blocks, Jesuits were to be found in every country arguing, instructing, consoling, animating the courage of the timid, inflaming the hearts of the tepid, and holding up the cross before the eyes of the dying.' W. Waterworth S. J., *Jesuits* (1852), p. 12.

frequent absences from Castlewood House, at a time when James II and thus Catholicism are becoming deeply unpopular in the country, only heighten the glamour that surrounds him and his faith in the eyes of young Harry, 'placed under a Popish Priest and Bred to that Religion' (I, 4). When the boy's 'Superiors are Engaged in Plots for the Restoration of King James II' (I, 5), Holt lets him into the secret of his escape route through a window that can be lowered by hidden machinery and a locked wardrobe where he keeps his disguises. The mature Henry Esmond's memories of Holt conform to the image of the Jesuit presented by Andrew Snape, the eighteenth-century High Church clergyman, who wrote: ''Tis one of the known and stale Artifices of that Politick Fraternity, to personate all Characters, to appear in every Shape and Dress, in order to promote their main Design, the Subversion of the Church of *England*.'[22]

When Castlewood House becomes the county's focal point for the proposed rising against King William and Queen Mary, Mr Holt comes 'to and fro, busy always', and little Harry longs to be a 'few inches taller, than he might draw a sword in his good cause' (I, 6). But the cause fails, Sir Thomas dies and his Viscountess is imprisoned, in the best rooms in Hexton Castle. It is at this point that Harry, left alone in the house with Captain Westbury and his troopers, comes into contact with 'Dick the Scholar' – the Whig, Richard Steele. His movement away from Catholicism, initiated by this new mentor, is represented as a process of maturation, mirroring that of the nation, as in Scott and Macaulay.[23] By now it is 1691 and Henry Esmond, as narrator, has brought the reader up to the point at which Rachel, the new Lady Castlewood, found him, aged twelve, 'busy over his great book', at the beginning of the novel. While Francis, the fourth Viscount, is away at court, Harry admires 'this fair young lady of Castlewood, her little daughter at her knee, and her domestics gathered round her, reading the Morning Prayer of the English Church' (I, 7). Rachel's Protestant chaplain, Dr Tusher, shows Harry that the 'prayers read were those of the Church of all ages'. Harry wants to be close to his 'mistress' and to 'think all things she did right'; he joins the rest of the household at prayers; and, within a couple of years, 'my Lady' makes a 'thorough convert'.

Dr Tusher is a lazy Anglican foil to the dynamic Jesuit, Father Holt. Throughout the nineteenth century, Catholics defended their rule of celibacy for the clergy by pointing to the priests' willingness to make sick

[22] A. Snape, *Second Letter to the Bishop of Bangor* (1717), cited in R. D. Tumbleson, *Catholicism in the English Protestant Imagination* (1998), p. 165.

[23] See pp. 116, 96 above.

calls during dangerous epidemics.[24] When smallpox comes to the village, it is Dr Tusher who rushes to Castlewood House to warn the family. In explaining to Lady Castlewood why he is not obliged to visit his parishioners if they are dying of the disease, he betrays worldly interests rather than spiritual principles:

'We are not in a Popish country: and a sick man doth not absolutely need absolution and confession,' said the Doctor. ''Tis true they are a comfort and a help to him when attainable, and to be administered with hope of good. But in a case where the life of a parish priest in the midst of his flock is highly valuable to them, he is not called upon to risk it (and therewith the lives, future prospects, and temporal, even spiritual welfare of his own family) for the sake of a single person, who is not very likely in a condition even to understand the religious message whereof the priest is the bringer – being uneducated, and likewise stupified or delirious by disease. If your Ladyship or his Lordship, my excellent good friend and patron, were to take it – '

'God forbid!' cried my Lord.

'Amen,' continued Doctor Tusher. 'Amen to that prayer, my very good Lord! for your sake I would lay my life down' – and, to judge from the alarmed look of the Doctor's purple face, you would have thought that that sacrifice was about to be called for instantly. (I, 8)

Far from being idealized, however, Jacobitism and Catholicism are also treated facetiously, as one would expect in Thackeray, and are represented as the domain of women – especially old women – and priests. The Dowager Lady Castlewood, locked away in Hexton Castle, repeatedly asks to be led out to execution, like Mary Queen of Scots, although there is never 'any thought of taking her painted old head off' (I, 6). In the novel's second Book, which treats of Henry Esmond's military life, she introduces him to Jacobite circles in London. This 'credulous old woman' has a 'hundred authentic stories of wondrous cures effected by the blessed King's rosaries, the medals which he wore, the locks of his hair, or what not' (II, 3). She is reminiscent of another old lady – the 'venerable' Mrs Skyring, whose house is full of Jacobite memorabilia, in Katharine Thomson's *The Chevalier*.[25]

Whereas Waverley was attracted by the romance of Jacobitism in early manhood, Esmond has already grown out of such feelings. He remembers a 'score' of the old Dowager's 'marvellous tales', believes as much as he chooses, and casts a cold eye on the Jacobite cause:

[24] See p. 84 above and J. H. Newman, *Apologia* (1967), p. 243. For an account of a sick call to a woman dying of cholera in London in 1849, see E. Price, *Sick Calls* (1855), pp. 175–8 (1st edition published 1850).

[25] K. Thomson, *Chevalier* (1844), I, 217. A wedding service is conducted on board ship in the novel by 'the *old* Romish chaplain' (II, 269, my emphasis).

After King James's death, the Queen and her people at St Germains – priests and women for the most part – continued their intrigues in behalf of the young Prince, James the Third, as he was called in France and by his party here (this Prince, or Chevalier de St George, was born in the same year with Esmond's young pupil, Frank, my Lord Viscount's son); and the Prince's affairs, being in the hands of priests and women, were conducted as priests and women will conduct them, artfully, cruelly, feebly, and to a certain bad issue. The moral of the Jesuits' story I think as wholesome a one as ever was writ: the artfullest, the wisest, the most toilsome, and dexterous plot-builders in the world – there always comes a day when the roused public indignation kicks their flimsy edifice down, and sends its cowardly enemies a-flying. (II, 4)

For the novel's first readers, in 1852, the description of the Jesuits as the old 'enemies' would have had a familiar contemporary ring to it.[26]

Book Three chronicles the end of Esmond's adventures in England, in which, 'without entering very eagerly into the controversy', he reverts to old family loyalties: 'It seemed to him that King James the Third was undoubtedly King of England by right' (III, 1). Father Holt tries to draw him back to the Catholic fold, but he chooses to remain faithful to the 'Church of his country'. His views on English politics and religion are somewhat jaundiced:

While the Tories, the October Club gentlemen, the High Church parsons that held by the Church of England, were for having a Papist king, for whom many of their Scottish and English leaders, firm churchmen all, laid down their lives with admirable loyalty and devotion; they were governed by men who had notoriously no religion at all, but used it as they would use any opinion for the purpose of forwarding their own ambition. The Whigs, on the other hand, who professed attachment to religion and liberty too, were compelled to send to Holland or Hanover for a monarch around whom they could rally. A strange series of compromises is that English History: compromise of principle, compromise of party, compromise of worship! (III, 5)

On his return from France, Henry, now Colonel Esmond, puts himself at the head of the 'little knot of fond conspirators' who keep the Jacobite flame alive (III, 7). The cause proves to be hopeless, however, and Charles Stuart a weak philanderer.[27] Esmond is present outside Kensington Palace when the herald-at-arms proclaims ' "GEORGE, by the Grace of God, of Great Britain, France, and Ireland, King, Defender of the Faith": And the people shouted

[26] See, for example, F. Trollope, *Father Eustace* (1847); M. H. Seymour, *Nature of Romanism* (1849); T. Carlyle, 'Jesuitism', *Works* (1884), XII, 360–401, and W. Waterworth's defence, in *The Jesuits* (1852). (Cf. also an earlier novel by J. F. Smith, *The Jesuit* (1832).)

[27] Stephen Bann comments that the 'predominant movement is not centripetal, but centrifugal. . . . the Young Pretender is, in effect, the *Deus ex machina* who speeds the plot to its conclusion. . . . It is precisely because Prince Charles has compromised Beatrix that Esmond is cured of his passion for her': *Clothing of Clio* (1984), p. 146.

God save the King!' (III, 13). Among the crowd is 'one sad face': 'poor Mr Holt's, who had slipped over to England to witness the triumph of the good cause; and now beheld its enemies victorious, amidst the acclamations of the English people'. As in *Redgauntlet*, there are to be no heroics at the end of *Henry Esmond*. Instead of being executed for treason, our half-hearted Jacobite hero emigrates. Jesuitism, in the chameleon shape of Father Holt, is denied the glory of martyrdom. Instead, it fulfils the desires of English Protestant culture by growing old, dying and being quietly buried:

> I saw him in Flanders after this, whence he went to Rome to the headquarters of his Order; and actually reappeared among us in America, very old, and busy, and hopeful. I am not sure that he did not assume the hatchet and mocassins there; and, attired in a blanket and war-paint, skulk about a missionary among the Indians. He lies buried in our neighbouring province of Maryland now, with a cross over him, and a mound of earth above him; under which that unquiet spirit is for ever at peace. (III, 13)

III

Jacobitism was mortally wounded at the Battle of Culloden (1746) and was declared dead on the accession of George III (1760). Meanwhile the story of anti-Catholicism became more complicated. After the 'Forty-five', when hostility towards Catholics had united the nation, there was a breakdown of consensus on the matter.[28] While an increasingly tolerant attitude developed among the élite over the subsequent thirty years, strong anti-Catholic feeling remained in the rest of the population, and particularly in the mid-1770s, when there was a hostile reaction to government policy in the colonies.[29] The very modest Catholic Relief Act of 1778 was enough to raise fears in Protestant minds that the constitution established in 1689 was in danger.[30] Those fears were played upon by the Protestant Association and its leader, Lord George Gordon, in the period leading up to the riots of June 1780.

Nineteenth-century historians and novelists looked back at the eighteenth century through the lens of the French Revolution, after which the

[28] C. Haydon, *Anti-Catholicism* (1993), pp. 161, 164.

[29] 'The unfortunate management of the War against the American Rebellion exercised a greater influence in stimulating the Riots than has been recognised', and that 'the sense of insecurity was intensified by a rumour everywhere rife, but never substantiated, that French and American agents were at work among the rioters': J. P. de Castro, *Gordon Riots* (1926), pp. vii–viii.

[30] 'It formally suspended the old restrictions on Popish priests and it was felt that this new licence might allow them to win converts. Above all, it aided a group which had long been represented by those in authority as dangerous and seditious.' C. Haydon, *Anti-Catholicism* (1993), p. 206.

world had changed for ever. When describing the hanging, drawing and quartering of a Jacobite rebel at Kennington, for example, following the slaughter at Culloden, Katharine Thomson commented that a 'scene worthy of the French Revolution was rehearsed'.[31] John Milner exaggerated not only the number of 'Protestant rioters' in London on 9 June 1780, but also their intentions: to 'exterminate' the Catholics, and to 'anticipate the horrors of Jacobinism in this country'.[32] Closer to home for Victorian commentators were the Chartist riots of 1839, themselves compared to events in Paris fifty years earlier by Carlyle, the historian of the French Revolution and a powerful influence upon Dickens.[33] Indeed, it has been suggested that Dickens himself, intent upon warning his readers of the dangers of mob rule, exaggerated the number of rioters and the extent of the damage that they caused.[34]

Whereas Carlyle, in *The French Revolution* (1837), sees the mob as the 'agent of a necessary historical process', Dickens sees it as the 'enemy of progress'.[35] In two other respects, however, Carlyle's vivid narrative does anticipate and possibly influence *Barnaby Rudge*. First, Carlyle combines an antipathy towards French Catholicism with sympathy for its persecuted priests. The 'Civil Constitution of the Clergy', he argues, 'divides France from end to end, with a new split, infinitely complicating all the other splits: – Catholicism, what of it there is left, with the Cant of Catholicism, raging on the one side, and sceptic Heathenism on the other; both, by contradiction, waxing fanatic.'[36] Having described the burning of an effigy of Pope Pius VI – 'our lath-and-gum Holiness' – in Paris, on 4 May 1791, he comments: 'In such extraordinary manner does dead Catholicism somerset and caper, skilfully galvanized.'[37] Yet, having described the infamous

[31] K. Thomson, *Chevalier* (1844), II, 230.

[32] J. Milner, *Letters to a Prebendary* (1800), p. 180. Gordon had only half the 100,000 supporters that Milner suggests.

[33] See T. Carlyle, *Works* (1884), XVI, 65.

[34] 'Dickens, writing in part to warn his own age of the dangers of popular action, portrayed the disturbances in *Barnaby Rudge*, the novel being dominated by pictures of a vast, surging mob, giving way to the basest passions, and the account of the troubles reaching its climax, predictably, with the dreadful scene at the distillery. Yet were the Gordon Riots really like that? In the enormous metropolis of London, only thirty-two private houses were destroyed or seriously damaged, whilst only eighty-one people in the City and in Middlesex received, and twenty-nine persons in Surrey claimed, compensation for damage to their property – a mere forty-seven receiving sums of £200 or more. The most costly damage was to Newgate and the King's Bench, but this was exceptional. . . . Were not the riots, in fact, far from being an outbreak of indiscriminate violence, circumscribed and very largely directed against a specific group in English society?' Haydon, *Anti-Catholicism* (1993), pp. 241–2.

[35] W. Oddie, *Dickens and Carlyle* (1972), p. 104.

[36] T. Carlyle, *Works* (1884), III, 289.

[37] Ibid., IV, 9.

drownings, or '*Noyades*', in the Loire at Nantes in December 1793, when ninety priests 'lie deep', he can comment that, 'simultaneously with the Tophet-black aspect, there unfolds itself another aspect, which one may call a Tophet-red aspect, the Destruction of the Catholic Religion'.[38] In *Barnaby Rudge*, Dickens invites his readers to sympathize with the persecuted Catholics of 1780, but without showing more than glimpses of their religious life, and then only the externals.

Secondly, Carlyle hates 'Anarchy as Death, which it is', and treats '*Sansculottism*' as 'madness', a 'World-bedlam': 'What, then, is this Thing called *La Révolution*, which, like an Angel of Death, hangs over France, noyading, fusillading, fighting, gun-boring, tanning human skins? . . . It is the Madness that dwells in the heart of men.'[39] During and after the 'Papal Aggression' crisis, anti-Catholic writers scored cheap points by commenting on the proximity of St George's Catholic Cathedral to Bedlam.[40] In *Barnaby Rudge* Dickens associates Bedlam, not with the Catholics, but with the anti-Catholic mob.[41]

Dickens had signed the original contract for *Barnaby Rudge: A Tale of the Riots of 'Eighty* in May 1836, when anti-Catholic feeling was strong.[42] Writing did not begin until 29 January 1841, although much of the research had been completed by then.[43] Publication followed rapidly, and the novel was serialized in *Master Humphrey's Clock* from February to November, at the same time that Ainsworth's *Guy Fawkes* was appearing in *Bentley's Miscellany*. Dickens's preface, like Ainsworth's, although less directly, appeals for religious toleration:[44]

That what we falsely call a religious cry is easily raised by men who have no religion, and who in their daily practice set at nought the commonest principles of right and wrong; that it is begotten of intolerance and persecution; that it is senseless, besotted, inveterate and unmerciful; all History teaches us. But perhaps we do not know it in our own hearts too well, to profit by even so humble an example as the 'No Popery' riots of Seventeen Hundred and Eighty.

However imperfectly those disturbances are set forth in the following pages, they are impartially painted by one who has no sympathy with the Romish Church,

[38] Ibid., IV, 365, 367. [39] Ibid., IV, 454; III, 187, 189, 390.

[40] 'It is a curious coincidence that, in London, the nearest public building to the Popish Cathedral is Bedlam; and those who study the Jesuit doctrines will allow, that there could scarcely be a more appropriate neighbourhood.' C. Sinclair, *Beatrice* (1852), I, xxxv–xxxvi.

[41] 'If Bedlam gates had been flung open wide, there would not have issued forth such maniacs as the frenzy of that night had made' (55).

[42] D. Walder, *Dickens and Religion* (1981), p. 95. [43] P. Ackroyd, *Dickens* (1990), pp. 323, 325.

[44] See p. 93 above. 'In the end, *Barnaby Rudge* is less concerned with expressing sympathy for the Catholics (although it does so), than with reminding Protestants that, as Locke said, mutual toleration signified the true church:' D. Walder, *Dickens and Religion* (1981), p. 96.

though he acknowledges, as most men do, some esteemed friends among the followers of its creed.

Dickens's distinction between his antipathy to Catholicism and his esteem for Catholic friends provides a clue to the structure of the novel, which begins in the private domain of the heart and soul, and then moves into the public world of politics. His reference to the lessons of history, to which his own novel will contribute, provides a clue to his plotting, in which both public and private histories of violent death press in upon the imagined present of the action.

In order to stir up 'intolerance and persecution' in the present, the Protestant Association looks back to the reign of 'Bloody Mary', whose name will later become a curse on the lips of the rioters:

> But when vague rumours get abroad, that in this Protestant association a secret power was mustering against the government for undefined and mighty purposes; when the air was filled with whispers of a confederacy among the Popish powers to degrade and enslave England, establish an inquisition in London, and turns the pens of Smithfield market into stakes and cauldrons; when terrors and alarms which no man understood were perpetually broached, both in and out of Parliament, by one enthusiast who did not understand himself, and bygone bugbears which had lain quietly in their graves for centuries, were raised again to haunt the ignorant and credulous; when all this was done, as it were, in the dark, the secret invitations to join the Great Protestant Association in defence of religion, life, and liberty, were dropped in the public ways, thrust under the house-doors, tossed in at windows, and pressed into the hands of those who trod the streets by night; when they glared from every wall, and shone on every post and pillar, so that stocks and stones appeared infected with the common fear, urging all men to join together blindfold in resistance of they knew not what, they knew not why; – then the mania spread indeed, and the body, still increasing every day, grew forty thousand strong. (37)

Dickens identifies precisely those Protestant anxieties, based on historical precedents, or 'bygone bugbears', that were to fuel anti-Catholic paranoia in his own generation: Queen Mary's fires at Smithfield; the work of the Inquisition; and the vaguely defined threat of Papal enslavement and the loss of that hard-won 'liberty' which John Bull will defend to the last.[45] Later in the chapter, Dennis the hangman's grotesque gloss on the lessons of English Reformation history exposes that paranoia for what it is. He fears that a Catholic take-over would ruin his trade: 'If these Papists gets

[45] A gentleman, known by some as 'a country gentleman of the true school' and by others as 'a genuine John Bull', will not believe that Barnaby has been an 'idiot' from birth, as his mother claims, but is a malingerer who needs flogging: 'He was warmly attached to church and state' (47).

into power, and begins to boil and roast instead of hang, what becomes of my work! . . . No Popery! I'm a religious man, by G– !'

The hangman's halter casts a long shadow across the plot of *Barnaby Rudge*, the dénouement of which follows a series of public executions (77). John Willet, landlord of the Maypole and one of several bad fathers in the novel, explains to his regulars that the mother of Hugh the ostler was 'hung when he was a little boy, along with six others, for passing bad notes – and it's a blessed thing to think how many people are hung in batches every week for that, and such-like offences, as showing how wide awake our government is' (11). Hugh gives his own version of events to John Chester, saying that, when he was a boy of six, they 'hung [his] mother up at Tyburn for a couple of thousand men to stare at' (23). During the riots, Hugh himself becomes a voyeur, when he and Dennis kidnap Emma Haredale, niece of the main Catholic character in the novel, and Dolly Varden, daughter of the worthy locksmith, Gabriel Varden. Dennis, who constantly gloats over his long tally of 'turning off' the condemned, says that he ought to be able to lift the fainting Emma by himself: 'I always like 'em to faint, unless they're very tender and composed. . . . I've lifted up a good many in my time' (59). When Dolly embraces Emma, the narrator implicates the male reader by asking, 'what mortal eyes could have avoided wandering to the delicate bodice, the streaming hair, the neglected dress, the perfect abandonment and unconsciousness of the blooming little beauty?': 'Not Hugh. Not Dennis.' Like Ainsworth in *Guy Fawkes*, Dickens exploits the ambiguous power of the pornography of violence.[46]

When the obsessively fastidious Chester, the worst of the novel's fathers, explains to his son why he should not marry Emma, he 'carelessly' plays with the idea of her being hanged:

> In a religious point of view alone, how could you ever think of uniting yourself to a Catholic, unless she was amazingly rich? You ought to be so very Protestant, coming of such a Protestant family as you do. . . . The very idea of marrying a girl whose father was killed, like meat! Good God, Ned, how disagreeable! Consider the impossibility of having any respect for your father-in-law under such unpleasant circumstances – think of his having been 'viewed' by jurors, and 'sat upon' by coroners, and of his very doubtful position in the family ever afterwards. It seems to me such an indelicate sort of thing that I really think the girl ought to have been put to death by the state to prevent its happening. But I tease you perhaps. (15)

[46] See p. 93 above.

So appalled is Edward Chester by his father's casual cruelty that he sits alone afterwards, 'with his head resting on his hand', in what appears to be a 'kind of stupor'.

While the murder of Reuben Haredale at the Warren, twenty-two years earlier, provides the mystery interest in the early chapters of the novel, tension between Catholic and Protestant is established through two old personal enemies: John Chester and Geoffrey Haredale, Reuben's brother, who has inherited the Warren. The antagonists first met in boyhood, at St Omer's, a Jesuit secular college in northern France, which Chester describes as 'a remarkably dull and shady seminary' (43). Whereas Haredale, being Catholic and 'of necessity educated out of England', was brought up there, Chester, 'being a promising young Protestant at that time, was sent to learn the French tongue from a native of Paris!' The Warren, redolent of Catholic concealment and a tale of murder, has an oppressive 'air of melancholy': 'Great iron gates, disused for many years, and red with rust, dropping on their hinges and overgrown with long rank grass, seemed as though they tried to sink into the ground, and hide their fallen state among the friendly weeds' (13). When Chester wants to meet Haredale and discuss how they can thwart the romance between Edward and Emma, he instals himself in the nearby Maypole, a Protestant stronghold as dozy as old England itself, built 'in the days of King Henry the Eighth' and, it is said, slept in by Queen Elizabeth (1). On John Willet's telling his customers that Chester has hired the large room for the meeting, Solomon Daisy assumes that violence will follow: 'speaking softly and with an earnest look', he says that Chester and Haredale are 'going to fight a duel in it' (11).

The outspoken and roughly dressed Haredale compares favourably with the smooth hypocrite, Chester. Yet Dickens, perhaps as with his esteemed Catholic friends, makes no attempt to explore his spirituality, dropping only occasional hints that he is a practising Catholic.[47] Rather, Dickens's interest is in the villainous, self-congratulating Protestant whose 'most refulgent smile' conceals a hatred of Haredale that tempts him to draw his sword 'fifty times' when he encounters him (27). Chester resists the temptation, however: '"You are the wise man's very last resource," he said, tapping the hilt of his weapon; "we can but appeal to you when all else is said and done"'. Smiling at this thought as he walks along, he is approached by a beggar, whom he allows to follow him before graciously dismissing him

[47] When John Willet visits the Warren at night, Haredale says: 'Gently with your light, friend. You swing it like a censer' (34). Later in the novel, Haredale's home is described as 'but another bead in the long rosary of his regrets' (61).

with a 'fervent blessing', which is 'as easy as cursing' and 'more becoming to the face'.

A sedan chair then takes Chester to the locksmith's house, where he charms Dolly's mother, Martha Varden, in order to persuade her that Emma Haredale should not marry his son, whom he maliciously describes as having 'no heart at all'. When Mrs Varden languishes over the thought that she might soon 'take an easy flight towards the stars' and leans her arm upon the 'Protestant Manual', 'as though she were Hope and that her Anchor', Chester assures her that it is his 'favourite book': '"How often, how very often in his early life – before he can remember" – (this clause was strictly true) – "have I deduced little easy moral lessons from its pages, for my dear son Ned!"' The Manual, in 'two volumes post octavo', is devoured by an ill-tempered Mrs Varden, along with a substantial breakfast in bed (4). It guides her in her literalist interpretation of the Bible: she regards John Willet with disfavour, believing that the 'publicans coupled with sinners in Holy Writ' are 'veritable licensed victuallers' (13). It shapes her theology: she chides Miss Miggs for 'speaking of angels in connection with your sinful fellow-beings', as we are 'mere worms and grovellers'. It fuels her anti-Catholic prejudices: she believes that 'to take a draught of small ale in the morning' is to observe a 'pernicious, irreligious, and Pagan custom, the relish whereof should be left to swine, and Satan, or at least to Popish persons, and should be shunned by the righteous as a work of sin and evil' (19).

In the early, domestic chapters of the novel, John Chester and Martha Varden represent themselves as 'staunch Protestants' in their handling of family affairs (21). In the novel's central chapters, which move into the public arena, they, like everybody else, are caught up in events leading to the Gordon Riots: they become politicized, 'true blue' Protestants. Simon Tappertit, Gabriel Varden's diminutive apprentice, epitomizes the kind of working-class militant who exploited the opportunity presented by the activities of the anti-Catholic Protestant Association in 1780. He mouths the clichés of the demagogue, as he explains the difference between his private and public personae to the kidnapped Dolly Varden: 'You behold in me, not a private individual, but a public character; not a mender of locks, but a healer of the wounds of his unhappy country' (59).

Following a five-year gap in the narrative, Dickens marks the transition from the private to the public in his treatment of events on 19 March 1780 (33). On the twenty-seventh anniversary of the murder of Reuben Haredale and, supposedly, his steward, Rudge, at the Warren, Solomon Daisy tells the company at the Maypole that he has just seen the ghost of Rudge – the 'likeness of a murdered man'. Later that evening, on a dark and stormy night, and as the murder mystery deepens, Lord George Gordon and his cronies arrive at the Maypole on their way to London (35). Suited in black and of 'Puritan's demeanour', Lord George is accompanied by John Grueby and by his secretary, the 'sly and slinking' Gashford, who, 'stricken by the magic of his eloquence in Scotland but a year ago, abjured the errors of the Romish church', and is later to be exposed by Haredale as a villain (43).

Gashford's energies are devoted to making new recruits and buoying up his leader through sycophancy. His account of a recent rally in Scotland introduces into the novel the kind of public bombast that is soon to be heard in the riots:

> When you spoke of a hundred and twenty thousand men across the Scottish border who would take their own redress at any time, if it were not conceded; when you cried 'Perish the Pope and all his base adherents; the penal laws against them shall never be repealed while Englishmen have hearts and hands' – and waved your own and touched your sword; and when they cried 'No Popery!' and you cried 'No; not even if we wade in blood', . . . ah! then I felt what greatness was indeed, and thought, When was there ever power like this of Lord George Gordon's!

When the 'deluded lord' has retired to bed, however, Grueby reminds Gashford of how things actually stand: 'Between Bloody Marys, and blue cockades, and glorious Queen Besses, and no Poperys, and Protestant associations, and making of speeches, . . . my lord's half off his head'; and the loyal English Protestants whom he addresses are 'very fond of spoons . . . and silver-plate in general, whenever area-gates is left open accidentally'. 'One of these evenings', he warns Gashford, 'when the weather gets warmer and Protestants are thirsty, they'll be pulling London down, – and I never heard that Bloody Mary went as far as *that*'.

Among Gashford's list of contributions to the Protestant Association, ranging from the 'United Link Boys, three shillings – one bad' to the 'anti-popish prisoners in Newgate', and from the 'Associated Remembers of Bloody Mary, half-a-guinea' to Tappertit's 'United Bull-dogs, half-a-guinea', is 'Mrs Varden's box (fourteenth time of opening), seven shillings and sixpence in silver and copper, and half-a-guinea in gold; and Miggs (being the saving of a quarter's wages), one-and-threepence'. Gabriel Varden, who refuses to join, 'remains in outer darkness'. Gabriel's damnation by Gashford reveals the Protestant Association's inversion of the novel's moral order, as represented in the values of the locksmith, whose name is that of a divine messenger and who is to act heroically during the riots. Dickens underlines the moral connection between the private contributor to the Association and the destructiveness of its work by focusing upon Martha Varden's collection box, 'painted in imitation of a very red-brick dwelling-house, with a yellow roof; having at top a real chimney, down which voluntary subscribers dropped their silver, gold, or pence, into the parlour; and on the door the counterfeit presentment of a brass plate, whereon was legibly inscribed "Protestant Association"' (41). Later, when Gabriel upbraids Tappertit for his part in the riots and warns him that the 'rope is round [his] neck', a 'crestfallen' Martha hides the box under her skirts, realizing that she has 'aided and abetted the growth of disturbances, the end of which it was impossible to foresee' (51).

With respect to the religious politics of 1780, Dickens's satire is directed against wrong-headed Protestantism. His treatment of the Catholics against whom the violence of the mob is directed is much sketchier, and is similar to his treatment of Haredale earlier in the novel: he shows sympathy for them, but carefully avoids going beyond the externals of their religion. When Haredale and Chester meet at Westminster Hall, and Sir John recalls their first meeting at St Omer's, the political issue of the day comes sharply into focus:

> 'Add to the singularity, Sir John,' said Mr. Haredale, 'that some of your Protestants of promise are at this moment leagued in yonder building, to prevent our having the surpassing and unheard-of privilege of teaching our children to read and write – here – in this land, where thousands of us enter your service every year, and to preserve the freedom of which, we die in bloody battle abroad, in heaps.' (43)

After Haredale is introduced by Sir John as a 'Catholic gentleman unfortunately – most unhappily a Catholic', Lord George announces in an agitated manner that he cannot talk to him, as they have nothing in common. Haredale's response – that they have much in common, 'all that the Almighty gave us, and common charity, not to say common sense' – is rebuffed, as Lord George announces, in words that go to the heart of many religious differences: 'I don't hear you, sir. . . . I can't hear you.' The crowd that harasses Haredale on his departure proceeds to 'giving Protestant knocks at the door of private houses, breaking a few lamps, and assaulting some stray constables', when it has finished with him.

The 'great stone' that Hugh throws at Haredale on the river steps of Westminster Hall presages the mob violence that is soon to be directed against Catholic houses and chapels. During the riots, Gashford looks down at the crowd from an upper window, trying to make out what they have been doing:

That they had been engaged in the destruction of some building was sufficiently apparent, and that it was a Catholic place of worship was evident from the spoils they bore as trophies, which were easily recognisable for the vestments of priests, and rich fragments of altar furniture. Covered with soot, and dirt, and dust, and lime; their garments torn to rags; their hair hanging wildly about them; their hands and faces jagged and bleeding with the wounds of rusty nails; Barnaby, Hugh, and Dennis hurried on before them all, like hideous madmen. (50)

The reader is shown the rioters and their 'trophies', illuminated by torch-light, while the chapel that they have destroyed remains out of sight, somewhere in the darkness. When the rioters systematically destroy the famous chapel at Moorfields on the Sunday evening, like 'mere workmen' with a 'certain task to do', they take away the 'very altars, benches, pulpits, pews, and flooring', and put 'priestly garments, images of saints, rich stuffs and ornaments, altar-furniture and household goods' on 'great fires in the fields' (52). While the violence is deplored, there is no sense of sacrilege in Dickens's description, and thus no acknowledgment of the mystery of the sanctuary that is violated.

Whereas the unhurried, workmanlike destruction of Moorfields chapel is briefly and coolly described, the hellish scenes at the Warren that follow are described with passionate intensity and extend over several pages. Dickens's descriptive prose is reminiscent of John Martin's 'apocalyptic sublime', in his depictions of hell:[48]

The besiegers being now in complete possession of the house, spread themselves over it from garret to cellar, and plied their demon labours fiercely . . . Men who had been into the cellars, and had staved the casks, rushed to and fro stark mad, setting fire to all they saw – often to the dresses of their own friends – and kindling the building in so many parts that some had no time for escape, and were seen, with drooping hands and blackened faces, hanging senseless on the window-sills to which they had crawled, until they were sucked and drawn into the burning gulf. (55)

Dickens draws attention to the 'exposure to the coarse, common gaze, over every little nook which usages of home had made a sacred place, and the destruction by rude hands of every little household favourite which old associations made a dear and precious thing'. For him, the destruction of a home, even a Roman Catholic home, *is* a sacrilegious act. The reader is asked to sympathize with a family left on the street with its furniture at midnight, after the carrier has taken fright at the fires in the area (61), but not with the worshippers whose chapels have been destroyed.

[48] See M. D. Paley, *Apocalyptic Sublime* (1986).

The section of the novel devoted to the riots concludes with a series of judicial punishments, through which Dickens explores the ironies of history. He reminds the reader that it was mainly the 'weakest, meanest, and most miserable' among the rioters who were hanged, adding that 'it was a most exquisite satire upon the false religious cry which had led to so much misery, that some of these people owned themselves to be Catholics, and begged to be attended by their own priests' (77). Dennis the hangman has to be supported by two men on his way to the gallows, while Hugh, who now knows that he is Sir John Chester's bastard son, curses his father before dying 'with a careless air'. Barnaby is condemned to death, but is reprieved at the last moment (79). Lord George Gordon is found not guilty of high treason ('Chapter the Last'). After seven years of remaining comparatively quiet, he is 'stimulated by some new insanity' to write a pamphlet attacking the Queen of France 'in very violent terms'. He converts to Judaism and dies in Newgate, missed by his fellow prisoners as one who, ironically, 'knew no distinction of sect or creed' in his alms-giving.

The dénouement of the private plot takes place when Haredale finds Sir John Chester amongst the ruins of the Warren, which Dickens again describes in terms of the sacred:

> The ashes of the commonest fire are melancholy things . . . How much more sad the crumbled embers of a home: the casting down of that great altar, where the worst among us sometimes perform the worship of the heart, and where the best have offered up such sacrifices, and done such deeds of heroism, as, chronicled, would put the proudest temples of old Time, with all their vaunting annals, to the blush! (81)

Sir John finds the scene 'very picturesque' and calmly takes snuff as Haredale accuses him of having 'set on Gashford to this work'. When he finally loses his self-control, however, and draws his sword against Haredale, he drops his mask and shows his 'hatred in his face'. Haredale's sword plunges through him 'to the hilt', and even now he tries to smile, as if remembering that an expression of hatred 'would distort his features after death'.

Like Scott's Pretender and Thackeray's Father Holt, Haredale dies in exile. Dickens's description of his escape to the Continent reflects the view, shared by Kingsley and most of their generation who called themselves Protestants, that monasticism is fundamentally unnatural and imprisoning:

> Repairing straight to a religious establishment, known throughout Europe for the rigour and severity of its discipline, and for the merciless penitence it exacted from those who sought its shelter as a refuge from the world, he took the vows which

thenceforth shut him out from nature and its kind, and after a few remorseful years was buried in its gloomy cloisters. (Chapter the Last)

Many readers of *Barnaby Rudge* have wondered why Dickens chose as a title the name of a character who flits in and out of the action rather than playing a central, shaping role in it. I believe that the answer lies in the supreme value that Dickens attaches to Barnaby's spirituality. This 'idiot boy' is closely in touch with 'nature and its kind':

It is something to look upon enjoyment, so that it be free and wild and in the face of nature, though it is but the enjoyment of an idiot. It is something to know that Heaven has left the capacity of gladness in such a creature's breast; it is something to be assured that, however lightly men may crush that faculty in their fellows, the Great Creator of mankind imparts it even to his despised and slighted work. Who would not rather see a poor idiot happy in the sunlight, than a wise man pining in a darkened jail! (25)

When Barnaby is himself imprisoned in a 'darkened jail', his relationship with God is shown to be open, unmediated and redemptive:

But the moon came slowly up in all her gentle glory, and the stars looked out, and through the small compass of the grated window, as through the narrow crevice of one good deed in a murky life of guilt, the face of Heaven shone bright and merciful . . . He, a poor idiot, caged in his narrow cell, was as much lifted up to God, while gazing on the mild light, as the freest and most favoured man in all the spacious city; and in his ill-remembered prayer, and in the fragment of the childish hymn, with which he sung and crooned himself asleep, there breathed as true a spirit as ever studied homily expressed, or old cathedral arches echoed. (73)

Like the simple country folk who lie in their 'narrow cells' in Gray's country churchyard, and like the children who sport on the green in Blake's *Songs of Innocence*, Barnaby Rudge holds the key to salvation in his childlike simplicity.[49] For Dickens, Barnaby is all that a confessing Catholic is not. Yet Catholicism is also, among other things, a religion of the heart. Ironically, like Lord George Gordon in this one respect, Dickens 'can't hear' this, perhaps because he is not listening.

[49] T. Gray, *Elegy Written in a Country Churchyard* (1751), line 15; W. Blake, 'The Ecchoing Green', line 9, in *Songs of Innocence* (1789); cf. Matthew 18.3, where Jesus says to his disciples: 'Verily I say unto you, Except ye be converted, and become as little children, ye shall not enter into the kingdom of heaven.'

PART II

Creeds and crises

CHAPTER 5

The fortress of Christianity

There is the strongest reason for believing, that as Judæa was chosen for the especial guardianship of the original Revelation; so has England been chosen for the especial guardianship of Christianity.
George Croly, *The Apocalypse of St John* (1827)[1]

I

As in Northern Ireland during the 'Troubles' of the last quarter of the twentieth century,[2] so, in the second quarter of the nineteenth, Catholic and Protestant interpretations of distant bloody histories lay behind each side's response to current crises in England. History pressed in upon those caught up in the conflicts which surrounded Catholic Emancipation (1829), Newman's conversion to Roman Catholicism (1845) and the Gorham crisis (1847–50), and many of the old arguments were rehashed again and again on both sides. New political realities, however, sharpened the debates, as the Church of England lost power and influence, while both Catholicism and Dissent gained ground. Each of these three major crises contributed to the climate of fear and suspicion that made the nation's misreading of 'Papal Aggression' possible in 1850. Each also revealed different facets of the three hundred years war between the old enemies, and each crystallized around a particular set of signs and symbols. In the case of Catholic Emancipation – a purely secular and political settlement, but one with enormous religious significance – it was the Protestant constitution itself that seemed to be in danger, as Fortress England, providentially chosen by God as the site of true reformed religion, was penetrated by a foreign power, when Catholics once again took their seats in Parliament.

Although only a minority of the Revd George Croly's contemporaries shared the extreme views expressed in *The Apocalypse of St John*, quoted

[1] G. Croly, *Apocalypse of St John* (1827), p. i.

[2] See M. Tanner, *Ireland's Holy Wars* (2001).

above and discussed later, millions were John Bullish in tendency: hostile to Catholicism, fearful of mass immigration, and deeply suspicious of those elected Catholic Irish Members of Parliament who were unable actually to take their seats in the House of Commons.[3] For Catholic Emancipation, which played an increasingly dominant role in British politics in the 1820s, was part of the much older Irish Question.[4]

Following the failure of a Catholic Relief Bill in 1821, the British Catholic Association had been formed by Daniel O'Connell ('the Liberator') in 1823, with provincial branches in 1824. O'Connell argued that Emancipation was a matter of political rights.[5] In 1825, Burdett's Catholic Relief Bill, draughted mainly with Irish needs in mind, was defeated in the House of Lords, and the government outlawed the Irish Catholic Association.[6] Meanwhile, anti-Catholicism intensified: militant Evangelical efforts over a broad front in the 1820s were called the 'Second Reformation', and James Edward Gordon, the founder of the more focused Reformation Society, was forwarding his campaign.[7]

When the Duke of Wellington became Prime Minister, in January 1828, he was still opposed to Catholic Emancipation, and his Home Secretary, Sir Robert Peel, was regarded as the leading anti-Catholic in the Commons. By this time, however, there was considerable pressure for reform, and the Ultra-Tories failed to rouse sufficient popular opposition in time to arrest the momentum. Dissenters were later to thank the Catholics for creating the climate that made the repeal of the Test and Corporation Acts possible, in February 1828. Twelve months later, plans to review the civil disabilities of Roman Catholics were announced in the King's Speech. Neither Wellington nor Peel had undergone a sudden conversion on the issue; both, in their

[3] See G. I. T. Machin, *Catholic Question* (1964), p. 5.

[4] 'The main concessions, including the final one of 1829, were only made because Catholic emancipation was mainly an Irish question . . . Catholic Ireland was a standing threat to the unity and security of the British Isles . . . The English Catholics wanted only the removal of their civil and political disabilities. But the Irish Catholics had a whole range of grievances of which religious exclusion was only one; these grievances were basically social and economic, and ultimately nationalist. Catholic emancipation, a final objective to the English Catholics, was likely to be only one step towards satisfying the Irish . . . unlike the English, they could bring political pressure to bear on the Government, and this fact eventually proved to be decisive': ibid., p. 11.

[5] In contrast, the English movement, led by traditionalists, was looking for integration within the existing system: see E. R. Norman, *English Catholic Church* (1984), p. 41.

[6] The 'staunchly anti-catholic Duke of York' stated that the 'Coronation Oath eternally prevented the royal assent being given to a Catholic relief bill, and implied that should he succeed George IV (as seemed likely at that time), such assent would never be given in his lifetime': G. I. T. Machin, *Catholic Question* (1964), p. 58.

[7] J. Wolffe, *Protestant Crusade* (1991), pp. 33–4. On two paintings of the period see *Imagining Rome*, ed. M. Liversidge and C. Edwards (1996), p. 49.

different ways, were political pragmatists. Peel's nickname was changed from 'Orange Peel' to 'Turncoat Peel'.

Following large majorities for the Bill in both the Commons, on 10 April 1829, and the Lords, on 13 April, the Duke of Wellington obtained the Royal Assent from a tearful George IV, having previously fought the only duel of his life over the matter, with the Earl of Winchilsea.[8] In return for an oath to 'abjure any intention to subvert the present Church Establishment as settled by law', Catholics could now be elected to Parliament, hold commissions in the army and navy, qualify for ministerial office and serve on lay corporations. Although the banning of Catholic religious orders at first caused dismay among the orders themselves, this part of the measure proved to be a dead letter.

The reaction of the British public was peaceful: there were no riots, and the British Catholic Association was quietly wound up. A widespread sense of anxiety remained, however. Gladstone wrote home from Christ Church, Oxford, reporting that the bedmakers

> seem to continue in a great fright, and mine was asking me this morning whether it would not be a very good thing if we were to give them [the Irish] a king and parliament of their own, and so to have no more to do with them. The old egg-woman is no whit easier, and wonders how Mr. Peel, who was always such a well-behaved man here, can be so foolish as to think of letting in the Roman catholics.[9]

The 'unthinking and the ignorant of all classes', John Morley adds drily, 'were much alike'. For many of the thinking and the educated, too, 1829 remained an even more important turning-point, politically and culturally, than 1832, the year of the first Reform Act.[10] It was in 1829 that both Edward Irving and Thomas Carlyle published essays entitled 'Signs of the Times', and that end-of-worldish prophecies abounded. As Mr Vincy says in *Middlemarch*, 'Some say it's the end of the world, and be hanged if I don't think it looks like it!' (36)

In the next section of this chapter, I examine some of the historical writing – by William Cobbett, Robert Southey and Charles Butler, among others – that appeared around 1825, when Catholic claims were being hotly debated across the nation. In section III I go on to consider how millenarian writers such as Robert Pollok, George Croly and Edward Irving, drawing

[8] See C. C. F. Greville, *Memoirs* (1875), I, 192–4.

[9] J. Morley, *Life of Gladstone* (1903), I, 53–4.

[10] 'Catholic Relief had a considerable psychological impact on Protestants and Catholics alike, and was symbolic of other important changes. It coincided with a period of ferment in evangelicalism and of new-found self-confidence among Roman Catholics, while the Irish influx was reaching a level which was forcing adaptation upon the Roman Church': J. Wolffe, *Protestant Crusade* (1991), p. 8.

upon history, used analogies of defence and attack in their anti-Catholic prophecies. The concept of Protestant England as the 'Fortress of Christianity' was central to popular millenarianism, which flourished after 1789, when modern defensive fortresses were built along the English coast, during the wars with France. The word 'castle', on the other hand, was less useful for anti-Catholics, for whom the whole project of Medievalism seemed highly suspect. The ambiguities surrounding one particular English castle – Arundel, itself on the south coast – are the focus of section IV, in which Catholic positions on reform are explored. The very different reactions of successive Dukes of Norfolk to unfolding political and religious developments, and the restoration of the castle, explain how Arundel, far from being seen as a sinister Catholic stronghold, came to be regarded as part of the shared heritage of Fortress England.

II

Sixty thousand copies of Cobbett's *History of the Protestant "Reformation," in England and Ireland* (1824–5) were sold in the first year. Gladstone wrote on the half-title of his copy of a later edition, in 1837: 'A "rollicking", impudent, mendacious book; most readable; with great art and felicity of narration . . . Here truly is a man master of his work, not servant of it.'[11] Ruskin referred to 'Cobbett's little History of the Reformation, the only true one ever written as far as it reaches', adding: 'I do not, of course, like his style, but the sum of my forty-four years of thinking on the matter . . . has led me to agree with Cobbett in all his main ideas, and there is no question whatever, that Protestant writers are, as a rule, ignorant and false in what they say of Catholics – while Catholic writers are as a rule both well-informed and fair.'[12]

Cobbett paid his respects to the 'usual fairness' of John Lingard, the Catholic historian (Letter X, para. 293), while himself presenting an account of the English 'Reformation' – for Cobbett always in quotation marks – that was biased towards Catholicism and that rehabilitated it within a redefined patriotic radical tradition.[13] Like the Catholic historian John Milner, whom he cites in the *History* (Letter IV, para. 123), Cobbett draws attention to the abusive language that is so often applied to Catholics (Letter I, para. 10). When himself in cursing mode, however, Cobbett the farmer describes Henry VIII – the 'first born son of the Reformation' – as 'the master-butcher,

[11] As the *History* has no pagination, references are by paragraph numbers in the main text. Gladstone's copy is at St Deiniol's Library, Hawarden (shelfmark I 52.78).
[12] *Works of Ruskin* (1903–12), XXXVII, 503, 507.
[13] See J. Wolffe, *Protestant Crusade* (1991), p. 15.

fat and jocose', who 'sat in his palace issuing orders for the slaughter' (Letter IV, para. 114). His idealized version of pre-Reformation Catholic England was to have a profound influence upon Victorian medievalism.[14] Referring to the row between Milner and Sturges over the history of Winchester, discussed earlier,[15] Cobbett defends the celibate clergy, arguing that, 'if WILLIAM OF WYKHAM had been a married man, the parsons would not now have had a COLLEGE at Winchester; nor would there have been a College either at Eton, Westminster, Oxford, or Cambridge, if the bishops, in those days, had been married men' (Letter IV, para. 124). A passage on the 'melancholy change which the "reformation" has produced' at St Cross, Winchester, has the freshness of Cobbett's characteristic eye-witness accounts that were to influence Carlyle's *Past and Present* (1843) and Ruskin's *Fors Clavigera* (1871–84), as well as Trollope's Barsetshire novels (Letter IV, para. 125).[16]

Like the Catholic historians of his generation, Cobbett strove to combat the impression, carefully cultivated by generations of Protestant historians, that Mary was a serial killer, whereas Elizabeth was innocent of Catholic blood. Once again, he critically examines the discourse that has become habitual in Protestant English culture, and substitutes for clichéd anti-Catholic jargon the earthy language of the English yeoman farmer. It will be his duty, he writes, to show that the 'mass of punishment' inflicted in Mary's reign has been 'monstrously exaggerated'; that the 'circumstances under which they were inflicted found more apology for the severity than the circumstances under which the Protestant punishments were inflicted'; that they were 'in amount as a single grain of wheat is to a whole bushel, compared with the mass of punishments under the Protestant Church, "as by law established"'; and that, 'be they what they might, it is a base perversion of reason to ascribe them to the principles of the Catholic religion' (Letter VIII, para. 223). Our 'deceivers', he adds, have taught us to talk of 'the reign of "BLOODY QUEEN MARY"; while they have taught us to call that of her sister, the "GOLDEN DAYS OF GOOD QUEEN BESS." They have taken good care never to tell us, that, for every drop of blood that Mary shed, Elizabeth shed a pint.' (In fact around three hundred Protestants were executed under Mary and around two hundred Catholics under Elizabeth.)

Like Milner, Cobbett argues that this Protestant tradition is hypocritical. Of *Foxe's Book of Martyrs* he writes, 'What a hypocrite . . . must that man

[14] See R. Williams, *Culture and Society* (1963), p. 37.
[15] See pp. 81–4 above. [16] Like the *History*, *Fors* is arranged in 'Letters' and paragraphs.

be, who pretends to believe in this Fox! [sic]' (Letter VIII, para. 248). At the deathbed of her sister Mary, he records, Elizabeth 'prayed God that the earth might open and swallow her, if she were not a true Roman Catholic'; yet 'it was not long before she began ripping up the bowels of her unhappy subjects, because they were Roman Catholics' (Letter IX, para. 260), killing more Catholics in one year than Mary killed Protestants in her whole reign (Letter IX, para. 269). Her '"virgin" propensity', he argues, 'led her to prefer that sort of intercourse with men, which [he] need not more particularly allude to' (Letter X, para. 295).

Cobbett weakens his case when he loses his temper, just as popular anti-Catholic writers lose their credibility when they abandon lucid argument for mindless fulminations against the Antichrist or the Whore of Babylon. In Letter XI, for example, he describes Elizabeth as 'this wicked woman' (para. 322) and 'this foul tyrant' (para. 340), and at the end, whips himself into this frenzied (and ludicrous) statement: 'Historians have been divided in opinion as to which was the worst man that England ever produced, her father, or Cranmer; but, all mankind must agree, that this was the worst woman that ever existed in England, or in the whole world, Jezebel herself not excepted' (para. 349). What is more, he rapidly revised his idealized view of the Catholic Middle Ages soon after the publication of his book.[17]

It was in 1824, the same year as Cobbett's *History*, that Robert Southey, formerly a Radical and now the Tory Poet Laureate, published his famous response to these troubled times in *The Book of the Church*, intended as a reply to the early volumes of Lingard's *History of England*.[18] Writing in the cool style of a refined Anglican, Southey begins with a classic, conservative defence of the Established Church, arguing that it brings material as well as spiritual benefits to the nation. Interestingly, however, he also comments on the need to 'arm the young *heart*' (my emphasis) with knowledge of this fact.[19] In establishing the Church of England, Southey argues, our forefathers rescued the nation from the 'corruptions' of the medieval Catholic Church, one of the earliest of which 'grew out of the reverence which was paid to the memory of departed Saints', at a time when 'the Monks and Clergy promoted every fantastic theory, and every vulgar superstition, that could be made gainful to themselves'. Monasticism and vows of celibacy encouraged an inward-looking, obsessive and unnatural mentality, so that

[17] This occurred in 1825, when Catholic leaders accepted a policy which was 'opposed to the strict principles of parliamentary reform': G. I. T. Machin, *Catholic Question* (1964), p. 8.

[18] J. P. Chinnici O. F. M., *English Catholic Enlightenment* (1980), p. 114.

[19] R. Southey, *Book of the Church* (1824), I, 1–2. Further quotations in this paragraph are from I, 289, 295, 305, 313, 315.

'enthusiasts, in order to obtain Heaven, spent their lives, not in doing good to others, but in inflicting the greatest possible quantity of discomfort and actual suffering upon themselves'. If auricular confession is considered 'in connexion with the celibacy of the Clergy, the cause will be apparent why the state of morals is generally so much more corrupt in Catholic than in Protestant countries'. Of all the 'corruptions of Christianity', however, 'there was none which the Popes so long hesitated to sanction' as the 'prodigious' and 'astonishing' doctrine of transubstantiation.

Southey's reading of Reformation English history flows from his perception that Roman Catholicism is fundamentally 'corrupt' – the term of abuse that was later to be used most frequently by Anglican bishops during the 'Papal Aggression' crisis.[20] Whereas chapter XIV, 'Queen Mary: The Persecution', contains numerous vivid accounts of burnings at the stake and canonizes Cranmer, chapter XV, 'Queen Elizabeth', passes over the 'bowellings' of Catholics in silence.[21] Southey ends, as he began, by applying the lessons of history to the present, representing the British constitution as the guarantor of national liberty and prosperity.[22]

Charles Butler's reply to Southey, entitled *The Book of the Roman-Catholic Church* (1825), was the product of almost half a century's experience in defending Catholicism against Anglican aggression and specifically refuting Anglican readings of history. A nephew of Alban Butler, the hagiographer, he had in 1791 become the first Catholic to be called to the Bar since the 'Glorious Revolution' of 1688, when England reaffirmed its Protestantism. As Secretary, from 1782, to national committees seeking the abolition of the Catholic penal laws, he had published many pamphlets and books, the most substantial of which was his *Historical Memoirs respecting the English, Irish, and Scottish Catholics, since the Reformation to the Present Time*, in four volumes (1819–21), an enlarged third edition of which appeared in 1822.

Adopting the tone of the defence lawyer in *The Book of the Roman-Catholic Church*, Butler asks whether Southey is 'sufficiently aware' that the Roman Catholics have 'sustained a defamation of three hundred years? – That, in consequence of it, an immense mass of prejudice was raised against them? That it yet retains its place in many uninstructed minds; and that it is not wholly eradicated from all the liberal and informed?'[23] In responding to Southey with a blow-by-blow defence of Rome and her doctrines, Butler frequently quotes Lingard, his co-religionist, and fills the gaps in his opponent's case, providing long sections on the persecution of Catholics during

[20] See p. 30 above. [21] Ibid., II, 141–251, 241, 252–312. [22] Ibid., II, 528.

[23] C. Butler, *Book of the Roman-Catholic Church* (1825), p. iv. Further quotations in this paragraph are from pp. 239–40, 243, 269, 271–2, 219.

the reign of Elizabeth – the frightful tortures in the Tower of London, Campion on the rack, the execution of Southwell, the pressing of Margaret Clitheroe. Being a barrister, he is also ready with a proposed settlement: 'Let protestants cease to reproach the roman-catholics with Mary's fires, and roman-catholics shall be equally silent on the sanguinary code of Elizabeth, and the savage executions under it.'

Butler's arguments against Southey were in turn answered by no fewer than ten Protestant writers. Among these, Henry Phillpotts, then Rector of Stanhope, later Bishop of Exeter, and always a controversialist, attempted to undermine Butler's credibility by challenging his sources, devoting Letter IV to 'Dr. Lingard. – His unfaithfulness in quotation'.[24] Joseph Blanco White, who had the distinction of abandoning first Catholic and then Anglican orders, accused Butler of lumping Protestants together with pagans.[25] George Townsend, an Anglican clergyman and editor of Foxe, described Butler's book as 'an entire failure', a view that would seem to be disproved by the sheer volume of Protestant writing that it provoked.[26]

In the face of such impassioned opposition, Butler published a vindication of *The Book of the Roman-Catholic Church* in 1826, which was answered by a further six Protestant books, including Southey's own vindication.[27] In terms of redrawing the contours of English Reformation history, however, Butler, together with Catholic historians like Lingard and Milner, failed to make much immediate impression on mainstream Protestant opinion. In Sharon Turner's preface to the first part of his *Modern History of England*, on the reign of Henry VIII (1826), he explained that he had read the discussions between his 'highly respected friends, Mr Southey and Mr Butler', that he did not want to upset his Catholic colleagues, but that new manuscript evidence convinced him that the Catholic clergy who were executed under Henry were indeed guilty of acts of insurrection and treason.[28] Four years later, Coleridge was to endorse Turner's work in forthright terms: 'The *most* honest of our English historians, and with no *superior* in industry and research, Mr. Sharon Turner, has labored successfully in detaching from the portrait of our first Protestant king the layers of soot and blood,

[24] H. Phillpotts, *Letters to Butler* (1825), pp. 101–12.

[25] J. B. White, *Letter to Butler* (1826), p. 43. Samuel Taylor Coleridge, author of *Aids to Reflection* (1825), wrote notes in his presentation copy of White's *Letter*, questioning some of Butler's evidence: British Library copy, shelfmark C.126.h.12, pp. 84–92. White also published *Practical and Internal Evidence against Catholicism* (1825).

[26] G. Townsend, *Accusations of History* (1825), p. 2.

[27] R. Southey, *Vindiciae Ecclesiae Anglicanae* (1826).

[28] S. Turner, *Modern History of England* (1826–9), I, v–x.

with which pseudo-Catholic hate and pseudo-Protestant candour had coated it.'[29]

Similarly, when Henry Soames, Rector of Shelley in Essex, published the fourth volume of his *History of the Reformation of the Church of England* in 1828, just a year before Catholic Emancipation, he focused upon Mary's cruelty, devoting around 600 pages to the 'History of the Reformation under Queen Mary' and only 150 to the 'History of the Reformation under Queen Elizabeth'.[30] Mary was 'far too narrow-minded for the government of a kingdom', he argues, whereas 'to the wisdom of Elizabeth's religious choice her native land has borne uninterrupted testimony ever since her auspicious occupation of its throne'.[31]

So Butler had not changed the hearts and minds of Protestant historians of the Reformation. His purpose, however, in engaging in a battle of the books over Reformation history was to gather ammunition for the debate over Catholic reform; and in terms of reform, he was soon to be on the winning side, in 1829.

III

In the early eighteenth century, Andrew Snape, a High Anglican clergyman, claimed that the Jesuits had always regarded the Church of England as the 'Grand Bulwark of the *Reformation*, and the Mound that keeps out the Inundation of *Popery* from once more overflowing all Europe'.[32] Similar claims were made one hundred years later by Richard Parkinson, Fellow of Christ's College, Cambridge, in his *The Church of England a Bulwark between Superstition and Schism* (1835). The Baptist, Hugh Stowell Brown, summarized the Establishment position in this way:

In parliamentary debates, in public speeches, in lectures, in sermons, in newspaper articles, it has been asserted, as if it were an indisputable axiom, a first principle, a self-evident truth, that a State Church is the Bulwark of Protestantism; that, were this barrier swept away, nothing would be left to protect us from the inroads and invasions of Popery, which would soon overrun the country and demolish all our Protestant institutions.[33]

[29] S. T. Coleridge, *Constitution of Church and State* (1976), p. 52 .

[30] H. Soames, *History of the Reformation* (1826–8), IV, 1–596, 597–740. [31] Ibid., pp. 595, 740.

[32] A. Snape, *Second Letter to the Bishop of Bangor* (1717), cited in R. D. Tumbleson, *Catholicism in the English Protestant Imagination* (1998), p. 165.

[33] H. S. Brown, *Bulwark of Protestantism* (1868), p. 4. Brown went on to argue that, 'In so far . . . as human agency can be a bulwark for Protestantism against the deadly, mighty, and subtle foe that is ever watching at her gates and assailing her defences, we may look to some other quarter than that of the State Church, which has so signally failed that it is, in reality, rather the way to Rome than anything else': ibid, pp. 14–15.

Meanwhile, a real bulwark against Catholicism maintained the *status quo* in Ireland, where only a quarter of the population was Protestant. Dublin Castle was run by a 'rabidly Protestant clique'.[34]

It would be mistaken, however, to regard the seventy-four martello towers that were built around the English coastline during the invasion scare of 1803 as Protestant England's defences against Continental Catholicism.[35] After all, ten years earlier, Fortress England had lowered the drawbridge in order to admit many foreign Catholic priests, after their expulsion from Revolutionary France. In 1803, John Bull thought of the external enemy as a Corsican upstart. Typical of the numerous broadsides published that year was a single sheet entitled 'People of the British Isles', which trumpeted that we 'have to defend and to maintain, such glorious privileges as collectively no other nation on earth can boast of possessing. We have a MAGNA CHARTA and a FREE PRESS; but above all, our glorious and invaluable Constitution, the admiration and the wonder of the world.'[36] Some of the latter-day prophets of the time did, however, choose to associate the new threat posed by Napoleon (who had re-opened the churches of France in April 1802) with the old threat of Romanism. In his *Hint to England* (1803), for example, Lewis Mayer quoted Daniel 7.11 to support his argument that,

> As the Dragon represents the ecclesiastical power of Rome, the speaking like a Dragon hath evidently an allusion to the threatenings and anathemas that have been thundered forth from Rome; and similar to the threatenings therefore of Pope Pius V. at the time of the invasion of Britain by the Spanish armada, Bonaparte now saith, My invincible armies shall pass over and subdue, ravage and plunder thee, and deliver thy inhabitants over to perpetual bondage; but the usurper should know that the fire from God, namely, defensive war, may destroy him, and prove those proud words are the certain indications of his approaching ruin . . .[37]

As usual, it is the constitution that is said to be in danger.[38]

After Waterloo, in the early years of a lengthy period of *pax Britannica*, the threat to the constitution posed by Catholicism replaced the threat of an invasion from France in the public mind. Thomas Burgess, the Bishop of St David's, expressed views that were shared by other senior Anglican clergy in his *Protestant's Catechism* (1818):

[34] P. Ziegler, *Melbourne* (1976), p. 92.

[35] See H. F. B. Wheeler and A. M. Broadley, *Napoleon* (1908), II, 150.

[36] See F. J. Klingberg and S. B. Hustvedt, eds, *Warning Drum* (1944), p. 198. Also see the British Library's collection of *Broadsides, etc., relating to the expected Invasion of England by Bonaparte* (pressmark 1851.c.3).

[37] L. Mayer, *Hint to England* (1803), p. 13.

[38] Ibid., p. 36 *et passim.*

Q. How would the re-admission of Papists to political power be *destructive* to our Protestant Constitution in Church and State? A. Papists could not be admitted into Parliament without *repealing* the principle of all the great constitutional statutes. And though all these statutes would not at once be repealed, the Constitution would, for the most part, in fact, and *wholly in principle*, cease to be Protestant.[39]

The defence of the constitution against 'Popery' was enshrined in the Parliamentary Oath:

I AB do swear that I do from my heart detest and abjure as impious and heretical that damnable doctrine and position, that Princes excommunicated or deprived by the Pope or any authority of the See of Rome, may be deposed or murdered by their subjects or any other whatsoever; and I do declare that no foreign prince, prelate, state or potentate hath or ought to have any jurisdiction, power, superiority, pre-eminence or authority, Ecclesiastical or Spiritual, within this realm.[40]

In 1829, however, Gilbert Chesnutt argued that no 'efficient security' was to be found in the 'imposition of Oaths and Declarations'.[41] In his view, the concession of Catholic claims amounted to 'no less than a radical and total change in the principles upon which the religious and civil institutions of a great nation are founded'.[42]

Talk of 'Protestant securities' against Rome – an obsession in the 1820s – reflects a deep sense of insecurity. With the general public becoming increasingly anxious about radical reforms at home and revolutionary movements abroad, England was receptive to representations of a world in which events reflected a cosmic battle between good and evil, the saved and the damned, us and them, Protestant and Catholic. Hence, in part, the popularity of the 'apocalyptic sublime' in English art, epitomized by John Martin's paintings, such as *The Fall of Babylon* (1819), *Belshazzar's Feast* (1821) and *The Destruction of Pompeii* (1822).[43] Millennialist readings of history that had flourished in England since the French Revolution of 1789, described by George Croly as the 'most stupendous transaction of Europe since the Christian Æra',[44] were now applied to the Catholic threat.

Croly (1780–1860), a giant of a man, had an unusual career, making his way as a poet and author in London for many years, before finally being presented with the living of St Stephen's, Walbrook, in the City

[39] T. Burgess, *Protestant's Catechism* (1818), p. 34.

[40] Cited in E. R. Norman, *English Catholic Church* (1984), p. 52. Cf. the Coronation Oath, p. 3 above.

[41] G. Chesnutt, *Protestant Securities* (1829), p. 8.

[42] Ibid., p. 1.

[43] See M. D. Paley, *Apocalyptic Sublime* (1986).

[44] G. Croly, *Poetical Works* (1830), I, iii.

of London in 1835.[45] An Ultra-Tory and an Orangeman, he refers to the 'coxcombery' of Butler in *Popery and the Popish Question* (1825) and, like his fellow Churchmen, bases his own argument upon the defence of Protestant 'liberty' against the 'yoke' of Catholicism.[46] The 'Papist', he believes, demands the 'unconditional surrender of the last security of the rights and lives of Protestantism'.[47] The 'Popedom' looks upon England as the 'great antagonist', and Ireland as the 'most devoted auxiliary', and now the 'note of Irish Papism' can be heard in the taverns of London.[48] The Catholic priest performs the Mass for money, claiming that 'what was but flour and water this moment', will, at the next, by the simple effect of his 'prayer and fingers', be the 'actual omnipotent Lord that reigneth in the heaven of heavens!'[49]

It is in the later stages of the book, however, that Croly's most characteristic note is struck, as he draws upon metaphors of invasion and defence:

> Can these things be forgotten, when we see Popery advancing to the gates of our Parliament and demanding to be let in? Shall we suffer it to march through those portals of the constitution, with spread ensigns, and in the undiminished pride of holy victory, with the banner of St Dominic, and the sword of St Bartholomew; or shall we not compel it to disarm before its approach, and require from it the common oath of all, to the safety of the commonwealth?[50]

Basing his views upon a fiercely patriotic idealism and a millenialist reading of history, Croly closes with a vision of England as providentially chosen for the defence of true religion:

> . . . if the worse come, let us have the high consolation . . . that we have not, in an unworthy fear, voluntarily abdicated the throne, on which England sits as the SOVEREIGN PROTECTOR OF PROTESTANTISM THROUGHOUT THE WORLD! that we have not flung open to the torrent of impiety and folly, of revolution and superstition, the portals of that Temple, whose wall our fathers "built in troublous times," and where their sons still see the visible presence of God's glory! But if trial should still come, let us take our stand, armed in a confiding faith, and look up to PROVIDENCE![51]

Two years later, Croly turned to the Book of Revelation for signs of the times, in *The Apocalypse of St John* (1827):

[45] See T. D. Gregg, *Christian Hero* (1861), pp. 13, 15; R. Herring, *A Few Personal Recollections* (1861), p. 4.
[46] G. Croly, *Popery* (1825), p. 131, 2–3. Cf. p. 11 above. [47] Ibid., p. 5.
[48] Ibid., pp. 23, 36. [49] Ibid., p. 54. [50] Ibid., p. 114. [51] Ibid., pp. 146–7.

It is fully predicted that there shall be a sudden revival of Atheism, superstition, and religious violence, acting upon the European nations until they are inflamed into universal war. All the elements of terror and ruin shall be roused; Protestantism persecuted; Popery, after a momentary triumph, utterly destroyed; in a general shock of kingdoms, consummated by some vast and palpable development of the Divine Power, at once protecting the Church, and extinguishing, in remediless and boundless devastation, infidelity and idolatry.[52]

It is in the preface to *The Apocalypse of St John* that Croly first makes the claim that is quoted in the epigraph to this chapter: 'There is the strongest reason for believing, that as Judæa was chosen for the especial guardianship of the original Revelation; so has England been chosen for the especial guardianship of Christianity.'[53] Croly's preface was to be separately reprinted several times, with added reflections upon the latest 'signs'. In 1828 it appeared as *The Englishman's Polar Star!!*, with a note pointing out that the disasters surrounding Canning's recent resignation as Prime Minister were a 'further confirmation of the opinions held forth in this Preface'.[54] Its main theme, however, is best summarized in the title by which it came to be known more widely – *England the Fortress of Christianity*. For Croly, England's divinely ordained constitution was once again under siege. The Tory government, in attempting to solve the Irish Question while also giving way to pressure for Catholic Relief in England, had opened a gate which had long been guarded, making Fortress England vulnerable to invasion by a foreign prince, the Antichrist of the seven hills. Adding to remarks that he had originally published 'on the eve of the year 1829', Croly writes:

The Bill of that calamitous year replaced the Roman Catholic in the Parliament, from which he had been expelled a century before, by the united necessities of religion, freedom, and national safety. The whole experience of our Protestant history had pronounced that evil must follow. *And it has followed.*

From that hour all has been changed. British legislation has lost its stability. England has lost alike her pre-eminence abroad, and her confidence at home. Every great institution of state has tottered. The Church in Ireland, bound hand and foot, has been flung into the furnace, and is disappearing from the eye. The Church in England is haughtily threatened with her share of the fiery trial.[55]

If the kind of xenophobic, anti-Catholic bombast that characterizes Croly's work reflects a deep sense of anxiety, so too does the sheer bulk

[52] G. Croly, *Apocalypse of John* (1827), p. 4.

[53] Ibid., p. i. This view had been not uncommon among some Protestants for generations.

[54] G. Croly, *Englishman's Polar Star* (1828), pp. 23–4.

[55] G. Croly, *England the Fortress of Christianity* (1839), p. 6. Croly's sustained interest in the essay is reflected in his frequent revisions. See, for example, the tightening of the phrasing in the opening paragraph in his *Historical Sketches, Speeches, and Characters* (1842), p. 1.

of Protestant epic poems of the period on apocalyptic and eschatological themes, as thousands of lines of blank verse are piled one upon the other to create a kind of heavenly firewall. Although Croly was known mainly as a poet in his own day, his verse made nothing like the impact of one of the best-sellers of the decade, *The Course of Time* (1827), by the young Scottish Secessionist, Robert Pollok.[56] The epic's narrator, an 'ancient bard of Earth', in paradise after the last judgment, offers a classic, uncritical Protestant view of the authority of Scripture:

> . . . the author, God himself;
> The subject, God and man, salvation, life
> And death – eternal life, eternal death –
> Dread words! whose meaning has no end, no bounds!
> Most wondrous Book! bright candle of the Lord![57]

Pollok regarded Church Establishment as harlotry – a vestige of a 'polluted' Catholic Christendom, prophesied in the Book of Revelation:

> . . . the scarlet-coloured Whore,
> Who on the breast of civil power reposed
> Her harlot head – the Church a harlot then,
> When first she wedded civil power – and drank
> The blood of martyred saints; whose priests were lords,
> Whose coffers held the gold of every land,
> Who held the cup of all pollutions full,
> Who with a double horn the people pushed,
> And raised her forehead, full of blasphemy,
> Above the holy God, usurping oft
> Jehovah's incommunicable names.[58]

At the Millennium, the 'Church and State, that long had held / Unholy intercourse, were now divorced'.[59] The 'bad years' that follow are characterized by the clergy's 'slumbering in the lap of civil power'. On the last day, all the 'scarlet troops' of Antichrist are condemned, 'Familiar most in hell, their dungeon fit'. In a long poem, structured in such a way that man's sinfulness can be dwelt upon repeatedly, Book Eighth describes the 'last pause of expectation' before the last judgment. Pollok reserves some of his strongest rhetoric for Roman Catholic religious. Of the monk, he writes:

[56] There were two 'Secessions' from the Established Presbyterian Church of Scotland in the eighteenth century (1733, 1761). The major 'Disruption' occurred in 1843, when the Free Church of Scotland was formed.

[57] R. Pollok, *Course of Time* (1867), pp. 25–6.

[58] Ibid., p. 127. Pollok echoes Revelation 17.1–6, 13.11.

[59] Ibid., p. 130. Further quotations in this paragraph are from pp. 141, 172, 190–1.

Unprofitable seemed, and unapproved
That day, the sullen, self-vindictive life
Of the recluse. With crucifixes hung,
And spells, and rosaries, and wooden saints,
Like one of reason reft, he journeyed forth . . .
On his own flesh inflicted cruel wounds;
With naked foot embraced the ice, by the hour
Said mass, and did most grievous penance vile;
And then retired to drink the filthy cup
Of secret wickedness, and fabricate
All lying wonders, by the untaught received
For revelations new. Deluded wretch!
Did he not know that the most Holy One
Required a cheerful life and holy heart?

The contemporary taste for this kind of verse helps to explain how Edward Irving, Minister of the Caledonian Chapel, Hatton Garden, became the most celebrated preacher in London in the 1820s. Having been intoxicated with his own success, he developed a strongly millenarian line in his three hour long services. Like his sermons, his books related biblical prophecy to the spiritual condition of Britain, which, like Croly, he regarded as a Protestant bulwark against both Catholicism and infidelity. In the preface to his *Babylon and Infidelity Foredoomed of God* (1826), for example, he explained that his discourse was

composed for the special instruction of the members of a society, called, THE CONTINENTAL SOCIETY, established in London some years ago, for the purpose of seeking the Lord, exercising carefulness, and labouring by all means for the spiritual condition of the Continent of Europe; distressed at present on the one hand, by the superstition of the Papists, and, on the other, by the Infidelity of those calling themselves Reformed.[60]

Britain alone, he argued, among the ten nations 'implicated with the Papacy', had been 'preserved from the six plagues' that had come upon the 'Papal nations' and from the 'thirty year's judgments which have come upon all the rest'.[61]

In Irving's view, the country's 'prosperity and glory, inward peace and outward power, during the last thirty years, have been due to our abjuration of the Papal superstition and idolatry'.[62] God's 'gracious providence' can be discerned in history:

[60] E. Irving, *Babylon and Infidelity* (1826), I, ix. [61] Ibid., II, 338.
[62] Ibid., II, 368. Further quotations in this paragraph are from pp. 370, 373, 378.

He brought his servant, William of Orange, from that family which had laboured most and suffered most in the cause of reformation, and exalted him to be the king of that nation, which, in like manner, had travailed through two centuries of distress in the same cause of her glory. Fit monarch for such a people. Fit husband for such a spouse.

Britain is the modern Israel, delivered out of ('Popish') slavery; but if we 'extend the same encouragement to the idolatry of Rome, as we do to the worship of the living and true God, then the anger of the Lord will burn against us as a fire'.

In June 1829, Irving stayed at Craigenputtoch and dined with his old friends the Carlyles. Irving had recently published an apocalyptic pamphlet entitled *The Signs of the Times*, which began with the proposition that, 'like Jerusalem, we have not known, and will not know, the time of our visitation'.[63] Writing shortly before the decision on Catholic Emancipation was taken, in April, Irving had expressed his deepest anxieties in the language of Daniel and the Revelation:

I, with fear and trembling, with much alarm, am waiting, with all the wise and understanding in this land, to see what will be the issue of these attempts towards a concordat, or treaty, or agreement, which they say are going on in high quarters. If God give us up to our own foolish mind in this matter, and bring us again into that family of hell, then observe you what a sign it will prove: for then, the ten horns of the beast shall be again brought under the dominion of the little horn; which hath not been since the Reformation: and God's testimony by Britain against the Papacy being by Britain relinquished, he will begin to testify after another sort, by judgments fiery and consuming.[64]

For his part, Carlyle had been uneasy about Irving since seeing him in Edinburgh the previous month, in May 1829.[65] In June, his own prophetic essay entitled 'Signs of the Times' appeared in the *Edinburgh Review* and began with the observation that it is 'no very good symptom either of nations or individuals, that they deal much in vaticination'.[66] Whereas Irving uncritically adopts the role of Old Testament prophet, Carlyle plays with the role, teasing the doom merchants: 'The "State in Danger" is a condition of things, which we have witnessed a hundred times; and as for the Church, it has seldom been out of "danger" since we can remember it.' The repeal of the Test Acts and of Catholic disabilities, he observes, has 'struck many of their admirers with an indescribable astonishment'. Our

[63] E. Irving, *Signs of the Times* (1829), p. 1. [64] Ibid., pp. 11–12.
[65] See F. Kaplan, *Thomas Carlyle* (1983), pp. 147–9.
[66] T. Carlyle, 'Signs of the Times', in *Works* (1884), XIII, 462. Further quotations in this paragraph are from XIII, 463, 464.

'worthy friends' mistook the 'slumbering Leviathan for an island', and, 'mooring under the lee, they had anchored comfortably in his scaly rind'. But now 'their Leviathan has suddenly dived under; and they can no longer be fastened in the stream of time':

At such a period, it was to be expected that the rage of prophecy should be more than usually excited. Accordingly, the Millennarians [sic] have come forth on the right hand, and the Millites on the left. The Fifth-monarchy men prophesy from the Bible, and the Utilitarians from Bentham . . . Left to themselves, they will the sooner dissipate, and die away in space.

Carlyle's rhetorical strategy is to satirize other kinds of prophecy in order to make space for his own, which he presents in a calm, commonsensical tone: 'It is the Age of Machinery, in every outward and inward sense of that word; the age which, with its whole undivided might, forwards, teaches and practises the great art of adapting means to ends.'[67] Whereas the apocalyptic sublimities of Pollok, Croly and Irving were of and for the 1820s, Carlyle's essay was one of the documents that set the agenda for the Victorian Age. Another was published the same year. Coleridge's concept of the national Church as the 'clerisy', in *On the Constitution of the Church and State* (1829), profoundly influenced figures such as Thomas Arnold, F. D. Maurice and W. E. Gladstone, and, together with *Aids to Reflection* (1825), shaped liberal Anglican thought for several generations.[68] While disagreeing with Irving's interpretation of the Revelation, however, Coleridge shares his hostility to Roman Catholicism. On a recent tour of Holland, Flanders, and up the Rhine as far as Bergen, Coleridge had made the following note:

Every fresh opportunity of examining the Roman Catholic religion on the spot, every new fact that presents itself to my notice, increases my conviction, that its immediate basis, and the true grounds of its continuance, are to be found in the wickedness, ignorance, and wretchedness of the many; and that the producing and continuing cause of this deplorable state is, that it is the interest of the Roman Priesthood, that so it should remain, as the surest, and in fact, only support of the Papal Sovereignty and influence against the civil powers, and the reforms wished for by the more enlightened governments, as well as by all the better informed and wealthier class of Catholics generally.

[67] Ibid., XIII, 465. George Eliot's own, retrospective analysis of a 'Mechanical Age' in *Felix Holt, The Radical* (1866), begins with her 'Author's Introduction', in which she comments that, 'till the agitation about the Catholics in '29, rural Englishmen had hardly known more of Catholics than of the fossil mammals'.

[68] See S. T. Coleridge, *Church and State* (1976), p. lxii. Further quotations in this paragraph are from pp. 140–1, 136, 51.

He describes the Middle Ages as the 'dark times, when the incubus of superstition lay heavy across the breast of the living and the dying; and when all the familiar "tricksy spirits" in the service of an alien, self-expatriated and anti-national priesthood were at work in all forms, and in all directions, to aggrandize and enrich a "kingdom of this world"'.

IV

Whereas Protestant commentators on Catholic Emancipation thought in terms of defence, Catholics used the language of liberation. Cardinal Wiseman later claimed that the 'year 1829 was to us what the egress from the catacombs was to the early Christians'.[69] When Ronald Knox preached a centenary sermon in 1929, he revised Croly's and Irving's parallel between Protestant England and Israel, telling his co-religionists that, before 1829, 'we suffered like Israel in Egypt'.[70] The concept of a 'Catholic stronghold' was therefore problematic in post-Reformation England, not only for Protestants, but also for Catholics themselves. After all, their sanctuaries had long been remote country houses and private chapels, rather than massive castles and awe-inspiring cathedrals.

The symbol of the ancient castle was deeply inscribed in the eighteenth-century Gothic tradition inherited by Ann Radcliffe, whose novel, *The Mysteries of Udolpho* (1794) centres upon Montoni's castle in Italy – a 'gloomy and sublime object' (II, 5), and the site of the vulnerable heroine's entrapment. 'Since Mrs Radcliffe's justly admired and successful romances', observed *The Critical Review* in 1795, 'the press has teemed with stories of haunted castles and visionary terrors'.[71] The Minerva Press and their rivals knew their markets, and no fewer than 62 novels published in Britain between 1794 and 1810 had the word 'castle' in the title.[72] This literary castle-building coincided with the French Revolutionary and Napoleonic Wars, which followed the destruction of the fortress that symbolized feudalism and *l'Ancien Régime* – the much hated Bastille.

Another Continental castle, but this time an idealized one, achieved widespread fame in the 1820s when Kenelm Henry Digby published and then expanded his study on chivalry – *The Broad Stone of Honour*

[69] N. P. S. Wiseman, *Religious and Social Position* (Dublin and London, 1864), p. 9.
[70] R. Knox, *Occasional Sermons* (1960), p. 165.
[71] J. Raven, *et al.*, *English Novel* (2000), I, 629. Cf. e.g. Dyer's *Grongar Hill* (1726) and Cowper's *The Task* (1785).
[72] This figure was arrived at by counting the revelant titles in ibid.

(1822–7).[73] The first version was subtitled 'Rules for the Gentlemen of England' (1822), and a second edition was called for in 1823. In the Prologue, Digby describes the work as a 'book of ensamples and doctrines, which I call *The Broad Stone of Honour*; seeing that it will be a fortress like that rock upon the Rhine where coward or traitor never stood, which bears this proud title, and is impregnable'.[74] Digby visited the castle of Ehrenbreitstein, opposite Coblentz, on the Rhine, while an undergraduate at Trinity College, Cambridge, where, following a night alone in King's College Chapel, he had resolved to become a knight.[75] The classic perpetual student, he stayed on in Cambridge for ten years after graduation, keeping in touch with a generation of brilliant liberal Protestant dons, including his tutor, Julius Hare, who wrote favourably on *The Broad Stone*.[76] When, however, Digby converted to Roman Catholicism in 1825, bought up all the copies he could, and produced a new, much larger version in 1826, Hare found that, though even finer, it could not be 'recommended without hesitation to the young': 'The very charm, which it is sure to exercise over them, heightens one's scruples about doing so. For in it the author has come forward as a convert and champion of the Romish Church, and as the implacable enemy of Protestantism.'[77]

In the first of four books – *Godefridus*, *Tancredus*, *Morus* and *Orlandus* – in the larger, Catholic version of *The Broad Stone*, Digby expresses the wish that his *magnum opus* 'may resemble one of those beautiful old cities of Spain, in which one finds everything'.[78] The instinct to include everything – which he shared with other writers of his day, including Coleridge – leads in his case to prolixity, and this helps to explain why he is now seldom read. He does, however, provide an outline in the Prologue to *Godefridus*, explaining that, in the first book he will 'endeavour to give a general idea of the views and principles respecting chivalry'; the second will contain a 'view of the religion and the discipline which belonged to chivalry in the heroic age of Christianity'; the third will consider the 'objections which have been urged by various sects of innovators against the principles and practice of the Christian chivalry'; and in the fourth, the 'main subject will be resumed, by giving a more detailed view of the virtues of the chivalrous

[73] Recent studies of 'medievalism' merely nod at Digby in passing references, thus distorting literary history. See, e.g., C. A. Simmons, *Reversing the Conquest* (1990), pp. 84, 103, 107; C. A. Simmons, ed., *Medievalism* (2001), p. 5.

[74] K. H. Digby, *Broad Stone of Honour* (1823), p. xiii.

[75] See B. Holland, *Memoir of Digby* (1919), pp. 9–10.

[76] J. C. and A. W. Hare, *Guesses at Truth* (1867), pp. 163–4.

[77] Ibid., p. 165.

[78] K. H. Digby, *Broad Stone of Honour* (1877), *Godefridus*, p. 65. Digby refers here to Victor Hugo's 'Les Orientales'.

character, when it was submitted to the genuine and all-powerful influence of the Catholic faith'.[79]

Digby believes that the Catholic Church in the ages of faith was a 'society in which not only all the wants of each man's own heart were fulfilled, but where alone all the principles which they so much desired to behold in action, of political justice and of the highest legislative wisdom, were realised and maintained in exercise'.[80] He seizes upon an important feature of Catholicism – that it is a religion of the heart – and relates this to chivalry. 'It is clear', he writes, that the 'philosophy of chivalry must be a philosophy of the heart; therefore it was prepared to admit the wisdom of those counsels which said, with St Augustin [sic], that "the intention makes the good work"; and with St Thomas, that "moral actions receive their form from the intention."'[81]

Like some other converts from Anglicanism,[82] Digby describes the faith of Catholics in terms of the security which issues from certainty: 'difficulties are removed; the night is past; a bright and everlasting day has dawned; there is an end of wandering and uncertainty, of doubt and disputation. All the articles of faith and all the truths of revelation are immoveably and definitely settled.'[83] The title-page vignette of the castle of Ehrenbreitstein on its great rock thus takes on a new meaning, as a symbol of chivalry, established upon the (Petrine) rock of the Catholic faith. Digby offers a lengthy analysis of castles and their significance in *Godefridus*, beginning with the quiet observation that, in the Middle Ages, the 'feudal nobility lived in castles, which, as may be inferred from the names of many, like Rochefort, Rochetaillée, Montfort, and others, were generally placed upon high rocks'.[84] His tour of European castles includes England, where he conjures up the archetypal native scene: 'What beauty in the terraced height, the ancestral grove of heroes dear to fame, the softly swelling hills, the ivy-mantled tower washed by the silver stream, the hoary cloister, and the level lawn!' Then, having moved from the individual features of particular castles to their shared characteristics and their meaning in relation to chivalry, he creates a vision which reminds us that knew Tennyson: 'No wonder that many of these castles, cresting high peaks of rock, should have seemed like the work of enchantment, when beheld at the rising of the sun, with their long exterior galleries glittering with the armour of those who kept watch, so that the whole seemed as if lighted up with living flames!'

[79] Ibid., p. 11. [80] *Godefridus*, pp. 158–9. [81] Ibid., p. 226. [82] See pp. 179–80 below.
[83] From *Morus*, cited in Holland, *Memoir of Digby* (1919), p. 34.
[84] *Godefridus*, p. 280. Further quotations in this paragraph are from pp. 352–3, 357–8.

Between *The Broad Stone of Honour* and *Idylls of the King* there appeared another castellated vision of the Middle Ages by one Julia Tilt, a society versifier whose main theme, frequently reiterated, is '*sic transit gloria mundi*'. Significantly, the title poem in *Arundel Castle, and other Poems* (1849), dedicated to the daughters of 'Her Grace the Duchess of Norfolk' and subscribed to by four Duchesses, four Marchionesses and eleven Countesses, is 'A Vision'. The narrator falls asleep, and dreams of Arundel Castle in the Catholic Middle Ages:

I looked; and lo, fair Arundel
Had faded into air;
The stately keep, the castle walls,
Alone were standing there.

I gazed into the vale below,
And by the water's side;
Where late I saw, the haunts of man,
An abbey flourished wide.[85]

The 'Empress Maud' (or Matilda) and the future King Henry II, who are fleeing from Stephen's 'rebel band', are welcomed to the castle by Queen Adeliza (or Alice), who announces that

'The rebel chief may hedge it round,
And lay it stone by stone,
Ere I give up my sovereign queen,
And send her forth to roam.'

Stephen's army lays seige to the castle for three weeks, after which Maud is allowed to move to 'some other tower'. As the dream fades, it is not masculine chivalry that is celebrated, but Queen Adeliza's feminine 'social virtues' and 'kindly deeds':

And dames as true, and maids as fair,
Still bless the Howard name;
And equally from rich and poor,
Respect and reverence claim.

The virtues of the lovely queen
Shine brighter still in them;
They're like the blossoms of the rose,
And she the parent stem.

[85] J. Tilt, *Arundel Castle* (1849), p. 2. Further quotations in this paragraph are from pp. 7, 10, 12. A second, revised edition appeared in 1852.

In reality, both the Howard line of descent and the history of Arundel Castle are extremely complex, not least with regard to Catholicism. On his accession to the throne, Henry II showed his gratitude by giving the castle, then in royal hands, to the Earl of Arundel, William de Albini, and his successors.[86] Five hundred years later, during the English Civil War, the castle was left 'almost a ruin' after General Waller had beseiged it in 1643–4.[87] According to Father Tierney, the Duke of Norfolk's chaplain from 1824, it was not until about 1720 that 'any attention appears to have been bestowed on these ruinous remains'. The castle's future as the principal seat of the modern Dukes of Norfolk was not secured, however, until the end of the eighteenth century. Charles, the scholarly and rather morose 10th Duke, who succeeded to the title in 1777, described himself as a 'whig Papist – a monster in nature'.[88] A practising Catholic, he believed that religion and politics should be kept separate, and resented the fact that, although head of the peerage, he was unable to play a major role in public life. (In 1764 he had published his *Considerations on the Penal Laws against Roman Catholics in England and the new-acquired Colonies in America*.) In 1783, he managed to secure the revenues from the renewal of the leases of the Strand estate in London for the restoration of Arundel Castle, through a private Act of Parliament.

The major restoration and reordering of the castle was actually carried out by his son, Charles, the 11th Duke, whose architectural skill and enthusiasm for the revival of Norman architecture place him in the first rank of Gothic builders of the period. Work began in 1791. Tierney records that the 'erection of the north-west front, which was begun in 1795, produced a manifest enlargement of the ancient edifice', and that the 'north-east wing, which contains the library, was commenced in 1801'.[89] William de Albini, the 3rd Earl, had been present at Runnymede, and his name inserted at the head of the great Charter. In 1806, the 11th Duke, who has been described as a 'feudal-republican',[90] began to build the famous Barons' Hall at Arundel, designed to 'commemorate the triumph of the barons over the tyranny of King John'.[91] It was sufficiently finished to be inaugurated on 15 June 1815, the same day as the Duchess of Richmond's ball in Brussels, before the battle of Waterloo. One nineteenth-century observer regarded the Barons' Hall as an example of dynastic medievalism:

[86] M. A. Tierney, *History of Arundel* (1834), p. 15. [87] Ibid., pp. 61–79.

[88] J. M. Robinson, *Dukes of Norfolk* (1995), p. 166. I am indebted to this study for the information on the Dukes of Norfolk that follows.

[89] M. A. Tierney, *History of Arundel* (1834), pp. 82–3.

[90] J. M. Robinson, *Dukes of Norfolk* (1995), p. 171.

[91] M. A. Tierney, *History of Arundel* (1834), p. 84.

> The late Charles Duke of Norfolk, in restoring his ancient Castle of Arundel to more than its former splendour, designed as the grand and characteristic feature of it, a Hall of such magnitude and embellishment as should rival those which are celebrated in every Gothic legend, and some of which remain in England to the present day.
>
> Upon the very site of one which had for the lapse of at least three hundred years been the scene of the baronial pomp and hospitality of his noble ancestors, and which had been reduced to ruin by the siege in 1643, he laid in 1806, the foundation of the present magnificent structure.[92]

The festival, this commentator felt, was 'intended to renew and exhibit (if upon this occasion only) the "pomp of older days"'.[93]

Ironically, it is the Catholic chaplain, Tierney, who refers to the 'six hundredth anniversary of the great foundation of English *liberty*', a term long ago appropriated by a Protestant Establishment in England.[94] Furthermore, the 11th Duke, who could hardly have been more different from his father, being a hard-drinking, generous extrovert, a friend of the Prince Regent, and who failed to toast the King at the festival, was known to posterity not only as the 'Drunken Duke', but also as the 'Protestant Duke'. He was an Anglican. Thus the Duke who built the Barons' Hall as a sign of dynastic continuity himself represented a break – one of several down the centuries – in the Howard family's strong links with the Catholic Church.

It was only with the accession in 1815 of Bernard Edward, the first of the devoutly Catholic Glossop Howards to become Duke of Norfolk, that another kind of restoration was achieved.[95] In 1824, the hereditary office of Earl Marshal, from which they had been excluded by their religion, was restored to the family, with the result that, irony of ironies, the nation's senior Catholic has presided at Protestant coronations in Westminster Abbey ever since. The 12th Duke's main aim in life, however, was Catholic Emancipation, and in 1820 he presented to the new King, George IV, a petition from himself and his co-religionists which combined a request for the repeal of the penal laws with a pledge of allegiance to the Crown. The robes that he wore when he finally took his seat in the House of Lords in 1829 are still used today by the Duke of Norfolk at the State Opening of Parliament.[96]

[92] Anon. ('J. D.'), 'An Account of the Festival', fol. 1, quoted by permission of His Grace the Duke of Norfolk.

[93] Ibid., fol. 5. [94] M. A. Tierney, *History of Arundel* (1834), p. 87.

[95] 'The Glossop Howards who inherited the dukedom in 1815, and who still hold it, have been the most consistently and strongly Catholic of the descendants of Henry Frederick, Earl of Arundel. For a hundred and fifty years they were faithful to an exiled dynasty and a proscribed faith': J. M. Robinson, *Dukes of Norfolk* (1995), p. 185.

[96] Ibid., pp. 189–92.

The 12th Duke was the epitome of the old English Catholic, reserved in manner and disapproving of any aspect of ritual or church building which smacked of Continental Ultramontane practice. His son, Henry Charles, who succeeded in 1842, took a similar anti-Ultramontane line during the 'Papal Aggression' crisis, going so far as to support Lord John Russell's Ecclesiastical Titles Bill in 1851. Indeed, so violent was his negative reaction to the restoration of the Catholic hierarchy, and to Wiseman and all his pomps, that he walked out of Arundel Castle one Sunday and began to worship in the Anglican parish church.[97]

Five years earlier, the high point of the 12th Duke's life had been a three-day visit to Arundel Castle made by Queen Victoria and Prince Albert, in December 1846. Special heaters were brought in to make the draughty corridors comfortable. Victoria and Albert enjoyed comparing Arundel with Windsor. With obvious amusement, the Queen recorded that she and the Duke nearly fell down on a steep path, but managed to pick themselves up. Clearly, the visit was a great success. Successive Dukes of Norfolk, in 1829 and in 1850–1, regarded themselves as the most loyal of Englishmen as well as Catholics.[98] Far from being a Catholic stronghold, or bulwark against Protestantism, the restored Arundel Castle, with its massive ramparts, proved to be more closely allied to the local parish church and to the Hanoverian monarchy than it was to the restored Catholic hierarchy. The drawbridge over the moat remained firmly down.

[97] Faber wrote to Newman on 28 January 1851, 'You must not be surprised if the Duke apostatises': *Faber, Poet and Priest* (1974), p. 223. In fact the Duke died a Catholic, in February 1856, 'receiving the last sacraments at the hands of Canon Tierney': J. M. Robinson, *Dukes of Norfolk* (1995), p. 202. In 1877 St Nicholas's, Arundel, was divided into Anglican and Roman Catholic sections, separated by a screen.

[98] The Duchess of Norfolk wrote to Queen Victoria criticizing the tone of the papal bull 'From Without the Flaminian Gate of Rome' (1850). The Queen replied that she saw 'no *real* danger' in the proceedings of Roman Catholic clergy: it was the Puseyites who worried her. See J. M. Robinson, *Dukes of Norfolk* (1995), p. 201.

CHAPTER 6

Out of the war of tongues

> I have not had any feeling whatever but one of joy and gratitude that God called me out of an insecure state into one which is sure and safe, out of the war of tongues into a realm of peace and assurance.
>
> John Henry Newman to Henry Bourne (13 June 1848)[1]

I

Like 1829, 1845 was a year of social and political crises in England. Six years of anti-Corn-Law agitation, during a series of bad harvests and a severe trade depression, finally came to a head when Sir Robert Peel succeeded in repealing the Corn Laws in November 1845, after yet another bad corn harvest. The main potato crop in Ireland, usually lifted in October and November, had been blighted by the fungus *Phytophthora infestans*, which came in from America. Over the following five years, more than a million Irish men, women and children either travelled in the opposite direction or emigrated to England, Scotland or Australia. Industrial unrest in England and the threat of Chartism also threatened national stability. It was at the end of April 1845 that Disraeli hurriedly finished his reform novel, *Sybil; or, The Two Nations*, in which early Victorian readers were invited to stare into the yawning gap between rich and poor.

1845 was also a year of religious crises. In February, W. G. Ward, a Tractarian, was degraded from his Oxford degrees for publishing pro-Roman Catholic views. In October, Newman's conversion to Roman Catholicism sent shockwaves through the Church of England and indeed the whole body politic. His long journey from Evangelicalism to Tractarianism and thence to Roman Catholicism fascinated his contemporaries, who had to wait for nineteen years to read the inside story in his *Apologia Pro Vita Sua*, but meanwhile could examine the book that he worked on right up to

[1] *Letters of Newman* (1961–77), XII, 218.

his conversion and published soon after it – *An Essay on the Development of Christian Doctrine*. If the Oxford Movement had suffered a setback in February, it now felt seriously weakened. The Catholic Church in England, on the other hand, gained in confidence as Newman and some of his followers were ordained into the priesthood.[2] The impact of the converts is reflected in the fact that both Newman and Manning were to become Cardinals, while Faber was regarded as more influential than either of them by most English Catholics.

The year began with a dire warning from the newly founded *British Protestant; or, Journal of the Religious Principles of the Reformation*:

> It seems to us abundantly plain that the impending and paramount controversy of the age will be between ROMANISM and PROTESTANTISM. The supremacy of Tradition – the Priest – and the Church is the essence of the one; and the supremacy of Scripture, as the only rule of faith, lies at the root of the other. These two points are the poles of the increasing movement, and towards the one or the other every party and church and ceremony seems rushing to its place.[3]

Wiseman's claim that, during the previous six years, fifty new Roman churches had been built in England alone, many of them large, was deeply disturbing and demanded a response from the *British Protestant*:

> The deadly doctrines of the Romish Church – deadly in all their formulas and under every variety of disguise – the progress and to some extent the popularity to which they have attained, and the extensive igorance of their real nature which prevails, especially among the lower and middle classes of society, render such publications as the present of the utmost importance. It will be our great design to promote the principles of the Reformation.[4]

There had been alarms earlier in the 1840s,[5] but now the Ward case threatened to split the Church of England. In January 1843, William George Ward, a Fellow and Tutor of Balliol, had argued in *The British Critic* that, 'if the duty of obeying our Church be grounded on our cognizance of its agreement with antiquity, it becomes the bounden duty of each one of us to examine carefully the force of *Roman* arguments drawn from Church

[2] The doubling of the number of Vicars Apostolic in 1840, which 'led to a great religious development', had also been an 'epoch' in the 'progress' of English Catholicism: N. P. S. Wiseman, *Religious Position of Catholics* (1864), p. 8.

[3] Anon., 'The Progress of Popery', *British Protestant*, January 1845, p. 1. The journal continued until 1864 as an organ of conservative Anglican Evangelicalism.

[4] Ibid., p. 3.

[5] See, e. g., W. Dodsworth, *Allegiance to the Church* (1841), p. 3; H. Rogers, 'Recent Developments of Puseyism', *Edinburgh Review*, October 1884, p. 310.

history'.[6] Then, in June 1844, he published his first book, in which his 'ideal of a Christian Church' bore a strong resemblance to the Church of Rome, which, be believed, had 'preserved in the main, and we have not, what is so inestimably precious, the high and true *idea* of a Church', adding that, 'on a great number of points', the 'English Church would act wisely in making Rome her model'.[7] When the Oxford authorities moved to strip Ward of his degrees, a pamphlet war began – an example of what Newman called the 'war of tongues' in the Church of England.[8] On 13 February 1845, a day of snow and ice, almost 1,200 of those eligible to vote gathered in the Ashmolean Theatre. *The Ideal of a Christian Church* was condemned by a vote of two to one, but what shocked those who attended was the size of the minority. Ward was degraded by a vote of 569 to 511.[9]

Only a month later, in March 1845, public attention focused upon another issue associated with 'Romanism', when Sir Robert Peel, the Tory Prime Minister, passed an act to raise the annual government grant to Maynooth, a college in Ireland for the training of Catholic priests, from £9,000 to £27,000, and to give £30,000 for capital expenditure on dilapidated buildings.[10] Peel's good will gesture was greeted with the inevitable 'No Popery' campaign in England[11] and the *British Protestant* was outraged.[12] The political fallout continued for months.

Ward's degradation and the Maynooth grant were often referred to in the autumn and winter of 1845, when the conversion of Newman and several of his followers was the subject of numerous tracts, articles in the religious press and published sermons. During a period of genuine alarm, High Churchmen such as William Sewell, who distanced himself from the Oxford Movement and was later described by Aubrey de Vere as suffering from 'Jesuit on the brain',[13] begged his fellow Churchmen to stay within the fold, in a sermon delivered to the University of Oxford in 1845 on

[6] W. G. Ward, 'Church Authority', *British Critic*, January 1843, p. 230.

[7] W. G. Ward, *Ideal of a Christian Church* (1844), pp. 53–4.

[8] The Ward case produced a 'rapid and copious stream from the press, which no former occasion [had] equalled'; 'The collective issue of six weeks gives nearly a pamphlet or paper *per diem*, and the pile continued to swell up to the very day of Convocation': Anon., 'Recent Proceedings at Oxford', *Christian Remembrancer*, April 1845, p. 519.

[9] Ward announced his engagement the next day. The couple were later married by Newman and converted to Catholicism. For fuller accounts of the Ward case, see O. Chadwick, *Victorian Church* (1966–70), I, 200–1, 207–11 and *From Bossuet to Newman* (1987), pp. 120–38.

[10] See O. Chadwick, *Victorian Church* (1966–70), I, 223–4, 236.

[11] See J. Wolffe, *Protestant Crusade* (1991), p. 1.

[12] Anon., 'Maynooth', *British Protestant*, February 1845, pp. 57–9.

[13] A. de Vere, *Recollections* (1897), p. 117.

5 November, the anniversary of the Gunpowder Plot, and in a strongly anti-Catholic novel.[14]

While Tractarians lamented the 'schism' caused by the converts to Rome for ecclesiastical reasons, they also felt a particularly acute sense of loss when Newman converted, because his personal influence had been so strong. The *Christian Remembrancer* adopted a funereal tone at the beginning of its long retrospective article on 'The Recent Schism': 'It is with deep pain that we commence some remarks on the secession of Mr. Newman, Mr. Ward, and Mr. Oakeley, with several others, from the communion of our Church.'[15] Newman was clearly Rome's most important new convert, as it was he who had captured the *hearts* of so many Anglicans who had heard and read him:

> To the first-mentioned name, of course, does the whole importance of this movement attach. Mr. Newman has, for the last ten years, had a position and influence which must make such a step on his part a heavy blow indeed to the Church. He has been loved, admired, looked up to. Over his circle of friends, and over that large ground which his authorship covered, his mind has irresistibly won. Everything that he has written has told. His books have made their way to persons' hearts. We can point to no one who has had the influence he has had amongst us; and now he has left us; has transferred himself and his name to another communion, and instead of being a witness for the English Church, become a witness against it.

In emphasizing that Newman's conversion is a crisis in the history of the Church of England, the author expresses the hope that Anglicans can turn it to good and strengthen their own personal commitment:

> To take a final leave of Mr Newman is a heavy task. His step was not unforeseen; but when it is come, those who knew him feel the fact as a real change within them, feel as if they were entering upon a fresh stage of their own life. May that very change turn to their profit, and discipline them by its hardness! It may do so, if they will use it so. Let nobody complain.

Accusations of schism and of 'Romish' superstition were, however, rebutted by some Tractarians who wanted to defend their tradition within the Established Church, post-Newman. Thomas William Allies, for example, the Rector of Launton in Oxfordshire, published *The Church of England cleared from the Charge of Schism, by the Decrees of the Seven Ecumenical Councils and the Tradition of the Fathers* in 1846.[16] Yet Allies himself was to convert to Rome, after the Gorham judgment of 1850.[17] In sharp contrast, militant Anglican Evangelicals could take a quite different line, arguing

[14] W. Sewell, *Plea of Conscience* (1845) and *Hawkstone* (1845). See also p. 187 below. Sewell publicly burned J. A. Froude's novel, *The Nemesis of Faith*, in 1849.

[15] Anon., 'The Recent Schism', *Christian Remembrancer*, January 1846, p. 167.

[16] This was a book of 'great value', according to the *Christian Remembrancer*, October 1846, p. 397.

[17] See p. 233 below.

that it was good to see the back of Newman and other 'wolves in sheep's clothing', and wishing that more would leave the Church of England.[18] Tractarianism's 'grossest superstitions', a leader writer in the *British Protestant* argued, were the result of the study of 'idolatrous' classical authors at Oxford, where 'pedantic habits of a college life' inspired an unhealthy 'clerical authority, and a dominion over the minds and consciences of the laity'.[19]

The impact of all these events of 1845 was most vividly described by William Gresley, Prebendary of Lichfield and a prolific writer on Church affairs:

> The Church of England has just passed through a process of fermentation, a fever has raged in her veins, a storm has troubled her atmosphere; and now that these symptoms have subsided, a great change is found to have taken place within her. The Church is not what she was. Parties within her have shifted their ground. New combinations have taken place. Opinions and doctrines which had almost become extinct are again recognised and widely established.[20]

Roman Catholics later came to see that their communion had also undergone a sea change. In 1930, Ronald Knox was to claim that, from the day of Newman's conversion, the Catholic Church in England had 'ceased to be what it was, a remnant, half-pitied, half-despised, to be hated perhaps, but certainly not feared': 'It has grown into a vigorous movement.'[21] At the time, however, *Dolman's Magazine*, a new Roman Catholic monthly, could not know this, and while it celebrated the number of conversions that followed Newman's, it carefully refrained from 'intruding within the sanctuary of private life by mentioning the names of converts, until they themselves shall have proclaimed their change of opinions to the public'.[22]

In this chapter I examine Newman's own ideas and feelings relating to his conversion to and membership of the Catholic Church. In section II, I argue that *An Essay of the Development of Christian Doctrine*, which is usually discussed in relation to Newman's thought, also draws upon his feelings about communities. In section III, his description of his conversion as a call from God 'out of the war of tongues into a realm of peace and assurance' is compared with the experience of other converts, and finally illustrated in Newman's novel, *Loss and Gain* (1848).

[18] Anon.,'Recent Converts', *British Protestant*, November 1845, p. 245.
[19] Ibid., pp. 245–6. [20] W. Gresley, *Real Danger of the Church* (1846), p. 3.
[21] R. A. Knox, *Occasional Sermons* (1960), p. 174.
[22] Anon., 'Catholic Monthly Intelligence', *Dolman's Magazine*, December 1845, p. 496. That month they were able to name Oakeley, Miles, Browne and Faber.

II

In the second of his 'Young England' trilogy of novels, *Sybil; or, The Two Nations*, published in May 1845, Disraeli announces his central theme in a stagey set piece that owes something to the Catholic version of Digby's *The Broad Stone of Honour* (1826) and something to the early poetry of Tennyson. Among the ruins of Marney Abbey, in 1837, the radical Stephen Morley explains to Charles Egremont that the new Queen reigns over two nations – 'THE RICH AND THE POOR': 'At this moment a sudden flush of rosy light, suffusing the grey ruins, indicated that the sun had just fallen; and, through a vacant arch that overlooked them, alone in the resplendent sky, glittered the twilight star' (II, 5). Sybil appears, 'apparently in the habit of a Religious', and her solo 'evening hymn to the Virgin' produces tones of 'almost supernatural sweetness'. The love plot – for both the young men are smitten – and the social problem plot are launched together.

During the conversations that precede this epiphany, Walter Gerard, a Catholic Chartist, laments the dissolution of the monasteries, for which the union workhouses seem but a poor substitute. His Cobbett-like reference to 'community' is taken up by Stephen Morley:

> As for community, . . . with the monasteries expired the only type that we ever had in England of such an intercourse. There is no community in England; there is aggregation, but aggregation under circumstances which make it rather a dissociating than a uniting principle.

For Disraeli the ambitious but mercurial Tory politician, it was perhaps safer to develop his theme of 'two nations' through Carlylean contrasts between medieval Catholicism and modern materialism than to emulate his friend, Lord John Manners, in associating himself with the Tractarians, who in the mid-1840s seemed an even greater threat to the Establishment than an emancipated Roman Catholic minority.[23] In Disraeli's view, the English were 'at all times a religious and Catholic people', but always 'anti-papal' (I, 3).[24] The Church of England had 'lost her sacred mission' and was failing to minister to a society in which godless and dissociating 'aggregation' was replacing Catholic community.

[23] See C. Whibley, *Lord John Manners* (1925), I, 63, 73, 130. In Disraeli's view, the Maynooth grant, which he attacked on the second reading, marked the end of Young England (I, 180–4). Lord John supported the scheme and longed for unity in the 'Church Universal'.

[24] The nineteen-year-old Disraeli had been so impressed by the 'inconceivably grand' Mass that he attended in Ghent that, at the elevation of the host, he 'flung [himself] on the ground'. See C. Hibbert, *Disraeli* (2004), p. 18.

Newman would have agreed. Not only did he preach community, he also practised it: as an Anglican, at Oriel College, Oxford, where he was a resident Fellow, and later at Littlemore; and as a Catholic at the Birmingham Oratory, of which he was the founder, in 1848, and where he died, in 1890. Intellectually, he was a lone voyager: he did not bother to read Möhler on development, for example, in a work published twenty years before his own *Essay*.[25] Emotionally, although he found aspects of community life irksome, he also found companionship and moral support in the male communities in which he lived throughout his adult life; and his experience of such communities influenced his thinking on the history of the Church and its doctrines.

When the 'war of tongues' was at its height in Oxford, during the row over his treatment of the Thirty-Nine Articles in *Tract XC* (1841), Newman's position as Vicar of St Mary's, the University church, became unsustainable. Having already built a new church out at Littlemore, which was part of his parish, he decided to 'set up a half College half monastery' there, a 'place of retirement' which could accommodate a curate and a teacher, and a small number of clergy who shared Newman's current dilemma and might be dissuaded from going over to Rome.[26] The contrast between the high table privileges of Oriel, which had been his 'own home',[27] and the monastic simplicity of Littlemore was complete. Newman had to explain himself to his bishop when the newspapers invented stories about cloisters.[28] By the summer of 1842 he could report that the cottages were full, but that 'men come and go; I have hardly one constant inmate, and in winter perhaps I may be left alone'.[29] The following year, he was planning to expand the accommodation. Even though he had been in a state of suspense throughout his time at Littlemore, those few years 'meant far more to him than all his years at Oxford';[30] and he was so fond of Ambrose St John, one of the young clergymen who joined him there and was to devote himself to his service for the next thirty years, that he asked to be buried in St John's grave at Rednal.

Newman found leaving Littlemore in 1846 such a 'very trying thing' that he 'could not help kissing [his] bed, and mantlepiece, and other parts of the

[25] O. Chadwick, *From Bossuet to Newman* (1987), p. 111.

[26] See I. T. Ker, *John Henry Newman* (1988), pp. 245–50, 271–8, 317–20.

[27] Having felt 'not quite at home' at Oriel to begin with, Newman was 'truly at home' in the years leading up until 1841. He left his 'own home' to move to Littlemore, where he attempted 'nothing ecclesiastical, but something personal': *Apologia* (1967), pp. 27, 76, 90, 161.

[28] Caroline Clive was also to make fun of the 'Monastery of Littlebitmore' in *Saint Oldooman* (1845).

[29] See I. T. Ker, *John Henry Newman* (1988), p. 249.

[30] Ibid., p. 319.

house'.[31] The strength of his emotional attachment to particular places and people is also reflected in his famous sermon on 'The Parting of Friends', delivered at St Mary's on the anniversary of the consecration of Littlemore church, on 25 September 1843, following his resignation. In looking back with the congregation, he speaks of their communal life together in terms of keeping the feast with merry hearts:

> It was a glad time when we first met here, – many of us now present recollect it; nor did our rejoicing cease, but was renewed every autumn, as the day came round. It has been a 'day of gladness and feasting, and a good day, and of sending portions one to another [Esther ix.19]'. We have kept the feast heretofore with merry hearts; we have kept it seven full years unto 'a perfect end [Compline]'; now let us keep it, even though in haste, and with bitter herbs, and with loins girded, and with a staff in our hand, as they who have 'no continuing city, but seek one to come [Heb.xiii.14]'.[32]

It was at Littlemore, on 9 October 1845, that the Italian Passionist, Father Dominic Barberi, received Newman into the Catholic Church, rather sooner than the new convert had anticipated. Newman stayed on there and was the last to leave, on the afternoon of 22 February 1846, having accepted Wiseman's offer of the old Oscott College (which Newman called 'Maryvale' or 'Mary Vale') near Birmingham, for the use of his community. Old Oscott was to be 'Littlemore continued'.[33] By 28 October 1846, Newman and St John were in Rome, where they were to undergo training at the College of Propaganda. Humiliating as aspects of his return to school must have been, England's greatest Christian apologist could write to Henry Wilberforce, on 13 December 1846: 'I was happy at Oriel, happier at Littlemore, as happy or happier still at Maryvale – and happiest here.'[34]

Better things were yet to come, however. Newman and St John were impressed by the Roman Oratory, which was reminiscent of an Oxford college. Wiseman's suggestion that their community should become Oratorians began to make sense, although it took eighteen months in all for Newman finally to decide on establishing a house dedicated to St Philip Neri in England.[35] He found himself in a different world as a Catholic priest, yet the Oratorians, with their musical tradition and emphasis upon friendship and community life, offered a vocation of which his former

[31] *Letters of Newman* (1961–77), XI, 132.
[32] J. H. Newman, *Sermons on Subjects of the Day* (1869), p. 399.
[33] *Letters of Newman* (1961–77), XI, 54. [34] Ibid., XI, 294.
[35] See *Newman the Oratorian* (1969), pp. 70–95.

Anglican ministry provided the 'spiritual matrix'.[36] Indeed, whereas the London Oratory was later to become Italianate in spirit, under the leadership of the Ultramontane Faber, Newman's Birmingham Oratory remained 'thoroughly English'.[37] It is a measure of Newman's charisma that some of the younger recruits simply followed him, rather than choosing the Oratory tradition *per se*.[38] Those who joined him at Maryvale learned that St Philip had 'formed a community, yet without vows and almost without rules', by 'forming in his disciples a certain character instead'.[39] In his Chapter Address of February 1848, Newman goes on to use the language of the heart: 'It was St Philip's object therefore, instead of imposing laws on his disciples, to mould them, as far as might be into living laws, or, in the words of Scripture, to write the law on their hearts'.[40]

Newman always kept the numbers down in what Oratorians traditionally called the 'Santa communità', one of the few Italian phrases which he cherished,[41] because, as he explained in another Address, the 'Congregation is to be the *home* of the Oratorian'.[42] In the remarkable passage that follows, he discerns helpful continuities between Roman sources of Oratorian tradition and Anglican sources of the Tractarian tradition that he and his fellow converts have left behind, as he quotes Keble's poem for the First Sunday of Lent in *The Christian Year*:

The Italians, I believe, have no word for home – nor is it an idea which readily enters into the mind of a foreigner, at least not so readily as into the mind of an Englishman. It is remarkable then that the Oratorian Fathers should have gone out of their way to express the idea by the metaphorical word *nido* or nest, which is used by them almost technically. . . . The Congregation, according to St Philip's institution, is never to be so large that the members do not know each other. They are to be 'bound together by that bond of love, which daily intercourse creates, and thereby all are to know the ways of each, and feel a reverence for "countenances of familiar friends."' Familiar faces, exciting reverence, daily intercourse, knowledge of each other's ways, mutual love, what is this but a description of home? As the Poet says

[36] Ibid., p. 68. Manning thought that Newman's progression from Anglican to Catholic was too smooth, and that *discontinuities* characterized the true conversion to Catholicism: see E. Norman, *English Catholic Church* (1984), p. 266.

[37] *Newman the Oratorian* (1969), p. 93. He founded the first Birmingham Oratory at Alcester Street, on 26 January 1849, preferring this provincial setting to London because it combined opportunities for missionary work with more time for intellectual pursuits than the capital would allow: ibid, pp. 99–100.

[38] Ibid., p. 106. [39] Ibid., p. 206.

[40] Cf. Wiseman's description of a jubilee in December 1850 as a 'period of reconsideration of themselves, and of remodelling, if I may so speak, of their very *heart*, and shaping and forming it more truly in the *mould* of the Gospel' (my emphases): see p. 45 above.

[41] *Newman the Oratorian* (1969), p. 118. [42] Ibid., p. 192.

> "Sweet is the smile of home; the mutual look
> Where hearts are of each other sure;
> Sweet all the joys that crowd the household nook
> The haunt of all affections pure."

Thus, Newman goes on to explain, the Oratorian is to have his own room – a 'nest' rather than a cell – and his own modest furniture, with his 'things about him, his books and little possessions': 'In a word, he is to have what an Englishman expresses by the distinctive word *comfort*. . . . Meanness, poverty, austerity, forlornness, sternness, are words unknown in an Oratorian House.'

Newman, who kissed his bed on leaving Littlemore and preached on the parting of friends, was highly sensitive to the emotional significance of domestic objects and the 'family' circle of friendship. When Faber and his 'dear monks',[43] the Brothers of the Will of God, or 'Wilfridians', decided that they wanted to join Newman and his Oratorians, Newman described the offer in terms that are uncannily close to those of a marriage settlement: 'Faber has offered himself and his to me, simply and absolutely – his house, his money, his all. The proposal came through Dr Wiseman.'[44] Eleven months later, in November 1848, he reported that they were 'now nearly 40 in family altogether'.[45] The following month he wrote to Bishop Ullathorne, explaining that an Oratory is not a mission and ought not to have a district attached to it: 'Its work is simply within its own homestead for those who choose to come, whether for the Sermons, for Confession, or for its exercises.'[46]

Soon after the London Oratory was established under Faber, in the Spring of 1849, the suggestion was made that Newman should move there, too, and thus have a more direct influence upon the national life. His negative reaction was expressed in terms of personal possessions and emotional bonds, in a letter of March 1849: 'It will not be an easy thing to move a man of 50 with a large library at his back, habituated and acclimated to the place, without a London House (to hold a library) to go to, with penitents and other ties, not to say those Fathers with whom he has lived, pulling him back.'[47] He seemed at last to be where he could 'live and die, having been for 10 years without what promised to be a home'.[48] Newman did

[43] *Faber, Poet and Priest* (1974), p. 134.

[44] To J. D. Dalgairns, from Maryvale, Perry Bar, Birmingham, 2 January 1848: *Letters of Newman* (1961–77), XII, 143. Unfortunately it was to be an unhappy marriage.

[45] Ibid., XII, 329.

[46] He added, 'in the *present* state of Birmingham, we wish . . . to undertake a mission, leaving the future to take care of itself': ibid., XII, 362–3.

[47] Ibid., XIII, 92. [48] Ibid., XIII, 108.

indeed live and die at the Birmingham Oratory, which moved to Egbaston in April 1852. In 1878, thirty years after speaking to the Fathers gathered at Maryvale, he gave a Chapter Address for the first time for many years, in which he explained why he had never liked a large Oratory: 'Twelve working Priests has been the limit of my ambition. One cannot love many at one time; one cannot really have many friends. An Oratory is a family and a home; a domestic circle, as the words imply, is bounded and rounded.'[49] Characteristically, Newman applied the same language to the Church as a whole. In his lectures on *Certain Difficulties felt by Anglicans in Catholic Teaching* (1850), he stated his wish that they should be 'lodged safely in the true home of [their] souls and the valley of peace'.[50]

Domestic and familial language was also applied to Church matters by Anglicans – including Tractarians – in the 1840s, when schism was thought of as a kind of divorce. As both clergy and laity could marry, divisions among families along sectarian lines were felt particularly keenly.[51] Above all, however, it was infidelity to 'Mother Church' which most disturbed commentators of the period. The true voice of conscience, William Sewell argued, would keep people faithful to our true 'mother', the Church of England.[52] William Gresley wrote of those who remained in the Established Church looking to her as their 'spiritual mother', their 'Nurse, who has fed them from their youth with spiritual food'.[53] Keble described the Church of England, not as a home, but as a mother, arguing that it is an 'undutiful thing' for a good Anglican to 'doubt whether she be our real Mother, who has ever professed to be so'.[54] William Bennett, the Tractarian Vicar of St Paul's, Knightsbridge, preaching on the recent 'schism' in November 1845, contrasted the 'circumcision of the heart' with what St Paul called 'the concision' (Philippians 3.2):

How should we feel *literally*, to take the knife, and cut down and up great wounds in the body of our *mother*? Thus shuddering should we feel, if, baptized in the holy fonts of our Church, we all at once should deny her baptism; if, kneeling at her altars for years and years, we deny her power to feed the flock with the Body and Blood of her Redeemer; if, ordained by the laying on of hands of her bishops, we

[49] *Newman the Oratorian* (1969), p. 387.

[50] J. H. Newman, *Lectures on certain Difficulties* (1850), p. 293.

[51] 'O my brethren, what concision of the spirit have I looked upon – in parents no longer worshipping together with their children – in children taught to look upon parents in anathema – in brethren and sisters – husband and wife, going divers ways to their houses of prayer, and at the threshold of their own dwelling, parting when they think of God': W. J. E. Bennett, *Schism of certain Priests* (1845), p. 12.

[52] W. Sewell, *Plea of Conscience* (1845), p. 39.

[53] W. Gresley, *Real Danger of the Church* (1846), p. 5.

[54] J. Keble, *Sermons* (1847), p. iii.

spurn that laying on of hands, and do despite to the Holy Ghost, so long believed and cherished. O fearful work of the '*concision!*' Fearful work of those who, while they imagine they are seeking light, are groping in darkness.[55]

In the Catholic lexicon, the word 'mother' had two distinct religious meanings. One of Faber's Oratory hymns begins, 'O Mary, Mother Mary! our tears are flowing fast', and on Thursday nights, Oratory prayers included the words 'we choose the ever-Blessed Virgin Mary for our Mother'.[56] Newman wrote to Mrs William Froude from Maryvale, on 16 June 1848, relating the largest of all communities, the Communion of Saints, to the Blessed Virgin Mary:

To know . . . that you are in the Communion of Saints – to know that you have cast your lot among all those Blessed Servants of God who are the choice fruit of His Passion – that you have their intercessions on high – that you may address them – and above all the Glorious Mother of God, what thoughts can be greater than these?[57]

The term 'mother' was also applied to the Catholic Church. Wiseman preached in 1850 on the Church as a 'kind, and merciful, and most loving mother',[58] and Newman, who lost his own mother in 1836, wrote in the *Apologia*: 'In spite of my affection for Oxford and Oriel, yet I had a secret longing love of Rome the Mother of English Christianity.'[59] It is this second meaning of 'mother' on which the climax of Newman's conversion novel, *Loss and Gain* (1848) turns, after Willis gives Charles Reding the kiss of peace:

'Farewell again; who knows when I may see you next, and where? may it be in the courts of the true Jerusalem, the Queen of Saints, the Holy Roman Church, the Mother of us all!' He drew Charles to him and kissed his cheek, and was gone before Charles had time to say a word.

. . . It seemed as if the kiss of his friend had conveyed into his own soul the enthusiasm which his words had betokened. He felt himself possessed, he knew not how, by a high superhuman power . . . He felt he was no longer alone in the world, though he was losing that true congenial mind the very moment he had found him. Was this, he asked himself, the communion of Saints? Alas! how could it be, when he was in one communion and Willis in another? 'O mighty Mother!' burst from his lips; he quickened his pace almost to a trot, scaling the steep ascents

[55] W. J. E. Bennett, *Schism of certain Priests* (1845), pp. 10–11. Bennett was later forced to resign his living at the time of the 'Papal aggression': see p. 19 above.

[56] M. Napier and A. Laing, eds, *London Oratory Centenary* (1984), p. 26; Anon., *Oratory Night Services* (1909), p. 29. See also pp. 234–5 below.

[57] *Letters of Newman* (1961–77), XII, 224.

[58] See p. 45 above. [59] J. H. Newman, *Apologia* (1967), p. 152.

and diving into the hollows which lay between him and Boughton. 'O mighty Mother!' he still said, half unconsciously; 'O mighty Mother! I come, O mighty Mother! I come; but I am far from home. Spare me a little; I come with what speed I may, but I am slow of foot, and not as others, O mighty Mother!' (II, 20)[60]

After walking two miles in a state of heightened excitement, the exhausted Reding wonders where these words came from and tells himself that 'enthusiasm is not truth'. The chapter ends with him saying, 'O mighty Mother!... Alas, I know where my heart is! but I must go by reason . . . O mighty Mother!'

Whereas the Communion of Saints exists in an eternal present, Mother Church operates in time and space, and, like any family, has a history and heredity. W. G. Ward argued in his controversial *Ideal of the Christian Church* (1844) that the 'miserable sin' of the Reformers was 'grievously to violate [the] *high sacredness of hereditary religion*'.[61] For Newman, more than for any other leading religious figure of the nineteenth century, in order to understand the history and mission of the Church 'family' one had to study the *Fathers* of the Church.[62] In 1849, he wrote to Henry Wilberforce from the Oratory at Alcester Street, Birmingham, stating that it was not the 'argument from unity or Catholicity' which immediately weighed with him in *An Essay on the Development of Christian Doctrine*, 'but from Apostolicity': 'If that book is asked, why does its author join the Catholic Church? The answer is, because it is the Church of St Athanasius and St Ambrose.'[63]

Although Newman had done preliminary work on the *Essay* in 1844, the serious writing began in January 1845. By June he was complaining of exhaustion, having completed more than half of the first draft.[64] Father Stanton described him standing at his high writing desk, 'completing and revising [the *Essay*] with the infinite care which was his wont' and seeming to 'grow ever paler and thinner'.[65] In August the final chapter was still causing him difficulties, but in September he let the book go to press, in his mind unfinished, so that it could be published before his reception into the Catholic Church. The printers were too slow, however, with the

[60] References to *Loss and Gain* by part and chapter number are from the edition of 1986 (see Bibliography, Primary Texts).

[61] W. G. Ward, *Ideal of a Christian Church* (1844), p. 74.

[62] Newman worked intensively on the Fathers from 1828 onwards: see pp. 64–7 above. As a Catholic he wrote, 'if we differ from the Fathers in a few things, Protestants differ in all': *Lectures on certain Difficulties* (1850), p. 296.

[63] *Letters of Newman* (1961–77), XIII, 78. Cf. *Lectures on certain Difficulties* (1850), pp. 298–9.

[64] See I. T. Ker, *John Henry Newman* (1988), pp. 299–301.

[65] W. Ward, *Life of Newman* (1912), I, 87.

result that the Advertisement, dated 'Littlemore, Oct. 6, 1845' – three days after his resignation from his Oriel Fellowship and three days before his reception – had to be followed by a Postscript announcing that, 'since the above was written, the author has joined the Catholic Church'. Seldom has the writing and publishing of a book so closely tracked a writer's spiritual journey, and seldom have the emotions associated with that journey been more carefully veiled.[66]

Newman refers to the 'home' of Christianity in the very first paragraph of his introduction. As we have seen, like Darwin in the *Origin of Species*, he peppers his opening sentences with the word 'facts'.[67] He argues that Christianity, a 'fact in the world's history', is 'no dream of the study or the cloister': 'It has from the first had an objective existence, and has thrown itself upon the great concourse of men. Its home is in the world; and to know what it is, we must seek it in the world, and hear the world's witness of it.'[68] The word 'concourse' means more than 'aggregation', containing as it does the sense of a shared humanity. It is one of many words with the prefix 'con-' or 'com-' (Lat. *cum-*, with, together) which Newman uses in the *Essay* when reflecting upon the *com*munal nature of Christianity and the development of doctrine. The cumulative, if unconscious effect of these words is to strengthen the sense of an emotional bond between the author and the Fathers of the Church. Like the members of the Oratory whom Newman addressed in 1848, they are all 'bound together by that bond of love, which daily intercourse creates'.

It is not until the final section of the introduction that Newman states the purpose of the *Essay* and outlines his theory of development:

> . . . from the nature of the human mind, time is necessary for the full *com*prehension and perfection of great ideas; and . . . the highest and most wonderful truths, though *com*municated to the world once for all by inspired teachers could not be *com*prehended all at once by the recipients, but, as received and transmitted by minds not inspired and through media which were human, have required only the longer time and deeper thought for their full elucidation. This may be called the *Theory of Developments*.[69]

In 1878, a year before he was elected Cardinal by Pope Leo XIII, Newman changed '*Theory of Developments*' to '*Theory of Development of Doctrine*', one of countless amendments that he made in the third, 'new edition' of the *Essay*. Newman not only tightened up the structure of the work, cutting

[66] Newman said of Scripture, 'its language veils our feelings while it gives expression to them': *Sermons on Subjects of the Day* (1869), p. 408.

[67] See p. 56 above.

[68] J. H. Newman, *Essay on Development* (1974), p. 69.

[69] Ibid., p. 90. The emphases on '*com* –' are mine.

and pasting extensively, but also 'left no word unexamined', with the result that 'stylistically and structurally it is a distinctly different book'.[70] Bearing this in mind as we come to examine variants in certain key passages of the *Essay*, we can be confident that Newman was fully aware of what he was retaining and what he was rejecting in 1878, after thirty-three years in the Catholic priesthood.

Here, first, is a single sentence from the third paragraph of the Introduction to the 1845 edition, in which Newman uses no fewer than seven 'con-' / 'com-' words, which enact in their own syntactic time, marked by the semi-colons, the kind of binding process (*cum-*, with, together) that he believes to be characteristic of the 'society of Christians' in historic time, through tradition:

> Till it is shown why we should view the matter differently, it is natural, or rather necessary, it is agreeable to our modes of proceeding in parallel cases, to consider that the society of Christians which the Apostles left on earth was of that religion to which the Apostles had converted them; that the external continuity of name, profession, and communion is a *primâ facie* argument for a real continuity of doctrine; that, as Christianity began by manifesting itself to all mankind, therefore it went on to manifest itself; and that the more, considering that prophecy had already determined that it was to be a power visible in the world and sovereign over it, characters which are accurately fulfilled in that historical Christianity to which we commonly give the name.[71]

Having then argued that Protestantism is 'not the Christianity of history', Newman goes on to emphasize the shared and public nature of doctrine, again using words suggestive of community and communion. What, he asks, in words that he allowed to stand in 1878, can be 'more conclusive than that the doctrine that was common to all at once was not really their own, but public property in which they had a joint interest, and proved by the concurrence of so many witnesses to have come from an apostolical source?'[72] He then rings another chime of eight 'con-' / 'com-' words (again retained in 1878) with reference to doctrine, as he corrects his own former statements 'in another publication':

> This may be considered as true. It may be true also, or shall here to assumed, for there will be an opportunity of recurring to the subject, that there is also a consensus in the Ante-nicene Church for the doctrines of our Lord's Consubstantiality and Coeternity with the Almighty Father. Let us allow that the whole circle of doctrines,

[70] J. H. Newman, *Essay on Development* (1949), pp. 433, 417. Inexplicably, the editor of this edition, C. F. Harrold, describes the edition of 1878 as the second (which appeared in 1846), when even the running-head of its Contents list identifies it as the third.

[71] J. H. Newman, *Essay on Development* (1845), pp. 2–3.

[72] Ibid., p. 8; cf. (1878), p. 11.

of which our Lord is the subject, was consistently and uniformly confessed by the Primitive Church, though not ratified formally in Council. But it surely is otherwise with the Catholic doctrine of the Trinity. I do not see in which sense it can be said that there is a consensus of primitive divines in its favour, which will not avail also for certain doctrines of the Roman Church which will presently come into mention. And this is a point which the writer of the above passages [i.e. himself] ought to have more distinctly brought before his mind and more carefully weighed; but he seems to have fancied that Bishop Bull proved the primitiveness of the Catholic doctrine concerning the Holy Trinity as well as concerning our Lord.[73]

Later in the *Essay*, he domesticates the phrase 'circle of doctrines' and claims that the Catholic doctrines are 'members of one family, and suggestive, or correlative, or confirmatory, or illustrative of each other'.[74]

One of Newman's most significant revisions was of a passage in chapter 1, 'On the Development of Ideas', in which he applies organic metaphors to the process of development among bodies of men, and strikingly anticipates Darwin in his use of the words 'germination', 'modifying' and 'evolution':

This process is called the development of an idea, being the germination, growth, and perfection of some living, that is, influential truth, or apparent truth, in the minds of men during a sufficient period. And it has this necessary characteristic, – that, since its province is the busy scene of human life, it cannot develope at all, except either by destroying, or modifying and incorporating with itself, existing modes of thinking and acting. Its development then is not like a mathematical theorem worked out on paper, in which each successive advance is a pure evolution from a foregoing, but it is carried on through individuals and bodies of men; it employs their minds as instruments, and depends upon them while it uses them.[75]

In 1878, Newman substituted for 'individuals and bodies of men' the phrase 'communities of men and their leaders and guides', drawing upon his long experience of community life and of leadership within the Catholic communion.[76]

My final example is a passage in the first edition which contains all the key terms that we have been considering: 'community', 'fact' and three 'con-' / 'com-' words. Newman is defining historical developments:

[73] Ibid., p. 11; cf. 1878 (p. 14). George Bull, later Bishop of St Davids, published his *Defensio Fidei Nicænæ* in 1685.

[74] J. H. Newman, *Essay on Development* (1845), p. 154. In this case Newman cut the whole paragraph in 1878, presumably because it repeated a point made earlier.

[75] Ibid., p. 37. [76] J. H. Newman, *Essay on Development* (1878), p. 38.

Another class of development may be called *historical*; I mean when a fact, which at first is very imperfectly apprehended except by a few, at length grows into its due shape and complete proportions, and spreads through a community, and attains general reception by the accumulation, agitation, and concurrence of testimony. . . . Thus by development the Canon of the New Testament has been formed. . . . Thus saints are canonized in the Church, long after they have entered into their rest.[77]

Here the word 'community' is applied to the life of the early Church which its members held in 'common' (Lat. *communis*), and expressed most fully in the 'communion', at which the eucharist was celebrated; and 'testimony' (Lat. *testis*, witness), which, when confirmed by others and accepted by the community over time, becomes 'tradition' (Lat. *tradere* to hand on).

Newman's intellectual grasp of Catholic tradition as a living, dynamic force, rather than the dead hand of a static power, informed his theory of development. With characteristic reserve, he also smuggled into the text of his *Essay* the emotional response of one who lived in and led small communities of faithful men to Catholicism as a 'family and a home' – a 'domestic circle'.

III

'From the time that I became a Catholic, of course I have no further history of my religious opinions.'[78] Having seized the attention of his readers at the beginning of the final chapter of his *Apologia Pro Vita Sua* (1864), entitled 'Position of my Mind since 1845', Newman goes on to explain himself: 'In saying this, I do not mean to say that my mind has been idle, or that I have given up thinking on theological subjects; but that I have had no variations to record, and have had no anxiety of heart whatever. I have been in perfect peace and contentment; I never have had one doubt.' Newman's sense of peace relates specifically to his 'religious opinions' here, and not to areas of pain in his life as a Catholic, of which there were many.[79] Nevertheless, the feeling that God had called him 'out of an insecure state into one which is sure and safe, out of the war of tongues into a realm of peace and assurance' (1848),[80] is one that he often repeated. Various elements of these statements of Newman's are also to be found in the testimony of other converts. His

[77] J. H. Newman, *Essay on Development* (1845), p. 49; cf. (1878), pp. 46–7.
[78] J. H. Newman, *Apologia* (1967), p. 215.
[79] On 17 July 1870 he wrote in a letter to Edward Husband concerning his conversion: 'I did not hope or long for any "peace or satisfaction", as you express it, for any illumination or success. I did not hope or long for any thing except to do God's will, which I feared not to do': *Letters of Newman* (1961–77), XXV, 160.
[80] See epigraph, p. 163 above.

attitude towards the Roman Church, for example, is one of *submission* to its authority – a term that was often applied by converts to their reception.[81] Similarly, his sense of *security* in 1848 echoes that of Kenelm Digby in the 1820s.[82] Above all, however, it is his sense of perfect *peace* that finds the strongest echoes elsewhere.[83] Faber, who was famous for his enthusiasm and his pious ejaculations, added this postscript to a letter written on the day of his reception: 'Peace – peace – peace!'[84] W. G. Ward wrote of the voice of the Church being 'as the Voice of God heard amidst the din of this restless and sinful world', and Wilfrid Ward later described the 'gain of peace and rest, and of much more which a Catholic only can understand', which placed his father 'beyond the reach of any feelings of regret from personal misunderstandings'.[85]

What aspects of Newman's experience, which is particularly well documented in his letters and diaries, offer clues to that which only a Catholic can understand? He described being at the College of Propaganda in Rome as a 'kind of dream': 'so quiet, so safe, so happy – as if I had always been here – as if there had been no violent rupture or vicissitude in my course of life – nay more quiet and happy than before'.[86] Newman's sense of *safety* is specifically Catholic, and refers to his salvation. In a letter of February 1849, he was to write: '*the* reason *why* I left the Anglican Church was that I thought salvation was not to be found in it'. His sense of *peace* and happiness is usually related to another aspect of Catholic spirituality – the sense of Christ's presence in the Blessed Sacrament. He wrote to Henry Wilberforce from Mary Vale, on 26 February 1846:

I am writing next room to the Chapel – It is such an incomprehensible blessing to have Christ in bodily presence in one's house, within one's walls, as swallows up all other privileges and destroys, or should destroy, every pain. To know that He is close by – to be able again and again through the day to go in to Him.

[81] Frederick Oakeley wrote his *Letter on Submitting to the Catholic Church: Addressed to a Friend* (1845) at Littlemore. Aubrey de Vere wrote in a letter of 1851, 'Such submission I regard as an act of obedience; and yet hardly of self-sacrifice. I firmly believe that in submitting to that authority on which Christ has set His seal, I but exchange a lawless freedom for a "glorious Liberty".' W. Ward, *Aubrey de Vere* (1904), p. 199. Cf. also H. G. Gill, *'Plain Reasons'* (1869).

[82] See p. 158 above.

[83] Ironically, it was on the grounds that Newman could *not* achieve peace within the Catholic fold that his *Essay on Development* was criticized by George Moberly, Head Master of Winchester College, in *Sayings of the Great Forty Days* (1846), p. xix.

[84] *Faber, Poet and Priest* (1974), p. 130.

[85] W. G. Ward, *Ideal of a Christian Church* (1844), p. 11; W. Ward, *William George Ward* (1889), p. 365. Compare Aubrey Beardsley, who wrote after his reception, 'It is such a rest to be folded after all my wandering': C. Snodgrass, *Aubrey Beardsley* (1995), p. 156.

[86] *Letters of Newman* (1961–77), XI, 294. Further quotations in this paragraph are from XIII, 59, XI, 129 (cf. XXV, 156), XI, 257, XII, 168.

Christ is present in the house, within the walls of the Oratorian 'home', just next door to the Fathers' own 'nests'. This is an important source of Newman's sense that his journey from Anglicanism to Catholicism had been 'from clouds and darkness into light', and 'out of shadows into truth'.

In Newman's novel, *Callista: A Sketch of the Third Century* (1856), the narrative drive is towards resolution, closure and peace, which are achieved through conversion, figured as an entry into a community of believers, and then martyrdom, from which that community benefits.[87] Waking on the morning of her martyrdom, Callista can say, 'I am ready; I am going home' (33). Miraculously, her body later lies under the open skies, untouched and uncorrupted: 'Her features have reassumed their former majesty, but with an expression of childlike innocence and heavenly peace' (35).

The 'sense of an ending'[88] is particularly strong in Victorian religious fiction, where the shape of the final chapters often reflects the novelist's position *vis-à-vis* Christian eschatology. Alternatively, as in the Anglican novels published after Newman's conversion which illustrated the perils of 'going over' to Rome, the conversion (or non-conversion or reconversion) endings are soteriological in emphasis. Elizabeth Missing Sewell's *Margaret Percival* (1847), edited by her anti-Catholic brother, William,[89] went through at least three editions in the first year of publication. Margaret steps back from the brink, rescued by Mr Sutherland, a High Church clergyman. In the final chapter, she at last achieves peace of mind:

> The English Church! yes, Margaret had now learnt to look to no other. The steady adherence to the line which Mr. Sutherland had marked out, avoiding all controversy, all practices which, without being sanctioned, had a tendency to Romanism, and the carrying out, in private, the principles of the Church, had at length brought its recompense. Margaret was no longer harassed by doubts.[90]

Whereas *Callista* was Newman's answer to Kingsley's *Westward Ho!*, his first novel, *Loss and Gain: The Story of a Convert* (1848), was his response to another conversion novel of the day – Elizabeth Harris's *From Oxford to Rome: and how it Fared with some who lately made the Journey, by a Companion Traveller* (1847). Harris was clearly well informed and knew where the flashpoints between the old enemies lay in the 1840s. Near the end of the novel, for example, she attacks what she regards as Catholic mistranslation

87 See pp. 74–6 above. 88 See F. Kermode, *Sense of an Ending* (1967).

89 William Sewell's own novel, *Hawkstone* (1845), addressed to a 'distracted and degraded England', was more strongly anti-Catholic. See also pp. 165–6 above.

90 E. M. Sewell, *Margaret Percival* (1847), II, 483.

of key biblical texts, citing 'And Jacob *worshipped the top of his staff!*'[91] Eustace, our hero, returns to Oxford in order to vote for Ward, having been persuaded that Puseyism is at least a step in the right direction. Disgusted by the result of the vote, Eustace and his sister Augusta are received into the Roman Catholic Church, thus offering Harris the opportunity to lament the breaking up of families – 'a tragic series, stretching out months long its melancholy dole of suffering; a soul-rending spectacle, repeated in its sorrowful fulness in each separate individual case, till among the forsaken studies, and the closed books, and in their place in the frequented chapel, the chief and father, and one devoted friend lingering by him to the last, were left alone'.

Harris also attacks another favourite target of the anti-Catholic writers of the 1840s, the corrupt monastery or convent, which, as usual, is the site of bad deaths.[92] When Eustace dies, the fanatical practices of the monastery 'household' are said to be inscribed upon his body, for the edification of his brother monks:

> . . . they had arrayed the body in the habilments of the grave, and displayed for the admiration of the household those secret severities of penance with which he had sought once to subdue and mortify the flesh, and in latter days, through their vivid suffering, to keep the mind from wandering into past thoughts. The coarse shirt of hair was exposed under the shroud; and the sharp iron cross, worn till it had wounded deep into the flesh, was shewn on the open breast. A scourge of cords, which had been discovered, stained with its cruel use, in a corner of the cell, was laid at his feet, with a worn book of devotions . . .[93]

His dying words indicate a reconversion, or 'perversion' in Catholic eyes: 'There is one God and one Mediator between God and man, Jesus Christ: and He is not an High Priest, that cannot be touched with the feeling of our infirmities.' Like her reference to the monastic 'household', Harris's description of poor Eustace, shivering in his chilly cell, touches a Catholic nerve in its denial of the peace which converts claimed to have found: he 'had asked for peace, and received a whirlwind'.

Newman's most effective weapon against these Anglican novels, which are unrelievedly earnest, is his humour. Take, for example, the discussion, early in *Loss and Gain*, about Evangelicalism as a religion of the heart.[94]

[91] E. F. S. Harris, *From Oxford to Rome* (1847), p. 244. Further quotations in this paragraph are from pp. 118–24, 139. (The third edition of Berington and Kirk's controversial *The Faith of Catholics*, in which this text was translated, was published in 1846: see pp. 62–3 above.)

[92] On Harris's treatment of nuns and convents, see pp. 215–16 below.

[93] Ibid., p. 177. Further quotations in this paragraph are from pp. 176, 162.

[94] See also p. 45 above.

The aptly named Freeborn believes that religion is a 'matter of the heart; no one could interpret Scripture rightly whose heart was not right' (I, 7). His argument is taken up at a solemn Evangelical tea-party, where the conversation is a 'dropping fire of serious remarks' (I, 17). The adherence of the discussants to the party line is reflected in Newman's amusingly naming them simply by numbers. No. 3 believes that faith is a 'feeling of the heart': 'We believe with the heart, we love from the heart, we obey with the heart; not because we are obliged, but because we have a new nature.' The Evangelicals' internal doctrinal disputes, however, are enough to put Reding off for life, and they therefore disappear from the novel.

In contrast, Reding's conversations with Willis, who is later to convert to Catholicism, that other 'religion of the heart', are enlivened with humour. When the hero learns that the building from which a blushing Willis emerges is not a Dissenting chapel, as he had thought, but a Catholic church, he comments that it '*is* a dissenting meeting, call it what you will, though not the kind of one I meant' (I, II). Reding believes that Willis's idea of the Established Church being 'one Church with the Roman Catholics' is a 'sheer absurdity', but Willis asks him not to talk like that:

> 'I feel all my heart drawn to the Catholic worship; our own service is so cold.' 'That's just what every stiff Dissenter says,' answered Charles; 'every poor cottager, too, who knows no better, and goes after the Methodists – after her dear Mr Spoutaway or the preaching cobbler. *She* says (I have heard them), "Oh, sir, I suppose we ought to go where we get most good. Mr So-and-so goes to my heart – he goes through me."' Willis laughed.

Willis's sense of humour and good-heartedness engage the reader's sympathy. 'O Reding', he says soon afterwards, following his reception, 'I'm so happy!' (I, 13).

Reding's transition from youth to manhood follows the death of his beloved father, a clergyman of the Church of England, to which the son now rededicates himself (I, 18). In time, however, the old doubts about the Articles return, and he confesses to his anxious young tutor, Carlton, that he loves their own Church more than he trusts her, but feels he can trust the Church of Rome (II, 8). As he and his friend Sheffield read for honours, a gap begins to open up between them. Whereas Sheffield seems to be content with what the 'perishable world' gives him, Reding's main characteristic is 'an habitual sense of the Divine Presence': 'a sense which, of course, did not insure uninterrupted conformity of thought and deed to itself, but still there it was – the pillar of the cloud before him and guiding him. He felt himself to be God's creature, and responsible to Him – God's possession, not his own' (II, 9).

As we have seen, he sets his face towards Rome, his 'mighty Mother', after Willis gives him the kiss of peace (II, 20). Like Newman, he does not think that he will be saved if he stays in the English Church (III, 5). Before finding certainty, however, he engages in a hilarious 'war of tongues' with a range of sectarians who try to capture him for themselves (III, 7–8). It is at the church of a Passionist House in the East End of London that a sense of community and of God's presence finally come together, and it is here that Reding can find peace. As in *Sybil*, Catholicism seems to offer an alternative to a modern world of godless and dissociating aggregation, in which 'two nations' are divided: 'rich and poor were mixed together – artisans, well-dressed youths, Irish labourers, mothers with two or three children – the only division being that of men and women' (III, 10). The priest is hardly audible, but the whole congregation moves 'all together', as if 'self-moved', like 'one vast instrument or Panharmonicon', until they all suddenly bow, and Reding understands:

> . . . it was the Blessed Sacrament – it was the Lord Incarnate who was on the altar, who had come to visit and to bless His people. It was the Great Presence, which makes a Catholic Church different from every other place in the world; which makes it, as no other place can be, holy.[95]

After Mass, he meets the kindly Superior: 'His *heart* beat, not with fear or anxiety, but with the thrill of delight with which he realized that he was beneath the shadow of a Catholic *community*, and face to face with one of its priests'; and within an hour he is sitting by himself, 'with pen and paper and his books, and with a cheerful fire, in a small cell of his new *home*' (my emphases).

The opening of the short final chapter is characteristically reserved, and it is the reserved sacrament in the tabernacle before him that brings the hero peace:[96]

> A very few words will conduct us to the end of our history. It was Sunday morning about seven o'clock, and Charles had been admitted into the communion of the Catholic Church about an hour since. He was still kneeling in the church of the Passionists before the Tabernacle, in the possession of a deep peace and serenity of mind, which he had not thought possible on earth. It was more like the stillness which almost sensibly affects the ears when a bell that has long been tolling stops,

[95] In the first edition, 'that Great Presence': *Loss and Gain* (1848), p. 382.

[96] On Newman's powerful response to the reserved sacrament in the tabernacle, even when an Anglican abroad, see I. T. Ker, *Catholic Revival in English Literature* (2003), pp. 19–22.

> or when a vessel, after much tossing at sea, finds itself in harbour. It was such as to throw him back in memory on his earliest years, as if he were really beginning life again. But there was more than the happiness of childhood in his heart; he seemed to feel a rock under his feet; it was the *soliditas Cathedræ Petri*. He went on kneeling, as if he were already in heaven, with the throne of God before him, and angels around, and as if to move were to lose his privilege. (III, II)

His prayers are interrupted by Willis, or Father Aloysius, in his 'dark Passionist habit', who was 'present through the whole', but now has to leave. Having received the young priest's blessing, Reding returns to a peaceful state. It is not through prayer, however, that he re-enters a realm of peace and assurance, in the closing words of the novel. It is through his absorption into the community of Catholics: 'and the new convert sought his temporary cell, so happy in the Present, that he had no thoughts either for the Past or the Future'.

CHAPTER 7

Authority on the rocks

> The great question to be solved is, *what is the authority to which as Christians we are bound to submit ourselves?*
>
> Edward Bellasis, *The Archbishop of Westminster* (1850)[1]

I

On the morning of Sunday 15 September 1850, less than four weeks before the 'Papal Aggression' crisis began, one of the country churches near Exeter was unusually crowded.[2] The sixty-three-year-old George Cornelius Gorham was 'reading himself in' as the new incumbent at Brampford Speke. Yet it was over three years since the Lord Chancellor had signified his willingness to present Gorham to the living, in August 1847. Bishop Henry Phillpotts, a litigious High Churchman, had refused to institute Gorham, on the grounds that he was unsound on the doctrine of baptismal regeneration. The intervening three years were the most divisive in the history of the Victorian Church of England and each new development in the Gorham case was widely reported in the press. Evangelical supporters of Gorham and the freedom of the individual conscience emerged victorious. His Tractarian opponents, already weakened by the loss of Newman, believed that the authority of the Church was undermined when the Judicial Committee of the Privy Council found for Gorham on appeal in March 1850.

Anti-Catholic feeling was strong at the time, even before the 'Papal Aggression' crisis of the autumn, not least because Gorham and his Evangelical supporters had attacked the 'Romanizing' tendencies of the Tractarians and had thus demonized Rome. Following the Gorham judgment, however, Catholics could retaliate by arguing that Queen Victoria was head

[1] E. Bellasis, *Archbishop of Westminster* (1850), p. 19.

[2] Anon., 'The Gorham Case', *Illustrated London News*, 21 September 1850, p. 242. Gorham had been inducted on 11 August.

of an Established Church in which it was possible for clergy to hold conflicting positions on a foundational doctrine and that therefore the Church of England condoned heresy.

Catholic claims to authority were based upon the Petrine rock of the Church of Rome, whose interpretation of both the Bible and tradition on major matters of faith and doctrine was regarded as infallible. Protestant claims were based upon the rock of the Scriptures – a narrower and, in an age of criticism, more vulnerable kind of authority. In 1833 Richard Towers, Prior of Ampleforth, argued that the Catholic rests his 'Rule of Faith' on the 'Testimony, on the Monuments, on the writings of every age, from the establishment of the Christian Religion to his own day', whereas the 'Protestant Rule of Faith is founded on the dead letter of the Scripture, as interpreted by the fanciful imagination, of every self conceited, illiterate, pretender to knowledge'.[3] Newman made a similar point twelve years later, in his *Essay on Development*, when he commented that 'in an age in which reason, as it is called, is the standard of truth and right, it is abundantly evident to any one, who mixes ever so little with the world, that, if things are left to themselves, every individual will have his own view of things, and take his own course; that two or three agree together today to part tomorrow; that Scripture will be read in contrary ways, and history will be analysed into subtle but practical differences'.[4] Hence the primacy of Church authority, as there can be 'no combination on the basis of truth without an organ of truth'.

Anglican replies to such arguments drew attention to the special 'commission' of the Established Church. William Sewell, for example, in his Gunpowder Plot sermon of November 1845 addressed to the University of Oxford, posed the 'real and only ultimate question to be decided in the present case': 'Which power in this land is entrusted with the charge of our souls by a Commission from Almighty God, the Bishop of Rome, or the Episcopate of our own Church?'[5] While Churchmen could indeed invoke history in this way and could celebrate the three hundredth anniversary of the Act of Uniformity (21 January 1549), they were vulnerable to the accusation that there was no uniformity in their interpretation of Anglican doctrine, as enshrined in the services of the Book of Common Prayer and the Articles of Religion. Gorham may have been exceptional in his tenacity during his extended battle with Bishop Phillpotts, but his antagonism towards his High Church brethren was far from unusual.

[3] *Second Letter to Comber* (1833), p. 43. [4] J. H. Newman, *Essay on Development* (1974), pp. 176–7.
[5] *Plea of Conscience* (1845), p. 7.

While a Canon of Durham Cathedral, Henry Phillpotts had made his name in a famous controversy with Charles Butler, then English Catholicism's most dynamic defender and controversialist.[6] Having next served as Bishop of Chester, Phillpotts was translated to Exeter, then a largely Evangelical diocese which was unused to disciplinarian bishops, in 1831. Although never an accomplished speaker, he was respected outside the diocese for his robust defence of the Church against 'error'. Inside the diocese, his abrasive approach to erring clergy made him unpopular.

The Revd Gorham had been in trouble before, in 1811, when Bishop Dampier of Ely examined him privately on baptismal regeneration before ordaining him. On that occasion, Gorham stood firm and the bishop backed down. After thirty-five years as an unbeneficed clergyman in a number of curacies he was presented to the vicarage of St Just, in the far west of Cornwall, in January 1846. When Gorham advertised for a curate who was 'free of Tractarian error', Bishop Phillpotts informed the successful candidate that his acceptance of the post was 'unfavourable to him', but did finally license him.[7] It was Gorham's request for a living nearer Exeter that led to his head-on collision with Phillpotts. The bishop examined Gorham on the 'soundness of [his] views of Christian doctrine, and, in particular, of Baptism, the foundation of all', for a total of fifty-two hours, between December 1847 and March 1848.[8] Under protest, the Vicar of St Just answered 149 questions on the subject of baptismal efficacy. Broadly speaking, for High Churchmen baptism was a matter of divine grace imparted sacramentally by a priest, but for Evangelicals a matter of an affirmation of faith by the person being baptized or, in infancy, by his or her supporters. Gorham's doctrinal position was Calvinistic but different from that of most Evangelicals, who nevertheless rallied to his cause.[9]

Gorham's published account of his ordeal by interview reads like an apology for Protestantism. The martyrs of the English Reformation, he believes, were not wholly successful in dispelling the 'cloudy errors of Popery'.[10] He records his dismay at having found 'Popish' church furniture and services at St Just. He can see no difference between Bishop Phillpotts's 'dogma' and the 'Popish doctrine of the unconditional regenerating efficacy

[6] See p. 146 above. [7] G. C. Gorham, ed., *Examination before Admission* (1848), p. v.

[8] Ibid., p. 15. On 'views', cf. Milman discussing various 'opinions' on doctrine in the early Church, whereas Newman argued that the great Christian doctrines were 'facts, not opinions': see p. 56 above.

[9] 'He believed that infants were not worthy recipients of baptism and that an act of "prevenient grace" was required to make them worthy, otherwise no spiritual grace was conferred in the Sacrament': P. J. Jagger, *Clouded Witness* (1982), p. 3.

[10] G. C. Gorham, ed., *Examination before Admission* (1848), p. xxii. Further quotations in this paragraph are from pp. 26, xxiv–xxv, 24, 13–14.

of that initiatory Sacrament'. Indeed, he accuses the bishop of drawing him towards the 'stream which is gliding into the gulf of Popery'. He considers that the 'errors, which the Oxford Tracts revived with mischievous popularity, are most dangerous, and effective in recruiting the ranks of the *Roman*-Catholic from (what is called in un-Protestant phraseology) the *Anglo*-Catholic Church, and of corrupting those who professedly remain in communion with her'.

Having failed to satisfy Phillpotts, who withheld his license, Gorham took his case to the Court of Arches at Canterbury, which found for the bishop on 2 August 1849. When Gorham then appealed to the Judicial Committee of the Privy Council, both Archbishops studied the appeal and approved it, although Bishop Blomfield of London disagreed. It was a committee of laymen, however, which actually voted on Gorham and upheld his appeal on 8 March 1850. When the judgment was announced, the face of William Goode, a leading Evangelical, shone with bliss, while the Tractarians Robert Wilberforce and James Hope (Scott) walked silently down the steps, their heads drooping.[11]

For High Churchmen, then, the Gorham case was about Church authority, invested in the episcopacy and the ecclesiastical courts. The overruling of a decision of the Court of Arches by lay members of the Privy Council was nothing short of Erastianism (the ascendancy of the State over the Church in ecclesiastical matters). For Evangelicals the case was about the Protestant identity of the Church of England and the need to arrest its apparent drift towards 'Popery'. Catholics looked on with a mixture of pity, bemusement and glee, and congratulated each other on belonging to a Church founded upon a rock. As a convert, Newman was particularly interested in the question of authority in the Roman and Anglican communions. He wrote to the Revd E. J. Phipps, Rector of Devizes, on 3 July 1848, inviting him to 'consider the vast difference between believing in a living authority, unerring because divine, in matters of doctrine, and believing none; – between believing what an external authority defines, and believing what we ourselves happen to define as contained in Scripture and the Fathers, where no two individuals define quite the same set of doctrines'.[12]

In the aftermath of the Gorham judgment, Newman's co-religionists reiterated his claim that Catholic authority was 'unerring because divine', as we will see in section II, where Roman and Tractarian commentators on Erastianism and heresy in the Church of England are compared. In section III

[11] See Baroness Bunsen's *Memoir of Bunsen*, cited in O. Chadwick, *Victorian Church* (1966–70), I, 262.
[12] *Letters of Newman* (1961–77), XII, 234.

I consider the impact of the case upon Victorian culture at mid-century, and particularly the early religious art of Millais and Holman Hunt and some of Ruskin's anti-Catholic writings.

II

Reviewing the series of crises that occurred in the middle of the nineteenth century, one might have expected that the years in which the most book titles with the word 'crisis' in them were published might have been those of the Chartist riots (1848), or the outbreak of the Crimean War (1854), or the Indian Mutiny (1857), years for which the British Library catalogue does indeed list no fewer than fifteen, twenty-four and twenty-five such titles respectively. In 1850, however, no fewer than thirty-two appeared, most of them in response to the 'Gorham crisis'.

Following the Privy Council judgment a great cry of anguish went up from the High Church party of the Church of England.[13] Bishop Phillpotts's own response was characteristically combative: he wrote an open letter, dated 20 March 1850, to John Bird Sumner, the Evangelical Archbishop of Canterbury, virtually accusing him of heresy.[14] An earlier public statement was that of Robert Wilberforce, Archdeacon of the East Riding, who preached on 'The Sacramental System' in Oxford on the Sunday after the announcement. He lamented the religious state of the nation: 'silent Churches, deserted altars, infrequent Eucharists, are but too plain a witness to the national unbelief'.[15] The 'Sacramental and Anti-sacramental systems', he affirmed, 'are two different religions, and to rest our hope of salvation on the one, is to say anathema to the other'. To deny the doctrine of baptismal grace is a 'virtual denial that Jesus Christ is come in the flesh'.

In his charge to the clergy of the East Riding for 1850 Wilberforce addressed the practical effect of the Gorham case, arguing that the judgment was a 'crisis' in the history of the Church of England, 'by which future times will decide whether she is a portion of Christ's Catholic Church, or a department of the secular government'.[16] If the Church 'subjects herself by voluntary act to the dictation of a lay-tribunal, she transfers to it that divine

[13] 'Baptismal regeneration became the watchword of the Tractarians': P. J. Jagger, *Clouded Witness* (1982), p. 16.

[14] 'To require as necessary to the efficacy of the Baptism of Infants that there be faith on the part of those who present them, is little short, if indeed short, of heresy': H. Phillpotts, *Letter to the Archbishop of Canterbury* (1850), p. 13.

[15] R. I. Wilberforce, *Sacramental System* (1850), p. 6. Further quotations in this paragraph are from pp. 14–15, 18.

[16] R. I. Wilberforce, *Practical Effect of the Gorham Case* (1850), p. 18.

THE GORHAM CONTROVERSY.—GREAT MEETING OF CLERGY AND LAITY, IN ST. MARTIN'S HALL, LONG-ACRE.

26. 'Great Meeting of the High Church Party', *The Illustrated London News*, 17, 438 (27 July 1850), 77.

authority, with which she herself claims to be invested'.[17] For the moment, with most of the bishops 'understood to have been united in demanding the abatement of this grievance', we, the clergy, must be patient and wait to see what happens.

While watching they could also pray. On 23 July a 'Great Meeting of the High Church Party' assembled at St Martin's Hall, Long Acre. 2,000 clergy and laity 'opposed to Mr. Gorham' crammed into the room (illustration 26) and a separate meeting, also packed, had to be held elsewhere for those who

[17] Ibid., p. 16.

could not get in.[18] Both Robert Wilberforce and his friend Henry Edward Manning, Archdeacon of Chichester and a future Cardinal Archbishop of Westminster, attended the main meeting. Of the three or four hundred conversions to Rome in the months that followed the Gorham judgment, the most significant was Manning's, who was to 'go over' on 6 April 1851. In Robert Wilberforce's case, having waited patiently and listened carefully to the pleas of friends and family, including his brother Samuel, Bishop of Oxford, for him to remain in the Church of England, he was to follow Manning into the Catholic Church on 1 November 1854.

Meanwhile the English Catholic Church, bruised by recent Protestant verbal assaults, seized the opportunity presented by an Establishment crisis. Nine days after the Gorham judgment was announced, Nicholas Wiseman preached in St George's Catholic Church, Southwark (soon to be a cathedral) on 'The Final Appeal in Matters of Faith'. His text sums up Catholicism's response to what it regarded as proof of Erastianism in the Church of England: '"MY KINGDOM IS NOT OF THIS WORLD" JOHN xviii.36.'[19] 'You are all aware', he said, 'that within these few days there has occurred an event calculated, according to the opinion of all men, in a most important, and, perhaps to us, in a most consoling way, to affect the state of Religion in this country, and, more particularly the position of that Establishment, which, under the name of the Church of England, is most especially connected with the religious and spiritual destinies of the great mass of the people.'[20]

Wiseman's aim in the sermon is to 'console' his congregation and the wider Catholic reading public by comparing the strength of Rome's authority, which is not of this world, with the weakness of Canterbury's, which is. Having sketched the history of the Church of England and its royal supremacy, he focuses upon the disagreement between the Archbishops and the Bishop of London over Gorham, and the reversing of the first, ecclesiastical judgment by a lay tribunal. The Church of England, he declares, has 'committed itself to a jurisdiction, which it now sees may, in its turn, commit her to a heresy'.[21] The nation may have invited the whole world to its door for the Great Exhibition, but its Church is in distress and Anglicans 'cannot in future speak of their Church, as if characterised by its baptism, or as entitled to any allegiance, because of any security which she affords her followers in regard to this vital point'. Catholics, he concludes, must pray for the 'multitudes, whom we believe, nay, whom we know, to be

[18] 'Great Meeting', *Illustrated London News*, 27 July 1850, p. 77.
[19] N. P. S. Wiseman, *Final Appeal in Matters of Faith* (1850), p. 1.
[20] Ibid., pp. 3–4. [21] Ibid., p. 27.

ensnared in the meshes of error and crooked doctrines', that they may be safely brought within the 'precincts' of Christ's true Church, to the 'bosom of their dear Mother, who stretches forth her hands to receive them'.[22]

Two months later, when the Pope had been safely returned to Rome, Wiseman delivered a lecture entitled *The Papal and Royal Supremacies Contrasted* at St George's. We Catholics, he rhapsodizes, have met here to rejoice that the 'successor of Peter has gone back in triumph, in the midst of the joy and of the tears of his subjects, to that seat of his spiritual jurisdiction which was allotted to him by the Providence of God, – by his Divine Master Himself'.[23] In contrast, following the Gorham judgment, the royal supremacy is becoming 'more and more a subject of anxiety, of doubt, of difficulty, and many are only considering in what way they can best escape from its influence'. Whereas in the Church of England a matter of doctrine has been decided by the Queen, on the advice of judges, every Catholic believes that the Pope 'holds his supremacy directly from God – from Christ; that he is His Vicar, His Representative on earth, the Head of the Church under Him; and therefore he obeys him as the Vicar of Christ – he obeys him as Christ Himself'. Although Protestants could argue that politics rather than providence was at work when the Pope returned to the Vatican, and that the whole basis of his Petrine authority was questionable, Wiseman's rhetoric again opens up a yawning gap between a Church that is not of this world and one that is.[24]

Later in the summer of 1850, as Anglicans continued to agonize over the Gorham judgment, Wiseman took a third opportunity to publish an address on the subject, two months before the restoration of the Catholic hierarchy. On the Sunday following the London meeting of the Anglican High Church party he delivered a discourse to a mixed audience of Catholics and Protestants at St John's Catholic Church in Salford.[25] His argument is simple: the Church of England is dead. Like a branch cut from a tree, however, it maintains for a while 'the outward semblance of life'. He then develops the analogy:

Not that I would insinuate, that before this the Church of England was a living branch of the true church: for I know in common with every intelligent man, that

[22] Ibid., pp. 38–9.

[23] N. P. S. Wiseman, *Papal and Royal Supremacies* (1850), p. 5. Further quotations in this paragraph are from pp. 7, 28, 40.

[24] Gladstone, whose copy of Wisman's lecture is annotated (St Deiniol's Library, Hawarden, classmark N 62/9), published a reply: *Remarks on the Royal Supremacy* (1850).

[25] The audience, who included Tractarians, paid a shilling a head to enter the 'magnificent edifice recently erected': N. P. S. Wiseman, *Discourse on the Gorham Controversy* (1850), p. 1.

for the last three hundred years, the Church of England has been a lifeless branch, and has formed no part whatever, of the true living Catholic Church of Christ. It has been a branch separated from the vine [John 15.4]. But this act has come as it were to set an outward and more visible seal to this unnatural state, by which it existed. It seems to be sent by the Almighty to open the eyes of people, and by showing them how this church is unable to grapple with error, to teach the value of our own more sacred ordinances.[26]

The reference to 'every intelligent man' is the mark of a skilled debater, as is his turning of its three hundredth anniversary against the Church of England.

A more poetical passage that follows bears the mark of a devout Catholic, in that Wiseman can describe the Church as beautiful:

Now if I look at any part of a tree I need not point out to you each part individually, I direct your eyes at once to its majesty – its luxuriantly leaved head, spreading on every side its gigantic branches and beautiful verdure, the trunk, magnificent in its proportions, gnarled roots, that have ploughed up the earth and come forth as if to seek light, from the depths in which they were buried. As beautiful and as united in its proportions as that tree, is the Catholic Church.[27]

Whereas Anglican bishops and poets – Wordsworth, Coleridge, Southey, Keble – tend to celebrate the organic relationship between Church and nation through such analogies, Wiseman's tree stands alone, like the tree in the parable of the mustard seed [Matthew 13.32].[28]

Meanwhile, with many High Churchmen teetering on the brink of conversion to Rome, Newman had been persuaded to prepare six *Lectures on certain Difficulties felt by Anglicans in Submitting to the Catholic Church*, which he delivered at the London Oratory between 9 May and 5 July 1850. Newman's 'impassioned eloquence' deeply impressed Aubrey de Vere, who attended them.[29] Drawing upon his own experience of losing faith in what he calls the 'movement of 1833' (the Oxford Movement), Newman argues that the Tractarians have no power and no future in a Church of England which is 'nothing more or less than an establishment, a department of government, or a function or operation of the state, – without a substance, – a mere collection of officials, depending on and living in the supreme civil power'.[30] The history of the royal supremacy is witheringly summarized: 'Elizabeth boasted that she "tuned its pulpits;" Charles forebade discussions

[26] Ibid., pp. 8–9. [27] Ibid., p.12.

[28] In 1856 Richard Simpson, the Liberal Catholic critic, parodied Anglican sermons which used the simile of a tree when addressing issues of the day such as baptismal regeneration, in a review of Browning's 'Bishop Blougram's Apology': *Richard Simpson as Critic* (1977), pp. 68–9.

[29] He also commented on their 'extreme subtlety': *Recollections of de Vere* (1897), p. 263.

[30] J. H. Newman, *Lectures on certain Difficulties* (1850), p. 7. Further quotations in this paragraph are from pp. 9, 21, 24–6, 80.

on predestination; George, on the Holy Trinity; Victoria allows differences on Holy Baptism.' The question raised by the Gorham case, he argues, was 'not what God had said, but what the English nation had willed and allowed'. Directly addressing his former Anglican brethren, Newman suggests that the doctrines of the Trinity and of eternal punishment are now under erasure in the Church of England. We Catholics, he says, long to convert you, in order to save your souls. Only the Catholic, he claims in his most appealing manner, has within him 'that union of external, with internal notes of God's favour, which sheds the light of conviction over his soul, and makes him both fearless in his faith, and calm and thankful in his hope'.

Newman's main strategy in the lectures is to show that the Tractarians have been effectively cut adrift by the Gorham judgment, as the purpose of their movement had been to maintain ecclesiastical authority, 'as opposed to the Erastianism of the State'.[31] Only the Catholic Church, he claims, is 'proof against Erastianism'; only she 'professes to be built upon facts, not opinions; on objective truths, not on variable sentiments; on immemorial testimony, not on private judgment; on convictions or discernments, not on conclusions'; only she is 'an organ and oracle, and nothing else, of a supernatural doctrine'.

One of Newman's most powerful analogies, that of the shipwreck, was a recurrent leitmotif in English Catholicism's response to Gorham. Wiseman, in the first Southwark sermon of 17 March, had shrewdly reversed John Bullish appeals to the providential claims of an 'island race', of the kind favoured by Charles Kingsley, by presenting the countries of Catholic Europe as rock-like. 'These various parts of the living Church of Christ', he comments, 'can in fact only look on the agitated Establishment of this country, as men can do from a firm and lofty shore, upon a frail and shattered bark, tossed upon the billows'.[32] Drawing upon one of the oldest analogies for the Church, he later modulates from 'bark' to 'ark', in an appeal to Anglicans to convert to Rome:

> Our blessed Redeemer . . . will pour forth at once the grace of complete and immediate conversion upon those who are wavering, who desire still to cling to this bark that is sinking beneath their feet; but who have not the courage at once generously to throw themselves into the bark of Peter, the only ark in which God is pleased to rescue from error and perdition.[33]

Newman, having argued in the first of his *Lectures on certain Difficulties* that the 'Establishment, whatever it be in the eyes of men, whatever its

[31] Ibid., p. 108. Further quotations in this paragraph are from pp. 163, 181, 182.
[32] N. P. S. Wiseman, *Final Appeal* (1850), p. 33. [33] Ibid., p. 38.

temporal greatness and its secular prospects, in the eyes of faith is a mere wreck', later embellishes Wiseman's analogy with more pointed Anglican reference:

The giant ocean has suddenly swelled and heaved, and majestically but masterfully snaps the cables of the small craft which lie upon its bosom, and strands them upon the beach. Hooker, Taylor, Bull, Pearson, Barrow, Tillotson, Warburton, and Horne, names mighty in their generation, are broken and wrecked before the power of a nation's will. One vessel alone can ride these waves, the boat of Peter, the ark of God.[34]

Wiseman takes up the theme again at the beginning of his third address, at Salford, on 28 July 1850. Whereas the Established Church of England had proved to be a 'barque exposed to be borne to and fro by every wind of doctrine', we who are 'placed upon the rock of Catholic faith' feel ourselves 'secure upon every point of doctrine'.[35] By this time the Tractarian brethren must have felt decidedly seasick. Gorham himself, writing from the vicarage at Brampford Speke in August 1851, made them feel no better when he protested against Bishop Phillpotts's convening of a diocesan synod and applied the same analogy to them:

May God deliver our Church from the perilous breakers, among which the reckless semi-Romish party has directed her! May He, in the pitifulness of His great mercy, 'bring her out of her distresses;' making 'the storm a calm, so that the waves thereof are still;' rendering her 'glad because she is *quiet*, and so bringing her to the desired haven!'[36]

III

Effective as the shipwreck analogy was, Wiseman's reference to the shepherd and the sheep (John 10.1–18) in his Southwark sermon of 17 March 1850 touched a more sensitive Anglican nerve. To Catholics, he reflects, it seems a 'most strange inconsistency . . . that this is the case of a lay tribunal reversing the judgment, or decision, of the Archbishop in his Court of Arches':

In fine, my Brethren, who after all represent the Church? Who are they whom Christ has appointed to watch over the law, over the interests, over the principles of His Church? Is it the sheep, or is it the shepherd? Is it the subject, or is it the ruler? Is it the Clergy – the ministering Clergy of the Church, and [sic] the Laity?[37]

[34] J. H. Newman, *Lectures on certain Difficulties* (1850), pp. 6, 23.
[35] N. P. S. Wiseman, *Discourse on the Gorham Controversy* (1850), p. 1.
[36] G. C. Gorham, *Exeter Synod* (1851), p. 20.
[37] N. P. S. Wiseman, *Final Appeal* (1850), pp. 22–3.

This was one of the most vexed questions dividing Tractarians from Evangelicals in the Church of England, as the Gorham case had shown and as Wiseman well knew. Sheep and shepherds, as well as baptismal signs of water and water vessels, figure in the iconography of early Pre-Raphaelite religious paintings which were conceived, executed and exhibited during and in the aftermath of the Gorham case.[38] Both Millais's *Christ in the House of His Parents (The Carpenter's Shop)* and Hunt's *A Converted British Family Sheltering a Christian Missionary from the Persecution of the Druids* were conceived while the Gorham case was before the Court of Arches, in the summer of 1849, completed the following winter, while the Judicial Committee of the Privy Council heard Gorham's appeal, and exhibited at the Royal Academy in May 1850, two months after the public announcement of the verdict for Gorham.[39]

Millais's original inspiration for a painting on 'Christ wounded in the House of His Friends' came from a sermon that he heard in Oxford in the summer of 1849 on Zechariah 13.6: 'And one shall say unto him, What are these wounds in thine hands? Then he shall answer, Those with which I was wounded in the house of my friends?'[40] Millais first showed the Rossettis a sketch or sketches for the painting on 1 November 1849. The sheep, symbolizing the laity or 'the people' in liturgical terms, are penned in and separated from the main actors in the foreground, whose positions are reminiscent of those of the clergy at a High Church celebration of Holy Communion or a Catholic Mass (illustration 27). As Millais began to paint in the early weeks of 1850 and interest in the pending Gorham judgment grew, the figure of the young John the Baptist on the right emerged as a key figure in the work. St John brings a bowl of water with which to bathe Christ's wound, thus prefiguring both his Baptism, indicated by the dove in the background, and his Passion, symbolized by the tools, the wood, the ladder and the unfinished basket, and recalled in the mixing of the water and the wine in the sacrament of the eucharist. Millais's painting, completed by 8 April 1850, would seem to support the view expressed by Robert Wilberforce in his Oxford sermon of 10 March on the 'Sacramental and Anti-sacramental systems', that to deny the doctrine of the baptismal

[38] See A. Grieve, 'The Pre-Raphaelite Brotherhood', *Burlington Magazine*, 1969, pp. 294–5. I am indebted to Grieve for his analysis of some of the iconography in these paintings.

[39] For chronologies and interpretations of these paintings see entries by Judith Bronkhurst and Malcolm Warner in A. Bowness, *et al.*, *Pre-Raphaelites* (1974), pp. 76–9.

[40] See W. H. Hunt, *Pre-Raphaelitism* (1905), I, 194–5. On Millais's and Hunt's original drawings for their paintings see A. Grieve, 'Style and Content in Pre-Raphaelite Drawings', in *Pre-Raphaelite Papers* (1984), pp. 23–43.

27. John Everett Millais, *Christ in the House of His Parents (The Carpenter's Shop)*, 1850.

regeneration of infants is a 'virtual denial that Jesus Christ is come in the flesh'.[41]

The 'Papist' flavour of all this High Anglican iconography contributed to the unpopularity of *Christ in the House* when it was exhibited in the Royal Academy in May 1850. The same was true of Hunt's submission on that year's Academy theme of 'An Act of Mercy', his *Converted British Family* (illustration 28), which was probably inspired partly by demonstrations in 1845 and 1848 against the Tractarians' adoption of the surplice.[42] Hunt himself recorded that he had read extensively in the 'early history of England', on Druidism and the missionaries who 'came to destroy the bloody creed'.[43] An early study for the painting which Hunt prepared in May 1849 already included a number of sacramental signs: the cross on the door, the pendent cruse of oil, the baptismal dish of water, and the bullrushes, suggestive of the story of the infant Moses in Egypt, itself a type of baptism.

Once he was working on canvas, in the summer of 1849, Hunt was confronted with a technical problem:

> As I worked out my composition it was apparent that the regulation size of the Academy canvas would not allow me to add to the central group a margin, most precious in my eyes, on which to paint from nature the landscape outside the hut, with the shallows of the river in front, by which the openness of the homestead on this side might be justified. I therefore gave up the wish to become the foremost student of the time, and developed my plans so that the composition should have the more justice done to it.[44]

The 'precious' foreground of the painting, which he painted at Homerton that August, is more than a meticulous representation of a clear river: it suggests the waters of baptism, while the addition of a fur loincloth on the smallest boy recalls John the Baptist and the baptismal regeneration of children, as in Millais's *Christ in the House*.[45]

When Millais's and Hunt's pictures were exhibited at the Royal Academy, in May 1850, they were mocked and vilified by the press. It was not until Ruskin was asked to defend their work against such attacks the following year that he became involved with the 'Pre-Raphaelites', a term which he

[41] 'The extreme youth of the Baptist and his intention of washing the wound of Christ can be related to the Tractarian emphasis on child baptism and regeneration through baptism': A. Grieve, 'The Pre-Raphaelite Brotherhood', *Burlington Magazine*, 1969, p. 294.

[42] See A. Grieve, 'Style and Content in Pre-Raphaelite Drawings', in *Pre-Raphaelite Papers* (1984), p. 37.

[43] W. H. Hunt, *Pre-Raphaelitism* (1905), I, 173. [44] Ibid.

[45] See A. Grieve, 'Style and Content in Pre-Raphaelite Drawings', in *Pre-Raphaelite Papers* (1984), pp. 28–30.

28. William Holman Hunt, *A Converted British Family Sheltering a Christian Missionary from the Persecution of the Druids*, 1850.

disliked. His writings on aesthetics had already influenced them; now his writings on religion were also to have their effect.

For Ruskin, the product of an Evangelical upbringing and still, at the age of thirty-one, keen to please his doting parents, the Gorham judgment inspired home thoughts from abroad.[46] On 31 March 1850 he read about Phillpotts's open letter to Archbishop Sumner in an English language newspaper at Avignon, on his way back to London from Venice, where he had spent the winter gathering masses of source material for *The Stones of Venice* (1851–3). Ruskin wrote in his diary:

> I read to-day in Galignani part of an acrimonious and of what I fear will become an indecent controversy between the Arch of Canterbury and the B of Exeter, respecting Infant Regeneration by Baptism. I am induced to set down what seems to me to be principles of right judgment in this case which a man of candour belonging by Education to neither party could hardly fail to acknowledge.[47]

The result was a draft of an 'Essay on Baptism', where Ruskin bases his argument upon the authority of Scripture rather than that of the Church and upon an understanding of Christianity as a faith to be embraced and lived out 'To-day', in the world, rather than disputed upon by opposing religious parties. The Angels of God, he suggests, must be surprised to see British Christians quarrelling over baptismal regeneration 'while half the world is unbaptized, and the other half blaspheming Christ'.[48] The 'great question for every man' is '"Whether he be Now serving God or not?"' and is not to do with his opinions on baptism. Ruskin does, however, offer the Evangelical a compromise position on baptismal regeneration. Just as we say that a person is dying, he suggests, so we can say that a person is 'converting', and that in God's eyes the beginning of that converting process – of which the individual and those around him or her may be unaware – could be as important as the moment of conversion itelf. That beginning 'may more properly be termed Regeneration'.[49]

Whereas the 'Essay on Baptism' remained unpublished in Ruskin's lifetime, his views on the shepherd and the sheep in relation to Church authority were published and widely discussed in 1851. On 3 March that year the first volume of *The Stones of Venice*, entitled 'The Foundations', was published. Here he vented his own and his father's anti-Catholic sentiments in appendices on 'Papal Power in Venice', 'Romanist Modern Art' and

46 On Ruskin's religion see M. Wheeler, *Ruskin's God* (1999).
47 *Diaries of Ruskin* (1936–9), II, 464.
48 *Works of Ruskin* (1903–12), XII, 574. Further quotations in this paragraph are from pp. 576, 583.
49 Cf. the position of Frederick Meyrick, an Evangelical controversialist, cited in P. J. Jagger, *Clouded Witness* (1982), p. 23.

'Romanist Decoration of Bases'.[50] Three days later the separate and supplementary *Notes on the Construction of Sheepfolds* appeared. This was a tract on the Church of England which disappointed a number of sheep-farmers who bought it, thinking that it was a technical manual. As if in response to Wiseman's question, the Evangelical Ruskin states unequivocally that 'all members of the Invisible Church become, at the instant of their conversion, Priests', and adds that the 'blasphemous claim on the part of the Clergy of being *more* Priests than the godly laity – that is to say, of having a higher Holiness than the Holiness of being one with Christ, – is altogether a Romanist heresy'. For Ruskin, as for Gorham, an Anglican clergyman is not a priest but a presbyter.[51]

On 7 May 1851 Ruskin defended the young Pre-Raphaelites in a letter to *The Times*. His comments were not entirely uncritical, however, and he was careful to clarify his own religious position: 'No one who has met with any of my writings', he wrote, 'will suspect me of desiring to encourage them in their Romanist and Tractarian tendencies.'[52] Soon after the publication of Ruskin's letter, Hunt started work on both *The Light of the World* and *The Hireling Shepherd*. In July he and Millais, who was working on his *Ophelia*, stayed at Ewell in Surrey together and discussed Ruskin's *Sheepfolds* and William Dyce's printed reply to Ruskin.[53] *The Hireling Shepherd*, exhibited at the Royal Academy in 1852, clearly reflects Hunt's engagement with the ecclesiastical debates of the day and his response to Ruskin's pamphlet. *Notes on the Construction of Sheepfolds* develops some of the themes in Ruskin's anti-sectarian 'Essay on Baptism':

> A man becomes a member of this Church only by believing in Christ with all his heart; nor is he positively recognizable for a member of it, when he has become so, by any one but God, not even by himself. Nevertheless, there are certain signs by which Christ's sheep may be guessed at. Not by their being in any definite Fold – for many are lost sheep at times; but by their sheep-like behaviour . . .[54]

Baptism, he argues, cannot be regarded as a 'sign of admission into the Visible Church', for 'half the baptized people in the world are very visible

[50] *Works of Ruskin* (1903–12), IX, 419–24, 436–40, 471–3. The 'basest' of all 'fatuities', Ruskin believes, 'is the being lured into the Romanist Church by the glitter of it, like larks into a trap by broken glass' (IX, 437). Further references in this paragraph are to XII, 537, 538.

[51] Cf. G. C. Gorham, ed., *Examination before Admission* (1848), p. 24.

[52] *Works of Ruskin* (1903–12), XII, 320.

[53] See J. G. Millais, *Life of Millais* (1899), i, 120–1. Dyce's reply focuses upon Ruskin's use of the term 'Visible' Church, which comes from Hooker rather than the Bible: W. Dyce, *Notes on Shepherds and Sheep* (1851). Hunt borrowed Coventry Patmore's copy of Hooker before going down to Ewell.

[54] *Works of Ruskin* (1903–12), XII, 528. Further quotations in this paragraph are from pp. 529, 530, 533–4, 548, 557, 558.

rogues'. We 'spend much time in arguing about efficacy of sacraments and such other mysteries; but we do not act upon the very certain tests which are clear and visible'. There is, in Ruskin's view, '*no such thing* as the Authority of the Church' in matters of doctrine, and 'ecclesiastical tyranny has, for the most part, founded itself on the idea of Vicarianism, one of the most pestilent of the Romanist theories, and most plainly denounced in Scripture'. The Church of England, 'paralyzed at its very heart by jealousies, based on little else than mere difference between high and low breeding', must heal the rift between the High Church party and the Evangelicals by 'keeping simply to Scripture', and must 'take her stand' against the Papacy, so that all Protestants can be 'united in one great Fold' and 'Anti-christ overthrown'.

In a letter of 1897 Hunt explained that his hireling shepherd (illustration 29) was a 'type' of 'muddle headed pastors who instead of performing their services to their flock – which is in constant peril – discuss vain questions of no value to any human soul'.[55] For the visitor to the Royal Academy in 1852 the message was clear: the 'hireling shepherd . . . careth not for the sheep' (John 10.13). Distracted by Catholicism, the seductive whore of Babylon 'arrayed in purple and scarlet colour' (Revelation 17.1–5), he allows the sheep to cross over the brook to feed on the corn of Rome and thus, like the lamb on the girl's knee eating unripe fruit, to go to their destruction.

Hunt's 'first object' in the painting, however, was to 'pourtray a real Shepherd and Shepherdess . . . sheep and absolute fields and trees and sky and clouds instead of the painted dolls with pattern backgrounds called by such names in the pictures of the period'.[56] Again, Ruskin's *Sheepfolds* pamphlet may have been an inspiration to Hunt, in both *The Hireling Shepherd* and *Our English Coasts, 1852 (Strayed Sheep)* (1853),[57] in offering realist detail in the working out of the meaning of the parable in John. He writes, for example, of 'Christ's Sheep' getting 'perpetually into sloughs, and snows, and bramble thickets', and at other times 'following each other on trodden sheepwalks, and holding their heads all one way when they see strange dogs coming'.[58]

In the same year that Hunt's *Strayed Sheep* was exhibited, Ruskin combined realism and allegorical interpretation in his analysis of St Mark's in the second volume of *The Stones of Venice*, entitled 'The Sea-Stories' (1853). One

[55] Cited in Judith Bronkhurst's entry on *The Hireling Shepherd* in A. Bowness, *et al.*, *The Pre-Raphaelites* (1974), p. 96.

[56] Letter of c. 1890s cited by Bronkhurst, ibid., p. 94.

[57] Ruskin refers to the 'white cliffs of England' in *Sheepfolds*: *Works of Ruskin* (1903–12), XII, 557.

[58] Ibid., XII, 534.

29. William Holman Hunt, *The Hireling Shepherd*, 1852.

of his main aims in chapter IV, 'St. Mark's', is to challenge the Catholic and Anglo-Catholic theory, promoted most energetically by Augustus Welby Pugin, that 'because all the good architecture that is now left is expressive of High Church or Romanist doctrines, all good architecture ever has been and must be so'.[59] This he achieves with ease, pointing out that whereas in the Middle Ages all the secular buildings around cathedrals were also in the Gothic style, today only the largest and most sacred buildings are left to us, with the result that we associate Gothic with medieval Catholicism. He also aims to explain why St Mark's, a Catholic cathedral built in the Byzantine style with Gothic and Renaissance additions, and often regarded as strange and alien by Protestant visitors to Venice, is one of the greatest buildings in the world. This he also achieves, but with much more imaginative effort.

Addressing his 'intelligent Protestant reader', Ruskin mobilizes a wide range of arguments. The Venetians, we must remember, had little respect for Rome and the papacy. Yes, the great religious painters were Catholics and we encounter unacceptable Catholic iconography in their works, but if we look at the 'heart of the work' we will find 'those deeper characters of it, which are not Romanist, but Christian, in the everlasting sense and power of Christianity'.[60] The 'incrusted' quality of St Mark's, which is built of brick faced with marble, is not a deceit of the kind that one might associate with Catholicism, but a brilliant and beautiful use of scarce materials imported from afar and overtly attached to the building with 'confessed rivets'. The fact that the despicable modern Venetian pays no heed to the building's message of divine judgment should not affect our assessment of the Venice of the doges, whose faithfulness and obedience made such a wonderful creation possible. St Mark's can be read as a 'Book Temple', in which the illiterate could learn the great truths of Scripture from the mosaics: 'Never had city a more glorious Bible.' And so on.[61]

Ruskin removed two sections of his analysis of 'Romanism' from the main body of the text of *The Stones of Venice* during drafting. He relegated to an appendix of volume II ('Proper Sense of the Word Idolatry') his argument that Protestants like himself should be careful of accusing Catholics of idolatry, when we are all guilty, in one way or another.[62] As we have seen, three days after the publication of volume II he also published his thoughts on the definition of the Church in a separate pamphlet, *Notes on the Construction of Sheepfolds*, which contains material originally intended to be

[59] Ibid., x, 120.
[60] Ibid., x, 125–6. Further quotations in this paragraph are from x, 93–5, 92, 129–30, 141.
[61] For a more detailed discussion of these themes, see M. Wheeler, *Ruskin's God* (1999), pp. 86–97.
[62] Ibid., x, 450–2.

part of the appendix on 'Romanist Modern Art' in volume I. Significantly, however, his thinking on baptism, made public in *Sheepfolds*, also informs his most subtle strategies in manœuvring the Protestant reader through St Mark's.

In order first to establish the contrast between northern and southern Gothic, and thus to play along with English Protestant prejudices, Ruskin famously leads the reader to the west front of an English cathedral with its 'great mouldering wall of rugged sculpture and confused arcades' before plunging him or her into the alleyways of Venice that lead to St Mark's square, out of which there rises a 'vision out of the earth', a 'treasure-heap, it seems, partly of gold, and partly of opal and mother-of-pearl'.[63] The 'Cleopatra-like' oriental luxury of the decoration is rescued from potential accusations of idolatry by Ruskin's emphasis upon the fact that the capitals of the pillars of variegated stones, 'rich with interwoven tracery', all begin and end with 'the Cross'. Ruskin then points out that modern Venice ignores the message of judgment that is embodied in St Mark's. The vendors of toys and caricatures sit on the foundations of its pillars and 'idle' middle-class Venetians 'lounge' in the cafés of the square, yet nobody pays the slightest attention to the 'images of Christ and His Angels' which 'look down upon it continually'.

Ostensibly in order to avoid entering the church 'out of the midst of the horror' of the hubbub, Ruskin invites the reader to turn southwards: 'let us turn aside under the portico which looks across the sea, and passing round within the two massive pillars brought from St Jean d'Acre, we shall find the gate of the Baptistery; let us enter there'.[64] This small external area adjacent to the Baptistery door (today fastened as a window) was of great personal significance to Ruskin, who drew it from all angles and seems to have regarded it almost as an interior.[65] He had long been aware that the sculpture of the Judgment of Solomon on the Ducal Palace directly faces the door of the Baptistery. As in his 'Essay on Baptism', where the 'unhappy parent' who comes to doubt the efficacy of baptism is warned of future divine judgment, the connection remains mute in 'St Mark's', as the 'heavy door' of the Baptistery 'closes behind us instantly, and the light and the turbulence of the Piazzetta are together shut out by it'.

Other reasons for Ruskin's taking the reader into the church via the Baptistery can be suggested. First, and most obviously, he can focus there upon the tomb of Doge Andrea Dandalo, 'a man early great among the great of Venice', which is an important 'touch-stone' in Ruskin's cultural history of the city. Secondly, he thus avoids entering through the north

[63] Ibid., x, 79–85. [64] Ibid., x, 85. [65] See J. Unrau, *Ruskin and St. Mark's* (1984), p. 127.

('Arabian') porch, as most nineteenth-century Venetians do in order to worship in their favourite side chapel, which is dedicated to the Virgin and vulgarly decorated with 'silver hearts'.[66] Thirdly, he softens the impact of the extraordinary interior of St Mark's upon the untrained English Protestant eye by repeating the familiar route taken by a visitor to a church at home – through the door, past the font, and into the main body of the building. That route symbolizes the passage of an individual soul into membership of 'Christ's holy Church' through baptism (Book of Common Prayer).

Like Gorham, Ruskin emphasizes the authority of Scripture, as he encourages the reader to 'look round at the room in which he [Dandalo] lies'.[67] The dimness of the place allows the eye of the Protestant beholder – both author and reader – to dwell, not on the Catholic altar, but on the surrounding biblical iconography, which is legimated through reference to England's greatest Puritan writer:

> The light fades away into the recess of the chamber towards the altar, and the eye can hardly trace the lines of the bas-relief behind it of the baptism of Christ: but on the vaulting of the roof the figures are distinct, and there are seen upon it two great circles, one surrounded by the "Principalities and power in heavenly places," [Ephesians 3.10] of which Milton has expressed the ancient division in the single massy line,
>
> "Thrones, Dominations, Princedoms, Virtues, Powers," [*Paradise Lost*, v, 601] and around the other, the Apostles; Christ the centre of both: and upon the walls, again and again repeated, the gaunt figure of the Baptist, in every circumstance of his life and death; and the streams of the Jordan running down between their cloven rocks; the axe laid to the root of a fruitless tree that springs up on their shore. "Every tree that bringeth not forth good fruit shall be hewn down, and cast into the fire." [Matthew 3.10] Yes, verily: to be baptized with fire, or to be cast therein; it is the choice set before all man. The march-notes still murmur through the grated window, and mingle with the sounding in our ears of the sentence of judgment, which the old Greek has written on that Baptistery wall. Venice has made her choice.
>
> He who lies under that stony canopy would have taught her another choice, in his day, if she would have listened to him; but he and his counsels have long been forgotten by her, and the dust lies upon his lips.

Ruskin chooses to spell to an endangered England the 'warning' which a now ruined Venice embodies here in the Baptistery, where, in contradistinction to Catholic doctrine, his Protestant interpretation of its iconography emphasizes choice and the convert's baptism by fire, in a passage crammed with favourite Evangelical types of baptism and appropriate biblical texts.

66 *Works of Ruskin* (1903–12), x, 91. 67 Ibid., x, 86–7.

The ground is thus prepared for the final, analytical section of this long chapter on St Mark's, where Ruskin attempts to 'form an adequate conception of the feelings of its builders, and of its uses to those for whom it was built', by showing the Protestant reader how to 'take some pains' to 'read all that is inscribed' in the building. By showing the reader first the atrium or portico reserved for 'unbaptized persons and new converts', and then the mosaics over the main entrance which the newly baptized saw when they were first permitted to pass into the church, Ruskin has him or her repeat in spirit the same journey completed earlier, through the Baptistery.[68] Yet again he asks the Protestant reader to attend to particular aspects of mosaics in which Christ is represented with appropriate texts such as 'I am the door' (John 10.9):

> Now observe, this was not to be seen and read only by the catechumen when he first entered the church; every one who at any time entered was supposed to look back and to read this writing; their daily entrance into the church was thus made a daily memorial of their first entrance into the spiritual Church; and we shall find that the rest of the book which was open for them upon its walls continually led them in the same manner to regard the visible temple as in every part a type of the invisible Church of God.

As in his 'Essay on Baptism', where Ruskin sought common, non-sectarian ground by emphasizing the 'converting *process*', the beginning of which 'may more properly be termed Regeneration', he now wants the Protestant reader to believe that the makers of these physical mosaics in a physical Catholic church of the twelfth and thirteenth centuries understood that baptism was an initiation into the spiritual Church, daily brought to mind in the process of re-entering the building. Therefore the mosaic of the first dome, he points out, 'which is over the head of the spectator as soon as he has entered by the great door (that door being the type of Baptism), represents the effusion of the Holy Spirit, as the first consequence and seal of the entrance into the Church of God'.

Unlike the first worshippers in St Mark's, Venetians in Ruskin's day focus upon the building's remarkable collection of 'stage properties of superstition', and never in his experience 'regard for an instant' the 'Scripture histories on the walls'.[69] Ruskin then underlines the point through a vivid contrast in which a familiar English rural scene is evoked – the shepherd and his sheep passing through the ruins of a monastery, three hundred years after its dissolution:

[68] Ibid., x, 135. [69] Ibid., x, 90–2.

> The beauty which it [St Mark's] possesses is unfelt, the language it uses is forgotten; and in the midst of the city to whose service it has so long been consecrated, and still filled by crowds of the descendants of those to whom it owes its magnificence, it stands, in reality, more desolate than the ruins through which the sheep-walk passes unbroken in our English valleys; and the writing on its marble walls is less regarded and less powerful for the teaching of men, than the letters which the shepherd follows with his finger, where the moss is lightest on the tombs in the desecrated cloister.

The claims of Catholic religion, in modern Venice as in England, are represented here as a dead letter, whereas the shepherd and his sheep, who need no sacred building or inscription or font, are alive and live in peace with their environment.

Later in life, Ruskin achieved a sympathy for and understanding of the Franciscan tradition, and even considered himself to be a tertiary of the order. He always retained his dislike of modern 'Romanism', however, notwithstanding his friendship with Cardinal Manning. Ironically, it was his vivid descriptions of St Mark's that gave an impetus to a revival of interest in Byzantine architecture in England, the zenith of which was John F. Bentley's Roman Catholic Cathedral at Westminster (1895–1902).[70]

[70] See ibid., x, li.

PART III

Cultural spaces

CHAPTER 8

Maiden and mother

Mary mother of maid and nun . . .

Gerard Manley Hopkins, 'Margaret Clitheroe'[1]

I

So far in this book I have focused upon the conflict between the old enemies in the first half of the nineteenth century, when two time-honoured patriarchies, one owing its allegiance to the Crown, the other to Rome, promoted their different versions of history and claims to ecclesiastical authority in England. In this final section I turn to the second half of the century, when English Catholicism gained in strength and confidence and when new cultural spaces opened up in which Catholic themes and symbols were reinterpreted. The old enemies continued their public battles, over Pope Pius IX's definition, for example, of the doctrines of the Immaculate Conception of the Blessed Virgin and Papal Infallibility. Meanwhile the inner conflict described by Joseph Henry Shorthouse in one of the epigraphs to this book, as between 'on the one side obedience and faith, on the other, freedom and the reason', expressed itself in new and often startling ways.

In section II of this chapter, I discuss the furore at mid-century surrounding convents, enclosed spaces dedicated to celibate women and the site of wild misreadings by anxious anti-Catholic writers, who drew upon old stories of sexual depravity in Continental 'nunneries'. So familiar was the association between convents and 'Jesuitical' control over women that two Protestant writers, Jemima Luke and Charlotte Brontë, could exploit its narrative potential in female spiritual autobiographies which examine the desires and delusions of celibate young women: *The Female Jesuit* (1851) and

[1] G. M. Hopkins, *Poems* (1967), p. 182.

Villette (1853). In subsequent decades, however, Catholic poets could celebrate rather than defend the figure of the professed nun as a daughter of the Virgin Mary. The identification of celibacy with holiness by Catholics and Tractarians, and resistance to this identification by the Protestant majority in England during the convents debate of the early 1850s, provide the background to section III, where I examine writings on Mary after the definition of her Immaculate Conception (1854), focusing particularly on the poetry of Gerard Manley Hopkins, Coventry Patmore and Alice Meynell.

II

In the first volume of his best-selling *History of England* (1849), Thomas Babington Macaulay stated that 'no part of the system of the old Church had been more detested by the reformers than the honour paid to celibacy'.[2] When Macaulay wrote this, three hundred years after the Reformation, honour was again being paid to celibacy by both Tractarians and Catholics in their attempts to re-establish monastic orders in England. Even more threatening to the English Protestant mind than the Catholic and Tractarian ideal of a celibate clergy was their ideal of sisterhoods, set apart from mainstream society, led by women and open to unspeakable abuses.

The convent, or 'nunnery', had provided settings for novels and romances since the early days of modern English fiction: examples include Mrs E. Slade's *Nunnery Tales; or, The Amours of Monks Priests and Nuns* (1743), Sarah Scott's *Millennium Hall* (1761), which describes a religious community of women, Joseph Trapp's *The Sprite of the Nunnery* (1796), Catharine Selden's *The English Nun* (1797), Charlotte Dacre's *Confessions of the Nun of St Omer* (1805), Sophia Frances's *The Nun of Misericordia* (1807), Edward Montague's *The Legends of a Nunnery* (1807), Thomas Rickman's *Atrocities of a Conventor; or, The Necessity of Thinking for Ourselves* (1808) and the anonymous *The Black Convent; or, A Tale of Feudal Times* (1819). The abused nun, punished by her mother superior, bullied by her sisters and raped or murdered by her confessor, had become a stock figure during the Revolutionary and Napoleonic Wars, when the flames of John Bull's xenophobic anti-Catholicism were fanned by escapist romance.[3]

We have already seen how Emma Robinson treats these themes in *Westminster Abbey* (1854), where Cardinal Wolsey and Sancgraal prey upon

[2] T. B. Macaulay, *History of England* (1856), I, 69.

[3] For negative comments on ill-fated nuns, see also Ann Radcliffe, *The Mysteries of Udolpho* (1794), III, II and Jane Austen, *Northanger Abbey* (1818), II, 2. The escaped nun became a stock figure in American anti-Catholic fiction: see S. M. Griffin, *Anti-Catholicism* (2004), pp. 27–61.

nuns.[4] Contemporary with this novel were Kingsley's attack on the Catholic ideal of celibacy in *Hypatia* (1853), where the heroine is killed by monks, and two Catholic responses to Kingsley, Wiseman's *Fabiola* (1855) and Newman's *Callista* (1856), where the heroines are presented as early Christian virgins and martyrs.[5] All these novels reflect the fact that in the early 1850s convents and female celibacy were the focus for fierce debate between the old enemies.

In 1830, as one anti-Catholic preacher informed his London congregation, there had been ten convents in England, but by 1850 there were fifty.[6] Several attempts were made in parliament to check this development and to introduce an inspection system. This was not enough for one anonymous pamphleteer, however, who demanded the '*suppression* of these dens of infamy'.[7] During the Crimean War, when Florence Nightingale recruited ten Catholic nuns and fourteen Anglican sisters, the furore subsided for a while, as the positive side of the convent movement was reported in the press.[8] Macaulay had already pointed out that the Catholic Church was better at channelling the skills and energies of women than was the Church of England.[9] In practice, however, to establish a 'good English convent' was a 'great problem', as Newman knew.[10]

Many anti-Catholic writers of the early 1850s referred to what they regarded as Catholicism's habitual suppression of intellectual freedom amongst its adherents. This had been one of Jules Michelet's themes in his hugely popular *Priests, Women, and Families* (1845), extracts from which were reprinted as *The Confessional and the Conventual System* (1850).[11] Catherine Sinclair comments in her preface to *Beatrice; or, The Unknown Relations* (1852): 'The object of Romanism is, entirely to subjugate the will and the intellect; therefore, as Niebuhr says of the Italians, their slavish subjection to the Church is "ghastly death".'[12] Convents were regarded as extreme examples of such 'subjugation'.[13] In Elizabeth Harris's *From Oxford*

[4] See above, p. 102. [5] See above, pp. 68–76. [6] J. B. Cartwright, *No Popery!* (1850), p. 155.

[7] See W. L. Arnstein, *Protestant versus Catholic* (1982), p. 62. [8] Ibid., p. 63.

[9] 'Even for female agency there is a place in her [the Church of Rome's] system. To devout women she assigns spiritual functions, dignities, and magistracies. . . . At Rome, the Countess of Huntingdon would have a place in the calendar as St Selina, and Mrs Fry would be foundress and first Superior of the Blessed Order of Sisters of the Gaols.' T. B. Macaulay, 'Von Ranke', in *Critical and Historical Essays* (1848), III, 241.

[10] Newman added, 'we want some persons of strong sense and wisdom to begin it. English people are so different from foreigners': letter to Mrs J. W. Bowden, Mary Vale, 4 March 1848, *Letters of Newman* (1961–77), XII, 175.

[11] The first edition of Michelet's *Du Prêtre, de la Femme, de la Famille* was published in Paris in 1845.

[12] C. Sinclair, *Beatrice* (1852), I, xvi.

[13] The word seems to have had a strong resonance in this context. Cf. e.g. T. H. Home, *Mariolatry* (1841), p. 5.

to Rome (1847), the Anglican novel that drew a reply from Newman in *Loss and Gain*,[14] Eustace's younger sister, Margaret, also converts to Catholicism and becomes a religious. In the convent, her friend, Sister Mary Beatrice, is said to have died '*unabsolved, and without the Sacraments* – and will be buried as a dog is buried'.[15] Margaret soon follows her, a brain-washed victim of a sinister system: 'The Nun has no mind – it is crushed, extinguished, annihilated in obedience.'[16] Catholic converts like Alice Meynell could make more sense of the monastic impulse, as it was precisely the 'hard old common path of submission and self-discipline' that had brought her to the 'gates' of the Catholic Church in the first place.[17] Obedience is at the heart of the Catholic faith, but nowhere more clearly than in the lives of those under vows. Hopkins, a Jesuit priest, was appalled by the story of a 'wretched being' who 'refused in church to say the words "and obey"' in the marriage service.[18] If it had been a Catholic wedding and he the priest, he 'would have let the sacrilege go no further'.

Whereas the subjugation of the intellect can perhaps be defended in the name of obedience, corruption in the convents is less easily explained away. As Michelet comments when citing documented examples of pregnant nuns, infanticide and the burial of dead babies in convents, 'When Lewis's "Monk" appeared, in 1796, people little expected to see that terrible novel outdone by a real history.'[19] In 1851 the British Society for Promoting the Religious Principles of the Reformation published an anonymous penny tract which begins with the statement that 'nunneries' are 'opposed to the principles of Christianity' and encourage 'cruelty, fraud, and deceit in the priests'.[20] R. P. Blakeney, an anti-Catholic writer who wished to abolish the 'conventual system' but had not himself experienced its rigours, insisted that the 'idle habits of monks and nuns increase the mere animal propensities, and produce great harm'.[21]

Bishop Ullathorne published an important tract on the rights of religious women in 1851,[22] and Cardinal Wiseman made a sustained effort to defend the system the following year, in response to M. Hobart Seymour's inflammatory lectures on 'nunneries' in the Assembly Rooms at Bath. Having been taken down in shorthand and then published, Seymour's first lecture, delivered on 21 April 1852, sold briskly. It paints a gloomy picture. The 'inner life of a nunnery', Seymour argues, 'is a life of monotony, wearisomeness,

[14] See above, pp. 181–2.
[15] E. F. S. Harris, *From Oxford to Rome* (1847), p. 265.
[16] Ibid., pp. 271–3.
[17] V. Meynell, *Alice Meynell* (1929), p. 42. Alice Meynell converted on 20 July 1868.
[18] Letter to Coventry Patmore, 24 September 1883, in G. M. Hopkins, *Selected Letters* (1990), p. 193.
[19] J. Michelet, *Priests, Women, and Families* (1846), p. 86.
[20] Anon., *Nunnery* (1851), p. 3.
[21] R. P. Blakeney, *Popery in its Social Aspect* (n.d.), p. 256.
[22] W. B. Ullathorne, *Plea for Religious Women* (1851).

disappointment, contention, bitterness, and despair'.[23] He cannot report the details of a 'solemn inquiry' into nunneries in Tuscany in 1782 as it revealed 'such a hideous system of long-continued immorality and vice among the nuns and their confessors, that the Pope of Rome was constrained to reform and re-model some of the nunneries'. His present aim is to awaken in his audience a 'sympathy for these poor and imprisoned females of our nunneries' and to encourage a 'more general movement throughout the country' in opposition to the conventual system.

Wiseman responded in an unpublished lecture delivered in the Catholic Chapel at Bath, on Sunday 23 May 1852, to which Seymour replied in turn on 7 June, again in the Assembly Rooms. This time Seymour moved into territory that Kingsley was to explore in *Hypatia* the following year:

> . . . in the beginning God made them male and female; in the beginning he made them man and wife; in the beginning he desired them to increase and multiply amidst the purity, and the innocence, and the holiness, and the happiness of Eden. But the Cardinal steps in with another and a different arrangement, and he would separate the man from the woman, and separate the woman from the man. The Church of Rome has adopted the principle that celibacy is more holy than marriage, and that married persons, as such, are not so holy as unmarried persons, as such. And, accordingly, it is held by many in the Church of Rome that the true atmosphere of religion is solitude and retirement; and that if we would attain to the highest flights of perfection, it must be in the cell of the hermit, or the cave of the anchorite; and as this would not be seemly or possible with women, so we must seek the loftiest flights of holiness, and the lowest depths of humility, in those women who retire to the silence, and the solitude, and the devotion of the cloister.[24]

Moving on from the bizarre picture of the portly Wiseman in Eden, Seymour underlines the supremacy of 'civil liberty and religious freedom', the 'glory of England', by saying that if ladies 'choose to dress themselves in a monastic fashion, black, white, and grey, with rosaries and crucifixes, it may all seem to us extremely silly, but we have no right to interfere; and any interference would be an infringement of their civil and religious rights.' But if 'young persons' are lured in and not allowed to leave, we have a right to intervene. Seymour then says of the bishop or cardinal who pronounces an 'awful anathema' on a nun who makes her final vows and on those who might help her escape: 'Let his name be blotted out from the book of the living, and not be written with the righteous.'

Wiseman's review of both Seymour's lectures in the *Dublin Review* was reprinted as a separate pamphlet. Like Newman in *Loss and Gain*, the

[23] M. H. Seymour, *Nunneries: A Lecture* (1852), p. 6. Further references in this paragraph are to pp. 18, 47.

[24] M. H. Seymour, *Convents or Nunneries* (1852), p. 8. Further references in this paragraph are to pp. 9, 18.

Cardinal uses humour as a weapon against Protestant arguments, thus making Seymour look silly, rather than the nuns. Replying, for example, to the claim that nuns are imprisoned in heavily fortified convents, he points out that 'not a single religious house in England, though belonging to what is called an enclosed order, is secured against easy escape, through windows or doors, should it be desired' and that 'almost every such house has grounds attached to it, in which the religious walk, without any enclosing wall; and a discontented nun might really run away at no greater risk, than a few scratches in getting through a hedge'.[25] Wiseman also has the statistics to hand which disprove the assertion that young girls are lured into convents against their will. Out of 325 professed nuns in England, he states, only five have been professed under the age of twenty-one. His clinching argument, however, is one that he used during the Gorham controversy, when contrasting the Erastianism of the Church of England with Catholicism's claim to authority that is 'not of this world':[26] 'The enemies of the religious state *assume*', he points out, 'that it is a mere human invention; we assert that [it] is the carrying out of a divine injunction.'

Most disturbing of all to Protestant England was the development of sisterhoods by the Tractarians, the enemy within.[27] Pusey (who encouraged his daughter's vocation to become an Anglican nun), John Mason Neale and Thomas Thelluson Carter all founded Anglican sisterhoods.[28] The celibate nun, like the Virgin Mary herself,[29] was regarded by many Protestants as a threat to family values, particularly if the nun were Anglican. In 1852 the Revd John Gladstone published an open letter on the case of Ann Maria Lane, 'now a Sister of Mercy, *against her Father's Wish*' [my emphasis] in Plymouth, in which he links abuses with the 'Popish doctrine of Confession and Absolution' and concludes that the Church of England 'must be purged of Popish leaven; and that it may be so, the bishops must be called to account, in some cases for their betrayal of, in others, for their carelessness concerning, Protestant truth'.[30]

[25] N. P. S. Wiseman, *Convents* (1853), p. 7. Further references in this paragraph are to pp. 37, 23.
[26] See above, p. 192.
[27] 'If we wish to find the real meaning of the Oxford Movement we must turn from the external controversies, to the hidden life and development of the sisterhoods': A. M. Allchin, *Silent Rebellion* (1958), p. 55.
[28] C. M. Engelhardt, 'The Paradigmatic Angel in the House', in A. Hogan and A. Bradstock, eds, *Women of Faith* (1998), p. 161.
[29] Ibid., p. 162.
[30] J. E. Gladstone, *Protestant Nunneries* (1853), pp. 39, 46.

30. Anon, 'Convent of the Belgravians', *Punch*, 19 (July–December 1850), 163.

Two years earlier *Punch* had published an article entitled 'Convent of the Belgravians' [illustration 30] which ends with the kind of awful pun that Victorian readers seem to have enjoyed: 'That the Anglican Convent, thus constituted, will lead to "perversions" there is no fear. Alas! the hard multitude will rather say that the Puseyite sisters are only playing at Roman Catholics, and the vile punster will remark that their Convent is more a Monkey-ry than a Nunnery.'[31] In the same year Kingsley poked fun at 'Popish' practices in his first novel, *Yeast* (1850). Argemone Lavington is encouraged by her Tractarian vicar to join what Kingsley calls a

[31] Anon., 'Convent of the Belgravians', *Punch*, 19 (July–December 1850), 163. ('Perversions' means conversions to Catholicism.) Ruskin wrote: 'Don't wear white crosses, nor black dresses, nor caps with lappets. Nobody has any right to go about in an offensively celestial uniform, as if it were more *their* business or privilege, than it is everybody's, to be God's servants': *Letter to Young Girls* (1876), p. 8, cited in L. H. Peterson, 'The Feminist Origins', in *Ruskin and Gender*, ed. D. Birch and F. O'Gorman (2002), p. 99.

'Church-of-England *béguinage*, or quasi-Protestant nunnery' in a neighbouring city, which the vicar visits in order to 'confess the young ladies' (10). Kingsley believes that the vicar is simply 'pampering the poor girl's lust for singularity and self-glorification', so he addresses Argemone directly, telling her that the angels laugh kindly 'at the rickety old windmill of sham-Popery which you have taken for a real giant'.

Kingsley's anxiety concerning repressed sexuality is an extreme version of a familiar phenomenon in early and mid-Victorian culture, most easily identifiable in the many images of nuns and convents in paintings of the period. Contemporary critiques of these works of art of the 1830s, 1840s and 1850s revealed what has been described as a 'mixture of repulsion and attraction' which 'mirrored the alloy of emotions that the religious issue evoked on its own'.[32] The same mixed feelings are to be found in the fiction of the period, as we have seen with Kingsley himself.[33] Many Protestant writers shared a disturbing sense of uncertainty concerning the 'unnatural' state of female celibacy in which nuns lived, separated from their parents and 'normal' family life. And the Jesuits were blamed. Kingsley and his contemporaries demonized the order as the greatest threat to the novice's security of body, mind and spirit, as Jesuits were associated with dissembling, lying and the manipulation of the emotions. Michelet, whose views were influential in this regard, argued in 1845 that 'nothing did greater service to the cause of the Jesuits than their constantly repeating that their austere founder had expressly forbidden them ever to govern the convents of women'.[34] This was true, he explains, 'as applied to convents generally, but false as regarded nuns in particular, and their special direction; they did not, indeed, govern them collectively, but they directed them individually'. What most disturbs Michelet is the fact that a bishop has said, 'We are Jesuits, all Jesuits', and nobody has contradicted him.[35]

Little wonder, then, that an anonymous documentary narrative entitled *The Female Jesuit* (1851) enthralled Victorian readers at the time of the 'Papal Aggression'. *The Female Jesuit* was written by Jemima Luke, née Thompson (1813–1906), the wife of Samuel Luke, a Congregational minister. Its vivid

[32] S. P. Casteras, 'Virgin Vows', in *Religion in the Lives of English Women*, ed. G. Malmgreen (1986), pp. 129–60 (p. 153). Casteras's examples include William Collins, *The World or the Cloister* (1843), Charles Eastlake, *A Visit to the Nun* (1846), Alexander Johnston, *The Novice* (1850), Charles Allston Collins, *Convent Thoughts* (1851), Alfred Elmore, *The Novice* (1852), John Callcott Horsley, *The Novice* (1856), John Everett Millais, *The Vale of Rest* (1859), Francis S. Walker, *The Convent Garden* (1878), Marie Spartali Stillman, *The Convent Lily* (1890), Arthur Hacker, *The Cloister or the World?* (1896), Edmund Blair Leighton, *Vows* (1906).

[33] See above, pp. 68–70, 105–10.

[34] J. Michelet, *Priests, Women, and Families* (1846), p. 39.

[35] Ibid., p. xxxii.

treatment of 'simulative hysteria' makes it a potential source not only for John Fowles's *The French Lieutenant's Woman* (1969) but also for Charlotte Brontë's *Villette* (1853), while its title has lured generations of anti-Catholic writers, including the Northern Ireland 'Loyalist', Ian Paisley, into citing it as a straight anti-Jesuit document.[36]

Jemima Luke explains that she and her husband were duped by a young woman called 'Marie', who presented herself as the innocent victim of Catholic treachery. Forced to become a nun, she falsely claimed, she had escaped from her convent on the Continent and needed shelter and protection, which the worthy couple duly offered. Ironically, in drawing upon the hackneyed plotting of an 'escaped nun' novel, Marie herself became the subject of a published narrative. For when it emerged that Marie was not who she said she was, and had been sent away, Jemima Luke decided to write an account of the fraud as a warning to others. Being a narrator who is about to describe a complex web of deception, the author is keen to emphasize in her preface that the 'startling assertion that "truth is stranger than fiction" has seldom been more fully verified than in the details of this volume'.[37] She cannot verify part of the Introduction and all of the Autobiography, which were 'furnished' by Marie, but '*All the remainder of the book is strictly and literally true*'.[38]

The key episodes in the Introduction and Autobiography could have been lifted directly from the anti-Catholic fiction of the period. In order to achieve verisimilitude in her story, Marie draws upon familiar Protestant fantasies, and is believed. In a 'Convent at I — ',

> On a Wednesday evening, Jan. 17th, 1849, between six and seven o'clock, while the nuns were at lecture, a young lady, who had been for seventeen years a pupil in the convents of this Order, and who for the last two years had been a postulant, entered the grotto. She knelt, and wept, and prayed in an agony of feeling, which He who searches the heart alone could fully estimate. She had been gradually but fully convinced of the errors of Romanism, and intensely longed for the light of God's truth and the liberty of His Gospel. She had looked forward for some time with increasing dread and disgust to the profession of a nun, yet she could see no escape.[39]

Travelling on the omnibus next day, Marie drops a little cross from her 'Catholic prayer-book'. The gentleman sitting opposite her picks up the cross, sees that she is a Roman Catholic and discusses with her the 'errors' of Romanism.[40]

[36] See www.acts2.com/thebibletruth/Jesuit_Oath.htm : Ian Paisley, 'The Jesuit Oath Exposed'.

[37] J. T. Luke, *Female Jesuit* (1851), p. iii. [38] Ibid., p. iv. [39] Ibid., p. 9. [40] Ibid., pp. 11–12.

Marie takes this gentleman's advice and seeks out a Protestant minister, Mr L —, who gives her a New Testament, formally a forbidden book to her. Mr and Mrs L — are 'greatly surprised' to find that, 'unaided by the Scriptures, and removed from Protestant books, or influence of any kind, she had detected the leading errors of Popery':

> The doctrine of transubstantiation had from the first horrified her, as a species of spiritual cannibalism: the worship of the Virgin and saints, and especially of the waxen images of the infant Jesus, had shocked her as idolatry: the daily repetition of scores of useless prayers, and the idle mummery of the public services, had been an insult to her understanding: the revolting questions of the confessional had outraged her modesty: the refusal of her confessors to permit her to read the Scriptures had awakened her suspicions: her naturally frank and upright mind had been disgusted, by the mystery and concealment which characterised all the movements of her Order; and her free spirit had risen in rebellion, against the spiritual slavery to which she had been condemned, as she had feared for life.[41]

Convinced of Marie's honesty, Mr and Mrs L — offer her 'an asylum in their house'.

Marie's 'Autobiography' then records that she was born in November 1825 in London, that her mother came from an old Catholic Yorkshire family, that her father was German and that she was taken to the Continent when she was seven years old.[42] Having described her convent education she comments: 'In each of the five convents in England there is the same adoration of the Virgin and the saints, and the same superstitious ideas are infused into the minds of the young.' She then describes her novitiate in Switzerland, her move to Nice on the orders of the Pope, and her entry to her order as a postulant in November 1846 at the age of twenty-one.

Part III, 'Sequel' is narrated by Jemima Luke, who describes Marie's employment as a governess by 'Mr and Mrs S —'.[43] Eventually Marie admits that she is an impostor: she is in fact the daughter of a surgeon. She was placed in the care of a guardian when both her parents died, and she entered a convent when she was 'thrown upon the world'. 'Growing tired of its discipline, and having heard of Mr L — in the north, as a man of benevolence and kindness, on coming to London, she determined to find him out.' Following her discovery, Marie emigrates. The narrative ends: 'It will surely be conceded, that the agent in so extraordinary a series of plots, has earned for herself the title *she assumed*, of "a Female Jesuit."'

Encouraged by the sale of almost four thousand copies of *The Female Jesuit* in under a year, Jemima Luke published under her own name *A Sequel*

[41] Ibid., pp. 21–2. [42] Ibid., p. 45. Further quotations in this paragraph are from pp. 59, 99.
[43] Ibid., p. 146. Further quotations in this paragraph are from pp. 391, 433.

to The Female Jesuit, containing her Previous History and Recent Discovery (1852). In the preface she explains that there would have been no need to publish the original book had *The Times* not rejected a letter containing an outline of Marie's 'intrigues', because the story was 'too extraordinary to be believed'.[44] In the past year, readers of *The Female Jesuit* have written to offer various solutions to Marie's case. As several 'literary and medical correspondents' regarded it as a 'singular mental developement, under the technical designation of "Simulative Hysteria"', she offers a few parallel cases that have been brought to her attention, such as that of a 'nun said to have escaped from the convent at Banbury, in 1851'. The most dramatic communication, however, came from Elizabeth Jobson, wife of a Wesleyan minister, who had met Marie, under the different assumed name of 'Lucy', in Manchester in 1847. Mrs and Mrs Jobson had responded positively to her request for help, just as Mr and Mrs Luke had.

From this point on in the *Sequel*, Thompson intersperses her narrative with attempts at interpretation. Is Marie's career as a fraudster to be explained in relation to nature or nurture, she wonders, as she frets over the fact that she and her husband have been able to obtain 'but few particulars' of Marie's childhood?[45] She pieces together Marie's graduation from the petty crimes of childhood to a period of two years as a governess, during which she perfected the art of faking the vomiting of blood, but was clearly not a 'common impostor'. Having later given up the idea of living with the Jobsons, Marie/Lucy lived in a surgeon's house in Manchester until her illness was shown to be feigned and she was evicted. She then persuaded a Manchester lady to go into partnership in expanding her boarding school before obtaining an introduction to a 'London convent' in November 1848. As what followed has already been described in *The Female Jesuit*, Jemima Luke moves on to events after June 1850.

Fleeing England when the Lukes discovered her deceit, Marie landed at Ostend after a difficult crossing, took the train to Ghent and sought admission at the 'English convent'. A few days later she reappeared in Brussels, where she approached a kindly old English Abbé and announced that she wished to be converted to Catholicism.[46] Her baptism in the church of Saint Gudule, Brussels, is, Jemima Luke believes, her '*third* conversion to popery within the space of six years'. She then latches on to a Mr and Mrs Seager, the third kindly married couple who are duped, during

[44] J. T. Luke, *Sequel to The Female Jesuit* (1852), p. iv. Further references in this paragraph are to pp. 1, 2, 8.
[45] Ibid., p. 34. Further references in this paragraph are to pp. 38–45.
[46] Ibid., pp. 55–6. Further references in this paragraph are to pp. 64, 66.

their temporary residence in Brussels. The narrator leaves Marie in their hands and passes over 'a period of nearly fifteen months, diversified with as romantic a series of conversions and persecutions, marriages and burials, tragic accidents and violent deaths, startling and strange reverses, as it has ever been the lot of the most enwrapt novel reader to meet with in the most entrancing novel'. Jemima Luke comments explicitly on the fact that Marie seems to be living out a novel or a popular stage play: 'The tale which the novelist is content to write, Marie acted. The farce which the stage-player condenses into an evening's amusement, she extended with unwearying activity and variety, and the most perfect imitation of nature, through all the details of her daily life.'

It is when Marie finally moves to Bonn and starts converting members of a Protestant congregation to Catholicism that she is finally tracked down and imprisoned for fraud.[47] Thompson records one more 'singular discovery' of 'what may surely be regarded as the master-piece of Marie's audacity and skill':

> It is the month of February 1851. London is unusually busy and excited, for the recent Papal Aggression, and the approaching Great Exhibition, were subjects of importance to all. Parliament is sitting, and these topics are debated with untiring interest from night to night.
>
> . . . Cardinal Wiseman is at his residence in Golden Square. He has much business to transact. A pile of letters meet for a Cardinal claims his attention, and among them is one from *Marie*. She writes from Bonn, and tells the Cardinal that she is a young English lady, once a Protestant, but now a Catholic: that her health is very delicate, and her life uncertain: that she is possessed of considerable property, and is anxious to leave it for the service of the Church; but as all her friends are Protestants, and would greatly object to such an arrangement, she cannot employ the family solicitor, and she ventures under these circumstances, to request that his Eminence will kindly recommend her to a Catholic lawyer.[48]

In the midst of all his other business, Wiseman wrote back to say that he had handed her request to his own solicitor and asked him to advise her. Where formerly a number of pious Protestants had taken Marie in as a refugee from a Catholic convent, the head of the Catholic Church in England now accepts her claim to be a convert to Catholicism.

Just because Marie may not have been a Jesuit, Jemima Luke insists, we are not to conclude that there are no female Jesuits. The central question of the narrative, however, is expressed in the title of its final chapter: 'What Is She?'[49] Why did Marie act as she did? A friend of Samuel Luke's, also

[47] Ibid., p. 108. [48] Ibid., pp. 138–9. [49] Ibid., pp. 194–207.

in the ministry, suggested 'a GIGANTIC EGOISM, which could not live without being the object of attention'. A lady reader of *The Female Jesuit* believed that the 'whole deception was suggested and carried through by Marie under the influence of "Simulative Hysteria", and not at the bidding of the Jesuits or of any other persons'. Significantly, the diagnoses that are put forward reflect the experience and world-views of the correspondents:

> Men of business pronounce her a swindler; medical men look upon it as an instance of "simulative hysteria;" literary men as an example of idosyncracy or monomania; many of the clergy believe her to be a Jesuit still; some pious persons consider it a case of Satanic possession; and it is interesting to observe how the track of thought which each class has been accustomed to pursue, directs or modifies the decision of Marie's case.

As 'no one theory appears to meet all the difficulties of the subject', Jemima Luke leaves the task of elucidating Marie's character to a 'more qualified hand', namely that of her husband, who observes that Marie was 'not in the ordinary use of the term an impostor – one simply living by her wits, and with no object beyond – that money was not her object, and that swindling was not her profession, is a legitimate conclusion from her whole course'. Rather, Marie had '*an intense delight in scheming*'.

Marie's term of imprisonment is coming to an end and, as Jemima Luke comments, the 'conclusion of her history is hidden in the pages of futurity'. Will she be 'spared, and changed, and saved?' she asks at the end. 'Who that contemplates the fearful alternative for guilt like hers will not breathe a wish for *such* a result?' Thus a narrative as ambiguous and opaque as any that one might encounter in the period is wrapped in the kind of conventional Evangelical sentiment that is expressed in Jemima Luke's well known hymn for children, 'I think, when I read that sweet story of old', first published in the *Sunday School Teacher's Magazine*:

> In that beautiful place He is gone to prepare
> For all who are washed and forgiven;
> And many dear children are gathering there,
> For of such is the kingdom of heaven.[50]

In *Villette*, published the year after Jemima Luke's *Sequel*, Charlotte Brontë, the daughter of an Irish Protestant clergyman, describes Madame Beck as a 'little Jesuit inquisitress' (26). In a reversal of *The Female Jesuit*, it is the Catholic host of the Protestant visitor who proves to be elusive and ambiguous. Lucy Snowe comments upon the way in which Madame

[50] *Methodist Hymn-Book* (1954), p. 760.

Beck's system of 'surveillance' works in Villette.[51] 'All these little incidents', she writes, 'taken as they fell out, seemed each independent of its successor; a handful of loose beads: but threaded through by that quick-shot and crafty glance of a Jesuit-eye, they dropped pendant in a long string, like that rosary on the prie-dieu' (34). In Villette 'they say' that Père Silas is a Jesuit (17) and Lucy refers to his 'Jesuit-slanders' against her (36), while Ginevra Fanshawe describes M. Paul as 'that tiger-Jesuit' (40).

In her school narrative Lucy frequently attacks the old familiar targets of anti-Catholic writing. 'Great pains were taken to hide chains with flowers', she writes: a 'subtle essence of Romanism pervaded every arrangement' (14). There, as elsewhere, 'the CHURCH strove to bring up her children robust in body, feeble in soul, fat, ruddy, hale, joyous, ignorant, unthinking, unquestioning' (14). She comments on the 'pamphlet in lilac' that M. Paul leaves in her desk for her: 'The voice of that sly little book was a honeyed voice'; 'Its appeal was not to intellect'; 'Portions of it reminded me of certain Wesleyan Methodist tracts I had once read as a child'; 'I smiled then over this dose of maternal tenderness, coming from the ruddy old lady of the Seven Hills' (36). She regards the Catholic Church as a 'tyrant', as 'tawdry, not grand', as 'grossly material', although she does not want M. Paul to lose his faith, just as he wants her to retain hers (36). The chapter in which she confronts Madame Beck as a potential rival in love is prefaced by Pauline commentary on the race that is set before us and references to Bunyan's Apollyon and Greatheart (38). 'Madame', she says with fervour, 'you are a sensualist.'

Lucy's exploration of the convent theme gives Brontë's critique of Catholicism more critical edge and contemporary relevance than these generalized attacks. Although the Pensionnat Heger was in the old quarter of Brussels, the building itself was only forty years old.[52] Part of the fictional Pensionnat Beck, however, is of 'ancient date' and its 'queerest little dormitories' were once 'nuns' cells' (8). There is a tradition that Madame Beck's house was a convent 'in old days': there were certain 'convent-relics, in the shape of old and huge fruit-trees', at the foot of one of which, beneath the mossy earth, can be seen a 'glimpse of slab, smooth, hard, and black' (12). The ghost of a nun is said to walk on 'some night or nights of the year', a story that Lucy at first dismisses as 'romantic rubbish'. 'The same hour which tolled curfew for our convent', she writes, 'which extinguished each lamp, and dropped the curtain round each couch, rang for the gay city about us the summons to festal enjoyment' (12). On the day of Madame

[51] 'Marie' writes of her fictitious convent, 'As with the Jesuits, a perfect system of espionage is maintained over every member of the community': J. T. Luke, *Female Jesuit* (1851), p. 5.

[52] See J. Barker, *Brontës* (1995), p. 379.

Beck's fête the girls are adorned in 'white muslin dress, and a blue sash (the Virgin's colours)', whereas Lucy chooses a dress in 'purple-gray', the colour of 'dun mist, lying on a moor in bloom' (14). In the long vacation, when M. Paul is on a pilgrimage to Rome and Lucy is almost alone in the school, she is driven to pour out her soul in Catholic confession (15). Had she gone to the priest's house, however, as he requested, he would have shown her 'all that was tender, and comforting, and gentle, in the honest Popish superstition': 'I might just now, instead of writing this heretic narrative, be counting my beads in the cell of a certain Carmelite convent on the Boulevard of Crécy, in Villette.'[53] Having left the church, Lucy faints 'near the Béguinage' (17).

When Lucy escapes from the school/convent on the night of the fête she is 'so close under the dungeon' that she can 'hear the prisoners moan' (38). Only by escaping the controlled environment of the pensionnat can Lucy witness Madame Beck's method of working through her 'secret junta' and a Jesuitical confusion between reality and illusion, fact and fantasy, truth and lies. In the park, she overhears Madame Walravens mention 'Justine Marie', the name of the dead nun (39). It turn out, however, that she is referring to a girl who joins the company, 'well-nourished, fair, and fat of flesh', and who is not 'her' nun. Similarly, Paul Emanuel, who was said to have departed for Guadaloupe, also appears in the flesh. Lucy returns to the school/convent to find the 'old phantom – the NUN' stretched on her bed. The 'long nun' proves, however, to be only a 'long bolster dressed in a long black stole, and artfully invested with a white veil'. Like so many features of Catholicism in Villette, the nun's vestments are no more than theatrical props: M. le Comte de Hamal and Miss Fanshawe have eloped (40).

The ground has now been cleared for the dénouement in Lucy's narrative. She describes her love for Paul Emanuel in pre-lapsarian terms, as they affirm their own religious beliefs in 'such moonlight as fell on Eden': 'Once in their lives some men and women go back to these first fresh days of our great Sire and Mother – taste that grand morning's dew – bathe in its sunrise' (41). Ironically, however, their consummation is that of celibates who meet under a chaste moon which signifies the risen Blessed Virgin Mary. For Lucy is 'penetrated' with Paul's 'influence' on the Feast of the Assumption, generally regarded in John Bull's England as one of the worst of Continental Catholicism's blasphemous additions to the Church's calendar.[54] Among

[53] This incident was based upon Brontë's own experience: see ibid., p. 424.

[54] For a modern feminist critique of the Assumption as a means of placing Mary 'outside the "feminine" realm of material time and bodily decay and into the "masculine" symbolic realm of eternal unchanged forms', see E. Bronfen, *Over Her Dead Body* (1992), p. 68.

M. Paul's last recorded words in his correspondence are a plea that she should 'remain a Protestant' (42). Lucy's overdetermined interpretation is that he is not a 'real Jesuit': 'He was born honest, and not false – artless, and not cunning – a freeman, and not a slave.' Lucy's Romantic hope is that the shackles of Catholic nurture can be thrown off and the paradise of Protestant nature regained.

Whereas *The Female Jesuit* presents itself as fact based upon fiction, *Villette* is fiction based upon fact. Like Dickens's *David Copperfield* and *Great Expectations*, it is a fictional autobiography which springs from personal experience of alienation. In drawing upon her own experiences at the Pensionnat Heger in Brussels, where she fell in love with M. Heger and was finally packed off home to Haworth following the intervention of his wife, Brontë reflected upon the parallels between girls' schools and convents. She adapted to her own purposes some of the recurrent motifs of contemporary writing, possibly including *The Female Jesuit*, that was hostile to the convent movement: 'surveillance' and the 'subjugation' of the mind; fraud and deceit; subterfuge and dressing up; monasticism as a threat to family values. Harriet Martineau's criticisms of *Villette* in a review and in private correspondence – that there was too much emphasis upon the need to be loved and that Brontë was bigoted in her anti-Catholicism – painfully exposed the psychic origins of the novel and effectively ended the friendship between the two writers.[55] Her broad treatment of the convent theme in the novel reflects both obsessions, and the tensions associated with them are only partially resolved through conventional Gothic play, in which the ghost of the dead nun is eventually shown to be no more than a male interloper in disguise. For Lucy's psychological and professional independence from Madame Beck's 'convent' at the end of the novel, signifying female freedom and Protestant rationalism, is gained at the expense of her living on the fringes of a Catholic community and in a state of false expectation concerning the safe return of her beloved Paul Emanuel, whose status as 'Jesuit' remains indeterminate, at least in this world.

III

By the 1870s, when English Catholicism had weathered the storms that accompanied the definition of the Immaculate Conception of the Blessed Virgin Mary (1854) and of Papal Infallibility (1870), Catholic poets were openly celebrating the conventual system and the special role of nuns in

[55] See J. Barker, *Brontës* (1995), pp. 718–20.

the life of the Church. In '"Sur Monique": A Rondeau by Couperin' (1875), Alice Meynell, the most important Catholic woman poet of the nineteenth century, makes a contemplative nun herself the object of contemplation:

> Quiet form of silent nun,
> What has given you to my inward eyes?
> What has marked you, unknown one,
> In the throngs of centuries
> That mine ears do listen through?[56]

At the end of the poem Meynell explores the nature of time and eternity in relation to the nun's role as intercessor:

> Sur Monique, remember me,
> 'Tis not in the past alone
> I am picturing you to be;
> But my little friend, my own,
> In my moment, pray for me.
> For another dream is mine,
> And another dream is true,
>   Sweeter even,
> Of the little ones that shine
> Lost within the light divine, –
> Of some meekest flower, or you,
>   In the fields of heaven.

Coventry Patmore described himself as a 'beggar by the Porch / Of the glad palace of Virginity' and was in love with Alice Meynell.[57] In his poem 'Deliciæ Sapientiæ de Amore', from *The Unknown Eros* (1877), he associates the nun's vocation with the Annunciation:

> And how the shining sacrificial Choirs,
> Offering for aye their dearest hearts' desires,
> Which to their hearts come back beatified,
> Hymn, the bright aisles along,
> The nuptial song,
> Song ever new to us and them, that saith,
> 'Hail Virgin in Virginity a Spouse!'
> Heard first below
> Within the little house
> At Nazareth;
> Heard yet in many a cell where brides of Christ

[56] From *Preludes* (1875): *Poems of Alice* (1923), p. 11.

[57] C. Patmore, *Poems* (1906), p. 330. Alice Meynell's husband, Wilfrid, eventually had to ask Patmore to end his association with them: see J. Badeni, *Slender Tree* (Padstow, 1981), pp. 115–16.

Lie hid, emparadised,
And where, although
By the hour 'tis night,
There's light,
The Day still lingering in the lap of snow.
(Book II, IX)[58]

As Alice Meynell, author of *Mary, the Mother of Jesus* (1912), also knew, the special calling of the nun is to be not only the bride of Christ, but also the daughter of Mary. In the words of Hopkins quoted in the epigraph to this chapter, Mary is 'mother of maid and nun'.[59]

Discussion of Mary leads naturally to reflection on the mystery of the Incarnation and of Christ as both fully man and fully God, creature and creator. Mary is both maiden and mother – mother of God and of man. She too can be described as both creature and creator in giving birth to the Saviour, although she herself is not the creator and saviour of the world. When Mary is taken as a role model in society, however, new complications arise. Whereas the imitation of Christ is generally regarded as a lifelong spiritual exercise in which all Christians fall short, however holy they may become, the imitation of Mary is seen as a social goal as well as a spiritual exercise. As Mary represents a patriarchal feminine ideal, in opposition to that other masculine 'idea of woman', the Magdalen, the imitation of Mary easily becomes a means of social control within a patriarchy.[60] Some women's writing of the nineteenth century recognized and challenged such thinking.[61]

Writing by and about women and Catholicism often returns to received polar oppositions, between Mary and Martha, the sisters of Lazarus, for example, who epitomize the contemplative and active ways in the writings of medieval mystics and in monastic tradition, and between two other Marys, the 'Madonna' and the 'Magdalen', who are figured as the virgin

[58] C. Patmore, *Poems* (1906), p. 333.

[59] Cf. Ernest Dowson's poem, 'Nuns of the Perpetual Adoration', first printed in the *Century Guild Hobby Horse* in 1891: 'And there they rest; they have serene insight / Of the illuminating dawn to be: / Mary's sweet Star dispels for them the night, / The proper darkness of humanity.' E. Dowson, *Poetical Works* (1967), p. 37.

[60] Typical of the starkly binary thinking that surrounds the theme is this: 'Together, the Virgin and the Magdalene form a diptych of Christian patriarchy's idea of woman. There is no place in the conceptual architecture of Christian society for a single woman who is neither a virgin nor a whore. Indeed, in Catholic countries the unmarried woman who has not take the veil is a pathetic figure of fun': M. Warner, *Alone of All Her Sex* (1985), p. 235.

[61] 'Women's writings testify to their ability to recognize the ideological conflicts in scripture that were suppressed in the patriarchal feminine ideal, and to interpret scripture as offering divinely sanctioned challenges to masculine authority': C. L. Krueger, *Reader's Repentance* (1992), p. 8.

and the whore of patriarchal tradition.[62] It is, of course, the 'Mother of God' who presides over all other female figures in Catholicism and it was Mary's multiple roles and attributes that made her an appealing subject for poetry as her cult gathered momentum in England after 1854.

Reports of Marian apparitions at Paris in 1830, La Salette in 1846, Lourdes in 1858, Pontmain in 1871, and Knock in 1879 encouraged a dramatic revival in devotion to the Madonna on the Continent, and later in Ireland, particularly among simple country people who had been displaced by industrialization.[63] In Protestant England the revival was noticeable but less dramatic. In the old days the four episcopal Vicars-General had worn Marian blue cassocks, symbolizing England's special claim to the title of 'Mary's Dowry',[64] but Marian statues were rare in pre-Victorian Catholic churches and, when they did start to appear, many 'old' Catholics disapproved.[65] In Ireland, too, there was little enthusiasm for the cult of Mary before the 1850s and 1860s. So the growth of the cult in England after the restoration of the hierarchy was unrelated to the arrival of the Irish 'hordes'.[66] Cardinal Wiseman's imports – processions, banners and statues in honour of Our Lady – originated in Spain, Italy and France. His successor, Cardinal Manning, was less enthusiastic, regarding the rosary as being not only unintelligible to non-Catholics, 'but by its perpetual repetition a stumbling block'.[67] Devotional books on Mary sold well in the later decades of the nineteenth century. Faber's *The Foot of the Cross; or, The Sorrows of Mary* (1858), for example, went through ten editions by 1886.

Most English Protestants maintained their sang-froid. Marian devotions, it was widely believed, did not appeal to the 'masculine intellect of England'.[68] Catherine Sinclair, author of *Beatrice*, hoped that the 'strong good sense of English minds may long continue to be their salutary protection against the Church of "Our Lady Star of the Sea!" a name much more fit for the Arabian Nights than for Christian teachers, but which is very attractive to young lovers of the imaginative and picturesque, as well as the whole gorgeous paraphernalia of Romish pageantry'.[69] Henry Hart Milman's view that the Virgin Mary had 'gradually supplanted' many

[62] See *Mysticism*, ed. F. C. Happold (1963); C. W. Atkinson, *et al.*, eds., *Immaculate and Powerful* (1985); S. Haskins, *Mary Magdalen* (1993); P. Norris, *Story of Eve* (1998).

[63] The Papacy and the clergy encouraged this development: see J. Singleton, 'The Virgin Mary and Religious Conflict', *Journal of Ecclesiastical History*, January 1992, pp. 16–17.

[64] See J. M. Robinson, *Dukes of Norfolk* (1995), p. 202.

[65] J. Singleton, 'The Virgin Mary and Religious Conflict', *Journal of Ecclesiastical History*, January 1992, pp. 19–20.

[66] Ibid., citing Emmet Larkin's research.

[67] Ibid., pp. 19–21, citing E. S. Purcell, *Life of Manning* (1896), II, 791.

[68] See Henry Soames, for example, p. 106 above.

[69] C. Sinclair, *Beatrice* (1852), I, ix.

local pagan deities[70] was echoed by Anna Jameson in her *Legends of the Madonna* (1852), where she argued that, from the fifth century, 'step by step the woman was transmuted into the divinity'.[71] Here was a clear example of Catholic idolatry, a blasphemy against the Scriptures. 'How can there be union with Rome', asked Walter Farquhar Hook in 1842, 'while she practically elevates the Virgin Mary into an idol?'[72]

Mariolatry was also regarded by most Protestants as a blasphemy against nature, God's second book. Catholics, in contrast, believed that the Blessed Virgin, born without spot of sin, helps us to grasp the mystery of the Incarnation. Mary's 'active' conception was like that of any other human being. At her 'passive' conception, however, when her soul was infused by God, 'grace was poured into her soul at the first instant of its being' through a special divine intervention.[73] 'Those who fear to call Mary the "Mother of God"', Coventry Patmore suggested, 'simply do not believe in the Incarnation at all'.[74] For Newman, who drew upon the Fathers of the Church, the Virgin Mary was the 'second and better Eve, who brought salvation into the world, as our first mother brought death'.[75] In answer to Pusey's criticism of the growth of Roman Catholic Mariolatry in England, Newman made a crucial distinction between faith and devotion:

I fully grant that *devotion* towards the Blessed Virgin has increased among Catholics with the progress of centuries; I do not allow that the *doctrine* concerning her has undergone a growth, for I believe that it has been in substance one and the same from the beginning.

By "faith" I mean the Creed and the acceptance of the Creed; by "devotion" I mean such religious honours as belong to the objects of our faith, and the payment of those honours.[76]

Interestingly, Ruskin blurred another Catholic distinction – that between devotion and worship – at a time when he regarded himself as a '"Catholic" of those Catholics, to whom the Catholic Epistle of St. James is addressed'.[77] In *Fors Clavigera*, Letter 41 (May 1874), written on his way to Italy, he argues that

[70] H. H. Milman, *History of Christianity* (1875), III, p. 97.

[71] A. Jameson, *Legends of the Madonna* (1852), p. xliii.

[72] W. F. Hook, *Perils of Idolatry* (1842), p. 20. Cf. H. Home, *Mariolatry* (1841), p. 6; J. B. Cartwright, *No Popery!* (1850), p. 45; R. F. Littledale, *Plain Reasons* (1886), p. 53.

[73] W. E. Addis and T. Arnold, *Catholic Dictionary* (1897), pp. 469–70.

[74] 'Knowledge and Science', in *The Rod, the Root and the Flower* (1895), cited in J. C. Reid, *Mind of Patmore* (1957), p. 146.

[75] *Callista* (12). Cf. W. E. Addis and T. Arnold, *Catholic Dictionary* (1897), p. 472.

[76] J. H. Newman, *Letter to Pusey* (1866), p. 28.

[77] *Works of Ruskin* (1903–12), XXIX, 92.

the worship of the Madonna has been one of [Catholicism's] noblest and most vital graces, and has never been otherwise than productive of true holiness of life and purity of character. I do not enter into any question as to the truth or fallacy of the idea; I no more wish to defend the historical or theological position of the Madonna than that of St Michael or St Christopher; but I am certain that to the habit of reverent belief in, and contemplation of, the character ascribed to the heavenly hierarchies, we must ascribe the highest results yet achieved in human nature, and that it is neither Madonna-worship nor saint-worship, but the evangelical self-worship and hell-worship . . . which have in reality degraded the languid powers of Christianity to their present state of shame and reproach.[78]

Ruskin understood the benign influence of 'Madonna-worship' upon the medieval craftsmen who had created the French and Italian architectural decoration that he so admired. Thomas William Allies, the Rector of Launton in Oxfordshire, offered his own reflections on the Virgin Mary in a letter from Venice to an unnamed correspondent and dated 5 August 1847:

It really seems to me that the more men dwell upon the Incarnation, the more they will associate the Blessed Virgin with our Lord, and the saints with Him and with her; they will not analyse and divide, but rather always seem to be touching the skirts of His robe of glory, in every one of those who have suffered and conquered in His name; and most of all in the Mother, who was and is so unspeakably near to Him. Thus the Protestant sees in her 'a dead woman worshipped'; the Catholic, the mother of all Christians; the Protestant sees in the saints 'deified sinners'; the Catholic, living members of His body, in whom His virtue now dwells without let of human corruption.[79]

Allies converted to Catholicism three years later, in October 1850.

It was later in 1850, on 8 December, the Feast of the Blessed Virgin's Conception, that Cardinal Wiseman preached a sermon at St George's Catholic Church, Southwark, in which he referred to 'that earnest, deep, affectionate devotion towards the blessed mother of our Lord and Saviour Christ Jesus which now more than any time deserves to be openly proclaimed and professed by Catholic mouth, when outrage and blasphemy are most loud in her dishonour, and in that, consequently, of Him to whom she gave birth'.[80] Four years later, on the same feast day, Wiseman was present in Rome when Pio Nono defined the Immaculate Conception of the Blessed Virgin. Pusey wrote:

78 Ibid., XXVIII, 82.
79 Thomas William Allies, *Journal in France* (1849), p. 162.
80 See p. 44 above.

The controversies about the Immaculate Conception are older than the Reformation, but have only just been decided. It has now been ruled that that doctrine was always matter of faith; yet it has only been formally received in the Roman Church, when it had tacitly made its way, and its once powerful opponents had ceased. The object of the decision was understood to be, *not* to settle controversies which had long expired, but to obtain the favour of the Blessed Virgin towards the Church of Rome by doing honour to her.[81]

Preaching on Rome's 'new dogma', Samuel Wilberforce, Bishop of Oxford, set out the classic Protestant argument that it was a blasphemy against both scripture and nature. The 'simple truth of God's word', he said, had been 'overlaid by papal error'.[82] This 'unlawful addition of a new article to the Creed' claimed that the nature with which Mary was born into this world was, 'from the first moment in which she began to exist, not that nature which all inherit who "naturally are engendered of the offspring of Adam", but another nature'. Rome, 'once the faithful city', had now become a 'harlot'.

Newman, who in 1849 wondered whether 'the Catholic people' called for the definition, nevertheless dedicated the church of the Birmingham Oratory 'under the title of the Immaculate Conception' in 1851, three years before Pius IX acted.[83] 'There is no burden at all', he was to write in the *Apologia* (1864), 'in holding that the Blessed Virgin was conceived without original sin; indeed, it is a simple fact to say, that Catholics have not come to believe it because it is defined, but that it was defined because they believed it'.[84] Although Newman took Mary as his confirmation name and made the Immaculate Conception the focal point of his teaching on her, he was careful to avoid excesses which he believed to be unEnglish.[85] Such scruples never troubled his exuberant brother Oratorian, Father Faber, however, who wrote in *The Foot of the Cross*, 'O how, after long meditation on the Immaculate Conception, love gushes out of every pore of our hearts when we think of that almost more than mortal queen, heartbroken, and with blood-stains on her hand, beneath the Cross!'[86] Faber did, however, go some way to answering Protestant objections to the dogma by arguing that 'Mary herself, though in a different way, needed redemption as much as we

[81] E. B. Pusey, *Eirenicon* (1865–70), I, 91. In response, Newman wrote his *Letter to Pusey* 'during nine days of intensive work preceding the Feast of the Immaculate Conception, 1865': *Mary* (2001), p. 42.

[82] S. Wilberforce, *Rome – Her New Dogma* (1855), p. 2. Further references in this paragraph are to pp. 5–6, 26.

[83] Letter to W. G. Ward: *Letters of Newman* (1961–77), XIII, 81–2.

[84] J. H. Newman, *Apologia* (1967), p. 228.

[85] See J. H. Newman, *Mary* (2001), pp. 36, 51.

[86] F. W. Faber, *Foot of the Cross* (n.d.), p. 63.

do, and received it in a more copious manner and after a more magnificent kind in the mystery of the Immaculate Conception'.[87]

In his *May Carols* (1857), Aubrey de Vere celebrated the providential moment in which the doctrine was defined:

A soul-like sound, subdued yet strong,
  A whispered music, mystery-rife,
A sound like Eden airs among
  The branches of the Tree of Life –

The Church had spoken. She that dwells
  Sun-clad with beatific light,
From Truth's unvanquished citadels,
  From Sion's Apostolic height,

Had stretched her sceptred hands, and pressed
  The seal of Faith, defined and known,
Upon that Truth till then confessed
  By Love's instinctive sense alone.[88]

De Vere also celebrated the 'polyvalent' nature of Mary.[89] As T. E. Bridgett put it in 1893, the 'relations of the Blessed Mary to Jesus Christ are manifold – for she might be considered as His daughter, His spouse, or His mother – so, too, are her relations manifold to the Church. She is its type, she is its noblest member, and she is its mother.'[90] In devotional terms, these various 'relations' are reflected in both her great suffering and her great joy.

In *The Foot of the Cross; or, The Sorrows of Mary*, published four years after the definition of the Immaculate Conception, Faber exploits the whole emotional range of Catholic tradition, from the homely comforts of sentimentality through to the violent paradoxes of casuistry. It is Jesus Christ whom we seek, he reminds his readers, but there is 'no time lost in seeking Him if we go at once to Mary; for He is always there; always at home'.[91] Yet Jesus, the 'joy of the martyrs, is the executioner of His mother': 'Twice over, to say the least, if not a third time also, did He crucify her, once by His Human Nature, once by His Divine, if indeed Body and Soul did not make two crucifixions from the Human Nature only.' Later, however, he also asserts that Mary was 'made as it were executioner in chief of her own beloved Son': 'The more tenderly she loved Him, the more fondly she

87 Ibid., p. 381.
88 III, XII, 'Sine Lave originali Concepta': A. de Vere, *May Carols* (1857), p. 110.
89 M. Warner, *Alone of All Her Sex* (1985), p. xxvi.
90 T. E. Bridgett, *England for Our Lady* (1893), p. 7.
91 F. W. Faber, *Foot of the Cross* (n.d.), p. 2. Further references in this paragraph are to pp. 10, 32.

clung to Him, the more willingly she bore her griefs, so much the deeper the iron of them entered into the soul of Jesus.'

When Faber wants the reader to enter imaginatively into the agonies of the Passion he tends to use the present tense. Mary 'must listen to the fierce singing of the scourges as they cleave the air, and count the stripes, and take into her heart the variety of deadly sickening sounds they made as they lit on this or that part of His Sacred Body'.[92] 'Think of the value of each drop of blood?' he writes. 'But why talk of drops? She is slipping in it.' He also frequently uses the present tense in his reflections upon the seven 'dolours' of Mary,[93] thus closing the gap between the reader and the historic events of the Passion. 'Suffering is dearer to the Saints than happiness', he asserts with reference to the first dolour, the prophecy of St Simeon, 'for the similitude of Christ has passed upon them': 'They have His tastes, His inclinations. They thirst for suffering, because there is something in it which is favourable to union with God.' Writing on the fourth dolour, meeting Jesus with the cross, Faber claims that Mary also 'assists in spirit at the Agony in the Garden, sees our Lord's Heart unveiled throughout, and feels in herself, and according to her measure, a corresponding agony'.

The climax of Mary's fifth dolour, the crucifixion, is described in the past tense: 'The two relations of Mother and Son were two no longer; they had melted into one. She knew that never had He loved her more than now.'[94] In a carefully controlled modulation from past to present, however, Faber associates Mary's loving participation in the Passion of her Son to her loving participation in the dying of the faithful:

> Mary assisted her Son to die in many mysterious ways. By His will, and in the satisfaction of her own maternal love, she has now assisted at the deathbeds of many millions. She has great experience by this time, if we might so speak, and is wonderfully skilled in the science of the last hour.[95]

Whereas for Protestants Mary is, in Allies's phrase, a 'dead woman worshipped', for Catholics she is 'mother of all Christians' in the eternal present.

If Mary's sufferings are explored through meditations upon the Passion, her joys are associated with the Annunciation and the Magnificat. Pope Pius IX himself encouraged Aubrey de Vere to publish his *May Carols*, which

[92] Ibid., p. 35. Further references in this paragraph are to pp. 39, 77, 207.

[93] The prophecy of Simeon, the flight into Egypt, the loss of Jesus for three days, the carrying of the cross, the crucifixion, the descent from the cross and the entombment. In 1880 Herbert Gribble designed an altar of black and white marble for the Chapel of the Seven Dolours in his London Oratory: see H. G. S. Bowden, ed., *Guide to the Oratory* (1893), p. 17.

[94] Ibid., p. 257.

[95] Ibid., p. 291.

appeared three years after the definition of the Immaculate Conception.[96] Here the emphasis falls upon the performative present of each carol and the future hope of the Christian who trusts in Mary:

> The childlike heart shall enter in;
> The virgin soul its God shall see: –
> Mother, and maiden pure from sin,
> Be thou the guide: the Way is He.
>
> The mystery high of God made Man
> Through thee to man is easier made:
> Pronounce the consonant who can
> Without the softer vowel's aid![97]

Twenty years later de Vere's friend, Coventry Patmore, engaged with some of the paradoxes associated with Mary's 'manifold' relations in *The Unknown Eros*. In 'The Child's Purchase' he emulates seventeenth-century poets such as Crashaw, himself a convert to Catholicism:

> Life's cradle and death's tomb!
> To lie within whose womb,
> There, with divine self-will infatuate,
> Love-captive to the thing He did create,
> Thy God did not abhor,
> No more
> That Man, in Youth's high spousal-tide,
> Abhors at last to touch
> The strange lips of his long-procastinating Bride;
> Nay, not the least imagined part as much!
> *Ora pro me!*[98]

For Patmore, Mary's manifold nature is to God what a prism is to light:

> My Lady, yea, the Lady of my Lord,
> Who didst the first descry
> The burning secret of virginity,
> We know with what reward!
> Prism whereby
> Alone we see
> Heav'n's light in its triplicity;
> Rainbow complex
> In bright distinction of all beams of sex,

[96] See W. Ward, *Aubrey de Vere* (1904), p. 234. Newman appointed 'In Epiphania' and 'Docens' to be sung at the Oratory each day in May: ibid., pp. 242–3.

[97] I, IV, 'Sancta Maria': de Vere, *May Carols* (1857), p. 7.

[98] II, XVII, 'The Child's Purchase': Patmore, *Poems* (1906), p. 356.

Shining for aye
In the simultaneous sky,
To One, thy Husband, Father, Son, and Brother,
Spouse blissful, Daughter, Sister, milk-sweet Mother;
Ora pro me!

The virgin mother is also the virgin bride of Christ:

Mildness, whom God obeys, obeying thyself
Him in thy joyful Saint, nigh lost to sight
In the great gulf
Of his own glory and thy neighbour light;
With whom thou wast as else with husband none
For perfect fruit of inmost amity;
Who felt for thee
Such rapture of refusal that no kiss
Ever seal'd wedlock so conjoint with bliss;
And whose good singular eternally
'Tis now, with nameless peace and vehemence,
To enjoy thy married smile,
That mystery of innocence;
Ora pro me!

The blessed virgin, a mortal woman, provides the means by which sinful man may approach God:

Sweet Girlhood without guile,
The extreme of God's creative energy;
Sunshiny Peak of human personality;
The world's sad aspirations' one Success;
Bright Blush, that sav'st our shame from shamelessness;
Chief Stone of stumbling; Sign built in the way
To set the foolish everywhere a-bray;
Hem of God's robe, which all who touch are heal'd;
To which the outside Many honour yield
With a reward and grace
Unguess'd by the unwash'd boor that hails Him to His face,
Spurning the safe, ingratiant courtesy
Of suing Him by thee;
Ora pro me!

So multi-faceted is Mary that her attributes crowd together in the lines that follow, as Patmore reflects upon those who deny Our Lady and are thus 'disemparadised':

Creature of God rather the sole than first;
Knot of the cord
Which binds together all and all unto their Lord;
Suppliant Omnipotence; best to the worst;
Our only Saviour from an abstract Christ
And Egypt's brick-kilns, where the lost crowd plods,
Blaspheming its false Gods;
Peace-beaming Star, by which shall come enticed,
Though nought thereof as yet they weet,
Unto thy Babe's small feet,
The Mighty, wand'ring disemparadised,
Like Lucifer, because to thee
They will not bend the knee;
Ora pro me!

By collapsing the period of time between her Son's death on the cross and her own death – 'Thereafter, holding a little thy soft breath' – Patmore dramatizes the Virgin's Assumption. It is as if earth cannot hold her:

In season due, on His sweet-fearful bed,
Rock'd by an earthquake, curtain'd with eclipse,
Thou shar'd'st the rapture of the sharp spear's head,
And thy bliss pale
Wrought for our boon what Eve's did for our bale;
Thereafter, holding a little thy soft breath,
Thou underwent'st the ceremony of death;
And, now, Queen-Wife,
Sitt'st at the right hand of the Lord of Life,
Who, of all bounty, craves for only fee
The glory of hearing it besought with smiles by thee!
Ora pro me!

Whereas Patmore's perspective is idealist, Alice Meynell's is realist. In *Mary, the Mother of Jesus* (1912), she contrasts the ways in which we think of Christ and his mother, and, crucially, the language which can be applied to them: 'We may set Him apart; as man indeed, but not as what the habits of our language would name "a man." His mother, on the contrary, is not only woman, but "a woman"; young, moreover, in the chief time of her office, and held to have been fitted for that office by one only quality – innocence.'[99] This innocence Meynell regards as the source of

[99] A. Meynell, *Mary* (1912), p. 29. (She was in her sixties when she wrote *Mary*.) Further references are to pp. 30, 35, 36. The search for the 'historical Jesus' in the late nineteenth and early twentieth centuries seems to have made little impact upon Meynell's devotional life. Alfred Schweitzer's *The Quest of the Historical Jesus* was published in 1906.

the 'little Virgin's' strength in governing the 'fierce ages and the passionate, the cruel ages during which man starved man to death and man put out the eyes of man for vengeance'. Down the centuries religious art has been produced 'in honour of chastity, of humility, of constancy, of self-denial and self-sacrifice, of compassion, of truth, of all that the race of man calls virtue'. For certain centuries of the history of European nations, she adds, 'art was almost entirely devoted to all the virtues, and all the virtues centred about a girl'.

Gerard Manley Hopkins, the unpublished laureate of the Blessed Virgin, would not have described her as a 'girl'. The poeticism of 'maid' far better conveys his idea of Mary, viewed from his position as a celibate Jesuit for whom the Immaculate Conception, defined twelve years before he was received into the Catholic communion by Newman, lay at the heart of his faith. Several of the poems in which Hopkins celebrated Our Lady could be described as 'May carols', like de Vere's. In 'Mary's month' some of them were hung up, along with others, in front of the statue of the Virgin at Stonyhurst, where Hopkins underwent part of his training and later taught. 'May Magnificat', for example, written in 1878, begins,

> May is Mary's month, and I
> Muse at that and wonder why:
> Her feasts follow reason,
> Dated due to season –
>
> Candlemas, Lady Day;
> But the Lady Month, May,
> Why fasten that upon her,
> With a feasting in her honour?
>
> Is it only its being brighter
> Than the most are must delight her?
> Is it opportunest
> And flowers finds soonest?
>
> Ask of her, the mighty mother:
> Her reply puts this other
> Question: What is Spring? –
> Growth in everything . . . [100]

Five years later Hopkins was again at Stonyhurst, where he now wrote his finest Marian poem, 'The Blessed Virgin compared to the Air we Breathe':

[100] G. M. Hopkins, *Poems* (1967), pp. 76–7.

Wild air, world-mothering air,
Nestling me everywhere,
That each eyelash or hair
Girdles; goes home betwixt
The fleeciest, frailest-flixed
Snowflake; that's fairly mixed
With, riddles, and is rife
In every least thing's life;
This needful, never spent,
And nursing element;
My more than meat and drink,
My meal at every wink;
This air, which, by life's law,
My lung must draw and draw
Now but to breathe its praise,
Minds me in many ways
Of her . . .
Mary Immaculate . . .

(1–17, 24)[101]

Hopkins had heard extracts from *The Foot of the Cross* read aloud in the refectory at Stonyhurst in March 1872.[102] The contrast between his technical and physical use of the word 'home' here, and Faber's domestic reference to Christ being 'always at home', reflects differences not only of personality, but also of tradition – in Hopkins's case, Jesuit, in Faber's, Oratorian.

Like Patmore, with whom he corresponded on many aspects of poetics, Hopkins was interested in the use of tense in religious verse that explores the nature of time and eternity. In 'Deliciæ Sapientiæ de Amore', from *The Unknown Eros* (1877), Patmore refers to the 'nuptial song' that was 'Heard first below / Within the little house / At Nazareth' and is 'Heard yet in many a cell where brides of Christ / Lie hid, emparadised'. In Hopkins's case, shifts from past to present tense often also move the focus from the external to the internal, and thence into the future:

Of her flesh he took flesh:
He does take fresh and fresh,
Though much the mystery how,
Not flesh but spirit now
And makes, O marvellous!
New Nazareths in us,
Where she shall yet conceive
Him, morning, noon, and eve . . .

(55–62)

[101] Ibid., pp. 93–4. [102] See A. Thomas S. J., *Hopkins the Jesuit* (1969), p. 231.

Hopkins much admired Patmore's poem, 'The Child's Purchase', in which the Blessed Virgin is compared to a prism, refracting for us the blinding light of God.[103] Hopkins writes:

> . . . this blue heaven
> The seven or seven times seven
> Hued sunbeam will transmit
> Perfect, not alter it.
> Or if there does some soft,
> On things aloof, aloft,
> Bloom breathe, that one breath more
> Earth is the fairer for.
> Whereas did air not make
> This bath of blue and slake
> His fire, the sun would shake,
> A blear and blinding ball
> With blackness bound, and all
> The thick stars round him roll
> Flashing like flecks of coal,
> Quartz-fret, or sparks of salt,
> In grimy vasty vault. (86–102)

Hopkins was also an alert reader of Ruskin, whose reflections on the 'firmament' in *Modern Painters* IV (1856) seem to have been such a rich source for him that one is tempted to describe it as 'Hopkinsesque':

> This, I believe, is the ordinance of the firmament; and it seems to me that in the midst of the material nearness of these heavens God means us to acknowledge His own immediate presence as visiting, judging, and blessing us. 'The earth shook, the heavens also dropped, at the presence of God.' [Psalms 68.8] 'He doth set His bow in the cloud', [Genesis 9.13] and thus renews, in the sound of every drooping swathe of rain, His promises of everlasting love. 'In them hath He set a *tabernacle* for the sun'; [Psalms 19.4] whose burning ball, which without the firmament would be seen but as an intolerable and scorching circle in the blackness of vacuity, is by that firmament surrounded with gorgeous service, and tempered by mediatorial ministries; by the firmament of clouds the golden pavement is spread for his chariot wheels at morning; by the firmament of clouds the temple is built for his presence to fill with light at noon; by the firmament of clouds the purple veil is closed at evening round the sanctuary of his rest; by the mists of the firmament his implacable light is divided, and its separated fierceness appeased into the soft blue that fills the depth of distance with its bloom, and the flush with which the mountains burn as they drink the overflowing of the dayspring. And in this tabernacling of the unendurable sun with men, through the shadows of the firmament, God would

[103] See N. White, *Hopkins* (Oxford, 1992), p. 353.

seem to set forth the stooping of His own majesty to men, upon the *throne* of the firmament.[104]

For Hopkins, however, it is Mary, whose colour is blue, who presides in the 'blue heaven' and who is the atmosphere in which he can breathe freely:

> So God was god of old:
> A mother came to mould
> Those limbs like ours which are
> What must make our daystar
> Much dearer to mankind;
> Whose glory bare would blind
> Or less would win man's mind.
> Through her we may see him
> Made sweeter, not made dim,
> And her hand leaves his light
> Sifted to suit our sight.
> Be thou then, O thou dear
> Mother, my atmosphere;
> My happier world, wherein
> To wend and meet no sin;
> Above me, round me lie
> Fronting my froward eye
> With sweet and scarless sky;
> Stir in my ears, speak there
> Of God's love, O live air,
> Of patience, penance, prayer:
> Worldmothering air, air wild,
> Wound with thee, in thee isled,
> Fold home, fast fold thy child.
> (103–126)

During a period of training at St Bueno's College in North Wales, Hopkins wrote the most famous celebration of the sacrificial calling of a nun in the English language, 'The Wreck of the Deutschland', which he dedicated

To the
happy memory of five Franciscan nuns
exiles by the Falck Laws
drowned between midnight and morning of
Dec. 7th, 1875[105]

104 *Works of Ruskin* (1903–12), VI, 113–14.
105 G. M. Hopkins, *Poems* (1967), p. 51. For a detailed analysis of the whole poem see M. Wheeler, *Death and the Future Life* (1990), pp. 340–66.

The following day, 8 December, was the Feast of the Immaculate Conception of the Blessed Virgin Mary:

> Jesu, heart's light,
> Jesu, maid's son,
> What was the feast followed the night
> Thou hadst glory of this nun? –
> Feast of the one woman without stain.
> For so conceivèd, so to conceive thee is done;
> But here was heart-throe, birth of a brain,
> Word, that heard and kept thee and uttered thee outright.
>
> Well, she has thee for the pain, for the
> Patience, but pity of the rest of them!
> Heart, go and bleed as a bitterer vein for the
> Comfortless unconfessed of them –
> No not uncomforted: lovely-felicitous Providence
> Finger of a tender of, O of a feathery delicacy, the breast of the
> Maiden could obey so, be a bell to, ring of it, and
> Startle the poor sheep back! is the shipwrack then a harvest,
> does tempest carry the grain for thee? (stanzas 30–1)

Like all the other Catholic poets and most of the Catholic prose writers discussed in this chapter, Hopkins was a convert for whom prayers for the conversion of 'Mary's dowry' had a special personal significance:

> Dame, at our door
> Drowned, and among our shoals,
> Remember us in the road, the heaven-haven of the reward:
> Our King back, Oh, upon English souls!
> Let him easter in us, be a dayspring to the dimness of us,
> be a crimson-cresseted east,
> More brightening her, rare-dear Britain, as his reign rolls,
> Pride, rose, prince, hero of us, high-priest,
> Our heart's charity's hearth's fire, our thoughts' chivalry's
> throngs' Lord. (stanza 35)

CHAPTER 9

Liberalism and dogma

> My battle was with liberalism; by liberalism I mean the anti-dogmatic principle and its developments.
>
> John Henry Newman, *Apologia Pro Vita Sua* (1864)[1]

I

Newman is referring to his former Oxford days, when, as a Tractarian, he opposed Anglican liberalism. In his view 1834, the second year of the Tracts, saw the 'commencement of the assault of Liberalism upon the old orthodoxy of Oxford and England', when Hampden published a pamphlet entitled *Observations on Religious Dissent*.[2] Newman makes it clear, however, that liberalism of the 1830s was very different from that of the 1860s, the subject of this chapter:

> I am not going to criticize here that vast body of men, in the mass, who at this time would profess to be liberals in religion; and who look towards the discoveries of the age, certain or in progress, as their informants, direct or indirect, as to what they shall think about the unseen and the future. The Liberalism which gives a colour to society now, is very different from that character of thought which bore the name thirty or forty years ago. Now it is scarcely a party; it is the educated lay world. . . . At present it is nothing else than that deep, plausible scepticism . . . the development of human reason, as practically exercised by the natural man.
>
> The Liberal religionists of this day are a very mixed body, and therefore I am not intending to speak against them.[3]

[1] J. H. Newman, *Apologia* (1967), p. 54.

[2] Ibid., pp. 61–2. (*Inter alia*, Hampden stated that, although he disliked Unitarian theology, he could not deny them the 'name of Christians', and that he favoured the removal of all religious tests at admission to the University of Oxford: R. D. Hampden, *Observations on Religious Dissent* (1834), pp. 21, 39. The 2nd edn (1834) added 'with particular reference to the Use of Religious Tests in the University' to the title. On Newman's opposition to the liberalism of Milman in 1840, see p. 64 above.)

[3] Ibid., pp. 233–4.

One of Newman's difficulties in 1864, the year of his confrontation with Kingsley, was that whereas he had been a conservative Anglican in 1834, he was now a liberal Catholic.

Newman felt obliged to add an extensive note on 'Liberalism' to the second edition of the *Apologia* (1865), as he had been asked to explain the term more fully. 'Merely to call it the Anti-dogmatic Principle', he wrote, 'is to tell very little about it'.[4] Now he defines liberalism as false liberty of thought:

> Liberty of thought is in itself a good; but it gives an opening to false liberty. Now by Liberalism I mean false liberty of thought, or the exercise of thought upon matters, in which, from the constitution of the human mind, thought cannot be brought to any successful issue, and therefore is out of place. Among such matters are first principles of whatever kind; and of these the most sacred and momentous are especially to be reckoned the truths of Revelation. Liberalism then is the mistake of subjecting to human judgment those revealed doctrines which are in their nature beyond and independent of it, and of claiming to determine on intrinsic grounds the truth and value of propositions which rest for their reception simply on the external authority of the Divine Word.[5]

Newman is not alone, however, in first defining liberalism by what it opposes. A century later, Leo Strauss argued that liberalism is 'understood here and now in contradistinction to conservatism', although in its original sense it means to practise the 'virtue of liberality'.[6] Echoing both Newman and Matthew Arnold, Strauss defines liberal education as 'education in culture or toward culture': 'The finished product of a liberal education is a cultured human being.'[7]

'Liberalism' is a notoriously slippery word, not only in the contexts of religion and education, but also of politics and social theory; and the modern student of history 'has to deal with both liberalism and liberalisms'.[8] In terms of party politics, the Liberals and their forbears, the Whigs, were in power for about two-thirds of the period between 1832 and 1885, and 'Gladstonian Liberalism' established itself as a major political force during W. E. Gladstone's first and second ministries (1868–74, 1880–5). As Lord Acton put it, 'the Whig governed by compromise, the Liberal begins the reign of ideas'.[9] Yet when Andrew Reid published *Why I am a Liberal* (1885), the ideas presented by the celebrities who contributed did not cohere into a body of political doctrine. Gladstone believed that the principle of Liberalism was 'trust in the people, qualified by prudence', whereas the

[4] Ibid., p. 254. [5] Ibid., pp. 255–6. [6] L. Strauss, *Liberalism* (1968), p. v. [7] Ibid., p. 3.
[8] E. K. Bramsted and K. J. Melhuish, eds. *Western Liberalism* (1978), p. xviii.
[9] See G. Watson, *English Ideology* (1973), p. 16.

principle of Conservatism was 'mistrust of the people, qualified by fear'.[10] Reid himself, on the other hand, defined the ideal Liberal as one who would 'love the approval of his own conscience more than the approval of the conscience of the people'. Browning's prefatory sonnet, entitled 'WHY I AM A LIBERAL', ends with a sestet in which Liberalism is associated with liberty and emancipation:

> But little do or can the best of us:
> THAT LITTLE IS ACHIEVED THROUGH LIBERTY.
> Who, then, dares hold – emancipated thus –
> His fellow shall continue bound? Not I,
> Who live, love, labour freely, nor discuss
> A brother's right to freedom. That is "Why."

The picture is further complicated by the fact that both Liberal and Conservative administrations worked within a political system that was based upon broadly 'liberal' constitutional principles. The English idea of government, or 'English ideology', has been defined as the 'idea of liberty expressed through parliamentary institutions'.[11]

This chapter explores the cultural space occupied by religious liberalism and liberalisms from the mid-1860s to the mid-1870s, when the old enemies fought over the question of Catholic dogma and Papal Infallibility. It is a crowded and complex space, not least because these years witnessed a post-Darwinian intellectual revolution, led by liberal thinkers, which produced a wide range of responses, some of them literary landmarks, including Newman's *Apologia* (1864) and *Essay in Aid of a Grammar of Assent* (1870), Arnold's *Culture and Anarchy* (1867–8) and *Literature and Dogma* (1871–3), Browning's *The Ring and the Book* (1868–9), George Eliot's *Middlemarch* (1871–2), Gladstone's *Vatican Decrees in their bearing of Civil Allegiance* (1874) and Manning's response of the same title (1875).

During this period, the two leading English converts to Catholicism, Manning and Newman, both of whom were to become Cardinals, in 1875 and 1879 respectively, were more famous than any of the Anglican bishops, and their views carried considerable weight, both nationally and internationally. While they shared a mistrust of Anglicanism, however, they held very different positions on Ultramontanism and the loosely knit movement known as Liberal Catholicism which opposed it. In the years leading

[10] A. Reid, ed., *Why I am a Liberal* (1885]), p. 13. Further quotations in this paragraph are from pp. 114, 11.
[11] G. Watson, *English Ideology* (1973), p. 1.

up to the Vatican Council which defined Papal Infallibility in 1870, these fault lines were exposed: the Ultramontane 'Infallibilists', Manning and W. G. Ward, editor of the *Dublin Review*, were opposed by the leading English Liberal Catholic 'anti-Infallibilists', Lord Acton, the historian, and Richard Simpson, the critic and editor, who was on the radical wing of the movement. Newman, a conservative Liberal Catholic 'inopportunist', who questioned the wisdom of defining the dogma at this time, disagreed with all of them at various times. Thus the protagonists in the debate over Papal Infallibility were aware that it was no longer possible to present the Church of England and the Catholic Church as utterly contrasting monoliths, one increasingly liberal, the other increasingly dogmatic. Both were internally divided into liberal and traditionalist groupings, although the proportions were different and the authority of the Church of Rome ensured the kind of doctrinal unity which Anglicanism never achieved.

The debate between liberalism and dogma unfolded over an extended period, before, during and after the Council – three phases which are treated in separate sections of this chapter: section II, from 1864 to February 1869, when Matthew Arnold and Robert Browning, both liberal Protestants, though with different religious backgrounds, wrote during the build-up to the Vatican Council in Rome and the reform of the franchise at home; section III, from March 1869 to the end of 1870, when Dr Cumming prophesied the end-time and Newman, whose *Grammar of Assent* was published during the Council itself, explored the nature of belief; and section IV, from 1871 to 1876, when, in the aftermath of the Council, Newman, Manning, Gladstone and Arnold all published major statements, to which the obscure part-time novelist, John Henry Shorthouse, added an important coda, in *John Inglesant* (1880).

II

Towards the end of 1864, on 6 December, Pope Pius IX sounded out the Cardinals who had gathered in Rome for a session of the Congregation of Rites about calling a General Council.[12] Two days later, the Vatican published an encyclical, *Quanta cura*, with an anti-progressive *Syllabus Errorum* attached to it – a Syllabus or list of eighty censured propositions, the last of which was particularly inflammatory in the view of anti-Catholic

[12] Of the fifteen cardinals, only one recommended defining Papal infallibility, and of thirty-two bishops consulted, only seven were for the definition: A. B. Hasler, *How the Pope became Infallible* (1981), p. 53.

journalists: 'The Roman Pontiff can and ought to be reconciled to, and come to terms with, progress, liberalism and modern civilization.'[13]

In England, 1864 had been a mixed year for Catholics. In the prolonged dispute between Newman and Kingsley, from January to June, Newman had emerged the clear winner. The fêting of Garibaldi in London in April had, however, been accompanied by renewed cries of 'No Popery'. In the same month, Acton had closed his liberal quarterly, *The Home and Foreign Review*, which Simpson edited, before the Vatican could do so. Now, in December, English Catholics were divided over the Syllabus. Acton was mortified, regarding the document as an attempt to suppress Liberal Catholicism.[14] Newman thought that it was untimely, writing privately to Ambrose St John, on 8 January 1865: 'It is difficult to know *what* HE [the Pope] *means* by his condemnation. The words "myth," "non interference," "progress," "liberalism", "new civilization" are undefined . . . (entre nous) . . . the advisers of the Holy Father seem determined to make our position in England as difficult as ever they can.'[15]

To English Ultramontanism, however, the Syllabus represented a golden opportunity for the faithful to renew their commitment to Papal authority. Manning, when Provost of Westminster, had delivered three lectures on the *Temporal Sovereignty of the Popes* (1860), in which he argued that it had been the 'root, and the productive and sustaining principle of Christian Europe', and that its dissolution would bring with it the 'dissolution of Christian Europe'.[16] Having later succeeded Wiseman as Archbishop of Westminster in April 1865, Manning was to act as 'chief whip' of the pro-Infallibility Ultramontane majority party at the Vatican Council.[17]

W. G. Ward published a series of essays in the *Dublin Review*, from July 1864 onwards, which he reprinted as *The Authority of Doctrinal Decisions* (1866) and dedicated to Manning. In his preface he acknowledges that there are different schools within the Catholic communion. While describing himself in the Preface as 'undoubtedly' of the 'Ultramontane school' and as one who holds 'most confidently, that the Pope's declaration ex cathedra is at once and ipso facto the Church's infallible teaching', he claims that 'hardly any part of these Essays is intended as an argument for Ultramontanism', as almost all his reasoning 'possesses the same force for Gallicans as for

[13] One Piedmontese paper declared that the Pope was against all things modern, including the steam engine. See Dom C. Butler, *Vatican Council* (1962), p. 55.

[14] See H. A. MacDougall O. M. I., *Acton–Newman Relations* (1962), pp. 97–8.

[15] *Letters of Newman* (1961–77), XXI, 378. Eleven years later, Newman was to comment publicly that not a word of the Syllabus was 'of the Pope's own writing': *Letter to Duke of Norfolk* (1875), pp. 150–1.

[16] H. E. Manning, *Temporal Sovereignty of the Popes* (1860), p. 2.

[17] See Dom C. Butler, *Vatican Council* (1962), p. 108.

Ultramontanes'.[18] Like Manning, however, Ward accuses what he calls 'unsound Catholics' with 'rebellion' and 'audacity'. His own view is that 'God teaches the Holy See, and the Holy See teaches the Church; it is Peter whose faith fails not, and who in his turn confirms his brethren'.[19] Newman was disturbed by Ward's extremism.[20]

Meanwhile, Anglicans had also disagreed among themselves in 1864. Two events had again raised the question of the authority of the Established Church on matters of biblical interpretation: the vindication by the judicial committee of the Privy Council of two contributors to *Essays and Reviews* (1860), Rowland Williams and H. B. Wilson, in February; and the deposing of John William Colenso, author of *The Pentateuch and Book of Joshua Critically Examined* (1862–79), from his see in Natal by Bishop Gray in June. (Colenso's trial was to be ruled illegal by the judicial committee in March 1865.)

To conservative Churchmen, such interference by a body which included laymen smacked not only of Erastianism, but also of liberalism, in the sense defined by Newman: 'that deep, plausible scepticism . . . the development of human reason, as practically exercised by the natural man'. Authority – external and infallible authority – was invoked in response to the threat of religious liberalism by men like Pusey, the leader of the Tractarians or 'Puseyites'. Pusey claimed in the first part of his *Eirenicon*, addressed to John Keble in 1865, that Anglicans have, 'equally with those in the Roman Church, infallible truth, as resting on infallible authority': 'We do not need the present agency of an infallible Church to assure us of the truth of what has been ruled infallibly. Nor, in fact, have Roman Catholics any more infallible authority for what they hold than we, seeing that it was ruled by the Church in past ages, to whom, so far, the present Church submits.'[21]

Keble, perhaps one of the few leading Anglicans who could still agree wholeheartedly with Pusey on this matter, died in 1866. Like Newman and Pusey, Keble was no ritualist, but many of their successors were.[22] Ritualism, with its emphasis upon the drama of the liturgy and the mystery of the sacraments, can be regarded as another reaction to liberal Anglicanism. The infallible authority of 'Scripture' and of the 'testimony of the rocks' may be questionable in the 1860s, but what Newman calls the 'external authority

[18] W. G. Ward, *Authority of Doctrinal Decisions* (1866), p. viii. Gallicanism, which originated in France, asserted the freedom of the Roman Catholic Church from the ecclesiastical authority of the Pope. Further quotations in this paragraph are from pp. 22, 70, 71, 96.

[19] Hans Küng argues that Peter, who denied the Lord three times, does not 'exactly convey an impression of infallibility': A. B. Hasler, *How the Pope became Infallible* (1981), p. 2.

[20] See H. A. MacDougall O. M. I., *Acton–Newman Relations* (1962), p. 112.

[21] E. B. Pusey, *Eirenicon* (1865–70), I, 96. [22] See ibid., III, 341.

of the Divine Word' is to be found here, now, in the real presence of the Eucharist.

To those Protestants, however, who rallied to the flag of the 'Scriptures' and 'liberty', handed down from the Reformation, such thinking was blasphemous. In 1866, Francis Close, the Evangelical Dean of Carlisle, launched a blistering attack entitled *"The Catholic Revival"; or, Ritualism and Romanism in the Church of England.* Like other bellicose Evangelical controversialists, such as George Croly or John Cumming, Close uses a declamatory style in which each comma or semi-colon marks a descent of the fist on the pulpit cushion. By 'Ritualism', he announces,

> we just now understand excessive Ritualism, or such an attention to outwards forms and ceremonies as appears to be inconsistent with the spirit and practice of 'that pure and Reformed part of Christ's Holy Catholic Church established in these Realms'; a Ritual which closely imitates, and sometimes exceeds that of the Church of Rome itself; a Ritual which is not a finality, which does not terminate in decorations, pomp, and ceremony, but which, by means of these, labours to introduce into our Church, if not 'the ROMAN OBEDIENCE,' certainly the Roman dogmas – doctrines – the very doctrines against which the Reformers of our Church protested, many of them sealing that protest with their blood![23]

For Close, 'dogma' is the opposite, not of liberalism, but of liberty. Ritualism, he continues,

> can now be regarded as nothing less than the exponent of a powerful and restless party in the Church; founded by the Tractarian School some thirty years since, still advancing, and recently assuming a confidence and affrontery calculated to arouse the most indifferent, summoning us to surrender even the very term PROTESTANT, and to accept with revent homage that which its leading advocates denominate as 'THE GREAT CATHOLIC REVIVAL.'

The 'surgeon's knife must remove the cancer', he concludes, or the cancer will 'pervade and destroy the body'.[24]

Matthew Arnold, who wrote a hilarious description of genuflection in Ritualist churches,[25] is as cool on the subject of the 'Tractarian School' and its development as Close is warm. In the *Cornhill* essays of the summer of 1867, later to be published in book form as *Culture and Anarchy* (1869), Arnold analyses the 'great middle-class Liberalism' that 'really broke the Oxford movement'.[26] Its cardinal points of belief, he explains, were

[23] F. Close, *"The Catholic Revival"* (1866), p. 5. [24] Ibid., p. 28.

[25] In *Literature and Dogma* he writes of one's 'progress to the altar' being 'almost barred by forms suddenly dropping as if they were shot in battle': *Dissent and Dogma* (1968), p. 361.

[26] M. Arnold, *Culture and Anarchy* (1960), p. 62.

'the Reform Bill of 1832, and local self-government, in politics; in the social sphere, free-trade, unrestricted competition, and the making of large industrial fortunes; in the religious sphere, the Dissidence of Dissent and the Protestantism of the Protestant religion'. Unlike Close's semi-colons, Arnold's calmly separates items in a list, thus allowing the irony of his critique of 'Philistinism' to remain understated.

This 'great middle-class Liberalism', Arnold continues, 'was the force which till only the other day seemed to be the paramount force in this country'.[27] Arnold and his contemporaries had, over the previous twelve months, witnessed the Hyde Park riots, Fenian atrocities on the mainland, the passing of the second Reform Act and anti-'Papist' riots in some of England's major cities, following William Murphy's notorious 'No Popery' lectures. These, Arnold believes, are signs that the 'great force of Philistinism' has been 'thrust into the second rank'. Democracy, a new and wholly different power, has 'suddenly appeared'. This 'modern spirit' and the 'anarchical tendency of our worship of freedom in and for itself, of our superstitious faith . . . in machinery, is becoming very manifest', as the Englishman's right to 'march where he likes, meet where he likes, enter where he likes, hoot as he likes, threaten as he likes, smash as he likes' is 'put into practice'.[28]

Newman argued the previous year that liberalism had lost definition over three decades and was now synonymous with the 'educated lay world'. Arnold believes that liberalism is weaker, but considers that this is partly Newman's fault. Having quoted the *Apologia*, Arnold asks who will estimate how much the 'currents of feeling created by Dr. Newman's movement' contributed to 'swell the tide of secret dissatisfaction which has mined the ground under the self-confident Liberalism of the last thirty years, and has prepared the way for its sudden collapse and supersession?'[29] 'It is in this manner', he adds with relish, 'that the sentiment of Oxford for beauty and sweetness conquers, and in this manner long may it continue to conquer!'

Two months after Arnold's last essay was published in the *Cornhill* in September 1868, Robert Browning's most ambitious poem – *The Ring and the Book* – began to appear in four monthly parts. Browning had returned from Florence in 1861 with the 'Old Yellow Book' he found on a market stall there, and had started work on the poem three years later.[30] Although the plot is centred upon the murder trial of Count Guido Franceschini in

[27] Ibid., pp. 62–3. [28] Ibid., p. 76. [29] Ibid., p. 63.

[30] See R. Browning, *Ring and the Book* (1971), pp. 11–12. Quotations in the main text are from this edition.

Rome in 1698, Browning also drew upon events in England in the 1860s. Like Ruskin, he was a narrow-eyed observer of Anglican squabbles, and the trials of Colenso and of Williams and Wilson provided him with useful sources. After fifteen years living in Catholic Italy, however, it was Pope Pius IX, or 'poor old "infallibility"' as he called him, who most intrigued Browning;[31] and it was Newman's triumph over Kingsley in the *Apologia* and the growing power of English Catholicism which seem to have spurred him to plan the poem in the summer of 1864 and start writing that autumn.[32]

Ever since he first corresponded with Elizabeth Barrett in 1845, Browning, with his background in South London Dissent, had said 'savage things about Popes and Imaginative religions',[33] and had associated the Oxford Movement with Roman Catholicism.[34] His unease with modern English 'Papists' is reflected in what the reviewer in *The Spectator* rightly described as 'extremely bad puns' in Book I, 'The Ring and the Book':[35]

> Go get you manned by Manning and new-manned
> By Newman and, mayhap, wise-manned to boot
> By Wiseman, and we'll see or else we won't!
>
> (I, 444–6)

While Catholic reviewers took exception to confusions over Molinism and the highly unorthodox doctrine put into the mouth of Pope Innocent XII, several critics and private correspondents commented on the 'grandeur' of the old Pope.[36]

Whereas there is no evidence that Innocent actually took any interest in the case, Browning's Pope has absolute power over the life and death of Guido and the four assassins he hired, being able to overturn the guilty verdict.[37] As he is close to death himself, however, being in 'grey ultimate

[31] See B. Litzinger, *Browning and the Babylonian Woman* (1962), pp. 14–16.

[32] See R. W. Goldsmith, *Relation of Browning's Poetry* (1958), p. 252.

[33] In the second letter he wrote her, on 13 January 1845, he playfully said that he would not share these things with her: *Brownings' Correspondence* (1984–98), X, 22.

[34] In a letter to Frederick Oldfield Ward of 18 February 1845, Browning described the enclosed manuscript of 'The Tomb at St. Praxed's (Rome, 15 –)', as a 'pet' of his, and 'just the thing for the time – what with the Oxford business, and Camden society and other embroilments': ibid., X, 83. Like Ruskin, however, Browning was also fascinated by and in some ways attracted to Roman Catholicism.

[35] *The Spectator*, 12 December 1868, pp. 1464–6; rpt. in B. Litzinger and D. Smalley, *Browning: The Critical Heritage* (1970), p. 290.

[36] See V. P. Anderson, *Browning as a Religious Poet* (1983), p. 48; E. A. Khattab, *Critical Reception* (1977), p. 52 ; Helena Faucit Martin to Browning, 6 March 1869, ALS Armstrong Browning Library, Baylor University, 69:45; Shirley Brooks to Browning, 30 March 1869, ALS Armstrong Browning Library, Baylor University, 69:61.

[37] See C. W. Hodell, *Old Yellow Book* (1908), p. 270.

decrepitude' (Book x, 'The Pope', 388), he will be answerable only to God for his decision. He is 'near the end; but still not at the end; /All till the very end is trial in life' (x, 1302–3). Even in the final lines of Book X, the Pope's final reflections on his own and Guido's deaths, before he dispatches his fatal decision, are upon the contrast between the 'flash' of the moment of insight or the death blow and the dreary in-between state of purgatory that will follow:

> So may the truth be flashed out by one blow,
> And Guido see, one instant, and be saved.
> Else I avert my face, nor follow him
> Into that sad obscure sequestered state
> Where God unmakes but to remake the soul
> He else made first in vain; which must not be.
> Enough, for I may die this very night
> And how should I dare die, this man let live?
>
> Carry this forthwith to the Governor!
>
> (x, 2126–34)

After the drama and lurid colours of the murder and the trial, Book x is crepuscular in mood and grey in tint, emphasizing a shift from the clash of opposing versions of the truth to the exploration on this 'sombre wintry day' (x, 211) of the liminal area between life and death and the disputed area between extremes, where the truth may actually lie, and where liberal thought has most to contribute towards its discovery. The possibility of Infallibility, to which reference has already been made in earlier books,[38] is now denied even by the Pope, as, sounding remarkably like Browning, he asks himself which of his predecessors was correct in his judgment on Pope Formosus: Stephen VII, who exhumed him and degraded his appointments at the grotesque 'cadaveric synod', or Theodore II, who decided to 'repope the late unpoped' (x, 109): 'Which of the judgments was infallible? / Which of my predecessors spoke for God?' (x, 150–1). Faced with having to make a judgment of his own, on a dead man walking, or latter-day Formosus,[39] Innocent can but say, like a true nineteenth-century man: 'I take His staff with my uncertain hand' (x, 164). But Browning seems to ignore the fact that Papal Infallibility is not in question in such judgments, being applied only to matters of faith and morals *ex cathedra*.

The most significant reference to Infallibility earlier in the poem is Guido's, when he refers to

[38] v, 1350; viii, 710; ix, 182.
[39] 'A mere dead man is Franceschini here, / Even as Formosus centuries ago' (x, 209–10).

. . . a vain attempt to bring the Pope
To set aside procedures, sit himself
And summarily use prerogative,
Afford us the infallible finger's tact
To disentwine your tangle of affairs . . .
(v, 1347–51)

Instead of 'disentwine', the Pope himself, at a crucial point in his reflections upon the nature of truth, uses a derivative of the word 'evolve', in the seventeenth-century sense of to 'disengage from wrappings, disclose gradually to view; to disentangle; to set forth in orderly sequence' (*evolve* 2, 1664, *OED*):

I have worn through this sombre wintry day,
With winter in my soul beyond the world's,
Over these dismalest of documents
Which drew night down on me ere eve befell, –
Pleadings and counter-pleadings, figure of fact
Beside fact's self, these summaries to-wit, –
How certain three were slain by certain five:
I read here why it was, and how it went,
And how the chief o' the five preferred excuse,
And how law rather chose defence should lie, –
What argument he urged by wary word
When free to play off wile, start subterfuge,
And what the unguarded groan told, torture's feat
When law grew brutal, outbroke, overbore
And glutted hunger on the truth, at last, –
No matter for the flesh and blood between.
All's a clear rede [record] and no more riddle now.
Truth, nowhere, lies yet everywhere in these –
Not absolutely in a portion, yet
Evolvable from the whole: evolved at last
Painfully, held tenaciously by me.
Therefore there is not any doubt to clear
When I shall write the brief word presently
And chink the hand-bell, which I pause to do.
Irresolute? Not I more than the mound
With the pine-trees on it yonder! Some surmise,
Perchance, that since man's wit is fallible,
Mine may fail here? Suppose it so – what then?
(My emphasis, x, 211–38)

In the 1830s, Carlyle had used the word 'evolve' in a new sense that is not specifically scientific: 'To bring out (what exists implictly or potentially):

e.g. to educe (order from confusion, light from darkness, etc.); to deduce (a conclusion, law, or principle) from the data in which it is involved; to develop (a notion) as the result of reflection or analysis; to work out (a theory or system) out of pre-existing materials' (*evolve* 5, 1831, *OED*).[40] The second example of this usage, however, is more relevant to *The Ring and the Book*, being from the Oxford lectures of Robert Hussey, Regius Professor of Ecclesiastical History, on *The Rise of the Papal Power* (1851), published during the 'Papal Aggression' crisis, where he argues that the dogma of 'Papal Supremacy' was originally no part of Christ's religion, and that, of the 'other new claims of authority which were gradually evolved from the doctrine of the Supremacy, it is to be observed, that *the right to dispose of temporal sovereignties* was assumed by Nicholas II (AD 1059), when he took upon himself to confirm the Duke of Calabria and Sicily in the possession of his dominions, on certain conditions, for which dominions the Duke swore fealty to the Pope'.[41]

Taken in this sense, which emphasizes development over time, the Pope's 'evolved at last' suggests parallels with Newman's *Essay on Development* (1845). That essay has been described as the 'theological counterpart' of the *Origin of Species* (1859),[42] and for Browning's first readers in the late 1860s, 'evolvable' and 'evolved' would have had a modern scientific resonance.[43] If the Pope's reflections on the truth being 'evolved *at last*' smacks of Newman on development and Darwin on evolution, however, his belief that truth lies, not 'absolutely in a portion', yet 'evolvable *from the whole*' anticipates Newman's argument on the cumulation of probabilities in the *Grammar of Assent* (1870), the work which he published during the Vatican Council.[44] The Pope's reflections are not simply Browning's, but represent a distillation of the kind of thinking that was current in the 1860s in the contested area between dogma and liberalism.

III

In 1970, the centenary year of Papal Infallibility, Hans Küng published a book which put him on a collision course with the Vatican: *Unfehlbar?:*

[40] 'An English Editor, endeavouring to evolve printed Creation out of a German printed and written Chaos' (*Sartor Resartus*).

[41] R. Hussey, *Rise of the Papal Power* (1851), pp. v, 172–3.

[42] See p. 56 above.

[43] 'To develop by natural processes from a more rudimentary to a more highly organized condition; to originate (animal or vegetable species) by gradual modification from earlier forms; in wider sense, to produce or modify by "evolution"' (*evolve* 6, 1832, *OED*). The first recorded usage of definition 6 is, like that of definition 5, from the early 1830s, in this case Lyell's *Principles of Geology*: 'The orang-outang, having been evolved out of a monad, is made slowly to attain the attributes and dignity of man.'

[44] See pp. 261–5 below.

Eine Anfrage.[45] Nine years later, in February 1979, Küng, the Director of the Institute for Ecumenical Research at the University of Tübingen, contributed an introductory essay on 'The Infallibility Debate – Where Are We Now?' to Father August Bernhard Hasler's controversial study, *How the Pope became Infallible: Pius IX and the Politics of Persuasion*. On 18 December 1979, the Vatican announced that Küng could 'neither be considered a Catholic teacher nor engage in teaching as such'.[46] Hasler commented, in a foreword to his second edition: 'Hardly anyone thought it could still happen, but it has: Papal Rome is once again branding as heretics those unwilling to believe in its Infallibility.' Hasler's main argument, which Küng endorsed, is that an elderly and sometimes mentally disturbed Pius IX used every available method of intimidation and obfuscation to ensure that the Council defined Papal Infallibility, against the better judgment of most of those who remained to vote on 18 July 1870, after the departure of the strongest anti-Infallibilists. In a nutshell, the Pope 'wanted no discussion'.[47]

After the definition, however, it was only in Germany that there was outright opposition to the Vatican decrees. Johann Joseph Ignaz von Döllinger, Professor of Church History at the University of Munich, who had led the attack against the Ultramontanes and who corresponded with Acton, Gladstone and Newman, was defrocked and excommunicated in 1871. Two years earlier, shortly before the opening of the Vatican Council, Acton had published an essay on Döllinger's *Der Papst und das Concil*, which had recently been published under the pseudonym *Von Janus* and was deeply anti-Infallibilist.[48] Whereas Newman could not believe that Döllinger wrote the essay, which he thought would do 'immense mischief', Acton's only regret was that the Infallibilists had not been hit harder. When Döllinger was excommunicated, the English Liberal Catholics mourned, while Manning and the Ultramontanes regarded it as proof that a Liberal Catholic was not a true Catholic.[49]

Acton travelled to Rome in September 1869, in the hope of preventing the definition, even though he had no official position at the Council. He reported regularly to Gladstone, who was Liberal Prime Minister at the time, and received reports from Simpson on developments in England and Ireland. A month after his arrival, Acton's remarkable account of the origins of the Council was published in *The North British Review*. He was not afraid to bring Liberal Party politics into his Liberal Catholic essay, stating that

[45] *Infallible?: An Enquiry*. [46] See A. B. Hasler, *How the Pope became Infallible* (1981), p. 29.
[47] Ibid., p. 79. [48] H. A. MacDougall O. M. I., *Acton–Newman Relations* (1962), p. 114.
[49] Ibid., p. 120.

The English Government was content to learn more and to speak less than the other Powers at Rome. The usual distrust of the Roman Court towards a Liberal ministry in England was increased at the moment by the measure which the Catholics had desired and applauded . . . Mr. Gladstone was feared as the apostle of those doctrines to which Rome owes many losses.[50]

His central argument, however, was that the Jesuits had driven the Infallibilists forward in Rome:

They were connected with every measure for which the Pope most cared; and their divines became the oracles of the Roman congregations. The papal infallibility had been always their favourite doctrine. Its adoption by the Council promised to give to their theology official warrant, and to their Order the supremacy in the Church. They were now in power; and they snatched their opportunity when the Council was convoked.[51]

Such a story would certainly have appealed to an English Protestant readership, for whom the Jesuit was the classic 'Papist' under the bed.

In the early days of the Council, *Punch* sniped at the unEnglish 'Papist' proceedings in Rome. They printed some verses entitled 'Fallibilis Infallibis', for example, accompanied by a seasonal cartoon showing the Pope and his bishops 'Sliding on Thin Ice'.[52] In a series of spoof dispatches from Rome by a Victorian Mrs Malaprop, Mrs Lavinia Ramsbotham informs Mr Punch that she has 'ascertained a great many things about the Roman Candlestick rights and cemeteries, which other Co-respondents of the daily papers are unable to get hold of'.[53] She is also an observant tourist: 'Oh! The antipathies of the place! Wonderful!! All old, every bit of it. And talk of Underground Railways in London!! Ah, you should see the Roman Currycombs made by the earliest Christian Marthas, who used to meet between 2 and 3 A.M. to sing hymns for fear of prosecution.' Even on Christmas Day 1869, *Punch* carried some doggerel entitled 'Look up your Latin', in which the bishops attending the Council 'found for their eloquence limited scope; / So few of the lot could talk Latin!'[54]

[50] Ibid., p. 289.

[51] J. E. E. D. Acton, *Essays on Freedom and Power* (1956), pp. 281–2. Hasler argues that the Jesuits' 'chief tools were the pulpit, the confessional, the press, and the religious associations': *How the Pope became Infallible* (1981), p. 58.

[52] The poem ends: 'And him who this great circle has swept through, / From zenith unto nadir of thought's zone, / Whose "true" to-day is next-day's most "untrue," / The Church has gathered here INFALLIBLE to own!' Anon., 'Fallibilis Infallibis', *Punch*, July–December 1869, pp. 242, 244–5.

[53] Anon., 'Rome and Ramsbotham', *Punch*, January–June 1870, p. 23.

[54] Anon., 'Look up your Latin', *Punch*, July–December 1869, p. 252.

The sixth stanza of 'Look up your Latin' refers to Dr John Cumming, the famous preacher and Minister of the National Scottish Church, Crown Court, Covent Garden:

> The chair by St. Peter so handsomely feed
> Many classical pontiffs have sat in;
> But 'twould be a scandal outrageous indeed
> If prelates should have to send north of the Tweed
> For CUMMING, to coach them in Latin.

Cumming's unsuccessful request, through Manning, to attend the Council had already been satirized by *Punch* in October 1869, in a ballad entitled 'Hey, Johnny Cumming!',[55] featuring his correspondence with 'Pawpie' and accompanied by a cartoon entitled 'NON POSS.!', with a smiling Pius IX saying to the stern Protestant preacher: '"No, dear Doctor Cumming – you may kiss my toe if you like, but you mustn't speechify"' (illustration 31).

Barred from the Council, Cumming applied his energy to prophecy, which was his forte.[56] It had long been believed in English millenarian circles that the end of time would occur between 1866 and 1873,[57] and the second Reform Act and the domestic disturbances around 1867 were interpreted as signs of the times. Expectation was further heightened by events in Rome, as Cumming reveals in *The Fall of Babylon Foreshadowed* (1870). 'Romanism', he declares, 'will be destroyed, not reformed.'[58] It had been calculated that Babylon would fall in 1868. 'If these calculations are just', Cumming concludes, 'we are on the eve of stupendous events. Great Babylon rapidly nears her destruction.'[59]

Nature and man seemed to conspire in providing signs of the end-time in Rome in the second half of 1870. The storm that broke over the city on 18 July was described by Bishop Ullathorne in his last letter from the Council:

> The great Session is over. The decree was voted by 533 'placets' to 2 'non placets' amidst a great storm. The lightning flashed into the aula, the thunder rolled over the roof, and glass was broken by the tempest in a window nearly over the pontifical throne and came rattling down. After the votes were given the Pope confirmed it at once, and immediately there was a great cheering and clapping from the bishops, and cheers in the body of St. Peter's. Then the 'Te Deum' began, the thunder forming the diapason.[60]

[55] Anon., 'Hey, Johnny Cumming!', *Punch*, July–December 1869, pp. 128–9. [56] See p. 24 above.
[57] See J. O. Waller, 'Christ's Second Coming', *Bulletin of the New York Public Library*, 1969, pp. 476–7.
[58] J. Cumming, *Fall of Babylon* (1870), p. 5. [59] Ibid., p. 473.
[60] Dom C. Butler, *Vatican Council* (1962), p. 416.

"NON POSS.!"

PAPA PIUS. "NO, DEAR DOCTOR CUMMING—YOU MAY KISS MY TOE IF YOU LIKE, BUT YOU MUSTN'T SPEECHIFY."

31. '"NON POSS.!" PAPA PIUS. "No, dear Doctor Cumming – you may kiss my toe if you like, but you mustn't speechify"', *Punch*, 57 (July–December 1869), 129.

The following day, war was declared between France and Prussia. This in turn meant that the French could no longer defend Rome, which was entered by the Italians on 20 September 1870. Pius IX became the 'Prisoner of the Vatican' after the loss of the Papal States and Italian unification was completed.

Towards the end of 1870, an excited Cumming published a new book in which he claimed that his earlier prophecies were coming to pass. In the preface to *The Seventh Vial* he asks: 'Is not the story of 1870 and the inspired prophecy of Patmos as echo and sound?'[61] He goes on to argue

[61] J. Cumming, *Seventh Vial* (1870), p. xv.

that the second spirit that comes out of the mouth of the beast (Revelation 13.1–7)[62] inspired the Pope to define Papal Infallibility:

> When the Pope rose to proclaim and accept the new attribute, as if Heaven were offended at such blasphemy, thunder and lightning, the most vivid and terrible, broke on St. Peter's, and one cardinal, overpowered by the scene, shouted 'Another Mount Sinai on earth!' The success of this spirit culminated in the decree of infallibility.[63]

For Ullathorne, the thunder contributed to a hymn of praise, and for one cardinal was a sign that the Pope, like Moses, could mediate between God and man. For commentators such as Cumming, however, it was a sign of God's wrath.

Cumming lists other signs that have followed. He describes the battle of Sedan and quotes *The Times* for 5 September: 'the earth has opened in Paris – a revolution has begun'.[64] Prussia, a Protestant nation, seems to have been chosen as the '"battle-axe" in the hand of God to strike the nations that have been at once the champions and the serfs of the Papacy; there seems to be in the van of her armies recently a Presence higher than human, and in her tactics a wisdom wholly denied to her French adversaries'. Unlike its predecessors, he explains, the seventh vial was to be 'universal, not local in its effects'. The Church of Rome has committed suicide and the Pope's signature to the dogma of his personal infallibility is the 'signature of his death-warrant'. All is not yet fulfilled, however, as Infallibility 'without even temporal power will still do tremendous mischief': 'We are not yet done with the Pope.'

Like Cumming, Newman remained in England during the Council and was disturbed by events in Rome. Unlike Cumming, however, he had received four invitations to attend, one of them from the Pope himself, and Newman's subtle contribution to the debate over liberalism and dogma proved to be of infinitely more value than Cumming's. One of a number of motives for staying at Edgbaston was the wish to complete *The Grammar of Assent*, a project which he had been turning over in his mind for three or four decades, but which now had an urgent relevance.

Aubrey de Vere records a discussion with Newman at the Oratory, when he called in on his way to Rome in 1870. Newman said:

[62] Ibid., p. 5. 'And I stood upon the sand of the sea, and saw a beast rise up out of the sea, having seven heads and ten horns, and upon his horns ten crowns, and upon his heads the name of blasphemy . . . '

[63] Ibid., p. 7. [64] Ibid., p. 14. Further quotations in this paragraph are from pp. 59, 61, 66, 194, 239.

'People are talking about the definition of the Papal Infallibility as if there were and could be but one such definition. Twenty definitions of the doctrine might be made, and of these several might be perfectly correct, and several others might be *exaggerated* and incorrect.' Every one acquainted with Newman's teaching was aware that he fully believed the doctrine – nay, that he had expressed that conviction in nearly every volume published by him subsequently to his conversion.[65]

Newman saw that the extreme Ultramontanes and their opponents were arguing over the limitations that might be placed upon Infallibility. He was concerned not only about the divisiveness of the definition, but also its timing. Pius IX had already defined the dogma of the Immaculate Conception in 1854. To press for this further dogma was, in Newman's view, premature: 'we are not ripe yet for the Pope's Infallibility', he wrote.[66] On 28 January 1870, he sent a letter to Bishop Ullathorne which he later described as 'one of the most passionate and most confidential letters' he ever wrote:

As to myself personally, please God, I do not expect any trial at all; but I cannot help suffering with the various souls which are suffering, and I look with anxiety at the prospect of having to defend decisions, which may be not difficult to my private judgment, but may be most difficult to maintain logically in the face of historical facts. What have we done to be treated, as the faithful never were treated before? When has definition of doctrine de fide been a luxury of devotion, and not a stern painful necessity? Why should an agressive [sic] insolent faction be allowed to 'make the heart of the just to mourn, whom the Lord hath not made sorrowful?' Why can't we be let alone, when we have pursued peace, and thought no evil [Psalm 34.14]?

Ironically, Newman's application of the terms 'insolent' and 'aggressive' to the Infallibilists among his co-religionists – including Manning and the Jesuits – echoes Lord John Russell's description of 'Papal Aggression' in 1850.[67] As usual, however, Newman is quick to emphasize his readiness to submit to authority – in this case, the highest authority:

If it is God's will that the Pope's infallibility should be defined, then it is His blessed Will to throw back 'the times and the moments' [Acts 1.7] of that triumph which He has destined for His Kingdom; and I shall feel I have but to bow my head to His adorable, inscrutable Providence.

On 15 February 1870, copies of Newman's letter were sent to London and to Döllinger at Munich, and on 6 April, three weeks after the publication of

[65] A. de Vere, *Recollections* (1897), p. 274.
[66] *Letters of Newman* (1961–77), xxv, 93. Further quotations in the paragraph are from pp. 117, 18–19.
[67] On 4 November 1850, Russell wrote to the Bishop of Durham, in a letter that was subsequently published, agreeing with his statement that the recent 'Papal Aggression' was 'insolent and insidious': see p. 12 above.

The Grammar of Assent, the *Standard* published the letter, to the delight of the Liberal Catholics.[68] Döllinger had written to Newman, thanking him for his copy of the *Grammar*, which he much admired. In his reply, Newman said that he supposed 'in all Councils there has been intrigue, violence, management, because the truth is held in earthen vessels [2 Corinthians 4.7]. But God over rules.'[69] Döllinger wrote to Acton condemning this letter, as it suggested that Newman would submit to the Council's decision, whatever it was. Döllinger proved to be right. Although Newman refused to believe that the definition would be made until it actually was, and continued to lament the divisiveness that the debate had caused, he was pleased that the terms of the definition were 'vague and comprehensive' and stated that he had 'no difficulty in admitting it'.[70] The real question for him was whether the definition came 'with the authority of an Ecumenical Council'.

As in so many of the Anglican controversies of the nineteenth century, the nub of the Infallibility debate within the Church of Rome was authority – Papal, Church and divine authority – in relation to dogma. Newman addressed this question in chapter v of *An Essay in Aid of a Grammar of Assent*, entitled 'Apprehension and Assent in the Matter of Religion', which begins with this statement:

> We are now able to determine what a dogma of faith is, and what it is to believe it. A dogma is a proposition; it stands for a notion or for a thing; and to believe it is to give the assent of the mind to it, as it stands for the one or for the other. To give a real assent to it is an act of religion; to give a notional, is a theological act. It is discerned, rested in, and appropriated as a reality, by the religious imagination; it is held as a truth, by the theological intellect. [71]

As we saw earlier, Newman described his conversion to Catholicism as a divine call into a state which is 'sure and safe', a 'realm of peace and assurance'.[72] Now, as his co-religionists engage in the kind of 'war of tongues' which he felt that he himself had escaped in 1845, Newman writes of the religious imagination *resting* in a dogma – a concept which contrasts strongly with the restless striving associated with the liberal Protestant mind and the will to be believe, analysed and in some ways epitomized by Matthew Arnold.

At the end of the chapter, and thus also of Part I of the *Grammar*, 'Assent and Apprehension', Newman emphasizes that it is the infallibility of the *Church* that lies at the heart of Roman Catholicism:

[68] See I. T. Ker, *John Henry Newman* (1988), p. 654. [69] *Letters of Newman* (1961–77), xxv, 85.
[70] Ibid., xxv, 164. [71] J. H. Newman, *Grammar of Assent* (1985), p. 98. [72] See chapter 6.

The word of the Church is the word of the revelation. That the Church is the infallible oracle of truth is the fundamental dogma of the Catholic religion; and 'I believe what the Church proposes to be believed' is an act of real assent, including all particular assents, notional and real; and, while it is possible for unlearned as well as learned, it is imperative on learned as well as unlearned. And thus it is, that by believing the word of the Church *implicitè*, that is, by leaving all that that word does or shall declare itself to contain, every Catholic, according to his intellectual capacity, supplements the shortcomings of his knowledge without blunting his real assent to what is elementary, and takes upon himself from the first the whole truth of revelation, progressing from one apprehension of it to another according to his opportunities of doing so.[73]

As always in Newman, a universal idea or principle comes alive in its application to the self, to the 'learned' Newman for whom real assent to the authority of the Church, as described in the Catechism, is 'imperative' and to the individual Catholic Newman – 'every Catholic' – who combines submission with development in the apprehension of revelation. Crucially, however, he does not individualize or personalize the 'oracle of truth', unlike W. G. Ward, for instance, for whom 'God teaches the Holy See, and the Holy See teaches the Church'.[74]

The distance between Newman's assent to dogma as a conservative Liberal Catholic and Protestant liberalism's commitment to free inquiry becomes even clearer in Part II of the *Grammar*, 'Assent and Inference', where he argues that the Catholic cannot describe himself as both a believer and an inquirer: 'He cannot be both inside and outside of the Church at once.'[75] 'I may be certain that the Church is infallible', he writes, 'while I am myself a fallible mortal; otherwise, I cannot be certain that the Supreme Being is infallible, until I am infallible myself.' Far from suppressing (fallible) individuality, however, as anti-Catholics often accused the Roman Church of trying to do, Newman believes that in 'these provinces of inquiry egotism is true modesty': 'In religious inquiry each of us can speak only for himself, and for himself he has a right to speak.' The 'Illative Sense', which he defines as the 'reasoning faculty, as exercised by gifted, or by educated or otherwise well-prepared minds', is a 'rule to itself, and appeals to no judgment beyond its own'. Morally and spiritually, however, an inner authority is at work: 'Our great internal teacher of religion is . . . our conscience.'

[73] Ibid., p. 102. [74] See p. 250 above.

[75] J. H. Newman, *Grammar of Assent* (1985), p. 125. Further quotations in this paragraph are from pp. 148, 248, 233, 251.

For Newman, the authority of Scripture is problematic. 'It is by no means self-evident', he states, 'that all religious truth is to be found in a number of works, however sacred, which were written at different times, and did not always form one book; and in fact it is a doctrine very hard to prove.'[76] An historical fact that cannot be denied, however, is that Judaism and Christianity, two 'wonderful creations', should 'span almost the whole course of ages, during which nations and states have been in existence, and should constitute a professed system of continued intercourse between earth and heaven from first to last amid all the vicissitudes of human affairs'.[77] Here is Newman's strongest example of the cumulation of probabilities leading towards certainty: 'This phenomenon again carries on its face, to those who believe in a God, the probability that it has that divine origin which it professes to have; and . . . this phenomenon, I say, of cumulative marvels raises that probability, both for Judaism and Christianity, in religious minds, almost to a certainty.' In his vision of universal history, Newman focuses upon the continuities of 'marvels' in the Judæo-Christian tradition – continuities which provide more evidence of the authority of the Church than of the Papacy.

IV

Among the many religious commentators of different persuasions who reviewed their positions on English Catholicism during and immediately after the Vatican Council, some took the opportunity to seek a better understanding between communions. Pusey, for example, looked for reconciliation when he addressed the third part of his *Eirenichon* to Newman in 1870, and referred to the old problem of language and misinterpretation. 'Sons of the same fathers', he writes, 'we must in time come to understand each other's language.'[78] The following year, Newman's bishop, Ullathorne, sought to set the record straight when he published a history of the restoration of the Roman Catholic hierarchy in 1850, which had been misinterpreted as 'Papal Aggression'. The 'long train of circumstances that brought about the establishment of that Hierarchy', he states in the preface, has 'never been understood; nor has the history of the negotiations with the Holy See, that brought it into existence, been hitherto given to the light

[76] Ibid., p. 244. [77] Ibid., p. 283.

[78] E. B. Pusey, *Eirenicon* (1865–70), III, 342. Catholic readers had in fact been angered by Pusey's book and Newman, in his published response, had told Pusey that he had discharged his olive branch 'as if from a catapult': see I. T. Ker, *John Henry Newman* (1988), pp. 578–81.

of open day'.[79] It was also in the 1870s that popular anti-Catholicism in England began to decline as a 'major public issue'.[80]

Among the liberal intelligentsia, the definition of Papal Infallibility confirmed the view that Pius IX had failed to meet the challenges of the intellectual revolution of the mid-nineteenth century and had retreated to a position of pre-Enlightenment absolutism.[81] Nevertheless, Newman, together with Bishop Butler, had a powerful influence upon Matthew Arnold's religious thought in general and his 'Essay towards a better Apprehension of the Bible', *Literature and Dogma* (*Cornhill Magazine*, 1871–3), in particular.[82] Arnold wrote to Newman on 29 November 1871: 'In all the conflicts I have with modern Liberalism and Dissent, and with their pretensions and shortcomings, I recognize your work; and I can truly say that no praise gives me so much pleasure as to be told (which sometimes happens) that a thing I have said reminds people, either in manner or matter, of *you*.'[83]

Like other Protestant admirers, Arnold recognized that Newman was a true Catholic in his beliefs and in his devotion to the Pope, but that his reservations about Papal Infallibility marked him out, not as a campaigning Liberal Catholic like Acton or Simpson, but as one who retained a characteristically English moderation and a recognition of the claims of the critical spirit.[84]

In *Literature and Dogma*, Arnold's dialogue is sometimes with Newman,[85] but more often with Ultramontane Catholicism. In his opposition to Catholic dogma and conservatism, Arnold offers a classic liberal defence of progress and modernity, as he addresses three crucial areas of Catholic belief: the theology of the Mass, miracles and Papal Infallibility.

On the institution of the Eucharist, Arnold takes an extremely individualistic Protestant position, disputing the Catholic claim that it 'makes the Church indispensable, with all her apparatus of an apostolical succession,

[79] W. B. Ullathorne, *Restoration of the Catholic Hierarchy* (1871), p. iii.

[80] D. G. Paz, *Popular Anti-Catholicism* (1992), p. 18.

[81] This was later argued by the historian of the Pontificate of Pius IX, who 'concluded that he failed to give the leadership necessary to meet the intellectual challenge of the day. [R. Aubert, *Pontificat de Pie IX* (1952), p. 500] The results of this default were reflected in English Catholicism. No official leader made any serious effort to face the difficulties introduced by the intellectual revolution': H. A. MacDougall O. M. I., *Acton–Newman Relations* (1962), p. 4.

[82] M. Arnold, *Dissent and Dogma* (1968), p. 457.

[83] *Letters of Newman* (1961–77), XXV, 440–1. Newman was 'much pleased' to have the letter.

[84] See O. Chadwick, *Victorian Church* (1966–70), II, 422.

[85] Having referred to his 'respect' for Dr Newman, he quotes him and then disagrees with him: '"Faith is, in its very nature, the acceptance of what our *reason* cannot reach, simply and absolutely upon testimony." But surely faith is in its very nature (with all deference be it spoken!) nothing of the kind; else how could Jesus Christ say to the Jews: "If I tell you the truth, why do ye not believe me?" [John 8.46]': M. Arnold, *Dissent and Dogma* (1968), pp. 312–13.

an authorised priesthood, a power of absolution': 'Yet, as Jesus founded it, it is the most anti-ecclesiastical of institutions, pulverising alike the historic churches in their beauty and the dissenting sects in their unloveliness; – it is the consecration of absolute individualism.'[86]

Nowhere was the gap between Protestant liberalism and Catholic dogma wider than on the question of miracles. For Newman, as we have seen, miracles, including ecclesiastical miracles, lay at the heart of his faith.[87] In *Literature and Dogma*, Arnold contrasts Roman Catholics, who 'fancy that Bible-miracles and the miracles of their Church form a class by themselves', and Protestants, who 'fancy that Bible-miracles, alone, form a class by themselves'.[88] For Arnold, the findings of modern biblical criticism threw into doubt some of the 'evidences' of Christianity to which most believers, Catholic and Protestant, clung. On Christ's comments on His resurrection, for example, in the farewell discourses, he writes: 'The miracle of the corporeal resurrection ruled the minds of those who have reported Christ's sayings for us; and their report, *how* he foretold his death, cannot always be entirely accepted.'[89] His conclusion is that 'dogmatic theology is, in fact, an attempt at both literary and scientific criticism of the highest order; and the age which developed dogma had neither the resources nor the faculty for such a criticism'.[90]

In his preface to the Popular Edition of *Literature and Dogma* (1883), Arnold added that to 'pass from a Christianity relying on its miracles to a Christianity relying on its natural truth is a great change': 'Probably the abandonment of the tie with Rome was hardly less of a change to the Christendom of the sixteenth century, than the abandonment of the proof from miracles is to the Christendom of to-day.'[91] Miracles, he concludes, 'have to go the same way as clericalism and tradition'.[92]

In the preface to the first edition (1873), where Arnold discusses Papal Infallibility in relation to the liberal shibboleths of progress and the search for truth, the prose crackles with satiric energy as Pius IX comes into view:

The idea of the infallible Church Catholic itself, as I have elsewhere said,[93] is an idea the most fatal of all possible ideas to the concrete, so-called infallible Church of Rome, such as we now see it. The infallible Church Catholic is, really, *the prophetic soul of the wide world dreaming on things to come*;[94] the whole human race, in its outward progress, discovering truth more complete than the parcel of truth any momentary individual can seize. Nay, and it is with the Pope himself as

[86] Ibid., p. 355. [87] See above, pp. 66, 76.
[88] M. Arnold, *Dissent and Dogma* (1968), p. 246. [89] Ibid., p. 318. [90] Ibid., p. 345.
[91] Ibid., p. 143. [92] Ibid., p. 244.
[93] See *St. Paul and Protestantism* (May 1870), in ibid., p.92. [94] Shakespeare, Sonnet 107.

with the Church Catholic. That amiable old pessimist in St. Peter's Chair, whose allocutions we read and call them impotent and vain, – the Pope himself is, in his idea, the very Time-Spirit taking flesh, the incarnate '*Zeit-Geist*'. O man, how true are thine instincts, how over-hasty thine interpretations of them![95]

Whereas Arnold's comments on Papal Infallibility formed only part of a broader critique of Catholicism, other commentators in the period following the Vatican Council, including Gladstone, Newman and Manning and Ward, still focused specifically on the definition of the dogma. In 1874, Gladstone was feeling particularly hostile towards Rome. The fall of his first Liberal ministry in February 1874 had only been postponed since the defeat of his beloved Irish University Bill the previous year. Gladstone believed that the Irish hierarchy who had condemned the Bill, and who had encouraged Irish members to vote against it, had acted on instructions from the Pope.[96] His resentment deepened when, on 7 September 1874, the Marquis of Ripon was received into the Catholic Church, having resigned as Lord President of the Council in Gladstone's cabinet the year before, in order to search his soul and study Newman. Gladstone visited Döllinger, now an Old Catholic,[97] in Munich that September and returned to England with a burning need to relieve his feelings in a 'political expostulation'. *The Vatican Decrees in their bearing on Civil Allegiance* was published, in the good old style of inflammatory pamphlets, on 5 November, and sold 145,000 copies in the first month, mainly in a sixpenny edition.

The Rome of the Middle Ages, Gladstone argues, 'claimed universal monarchy', and the modern Church of Rome has 'abandoned nothing, retracted nothing'.[98] Writing as the supreme parliamentarian, he claims that, whereas in the 'national Churches and communities of the Middle Ages, there was a brisk, vigorous, and constant opposition which stoutly asserted its own orthodoxy', this same opposition has been 'put out of court, and judicially extinguished within the Papal Church, by the recent decrees of the Vatican'. Gladstone quotes Newman's letter to Ullathorne in order to demonstrate that the 'first living theologian now within the Roman Communion' was against the Infallibilists in 1870, and he describes the excommunication of Döllinger in 1871 as 'moral murder'.

[95] M. Arnold, *Dissent and Dogma* (1968), pp. 161–2.

[96] See *Letters of Newman* (1961–77), XXVII, xiii.

[97] The Old Catholics were a sect organized in German speaking countries to oppose Papal Infallibility. They believed that the modern Papacy was distorting Catholicism's true historic identity.

[98] The quotations from Gladstone in this paragraph are from J. H. Newman and W. E. Gladstone, *Newman and Gladstone* (1962), pp. 11, 14–15, 21.

32. 'A November Cracker', *Punch*, 67 (July–December 1874), 227.

At the heart of the document, as in so many of Gladstone's speeches and tracts, lies a question: 'England is entitled to ask, and to know, in what way the obedience required by the Pope and the Council of the Vatican is to be reconciled with the integrity of civil allegiance?'[99] In Gladstone's view, Manning's *Three Lectures on the Temporal Sovereignty of the Popes* prove that the two cannot be reconciled. 'Too commonly', he concludes, 'the spirit of the neophyte is expressed by the words which have become notorious: "a Catholic first, an Englishman afterwards".'

While *Punch* printed a number of squibs and cartoons at Gladstone's expense (illustration 32), no fewer than twenty serious responses to his expostulation were published, including Manning's official reply as

[99] Quotations in this paragraph are from ibid., pp. 43, 51, 56, 61.

Archbishop of Westminster, which often reads like a report from head office, headed by an executive summary:

1. That the Vatican Decrees have in no jot or tittle changed either the obligations or the conditions of civil allegiance.
2. That the civil allegiance of Catholics is as undivided as that of all Christians, and of all men who recognize a Divine or natural moral law.
3. That the civil allegiance of no man is unlimited; and therefore the civil allegiance of all men who believe in God, or are governed by conscience, is in that sense divided.
4. In this sense, and in no other, can it be said with truth that the civil allegiance of Catholics is divided. The civil allegiance of every Christian man in England is limited by conscience and the law of God; and the civil allegiance of Catholics is limited neither less nor more.[100]

Newman's open *Letter to the Duke of Norfolk* (1875) was widely regarded as the best response.[101] Gladstone's real difficulty, he suggested, was not the existence of a Pope, 'but of a Church, which is his aversion'.[102] As God has sovereignty, 'though He may be disobeyed or disowned, so has His Vicar upon earth; and further than this, since Catholic populations are found everywhere, he ever will be in fact lord of a vast empire'. In reply to the question whether Catholics can obey the Queen and yet obey the Pope, Newman coolly answers that he gives an 'absolute obedience to neither', and that if ever this 'double allegiance' pulled him in contrary ways, he would decide the issue 'on its own merits'.

Having further applied the Illative Sense to a defence of Papal Infallibility and a review of the troubled history of English Catholicism, up to the present, Newman concludes with a statement which illustrates his own dictum, that in 'these provinces of inquiry egotism is true modesty':

I say there is only one Oracle of God, the Holy Catholic Church and the Pope as her head. To her teaching I have ever desired all my thoughts, all my words to be conformed; to her judgment I submit what I have now written, what I have ever written, not only as regards its truth, but as to its prudence, its suitableness, and its expedience.

[100] H. E. Manning, *Vatican Decrees* (1875), p. 3. See also Manning's *Four Great Evils of the Day* (1871) and *True Story of the Vatican Council* (1877); and J. T. F. Mitford, Earl of Redesdale, *Infallible Church* (1875).

[101] William Froude described *A Letter addressed to His Grace the Duke of Norfolk on occasion of Mr. Gladstone's Recent Expostulation* as 'magnificent writing – it seems to me almost the best thing Newman ever wrote': *Letters of Newman* (1961–77), XXVII, xv.

[102] J. H. Newman, *Letter to the Duke of Norfolk* (1875), p. 100. Further quotations in this paragraph are from pp. 110, 125, 203.

In the *Grammar of Assent* Newman had stated that the 'fundamental dogma of the Catholic religion' was that 'the Church is the infallible oracle of truth', but had not individualized or personalized the oracle.[103] Now, through a masterly use of the word 'and' in the first sentence quoted here, he both denies the anti-Infallibilist position of Döllinger and the Old Catholics, affirming that the Church and its head are 'one', and affirms the primacy of the authority of the Church – 'Church' being the first and thus dominant term.

A year later, in 1876, Joseph Henry Shorthouse completed *John Inglesant*, the historical novel that he had been working on for the previous ten years, in a house only a 'stone's throw' from Newman's Oratory at Edgbaston.[104] Shorthouse, a wealthy semi-invalid who worked in the family chemical business and wrote in the evenings, was a High Anglican Platonist. He took a lively interest in the religious controversies of the day and corresponded with Arnold after hearing him lecture in 1871. When his privately printed novel was published by Macmillan in 1881, having been discovered by Mary ('Mrs Humphry') Ward, it became a literary phenomenon. Gladstone was photographed holding a copy and, having invited Shorthouse to a reception at Downing Street, beamed at him when he entered the room.

John Inglesant comes from a line of 'Papists at heart' who conformed to the Church of England during and after the English Reformation. Like Henry Esmond before him, he is tutored by a Jesuit, who grooms him as a pro-Catholic Anglican, so that he can serve as a mediator between Canterbury and Rome in the years leading up to the Civil War. (The Revd J. H. Smith pointed out that Shorthouse was 'originally a member of the Society of Friends, and also that *John Inglesant* is to a large extent a satire, and not altogether a fair one, on the principles and system of the Jesuits'.[105]) As a supporter of King Charles, Inglesant narrowly escapes execution, before travelling to Italy in search of his brother's murderer. At Rome he is impressed by the majesty of St Peter's, just being completed, but maintains his idealized view of the Church of England. The description of a Conclave in the Vatican can be read as implicit commentary upon the Papal Infallibility debate which unfolded as the novel was being written, both in its atmosphere – 'mysterious', 'confined', 'wearisome', 'darkened' – and its politics: 'The assembly was divided into different parties, each day by day

[103] See p. 264 above.
[104] S. Shorthouse, ed., *Life of Shorthouse* (1905), I, 74. Further quotations in this paragraph are from pp. 83–90, 113.
[105] Ibid., I, xv.

intriguing and manoeuvring, by every art of policy and every inducement of worldly interest, to add to the number of its adherents'.[106]

Having found and forgiven his brother's murderer, Inglesant – 'England's saint', who is mistaken for St George – stands trial for Molinism, being both a Benedictine novice and 'an accredited agent of the Queen Mother of England' – Henrietta Maria.[107] The novel ends with the hero offering his reflections upon Catholicism and Anglicanism, based on an unusually wide experience of both traditions in the seventeenth century. 'Mankind will do wrong', he says,

> if it allows to drop out of existence, merely because the position on which it stands seems to be illogical, an agency by which the devotional instincts of human nature are enabled to exist side by side with the rational. The English Church, as established by the law of England, offers the supernatural to all who choose to come. It is like the Divine Being Himself, whose sun shines alike on the evil and on the good. Upon the altars of the Church the divine presence hovers as surely, to those who believe it, as it does upon the splendid altars of Rome.[108]

He then waxes Arnoldian, arguing that Catholicism has reduced 'spiritual and abstract truth to hard and inadequate dogma', and 'terror and superstition are the invariable enemies of culture and progress'.

John Inglesant's most important statement on Rome and Canterbury offers telling commentary not only upon the subject of this chapter, but also of this book as a whole and Newman's central place in it:

> This is the supreme quarrel of all . . . This is not a dispute between sects and kingdoms; it is a conflict within a man's own nature – nay, between the noblest parts of man's nature arrayed against each other. On the one side obedience and faith, on the other, freedom and the reason. What can come of such a conflict as this but throes and agony?[109]

[106] J. H. Shorthouse, *John Inglesant* (1911), pp. 339–40.
[107] Ibid., p. 416. [108] Ibid., p. 442. [109] Ibid., p. 441.

CHAPTER 10

Painful epiphanies

> Those passion-flowers must blossom, to the last.
> Purple they bloom, the splendour of a King:
> Crimson they bleed, the sacrament of Death:
> About our thrones and pleasaunces they cling,
> Where guilty eyes read, what each blossom saith.
>
> Lionel Johnson, 'To a Passionist' (1895)[1]

I

Catholicism's much criticized emphasis upon pain and guilt found new and often extreme forms of cultural expression during the 1890s, in what has come to be known as 'the Decadence'. As in France, where Huysmans, Verlaine, Rimbaud and Villiers de L'Isle Adam were all converts to Catholicism, one of the defining features of the Decadence in England was the number of converts in its circle of writers and artists, including Frederick Rolfe, John Gray, Lionel Johnson, Ernest Dowson, Robert Ross, André Raffalovich, Aubrey Beardsley (shortly before his death), Oscar Wilde (on the very point of death) and Lord Alfred Douglas (in 1911).[2] As most of these figures were gay, it is not difficult to see how anti-Catholicism now became more explicitly anti-homosexual than in Charles Kingsley's day, when the homophobic Rector of Eversley referred darkly to the unmanliness of celibate Catholic priests.[3] For the converts themselves, Catholicism was both aesthetically attractive and overtly welcoming to the marginalized and the confused.

The 'Naughty Nineties' represent a discrete and clearly defined phase in English cultural history. In 1890 Oscar Wilde met Bosie (Lord Alfred Douglas) at Oxford and published *The Picture of Dorian Gray*, and John

[1] L. Johnson, *Poems* (1895), p. 70.
[2] See E. Hanson, *Decadence and Catholicism* (1997), pp. 10–14. I am indebted to Hanson's study.
[3] 'Newman was often described in feminine terms': ibid., p. 254.

Gray, who answered to the name of Dorian when among friends, was received into the Catholic Church, only to behave even more outrageously than before. On Gray's ordination to the priesthood in 1901, however, he distanced himself from his *fin de siècle* phase.[4] Ernest Dowson and Lionel Johnson both converted in 1891 and died in 1900 and 1902 respectively. Wilde and Beardsley both died in 1900, as the Victorian Age drew to a close with the death of Ruskin and that of the Queen herself the following year.

During the 1890s the era of the convert cardinals also came to an end and the revived Catholic Church in England came to maturity. In 1890 Newman was buried in the grave of his friend Ambrose St John at Rednal[5] and in 1892 Herbert Vaughan, a cradle Catholic from a large family of priests and nuns, succeeded Manning as Cardinal Archbishop of Westminster.[6] In 1893 Cardinal Vaughan joined his bishops for the 'Solemnity of the Renewal of England's Dedication to the Blessed Virgin Mary and to St Peter, Prince of the Apostles' and to hear the Revd T. E. Bridgett of the Congregation of the Most Holy Redeemer preach on 'England for Our Lady'. 'What we now ask', Bridgett said, 'and what the Catholic Church throughout England is about to ask, is that our dear Lady would enter again into full possession of her ancient Dowry.'[7]

Marian devotions had helped to soften the harshness of earlier Catholic teaching, while increased centralization under Pius IX had been accompanied by a greater emphasis upon the Incarnation than the Atonement and upon the God of love rather than of fear.[8] As the power of the Established Church of England and the idea of the confessional state declined, Catholicism underwent what has been described as a 'reflowering': 'It was the hierarchy's achievement that by the death of Manning, the Catholic community had become less inward looking, better educated, more socially aware and more loyally Roman that it was at any earlier period in the century.'[9] That relations between the Catholic Church and the British Establishment were much improved was evident in the Queen's gratified acceptance of an envoy from Pope Leo XIII at her Jubilee celebrations in

[4] Gray bought up copies of his *Silverpoints* in order to remove any reminder that he was once a decadent. See J. Adams, *Madder Music* (2000), p. 99.

[5] One of Newman's most brilliant converts, Gerard Manley Hopkins, died the previous year.

[6] Francis Thompson wrote to Katharine Tynan, 'We have exchanged a genius for a satisfied mediocrity': B. M. Boardman, *Between Heaven and Charing Cross* (1988), p. 178.

[7] T. E. Bridgett, *England for Our Lady* (1893), pp. 21–2.

[8] See V. A. McClelland, 'The Formative Years, 1850–92', in *From Without the Flaminian* Gate, ed. McClelland and M. Hodgetts (1999), p. 8.

[9] Ibid., pp. 13, 17.

1887 and her subsequent dispatch of the Duke of Norfolk to assure the Pope of her 'sincere friendship and unfeigned respect and esteem'.[10] Although Ritualism within the Church of England in the late 1890s caused some of the worst 'No Popery' riots since 'Papal Aggression', and some forms of Roman Catholic devotion were anathema to John Bull, Newman's widely read *Apologia* (1864) helped educated Protestants to see that Catholic priests could be warm hearted human beings, and even English gentlemen.[11]

Perhaps the most powerful symbol of English Catholicism's new status, prior to the building of Westminster Cathedral,[12] was Herbert Gribble's London Oratory church on the Brompton Road, consecrated on 25 April 1884, when Cardinal Manning preached to sixteen bishops and 250 priests, and completed in 1895.[13] The church's Italian Baroque was propagandist, proclaiming Ultramontanism and spurning Puginesque neo-Gothic, which for many Oratorians had overtones of the Oxford Anglicanism from which they had converted.[14] In its florid difference it contributed to a florid phase of English culture that was based upon Continental models, albeit mainly French rather than Italian in the literature; and both the Oratory church and the Decadence can be read as critiques of bourgeois Protestant Victorianism. In *Helbeck of Bannisdale* (1898), by Matthew Arnold's niece, Mary (Mrs Humphry Ward), Laura Fountain remembers a saying of her father's as to the Oratory's 'vicious Roman style' – the 'tomb of the Italian mind' (v, 1).

The Oratorians reached out to both the upper and lower classes. Their mission to the poor was an extension of 'family life', drawing Catholics who were lost in what Disraeli's Stephen Morley called the 'aggregation' of Victorian towns and cities into the 'community' that is the Catholic Church.[15] Faber's vocation in the late 1840s and 1850s, for example, was largely defined by his ministry to the poor, and he regarded the 'holy poverty' of the Catholic clergy as a form of Christian solidarity with their flocks.[16] The Irish-born population of England grew from 290,891 in 1841 to 519,959 in 1851 – three-quarters of the Roman Catholics in the country.[17]

[10] See O. Chadwick, *Victorian Church* (1966–70), II, 406–7.

[11] See ibid., II, 355, 408, 415.

[12] The main structure of this Byzantine basilica was begun in 1896 and used for the funeral of its founder, Cardinal Vaughan, in 1903. The cathedral was consecrated in 1910.

[13] See *The London Oratory Centenary, 1884–1984*, ed. M. Napier and A. Laing (1984), pp. 7, 44.

[14] R. O'Donnell, 'The Architecture of the London Oratory Churches', in ibid., pp. 21–47.

[15] On *Sybil; or, The Two Nations* (1845) see above, p. 168.

[16] Faber dedicated the first volumes of his 42 volume series of *The Saints and Servants of God* to the 'Secular Clergy of the Catholic Church in England, the Successors and Spiritual Children of Generations of Martyrs, who, by their Cheerfulness in Holy Poverty, their Diligence in Obscurity and under Oppression . . . have Preserved to their Country . . . the unfailing Light of the Catholic Faith': *Life of Saint Philip Neri* (1847), I, iii.

[17] See D. G. Paz, *Popular Anti-Catholicism* (1992), p. 51.

Some of the 'hapless Irish Celts' over whom the 'Romish priesthood' ruled, in Charles Kingsley's inflammatory phrase,[18] filled the first makeshift church of the London Oratory, bringing large numbers of fleas along with their spiritual needs.[19] Most of the eighteen thousand children running wild in the streets of London were Irish. In promoting the Oratory's Catholic Ragged School, Faber expressed his concern that Catholic children were being 'tempted to renounce their religion by the rewards held out to their parents and themselves if they attend the protestant Ragged Schools'.[20]

At the same time, the gentlemanly Oratorians had a particular appeal to the upper classes, and by the end of the century their splendid new church was attracting many of the wealthier members of London's Catholic community. The Oratory's best-known society priest was Father Sebastian Bowden (1836–1920), who drew many converts into the Catholic Church and who served as confessor and confidant to several key figures of the Decadence. An old Etonian and former Guards officer, Henry George ('Sebastian') Bowden was the nephew of Mrs Elizabeth Bowden, the convert widow of J. W. Bowden, Newman's Oxford friend. This soldier-priest was Provost of the London Oratory from 1889 to 1892 and again from 1904 to 1907. Maurice Baring described him as a 'sensible Conservative, a patriot, a fine example of an English gentleman in mind and appearance; a prince of courtesy, and a saint', and regarded his acquaintance with him, and the friendship and sympathy he gave him, as the 'greatest privilege bestowed on [him] by Providence'.[21]

Insight into both the author and the Oratory is provided by Father Sebastian's guidebook of 1893, in which he draws attention to notices in the nave inviting subscriptions 'for the completion of the Church and its decorations'.[22] The façade, he explains, 'now in course of erection, is much in need of funds. The outer dome is not begun, and the nave, transepts, and interior dome are still in their unrelieved whitewash.' Nevertheless, 'thousands of persons' visit the Oratory and 'much inconvenience is caused' by the disregard of a notice asking for silence.[23] 'Visitors should remember', he continues, 'that they are within the precincts of a building which is to a Catholic, by reason of the Eucharistic Presence the most sacred spot on earth.' His aim is to provide not only a description of the church but

[18] 'What, then, does Dr. Newman Mean?' (1864), in J. H. Newman, *Apologia* (1967), p. 384.

[19] Faber wrote in 1849: 'Brother Chad catches them by handfulls [sic]. The Irish are swamping us; they are rude and unruly and after many complaints the Catholic tradesmen are leaving us.' R. Chapman, *Father Faber* (1961), p. 233.

[20] *Faber, Poet and Priest* (1974), p. 225.

[21] M. Baring, *Puppet Show of Memory* (1922), p. 396. See also H. G. S. Bowden, *Spiritual Teaching* (1921).

[22] H. G. S. Bowden, ed., *Guide to the Oratory* (1893), p. 106.

[23] Ibid., p. 5.

also 'such information regarding Catholic doctrines and devotions as is constantly demanded by visitors'.[24] For Protestant readers, this was perhaps the first time that they had been offered such clear information by a Catholic priest.

Father Sebastian's description of the Mass would have been particularly arresting, linking as it does the guilt of sinners with the need to repeat the sacrifice of Calvary, and thus directly contradicting all shades of Protestant teaching and practice:

> . . . all the sacrifices of earth's best gifts, and even those of Israel, could not offer to God the worship which was His due, nor satisfy for the guilt contracted by sin. Both these ends were, however, accomplished by the sacrifice of the Cross, and in the Mass, by which, under the sacramental species of bread and wine, this sacrifice is mystically repeated, Jesus Christ is offered again as the Lamb slain in propitiation for our sins, and in adoration of the supreme Majesty of the One Eternal God.
>
> In the Mass, as on the Cross, Christ is both Priest and Victim; the human celebrant acts in His person, and consecrates in His name.[25]

One of the many sinners who sought Father Sebastian's counsel at the Oratory House[26] was Oscar Wilde, on 14 April 1878. Wilde was then an undergraduate at Oxford and was suffering from syphilis. Bowden wrote to him the following day, but on the Thursday, when he was due to return for further discussion, and possibly reception into the Catholic Church, Wilde backed out, sending a large bunch of lilies in his stead.[27] Much later, during Wilde's trial in 1895, Bosie's Protestant mother turned to Father Sebastian for help in finding a steady friend for her son – More Adey, a thirty-eight-year-old Catholic and a 'discreet' homosexual.[28] Two years later, on his release from prison, Wilde applied to the Jesuits at Farm Street asking for a six-month retreat. When he received a reply saying that retreats could not be arranged at such short notice, he burst into tears. His travelling bags for the Continent were initialled S. M., for Sebastian Melmoth, a pseudonym which honoured his favourite martyr and his great-uncle's solitary wanderer.[29]

Meanwhile Ernest Dowson had been received into the Catholic Church at the London Oratory, on 25 September 1891, after which he attended

[24] Ibid., p. v.

[25] Ibid., p. 40. Such 'priestcraft' would have seemed as alien to Protestant visitors as some of the more recent Ultramontane developments, such as the devotion of the Sacred Heart, and the exposure, on Tuesdays, of a wax figure of St Philip Neri, 'in a recumbent posture, clad in sacred vestments and containing a relic of his body' (pp. 25–6).

[26] The Oratory House was completed in 1854.

[27] See R. Ellmann, *Oscar Wilde* (1987), pp. 90–1.

[28] See R. Croft-Cooke, *Bosie* (1963), pp. 129–30.

[29] See R. Ellmann, *Oscar Wilde* (1987), pp. 495, 491.

church frequently, if not regularly,[30] and Aubrey Beardsley had been inspired by the building, declaring it to be 'the only place in London where one can forget that it is Sunday'.[31] One of the illustrations for his semi-pornographic novella, *Under the Hill*, published in *The Savoy* in 1896, was entitled *The Ascension of Saint Rose of Lima* (illustration 33). Taking the famous Madonna of Victory in the Oratory as his model for the Virgin Mary, Beardsley ambiguously combines a playful homosexuality with a desire for spiritual transcendence – two poles of his own experience.[32]

I will be discussing the Decadents in more detail later in this chapter, where I examine two features of Catholicism that are highlighted by Father Sebastian in his guidebook to the Oratory. First, he informs the reader that the plans for the church, drawn up in 1880, had included space for no fewer than twenty confessionals, and that confessions are heard whenever the Church is open.[33] In section II I focus upon some of the painful epiphanies that feature in the writings of the Decadence, in which shame and guilt are related to the pain and humiliation of Christ's Passion. Secondly, Father Sebastian asks the reader to remember that Catholicism is 'essentially a ceremonial religion': 'We are not pure spirits, but spirit and matter, or body and soul; and we naturally therefore express by outward acts our interior thought or feeling.'[34] In section III I show how aspects of Decadent aesthetics are informed by a sense of sin and a longing for purity, both of which are dramatized for some writers in Catholic ceremonial.

II

Meditations upon Christ's wounds are familiar aspects of Catholic devotion. On a Friday night in the Oratory, the procession, moving from the first to the second Station of the Cross, would sing:

Sancta Mater, istud agas,	Holy Mother, grant my prayer,
Crucifixi fige plagas	And make me in my heart to bear
Cordi meo valide.	The wounds of Jesus crucified.[35]

[30] See J. Adams, *Madder Music* (2000), p. 59. Adams describes Dowson as the 'archetypal decadent poet, combining all the characteristics which made up the decadent temperament: strange delights, sexual promiscuity and wild entertainments co-existing with classical scholarship and devotion to the Catholic Church to which Dowson converted as so many decadents did' (ix–x).

[31] A. Beardsley, *Under the Hill* (1904), p. 63. It was specifically Protestant Sundays in London that Dickens had satirized as dark and dreary in *Little Dorrit* (1855–7), chapter 3.

[32] See C. Snodgrass, *Aubrey Beardsley* (1995), p. 24. Snodgrass points out that the Oratory church also contained a statue of St Rose of Lima, depicted in a heavily pleated dress and veil.

[33] H. G. S. Bowden, ed., *Guide to the Oratory* (1893), pp. 3, 9. [34] Ibid., p. 6.

[35] Anon., *Oratory Night Services* (1909), p. 32.

33 Aubrey Beardsley, *The Ascension of Saint Rose of Lima*, *The Savoy*, 2 (April 1896), 189.

Christ's wounds are healing and saving:

> Strength and protection may His Passion be,
> O blessed Jesus, hear and answer me;
> Deep in Thy Wounds, Lord, hide and shelter me,
> So shall I never, never part from Thee.[36]

They are therefore beautiful, as Lionel Johnson's evocation of embroidered passion flowers, quoted in the epigraph to this chapter, suggests: 'Purple

[36] No. 42, 'Anima Christi', in F. W. Faber, *et al.*, *Oratory Hymns* (1909), p. 45.

they bloom, the splendour of a King: / Crimson they bleed, the sacrament of Death.'

John Gray draws upon this rich tradition in *Silverpoints* (1893). In 'Mon Dieu m'a dit: . . .' he writes:

God has spoken: Love me,
son, thou must; Oh see
My broken side; my heart,
its rays refulgent shine;
My feet, insulted, stabbed,
that Mary bathes with brine
Of bitter tears; my sad arms,
helpless, son, for thee;
With thy sins heavy; and my hands;
thou seest the rod;
Thou seest the nails, the sponge,
the gall; and all my pain
Must teach thee love, amidst a world
where flesh doth reign,
My flesh alone, my blood,
my voice, the voice of God.
Say, have I not loved thee,
loved thee to death,
O brother in my Father,
in the Spirit son?
Say, as the word is written,
is my work not done?
Thy deepest woe have I not sobbed
with struggling breath?
Has not thy sweat of anguished nights
from all my pores in pain
Of blood dripped, piteous friend,
who seekest me in vain? [37]

In 'The Crucifix', which is dedicated to Ernest Dowson, Gray moves beyond reflection upon shame and guilt ('With thy sins heavy'), expressed through the 'I / thou' of colloquy with God, in order to focus explicitly upon the beauty of an artistic *representation* of Christ, through which 'one' almost sees the agony of the cross:

A gothic church. At one end of an aisle,
Against a wall where mystic sunbeams smile
Through painted windows, orange, blue, and gold,

[37] J. Gray, *Silverpoints* (1893), p. 27.

The Christ's unutterable charm behold.
Upon the cross, adorned with gold and green,
Long, fluted golden tongues of sombre sheen,
Like four flames joined in one, around the head
And by the outstretched arms, their glory spread.
The statue is of wood; of natural size;
Tinted; one almost sees before one's eyes
The last convulsion of the lingering breath.
'Behold the man!' Robust and frail. Beneath
That breast indeed might throb the Sacred Heart.
And from the lips, so holily dispart,
The dying murmur breathes 'Forgive! Forgive!'
O wide-stretched arms! 'I perish, let them live.'
Under the torture of the thorny crown,
The loving pallor of the brow looks down
On human blindness, on the toiler's woes;
The while, to overturn Despair's repose,
And urge to Hope and Love, as Faith demands,
Bleed, bleed the feet, the broken side, the hands.
A poet, painter, Christian, – it was a friend
Of mine – his attributes most fitly blend –
Who saw this marvel, made an exquisite
Copy; and, knowing how I worshipped it,
Forgot it, in my room, by accident.
I write these verses in acknowledgment.[38]

Gray scandalously invites the late Victorian reader to behold, not the 'offence', or 'scandal' of the cross (Galatians 5.11), but the 'unutterable charm' of the Christ, with its suggestion of lovable attractiveness and graceful manners; and, in an age of Protestant anxiety concerning Catholic idolatry, he confesses to worshipping a copy of a wooden statue.

Whereas Gray celebrates the way in which a beautiful crucifix can reveal the 'charm' of Christ crucified, the story that he inspired, Wilde's *The Picture of Dorian Gray* (1890), turns upon the idea of beauty as a form of concealment or burial of sin and shame. The portrait of Dorian becomes a 'monstrous and loathsome thing', as it externalizes the corruption within him (8). The covering that he chooses for the picture in the old school-room is associated with Continental Catholicism:

As the door closed, Dorian put the key in his pocket, and looked round the room. His eye fell on a large purple satin coverlet heavily embroidered with gold, a splendid piece of late seventeenth-century Venetian work that his grandfather

[38] Ibid., p. 23.

> had found in a convent near Bologna. Yes, that would serve to wrap the dreadful thing in. It had perhaps served often as a pall for the dead. Now it was to hide something that had a corruption of its own, worse than the corruption of death itself – something that would breed horrors and yet would never die. What the worm was to the corpse, his sins would be to the painted image on the canvas. They would mar its beauty, and eat away its grace. They would defile it, and make it shameful. And yet the thing would still live on. It would be always alive. (10)

It is in hell that 'their worm dieth not, and the fire is not quenched' (Mark 10.44, 46, 48) and Dorian's is a living hell. Whereas the Catholic pall symbolizes sacramental purification of that which will turn to dust, Dorian's 'coverlet' conceals corruption that will never die; and whereas the penitents in church bring their sins into the light of grace by 'whispering through the worn grating the true story of their lives', Dorian can only 'look with wonder at the black confessionals, and long to sit in the dim shadow of one of them' (11).

The word 'shame' ('They would defile it, and make it shameful') had a special weight for Wilde, and for other Decadent homosexuals in the 1890s, as it became 'virtually a synonym for sexual acts between men'.[39] Like other Decadents, Wilde could find 'grace in the depths of shame and sainthood in the heart of the sinner',[40] and he expressed this most powerfully in *The Ballad of Reading Gaol* (1898). Whereas *Dorian Gray* examines the corruption that can result from the pursuit of pleasure, *Reading Gaol* describes the effects of physical and spiritual pain, and affirms the possibility of redemption. In his other prison narrative, *De Profundis* (1905), Wilde declared that 'Pain, unlike Pleasure, wears no mask.'[41] Christ, we recall, was naked upon the cross.

Wilde's opening description in *Reading Gaol* of the guardsman dressed in 'shabby grey', who walks amongst the 'Trial Men', associates him with the sacrament of the Eucharist, and thus with the Last Supper and the Passion:

> He did not wear his scarlet coat,
>   For blood and wine are red,
> And blood and wine were on his hands
>   When they found him with the dead,
> The poor dead woman whom he loved,
>   And murdered in her bed.[42]

[39] E. Hanson, *Decadence and Catholicism* (1997), p. 86.

[40] Ibid., p. 7. Hanson adds: '*Shame* is only the most polished gem in a constellation of synonyms, each with its own nuance – words like *degradation, humiliation, disgrace, blame, guilt, remorse, ruin, pariah, outcast*', p. 99.

[41] O. Wilde, *Complete Works* (1966), p. 920.

[42] Ibid., p. 843. Further quotations in this paragraph are from pp. 845, 850.

As in *Dorian Gray*, Wilde creates a hell on earth. The narrator walks 'with other souls in pain, / Within another ring', and from this Dantean hell the guardsman can catch glimpses of heaven – 'that little tent of blue / Which prisoners call the sky'. The raw ugliness of the 'hideous' execution shed and its paraphernalia is explicitly associated with Christ's Passion in the reference to the 'kiss of Caiaphas'. The night before the execution, prisoners kneel in prayer: 'And bitter wine upon a sponge / Was the savour of Remorse.'

Whereas Dorian Gray's sins are covered by an ornate Venetian pall, the guardsman's pall is quicklime:

> Deep down below a prison-yard,
>   Naked for greater shame,
> He lies, with fetters on each foot,
>   Wrapt in a sheet of flame![43]

Yet it is in the very depths of shame that grace blossoms. The guardsman's grave will have nothing sown in it for three years:

> They think a murderer's heart would taint
>   Each simple seed they sow.
> It is not true! God's kindly earth
>   Is kindlier than men know,
> And the red rose would but blow more red,
>   The white rose whiter blow.
> Out of his mouth a red, red rose!
>   Out of his heart a white!
> For who can say by what strange way,
>   Christ bring His will to light,
> Since the barren staff the pilgrim bore
>   Bloomed in the great Pope's sight?

In the original legend, Tannhäuser is supported by Our Lady but is refused papal absolution until the budding of his staff after three days – a sign of divine grace. Swinburne's and Beardsley's treatments of the legend, in 'Laus Veneris' (1866) and *Under the Hill* (1896), are deeply ambiguous,[44] and Coleridge's Catholic references in 'The Ancient Mariner' (1798) are an aspect of that literary ballad's medievalism. The spirituality of Wilde's poem is, however, unambiguously Catholic. The guardsman lacks a 'requiem that

[43] Ibid., p. 855.
[44] The legend was also adapted by Wagner, in his opera of 1843–4. Wilde's description of the gaol as a hell on earth probably owed something to Swinburne. See M. Wheeler, *Death and the Future Life* (1990), pp. 210–14.

might have brought / Rest to his startled soul', and the other prisoners lack the comfort of a Catholic tradition of flower symbolism:

> So never will wine-red rose or white,
> Petal by petal, fall
> On that stretch of mud and sand that lies
> By the hideous prison-wall,
> To tell the men who tramp the yard
> That God's Son died for all.[45]

'Pardon', a loaded term in the context of a prison, also has Catholic overtones:

> Ah! happy they whose hearts can break
> And peace of pardon win!
> How else may man make straight his plan
> And cleanse his soul from Sin?
> How else but through a broken heart
> May Lord Christ enter in?[46]

The guardsman shares a hope of heaven with the penitent thief who endured the shame of the cross alongside Christ:

> And he of the swollen purple throat,
> And the stark and staring eyes,
> Waits for the holy hands that took
> The Thief to Paradise;
> And a broken and a contrite heart
> The Lord will not despise [Psalm 51.17].

In the light of such passages from *Reading Gaol*, it is easy to see why Catholic literary historians have tried to reclaim Wilde from those critics who present him as a hedonistic gay martyr. His imprisonment was described in the 1930s, for example, as a 'personal blessing in the effect it had both on his subsequent life and upon his art'; and more recently the central theme of his late works was said to have 'precious little to do with the role of "sexual liberator" which posterity has thrust upon him, and everything to do with the Christian penitent seeking forgiveness'.[47] Like the Man of Sorrows, the Wilde of *De Profundis* and *Reading Gaol* was acquainted with grief (Isaiah 53.3). He experienced something of the

[45] O. Wilde, *Complete Works* (1966), p. 856.
[46] Ibid., p. 859.
[47] C. Alexander S. J., *Catholic Literary Revival* (1935), p. 98; J. Pearce, 'The Catholic Literary Revival', in *From Without the Flaminian Gate*, ed. V. A. McClelland and M. Hodgetts (1999), p. 303.

melancholy and loneliness associated with the Passion, as understood in Catholic tradition.

Although Lionel Johnson's credentials as a Decadent have been questioned, he has been described as 'not untouched by that strange sickness, the *maladie de siècle*'.[48] Johnson's Romantic and Catholic sensibility is attuned to the melancholy associated with those who worship the Man of Sorrows. He detects this strain of melancholy in Newman, for example –

> The freedom of the living dead;
> The service of a living pain:
> He chose between them, bowed his head,
> And counted sorrow, gain . . .

and in the Irish:

> Long Irish melancholy of lament!
> Voice of the sorrow, that is on the sea:
> Voice of that ancient mourning music sent
> From Rama childless: the world wails in thee.[49]

Whereas in Gray's 'The Crucifix' the '*mystic sunbeams smile / Through painted windows, orange, blue, and gold*',[50] in Johnson's 'The Church of a Dream' stained glass, exposed to the elements, signifies the melancholy of the religious life:

> Sadly the dead leaves rustle in the whistling wind,
> Around the weather-worn, gray church, low down the vale:
> The Saints in golden vesture shake before the gale;
> The glorious windows shake, where still they dwell enshrined;
> Old Saints, by long dead, shrivelled hands, long since designed:
> There still, although the world autumnal be, and pale,
> Still in their golden vesture the old saints prevail;
> Alone with Christ, desolate else, left by mankind.
> Only one ancient Priest offers the Sacrifice,
> Murmuring holy Latin immemorial:
> Swaying with tremulous hands the old censer full of spice,
> In gray, sweet incense clouds; blue, sweet clouds mystical:
> To him, in place of men, for he is old, suffice
> Melancholy remembrances and vesperal.[51]

For Johnson, melancholy is associated with loneliness: the old saints portrayed in the windows are 'alone with Christ', and inside the church 'Only

[48] C. Alexander S. J., *Catholic Literary Revival* (1935), p. 137.
[49] L. Johnson, 'In Falmouth Harbour' (1887) and 'To weep Irish' (1893), in *Poems* (1895), pp. 9, 41.
[50] See pp. 280–1 above.
[51] Ibid., pp. 84–5. The poem was written in 1888 and dedicated to Bernhard Berenson.

one ancient Priest offers the Sacrifice.' His poem entitled 'The Dark Angel' ends:

> Do what thou wilt, thou shalt not so,
> Dark Angel! triumph over me:
> *Lonely, unto the Lone I go;*
> *Divine, to the Divinity.*[52]

At the heart of Catholic teaching on the individual Christian journey lies the recognition that the soul is finally alone and will go to judgment alone. 'How entirely *alone* we must be at death, our soul on our lips', Father Sebastian Bowden of the Oratory counsels: 'to go before God and into Eternity with no human help'.[53] Passiontide, he teaches, is to be used for one purpose: 'to make you realize Our Lord's Death; and use that Death for yourself: apply it to yourself immediately. Think what your own death will be. When you die you will be able to speak very little: you will hardly hear or see anyone: in darkness and *absolute solitude* your soul will have to pass to God.' In life, as in death, we are alone, and 'feelings of loneliness', Father Sebastian believes, '*cannot be altered*': 'We are made so, and so must remain.'

For some writers of the Decadence it was the experience of spiritual and intellectual loneliness, epitomized in Gethsemane, that made monasticism so fascinating. Wilde, frustrated in his desire to go into retreat after his release from prison in 1897, responded approvingly the following year when he heard that Huysmans had entered a monastery.[54] Ernest Dowson's first religious poem was entitled 'Nuns of the Perpetual Adoration' (1891).[55] When Dowson visited a Carthusian monastery, the same year, he was impressed by the fact that the monks had separate cells where they even took their meals alone, and had only one hour a week for conversation together:[56]

> Through what long heaviness, assayed in what strange fire,
> Have these white monks been brought into the way of peace,
> Despising the world's wisdom and the world's desire,
> Which from the body of this death bring no release?
>
> ('Carthusians')[57]

52 Ibid., p. 69. The poem is dated 1893.

53 H. G. S. Bowden, *Spiritual Teaching* (1921), p. 63. Further quotations in this paragraph are from pp. 21, 88. (Cf. also pp. 34, 106, and J. H. Newman, *Dream of Gerontius*, in *Verses* (1869), pp. 356–61. In *Helbeck of Bannisdale*, Laura Fountain's letter explaining her suicide includes the sentence, 'How horrible it is that death is so *lonely!*' (v, 4). She cannot bring herself to convert to Catholicism.)

54 See V. A. McClelland and M. Hodgetts, eds., *From Without the Flaminian Gate* (1999), p. 304.

55 E. Dowson, *Poetical Works* (1967), pp. 36–7.

56 See J. Adams, *Madder Music* (2000), p. 51.

57 E. Dowson, *Poetical Works* (1967), p. 97. The poem was first published in *Decorations* (1899).

More than other orders, the Carthusians are called to 'dwell alone with Christ':

> It was not theirs with Dominic to preach God's holy wrath,
> They were too stern to bear sweet Francis' gentle sway;
> Theirs was a higher calling and a steeper path,
> To dwell alone with Christ, to meditate and pray.

Paradoxically, the way of the 'dolorous Cross' is sweet:

> O beatific life! Who is there shall gainsay,
> Your great refusal's victory, your little loss,
> Deserting vanity for the more perfect way,
> The sweeter service of the most dolorous Cross.[58]

In all the works discussed in this section there comes a moment of epiphany, when the writer recognizes in some aspect of the Passion both the type and the potential cure of his own spiritual pain, or loneliness, or melancholy. Poised in the 1890s between Pater and Joyce, each has 'to pay a heavy price in suffering, to risk his immortal soul, and to be alone'.[59]

III

Decadent excess, in the form of illicit sex, or drink and drugs, or extragavant dress and behaviour, is often accompanied by a strong sense of sin and a longing for purity of the kind that Dowson recognized in the Carthusians – 'The sweeter service of the most dolorous Cross'. These contrary impulses can find expression and resolution in the drama of the Mass, where all the senses, including touch and smell, are engaged in the worshipper's response to what Johnson describes as the 'gray, sweet incense clouds; blue, sweet clouds mystical'. For several writers of the Decadence – a concerted assault upon bourgeois Victorian culture, with its Protestant emphasis upon words and the Word – the sensuousness of Catholic ceremonial exercised a strong Romantic appeal.

During Wilde's time at Trinity College, Dublin, which felt threatened in the early 1870s by the rise of the Catholic University founded by Newman, he was attracted to the forms of Catholicism while maintaining his Protestant identity – a position which has been related to other 'self-contradictory inclinations' in him.[60] Then at Oxford he filled his college rooms with Catholic artifacts and was tempted to follow Hunter Blair into the Roman Church, after his friend attended the ceremony at which Manning was

[58] Ibid., p. 98. [59] F. Kermode, *Romantic Image* (2002), p. 4.
[60] See R. Ellmann, *Oscar Wilde* (1987), p. 32.

created Cardinal in Rome, in 1875. According to Frank Harris, Wilde liked the 'perfume of belief'.[61]

For Dorian Gray, the attraction of Catholic ritual is coolly visual:

It was rumoured of him once that he was about to join the Roman Catholic communion; and certainly the Roman ritual had always a great attraction to him. The daily sacrifice, more awful really than all the sacrifices of the antique world, stirred him as much by its superb rejection of the evidence of the senses as by the primitive simplicity of its elements and the eternal pathos of the human tragedy that it sought to symbolise. He loved to kneel down on the cold marble pavement, and watch the priest, in his stiff flowered vestment, slowly and with white hands moving aside the veil of the tabernacle, or raising aloft the jewelled lantern-shaped monstrance with that pallid wafer that at times, one would fain think, is indeed the "*panis cælestis*," the bread of angels, or, robed in the garments of the Passion of Christ, breaking the Host into the chalice, and smiting his breast for his sins. The fuming censers, that the grave boys, in their lace and scarlet, tossed into the air like great gilt flowers, had their subtle fascination for him.(11)

This visual emphasis, and Dorian's conception of the Catholic Church as a body that has wasted away by fasting and flagellation, prepare us for the gestures of concealment and covering in the story that I discussed earlier:

He had a special passion, also, for ecclesiastical vestments, as indeed he had for everything connected with the service of the Church. In the long cedar chests that lined the west gallery of his house he had stored away many rare and beautiful specimens of what is really the raiment of the Bride of Christ, who must wear purple and jewels and fine linen [Exodus 26.1], that she may hide the pallid macerated body that is worn by the suffering that she seeks for, and wounded by self-inflicted pain. (11)

In contrast, all the senses are at work in Ernest Dowson's evocation of Benediction, the most popular rite in late Victorian Catholicism,[62] in 'Benedictio Domini', which first appeared in *Verses* (1896) and was dedicated to Selwyn Image:

Without, the sullen noises of the street!
  The voice of London, inarticulate,
Hoarse and blaspheming, surges in to meet
  The silent blessing of the Immaculate.

Dark is the church, and dim the worshippers,
  Hushed with bowed heads as though by some old spell,
While through the incense-laden air there stirs
  The admonition of a silver bell.

[61] See ibid., pp. 51–2. [62] See M. Heimann, *Catholic Devotion* (1995), p. 46.

Dark is the church, save where the altar stands,
Dressed like a bride, illustrious with light,
Where one old priest exalts with tremulous hands
The one true solace of man's fallen plight.

Strange silence here: without, the sounding street
Heralds the world's swift passage to the fire:
O Benediction, perfect and complete!
When shall men cease to suffer and desire?[63]

Catholicism's role *contra mundum* is explored here through a series of oppositions – between the noises of the street 'without', in both senses, and the silence of the blessing; the surging movement of the street and the spellbound stillness of the bowed heads in the church; the profanity of the 'blaspheming' outside and the sacredness of the 'Immaculate' inside; the inarticulateness of the 'voice of London' and the 'admonition of a silver bell'; the darkness of the congregation and the light of the altar; the elevation of the Host and the fallen plight of man – culminating in that of the closing lines: 'O Benediction, perfect and complete! / When shall men cease to suffer and desire.'

Francis Thompson was friendly with the Decadents, but was himself extreme only in his down-and-out period, when he literally lay in the gutter but looked at the stars. In contrast to Dowson and his church interior, Thompson, in his 'Orient Ode' published the following year, maps the mystery of Benediction onto nature and the glories of the rising and setting sun:

Lo, in the sanctuaried East,
Day, a dedicated priest
In all his robes pontifical exprest,
Lifteth slowly, lifteth sweetly,
From out its Orient tabernacle drawn,
Yon orbèd sacrament confest
Which sprinkles benediction through the dawn;
And when the grave procession's ceased,
The earth with due illustrious rite
Blessed, – ere the frail fingers featly
Of twilight, violet-cassocked acolyte,
His sacerdotal stoles unvest –
Sets, for high close of the mysterious feast,
The sun in august exposition meetly
Within the flaming monstrance of the West.[64]

[63] E. Dowson, *Poetical Works* (1967), p. 48. [64] F. Thompson, *Works* (1913), II, 21.

So impressive were Thompson's *New Poems* (1897), which included 'Orient Ode', that Protestant reviewers accused Thompson of being 'deliberately put forward by the Catholic intelligentsia'.[65]

It is also characteristic of the Decadence that the artistic possibilities of ugliness should be exploited as well as those of beauty, including in the area of Catholic ceremonial. Ironically, it was the very crudity of the Mass celebrated by an untidy priest with a congregation of six peasant women in a scruffy church in the French countryside that persuaded the twenty-three-year-old John Gray that Catholicism was the one true religion.[66] In contrast, the novelist George Moore, an Irish cradle Catholic who publicly admitted to a loathing of both the country and religion of his birth, found confirmation of his scepticism in such scenes, which he drew upon in his provocatively naturalistic fiction.[67]

Moore could not think of Ireland without a 'sensation akin to nausea'. He loved the English, however, with a love that was 'foolish – mad, limitless': 'England is Protestantism, Protestantism is England. Protestantism is strong, clean, and westernly, Catholicism is eunuch-like, dirty, and Oriental.' In his private life, Moore would attend an English parish church on a Sunday, more for what it represented than for Anglican worship. In his fiction, however, it was Irish Catholicism that inspired him. In *A Drama in Muslin: A Realistic Novel* (1886), for example, Moore describes a peasant Mass in a barn of a church from the point of view of well-bred young ladies. The priest is

> a large fat man, whose new, thick-soled boots creaked terribly as he ascended the steps of the altar. He was preceded by two boys dressed in white and black surplices. They rang little brass bells furiously, and immediately a great trampling of feet was heard. The peasants came, coughing and grunting with monotonous, animal-like voices; and the sour odour of cabin-smoked frieze arose, and was almost visible in the great beams of light that poured through the eastern windows; and whiffs of unclean leather, mingled with the smell of a sick child, flaccid as the prayer of the mother who grovelled, beating her breast, before the third Station of the Cross; and Olive and May, exchanging looks of disgust, drew forth cambric pocket-handerchiefs, and in unison the perfumes of white rose and eau d'opoponax evaporated softly.[68]

Whereas Wilde is attracted to the 'perfume of belief' and Dowson delights in the sweet 'incense-laden air' of an urban church, Moore reverses the

[65] See B. M. Boardman, *Between Heaven and Charing Cross* (1988), p. 272.
[66] See B. Sewell, *In the Dorian Mode* (1983), p. 11.
[67] Moore published this statement in *Time* in October 1887. See A. Frazier, *George Moore* (2000), p. 162. Frazier claims that 'no memorial exists of any period in which George Moore believed in God' (p. 18).
[68] G. Moore, *Drama in Muslin* (1886), p. 70. (Chapter 4.)

olfactory categories, substituting man-made fragrancies for the odour of sanctity and cambric pocket-handerchiefs for the 'fine linen' of the altar.

The faith of the heroine, Alice, has burst 'like a wind-filled bladder under the point of a pin' after reading Shelley and Darwin, and her response to the Mass is more cerebral than physical:

> The mumbled Latin, the by-play with the wine and water, the mumming of the uplifted hands, were so appallingly trivial, and, worse still, all realisation of the idea seemed impossible to the mind of the congregation. Passing by, without scorn, the belief that the white wafer the priest held above his head, in this lonely Irish chapel, was the Creator of the twenty millions of suns in the Milky Way, she mused on the faith as exhibited by those who came to worship, and that which would have, which must have, inspired them, were Christianity now, as it once was, a burning, a vital force in the world.[69]

Moore challenges the assumption that, for the peasantry, the Mass provides a strangely beautiful if half understood interlude in the weekly round of sweaty toil, by emphasizing what he regards as the ugliness and absurdity of the rite, as celebrated in rural Ireland. For other Decadents, ugliness is associated with the 'Tempter'. Whereas Francis Thompson is pursued by the Hound of Heaven,[70] Lionel Johnson is 'reached' by the Dark Angel:

> Dark Angel, with thine aching lust
> To rid the world of penitence:
> Malicious Angel, who still dost
> My soul such subtile violence!
>
> Because of thee, no thought, no thing,
> Abides for me undesecrate:
> Dark Angel, ever on the wing,
> Who never reachest me too late![71]

The Dark Angel's method of perverting goodness is to destroy beauty through desire, thus desecrating what is holy:

> When music sounds, then changest thou
> Its silvery to a sultry fire:
> Nor will thine envious heart allow
> Delight untortured by desire.
>
> Through thee, the gracious Muses turn
> To Furies, O mine Enemy!

[69] Ibid.

[70] 'I fled Him, down the nights and down the days': 'The Hound of Heaven', in F. Thompson, *Works* (1913), I, 107–13.

[71] L. Johnson, *Poems* (1895), p. 67. 'The Dark Angel' is dated 1893.

And all the things of beauty burn
With flames of evil ecstasy.

More disturbing still is the Dark Angel's destruction of *natural* beauty:

Within the breath of autumn woods,
Within the winter silences:
Thy venomous spirit stirs and broods,
O Master of impieties!

The ardour of red flame is thine,
And thine the steely soul of ice:
Thou poisonest the fair design
Of nature, with unfair device.

Apples of ashes, golden bright;
Waters of bitterness, how sweet!
O banquet of a foul delight,
Prepared by thee, dark Paraclete![72]

And the poem ends with Johnson's version of a vision of judgment, in which he articulates the Catholic's fear of damnation and hope of salvation:

I fight thee, in the Holy Name!
Yet, what thou dost, is what God saith:
Tempter! should I escape thy flame,
Thou wilt have helped my soul from Death:

The second Death, that never dies,
That cannot die, when time is dead:
Live Death, wherein the lost soul cries,
Eternally uncomforted.

Dark Angel, with thine aching lust!
Of two defeats, of two despairs:
Less dread, a change to drifting dust,
Than thine eternity of cares.

Do what thou wilt, thou shalt not so,
Dark Angel! triumph over me:
Lonely, unto the Lone I go;
Divine, to the Divinity.[73]

The dramatic turn in the final stanza is achieved through an assertion of the single self or soul, which may fall into sin in this world, but is created in the image of God and intended, through grace, to return to

[72] Ibid., p. 68. [73] Ibid., p. 69.

God. Like Hopkins, Johnson regards the physical beauty and vigour of male youth as signs of grace. In place of the autumnal melancholy and loneliness associated with age in 'The Dream of a Church', his poem 'A Dream of Youth', written a year later and dedicated to Lord Alfred Douglas, attempts to sanctify the homoerotic attractions of the acolytes in a Catholic procession that takes place in the freshness of an early morning of spring or summer:

> With faces bright, as ruddy corn,
> Touched by the sunlight of the morn;
> With rippling hair; and gleaming eyes,
> Wherein a sea of passion lies;
> Hair waving back, and eyes that gleam
> With deep delight of dream on dream;
> With full lips, curving into song;
> With shapely limbs, upright and strong:
> The youths on holy service throng.
>
> Vested in white, upon their brows
> Are wreaths fresh twined from dewy boughs:
> And flowers they strow along the way,
> Still dewy from the birth of day.
> So, to each reverend altar come,
> They stand in adoration: some
> Swing up gold censers; till the air
> Is blue and sweet, with smoke of rare
> Spices, that fetched from Egypt were.[74]

Whereas for George Moore the word 'Oriental' is a negative term describing Catholicism, together with 'eunuch-like' and 'dirty', Johnson's reference to Egypt at once affirms the exotic in his description of the censing and hints at Catholicism's origins in Judaism and the Exile. The awkward rhythm of 'some / Swing up gold censers' creates tension in a poem that refers to a 'sea of passion' in the shapely acolytes' 'gleaming eyes', 'passion' in this sense being a negative term for the Catholic Church.

As we have seen, Father Sebastian Bowden of the London Oratory offered a sympathetic ear to troubled young men in the 1880s and 1890s and helped Bosie and his mother during Oscar Wilde's trial.[75] He was wary of excess, however, teaching that the Catholic Church had 'always blessed the enjoyment of beauty', and that it is 'impossible that such things should separate you from God *unless they are enjoyed to excess*' (my emphasis).[76] He was

[74] Ibid., pp. 54–5. 'A Dream of Youth' was written in 1889. [75] See pp. 276–7 above.
[76] H. G. S. Bowden, *Spiritual Teaching* (1921), p. 72.

deeply disturbed by the excesses of French naturalism, which he regarded as 'simply degrading – lowering everything high and good in human love and affection: pure passion and no real love – in fact, filthy'.[77] The title of 'A Dream of Youth' reflects the fact that in the 1890s these polarities – of passion and love, and of beauty and filth – can be dissolved only in the idealized worlds of art and of dreams, where Catholicism's declared universality can include the 'Love that dare not speak its name'.[78]

Dorian Gray, who so admires the ritual of the Catholic Church from which he is excluded, is nevertheless disturbed by parallels between modern literary confessions and traditional Catholic writings. Having covered up his portrait in the old schoolroom, he sadly recalls the 'stainless purity of his boyish life' and can only hope that 'some love might come across his life, and purify him, and shield him from those sins that seemed to be already stirring in spirit and in flesh' (10). Withdrawing to the library, he picks up a book from the 'little pearl-coloured octagonal stand, that had always looked to him like the work of some strange Egyptian bees that wrought in silver'. Lord Henry Wotton has sent over a yellow book written in the style that characterizes the work of 'some of the finest artists of the French school of *Symbolistes*':

> It was a novel without a plot, and with only one character, being, indeed, simply a psychological study of a certain young Parisian, who spent his life trying to realise in the nineteenth century all the passions and modes of thought that belonged to every century except his own . . . There were in it metaphors as monstrous as orchids, and as subtle in colour. The life of the senses was described in the terms of mystical philosophy. One hardly knew at times whether one was reading the spiritual ecstasies of some mediæval saint or the morbid confessions of a modern sinner. It was a poisonous book. The heavy odour of incense seemed to cling about its pages and to trouble the brain.

For Dorian, who is described as a 'lad', the conflation of the sacred and the profane is 'poisonous'. Yet this is precisely what Johnson strives for in 'A Dream of Youth'. Such 'jarring paradoxes of the fin de siècle' are epitomized in the life and art of Aubrey Beardsley, who was both a devout Catholic convert and a notorious 'pornographer'.[79] Yeats believed that Beardsley's

[77] Ibid., p. 73.

[78] The line is Lord Alfred Douglas's, from his poem 'Two Loves'. *Punch*'s response to Aubrey Beardsley is entitled 'From the Queer and Yellow Book': *Punch*, January–June 1895, p. 58. In 1896 the *Retrospective Review* referred to John Gray's 'hot-house erotics': see R. K. R. Thornton, *Decadent Dilemma* (1983), p. 49. Ellis Hanson comments that today the Catholic Church 'may well be the world's most homophobic institution, but it also may well be the world's largest employer of lesbians and gay men': E. Hanson, *Decadence and Catholicism* (1997), p. 26.

[79] See C. Snodgrass, *Aubrey Beardsley* (1995), p. 6.

conversion was sincere, but that 'so much of impulse as could exhaust itself in prayer and ceremony, in formal action and desire, found itself mocked by the antithetical image'.[80] Yet something of the genius of Catholicism is here, as in the ministry of Father Sebastian Bowden. At its best, the Catholic Church could absorb the shocks that accompany extreme emotional responses to the ambiguities of human existence.

In the early nineteenth century, when Catholics formed a small minority in England, the conflict between the old enemies had been largely political, as pressure built up for Catholic relief and for Emancipation. At mid-century the description of the restoration of the Catholic hierarchy as 'Papal Aggression' reflected visceral anxieties concerning the Other, expressed culturally in forms such as historical fiction and paintings. A troubled and internally divided Established Church of England now felt threatened by the growth and ambition of its old enemy on the margins of national life. By the end of the century the restored English Catholic Church had moved closer to the centre of that life, but had retained an identity which had a special appeal to those who felt themselves to be on the margins of late Victorian culture. Writers of the Decadence could not only find aesthetic and spiritual solace in such a Church, but could use Catholic language and symbols to intensify their scandalous impact upon a largely middle-class Protestant readership, waving the red rag of 'Popery' in the astonished face of John Bull. In 1881 John Henry Shorthouse had identified a universal inner conflict between the 'noblest parts of man's nature arrayed against each other' – on the 'Catholic' side, obedience and faith, on the 'Protestant' side, freedom and the reason.[81] In the 1890s, the Decadence prepared the ground for the century that was to follow by exposing the paradoxes and contradictions that lay on both sides of the divide between the old enemies.

[80] Ibid., p. 157. [81] See p. 272 above.

Bibliography

Place of publication is London unless otherwise stated.

PRIMARY TEXTS

Abbot, C. *The Fifth of November; or, Protestant Principles Revived, in Memory of the Glorious Revolution by King William III, including a correct and authentic Copy of a Speech on the Roman Catholic Relief Bill, delivered May 24, 1813* (1814).

Acton, J. E. E. D., Baron. *Essays on Freedom and Power*, ed. G. Himmelfarb, Meridian series (1956).

See also MacDougall O. M. I., H.A. (Secondary texts).

Acton, W. *Prostitution considered in its Moral, Social, and Sanitary Aspects, in London and other large Cities, with Proposals for the Mitigation and Prevention of its attendant Evils* (1857).

Addis, W. E. and T. Arnold. *A Catholic Dictionary, containing some Account of the Doctrine, Discipline, Rites, Ceremonies, Councils, and Religious Orders of the Catholic Church*, 5th edn (1897).

Ainsworth, W. H. *Guy Fawkes; or, The Gunpowder Treason: An Historical Romance*, Author's Copyright Edn (1878).

The Tower of London: A Historical Romance, new edn (1903).

See also Axon, W. E. A.; Worth, G. J. (Secondary texts).

Allies, T. W. *The Church of England cleared from the Charge of Schism, by the Decrees of the Seven Ecumenical Councils and the Tradition of the Fathers*, 2nd edn enlarged (Oxford, 1848).

Journal in France in 1845 and 1848, with Letters from Italy in 1847, of Things and Persons concerning the Church and Education (1849).

A Life's Decision (1880).

Per Crucem ad Lucem: The Result of a Life, 2 vols. (1879).

The Royal Supremacy, viewed in Reference to the Two Spiritual Powers of Order and Jurisdiction (1850).

St Peter, his Name and his Office, as set forth in Holy Scripture (1852).

The See of St Peter, the Rock of the Church, the Source of Jurisdiction, and the Centre of Unity (1850).

Allnatt, C. F. B. *Cathedra Petri: The Titles and Prerogatives of St Peter, and of His See and Successors, as described by the Early Fathers, Ecclesiastical Writers, and Councils of the Church*, 2nd edn, rev. and enlarged (1879).
Amherst S. J., W. J. *The History of Catholic Emancipation and the Progress of the Catholic Church in the British Isles (chiefly in England) from 1771 to 1820*, 2 vols. (1886).
Anon. *"1745": A Tale* (1859).
Anon. ('J. D.'). 'An Account of the Festival held in the Barons' Hall of Arundel Castle on the Fifteenth day of June 1815: Written June 16th', Arundel Castle MS.
Anon. 'The Address of the Synod on the Queen's Colleges', *The Illustrated London News*, 17, 447 (21 September 1850), 250.
Anon. 'Addresses to the Crown', *The Illustrated London News*, 17, 459 (14 December 1850), 446.
Anon. 'Addresses to the Crown (from the "Times")', in Anon., *The Roman Catholic Question* [see below], 11th series, pp. 1–7.
Anon. 'Allies on the Papal Supremacy', *The Christian Remembrancer*, 20, 71 (July–December 1850), 36–83.
Anon. *The Ancient Creed of the Reformed Churches, contrasted with the Modern Creed of Rome* (1851).
Anon. *Awful Disclosures of Maria Monk, and the Startling Mysteries of a Convent Exposed!* (Philadelphia, PA, n.d.).
Anon. 'The Bishop of London's Charge, at St Paul's Cathedral, Nov. 2, 1850', in Anon., *The Roman Catholic Question* [see below], 2nd series, pp. 1–13.
Anon. *Bishop Jewel's Challenge* (1851).
Anon. *The Black Convent; or, A Tale of Feudal Times*, 2 vols. (1819).
Anon. ('Captain Rock'). *Captain Rock in Rome*, 2 vols. (1833).
Anon. 'Catholic Monthly Intelligence', *Dolman's Magazine*, 2, 10 (December 1845), 496–510.
Anon. *Cats let out of a Scarlet Bag; or, Catholic Pusses caught in a Protestant Priest-Trap; or, what Protestants will have to believe, if they become Catholics* (Norwich, 1824).
Anon. *A Complete Description of St George's Cathedral, and Hand-Book to The Catholic Antiquities of Southwark* (1851).
Anon. *The Confession of Faith; The Larger and Shorter Catechisms, with the Scripture-Proofs at Large; Together with the Sum of Saving Knowledge, (Contained in the Holy Scriptures, and held forth in the said Confession and Catechisms,) and Practical Use thereof. . . .* (Edinburgh, 1852).
Anon. 'Convent of the Belgravians', *Punch*, 19 (July–December 1850), 163.
Anon. 'The Court at Osborne', *The Illustrated London News*, 17, 453 (2 November 1850), 343.
Anon. ('A Presbyter of the Reformed Catholic Church in England.') *The Constitutional Assemblies of the Clergy the Proper and only Effectual Security of the Established Church* (1829).
Anon. *Coronation Oath* (1851).

Anon. *The Crown of England Subject to None but God* (1851).
Anon. 'The Downfall of the Russell Administration', *The Illustrated London News*, 18, 472 (1 March 1851), 169–70.
Anon. 'Dr. Cumming on the Romish Aggression', in Anon., *The Roman Catholic Question* [see below], 2nd series, pp. 13–16.
Anon. *Ellmer Castle: A Roman Catholic Story of the Nineteenth Century* (Dublin, 1827).
Anon. *England's Fall is Babylon's Triumph: An Original Interpretation of the Apocalypse, with a special reference to the Greek Church* (1855).
Anon. *Extinction of Protestantism: Speeches of Lord Arundel, M.P., and Mr. Plumptre, M.P.* (1851).
Anon. 'Fallibilis Infallibis', *Punch*, 57 (July–December 1869), 242, 244–5.
Anon. 'Father Oswald, a Catholic Story', *Edinburgh Review*, 77, 156 (April 1843), 482–500.
Anon. *Father Oswald: A Genuine Catholic Story* (1842).
Anon. *The Form and Order of the Service that is to be Performed, and of the Ceremonies that are to be Observed, in the Coronation of Her Majesty Queen Victoria, in the Abbey Church of St Peter, Westminster, on Thursday, the 28th of June 1838* (1838).
Anon. 'From the "Daily News", Nov. 21, 1850', in Anon., *Roman Catholic Question* [see below], 5th series, pp. 14–15.
Anon. 'From the Queer and Yellow Book', *Punch*, 108 (January–June 1895), 58.
Anon. 'The Gorham Case', *The Illustrated London News*, 17, 441 (17 August 1850), 135; 17, 447 (21 September 1850), 242.
Anon. *Gorham* v. *Bishop of Exeter: The Judgment of the Judicial Committee of Privy Council, Delivered March 8, 1850, Reversing the Decision of Sir H. J. Fust* (1850).
Anon. 'The Great Exhibition', *The Illustrated London News*, 18, 481 (3 May 1851), 343–51.
Anon. 'Great Industrial Exhibition of 1851', *The Illustrated London News*, 17, 456 (23 November 1850), 399.
Anon. 'Great Meeting of the High Church Party', *The Illustrated London News*, 17, 438 (27 July 1850), 77–78.
Anon. 'Hey, Johnny Cumming!', *Punch*, 57 (July–December 1869), 128–9.
Anon. *A History of Popery, Containing an Account of the Origin, Growth, and Progress of the Papal Power, its Political Influence in the European States-System, and its Effects on the Progress of Civilization. To which are added, an Examination of the Present State of the Romish Church in Ireland, a Brief History of the Inquisition, and Specimens of Monkish Legends* (1838).
Anon. ('A Member of the Middle Temple'). *Is Papal Supremacy Recognized by the Law of England?; or, Is the Papal Hierarchy Legal?* (1851).
Anon. 'The Jewish Question', *The Illustrated London News*, 17, 439 (3 August 1850), 93–4.
Anon. ('A Protestant'). *A Letter to the Rev. Edward Stanley, A.M., Rector of Alderley, in Reply to his Address in Favour of the Roman Catholics* (1829).

Anon. 'Letters and Replies from Bishops of the Established Church', in Anon., *The Roman Catholic Question* [see below], 8th series, pp. 2–11.
Anon. 'Look up your Latin', *Punch*, 57 (July–December 1869), 252.
Anon. 'Lord John Russell and the Pope: To the Right Rev. the Bishop of Durham', in Anon., *The Roman Catholic Question* [see below], 1st series, p. 8.
Anon. 'Maynooth', *The British Protestant*, 2 (February 1845), 57–9.
Anon. 'Mr. Roebuck's Letter to Lord John Russell', in Anon., *The Roman Catholic Question* [see below], 9th series, pp. 1–3.
Anon. *Murderous Effects of the Confessional* (1851).
Anon. 'The New Roman Catholic Bishops', *The Illustrated London News*, 17, 453 (2 November 1850), 341–2.
Anon. ('C.'). 'The No-Popery Agitation', *The Christian Reformer; or, Unitarian Magazine and Review*, n.s. 7 (January–December 1851), 1–8.
Anon. *The Nunnery; or, Popery Exposed in her Tyranny* (1851).
Anon. *Oratory Night Services* (1909).
Anon. 'The Papal Aggression', *The Illustrated London News*, 17, 454 (9 November 1850), 357–8; 17, 459 (14 December 1850), 457–8.
Anon. 'The Papal Vindication', *The Illustrated London News*, 17, 456 (23 November 1850), 397–8.
Anon. *The Persecuting and Sanguinary Spirit of the Church of Rome* (1851).
Anon. ('John Bull'). 'A Plain Appeal to the Common Sense of all the Men and Women of Great Britain and Ireland' [1 November 1850], in Anon., *The Roman Catholic Question* [see below], 4th series, pp. 1–12.
Anon. 'Pontifical News', *Punch*, 19 (July–December 1850), 182.
Anon. ('Erinaceus'). *The Popish Divan; or, Political Sanhedrin: A Satirical Poem* (1809).
Anon. *Popish Persecution; or, A Sketch of the Penal Laws. No. I – Henry VIII. – Elizabeth*, The Clifton Tracts, Historical Library series (1851).
Anon. *The Priest in Absolution: A Manual for such as are Called unto the Higher Ministries in the English Church*, 2 pts (1866–73).
Anon. 'The Progress of Popery', *The British Protestant*, 1 (January 1845), 1–3.
Anon. *Progress of Truth in Ireland* (1851).
Anon. 'The Queen's Speech', *The Illustrated London News*, 18, 468 (8 February 1851), 106–7.
Anon. *The Real Causes of "The Papal Aggression" Considered, in a Statement respectfully Presented to the Lord Bishop of Gloucester and Bristol, by Certain of his Clergy, in Lent, 1851* (1851).
Anon. 'The Real Danger of the Church in England', in Anon., *The Roman Catholic Question* [see below], 1st series, pp. 12–13.
Anon. 'Recent Converts to the Church of Rome', *The British Protestant*, 11 (November 1845), 245–8.
Anon. 'Recent Proceedings at Oxford', *The Christian Remembrancer*, 9, 48 (April 1845), 519–79.
Anon. 'The Recent Schism', *The Christian Remembrancer*, 11, 56 (January 1846), 167–218.

Anon. ('A Layman'). *The Rejection of Catholic Doctrines attributable to the Non-realisation of Primary Truths, exemplified in Letters to a Friend on Devotion to the Blessed Virgin, the Angels, and Saints* (1878).

Anon. 'The Rev. Mr. Bennett's Letter to Lord John Russell', in Anon., *The Roman Catholic Question* [see below], 9th series, pp. 3–13.

Anon. *The Roman Catholic Question: A Copious Series of important Documents, of Permanent Historical Interest, on the Re-establishment of the Catholic Hierarchy in England, 1850–1* (1851).

Anon. 'Roman Catholic Synod', *The Illustrated London News*, 17, 443 (24 August 1850), 167.

Anon. ('A Barrister'). *Roman Catholicism: Being an Historical and Legal Review of its Past Position and Present Claims in England* (1851).

Anon. 'Rome and Ramsbotham', *Punch*, 58 (January–June 1870), 23.

Anon. *"Rome's Recruits": A List of Protestants who have become Catholics since the Tractarian Movement, Re-printed, with numerous additions and corrections, from "The Whitehall Review" of September 28th, October 5th, 12th, and 19th, 1878* (1878).

Anon. 'Royal Italian Opera', *The Illustrated London News*, 17, 443 (24 August 1850), 175.

Anon. 'Sacramental Confession', *The British Critic, and Quarterly Theological Review*, 33, 46 (April 1843), 296–347.

Anon. ('The Author of 'Quousque'). *Secession to Rome*, new and improved edn (1874).

Anon. *Secessions to Rome no Novelty, nor Ground for Alarm; but a Natural Result of an Ecclesiastical Movement or Revival in the Church* (1846).

Anon. *The Spectre of the Vatican; or, The Efforts of Rome in England since the Reformation* (n.d.).

Anon. 'Speech of the Very Rev. the Dean of Bristol (Dr. Gilbert Elliott), at a Meeting of Clergymen, held at Bristol, 6th November, 1850', in Anon., *The Roman Catholic Question* [see above], 4th series, pp. 12–16.

Anon. *Statistics of Popery in Great Britain and the Colonies, illustrated with a Map shewing the Situation of each Roman Catholic Chapel, College, and Seminary, throughout England, Scotland, and Wales, reprinted from "Fraser's Magazine" for March and April, 1839*, 6th edn (1839).

Anon. *Suggestions for a Practical Use of the Papal Aggression: A Letter to the Lord Bishop of Manchester, by one of his Clergy* (1851).

Anon. 'To Englishmen, Irishmen, and Scotchmen', in Anon., *Roman Catholic Question* [see above], 5th series, p. 2.

Anon. 'Tractarianism and Mr. Ward', *Dolman's Magazine*, 1, 1 (March 1845), 73–81.

Anon. *The Virgin Mary, a Married Woman* (1869).

Anon. ('MacGauran'). *Walter Clayton: A Tale of the Gordon Riots*, 3 vols. (1844).

Anon. *What is Romanism?*, 26 pts (1849–51).

Armstrong, G. *Infallibility not Possible; Involuntary Error not Culpable; in Letters, Originally Occasioned by the Celebrated Controversy, in Dublin, between Messrs. Pope and Maguire; and now, still more Urgently than then, Addressed to the*

Attention of Roman Catholics, and of Protestants of every Denomination, in the United Empire, 2nd rev. edn (London and Bristol, 1851).

Arnold, M. *Culture and Anarchy*, ed. J. D. Wilson, Landmarks in the History of Education series, ed. J. D. Wilson and F. A. Cavenagh (Cambridge, 1960).

Dissent and Dogma, ed. R.H. Super, Complete Prose Works of Matthew Arnold, 6 (Ann Arbor, MI, 1968).

Arnold, T. *The Christian Duty of Granting the Claims of the Roman Catholics, with a Postscript in Answer to the Letters of the Rev. G. S. Faber, printed in the St James's Chronicle* (Oxford and London, 1829).

Introductory Lectures on Modern History, delivered in Lent Term, MDCCCXLI, 2nd edn (1843).

Arthur, W. *The Modern Jove: a Review of the Collected Speeches of Pio Nono*, 2nd edn (1873).

Austen, J. *The History of England from the Reign of Henry the 4th to the Death of Charles the Ist*, intro. A. S. Byatt, ed. D. Le Faye (Chapel Hill, NC, 1993).

Baines, P. A. *A Defence of the Christian Religion, &c. in a Series of Letters addressed to Charles Abel Moysey, D.D., Archdeacon of Bath*, new edn (1825).

Baker, H. W., *et al.*, eds. *Hymns Ancient and Modern, for use in the Services of the Church, with accompanying Tunes*, Standard edn (1916).

Balthasar, H. U. von. *The Glory of the Lord: A Theological Aesthetics*, 7 vols. (Edinburgh, 1982–91).

Barnes, R. *The Papal Brief considered with reference to the Laws of England* (1850).

Barwick, E. *A Treatise on the Church, chiefly with respect to its Government: in which the Divine Right of Episcopacy is Maintained, the Supremacy of the Bishop of Rome proved to be Contrary to the Scriptures and Primitive Fathers; and the Reformed Episcopal Church in England, Ireland, and Scotland, proved to be a Sound and Orthodox Part of the Catholic Church. Compiled from the most Eminent Divines*, 2nd edn (Belfast, 1813).

Baxter, R. *A Key for Catholicks, To Open the Jugling of the Jesuits, and Satisfie all that are but truly Willing to Understand, whether the Cause of the Roman or Reformed Churches be of God; and to leave the Reader utterly Unexcusable that after this will be a Papist* (1659).

Beal, W. *An Analysis of Palmer's* Origines Liturgicæ; or, Antiquities of the English Ritual; *and of his* Dissertation on Primitive Liturgies, *for the use of Students at the Universities, and Candidates for Holy Orders, who have read the Original Work* (Cambridge and London, 1850).

Beardsley, A. *The Letters of Aubrey Beardsley*, ed. H. Maas, J. L. Duncan and W. G. Good (1970).

Under the Hill and other Essays in Prose and Verse, with Illustrations (London and New York, 1904).

See also Snodgrass, C. (Secondary texts).

Bellarmine, R. F. R. *A Fair and Calm Consideration of the modern Controversy concerning Justification, as it is Explained in the Five Books of Cardinal Bellarmine*, vol. 1 (Oxford, 1850).

Bellasis, E. *The Anglican Bishops versus the Catholic Hierarchy: A Demurrer to Farther Proceedings* (1851).

The Archbishop of Westminster: A Remonstrance with the Clergy of Westminster, from a Westminster Magistrate (1850).

Memorials of Mr. Serjeant Bellasis (1800–1873) (1893).

Bennett, W. J. E. *The Church, the Crown, and the State, their Junction or their Separation: Considered in two Sermons, bearing reference to the Judicial Committee of the Privy Council* (1850).

The Schism of certain Priests and Others, lately in Communion with the Church: A Sermon . . . preached at St Paul's on the 22nd Sunday after Trinity, 1845 (1845).

Berington, J. *An Address to the Protestant Dissenters, who have lately Petitioned for a Repeal of the Corporation and Test Acts* (Birmingham, 1787).

An Essay on the Depravity of the Nation, with a view to the Promotion of Sunday Schools, &c. of which a more Extended Plan is Proposed (Birmingham, 1788).

A Literary History of the Middle Ages; comprehending an Account of the State of Learning, from the Close of the Reign of Augustus, to its Revival in the Fifteenth Century (1814).

The Rights of Dissenters from the Established Church, in relation, principally, to English Catholics (Birmingham, 1789).

The State and Behaviour of English Catholics, from the Reformation to the Year 1781, with a View of their Present Number, Wealth, Character, &c., 2nd edn (1781).

Berington, J. and J. Kirk, compilers. *The Faith of Catholics, confirmed by Scripture, and attested by the Fathers of the Five First Centuries of the Church* (1813).

The Faith of Catholics on Certain Points of Controversy, confirmed by Scripture, and attested by the Fathers of the Five First Centuries of the Church, 2nd edn (1830).

The Faith of Catholics, on Certain Points of Controversy, confirmed by Scripture, and attested by the Fathers of the Five First Centuries of the Church, 3rd edn, rev. and enlarged by J. Waterworth, 3 vols. (1846).

Beste, H. D. *Four Years in France; or, Narrative of an English Family's Residence there during that Period; preceded by some Account of the Conversion of the Author to the Catholic Faith* (1826).

Bicheno, J. *The Signs of the Times; or, The Overthrow of the Papal Tyranny in France, the Prelude of Destruction to Popery and Despotism; but of Peace to Mankind* (1793).

Bingham, J. *Origines Ecclesiasticæ: The Antiquities of the Christian Church, with Two Sermons and Two Letters on the Nature and Necessity of Absolution, reprinted from the Original Edition, MDCCVIII–MDCCXXII, with an Enlarged Analytical Index*, 2 vols. (1850).

Blakeney, R. P. *Manual of Romish Controversy: Being a complete Refutation of the Creed of Pope Pius* IV, 27th thousand (Edinburgh and London, n.d.).

Popery in its Social Aspect: Being a Complete Exposure of the Immorality and Intolerance of Romanism, 6th thousand (Edinburgh, n.d.).

Blomfield, A., ed. *A Memoir of Charles James Blomfield, D.D., Bishop of London, with Selections from his Correspondence*, 2 vols. (1863).
Blomfield, C. J. See Blomfield, A.
See also Anon., 'The Bishop of London's Charge'.
Bosanquet, S. *The Romanists, the Established Church, and the Dissenters: Substance of a Speech at a Public Meeting at Monmouth, on the late Papal Aggression, on the 19th of November, 1850; together with an Explanatory Letter to the Rev. Robert Jackson* (London and Monmouth, 1851).
Bowden, H. G. S. ed., *Guide to the Oratory, South Kensington, with Explanations and Plates* (1893).
Spiritual Teaching of Father Sebastian Bowden of the London Oratory, consisting of Counsels on various Subjects, Notes of Addresses and Letters, with short Introductory Memoir, ed. Fathers of the Oratory (1921).
See also Baring, M. (Secondary texts).
Bowden, J. E. *The Life and Letters of Frederick William Faber*, 5th edn (n.d.).
Bowyer, G. *The Cardinal Archbishop of Westminster and the New Hierarchy* (1851).
Observations on the Arguments of Dr. Twiss respecting the New Roman Catholic Hierarchy (1851).
The Roman Documents relating to the New Hierarchy: with an Argument, 4th edn, with additions (1850).
Bradford, J. *Writings of the Rev. John Bradford, Prebedendary of St Paul's, and Martyr, A.D. 1555* (1827).
Brenan, M. *Papal Dominion Incompatible with the Natural Rights and Positive Laws of Christian Nations* (1828).
Brewster, D. *More Worlds than One: The Creed of the Philosopher and the Hope of the Christian* (1854).
Bricknell, W. S. *The Judgment of the Bishops upon Tractarian Theology: A Complete Analytical Arrangement of the Charges delivered by the Prelates of the Anglican Church, from 1837 to 1842 inclusive; so far as they relate to the Tractarian Movement, with Notes and Appendices* (Oxford, 1845).
Bridgett, T. E. *England for Our Lady: A Sermon preached . . . before His Eminence Cardinal Vaughan, Archbishop of Westminster, and the Bishops of England, in the Church of the Oratory, London, June 29th, 1893, on the Occasion of the Solemnity of the Renewal of England's Dedication to the Blessed Virgin Mary and to St Peter, Prince of the Apostles* (1893).
Brinckman, A. *The Controversial Methods of Romanism* (1888).
See also Moore, T. and A. Brinckman.
Bristow, C. *The Roman Catholic and Protestant Churches proved to be nearer Related to each other than most Men Imagine* (Dublin, 1841).
British Protestant, The; or, Journal of the Religious Principles of the Reformation (1845–64).
British Quarterly Review, The (1845–86).
Brogden, J. *Catholic Safeguards against the Errors, Corruptions, and Novelties of the Church of Rome; beings Discourses and Tracts, selected from the Works of Eminent*

Divines of the Church of England, who lived during the Seventeenth Century, 3 vols. (1851).

Brontë, C. ('Currer Bell'). *Villette*, new edn (1904).

Broughton, W. G. *A Letter to the Right Rev. Nicholas Wiseman, D.D., by the Bishop of Sydney, Metropolitan of Australasia, together with the Bishop's Protest, March the 25th, 1843, and the Resolutions of a special Meeting of the Standing Committee of the Sydney Diocesan Committee of the Societies for Promoting Christian Knowledge and the Propagation of the Gospel in Foreign Parts, held on the A Letter to the Right Rev. Nicholas Wiseman, D.D., 15th day of March, 1851* (1852).

Brown, H. S. *The Bulwark of Protestantism* (Manchester, 1868).

Brown, R. *Popery: A Gigantic Swindle of the Devil; Designed by Him to Enslave Mankind and Destroy Souls* (1894).

Browning, R. *The Poems*, ed. J. Pettigrew and T. J. Collins, 2 vols. (Harmondsworth, 1981).

The Poetical Works of Robert Browning, vol. IV, ed. I. Jack, R. Fowler and M. Smith (Oxford, 1991).

The Ring and the Book, ed. R. D. Altick (Harmondsworth, 1971).

See also Irvine, W. and P. Honan; Jack, I.; Khattab, E. A.; Korg, J.; Litzinger, B.; Litzinger, B. and D. Smalley (Secondary texts).

Browning, R. and E. B. *The Brownings' Correspondence*, ed. P. Kelley and R. Hudson, 14 vols. (Winfield, KS, 1984–98).

Bull, G. *Defensio Fidei Nicænæ, ex Scriptis, quæ Extant, Catholicorum Doctorum, qui intra tria prima Ecclesi Christian Secula Floruerunt, etc.* (Oxford, 1685).

Bulwark, The; or, Reformation Journal : In Defence of the true Interests of Man and of Society, especially in reference of the Religious, Social, and Political bearings of Popery (Edinburgh and London, 1851–87).

Burgess, T. *The Protestant's Catechism, in which it is clearly Proved, that the Ancient British Church existed several Centuries before Popery had any Footing in Great Britain; and many important Facts illustrative of the Fallacy of the Popish Doctrine of Infallibility, never before Published* (York, 1817).

The Protestant's Catechism on the Origin of Popery, and on the Grounds of the Roman Catholic Claims; to which are prefixed, The Opinions of Milton, Locke, Hoadley, Blackstone, and Burke: with a Postscript on the Introduction of Popery into Ireland by the Compact of Henry II. and Pope Adrian, in the Twelfth Century (1818).

Burgon, J. W. *Letters from Rome to Friends in England* (1862).

Burton, E. *History of the Christian Church, from the Ascension of Jesus Christ, to the Conversion of Constantine* (1836).

Butler, A. *The Lives of the Fathers, Martyrs, and other principal Saints*, ed. F. C. Husenbeth, Illuminated edn, 2 vols. (London and Dublin, 1883–6).

Butler, C. *The Book of the Roman-Catholic Church, in a Series of Letters addressed to Robt. Southey, Esq. LL.D. on his "Book of the Church"* (1825).

An Historical and Literary Account of the Formularies, Confessions of Faith, or Symbolic Books, of the Roman Catholic, Greek, and principal Protestant Churches. By the author of the Horæ Biblicæ, and intended as a Supplement to that Work . . . To which are added four Essays. I. A succinct Historical Account of the Religious Orders of the Church of Rome. II. Observations on the Restriction imposed by the Church of Rome on the general Reading of the Bible in the Vulgar Tongue. III. The Principles of Roman-Catholics in regard to God and the King, first published in 1684 . . . IV. On the Reunion of Christians (1816).

Historical Memoirs respecting the English, Irish, and Scottish Catholics, since the Reformation to the Present Time, 4 vols. (1819–21).

Vindication of "The Book of the Roman Catholic Church", against the Reverend George Townsend's "Accusations of History against the Church of Rome": . . . *With copies of Doctor Phillpott's Fourth Letter to Mr. Butler, Containing a Charge against Dr. Lingard; and of a Letter of Doctor Lingard to Mr. Butler, in Reply to the Charge* (1826).

Butler, W. A. *Letters on Romanism, in Reply to Mr. Newman's Essay on Development*, ed. T. Woodward (Cambridge, 1854).

Butt, M. M., later Sherwood. *The Nun* (1833).

Cahill, D. W. *The Rev. Dr. Cahill's First [-Fifth] Letter to Lord John Russell, relative to his Lordship's Epistle to the Bishop of Durham*, 5 pts (Manchester, 1850).

Campbell, G. D. *The Twofold Protest: A Letter from the Duke of Argyll to the Bishop of Oxford* (1851).

Campbell, J. *The Conquest of England: Letters to the Prince Consort on Popery, Puseyim, Neology, Infidelity, and the Aggressive Policy of the Church of Rome* (1861).

Capes, J. M. *Reasons for Returning to the Church of England* (1871).

To Rome and Back (1873).

Carlyle, T. *Works*, Edition de Luxe, 20 vols. (Boston, 1884).

See also Kaplan, F.; Oddie, W. (Secondary texts).

Carter, T. T. *The Roman Question* (1890).

Cartwright, J. B. 'Tradition', in *The Errors of Romanism: Six lectures on the Errors of the Church of Rome, delivered in Sir George Wheeler's Chapel . . . London . . . Lent 1838, by A. MacCaul, et al.* (1838).

No Popery!: A Course of Eight Sermons, preached at the Episcopal Jews' Chapel, Palestine Place, Bethnal Green (1850).

Casaubon, I. *Anglican Catholicity Vindicated against Roman Innovations: in the Answer of Isaac Casaubon to Cardinal Perron, reprinted from the Translation published by Authority in 1612, with an Introduction, Table of Contents and full Index*, by W. R. Whittingham and H. Harrison (Baltimore, MD, and New York, 1875).

Challoner, R. *The Garden of the Soul: A Manual of Spiritual Exercises and Instructions for Christians who, living in the World, aspire to Devotion*, new edn (n.d.).

Martyrs to the Catholic Faith: Memoirs of Missionary Priests and other Catholics of both Sexes that have Suffered Death in England on Religious Accounts from the Year 1577 to 1684, 2 vols. (Edinburgh, 1878).

Memoirs of Missionary Priests, as well Secular as Regular; and of other Catholics, of both Sexes, that have suffered Death in England, on Religious Accounts, from the Year of our Lord 1577, to 1684 . . ., 2 parts (1741–2).

See also Burton, E. H.; Duffy, E. (Secondary texts).

Chambers, R. *History of the Rebellion in Scotland in 1745, 1746*, 2 vols., Constable's Miscellany of Original and Selected Publications in the various Departments of Literature, Science, & the Arts series, vols. xv–xvi, new edn (Edinburgh and London, 1828).

Vestiges of the Natural History of Creation (1844).

Champneys, B. *Memoirs and Correspondence of Coventry Patmore*, 2 vols. (1900).

Chesnutt, G. *Protestant Securities Considered: in which the Insufficiency of all Securities hitherto Offered to the Consideration of Parliament is set forth, and New Securities Proposed* (1829).

Chettle, J. *Protestant Objections, against the Romish Doctrine of Transubstantiation, addressed to the Rev. F. Martyn, and to the Inhabitants of Walsall and Bloxwich*, 2nd edn (Walsall, 1827).

Chiniquy, C. *Fifty Years in the Church of Rome* (Chicago, IL, 1885).

Christian Remembrancer, The: A Quarterly Review (1840–68).

Churton, E. *The Early English Church* (1840).

Ciocci, R. *A Narrative of Iniquities and Barbarities practised at Rome in the Nineteenth Century*, 3rd edn (1844).

Clifton Tracts, The, published by the Brotherhood of St Vincent of Paul, nos. 1–69, extra series, nos. 1, 2 (1851–4).

Clive, C. *Saint Oldooman: A Myth of the Nineteenth Century, contained in a Letter from the Bishop of Verulanum to the Lord Drayton* (1845).

Close, F. *"The Catholic Revival"; or, Ritualism and Romanism in the Church of England, Illustrated from "The Church and the World": A Paper read at the Annual Meeting of "The Evangelical Union for the Dioceses of Carlisle", printed and published at their Request* (London and Carlisle, 1866).

The Restoration of Churches is the Restoration of Popery, Proved and Illustrated from the Authentic Publications of the "Cambridge Camden Society": A Sermon, preached in the Parish Church, Cheltenham, on Tuesday, November 5th, 1844, 2nd edn (1844).

Clough, A. H. *The Poems of Arthur Hugh Clough*, ed. A. L. P. Norrington (1968).

Cobbett, W. *A History of the Protestant Reformation in England and Ireland; showing how that Event has Impoverished the main Body of the People in those Countries; and containing a List of the Abbeys, Priories, Nunneries, Hospitals, and other Religious Foundations in England, and Wales, and Ireland, Confiscated, Seized on, or Alienated, by the Protestant "Reformation" Sovereigns and Parliaments, in a series of Letters addressed to all Sensible and Just Englishmen*, 2 vols. (1829).

See also Sambrook, J. (Secondary texts).

Cobbin, I. *The Book of Popery: A Manual for Protestants, descriptive of the Origin, Progress, Doctrines, Rites, and Ceremonies of the Papal Church*, Popular Elementary Works series (1840).

Colenso, J. W. *The Pentateuch and Book of Joshua Critically Examined*, 7 pts (1862–79).

Colenso, W. 'A few Remarks on the hackneyed Quotation of "Macaulay's New Zealander"', in his *Three Literary Papers, read before the Hawke's Bay Philosophical Institute, during the Session of 1882*, 3 (Napier, NZ, 1883).

Coleridge, S. T. *On the Constitution of the Church and State*, ed. J. Colmer, *The Collected Works of Samuel Taylor Coleridge*, gen. ed. K. H. Coburn, vol. x, Bollinger series, 75 (1976).

Colquhoun, J. C. *Speech of Mr. Colquhoun, at Exeter Hall, on Friday, March 11, 1836, upon the Subject of the Maynooth College Grant* (1836).

Comber, T. *A Letter to the Duke of Sussex, and to the Duke of Gloucester, on the Line of Conduct adopted by their Royal Highnesses in the late Debate, on the Roman Catholic Questions, June 9th and 10th, 1828* (1828).

Connelly, P. *The Coming Struggle with Rome, not Religious but Political; or, Words of Warning to the English People*, 5th edn (1852).

Domestic Emancipation from Roman Rule in England: A Petition to the Honourable House of Commons, with Notes (1852).

Reasons for Abjuring Allegiance to the See of Rome: A Letter to the Earl of Shrewsbury (1852).

Cooper, C. P. *The Common Law and the Pope's Apostolic Letters of Sept. 1850: Extracts of some Letters addressed in the month of December last to Charles Purton Cooper*, 3rd edn (1851).

The Government and the Irish Roman Catholic Members, 6th edn (1851).

The Late Edict of the Court of Rome: Lord Beaumont's Letter to Lord Zetland, 6th edn (1850).

The Pope's Apostolic Letters of September, 1850: Letter to an English Roman Catholic Peer, 6th edn (1851).

The Pope's Brief of September, 1850: Notes on some Conclusions arrived at 15th–25th January, 1851, in several Conferences between certain Roman Catholic Priests and a Queen's Counsel, 5th edn (1851).

So much of the Pope's Apostolical Letters of September, 1850, as relates to the Creation of an Archbishopric and Twelve Bishoprics . . ., 6th edn (1851).

The Ultra Party amongst the English Roman Catholics: Extract of a Letter, dated 15th February, 1851, from an English Roman Catholic Peer to Mr. Purton Cooper, 4th edn (1851).

Coplestone, E. *Essential Difference between the Church of England and the Church of Rome, illustrated in Two Sermons preached Nov. 15, 1840, at St Paul's Church, Newport, on the Opening of an Enlarged Popish Chapel in that Town, and in a Pastoral Address, on Roman Catholic Errors*, 3rd edn (1845).

Cox, J. E. ed. *Protestantism contrasted with Romanism by the Acknowledged and Authentic Teaching of each Religion*, 2 vols. (1852).

Craig, E. *A Second Friendly Address to Roman Catholics, occasioned by "The Answer of the Rev. Mr. M'Kay," intended as a Summary View of the Controversy between Protestants and Roman Catholics* (Edinburgh, 1831).

Crapsey, A. S. *The Last of the Heretics* (New York, NY, 1924).

Crawford, W. S. *Correspondence between Wm. Sharman Crawford, Esq., M.P. for Rochdale, and the Rev. Dr Molesworth, Vicar of the Same, on the Papal Aggression, and on the Spiritual Liberties and Temporal Rights of the Established Church* (1851).

Creevey, T. *The Creevey Papers: A Selection from the Correspondence & Diaries of the late Thomas Creevey, M.P., born 1768 – died 1838*, ed. H. Maxwell, 2 vols. (1903).

Croly, G. *The Apocalypse of St John; or, Prophecy of the Rise, Progress, and Fall of the Church of Rome; the Inquisition; the Revolution of France; the Universal War; and the Final Triumph of Christianity: being a New Interpretation* (1827).

Divine Providence; or, The Three Cycles of Revelation, showing the Parallelism of the Patriarchal, Jewish, and Christian Dispensations: Being a New Evidence of the Divine Origin of Christianity (1834).

England the Fortress of Christianity: The Preface to "The New Interpretation of the Apocalypse", 17th thousand (1839).

England, Turkey, and Russia: A Sermon, preached on the Embarkation of the Guards for the East, in the Church of St Stephen, Walbrook, February 26, 1854 (1854).

The Englishman's Polar Star!!; or, A deeply Interesting, and highly Important View of Unquestioned Historical Facts, as Connected with the Honour and Safety of the British Empire: being a Preface to a New Interpretation of the Apocalypse of St John, with Introductory and Concluding remarks, by R. H. M. (Preston, 1828).

The French Revolution, of 1848: A Sermon, preached in the Church of St Stephen, Walbrook (1848).

Historical Sketches, Speeches, and Characters (1842).

National Defence Essential to National Safety, and Justified by Holy Scripture: A Sermon, Preached in the Church of St Stephen, Walbrook, before Lieutenant-Colonel Wilson, and the Staff of the Royal London Militia (1853).

National Diseases – Divine Judgments: A Sermon, on the Approach of the Cholera, preached in the Church of St Stephen, Walbrook, in consequence of the Bishop of London's Pastoral Letter to his Clergy (1848).

The Poetical Works of the Rev. George Croly, 2 vols. (1830).

Popery and the Popish Question: being an Exposition of the Political and Doctrinal Opinions of Messrs. O'Connell, Keogh, Dromgole, Gandolphy, &c., &c. (1825).

The Popish Primacy: Two Sermons on the Conversions to Popery, and the Coming Trial of Nations, preached in St Peter's Church, Brighton, September, 1850 (1850).

The Reformation a Direct Gift of Divine Providence: A Sermon, preached in St Paul's Cathedral, on Monday, October 8, 1838, on the First Day of the Visitation of the Right Hon. and Right Rev. Charles James, Lord Bishop of London (1838).

Salathiel: A Story of the Past, the Present, and the Future, 3 vols. (1828).

A Sermon on the Death of the Duke of Wellington; Preached in the Church of St Stephen, Walbrook, September 19, 1852 (1852).

'Speech of the Rev. Dr. Croly at the Protestant Meeting at Hackney', *Protestant Magazine*, n.s. 13, 2 (February 1851), 49–54.

The Spread of the Gospel the Safeguard of England!: A Sermon, preached in St Stephen's Church, Walbrook, on Sunday, October 4, 1835, being the Tercentenary of the Translation of the whole Bible into the English Language, 3rd edn (1835).

Tales of the Great St Bernard, 3 vols. (1828).

The Universal Kingdom: A Sermon, preached at the Request of the Protestant Association of London, May 4, 1843 (1843).
See also Herring, R.
Cumming, J. *The Fall of Babylon Foreshadowed in her Teaching, in History, and in Prophecy* (1870).
The Seventh Vial; or; The Time of Trouble Begun, as shown in the Great War, the Dethronement of the Pope, and other Collateral Events (1870).
See also Anon., 'Dr. Cumming'.
See also Ellison, R. H. and C. M. Engelhardt (Secondary texts).
Cuninghame, W. *The Church of Rome the Apostasy, and the Pope the Man of Sin and Son of Perdition of St Paul's Prophecy, in the Second Epistle to the Thessalonians; with an Appendix, containing an Examination of the Rev. W. Burgh's Attempt to Vindicate the Papacy from these Charges . . .* , 2nd edn (1845).
Cusack, M. F. C. *Life Inside the Church of Rome* (1889).
Dacre, C. *Confessions of the Nun of St Omer: A Tale*, 3 vols. (1805).
Daly, W. *Thoughts, suggested by the Papal Aggression: in a Series of Discourses, with Notes and Observations: Forming a Sequel to "Thoughts chiefly on the subject of Differences among Christians"* (Dublin, 1852).
Daniel, G. *Merrie England in the Olden Time*, 2 vols. (1842).
Darby, J. N. *Analysis of Dr. Newman's Apologia Pro Vita Sua: with a Glance at the History of Popes, Councils, and the Church* (1866).
The Collected Writings of J. N. Darby, ed. W. Kelly, 34 vols. (Sunbury, PA, 1971–2).
Darwell, L. *The Catholic Church, together with the True Bible and Apostolical Tradition, Evangelical Doctrine, a Fixed Creed, Apostolical Succession and Jurisdiction; or, The Romish Scheme, with an Interpolated Canon, Men's Traditions, a Corrupt Doctrine, a Still-Developing Creed, with no True Mission, no Succession, no Jurisdiction: Which? Consisting of Certain Remarks on a Tract entitled "The Catholic Church, &c. or the Anglican Church, &c. – Which?"* (1851).
Christ's Ministers to give Attendance to Reading, Exhortation, and Doctrine: A Sermon, Preached at Church Stretton, May 13, MDCCCXL, upon Occasion of the Primary Visitation of the Venerable William Vickers . . . (1841).
The Church of England a True Branch of the Holy Catholic Church: A Reply to the Remarks of the Rev. Eugene Egan on the Pamphlet entitled, "The Catholic Church, or the Romish Schism, Which?" (London and Shrewsbury, 1853).
Darwin, C. *The Origin of Species by Means of Natural Selection* (1902).
See also Beer, G.; Desmond, A. and J. Moore (Secondary texts).
Davies, E. *et al. The Churchman Armed against the Errors of the Time, by "The Society for the Distribution of Tracts in Defence of the United Church of England and Ireland, as by Law Established"*, 3 vols. (1814).
Day, S. P. D. *Monastic Institutions: Their Origin, Progress, Nature, and Tendency* (Dublin and London, 1844).
Dearden, H. W. *Modern Romanism Examined*, 2nd edn (1899).
Denison, G. A. *The Kingdom of the Truth: A Sermon, preached in the Church of St Peter, Plymouth, on Monday, October 7, 1850* (1850).

De Vere, A. *English Misrule and Irish Misdeeds: Four Letters from Ireland, addressed to an English Member of Parliament*, 2nd edn (1848).
The Infant Bridal, and other Poems (1864).
May Carols (1857).
Recollections of Aubrey de Vere (New York and London, 1897).
The Search after Proserpine, Recollections of Greece, and other Poems (Oxford and London, 1843).
The Waldenses; or, The Fall of Rora: A Lyrical Sketch, with other Poems (Oxford and London, 1842).
See also Ward, W. (Secondary texts).
De Vere, S. E. *Is the Hierarchy an Aggression?* (1851).
Dickens, C. *Barnaby Rudge: A Tale of the Riots of 'Eighty* (n.d.).
David Copperfield, ed. N. Burgis, Clarendon Dickens (Oxford, 1981).
See also Ackroyd, P; Oddie, W.; Walder, D. (Secondary texts).
Digby, K. H. *The Broad Stone of Honour; or, Rules for the Gentlemen of England*, 2nd edn (1823).
The Broad Stone of Honour; or, The True Sense and Practice of Chivalry, in four parts: Godefridus, Tancredus, Morus, Orlandus (1877).
Compitum; or, The Meeting of the Ways at the Catholic Church, 3 vols. 2nd edn (1851–3).
Mores Catholici, 11 vols. (1831–42).
The Temple of Memory (1874).
See also Holland, B. (Secondary texts).
Disraeli, B. *Sybil; or, The Two Nations*, intro. W. Sichel, World's Classics series (1926).
See also Hibbert, C. (Secondary texts).
Dodd, C. *Dodd's Church History of England, from the Commencement of the Sixteenth Century to the Revolution in 1688*, with notes, additions, and a continuation by the Rev. M. A. Tierney, 5 vols. (1839–43).
Dodsworth, W. *Allegiance to the Church: A Sermon* (1841).
Dolmans Magazine and Monthly Magazine of Criticism (1845–9).
Douglas, A. *Poems* (Paris, 1896).
See also Croft-Cooke, R. (Secondary texts).
Dowson, E. *The Poetical Works of Ernest Dowson*, ed. D. Flower, 3rd edn (1967).
See also Adams, J. (Secondary texts).
Dyce, W. *Notes on Shepherds and Sheep: A Letter to John Ruskin, Esq., M.A.* (1851).
Ecclesiologist, The (Cambridge and London, 1841–68).
Eden, R. *The Pope's Movement: How Government may be Expected to Deal with it* (1851).
Eliot, E. G., Earl of St Germans. *Reasons for Not Signing an Address to Her Majesty, on the Subject of the Recent So-Called Papal Aggression*, 2nd edn (1850).
Eliot, G. See Evans, M.
Elliott, G. See Anon., 'Speech of the Dean of Bristol'.
Empson, W. 'The Papal States', *Edinburgh Review*, 86, 174 (October 1847), 494–8.

Evans, M. ('George Eliot'). *Felix Holt, the Radical*, ed. P. Coveney, Penguin English Library (Harmondsworth, 1972).
Middlemarch, ed. D. Carroll, Clarendon Edn of the Novels of George Eliot, ed. G. S. Haight (Oxford, 1986).
See also Beer, G. (Secondary texts).
Faber, F. W. *All For Jesus; or, The Easy Ways of Divine Love* (1853).
Faber, Poet and Priest: Selected Letters by Frederick William Faber, 1833–1863, ed. R. Addington (Cowbridge and Bridgend, 1974).
The Foot of the Cross; or, The Sorrows of Mary (n.d.).
Growth in Holiness; or, The Progress of the Spiritual Life, 2nd edn (1855).
The Life of Saint Philip Neri, Apostle of Rome, and Founder of the Congregation of the Oratory, 2 vols. Saints and Servants of God series (1847).
ed., *The School of S. Philip Neri, from the Italian* (1850).
See also Bowden, J. E.
See also Chapman, R. (Secondary texts).
Faber, F. W., *et al.*, eds. *The Saints and Servants of God*, 42 vols. (1847–56).
Faber, F. W., J. H. Newman, E. Caswall, *et al. Oratory Hymns* (1909).
Faber, G. S. *The Difficulties of Romanism in respect of Evidence; or, The Peculiarities of the Latin Church evinced to be Untenable on the Principles of Legitimate Historical Testimony*, 3rd edn (1853).
Fillan, A. D. *Stories, Traditionary and Romantic, of the two Rebellions in Scotland in 1715 and 1745* (1849).
Finch, G. *The Sketch of the Romish Controversy*, 2 vols. (1850).
Finlason, W. F. *The Catholic Hierarchy Vindicated by the Law of England* (1851).
Fish, H. *Chapters on the Teaching of the Roman Church, proving it to be Unscriptural, Absurd, and Scandalous* (1853).
Fitzgerald, W. See Whitaker, W.
Fleming, R. *An Epistolary Discourse containing a new Resolution and Improvement of the Grand Apocalyptical Question, concerning the Rise and Fall of Rome Papal* (1848).
Fleury, C. *The Ecclesiastical History of M. L'Abbé Fleury* ed. J. H. Newman, 3 vols. (Oxford and London, 1842–4).
Fox, S. *Monks and Monasteries: Being an Account of English Monachism*, The Englishman's Library series, 28 (1845).
Foxe, J. *The Acts and Monuments of John Foxe: With a Life of the Martyrologist, and Vindication of the Work by George Townsend*, 8 vols. (1843–9).
The Book of Martyrs, rev. W. Bramley-Moore (London and New York, 1873).
Popery as it Was and Is: The Suffolk Martyrs, carefully Compiled, without Abridgment, from "Fox's Book of Martyrs", ed. A. J. Green (Sudbury, 1851).
The Sussex Martyrs: Their Examinations and Cruel Burnings in the Time of Queen Mary . . . Extracted from Foxe's "Acts and Monuments", ed. M. A. Lower (Lewes and London, 1851).
Frances, S. *The Nun of Misericordia; or, The Eve of All Saints: A Romance*, 4 vols. (1807).

Frohschammer, J. *The Romance of Romanism: A Discovery & a Criticism* (Edinburgh, 1878).

Froude, J. A. 'England's Forgotten Worthies', in *Essays in Literature & History*, Everyman's Library, 13 (London and Toronto, 1906).

Lectures on the Council of Trent, delivered at Oxford, 1892–3 (London and Bombay, 1896).

The Nemesis of Faith, intro. R. Ashton (1988).

'The Oxford Counter-Reformation', in *Short Studies on Great Subjects*, 5 vols. (1907), v, 166–260.

'The Philosophy of Catholicism', in *Short Studies on Great Subjects*, 5 vols. (1920), I, 163–75.

See also Rowse, A. L. (Secondary texts).

Fuller, T. *The Church-History of Britain, from the Birth of Jesus Christ, untill the year 1648* (1655).

Fullerton, G. *Ellen Middleton: A Tale*, 3 vols. (1844).

Grantley Manor: A Tale, 3 vols. (1847).

Fullwood, F. *Roma Ruit: The Pillars of Rome Broken: wherein all the Several Pleas for the Pope's Authority in England, with all the Material Defences of them, as they have been Urged by Romanists from the Beginning of our Reformation to this Day, are Revised and Answered. To which is subjoined A Seasonable Alarm to all Sorts of Englishmen, against Popery, both from their Oaths and their Interests*, new edn, rev. C. Hardwick (Cambridge and London, 1847).

Galton, A. *The Message and Position of the Church of England: Being an Enquiry into the Claims of the Mediaeval Church, with an Appendix on the Validity of Roman Orders, with a Preface on the Royal Supremacy by J. Henry Shorthouse, Author of 'John Inglesant' etc.* (1899).

Gautrelet, F. X. *A Manual of Devotion to the Sacred Heart of Jesus* (1855).

Geddes, A. *A Modest Apology for the Roman Catholics of Great Britain: Addressed to all Moderate Protestants; Particularly to the Members of both Houses of Parliament* (1800).

Gibbes, P. ('A Lady'). *Friendship in a Nunnery; or, The American Fugitive: Containing a full Description of the Mode of Education and Living in Convent Schools, both on the low and high Pension; the Manners and Characters of the Nuns; the Arts practised on Young Minds; and their Baneful Effects on Society at large*, 2 vols. (1778).

Gibbon, E. *The History of the Decline and Fall of the Roman Empire*, new edn, 4 vols. (1828–9).

The History of the Decline and Fall of the Roman Empire, ed. H. H. Milman, 12 vols. (1838–9).

Gill, H. G. *"Plain Reasons Why I Submitted to the Catholic Church"* (1869).

The Pretensions of Ultramontanism (1873).

Gillies, A. C. *Popery Dissected; its Absurd, Inhuman, Unscriptural, Idolatrous and Antichristian Assumptions, Principles and Practices Exposed from its own Standard Works: Being a Series of Unanswered Letters addressed to the R.C. Bishop of Arichat, N.S.* (Pictou, NS, 1874).

Gillow, J. *A Literary and Biographical History or Bibliographical Dictionary of the English Catholics, from the Breach with Rome, in 1534, to the Present Time* (1885).

Gladstone, J. E. *Protestant Nunneries; or, "The Mystery of Iniquity" Working in the Church of England: A Letter to Sir Culling E. Eardley, Bart. concerning Ann Maria Lane, now a Sister of Mercy, against her Father's Wish, in Miss Sellon's Institution at Eldad, Plymouth; with an Introductory Letter, and Sir Culling E. Eardley's Reply* (1853).

Gladstone, W. E. 'Ward's *Ideal of a Christian Church*', *Quarterly Review*, 75 (1844–5), 149–200.

Remarks on the Royal Supremacy, as it is Defined by Reason, History, and the Constitution: A Letter to the Lord Bishop of London (1850).

See also Newman, J. H. and W. E. Gladstone.

See also Morley, J. (Secondary texts).

Goode, W. *The Divine Rule of Faith and Practice; or, A Defence of the Catholic Doctrine that Holy Scripture has been, since the Times of the Apostles, the Sole Divine Rule of Faith and Practice to the Church: against the Dangerous Errors of the Authors of the Tracts for the Times and the Romanists . . .*, 2nd edn, 3 vols. (1853).

A Letter to the Bishop of Exeter: Containing an Examination of his Letter to the Archbishop of Canterbury, 3rd edn (1850).

Rome's Tactics; or, A Lesson for England from the Past: showing that the great Object of Popery since the Reformation has been to Subvert and Ruin Protestant Churches and Protestant States by Dissensions and Troubles caused by Disguised Popish Agents: with a brief Notice of Rome's Allies in the Church of England (1867).

Gordon, H. *The Present State of the Controversy between the Protestant and Roman Catholic Churches* (1837).

Gorham, G. C. ed., *Examination before Admission to a Benefice by the Bishop of Exeter, followed by Refusal to Institute, on the Allegation of Unsound Doctrine respecting the Efficacy of Baptism* (1848).

The Exeter Synod: A Letter to the Bishop of Exeter on the Diocesan Synod Convened by his Lordship June 25, 1851, with an Appendix containing Lord John Russell's Speech in Parliament on the Synod, the Archibishop of Canterbury's Letter on the Same, and Two Protests of One Hundred and Eleven Clergymen, 2nd edn (London and Exeter, 1851).

Grattan, H. *et al. Necessity for Universal Toleration, exemplified in the Speeches on the Catholic Question in 1805 and 1808, by Mr. Henry Grattan, M.P., Lord Hutchinson, K.B., the Earl of Moira, the Bishop of Norwich; and in the sentiments of Sir J. C. Hippisley . . . Carefully Revised and Corrected from authentic MSS. To which are subjoined a succinct Expression of the Bishop of Llandaff's Opinion . . . and the last French Imperial Decrees respecting the Papal Dominions. With preliminary Observations, by a Protestant Layman* (1808).

Gray, J. *Silverpoints* (1893).

See also Sewell, B. (Secondary texts).

Gregg, T. D. *The Christian Hero Triumphant in his Fall: A Funeral Sermon on Croly, the Historian, the Moralist, the Poet, the Divine* (1861).

Gregg, T. D. and T. Maguire. *Authenticated Report of the Discussion between the Rev. T. D. Gregg, and the Rev. Thomas Maguire*, Church Edn (Dublin, 1839).

Gresley, W. *The Real Danger of the Church of England* (1846).

Greville, C. C. F. *The Greville Memoirs: A Journal of the Reigns of King George IV and King William IV*, ed. H. Reeve, 4th edn, 3 vols. (1875).

Grey, C., Earl. *Earl Grey's Circular: (A Memento)* (1849).

Griffith, E. *Worlds!! Above and Below, with Illustrative Diagrams, showing the Relative Positions of the Universal Kingdoms of our God, as Revealed in the Holy Scriptures* (London and South Norwood, 1885).

Halévy, J. F. F. *The Jewess: Complete Book of Lyrics and Words*, trans. and ed. P. Pinkerton (1900).

Hall, H. E. *Leadership not Lordship: A Series of Short Instructions on the Roman Question, with a hitherto Unpublished Letter on the Subject by the late Canon Liddon* (1892).

Hall, W. J. *The Doctrine of Purgatory and the Practice of Praying for the Dead as Maintained by the Romish Church, Examined* (1843).

Hallam, H. *The Constitutional History of England from the Accession of Henry VII to the Death of George II*, 2 vols. (1827).

Hamilton, A. M. *The Irishwoman in London: A Modern Novel*, 3 vols. (1810).

Hampden, R. D. *Observations on Religious Dissent* (Oxford and London, 1834).

Hare, J. C. and A. W. *Guesses at Truth* (1867).

Harris, E. F. S. *From Oxford to Rome: and how it Fared with some who lately made the Journey, by a Companion Traveller*, rev. edn (1847).

Rest in the Church (1848).

Hawkins, E. *Stand Fast in the Faith: A Sermon, Preached in Curzon Chapel, May Fair, on Sunday, Nov. 3, 1850* (1850).

Haydock, G. L., ed. *The Holy Bible: Translated from the Latin Vulgate . . . with Notes and Illustrations, Compiled by the Rev. George Leo Haydock. Embellished with Beautiful Engravings. The Text carefully Verified, and the Notes Revised, with the Sanction of the Ecclesiastical Dignitaries in Great Britain, by the Very Rev. F. C. Husenbeth, D.D. . . .* (1852).

Herring, R. *A Few Personal Recollections of the late Rev. George Croly, LL.D., Rector of St Stephen's, Walbrook, with Extracts from his Speeches and Writings* (1861).

Heygate, W. E. *Catholic Antidotes* (1858).

Hillier, G. *A Day at Arundel: Comprising the Antiquities of the Castle, Ecclesiastical Associations, and Neighbouring Beauties*, 2nd edn (1851).

Hislop, A. *The Two Babylons; or, The Papal Worship proved to be the Worship of Nimrod and his Wife. With Sixty-one Woodcut Illustrations from Nineveh, Babylon, Egypt, Pompeii, &c*, 3rd edn (Edinburgh and London, 1862).

Hogan, W. *Auricular Confession and Popish Nunneries* (London and Liverpool, 1846).

Hogg, J., ed. *The Jacobite Relics of Scotland: Being the Songs, Airs, and Legends, of the Adherents to the House of Stuart* (Edinburgh and London, 1819).

Holcroft, T. *A Plain and Succinct Narrative of the Gordon Riots, London, 1780*, ed. G. G. Smith (Atlanta, GA, 1944).

Home, T. H. *Mariolatry; or, Facts and Evidences demonstrating the Worship of the Blessed Virgin Mary by the Church of Rome*, 2nd edn (1841).

Hook, W. F. *Auricular Confession: A Sermon* (1848).

Perils of Idolatry: A Sermon (London and Leeds, 1842).

Hooker, R. *The Works of that Learned and Judicious Divine, Mr. Richard Hooker, with an Account of his Life and Death by Isaac Walton*, ed. J. Keble, 3 vols. (Oxford, 1836).

Hopkins, G. M. *The Poems of Gerard Manley Hopkins*, 4th edn, ed. W. H. Gardner and N. H. MacKenzie (Oxford and New York, 1967).

Selected Letters, ed. C. Phillips (Oxford, 1990).

See also Thomas S. J., A.; White, N. (Secondary texts).

Hopkins, J. H. *"The End of Controversy," Controverted: A Refutation of Milner's "End of Controversy," in a Series of Letters addressed to the Most Reverend Francis Patrick Kenrick, Roman Catholic Archbishop of Baltimore*, 2nd edn, 2 vols. (New York, 1855).

Hopkins, S. *The Religious and Political Evils of Catholicism; or, The Protestant Interests in Danger* (1829)

Horne, T. *The Religious Necessity of the Reformation Asserted, and the Extent to which it was Carried in the Church of England Vindicated, in Eight Sermons Preached before the University of Oxford, in the Year MDCCCXXVIII, at the Lecture Founded by the late Rev. John Bampton* (Oxford, 1828).

Howard, H. G. F., Earl of Arundel and Surrey. *A few Remarks on the Social and Political Condition of British Catholics* (1847).

Hull, W. W. *A Statement of some Reasons for Continuing to Protestants the whole Legislature of Great Britain and Ireland: In Reply to the Considerations of the Reverend John Davison, B.D., late Fellow of Oriel College, Oxford* (London and Oxford, 1829).

Hunt, W. H. *Pre-Raphaelitism and the Pre-Raphaelite Brotherhood*, 2 vols. (London and New York, 1905).

Husenbeth, F. C. *The Convert Martyr: A Drama in Five Acts, arranged from "Callista" by Permission of its Author, The Rev. J. H. Newman, D.D.* (1857).

Defence of the Creed and Discipline of the Catholic Church, against the Rev. J. Blanco White's "Poor Man's Preservative against Popery", with Notice of every thing Important in the same Writer's "Practical and Internal Evidence against Catholicism" (1826).

The Life of the Right Rev. John Milner, D.D., Bishop of Castabala, Vicar Apostolic of the Midland District of England, F.S.A. London, and Cath. Acad. Rome (Dublin, 1862).

See also Butler, A.; Haydock, G. L.; Milner, J.

Hussey, R. *The Great Contest: A Sermon preached in the Cathedral of Christ Church, on Easter Day, April 11, 1841* (Oxford and London, 1841).

The Rise of the Papal Power traced in Three Lectures (Oxford, 1851).

Huxley, T. H. *Lectures & Lay Sermons*, intro. O. Lodge, Everyman's Library series, ed. E. Rhys (London and New York, 1910).

Science and Hebrew Tradition: Essays (1904).

Irons, W. J. *The Judgments on Baptismal Regeneration: I. The Church-Court of Arches. II. The State-Court of Privy Council. III. The Present English Bishops. IV. The Present Scottish Bishops . . . ; A Discourse on Heresy and Open Questions is prefixed, by William J. Irons, B.D, Vicar of Brompton* (1850).

Notes of the Church: A Sermon preached on the Twenty-second Sunday after Trinity, at Brompton Church (1845).

The Present Crisis in the Church of England: Illustrated by a Brief Inquiry as to the Royal Supremacy . . . (1850).

Irving, E. *Babylon and Infidelity Foredoomed of God: A Discourse on the Prophecies of Daniel and the Apocalypse, which relate to these Latter Times, and until the Second Advent*, 2 vols. (Glasgow, 1826).

The Signs of the Times (1829).

Irwin, A. *Roman Catholic Morality, as Inculcated in the Theological Class-Books used in Maynooth College* (Dublin and London, 1836).

Jameson, A. *Legends of the Madonna, as Represented in the Fine Arts, forming the Third Series of Sacred and Legendary Art* (1852).

Legends of the Monastic Orders, as Represented in the Fine Arts (1850).

Sacred and Legendary Art, new edn, 2 vols. (1891).

Sisters of Charity, Catholic and Protestant, Abroad and at Home (1855).

Jenkins, R. C. *The Privilege of Peter and the Claims of the Roman Church confronted with the Scriptures, the Councils, and the Testimony of the Popes themselves* (1875).

Romanism: A Doctrinal and Historical Examination of the Creed of Pope Pius IV (1882).

Jerrold, B. *The Life of George Cruikshank, in Two Epochs*, new edn (1882; rpt. Chicheley, 1971).

Johnson, L. *The Gordon Riots*, Historical Papers series, 12 (1893).

Poems (London and Boston, 1895).

Keary, W. *The Comments of the Early Fathers of the Church on Scripture, opposed to the Modern Interpretations of Rome* (1830).

A Common-Place Book to the Fathers, containing a Selection of Passages from the Primitive Writers, Opposed to the Tenets of Romanism (1828).

A Continuation of the Ampleforth Discussion, comprising a Reply to the Letter of the Rev. Richard Towers (1833).

Keble, J. *The Case of Catholic Subscription to the Thirty-Nine Articles Considered: with especial Reference to the Duties and Difficulties of English Catholics in the Present Crisis: in a Letter to the Hon. Mr. Justice Coleridge* (1841).

The Christian Year: Thoughts in Verse for the Sundays and Holydays throughout the Year, . . . with Twenty-Four Illustrations by Fr. Overbeck, reproduced in permanent Photography (1875).

Church Matters in MDCCCL, No. 1 – Trial of Doctrine, No. 2 – A Call to Speak Out (1850).
A few very plain Thoughts on the proposed Admission of Dissenters to the University of Oxford (Oxford, 1854).
The First Edition of Keble's Christian Year, being a Facsimile of the Editio Princeps published in 1827, with a Preface by the Bishop of Rochester [E. S. Talbot], and a List of Alterations made by the Author in the Text of later Editions (1897).
Heads of Consideration on the Case of Mr. Ward (Oxford and London, 1845).
Sermons, Academical and Occasional (Oxford, 1847).
See also Newman, J. H., E. B. Pusey and J. Keble, *et al.*
Kennedy, G. *The Decision; or, Religion must be All, or is Nothing* (Edinburgh, 1821).
Father Clement: A Roman Catholic Story (Edinburgh, 1823).
The Works of Grace Kennedy, Author of "The Decision", 6 vols. (Edinburgh, 1827).
Kennedy, J. *The Impregnable Fortress* (Bath, 1852).
Kerns, T. *Analogia; or, Brief Notes of Pagan Idolatry and the Church of Rome*, 2nd edn, enlarged (Sheffield, 1851).
Kingsley, C. *Alton Locke, Tailor and Poet*, ed. H. Van Thal, intro. D. Lodge, First Novel Library series (1967).
Hypatia (1903).
Westward Ho! (London and Glasgow, n.d.).
'Why should we Fear the Romish Priests?', *Fraser's Magazine*, 37 (April 1848), 467–74.
Yeast: A Problem, Works of Charles Kingsley, 2 (1879).
See also Kingsley, F.; Ogilvie, G. S.
See also Chitty, S.; Morris, K. L. (Secondary texts).
Kingsley, F., ed. *Charles Kingsley: His Letters and Memories of His Life*, 2 vols., 2nd edn (1877).
Klingberg, F. J. and S. B. Hustvedt, eds. *The Warning Drum: The British Home Front Faces Napoleon: Broadsides of 1803* (Berkeley and Los Angeles, CA, 1944).
Knox, R. A. *Occasional Sermons of Ronald A. Knox*, ed. P. Caraman S. J. (1960).
Laing, F. H. *Catholic the same in Meaning as Sovereign, in which the genuine Nature of Catholicity is relieved from the false Notions with which the Protestant Usurpation of the Word has Embarassed it* (1848).
Leger, W. N. St. *Gorham v. the Bishop of Exeter, Protestant v. Catholic, Heresy v. Truth: "A Tract for the Times"* (1850).
Leslie, C. *A Short Method with the Romanists; or, The Claims and Doctrines of the Church of Rome Examined, in a Dialogue between a Protestant and a Romanist* (Edinburgh, 1835).
Lewis, M. *The Monk: A Romance*, ed. H. Anderson (1973).
Lewis, T. *The History of Hypatia, a most Impudent School-Mistress of Alexandria, Murder'd and Torn to Pieces by the Populace: In Defence of Saint Cyril and the Alexandrian Clergy from the Aspersions of Mr. Toland* (1721).
Liddon, H. P. *Life of Edward Bouverie Pusey, Doctor of Divinity, Canon of Christ Church, Regius Professor of Hebrew in the University of Oxford*, ed. J. O.

Johnston and R. J. Wilson, 2nd edn, 4 vols. (London and New York, 1893).
Lingard, J. *A Collection of Tracts, on Several Subjects, connected with the Civil and Religious Principles of Catholics* (1826).
The History and Antiquities of the Anglo-Saxon Church; containing an Account of its Origin, Government, Doctrines, Worship, Revenues, and Clerical and Monastic Institutions, 2 vols. (1845).
A History of England from the First Invasion by the Romans to the Revolution in 1688, 8 vols. (1819–30).
('by a Catholic') *A New Version of the Four Gospels, with Notes, Critical and Explanatory* (1836).
Observations on the Laws and Ordinances which exist in Foreign States, relative to the Religious Concerns of their Roman Catholic Subjects (1851).
Tracts occasioned by the Publication of a Charge, delivered to the Clergy of the Diocese of Durham, by Shute, Bishop of Durham, in 1806 (Newcastle, 1813).
See also Talbot, W.; White, T.
See also Halle, M. ('M. Haile') and E. Bonney (Secondary texts).
Littledale, R. F. *The Petrine Claims: A Critical Inquiry* (1889).
Plain Reasons against Joining the Church of Rome, rev. edn, 40th thousand (1886).
Lockhart, J. G. *The Life of Sir Walter Scott*, Edinburgh Edn, 10 vols. (Edinburgh, 1902–3).
Lord, J. *The Vatican and St James's; or, England, Independent of Rome: with Introductory Remarks on Spiritual and Temporal Power* (1851).
Luke, J. T. *The Female Jesuit: or, The Spy in the Family* (1851).
A Sequel to The Female Jesuit, containing her Previous History and Recent Discovery, 2nd thousand (1852).
Macaulay, T. B. *Essays and Lays of Ancient Rome*, Popular Edn (1899).
The History of England from the Accession of James II, 4 vols. (Philadelphia, PA, 1856).
'Von Ranke', *Edinburgh Review*, 72, 145 (October 1840), 227–58; rpt. in *Critical and Historical Essays, contributed to the Edinburgh Review*, 5th edn, 3 vols. (1848), III, 207–54.
See also Moncrieff, J.; Otto, G.; Trevelyan, G. O.
MacCabe, W. B. *A Catholic History of England: Part I, England, its Rulers, Clergy, and Poor, before the Reformation, as described by the Monkish Historians*, 3 vols. (1847–54).
McNeile, H. *The Church Establishment: Speech of the Rev. Hugh McNeile, in Defence of the Established Church, at the Second Annual Meeting of the Protestant Association, at Exeter Hall, May 10, 1837*, 3rd edn (1839).
Popery Theological, Another Challenge!: Reply of the Rev. Hugh McNeile, Rector of Albury, Surrey, to the Rev. Joseph Sidden, Roman Catholic Priest, Sutton Park, Surrey (1829).
Madden, R. R. *The History of the Penal Laws enacted against Roman Catholics; the Operation and Results of that System of Legalized Plunder, Persecution, and Proscription; Originating in Rapacity and Fraudulent Designs, concealed under*

False Pretences, Figments of Reform, and a Simulated Zeal for the Interests of True Religion (1847).

Maguire, J. F. *Pius the Ninth*, new edn, rev. J. L. Patterson (1878).

Malet, W. W. *The Olive Leaf: A Pilgrimage to Rome, Jerusalem, and Constantinople, in 1867, for the Reunion of the Faithful* (1868).

Manning, H. E. *A Charge delivered at the Ordinary Visitation of the Archdeaconry of Chichester, in July, 1849* (1849).

The Four Great Evils of the Day, 2nd edn (1871).

A Pastoral Letter to the Clergy and Laity of the Diocese of Westminster (1865).

Temporal Sovereignty of the Popes: Three Lectures . . . delivered in the Church of St Mary of the Angels, Bayswater (1860).

The True Story of the Vatican Council (1877).

Vatican Decrees in their bearing on Civil Allegiance (1875).

See also Purcell, E. S.

See also McClelland, V. A. (Secondary texts).

Marriott, W. B. *The Testimony of the Catacombs and of other Monuments of Christian Art, from the Second to the Eighteenth Century, concerning Questions of Doctrine now disputed in the Church* (1870).

Marsh, H. *A Comparative View of the Churches of England and Rome* (Cambridge, 1814).

Mayer, L. *A Hint to England; or, A Prophetic Mirror: Containing an Explanation of Prophecy that relates to the French Nation, and the Threatened Invasion, proving Bonaparte to be the BEAST that arose out of the Earth, with Two Horns like a Lamb, and spake as a Dragon, whose Number is 666 . . . Rev. xiii* (1803).

Meek, R. *Reasons for Attachment and Conformity to the Church of England* (1831).

Melvill, H. *Protestantism and Popery, extracted, by permission, from his second volume of Sermons, published by RIVINGTONS, London, 1838* (1839).

Mendham, J. *The Literary Policy of the Church of Rome Exhibited, in an Account of her Damnatory Catalogues or Indexes, both Prohibitory and Expurgatory, with Various Illustrative Extracts, Anecdotes, and Remarks*, 2nd edn (1830).

Merle D'Aubigné, J.-H. *History of the Great Reformation of the Sixteenth Century in Germany, Switzerland, &c.*, trans. D. Walther, 3 vols. (1838–41).

The Reformation in England, ed. S. M. Houghton, 2 vols. (1962).

Methodist Hymn-Book, The, with Tunes (1954).

Meynell, A. *Mary, the Mother of Jesus: An Essay*, illus. R. A. Bell (1912).

The Poems of Alice Meynell, complete edn (1923).

See also Meynell, V.

See also Badeni, J. (Secondary texts).

Michelet, J. *The Confessional and the Conventual System: Extracted from Michelet's "Priests, Women, and Families"*, Hints to Romanizers series, 1 (1850).

The People, trans. C. Cocks, 3rd edn (1846).

Priests, Women, and Families, trans. C. Cocks, 11th edn (1846).

Miles, M. M, ed. *Maiden & Mother: Prayers, Hymns, Songs and Devotions to Honour the Blessed Virgin Mary throughout the Year* (2001).

Millais, J. E. See Millais, J. G.

Millais, J. G. *The Life and Letters of Sir John Everett Millais, President of the Royal Academy*, 2 vols. (1899).

Miller, H. *The Testimony of the Rocks; or, Geology in its Bearings on the Two Theologies, Natural and Revealed*, 28th thousand (Edinburgh and London, 1862).

Miller, J. C. *"Subjection; No, not for an Hour:" A Warning to Protestant Christians, in behalf of the "Truth of the Gospel," as now Imperilled by the Romish Doctrines and Practices of the Tractarian Heresy: Being the Substance of a Sermon preached in Saint Martin's Church, Birmingham, on Sunday Evening, September 8, 1850*, 3rd edn (1850).

Mills, A. J. F. *The History of Riots in London in the year 1780, commonly called the Gordon Riots* (1883).

Milman, A. *Henry Hart Milman, D.D., Dean of St Paul's: A Biographical Sketch* (1900).

Milman, H. H. *The History of Christianity, from the Birth of Christ to the Abolition of Paganism in the Roman Empire*, new and rev. edn, 3 vols. (1875).

History of Latin Christianity, including that of the Popes, to the Pontificate of Nicolas V, 4th edn, 9 vols. (1883).

See also Gibbon, E.; Milman, A.

Milner, J. *The End of Religious Controversy, in a Friendly Correspondence between a Religious Society of Protestants, and a Roman Catholic Divine, Addressed to the Right Rev. Dr. Burgess, Lord Bishop of St David's, in Answer to his Lordship's* Protestant Catechism, 3 parts (1818).

The History Civil and Ecclesiastical, and Survey of the Antiquities of Winchester, 2 vols. (Winchester, 1798–9).

The History and Survey of the Antiquities of Winchester, with supplementary Notes; also a Biographical Memoir by F. C. Husenbeth, 3rd edn, 2 vols. (Winchester, 1839).

Letters to a Prebendary: Being an Answer to Reflections on Popery, by the Rev. J. Sturges, LL. D., Prebendary and Chancellor of Winchester, and Chaplain to his Majesty; with Remarks on the Opposition of Hoadlyism to the Doctrines of the Church of England, and on Various Publications, occasioned by the late Civil and Ecclesiastical History of Winchester (Winchester, 1800).

A Vindication of the End of Religious Controversy, from the Exceptions of the Right Rev. Dr. Thomas Burgess, Bishop of St David's, and the Rev. Richard Grier, A.M., Vicar of Templebodane . . . in Letters to a Catholic Convert (1822).

See also Husenbeth, F. C.

See also Couve de Murville, M. N. L.; Patten, B. (Secondary texts).

Mitford, J. T. F., Earl of Redesdale. *The Infallible Church and the Holy Communion of Christ's Body and Blood: Correspondence between Lord Redesdale and Cardinal Manning in the "Daily Telegraph"* (1875).

Moberly, G. *The Sayings of the Great Forty Days, between the Resurrection and Ascension, regarded as the Outlines of the Kingdom of God, in Five Discourses: with an Examination of Mr. Newman's Theory of Developments*, 3rd edn (London and Winchester, 1846).

Moncrieff, J. 'Macaulay's *History of England*', *Edinburgh Review*, 90, 181 (July 1849), 249–92.

Monro, E. *Reasons for Feeling Secure in the Church of England: A Letter to a Friend, in Answer to Doubts Expressed in Reference to the Claims of the Church of Rome* (1850).

Monsell, J. S. B. *The Parish Hymnal, after the Order of the Book of Common Prayer* (1873).

Montague, E. *The Legends of a Nunnery: A Romance*, 4 vols. (1807).

Montgomery, R. *The Omnipresence of the Deity: A Poem* (1828).

Moore, G. *A Drama in Muslin: A Realistic Novel*, Vizetelly's One-Volume series, 15 (1886).

Evelyn Innes (1898).

Sister Teresa (1901).

See also Frazier, A. (Secondary texts).

Moore, T. and A. Brinckman. *The Anglican Brief against Roman Claims* (1893).

Morison, J. *Homilies for the Times; or, Rome and her New Allies: A Plea for the Reformation* (1841).

Morris, F. O. ('A Yorkshire Clergyman'). *A Letter to Archdeacon Wilberforce on Supremacy* (London and York, 1854).

Neale, J. M. *Ayton Priory; or, The Restored Monastery* (Cambridge and London, 1843).

A few Words of Hope on the present Crisis of the English Church (1850).

A History of the Holy Eastern Church: The Patriarchate of Alexandria (1847).

The Lewes Riot: Its Causes and Consequences (1857).

See also Towle, E. A. (Secondary texts).

Neander, J. A. W. *The History of the Christian Religion and Church during the Three First Centuries*, trans. H. J. Rose, 2 vols. (1831–41).

History of the Planting and Training of the Christian Church by the Apostles, with the Author's Final Additions; also, his Antignostikus; or, Spirit of Tertullian, trans. J. E. Ryland, 2 vols. (1859).

Newdegate, C. N. *Monastic and Conventual Institutions* (Edinburgh, 1866).

Newland, H. *Whom has the Pope Aggrieved?* (1850).

Newman, J. H. *Apologia Pro Vita Sua: Being a Reply to a Pamphlet entitled "What, then, does Dr. Newman mean?"* (1864).

Apologia Pro Vita Sua: Being a History of his Religious Opinions, ed. M. J. Svaglic (Oxford, 1967).

The Arians of the Fourth Century, 3rd edn (1871).

Callista: A Sketch of the Third Century (1876).

Christ upon the Waters: A Sermon preached in substance at St Chad's, Birmingham, on Sunday, October 27, 1850, on occasion of Establishment of the Catholic Hierarchy in this Country (Birmingham and London, 1850).

Discussions and Arguments on Various Subjects (1872).

An Essay in Aid of a Grammar of Assent, ed. I. T. Ker (Oxford, 1985).

An Essay on the Development of Christian Doctrine (1845).

An Essay on the Development of Christian Doctrine, new edn (1878).

An Essay on the Development of Christian Doctrine, ed. C. F. Harrold (New York, 1949).

An Essay on the Development of Christian Doctrine: The Edition of 1845, ed. J. M. Cameron (Harmondsworth, 1974).

Essays, Critical and Historical, 2nd edn, 2 vols. (1872).

The Idea of a University, Defined and Illustrated: I. In Nine Discourses delivered to the Catholics of Dublin, II. In Occasional Lectures and Essays addressed to the Members of the Catholic University, ed. I. T. Ker (Oxford, 1976).

Lectures on certain Difficulties felt by Anglicans in Submitting to the Catholic Church, 2nd edn (1850).

Lectures on the Present Position of Catholicism in England (1850).

A Letter Addressed to His Grace the Duke of Norfolk on occasion of Mr. Gladstone's Recent Expostulation (1875).

A Letter to the Rev. E.B. Pusey, D.D. on his recent Eirenicon, 2nd edn (1866).

The Letters and Diaries of John Henry Newman, ed. C. S. Dessain *et al.*, 31 vols. (London and Oxford, 1961–77).

Loss and Gain (1848).

Loss and Gain: The Story of a Convert, ed. A. G. Hill, World's Classics series (Oxford and New York, 1986).

Mary: The Virgin Mary in the Life and Writings of John Henry Newman, ed. P. Boyce (Leominster and Grand Rapids, MI, 2001).

Newman The Oratorian: His unpublished Oratory Papers, with an Introductory Study on the Continuity between his Anglican and his Catholic Ministry, ed. P. Murray O.S.B. (Dublin, 1969).

Sermons bearing on Subjects of the Day, new edn (1869).

The Theological Papers of John Henry Newman on Biblical Inspiration and on Infallibility, ed. J. D. Holmes (Oxford, 1979).

Verses on Various Occasions, 2nd edn (1869).

See also Fleury, C.

See also Dessain, C. S.; Gilley, S.; Ker, I. T.; Lash, N.; MacDougall O. M. I., H.A.; Trevor, M.; Turner, F. M.; Ward, W. (Secondary texts).

Newman, J. H., E. B. Pusey, J. Keble, *et al. Tracts for the Times*, 1–90 (London and Oxford, 1833–41).

Newman, J. H. and W. E. Gladstone. *Newman and Gladstone: The Vatican Decrees*, intro. A. S. Ryan (Notre Dame, IN, 1962).

Newton, T. *Dissertations on the Prophecies, which have been remarkably fulfilled, and at this time are fulfilling in the world*, 3 vols. (1754–8).

Nicholson, A. *A Reply to Cardinal Manning's recently acknowledged Essay entitled "Dr. Nicholson's Accusation of the Archbishop of Westminster"* (1878).

Niebuhr, B. G. *The History of Rome*, trans. J. C. Hare, C. Thirwall and W. Smith, 3 vols. (Cambridge, 1828–32).

Northcote, J. S. 'The Roman Catacombs', *Rambler* (10 June–12 August 1848), 124–7, 222–5, 246–9, 340–3.

Oakeley, F. *A Letter on Submitting to the Catholic Church: Addressed to a Friend*, 2nd edn (1845).

The Youthful Martyrs of Rome: A Christian Drama, adapted from "Fabiola; or, The Church of the Catacombs" (1856).
O'Conor, C. ('Columbanus'). *Columbanus No. VII: The Gallican Liberties, Indispensable Securities for the Constitutional Government of the Irish Catholic Church* (London and Dublin, 1816).
Ogilby, J. D. *The Catholic Church in England and America: Three Lectures . . .* (New York and Philadelphia, PA, 1844).
Ogilvie, G. S. *Hypatia: A Play in Four Acts, founded on Charles Kingsley's Novel* (1894).
O'Hagan, T. *The Address of the Catholic Archbishops and Bishops of Ireland, to their beloved Flocks, on the Penal Enactment with which the Catholics of England and Ireland are Threatened. To which is added an Appendix, containing the Opinion of Thomas O'Hagan, Esq., Q.C.* (Dublin, 1851).
O'Meara, K. *Thomas Grant: First Bishop of Southwark*, 2nd edn (1886).
Oliphant, M. 'Madonna Mary', *Good Words*, 7 (1866), 1–17, 72–88, 145–60, 217–31, 289–303, 361–77, 433–46, 505–22, 577–91, 649–62, 721–36, 793–806.
Oliver, S. *Dr. Newman's Apologies: Being Four Letters addressed to the Editors of "The Undergraduate's Journal"* (Oxford and Cambridge, 1878).
Otto, G. *The Life and Letters of Lord Macaulay*, new impression (1899).
Ouseley, G. *Old Christianity against Papal Novelties, including a Review of Dr. Milner's "End of Controversy"* (Philadelphia, PA, 1842).
Oxenham, F. N. *The Validity of Papal Claims: Five Lectures delivered in Rome* (1897).
Pagani, J. B. *The Pillar and Foundation of Truth* (Prior Park, 1840).
Paley, W. *A View of the Evidences of Christianity, in Three Parts*, 12th edn, 2 vols. (1807).
Palmer, W. *Origines Liturgicæ; or, Antiquities of the English Ritual, and a Dissertation on Primitive Liturgies*, 4th edn, 2 vols. (1845).
A Supplement to the First Three Editions of the Origines Liturgicæ, *comprising Additions made in the Fourth Edition* (1845).
Patmore, C. *Poems*, intro. B. Champneys (1906).
See also Reid, J. C. (Secondary texts).
Patterson, J. W. *The Church of England* versus *The Roman Church in England: A Plain Answer to the Perversions of a Pervert* (1872).
Perceval, A. P. *The Roman Schism Illustrated, from the Records of the Catholic Church* (1836).
Phillipps, A. L. [later de L.]. *A Letter to the . . . Earl of Shrewsbury . . . on the Re-establishment of the Hierarchy of the English Catholic Church, and the present Posture of Catholic Affairs in Great Britain* (1850).
See also Purcell, E. S.
See also Pawley, M. (Secondary texts).
Phillips, E., ed. *An Octave of Catholic Prayers for the Eventful Year 1851* (1851).
Phillpotts, H. *Acts of the Diocesan Synod, held in the Cathedral Church of Exeter, by Henry, Lord Bishop of Exeter, on Wednesday, Thursday, and Friday, June 25, 26, 27, of the Year of our Lord 1851, By Authority* (1851).
A Letter to the Archbishop of Canterbury from the Bishop of Exeter (1850).

Letters to Charles Butler, Esq. on the Theological Parts of his Book of the Roman Catholic Church, with Remarks on certain Works of Dr. Milner, and Dr. Lingard, and on some Parts of the Evidence of Dr. Doyle before the two Committees of the Houses of Parliament (1825).

See also Shutte, R. N.

Pollok, R. *The Course of Time: A Poem*, 26th edn (Edinburgh and London, 1867).

Poole, G. A. *The Testimony of Saint Cyprian against Rome: An Essay towards Determining the Judgment of Saint Cyprian touching Papal Supremacy* (London and Edinburgh, 1828).

Pope, R. T. P. *Roman Misquotation; or, Certain Passages from the Fathers, adduced in a Work entitled "The Faith of Catholics," &c. brought to the Test of the Originals, and their Perverted Character Demonstrated* (1840).

Powell, H. T. *Strictures on Dr. Arnold's Pamphlet, entitled, "The Christian Duty of Granting the Claims of the Roman Catholics"* (1829).

Price, B. 'The Anglo-Catholic Theory', *Edinburgh Review*, 94, 192 (October 1851), 527–57.

'Tracts for the Times – Number Ninety', *Edinburgh Review*, 73, 147 (April 1841), 271–97.

Price, E. *Sick Calls: From the Diary of a Missionary Priest, mostly Republished from "Dolman's Magazine"*, 2nd edn (1855).

Pugin, A. W. N. *Contrasts; or, A Parallel between the Noble Edifices of the Middle Ages, and Corresponding Buildings of the Present Day* (1836).

An Earnest Address, on the Establishment of the Hierarchy (1851).

Purcell, E. S. *Life of Cardinal Manning, Archbishop of Westminster*, 2 vols. (London and New York, 1896).

Life and Letters of Ambrose Phillipps de Lisle (1900).

Pusey, E. B. *The Church of England, a portion of Christ's one Holy Catholic Church, and a means of restoring visible unity: An Eirenicon. Part 1. A letter to the author of "The Christian Year." Sixth thousand.-Part 2. First Letter to . . . J. H. Newman . . . in explanation chiefly in regard to the reverential love due to the ever-blessed Theotokos, and the doctrine of her Immaculate Conception; with an analysis of Cardinal de Turrecremata's work on the Immaculate Conception.-Part 3. Is healthful Reunion impossible? A second letter to . . . J. H. Newman, etc.*, 3 parts (Oxford, 1865–70).

See also Liddon, H. P.; Newman, J. H., J. Keble, *et al.*

Pyer, J. *Russell* v. *Wiseman; or, Reason* v. *Opinion: An Appeal to the Lords* (1851).

Radcliffe, A. *The Italian; or, The Confessional of the Black Pentitents: A Romance*, 3 vols. (1797).

Ranke, von L. *The Ecclesiastical and Political History of the Popes of Rome during the Sixteenth and Seventeenth Centuries*, trans. S. Austin, 3 vols. (1840).

History of the Reformation in Germany, 2nd edn, trans. S. Austin, 3 vols. (1845).

Reade, C. *The Cloister and the Hearth* (1861).

Reeve, J. *A General History of the Christian Church, from its First Establishment to the Present Century* (Dublin, 1851).

A Short View of the History of the Christian Church, from its First Establishment to the Present Century, 3 vols. (Exeter and London, 1802).

Reid, A. ed. *Why I am a Liberal: Being Definitions and Personal Confessions of Faith by the Best Minds of the Liberal Party* (1885).

Rickman, T. C. ('A Citizen of the World'). *Atrocities of a Convent; or, The Necessity of Thinking for Ourselves, exemplified in the History of a Nun*, 3 vols. (1808).

Robinson, E. *Westminster Abbey; or, The Days of the Reformation*, Routledge's Railway Library series (1859).

Whitefriars; or, The Days of Charles the Second: An Historical Romance, 3rd edn, 3 vols. (1844).

Rogers, H. 'Puseyism, or the Oxford Tractarian School', *Edinburgh Review*, 77, 156 (April 1843), 501–62.

'Recent Developments of Puseyism', *Edinburgh Review*, 80, 162 (October 1844), 310–75.

'Ultramontane Doubts', *Edinburgh Review*, 93, 190 (April 1851), 535–78.

Ruskin, J. *The Diaries of John Ruskin*, ed. J. Evans and J. H. Whitehouse, 3 vols. (Oxford, 1936–9).

The Works of John Ruskin, ed. E. T. Cook and A. Wedderburn, 39 vols. (London and New York, 1903–12).

Russell, C. A. *The Catholic in the Workhouse: Popular Statement of the Law as it Affects him, the Religious Grievances it Occasions, with Practical Suggestions for Redress* (London and Dublin, 1859).

Russell, J. See Walpole, S.

Russell, O. *The Roman Question: Extracts from the Despatches of Odo Russell from Rome, 1858–1870*, ed. N. Blakiston (1962).

Sall, A. *True Catholic and Apostolic Faith maintained in the Church of England, to which is prefixed a Sermon preached by him at Christ Church, Dublin . . . and a Preface shewing the Reasons for Deserting the Communion of the Roman Church, and Embracing that of the Church of England*, new edn, rev. and ed. J. Allport (London and Nottingham, 1840).

Salmon, G. *The Infallibility of the Church: A Course of Lectures delivered in the Divinity School of the University of Dublin* (1888).

Sanford, G. B. *The Reasons of a Romanist Considered: A Letter to the Hon. and Rev. George Spencer, containing Remarks upon the Reasons which he assigned, in a Sermon preached at Manchester, as the Cause of his Conversion* (Oxford, 1840).

Schulte, J. *Roman Catholicism, Old and New, from the Standpoint of the Infallibility Doctrine* (Toronto, 1876).

Scott, T. ed. *The Holy Bible*, with Explanatory Notes, Practical Observations, and Copious Marginal References, 9th edn, 6 vols. (1825).

Scott, W. *Redgauntlet: A Tale of the Eighteenth Century*, Everyman's Library series (New York, 1906).

Waverley, ed. A. Hook (Harmondsworth, 1972).

See also Johnson, E. (Secondary texts)

Scudamore, W. E. *England and Rome: A Discussion of the Principal Doctrines and Passages of History in Common Debate between the Members of the Two Communions* (1855).

Letters to a Seceder from the Church of England to the Communion of Rome (1851).

Selden, C. *The English Nun: A Novel* (1797).

Sewell, E. M. *Margaret Percival*, ed. W. Sewell, 2 vols., 3rd edn (1847).

Sewell, W. *Hawkstone: A Tale of and for England in 184–*, 2 vols. (1845).

The Plea of Conscience for Seceding from the Catholic Church to the Romish Schism in England: A Sermon preached before the University of Oxford, November 5, 1845, . . . to which is prefixed an Essay on the Process of Conscience, 2nd edn (Oxford, 1845).

See also Sewell, E. M.

Seymour, M. H. *Convents or Nunneries: A Lecture in Reply to Cardinal Wiseman, delivered at the Assembly Rooms, Bath, on Monday, June 7, 1852, reported in Short-hand* (Bath and London, 1852).

Mornings Among the Jesuits at Rome (1849).

The Nature of Romanism, as exhibited in the Missions of the Jesuits and other Orders, . . . delivered before the Young Men's Christian Association, in Exeter Hall, December 18, 1849, Robert C. L. Bevan in the Chair (1849).

Nunneries: A Lecture delivered at the Assembly Rooms, Bath, on Wednesday, April 21, 1852, reported in Short-hand, 3rd thousand (Bath and London, 1852).

Shaw, T. H. *The McPhersons; or, Is the Church of Rome making progress in England? To which is added England's Glory! the Roll of Honour* (1879).

Shorthouse, J. H. *John Inglesant: A Romance* (1911).

See also Galton, A.; Shorthouse, S.

Shorthouse, S. ed. *Life, Letters, and Literary Remains of J. H. Shorthouse*, 2 vols. (London and New York, 1905).

Shrewsbury, Earl of, *et al. Declaration of the Roman Catholic Laity of England* (1851).

Shutte, R. N. *The Life, Times, and Writings of the Right Rev. Dr. Henry Phillpotts, Lord Bishop of Exeter*, vol. 1 only (1863).

Simpson, R. *Edmund Campion: A Biography* (London and Edinburgh, 1867).

Richard Simpson as Critic, ed. D. Carroll, Routledge Critics series, ed. B. C. Southam (1977).

Sinclair, C. *Beatrice; or, The Unknown Relations*, 3 vols. (1852).

Sinclair, J. *An Essay on Papal Infallibility* (1850).

Skeats, H. S. *A History of the Free Churches of England, from A.D. 1688–A.D. 1851* (1868).

History of the Free Churches of England . . . From the Reformation to 1851 . . . With a continuation to 1891 by C. S. Miall (1891).

Slade, Mrs E. *Nunnery Tales; or, The Amours of Monks and Nuns* (1743).

Smith, C. *An Inquiry into the Catholic Truths hidden under Certain Articles of the Creed of the Church of Rome*, 4 pts (1844–56).

Smith, J. F. *The Jesuit*, 3 vols. (1832).

Snape, A. *A Second Letter to the Lord Bishop of Bangor, in Vindication of the Former* (1717).

Soames, H. *The Anglo-Saxon Church: Its History, Revenues and General Character*, 3rd edn (1844).

Elizabethan Religious History (1839).

The History of the Reformation of the Church of England, 4 vols. (1826–8).

Reasons for Opposing the Romish Claims (1829).

Southey, R. *The Book of the Church*, 2 vols. (1824).

Essays, Moral and Political, 2 vols. (1832).

Sir Thomas More; or, Colloquies on the Progress and Prospects of Society, 2 vols. (1829).

Vindiciae Ecclesiae Anglicanae: Letters to Charles Butler, Esq. comprising Essays on the Romish Religion and Vindicating The Book of the Church (1826).

Spielmann, M. H. *The History of "Punch"* (1895).

Stanford, C. S. *A Handbook of the Romish Controversy: being a Refutation in Detail of the Creed of Pope Pius the fourth, on the Grounds of Scripture and Reason: with an Appendix and Notes*, new edn, 27th thousand (Dublin, 1864).

Stanley, A. P. 'The Gorham Controversy', *Edinburgh Review*, 92, 185 (July 1850), 263–92.

Historical Memorials of Westminster Abbey (1868).

Sturges, J. *Reflections on the Principles and Institutions of Popery, with reference to Civil Society and Government, especially that of this Kingdom; occasioned by the Rev. John Milner's History of Winchester* (Winchester, 1799).

Sumner, C. R. See Sumner, G. H.

Sumner, G. H. *Life of Charles Richard Sumner, D.D., Bishop of Winchester, and Prelate of the Most Noble Order of the Garter, during a Forty Years' Episcopate* (1876).

Tait, W. *The Bible or Rome?: A Question for 1852*, 4th edn (London and Wakefield, 1852).

Talbot, W. ('Christianus'). *The Protestant Apology for the Roman Catholic Church; or, the Orthodoxy, Purity, and Antiquity of her Faith and Principles Proved, from the Testimony of her most Learned Adversaries, by Christianus. Prefixed is an Introduction, concerning the Nature, present State, and true Interests of the Church of England; and on the Means of Effecting a Reconciliation of the Churches: with Remarks on the false Representations, repeated in some late Tracts, of several Catholic Tenets, particularly the Supremacy of the See of Rome, by Irenæus* [J. Lingard] (Dublin, 1809).

Tayler, W. *Popery: Its Character and its Crimes* (1847).

Thackeray, W. M. 'A Dream of Whitefriars', *Punch*, 19 (July–December 1850), 184–5.

The History of Henry Esmond, ed. J. Sutherland and M. Greenfield, Penguin English Library series (1970).

Thompson, F. *The Works of Francis Thompson*, 3 vols. (1913).

Thompson, J. See Luke, J. T.

Thomson, K. *The Chevalier: A Romance of the Rebellion of 1745*, 3 vols. (1844).

Memoirs of the Jacobites of 1715 and 1745, 3 vols. (1845).

Tickell S. J., G., trans. *Month of the Sacred Heart of Jesus* (1858).

Tierney, M. A. *The History and Antiquities of the Castle and Town of Arundel, including the Biography of its Earls, from the Conquest to the Present Time* (1834).

A Letter to the Very Rev. G. Chandler, D.C.L., Dean of Chichester, and Rector of All Souls', Langham Place, etc., etc., containing some Remarks on his Sermon, Preached in the Cathedral Church of Chichester, on Sunday, October 15th, 1843, "On the Occasion of Publicly Receiving into the Church a Convert from the Church of Rome" (1844).

See also Dodd, C.

Tilt, J. *Arundel Castle, and other Poems* (1849).

Historical Ballads, illustrative of the History of England, with other Poems, 6th edn, with additions (1852).

Poems and Ballads (1847).

Todd, J. H. *The Search after Infallibility: Remarks on the Testimony of the Fathers to the Roman Dogma of Infallibility* (London and Dublin, 1848).

Toland, J. *Hypatia; or, The History of a most beautiful, most vertuous, most learned, and every way accomplish'd Lady; who was torn to Pieces by the Clergy of Alexandria, to gratify the Pride, Emulation, and Cruelty of their Archbishop, commonly but undeservedly stiled St Cyril* (1753).

Tonna, C. E. *The Rockite: An Irish Story* (1829).

Towers, R. *A Second Letter to the Rev. Thos. Comber, Rector of Oswaldkirk, in Reply to his and other Charges* (York, 1833).

Townsend, G. *The Accusations of History against the Church of Rome Examined, in Remarks on many of the Principal Observations in the Work of Mr. Charles Butler, entitled the "Book of the Roman Catholic Church"* (1825).

See also Foxe, J.

Trapp, J. *The Sprite of the Nunnery: A Tale from the Spanish*, 2 vols. (1796).

Trevelyan, G. O. *The Life and Letters of Lord Macaulay*, new impression (1899).

Trollope, A. *Clergymen of the Church of England*, intro. R. apRoberts (Leicester, 1974).

Trollope, F. *Father Eustace: A Tale of the Jesuits*, 3 vols. (1847).

Tupper, M. F. *Three Hundred Sonnets* (1860).

Turner, S. *The Modern History of England: Pt I, The History of the Reign of Henry the Eighth, comprising the Political History of the Commencement of the English Reformation; Pt II, The History of the Reigns of Edward the Sixth, Mary, and Elizabeth* (1826–9).

Turnour, E. J. *"Utrum Horum Mavis Accipe!": Tierney's Libel Exposed, in a Comparison of the Words, and Actions, of Protestant, and Papist, Ministers of the Gospel, Extracted from the Address to the Reader in the Sixth Volume of Sermons* (1831).

Twiss, T. *The Letters Apostolic of Pope Pius IX., considered with Reference to the Law of England and the Law of Europe* (1851).

Tyler, J. E. *The Image-Worship of the Church of Rome proved to be Contrary to Holy Scripture, and the Faith and Discipline of the Primitive Church, and to*

involve Contradictory and Irreconcilable Doctrines within the Church of Rome itself (1847).
Tyrrell S. J., G. *External Religion: Its Use and Abuse* (1899).
See also Sagovsky, N. (Secondary texts).
Ullathorne, W. B. *The Döllingerites, Mr. Gladstone, and Apostates from the Faith* (1875).
From Cabin-Boy to Archbishop (1891).
History of the Restoration of the Catholic Hierarchy in England (1871).
Letters, ed. the nuns of St Dominic's Convent, Stone (1892).
Notes on the Education Question (1857).
A Pastoral Letter to the Faithful of the Diocese of Birmingham (Birmingham, 1870).
A Plea for the Rights and Liberties of Religious Women, with reference to the Bill proposed by Mr. Lacy (1851).
'To the Editor of the Times', in Anon., *The Roman Catholic Question* [see above], 1st series, pp. 6–7.
See also C. Butler, *The Vatican Council* (Secondary texts).
Urwick, W. See Vaughan, K. and W. Urwick.
Varnier, J. J. ('Father Felix'). *Why I Left the Communion of the Church of Rome; or, A Narrative of Inquiries regarding the Grounds of Roman Catholicism* (Calcutta, 1860).
Vaughan, H. *Address on the Re-dedication of Catholic England to Blessed Peter* (1893).
Letters of Herbert Vaughan to Lady Herbert of Lea, 1867 to 1903 (1942).
The Reunion of Christendom (1897).
ed., *The Young Priest: Conferences on the Apostolic Life* (1904).
See also McCormack, A.; Snead-Cox, J. G. (Secondary texts).
Vaughan, K. and W. Urwick. *The Papacy and the Bible: A Controversy* (Manchester and London, 1874).
Victoria, H. M. Queen. *The Letters of Queen Victoria: A Selection from Her Majesty's Correspondence between the Years 1837 and 1861*, ed. A. C. Benson and Viscount Esher, 3 vols. (1908).
See also Anon., *The Form and Order of the Service.*
Walpole, S. *The Life of Lord John Russell*, 2 vols. (1889).
Ward, J. C. ('Οὔτις'). *"The Jesuits" Reviewed*, Pt 1 (1852).
Ward, M. A. *Helbeck of Bannisdale* (1898).
Ward, W. G. *The Authority of Doctrinal Decisions which are not Definitions of Faith, considered in a short Series of Essays reprinted from "The Dublin Review"* (1866).
'Church Authority', *The British Critic, and Quarterly Theological Review*, 33, 45 (January 1843), 202–33.
The Ideal of a Christian Church, considered in Comparison with existing Practice, containing a Defence of certain Articles in the British Critic *in Reply to Remarks on them in Mr. Palmer's 'Narrative'* (1844).
See also Anon.,'Tractarianism and Mr Ward'.
See also Ward, W. (Secondary texts).

Warter, J. W. *A Plain Protestant's Manual; or, Certain Plain Sermons on the Scriptures, the Church, and the Sacraments, &c. &c. .&c. in which the Corruptions of the Romish Church are Evidently Set Forth* (1851).

Waterworth, J. *An Examination of the Evidence adduced by Mr. Keary, against the Authenticity or Validity of Certain Passages from the Fathers, contained in the "Faith of Catholics, on Certain Points of Controversy, compiled by Rev. Jos. Berington and Rev. John Kirk"* (1834).

See also Berington, J. and J. Kirk.

Waterworth S. J., W. *The Jesuits; or, An Examination of the Origin, Progress, Principles, and Practices of The Society of Jesus; with Observations on the Leading Accusations of the Enemies of the Order* (1852).

Origin and Developments of Anglicanism; or, A History of the Liturgies, Homilies, Articles, Bibles, Principles and Governmental System of the Church of England (1854).

See also Ward, J. C.

Webster, W. *Some Features of Modern Romanism, with an Appendix, bringing the Work up to July 1898* (1898).

Welsh, D. *Elements of Church History, vol. I, comprising the External History of the Church during the First Three Centuries* (Edinburgh, 1844).

Westerton, C. *Popish Practices at St Paul's, Knightsbridge*, 4th edn (1854).

Whately, R. *The Errors of Romanism traced to their Origin in Human Nature* (1830).

Essays on the Errors of Romanism, having their Origin in Human Nature, 3rd series, 5th edn rev. (1856).

Whewell, W. *Of the Plurality of Worlds: An Essay* (1853).

Whitaker, W. *A Disputation on Holy Scripture, against the Papists, especially Bellarmine and Stapleton*, trans. and ed. W. Fitzgerald, Parker Society series (Cambridge, 1849).

White, J. B. *A Letter to Charles Butler, Esq. on his Notice of the "Practical and Internal Evidence against Catholicism"* (1826).

Practical and Internal Evidence against Catholicism, with Occasional Strictures on Mr. Butler's Book of the Roman Catholic Church: in Six Letters, Addressed to the Impartial among the Roman Catholics of Great Britain and Ireland (1825).

White, T. *Sermons for the Different Sundays, and Principal Festivals of the Year; with a few Additional Sermons on various Important Subjects*, selected and arranged by J. Lingard, 2 vols. (1828).

Wilberforce, R. I. *An Inquiry into the Principles of Church-Authority; or, Reasons for Recalling my Subscription to the Royal Supremacy*, 2nd edn (1854).

The Practical Effect of the Gorham Case: A Charge, to the Clergy of the East Riding, delivered at the Ordinary Visitation, A.D. 1850 (1850).

The Sacramental System: A Sermon preached at St Mary's Church, before the University of Oxford, on Sunday, March 10th, 1850 (1850).

Wilberforce, S. *Rome – Her New Dogma and our Duties: A Sermon, preached before the University, at St Mary's Church, Oxford, on the Feast of the Annunciation of the Blessed Virgin Mary, 1855* (Oxford and London, 1855).

Wilberforce, W. *Practical View of the Prevailing Religious System of Professed Christians, in the Higher and Middle Classes in this Country, Contrasted with Real Christianity* (Dublin, 1797).

Wilde, O. *Complete Works*, gen. ed. J. B. Foreman, intro. V. Holland, new edn (London and Glasgow, 1966).

See also Ellmann, R. (Secondary texts).

Wiseman, N. P. S., 'Address of the Catholics of England to her Majesty', in Anon., *The Roman Catholic Question* [see above], 4th series, p. 16.

An Appeal to the Reason and Good Feeling of the English People on the Subject of the Catholic Hierarchy (1850).

On Compromises of Truth in Religious Teaching: A Lecture delivered in St George's Catholic Cathedral, Southwark, on the Evening of Sunday, June 22nd (1851).

Convents: A Review of Two Lectures on this Subject, by the Rev. M. Hobart Seymour, embodying the Substance of a Lecture delivered at the Catholic Chapel, Bath, on Sunday, May 23, 1852, from the Dublin Review for December, 1852 (1853).

A Discourse delivered at St John's, Catholic Church, Salford, on Sunday, July 28th, 1850, on the Gorham Controversy (1850).

Essays on Various Subjects, with a Biographical Introduction by the Rev. Jeremiah Murphy (1888).

Fabiola; or, The Church of the Catacombs, Popular Edn (New York and Philadelphia, PA, n.d.)

The Final Appeal in Matters of Faith: A Sermon preached in St George's Catholic Church, Southwark, on Sunday, the 17th of March, 1850 (1850).

Lectures on the principal Doctrines and Practices of the Catholic Church, delivered at St Mary's Moorfields, during the Lent of 1836, rev. edn, 2 vols. (n.d.).

The Papal and Royal Supremacies Contrasted: A Lecture delivered in St George's Catholic Church, Southwark, on Sunday, the 12th of May, 1850 (1850).

'Pastoral', in Anon., *The Roman Catholic Question* [see above], 1st series, pp. 4–6.

The Real Presence of the Body and Blood of Our Lord Jesus Christ in the Blessed Eucharist, proved from Scripture in eight lectures delivered in the English College, Rome, Centenary Edn, intro. J. M. T. Barton (1934).

The Religious and Social Position of the Catholics in England: An Address Delivered to the Catholic Congress of Malines, August 21, 1863, translated from the French (1864).

'A Sermon preached at St George's Catholic Church', in *The Roman Catholic Question* [see above], 10th series, p. 4.

Sermons preached on various Occasions (Dublin, 1889).

The Social and Intellectual State of England, compared with its Moral Condition: A Sermon delivered in St John's Catholic Church, Salford, on Sunday, July 28th 1850 (1850).

Three Lectures on the Catholic Hierarchy: Lecture I, Delivered in St George's, Southwark, on Sunday, December 8th, 1850; II, 15th; III, 22nd (1850).

Words of Peace and Justice addressed to the Catholic Clergy and Laity of the London District, on the Subject of Diplomatic Relations with the Holy See, 2nd edn (1848).

See also Oakeley, F.

See also Gwynn, D. R.; Jackman, S. W.; Ward, W. (Secondary texts).

Wood, T. *The Origin, Learning, Religion, and Customs of the Ancient Britons: with an Account of the Introduction of Christianity into Britain, and the Idolatry and Conversion of the Saxons; Remarks on the Errors and Progress of Popery; Considerations on the Christian Church, its Foundation, Superstructure, and Beauty, &c., &c.* (1846).

Woodgate, H. A. *Anomalies in the English Church no Just Ground for Seceding; or, The Abonormal Condition of the Church considered with Reference to the Analogy of Scripture and of History* (Oxford, 1857).

Wordsworth, C. *Letters to M. Gondon, author of "Mouvement Religieux en Angleterre," "Conversion de Soixante Ministres Anglicans," &c. &c. &c. on the Destructive Character of the Church of Rome, both in Religion and Policy* (1847).

ed., *The New Testament of Our Lord and Saviour Jesus Christ, in the Original Greek, with Introductions and Notes*, new edn, 2 vols. (1877).

Sequel to Letters to M. Gondon, on the Destructive Character of the Church of Rome, both in Religion and Polity (1848).

Wordsworth, W. *Poems*, ed. J. O. Hayden, 2 vols. (Harmondsworth, 1977).

Wright, C. *The History and Description of Arundel Castle, Sussex, the Seat of His Grace The Duke of Norfolk, with an Abstract of the Lives of the Earls of Arundel, from the Conquest to the Present Time . . .* , 2nd edn (1818).

Wright, G. N. *Life and Campaigns of Arthur, Duke of Wellington, K.G.*, 4 vols. (1841).

Wright, T. *St Patrick's Purgatory: An Essay on the Legends of Purgatory, Hell and Paradise current during the Middle Ages* (1844).

Wylie, J. A. *Rome and Civil Liberty; or, The Papal Aggression in its Relation to the Sovereignty of the Queen and the Independence of the Nation* (London and Edinburgh, 1864).

Yonge, C. M. *Womankind* (1876).

SECONDARY TEXTS

Ackroyd, P. *Albion: The Origins of the English Imagination* (2002).

Dickens (1990).

London: The Biography (2000).

Adams, J. *Madder Music, Stronger Wine: The Life of Ernest Dowson, Poet and Decadent* (London and New York, 2000).

Adams, J. E. *Dandies and Desert Saints: Styles of Victorian Manhood* (Ithaca, NY, and London, 1995).

Addington, R. *The Idea of the Oratory* (1966).

Alexander S. J., C. *The Catholic Literary Revival: Three Phases in its Development from 1845 to the Present* (Milwaukee, WI, 1935).
Allchin, A. M. *The Silent Rebellion: Anglican Religious Communities, 1845–1900* (1958).
Altholz, J. L. *The Religious Press in Britain, 1760–1900*, Contributions to the Study of Religion series, ed. H. W. Bowden, 22 (New York, 1989).
Altick, R. D. *The English Common Reader: A Social History of the Mass Reading Public, 1800–1900* (1957; rpt. Chicago and London, 1963).
Anderson, V. P. *Robert Browning as a Religious Poet: An Annotated Bibliography of the Criticism* (Troy, NY, 1983).
Anson, P. F. *The Call of the Cloister: Religious Communities and Kindred Bodies in the Anglican Communion* (1956).
Apostolos-Cappadona, D. *Dictionary of Women in Religious Art* (Oxford, 1998).
Arnstein, W. L. *Protestant versus Catholic in Mid-Victorian England: Mr. Newdegate and the Nuns* (Columbia, MO, and London, 1982).
Atkinson, C. W., C. H. Buchanan and M. R. Miles, eds. *Immaculate and Powerful: The Female in Sacred Image and Society Reality*, Harvard Women's Studies in Religion series (1985; rpt. Wellingborough, 1987).
Aubert, R. *Le Pontificat de Pie IX* (Paris, 1952).
Auerbach, J. A. *The Great Exhibition of 1851: A Nation on Display* (New Haven, CT, and London, 1999).
Aveling, J. C. H. *The Handle and the Axe: The Catholic Recusants in England from Reformation to Emancipation* (1976).
Aveling, J. C. H., D. M. Loades and H. R. McAdoo, *Rome and the Anglicans: Historical and Doctrinal Aspects of Anglican-Roman Catholic Relations* (Berlin, 1982).
Axon, W. E. A. *William Harrison Ainsworth: A Memoir* (1902).
Badeni, J. *The Slender Tree: A Life of Alice Meynell* (Padstow, 1981).
Baker, J. E. *The Novel and the Oxford Movement*, Princeton Studies in English series, 8 (Princeton, NJ, 1932).
Bann, S. *The Clothing of Clio: A Study of the Representation of History in Nineteenth-century Britain and France* (Cambridge, 1984).
Baring, M. *The Puppet Show of Memory* (1922).
Barker, J. *The Brontës* (1994; rpt. 1995).
Barmann, L. *Baron Friedrich von Hügel and the Modernist Crisis in England* (Cambridge, 1972).
Battiscombe, G. *John Keble: A Study in Limitations* (1963).
Bebbington, D. *Evangelicalism in Modern Britain: A History from the 1730s to the 1980s* (1989).
Beck, G. A., ed. *The English Catholics, 1850–1950: Essays to Commemorate the Centenary of the Restoration of the Hierarchy of England and Wales* (1950).
Beer, G. *Darwin's Plots: Evolutionary Narrative in Darwin, George Eliot and Nineteenth-Century Fiction* (1983; rpt. 1985).
Bellenger, Dom A. *The French Exiled Clergy in the British Isles after 1789* (Bath, 1986).

Bergonzi, B. *A Victorian Wanderer: The Life of Thomas Arnold the Younger* (Oxford, 2003).

Bernstein, S. *Confessional Subjects: Revelations of Gender and Power in Victorian Literature and Culture* (Chapel Hill, NC, and London, 1997).

Blom, F., *et al. English Roman Catholic Books, 1701–1800: A Bibliography* (Aldershot, 1996).

Boardman, B. M. *Between Heaven and Charing Cross: The Life of Francis Thompson* (New Haven, CT, and London, 1988).

Bossy, J. *The English Catholic Community 1570–1850* (1975).

Bowness, A. *et al. The Pre-Raphaelites* (1974).

Bradley, I. *Abide with Me: The World of Victorian Hymns* (1997).

Bradstock, A. S., *et al.*, eds. *Masculinity and Spirituality in Victorian Culture* (Basingstoke and New York, 2000).

Bramstead, E. K. and K. J. Melhuish, eds. *Western Liberalism: A History in Documents from Locke to Croce* (London and New York, 1978).

Brendon, P. *Hurrell Froude and the Oxford Movement* (1974).

Bronfen, E. *Over Her Dead Body: Death, Femininity and the Aesthetic* (Manchester, 1992).

Brooke, J. H. *Science and Religion: Some Historical Perspectives*, Cambridge History of Science series, ed. G. Bassala (Cambridge, 1991).

Brown, R. *Church and State in Modern Britain, 1700–1850* (1991).

Burrow, J. W. *A Liberal Descent: Victorian Historians and the English Past* (Cambridge, 1981).

Burton, E. H. *The Life and Times of Bishop Challoner (1691–1781)*, 2 vols. (1909).

Butler, Dom C. *The Life & Times of Bishop Ullathorne, 1806–1889*, 2 vols. (1926).

The Vatican Council, 1869–1870: Based on Bishop Ullathorne's Letters, ed. C. Butler, Fontana Library series (1962).

Butler, M. *Romantics, Rebels and Reactionaries: English Literature and its Background, 1760–1830* (Oxford and New York, 1981).

Casteras, S. P. 'Virgin Vows: The Early Victorian Artists' Portrayal of Nuns and Novices', in *Religion in the Lives of English Women, 1760–1930*, ed. G. Malmgreen (Bloomington, IN, 1986), pp. 129–60.

Chadwick, O. *From Bossuet to Newman*, 2nd edn (Cambridge, 1987).

The Victorian Church, Ecclesiastical History of England, ed. J. C. Dickinson, vols. VII–VIII (1966–70).

Chapman, R[aymond]. *Faith and Revolt: Studies in the Literary Influence of the Oxford Movement* (1970).

Chapman, R[onald]. *Father Faber* (1961).

Chinnici O. F. M., J.P. *The English Catholic Enlightenment: John Lingard and the Cisalpine Movement, 1780–1850* (Shepherdstown, VA, 1980).

Chitty, S. *The Beast and the Monk: A Life of Charles Kingsley* (1974).

Clark, J. C. D. *English Society, 1660–1832: Religion, Ideology and Politics during the Ancien Regime*, 2nd edn (Cambridge, 2000).

Colley, L. *Britons: Forging the Nation, 1707–1837* (New Haven, CT, 1992).

Couve de Murville, M. N. L. *John Milner* (Birmingham, 1986).

Croft-Cooke, R. *Bosie: The Story of Lord Alfred Douglas, His Friends and Enemies* (1963).
Cruickshanks, E. and J. Black, eds. *The Jacobite Challenge* (Edinburgh, 1988).
Culler, A. D. *The Victorian Mirror of History* (New Haven, CT, and London, 1985).
Cunningham, V. *Everywhere Spoken Against: Dissent in the Victorian Novel* (Oxford, 1975).
Daiches, D. *Charles Edward Stuart: The Life and Times of Bonnie Prince Charlie* (1973).
De Castro, J. P. *The Gordon Riots* (1926).
Desmond, A. and J. Moore, *Darwin* (New York, 1992).
Dessain, C. S. *John Henry Newman* (1966).
Dobson, M. and N. J. Watson. *England's Elizabeth: An Afterlife in Fame and Fantasy* (Oxford and New York, 2002).
Duffy, E., ed. *Challoner and his Church: A Catholic Bishop in Georgian London* (1981).
Ebertshäuser, C. H., *et al. Mary: Art, Culture, and Religion through the Ages*, trans. P. Heinegg (New York, 1998).
Edwards, F. *The Jesuits in England, from 1580 to the Present Day* (Tunbridge Wells, 1985).
Ellison, R. H. and C. M. Engelhardt, 'Prophecy and Anti-Popery in Victorian London: John Cumming Reconsidered', *Victorian Literature and Culture* (2003), 373–89.
Ellmann, R. *Oscar Wilde* (1987).
Engelhardt, C. M. 'The Paradigmatic Angel in the House: The Virgin Mary and Victorian Anglicans', in Hogan, A. and A. Bradstock, eds, *Women of Faith in Victorian Culture* (London and New York, 1998), pp. 159–71.
'Victorian Masculinity and the Virgin Mary', in Bradstock, A. S., *et al.*, *Masculinity and Spirituality in Victorian Culture* (Basingstoke and New York, 2000), pp. 44–57.
See also Ellison, R. H.
Fitzsimons, J., ed. *Manning: Anglican and Catholic* (1951; rpt. Westport, CT, 1979).
Fleishman, A. *The English Historical Novel: Walter Scott to Virginia Woolf* (Baltimore, MD, and London, 1971).
Fletcher, I., ed. *Decadence and the 1890s*, Stratford-upon-Avon Studies series, ed. M. Bradbury and D. Palmer, 17 (1979).
Frazier, A. *George Moore, 1852–1933* (New Haven, CT, and London, 2000).
Gill, S. *Women and the Church of England: From the Eighteenth Century to the Present* (1994).
Gilley, S. 'The Garibaldi Riots of 1862', *Historical Journal*, 16, 4 (1973), 697–732.
Newman and his Age (1990).
'The Roman Catholic Church in England, 1780–1940', in *A History of Religion in Britain: Practice and Belief from Pre-Roman Times to the Present*, ed. S. Gilley and W. J. Sheils (Oxford, 1994).

Goldsmith, R. W. *The Relation of Browning's Poetry to Religious Controversy, 1833–1868*, unpublished dissertation, University of North Carolina (1958).

Grieve, A. 'The Pre-Raphaelite Brotherhood and the Anglican High Church', *Burlington Magazine*, 111 (1969), 294–5.

'Style and Content in Pre-Raphaelite Drawings 1848–50', in *Pre-Raphaelite Papers*, ed. L. Parris (1984), pp. 23–43.

Griffin, S. M. *Anti-Catholicism and Nineteenth-Century Fiction* (Cambridge, 2004).

Gwynn, D. R. *Cardinal Wiseman* (1929).

Father Dominic Barberi (1947).

Father Luigi Gentili and his Mission, 1801–1848 (Dublin, 1951).

Lord Shrewsbury, Pugin, and the Catholic Revival (1946).

The Second Spring, 1818–1852: a Study of the Catholic revival in England (1942).

Hale, R. *Canterbury and Rome, Sister Churches: A Roman Catholic Monk reflects upon Reunion in Diversity* (1982).

Hall, D., ed. *Muscular Christianity: Embodying the Victorian Age* (Cambridge, 1994).

Halle, M. ('M. Haile') and E. Bonney. *Life and Letters of John Lingard, 1771–1851* (1911).

Hallman, D. *Opera, Liberalism, and Anti-Semitism in Nineteenth-Century France: The Politics of Halévy's* La Juive, Cambridge Studies in Opera series, ed. A. Groos (Cambridge, 2002).

Hanson, E. *Decadence and Catholicism* (Cambridge, MA, and London, 1997).

Haskins, S. *Mary Magdalen: Myth and Metaphor* (1993).

Hasler, A. B. *How the Pope became Infallible: Pius IX and the Politics of Persuasion*, intro. H. Küng, trans. P. Heinegg (Garden City, NY, 1981).

Hayden, J. O., ed. *Scott: The Critical Heritage*, Critical Heritage series, ed. B. C. Southam (1970).

Haydon, C. *Anti-Catholicism in Eighteenth-Century England, c. 1714–80: A Political and Social Study* (Manchester and New York, 1993).

'The Gordon Riots in the English Provinces', *Historical Research*, 63 (1990), 354–9.

Healy, P. 'Man Apart: Priesthood and Homosexuality at the End of the Nineteenth Century', in A. S. Bradstock, *et al.*, eds, *Masculinity and Spirituality in Victorian Culture* (Basingstoke and New York, 2000), pp. 100–15.

Heimann, M. *Catholic Devotion in Victorian England* (Oxford, 1995).

Herbermann, C. G., *et al.*, eds. *The Catholic Encyclopedia*, 16 vols. (New York, 1913–14).

Hibbert, C. *Disraeli: A Personal History* (2004).

King Mob: The Story of Lord George Gordon and the Riots of 1780 (1958).

Hickey, J. V. *Urban Catholics: Urban Catholicism in England and Wales from 1829 to the Present Day* (1967).

Hilliard, D. '"UnEnglish and Unmanly": Anglo-Catholicism and Homosexuality', *Victorian Studies*, 25 (1981–2), 181–210.

Hilton, B. *The Age of Atonement: The Influence of Evangelicalism on Social and Economic Thought, 1795–1865* (Oxford, 1988).

Hodell, C. W. ed. and trans. *The Old Yellow Book: Source of Browning's 'The Ring and the Book', in complete Photo-reproduction* (Washington, DC, 1908).

Hogan, A. 'Angel or Eve?: Victorian Catholicism and the Angel in the House', in Hogan, A. and A. Bradstock, eds. *Women of Faith in Victorian Culture* (London and New York, 1998), pp. 91–100.

Hogan, A. and A. Bradstock, eds. *Women of Faith in Victorian Culture: Reassesssing the Angel in the House* (London and New York, 1998).

Holland, B. *Memoir of Kenelm Henry Digby* (1919).

Holmes, J. D. *More Roman than Rome: English Catholicism in the Nineteenth Century* (London and Shepherdstown, WV, 1978).

Hornsby-Smith, M. P. *Roman Catholic Beliefs in England: Customary Catholicism and Religious Authority in England* (Cambridge, 1991).

Irvine, W. and P. Honan. *The Book, the Ring, and the Poet: A Biography of Robert Browning* (1974).

Jackman, S. W. *Nicholas Cardinal Wiseman: A Victorian Prelate and his Writings* (Dublin and Gerrards Cross, 1977).

Jack, I. *Browning's Major Poetry* (Oxford, 1973).

Jagger, P. J. *Clouded Witness: Initiation in the Church of England in the Mid-Victorian Period, 1850–1875*, Pittsburgh Theological Monographs series, ed. D. Y. Hadidian, n.s. 1 (Allison Park, PA, 1982).

Jantzen, G. M. *Becoming Divine: Towards a Feminist Philosophy of Religion* (Manchester, 1998).

Jay, E. *The Religion of the Heart: Anglican Evangelicalism and the Nineteenth Century Novel* (Oxford, 1979).

Jenkyns, R. *Dignity and Decadence: Victorian Art and the Classical Inheritance* (1991).

Johnson, E. *Sir Walter Scott: The Great Unknown* (New York, 1970).

Kaplan, F. *Thomas Carlyle: A Biography* (Ithaca, NY, 1983).

Ker, I. T. *The Catholic Revival in English Literature, 1845–1961: Newman, Hopkins, Belloc, Chesterton, Greene, Waugh* (Leominster, 2003).

John Henry Newman: A Biography (Oxford and New York, 1988).

Kermode, F. *Romantic Image* (1957; rpt. London and New York, 2002).

The Sense of an Ending: Studies in the Theory of Fiction (New York, 1967).

Khattab, E. A. *The Critical Reception of Browning's* The Ring and the Book*: 1868–1889 and 1951–1968*, Salzburg Studies in English Literature series, Romantic Reassessment, ed. J. Hogg, 66 (Salzburg, 1977).

Kiely, R. *The Romantic Novel in England* (Cambridge, MA, 1972).

Korg, J. *Browning and Italy* (Athens, OH, and London, 1983).

Kreuger, C. L. *The Reader's Repentance: Women Preachers, Women Writers and Nineteenth-Century Social Discourse* (Chicago, IL, and London, 1992).

Küng, H. *Unfehlbar?: Eine Anfrage* (Zürich, 1970).

Lake, P., with M. Questier. *The Antichrist's Lewd Hat: Protestants, Papists and Players in Post-Reformation England* (New Haven, CT, and London, 2002).

Landry, D. *The Muses of Resistance: Labouring-Class Women's Poetry in Britain, 1739–1796* (Cambridge, 1990).

Lankewish, V. A. 'Love Among the Ruins: The Catacombs, the Closet, and the Victorian "Early Christian" Novel', *Victorian Literature and Culture*, 28 (2000), 239–73.

Litzinger, B. *Robert Browning and the Babylonian Woman*, Baylor Browning Interests series, 19 (Waco, TX, 1962).

Litzinger, B. and D. Smalley, eds. *Browning: The Critical Heritage*, Critical Heritage series, ed. B. C. Southam (1970).

Liversidge, M. and C. Edwards, eds. *Imagining Rome: British Artists and Rome in the Nineteenth Century* (1996).

McClelland, V. A. *Cardinal Manning: His Public Life and Influence, 1865–1892* (1962).

McClelland, V. A. and M. Hodgetts, eds. *From Without the Flaminian Gate: 150 Years of Roman Catholicism in England and Wales 1850–2000* (1999).

McCormack, A. *Cardinal Vaughan: The Life of the Third Archbishop of Westminster* (1966).

MacDougall O. M. I., H.A. *The Acton–Newman Relations: The Dilemma of Christian Liberalism* (New York, 1962).

Machin, G. I. T. *The Catholic Question in English Politics, 1820 to 1830* (Oxford, 1964).

Martin, J. R. *Baroque*, Style and Civilization series, ed. J. Fleming and H. Honour (1977).

Mason, M. *The Making of Victorian Sexuality* (Oxford and New York, 1994).

Mews, S., ed. *Modern Religious Rebels* (1993)

Meynell, V. *Alice Meynell: A Memoir* (1929).

Monod, P. K. *Jacobitism and the English People, 1688–1788* (Cambridge, 1989).

Morgan, S. *A Passion for Purity: Ellice Hopkins and the Politics of Gender in the late-Victorian Church*, CCSRG monograph series, 2 (Bristol, 1999).

Morley, J. *The Life of William Ewart Gladstone*, 3 vols. (London and New York, 1903).

Morris, K. L. 'John Bull and the Scarlet Woman: Charles Kingsley and Anti-Catholicism in Victorian Literature', *Recusant History*, 23, 2 (1996–7), 190–218.

Murray, P. and L. *The Oxford Companion to Christian Art and Architecture* (Oxford and New York, 1996).

Napier, M. and A. Laing, eds. *The London Oratory Centenary, 1884–1984* (1984).

Neill, S. *The Interpretation of the New Testament, 1861–1961*, The Firth Lectures series, 1962 (Oxford, 1966).

Newsome, D. *The Parting of Friends: A Study of the Wilberforces and Henry Manning* (1966).

The Convert Cardinals (1993).

Nockles, P. B. *The Oxford Movement in Context: Anglican High Churchmanship, 1760–1857* (Cambridge, 1994).

Norman, E. R. *Anti-Catholicism in Victorian England*, Historical Problems: Studies and Documents series, ed. G. R. Elton, 1 (1968).

The English Catholic Church in the Nineteenth Century (Oxford, 1984).

Roman Catholicism in England from the Elizabethan Settlement to the Second Vatican Council (Oxford and New York, 1985).
The Victorian Christian Socialists (Cambridge, 1987).
Norris, P. *The Story of Eve* (Basingstoke and Oxford, 1998).
O'Brien, C. 'When Radical meets Conservative: Godard, Delannoy and the Virgin Mary', *Literature & Theology*, 15, 2 (2001), 174–86.
O'Brien, S. '*Terra Incognita*: The Nun in Nineteenth-Century England', *Past and Present*, 121 (1988), 110–40.
Oddie, W. *Dickens and Carlyle: The Question of Influence* (1972).
Paisley, I. 'The Jesuit Oath Exposed' (www.acts2.com/thebibletruth/Jesuit_Oath.htm)
Paley, M. D. *The Apocalyptic Sublime* (New Haven, CT, and London, 1986).
Parker, R., ed. *The Oxford Illustrated History of Opera* (Oxford and New York, 1994).
Patten, B. *Catholicism and the Gothic Revival: John Milner and St Peter's Chapel, Winchester*, Hampshire Papers, 21 (Winchester, 2001).
Pawley, B. and M. *Rome and Canterbury through Four Centuries: A Study of the Relations between the Church of Rome and the Anglican Churches, 1530–1981*, rev. edn (London and Oxford, 1981).
Pawley, M. *Faith and Family: The Life and Circle of Ambrose Phillipps de Lisle* (Norwich, 1993).
Paz, D. G. *Popular Anti-Catholicism in Mid-Victorian England* (Stanford, CA, 1992).
Peterson, L. H. *Victorian Autobiography: The Tradition of Self-Interpretation* (New Haven, CT, 1986).
Pickering, W. S. F. *Anglo-Catholicism: A Study in Religious Ambiguity* (1989).
Pittock, M. *Poetry and Jacobite Politics in Eighteenth-Century Britain and Ireland* (Cambridge, 1994).
Prickett, S. *Origins of Narrative: The Romantic Appropriation of the Bible* (Cambridge, 1996).
Purbrick, L. *The Great Exhibition of 1851: New Interdisciplinary Essays*, Texts in Culture series, ed. J. Wallace and J. Whale (Manchester and New York, 2001).
Quinn, D. *Patronage and Piety: The Politics of English Roman Catholicism, 1850–1900* (1992).
Ralls, W. 'The Papal Aggression of 1850: A study in Victorian anti-Catholicism', in *Religion in Victorian Britain*, ed. G. Parsons, 4 vols. (Manchester, 1988), IV, 115–34.
Ranchetti, M. *The Catholic Modernists: A Study of the Religious Reform Movement, 1864–1907*, trans. I. Quigly (1969).
Raven, J., *et al. The English Novel, 1770–1829: A Bibliographical Survey of Prose Fiction Published in the British Isles*, 2 vols. (Oxford, 2000).
Reed, J. S. *The Glorious Battle: The Cultural Politics of Victorian Anglo-Catholicism* (Nashville, TN, and London, 1996).
Reed, S. *Women and the Church of England* (1994).
Reid, J. C. *The Mind and Art of Coventry Patmore* (1957).

Rhodes, R. W. *The Lion and the Cross: Early Christianity in Victorian Novels* (Columbus, OH, 1995).
Robinson, J. M. *The Dukes of Norfolk*, rev. edn (Chichester, 1995).
Rowell, G. *The Vision Glorious: Themes and Personalities of the Catholic Revival in Anglicanism* (Oxford, 1983).
Rowlands, J. H. L. *Church, State and Society: The Attitudes of John Keble, Richard Hurrell Froude and John Henry Newman, 1827–1845* (Worthing, 1989).
Rowse, A. L. *Froude the Historian: Victorian Man of Letters* (Gloucester, 1987).
Sagovsky, N. *'On God's Side': A Life of George Tyrrell* (Oxford, 1990).
Said, E. W. *Orientalism* (New York, 1978).
Sambrook, J. *William Cobbett*, Routledge Author Guides series, ed. B. C. Southam (London and Boston, 1973).
Sanders, A. *The Victorian Historical Novel, 1840–1880* (London and Basingstoke, 1978).
Schiefelbein, M. E. *The Lure of Babylon: Seven Protestant Novelists and Britain's Roman Catholic Renewal* (Macon, GA, 2001).
Sewell, B. *In the Dorian Mode: A Life of John Gray, 1866–1934* (Padstow, 1983).
Sewell, D. *Catholics: Britain's Largest Minority* (2001).
Shell, A. *Catholicism, Controversy and the English Literary Imagination, 1558–1660* (Cambridge and New York, 1999).
Singleton, J. 'The Virgin Mary and Religious Conflict in Victorian Britain', *Journal of Ecclesiastical History*, 43, 1 (January 1992), 16–34.
Snead-Cox, J. G. *The Life of Cardinal Vaughan* (1910).
Snodgrass, C. *Aubrey Beardsley: Dandy of the Grotesque* (New York and Oxford, 1995).
Stevenson, J. *Popular Disturbances in England, 1700–1832*, 2nd edn (1992).
Strauss, L. *Liberalism Ancient and Modern* (New York and London, 1968).
Stromberg, R. N. *Religious Liberalism in Eighteenth-Century England* (Oxford, 1954).
Strong, R. C. *And when did you last See your Father?: The Victorian Painter and British History* (1978).
Sussman, H. L., ed. *Victorian Masculinities: Manhood and Masculine Poetics in Early Victorian Literature and Art* (Cambridge, 1995).
Sutherland, J. *The Longman Companion to Victorian Fiction* (1988).
Swift, R. 'Guy Fawkes Celebrations in Victorian Exeter', *History Today*, November 1981, pp. 5–9.
Tallett, F. and N. Atkin, eds. *Catholicism in Britain and France since 1789* (1996).
Tanner, M. *Ireland's Holy Wars: The Struggle for a Nation's Soul, 1500–2000* (London and New Haven, CT, 2001).
Tanner, T. *Venice Desired* (Oxford, 1992).
Thomas S.J., A. *Hopkins the Jesuit: The Years of Training* (1969).
Thornton, R. K. R. *The Decadent Dilemma* (1983).
Tillotson, G. and D. Hawes, eds. *Thackeray: The Critical Heritage*, Critical Heritage series, ed. B. C. Southam (London and New York, 1968).

Toon, P. *Evangelical Theology 1833–1856: A Response to Tractarianism*, Marshalls Theological Library series (London and Atlanta, GA, 1979).
Towle, E. A. *John Mason Neale, D.D.: A Memoir* (1907).
Trevor, M. *Newman: The Pillar of Cloud* (1962).
Newman: Light in Winter (1962).
Tumbleson, R. D. *Catholicism in the English Protestant Imagination: Nationalism, Religion, and Literature, 1660–1745* (Cambridge, 1998).
Turner, F. M. *John Henry Newman: The Challenge to Evangelical Religion* (New Haven, CT, and London, 2002).
Unrau, J. *Ruskin and St Mark's* (1984).
Vance, N. *Sinews of the Spirit: The Ideal of Christian Manliness in Victorian Literature and Religious Thought* (Cambridge, 1985).
The Victorians and Ancient Rome (Oxford, 1997).
Walder, D. *Dickens and Religion* (1981).
Wallis, F. H. *Popular Anti-Catholicism in mid-Victorian Britain*, Texts and Studies in Religion series, 60 (Lewiston, 1993).
Ward, B. *The Dawn of the Catholic Revival in England, 1781–1803*, 2 vols. (1909).
The Eve of Catholic Emancipation: Being the History of the English Catholics during the first Thirty Years of the Nineteenth Century, 3 vols. (1911–12).
The Sequel to Catholic Emancipation: The Story of the English Catholics continued down to the Re-Establishment of their Hierarchy in 1850, 2 vols. (1915).
Ward, W. *Aubrey de Vere: A Memoir, based on his Unpublished Diaries and Correspondence* (1904).
The Life and Times of Cardinal Wiseman, 5th edn, 2 vols. (1899).
The Life and Times of John Henry Cardinal Newman, based on his Private Journals and Correspondence, 2 vols. (1912).
William George Ward and the Oxford Movement (1889).
Warner, M. *Alone of All Her Sex: The Myth and Cult of the Virgin Mary* (1976; rpt. 1985).
Watson, G. *The English Ideology: Studies in the Language of Victorian Politics* (1973).
Wheeler, H. F. B. and A. M. Broadley. *Napoleon and the Invasion of England: The Story of the Great Terror* (London and New York, 1908).
Wheeler, M. *Death and the Future Life in Victorian Literature and Theology* (Cambridge, 1990).
Ruskin's God (Cambridge, 1999).
Whibley, C. *Lord John Manners and his Friends*, 2 vols. (Edinburgh and London, 1925).
White, N. *Hopkins: A Literary Biography* (Oxford, 1992).
Williams, R. *Culture and Society, 1780–1950* (1958; rpt. Harmondsworth, 1963).
Wolff, R. L. *Gains and Losses: Novels of Faith and Doubt in Victorian England* (1977).
Wolffe, J. *The Protestant Crusade in Great Britain, 1829–1860* (Oxford, 1991).
Worth, G. J. *William Harrison Ainsworth*, Twayne's English Authors series, ed. S. E. Bowman, 138 (New York, 1972).
Ziegler, P. *Melbourne: A Biography of William Lamb, 2nd Viscount Melbourne* (1976).

Index